Study Guide

for use with

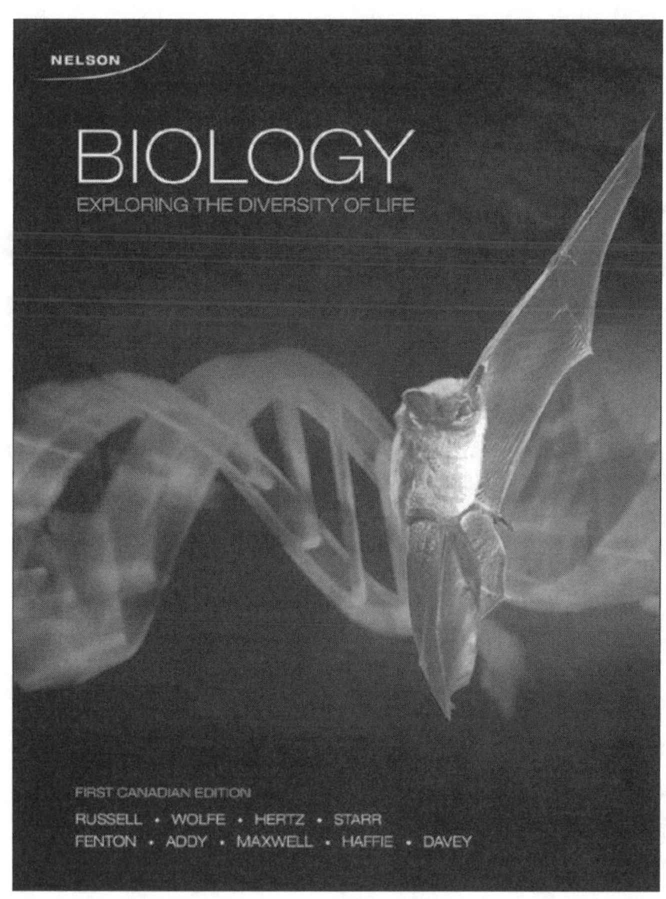

Prepared by COLIN MONTPETIT
UNIVERSITY OF OTTAWA

JULIE SMIT
UNIVERSITY OF WINDSOR

WENDY KEENLEYSIDE
UNIVERSITY OF GUELPH

NELSON EDUCATION

NELSON EDUCATION

Study Guide for use with *Biology: Exploring the Diversity of Life*, First Canadian Edition

Peter J. Russell, Stephen L. Wolfe, Paul E. Hertz, Cecie Starr, M. Brock Fenton, Heather Addy, Denis Maxwell, Tom Haffie, Ken Davey

Prepared by Colin Montpetit, Julie Smit, and Wendy Keenleyside

Vice President, Editorial Director:
Evelyn Veitch

Editor-in-Chief, Higher Education:
Anne Williams

Executive Editor:
Paul Fam

Senior Marketing Manager:
Sean Chamberland

Managing Editor, Development:
Alwynn Pinard

Content Production Manager:
Christine Gilbert

Proofreader:
Holly Dickinson

Production Coordinator:
Ferial Suleman

Design Director:
Ken Phipps

Managing Designer:
Franca Amore

Printer:
Edwards Brothers

Printed and bound in
The United States of America
1 2 3 4 12 11 10 09

For more information contact Nelson Education Ltd., 1120 Birchmount Road, Toronto, Ontario, M1K 5G4. Or you can visit our Internet site at http://www.nelson.com

ISBN-13: 978-0-17-647474-4
ISBN-10: 0-17-647474-9

Table of Contents

Unit One Setting the Stage

1 Light and Life ... 1
2 Origins of Life ... 18
3 Selection, Biodiversity, and Biosphere ... 41

Unit Two Energy: Process and Facilitation

4 Energy and Enzymes ... 56
5 Membranes and Transport ... 75
6 Cellular Respiration ... 94
7 Photosynthesis ... 112
8 Cell Communication ... 130

Unit Three Genes

9 Cell Cycles ... 144
10 Genetic Recombination ... 157
11 Mendel, Genes, and Inheritance ... 178
12 Genes, Chromosomes, and Human Genetics ... 190

Unit Four DNA and Gene Expression

13 DNA Structure, Replication, and Organization ... 205
14 Gene Structure and Expression ... 223
15 Control of Gene Expression ... 246
16 DNA Technologies and Genomics ... 266

Unit Five Evolution and Classification

17 Microevolution: Genetic Changes within Populations ... 281
18 Species ... 294
19 Evolution and Classification ... 310
20 Darwin, Fossils, and Developmental Biology ... 324

Unit Six Diversity of Life

21 Prokaryotes ... 340
22 Viruses, Viroids, and Prions: Infectious Biological Particles ... 355
23 Protists ... 365
24 Fungi ... 381
25 Plants ... 399
26 Protostomes ... 418
27 Deuterostomes: Vertebrates and Their Closest Relatives ... 431
28 The Plant Body ... 446
29 Transport in Plants ... 456

Unit Seven Systems and Process—Plants
30 Reproduction and Development in Flowering Plants 466
31 Control of Plant Growth and Development 480

Unit Eight Systems and Process—Animals
32 Introduction to Animal Organization and Physiology 492
33 Information Flow: Nerves, Ganglia, and Brains 504
34 Sensory Systems 526
35 The Endocrine System 543
36 Muscles, Skeletons, and Body Movements 562
37 The Circulatory System 574
38 Animal Reproduction 590
39 Animal Development 602
40 Animal Behaviour 616

Unit Nine Life Processes—an Integrated View
41 Plant and Animal Nutrition 630
42 Gas Exchange: The Respiratory System 647
43 Regulating the Internal Environment 662
44 Defences against Disease 676

Unit Ten Ecology
45 Population Ecology 691
46 Population Interactions and Community Ecology 706
47 Ecosystems 724

Unit Eleven Biology in Action
48 Conservation of Biodiversity 735
49 Domestication 747

CASE STUDIES C-1
ANSWERS A1

USING THIS STUDY GUIDE

Welcome to the Study Guide for use with *Biology: Exploring the Diversity of Life*, First Canadian Edition. This study resource has been dramatically adapted to meet the needs of Introductory Biology students to better prepare you for the content of your course and to help you make connections to the text's content and beyond. The Study Guide is organized into three distinct parts: study resources by chapter, a section of case studies and finally, an answers section which will allow you to check your work.

Study Resources by Chapter
Each chapter of this section corresponds to the chapter in your textbook. Within these chapters you will find the following:

Chapter Highlights: A quick overview of the subject matter in a bulleted list.
Study Strategies and Learning Outcomes: The study strategies are designed to provide insight into how best to approach the chapter content and study effectively while the learning outcomes address what you should know having completed your review of the chapter.
Topic Map: Each chapter contains one of these maps to provide you with a visual understanding of how the content within the chapter fits together.
Self-Test Questions: There are at least twenty of these multiple-choice questions towards the end of the chapter designed to test you on the content of the entire chapter. Answers to these questions are provided within the Answers section at the end of the Study Guide.
Integrating and Applying Key Concepts: These represent "big-picture" or "real-life" applications of the concepts presented in the chapter. Some guidance as to how to approach these concepts is provided within the Answers section.
Optional Activities: Some additional activities you may choose to pursue should you wish to further your understanding of the content or explore a particular area of interest.

Within the body of these chapters you will find various exercises designed to help familiarize yourself with the material such as: Section Review (fill-in-the-blank style questions), Labelling, True/False Questions, Matching, Complete the Table, Short Answer Questions and Building Vocabulary. The answers to all questions within these chapters are available in the Answers section at the back of the Study Guide.

Case Studies
Refer to the introduction to the Case Studies section (pg. C-1) for more information on this section. This is an exciting addition to the Study Guide and will help you better connect with the content within the text and your course.

Answer Section
At the very end of the Study Guide the answers to the questions noted above are contained and are organized as they appear within the chapter.

For more study aids and materials related to your text visit www.biologyedl.nelson.com. Here you will find flashcards, weblinks, crosswords, and a sample "Purple Pages" module that links back to the *Chemical and Physical Foundations of Biology* references pages from your text. The full modules are available through your access to CengageNOW—your online key to study success—please visit your text website for more details.

1 Light and Life

CHAPTER HIGHLIGHTS

- Light is the visible region of the sun's electromagnetic radiation.
  - It consists of particles called photons and travels as a wave. The energy of a photon is inversely proportional to the wavelength.
  - Pigment molecules have a molecular structure that allows them to absorb light energy.
  - Colour results from transmission-specific wavelengths of light.
- Light is a source of energy for many life forms.
  - Absorption of light energy by photosynthetic pigments is critical for life on Earth.
  - Photosynthetic organisms use the absorbed energy to drive energy-requiring biosynthetic processes.
- Light is a source of information for most life forms.
  - Rhodopsin is the most common photoreceptor for light sensing.
  - Light-sensing structures vary in structure, from the simple eyespot of *Chlamydomonas reinhardtii* to the extremely complex single-lens "camera eye" of humans.
  - The evolution of the image-forming eyes required the simultaneous evolution of a neural system or brain to process the images.
- Short-wavelength ultraviolet (UV) radiation possesses enough energy to damage biological molecules.
  - Photosynthetic structures are constantly being damaged and require constant repair. UV-induced thymine dimer formation in DNA can cause mutations and cancer. The skin pigment melanin absorbs UV light and protects skin cells.

- Light is important for the behaviour and ecology of all organisms.
  - The circadian rhythm is found in all life forms and depends on the sun's 24-hour cycle of light/dark but is set by the organism's biological clock.
  - Light-sensing organisms use colour/transmitted light to attract, warn, or hide.
- Some animals are adapted to live in the dark.
  - Some can see at very low light levels.
  - The blind mole rat is descended from predecessors with functional eyes and has a functional biological clock.
  - Ecological light pollution is excessive artificial light at night and can harm night-dwelling organisms.
- Bioluminescent organisms produce their own light for communication, warning, or camouflage or to attract a mate.

STUDY STRATEGIES AND LEARNING OUTCOMES

- Remember that the goal of this chapter is to understand the nature of light and to appreciate how it impacts all life on Earth. You should first be able to describe the nature of light and the relationship between wavelength and energy and remember the wavelength range of the visible region of the spectrum, as well as which colours are at the two ends and which of these are closer to the UV and infrared regions of the spectrum.
- Make sure you understand how light interacts with matter. What happens when a photon is absorbed or when light of a particular wavelength is transmitted? These processes underlie all of the effects of light on living organisms.
- Learn about retinal, rhodopsin, and the various levels of complexity of light-sensing units. The more basic the unit, the more basic the information and response. Once you understand how organisms can sense light/colour, then you can learn about the various roles light sensing plays in the biology of Earth.
- Take one section at a time and then work through the companion section(s) in the study guide.

By the end of this chapter, you should be able to

- Describe the physical nature of light and the relationship between wavelength and energy and identify the wavelength range of the visible region of the spectrum as well as which colours are at the two ends and which of these are closer to the ultraviolet and infrared regions of the spectrum
- Describe the relationship between photosynthesis and the growth of various life forms: plants as well as life forms that perform cellular respiration
- Describe the various nonphotosynthetic light-absorbing pigments—those used for information. Be aware of the various levels of complexity of light-sensing units: the more basic the unit, the more basic the information and response.
- Describe the various roles light sensing plays in the biology of Earth

TOPIC MAP

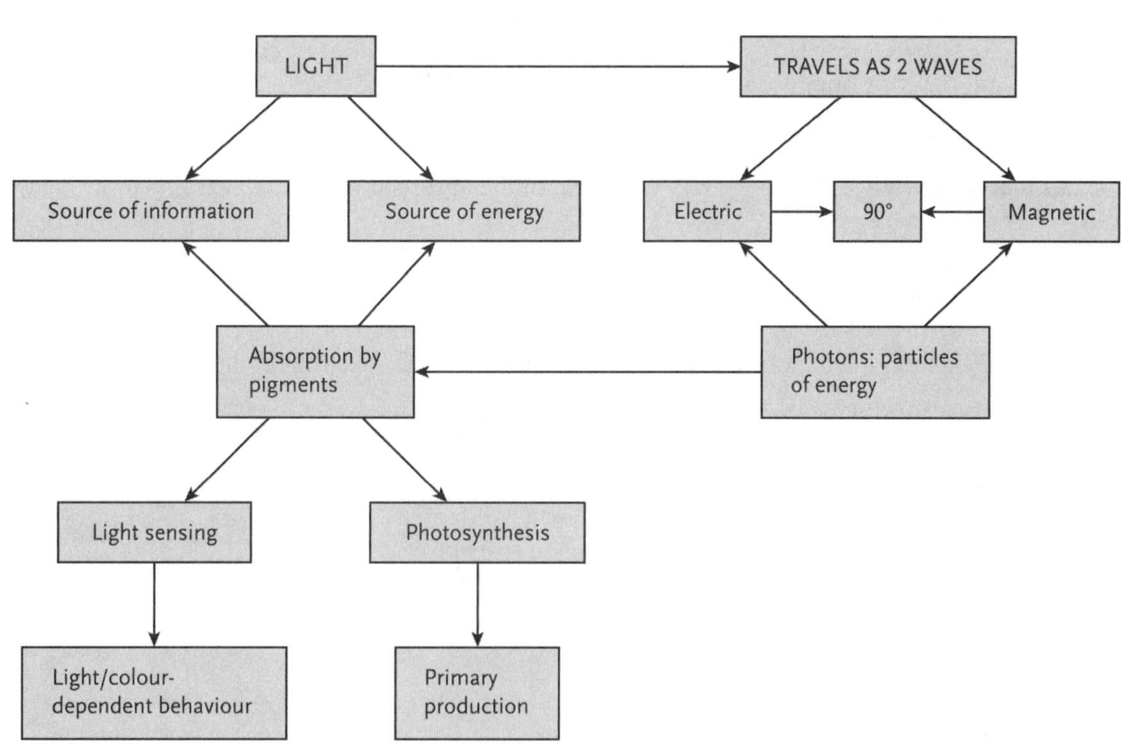

INTERACTIVE EXERCISES

1.1 The Physical Nature of Light [pp. 2–5]

Section Review (Fill-in-the-Blanks)

Light can be used as a source of (1) ____________ or (2) ________________ about the environment. The sun's energy is given off as (3) ______________ _________________ and comprises two waves, (4) _______________ and (5) __________________, oriented at _____________ degrees to each other. Light also behaves as a stream of energy particles or (6) _____________________. Although these possess no mass, they each contain a discrete amount of energy, which is (7) ___________________ related to its wavelength. When the photons hit matter, they have three possible fates: they can be (8) __________________ by the matter, (9) __________________ through the matter, or (10) ________________ by the matter. It is only in the latter case that the matter can actually use the energy of the photons. Molecules that are able to do this are called (11) ______________, and they are characterized by the presence of a (12) ______________________ system. Each absorbed photon promotes a single (13) ________________ from its lowest energy level, called its (14) _______________ state, to a higher energy, excited state. In order to be absorbed, the difference in energy between the two states must be exactly (15) ______________ that of the absorbed photon.

Sequence

16. Arrange the following components of the electromagnetic spectrum in the correct sequence, from short wavelength to long wavelength.

A. Infrared

B. Violet

C. UV

D. Red

____ ____ ____ ____

Labelling

17. Label the following diagram of light absorption by an excitable electron of a pigment molecule.

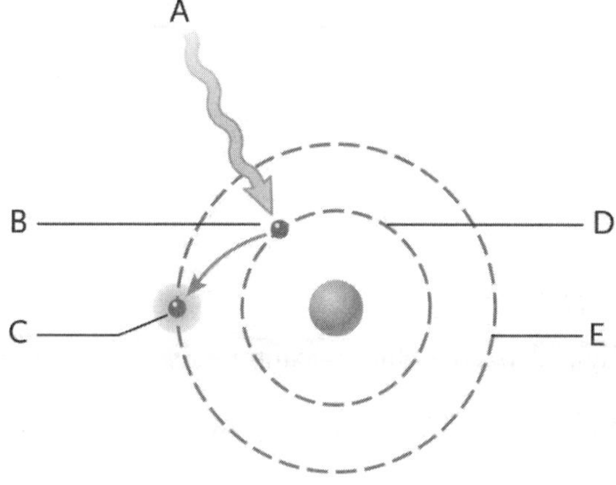

True/False

Mark if the statement is true or false. If the statement is false, provide the correct term and an explanation in the line below each statement.

18. _________ The energy of light is directly proportional to its wavelength.

____

____

19. _________ Sunlight results from the fusion of helium nuclei.

____

____

20. _________ Colour results from the inability of matter to absorb regions of the visible spectrum.

____

____

21. _________ The differences between the ground state and excited states of chlorophyll's light-absorbing electrons equal the energy of red and green wavelengths.

____

____

Matching

Match each of the following terms with its correct definition or descriptor.

22.	____	Accessory pigments	A.	A plot of wavelength versus effectiveness of a light-dependent biological process
23.	____	Electromagnetic radiation	B.	Molecules possessing alternating double and single bonds
24.	____	Ultraviolet radiation	C.	Absorb light between blue and red wavelengths for photosynthesis
25.	____	Pigment	D.	A light-dependent proton pump in the membrane of certain prokaryotes
26.	____	Action spectrum	E.	Electric and magnetic waves containing photons
27.	____	Bacteriorhodopsin	F.	Region of the electromagnetic spectrum below 400 nm wavelengths

1.2 Light as a Source of Energy [pp. 5–6]

1.3 Light as a Source of Information [pp. 6–10]

Section Review (Fill-in-the-Blanks)

An electron in its excited state is a source of (28) _________________ energy, and in photosynthesis, this is used to make energy-rich molecules such as (29) ________________ and (30) ________________. These then provide the energy for the conversion of CO_2 to (31) ______________________ and other cellular constituents. In contrast to classical photosynthetic organisms, certain prokaryotes use a light-dependent proton pump called (32) ______________ to generate a proton gradient; this gradient is then used to drive the formation of (33) _________________ by an ATP synthase in the membrane. In light-sensing organisms, light is used as a source of (34) ____________________. The most common photoreceptor in these organisms is (35) __________________. Upon absorption of light energy, the photoreceptor pigment (36) _________________ triggers signalling events that allow these organisms to respond to the light. In some light-sensing photosynthetic organisms, photoreceptors mediate (37) _______________, that is, swimming toward a light source. In plants, another type of photoreceptor, (38) ___________________, is present in the cytoplasm and triggers the developmental process of (39) __________________ when seedlings are exposed to light. In light-sensing organisms, light-sensing units vary from the most simple, the (40) __________________ of certain algae such as *C. reinhardtii,* to the image-forming (41) ____________________ eyes of insects and crustaceans and, finally, the (42) __________________ eyes of most vertebrates.

Matching

Match each of the following terms with its correct definition or descriptor.

43.	____	Phototaxis	A.	Light-sensing unit in compound eyes
44.	____	Photomorphogenesis	B.	Swimming in a light-dependent direction
45.	____	Photoreceptor	C.	Pigment molecule in rhodopsin
46.	____	Phytochrome	D.	Simple eye containing *ca.* 100 photoreceptor cells lining a cup or pit
47.	____	Retinal	E.	Back of single-lens eye containing photoreceptors
48.	____	Rhodopsin	F.	Light-sensing system
49.	____	Retina	G.	Protein associated with retinal in rhodopsin
50.	____	Ommatidia	H.	Most common photoreceptor
51.	____	Ocellus	I.	Photoreceptor in the cytosol of all plants
52.	____	Opsin	J.	Development of seedlings triggered by exposure to light

Complete the Table

53. Complete the following table.

Type of eye	*Type of organism*	*Characteristic property*
Ocellus	*Planaria*	A.
B.	Some invertebrates/most vertebrates	Image-forming "camera" eye contains cornea, lens, and retina
Compound	C.	Contains hundreds to thousands of visual units

Short Answer

Provide an explanation of the following:

54. How do eyes differ from eyespots?

____

____

55. How do eyespots differ from the ocelli of planaria?

____

____

56. What is the advantage of a compound eye over a single-lens eye?

____

____

57. What are the two initial effects of light absorption by retinal?

____

____

1.4 Light Can Damage Biological Molecules [pp. 11–13]

Section Review (Fill-in-the-Blanks)

The visible portion of the electromagnetic spectrum spans wavelengths from (58) _______________ nm at the short end to the longer wavelengths of (59) _________________ nm. One reason that life on Earth has evolved to use only this portion of the spectrum is that most of the wavelengths outside this range are (60) _______________ by molecules in Earth's atmosphere: the shorter wavelengths by the (61) _____________ of Earth's upper atmosphere and the longer ones by (62) ________________ vapour and (63) _______________. Another reason is that wavelengths in the (64) ________________ region possess enough energy to break the chemical bonds of molecules. The longer wavelengths of the (65) ________________ region of the spectrum do not have enough energy to excite the (66) ________________ of pigment molecules and, in addition, are absorbed by the (67) ____________________ molecules that comprise the major molecule in all life forms. While critical to life, visible light can also cause (68) _____________. In order to compensate for this, photosynthetic organisms have highly efficient (69) _______________ mechanisms for their photosystems. In addition, accessory pigments such as (70) __________________ absorb excess light and dissipate it as (71) ________________. In animals, the most serious damage from exposure to UV light occurs in DNA, where absorption of light results in the formation of (72) ___________________, and these can, in turn, lead to mutations and cancer. (73) _________________ is a protective pigment in the skin of humans that preferentially absorbs these wavelengths.

True/False

Mark if the statement is true or false. If the statement is false, justify your answer in the line below each statement.

74. _________ Wavelengths shorter than 400 nm do not contain enough energy to excite pigment electrons.

____

____

75. _________ The ozone layer of Earth's atmosphere absorbs the majority of UV radiation from the sun.

____

____

76. _________ Dark-skinned people living in sun-poor regions must rely on vitamin D–supplemented foods to prevent vitamin D deficiencies.

____

____

1.5 Role of Light in Ecology and Behaviour [pp. 13–18]

Section Review (Fill-in-the-Blanks)

Because of Earth's 24-hour day-night cycle, certain physiological and behavioural phenomena exhibit a 24-hour (77) _______________ rhythm. Although set by the external light environment, this rhythm does not rely on (78) _______________ changes in light but on the so-called (79) ___________ clock. In humans, this clock resides in the suprachiasmatic region of the (80) _______________. It receives information on the external light from the (81) ______________ nerve and, in turn, alters physiological and behavioural activities through changes in the levels of the hormone (82) _______________. Also related to the day-night cycle of Earth's rotation around the sun are the (83) _____________ cycles of plants and animals. These include (84) _______________ and dormancy in the former and migration and (85) ______________ of the latter. Unabsorbed light within the visible spectrum can also play a role in an organism's biology as this gives the organism its (86) ________________. (87) _______________ is a strategy to avoid detection by other animals and is based (88) _______________ and (89) ________________: an animal that has a similar (90) _______________ to its background and who remains still will be hard to detect. In contrast, animals may use (91) __________________ as a warning to other animals, while flowers use shape, smell, and (92) _____________ to attract (93) _________________. Artificial light can give rise to the phenomenon of (94) _____________________, which, in turn, can have adverse effects on (95) _____________ animals.

Labelling

96. Label the following diagram of the components associated with circadian rhythm in humans.

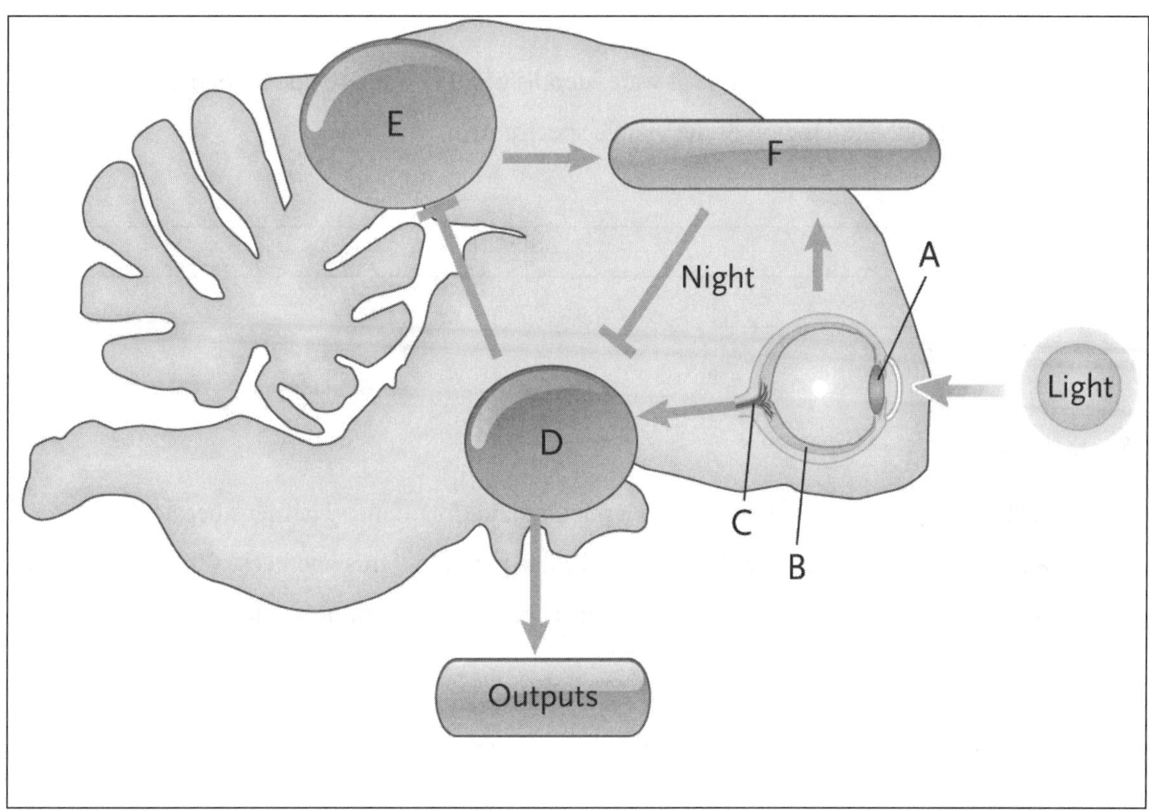

True/False

Mark if the statement is true or false. If the statement is false, justify your answer in the line below each statement.

97. ______ Only vertebrates display circadian rhythm.

____

____

____

98. ______ The pineal gland releases melatonin at night.

____

____

____

99. ______ Seasonal rhythms are set by the biological clock.

____

____

____

100. ______ Red algae are able to live at water depths of 30 m or more due to their ability to absorb light in the red region of the visible spectrum.

____

____

____

Short Answer

101. The "butterfly bush" (*Buddleia* spp.) is a popular addition to many gardens because it attracts so many different butterflies, as well as hummingbirds and bees. What aspect(s) of this flowering shrub would attract certain species of pollinators but not others?

____

____

102. Explain how, in the past, natural selection selected for the rare dark-coloured variants of the peppered moth.

____

____

1.6 Life in the Dark [p. 18]

1.7 Organisms Making Their Own Light: Bioluminescence [pp. 18–20]

Section Review (Fill-in-the-Blanks)

Some animals possess a (103) ________________ system but cannot see. One such animal is the blind (104) ________________, which lives in total (105) ________________. Although unable to see, because the (106) ______________, optic nerve, and suprachiasmatic nucleus remain functional, the animal can still detect (107) _____________, allowing it to set or entrain its (108) ________________ clock and establish a (109) ________________ rhythm. The phenomenon of (110) ________________ is based on the ability of certain life forms to produce light. The reaction is initiated by ATP, which provides the energy to drive a delocalized (111) _______________ of a substrate molecule from a ground state to a(n) (112) _______________ state. It then falls back to a ground state, and the potential energy is released as a (113) ____________ of light. This light is used for a variety of purposes, including attracting either a (114) _______________ or (115) ________________, camouflage, communication, and (116) _________________. The latter behaviour is exhibited by dinoflagellates.

Matching

Match each of the following terms with its correct definition or descriptor.

117.	____	Dinoflagellates	A.	Bioluminescence caused by fungi growing on rotten wood
118.	____	Retinitis pigmentosa	B.	Density-dependent behaviour of some species of bacteria
119.	____	Foxfire	C.	Bioluminescent algae
120.	____	Quorum sensing	D.	Age-related loss of vision

SELF-TEST

1. Which of the following is not true of light with a wavelength of 400 nm? [p. 3]

   a. Higher energy than light of 700 nm
   b. Blue region of the spectrum
   c. Closer to the infrared region than the UV region of the spectrum
   d. Contains photons of energy

2. Which of the following is not true of pigments? [pp. 3–4]

   a. Alternating single and double bonds
   b. Possess colour corresponding to the colour of absorbed light
   c. Possess delocalized electrons that absorb light energy and move to an excited state
   d. Characterized by a conjugated system

3. Chlorophyll is [pp. 3–5]

   a. Capable of absorbing light in the green region of the spectrum
   b. Involved in vision
   c. Capable of absorbing light in the green and yellow regions of the spectrum
   d. Capable of absorbing light in the blue and red regions of the spectrum

4. Which of the following is not true of photosynthesis? [pp. 5–6]

   a. Involves capture of light energy
   b. In plants, it is accompanied by the release of oxygen
   c. In plants, it provides the energy to convert oxygen to carbon dioxide
   d. Provides the energy for biosynthesis of high-energy molecules

5. Bacteriorhodopsin does not have the following characteristic: [p. 6]

   a. Found in certain prokaryotes
   b. Absorbs light energy
   c. Acts as a light-dependent proton pump
   d. Involved in certain types of photosynthesis

6. Which of the following is not true of retinal? [pp. 6–7]

   a. Common to bacteriorhodopsin and rhodopsin
   b. Changes shape when it absorbs light energy
   c. Contains a conjugated system
   d. Only found in organisms with rhodopsin

7. Which of the following structures is not associated with vision? [pp. 8–9]

   a. The ocellus
   b. A nervous system
   c. *C. reinhardtii* eyespot
   d. An ommatidium

8. Which of the following are associated with the human eye? [pp. 9–10]

   a. Ommatidia
   b. Accessory pigments
   c. Rods and cones
   d. Ocelli

9. The following is associated with photomorphogenesis: [p. 8]

   a. Phytochrome
   b. Phototaxis
   c. Flagella
   d. Retina

10. Ionizing radiation is associated with [p. 11]

    a. Infrared light
    b. Wavelengths of 700 nm or more
    c. Wavelengths in the UV region of the spectrum
    d. Absorption by atmospheric water vapour

11. Which of the following statements about electromagnetic radiation is false? [pp. 11–13]

   a. Infrared light is absorbed by water
   b. Only one portion of the UV region of the spectrum is absorbed in Earth's upper atmosphere
   c. UV light causes irreversible thymine dimers in DNA
   d. Melanin preferentially absorbs UV light

12. Which of the following statements about the circadian rhythm is false? [pp. 13–14]

   a. Exhibits a direct association with exposure to sunlight
   b. Related to the seasonal flowering of trees
   c. Found in all types of organisms
   d. Is the physiological basis of jet lag

13. Which of the following structures is not involved in the human circadian rhythm? [p. 14]

   a. Retina
   b. Melanin
   c. Suprachiasmatic nucleus
   d. Pineal gland

14. Which of the following statements about carotenoids is true? [p. 12]

   a. They protect the photosynthetic apparatus from light-induced damage
   b. They are the light-sensing pigments in the human retina
   c. They are responsible for photomorphogenesis
   d. They are the hormones responsible for the human circadian rhythm

15. Which of the following statements about peppered moths is correct? [pp. 14–15]

   a. The common form is dark (pepper coloured)
   b. It uses camouflage to avoid predation
   c. It almost became extinct during the Industrial Revolution
   d. It uses its bright colour to warn off predators

16. Which of the following is not dependent on colour perception? [pp. 14–17]

   a. Camouflage
   b. Attracting mates
   c. Warning off predators
   d. Phototaxis

17. The following is associated with the growth of red algae at greater depths in oceans: [p. 17]

   a. The production of carotenoid accessory pigments
   b. The higher energy and penetrating ability of red light
   c. The production of phytoerythrin to absorb blue light
   d. The attenuation of short-wavelength light, leaving only red light at greater depths

18. Which of the following statements about mole rats is false? [p. 18]

    a. They have eyes but are blind
    b. They live most of their lives in the dark
    c. Their biological clock is functional
    d. None of the above

19. Which of the following statements about bioluminescence is false? [pp. 18–20]

    a. It describes the excess nighttime light of urban environments
    b. It is an energy-requiring biological process
    c. It can be used to attract mates
    d. It is a form of communication among bacteria

20. The following is not an age-related vision problem: [p. 20]

    a. Phototaxis
    b. Macular degeneration
    c. Retinitis pigmentosa
    d. Retinal degeneration

INTEGRATING AND APPLYING KEY CONCEPTS

1. Explain why Darwin had difficulties in explaining the evolution of the eye and how the evolution of the eye can, in fact, be explained based on natural selection. How, then, would you explain the "backwards" evolution of the mole rat's eyes?

2. Why are carotenoids, chlorophyll, and phytochromes all critical to life on Earth?

3. Trace the path of the different wavelengths of the electromagnetic spectrum from fusion of the hydrogen nuclei to 30 m below the surface of Earth's bodies of water, explaining what happens to those wavelengths that do not penetrate to the 30 m depth of a body of water and why.

OPTIONAL ACTIVITIES

1. Mothers often try to convince their children to eat their carrots by telling them that carrots are good for vision. The basis for this statement is the presence of the pigment beta-carotene in carrots. Research how beta-carotene relates to a person's ability to see and why diet-related blindness is a significant public health problem in low-income countries.

2. In temperate countries such as Canada and the United States, the yellows and reds of deciduous trees signal the onset of fall. While one might conclude that this is just a manifestation of a decrease in the sun's intensity related to circadian rhythms, there is, in fact, considerable debate among scientists as to the reason(s) for these changes. Research some of the theories that are currently being investigated.

2 Origins of Life

CHAPTER HIGHLIGHTS

- Living and nonliving matter share the same chemical properties and obey the same laws of chemistry and physics but are distinguished by the seven unique properties of life.
- The smallest unit of life is the cell. Cells only arise from preexisting cells, and if broken, they lose all properties of life. Some organisms are unicellular, while others are multicellular and may have cells that are specialized or differentiated.
- The Earth and our solar system were formed ~4.6 billion years ago, and for at least 700 million years, our planet was devoid of life.
    - Oparin and Haldane theorized that life evolved after the abiotic or spontaneous synthesis of organic chemicals; the Miller–Urey experiment provides experimental support for this theory.
    - These reactions are dependent on the reducing conditions of primitive Earth and would have been followed by the gradual polymerization of these molecules and the formation of membrane-bound vesicles or protobionts.
    - The first life forms are thought to have existed in an "RNA world," where RNA molecules were both the molecules of information storage and the biological catalysts.
    - Alternatively, the theory of panspermia proposes that life was seeded from an extraterrestrial source.
- The earliest life forms were anaerobic prokaryotes. A critical transition occurred with the evolution of oxygenic photosynthetic bacteria. These bacteria used water as a source of electrons, allowing them to become the dominant form of life on Earth and resulting in the gradual accumulation of oxygen in Earth's atmosphere.
    - Increasing oxygen levels led to the evolution of bacteria that performed aerobic respiration, a far more energy-efficient form of metabolism.
- Eukaryotes are believed to have evolved with the in-folding of the plasma membrane of a prokaryotic ancestor, giving rise to the endomembrane system; the endosymbiosis of an aerobic respiratory bacterium, giving rise to the mitochondrion; and, subsequently, the

endosymbiosis of an oxygenic photosynthetic bacterium, giving rise to the chloroplasts of algae and plants.

- Other critical evolutionary changes in eukaryotes include the development of a cytoskeleton and, on several occasions, multicellularity, ultimately giving rise to the fungi, animals, plants, and multicellular algae.

STUDY STRATEGIES AND LEARNING OUTCOMES

- This chapter is relatively simple and mainly requires imagination to picture the prevailing conditions and evolutionary transitions believed to have led to modern-day life on Earth. Remember that the goal of this chapter is to understand the conditions on abiotic Earth and how these may have allowed for the spontaneous evolution of organic molecules, the polymerization of these monomers to yield the critical macromolecules of cells, and the assembly of membrane-bound vesicles leading ultimately to the first prokaryotic cell. From here, you need to understand how the metabolism of prokaryotes evolved, leading to the predecessor of the mitochondrion and chloroplast and the evolution of the first eukaryotic cells.
- Make sure you know the key dates and sources of evidence for the various evolutionary transitions, the names of the key hypotheses, and people who have contributed information or theories for the evolution of life on Earth.
- Also make sure that you understand the basic similarities and essential differences between prokaryotic and eukaryotic cells. Make sure you can draw a typical animal cell, identifying the basic cellular structures and the functions of each.
- Take one section at a time and then work through the companion section(s) in the study guide.

By the end of this chapter, you should be able to

- Name the seven characteristics of life and describe the cell theory
- Describe the prevailing conditions on abiotic Earth and the theories on how the first cell evolved; you should also know what the theory of panspermia proposes
- Describe how the metabolism of prokaryotes evolved, leading to the predecessor of the mitochondrion and chloroplast and the evolution of the first eukaryotic cells

- Identify the dates and sources of evidence for the various evolutionary transitions and the names of the key hypotheses and people who have contributed information or theories for the evolution of life on Earth
- Understand the basic similarities and essential differences between prokaryotic and eukaryotic cells
- Draw a typical animal cell, identifying the basic cellular structures as well as the functions of each

TOPIC MAP

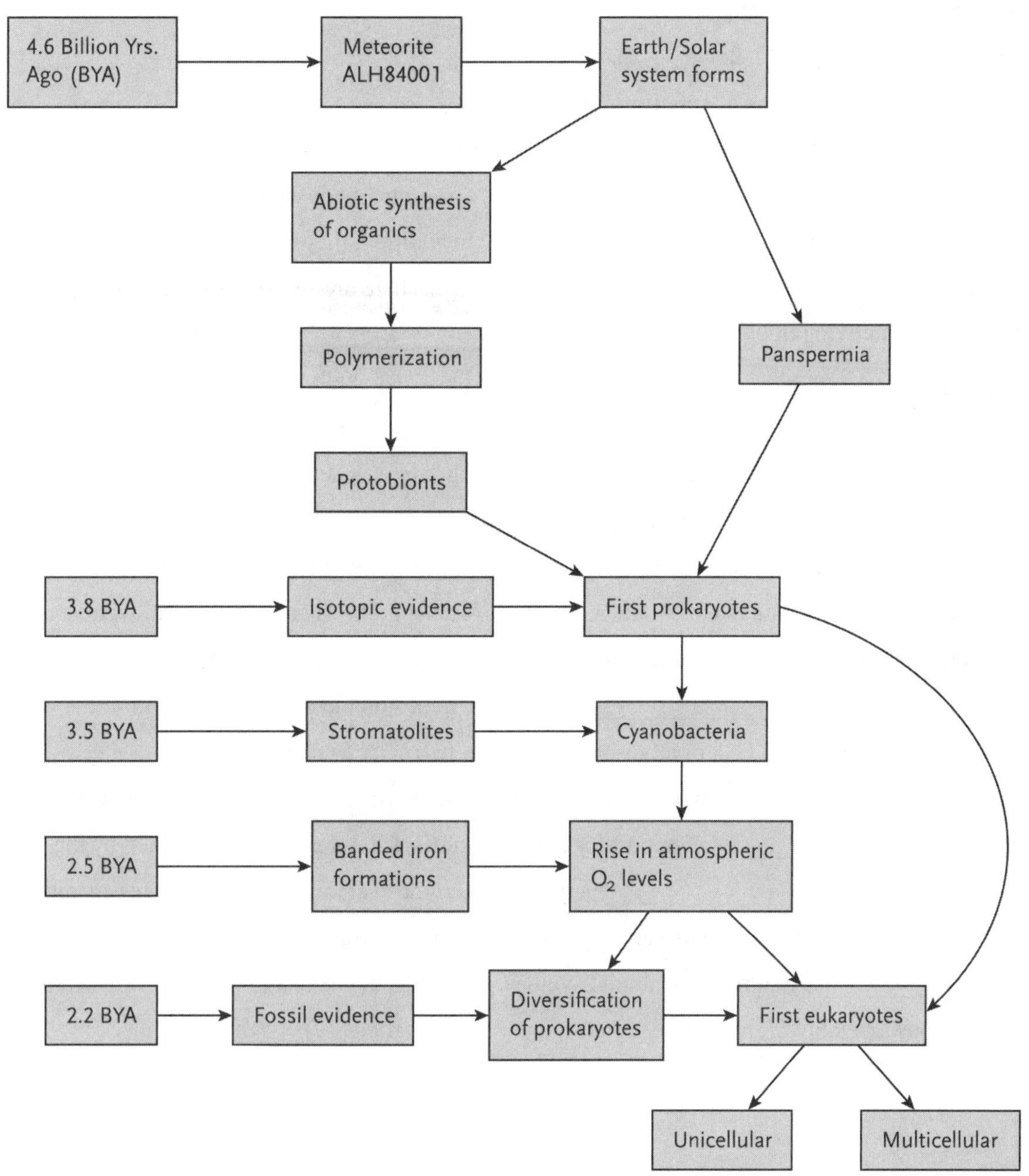

4.6 Billion Yrs. Ago (BYA)
Meteorite ALH84001
Earth/Solar system forms
Abiotic synthesis of organics
Polymerization
Protobionts
Panspermia
3.8 BYA
Isotopic evidence
First prokaryotes
3.5 BYA
Stromatolites
Cyanobacteria
2.5 BYA
Banded iron formations
Rise in atmospheric O_2 levels
2.2 BYA
Fossil evidence
Diversification of prokaryotes
First eukaryotes
Unicellular
Multicellular

INTERACTIVE EXERCISES

Why It Matters [pp. 23–24]

2.1 What Is Life? [pp. 24–25]

2.2 The Chemical Origins of Life [pp. 25–29]

Section Review (Fill-in-the-Blanks)

At the molecular level, living and nonliving (abiotic) things are made of the same (1) ______________ and (2) __________________. In addition, they both follow the same laws of (3) ________________ and (4) ________________. There are seven critical characteristics that distinguish living things: they display (5) ________________, harness and utilize (6) ____________, reproduce, respond to (7) ______________, exhibit homeostasis as well as growth and (8) _______________, and (9) _______________ in order to become better adapted to their environment. The organization of living things is described by the (10) _______________ theory. When Earth first formed, (11) _______________ billion years ago, there was no life present. It took approximately (12) __________ million years for primordial Earth to cool to temperatures that would support life. The conditions on abiotic Earth were very different from those of the present day and included high amounts of five gases: (13) ____. Based on the conditions of primordial Earth, which included the lack of atmospheric oxygen, two scientists named (14) ___________ and (15) ________________ independently proposed that organic molecules could have formed (16) ____________________. Support for this hypothesis was subsequently provided by the (17) __________________________ experiment. From those abiotically synthesized organic monomers, simple polymers have been hypothesized to have formed on solid surfaces such as (18) ___________. Assuming the abiotic synthesis of lipids, the next step would have been the spontaneous assembly of membrane vesicles, in structures called (19) ________________.

True/False

Mark if the statement is true or false. If the statement is false, justify your answer in the line below each statement.

20. ____________ One of the tenets of the cell theory is that when cells are broken open, they lose the properties of life.

____

____

21. ____________ Meteorite ALH84001 is the subject of much interest because it may have brought life to Earth.

____

____

22. ____________ Ozone was present on primordial Earth and was a critical component of the Miller–Urey experiment.

____

____

23. ____________ Deep sea vents emit heated water and high amounts of ammonia and methane and so may have been the site of the evolution of organics and early life.

____

____

Matching

Match each of the following terms with its correct descriptor or relevance to the origins of life on Earth.

24.	____	Homeostasis	A.	Spontaneously formed membrane-bound vesicles that may have been the predecessors to the first cells
25.	____	Oparin–Haldane	B.	The name for the experiment that replicated conditions on early Earth and demonstrated the abiotic synthesis of organic molecules
26.	____	Miller–Urey	C.	The name for the hypothesis that organic molecules were synthesized abiotically on primordial Earth
27.	____	Clay	D.	The process of maintaining constant internal conditions
28.	____	Protobionts	E.	A solid material that can store potential energy and may have been the site of polymerization of abiotically synthesized organic monomers

Sequence

29. Arrange the following proposed steps for the origin of life on Earth.

A. Aggregation of complex organic molecules inside a membrane-bound protobiont
B. Abiotic synthesis of organic molecules
C. Assembly of complex organic macromolecules
D. Condensation of interstellar gases

____ ____ ____ ____

Complete the Table

30. Complete the following table.

Characteristic of life	*Description*
A.	Requires energy to maintain and the cell as the fundamental unit
Harness and utilize energy	B.
C.	Generation of more of the same kind
D.	Changes in external conditions trigger adjustments in structure, function, and behaviour
Homeostasis	E.
Growth and development	F.
G.	Changes over generations to become better adapted to the environment

Short Answer

31. Why do scientists believe that the ancient atmosphere lacked oxygen and was a reducing environment?

____

____

32. Describe the features that viruses share with living things.

____

____

33. Discuss the results of the original Miller–Urey experiment and subsequent improvements.

____

____

2.3 The Origins of Information and Metabolism [pp. 29–31]

Section Review (Fill-in-the-Blanks)

All life on Earth exhibits the same flow of information, from (34) _______________ to (35) _________________ to (36) _______________. In considering how life evolved from protobionts, one must be able to answer how this flow evolved given that the end product, (37) _________________, catalyzes each step of the process. A likely answer came with the discovery of (38) __________________ by Thomas Czech. Like proteins, these molecules are able to fold into specific (39) _____________, which, in turn, determine their function. Early life is therefore theorized to have used (40) ______________ to serve both as the source of molecular (41) ____________________ and as a biological (42) ___________________. In this primordial "RNA world," it has been hypothesized that some RNA molecules began to synthesize simple (43) ______________________. Such cells would have had a (44) ___________________ advantage since (45) ________________ are far stronger catalysts than ribozymes and have a far greater structural (46) _________________. This change would have been followed by the evolution of (47) _______________. Selection would have favoured this molecule for information storage because the presence of deoxyribose makes this molecule more chemically (48) _________________, as does the replacement of uracil with (49) _________________. In addition, the (50) ______________________-stranded structure would allow cells to use the (51) _________________ strand to repair mutations. While the information system of cells was evolving, so was the cellular (52) __________________, which probably began with simple oxidation– (53) ________________ reactions and eventually became more complex with multistep (54) _______________ reactions allowing for stepwise energy release.

Short Answer

55. Discuss the connection between Thomas Czech and the proposed "RNA world."

____

____

56. Why are proteins better molecules for catalyzing chemical reactions than ribozymes?

____

____

57. Compare the structure of DNA with that of RNA.

____

____

58. Describe the process of evolutionary change.

____

____

True/False

Mark if the statement is true or false. If the statement is false, justify your answer in the line below each statement.

59. ____________ The catalyst for the polymerization of modern-day proteins is a ribozyme.

____

____

60. ____________ Because DNA contains the molecular instructions for the cell and also directs its own replication, purified chromosomes would be able to direct the synthesis of a new cell.

____

____

61. ___________ If RNA molecules were normally double-stranded, they would not be able to function as ribozymes.

____

____

62. ____________ Cellular respiration was probably one of the earliest methods of energy generation.

____

____

63. ____________ L1 ligase is an in vitro–generated ribozyme that can catalyze the polymerization of RNA from its monomeric building blocks.

____

____

2.4 Early Life [pp. 32–35]

Section Review (Fill-in-the-Blanks)

Indirect isotopic evidence based on the (64) ________________ composition of ancient rocks dates the first life forms to (65) ________________ billion years ago. More direct fossil evidence of early life comes from structures called (66) __________________ that date the appearance of photosynthetic bacteria called (67) __________________ to (68) ___________ billion years ago. In contrast to the idea that life evolved spontaneously after the abiotic synthesis of organics, (69) ___________________ postulates that early life on Earth had an extraterrestrial origin. Support for this theory comes from two observations: (i) the extremely short timeframe between when Earth had (70) ________________ to a level appropriate for life and the earliest evidence for life and (ii) the evidence that structures called (71) ____________________ that are formed by some prokaryotes can be revived to active growth after exposure to the extreme conditions typical of space as well as to extremely long periods of (72) ___________________. Regardless of how those first life forms arose on primordial Earth, the first forms were the simple (73) _____________________. Although the lack of a (74) _____________ and other features makes these cells strikingly different from eukaryotes, they possess certain features that are universal among all cells on Earth. They are all surrounded by a (75) ___________________, which separates the external environment from the (76) ________________ of the cell. In prokaryotes, the former is also the site of energy generation via the (77) ____________________ of certain molecules or by harnessing the energy of sunlight through (78) ___________________. In contrast, in eukaryotes, these energy-harvesting processes take place in (79) __________________ called mitochondria and chloroplasts, respectively. Other characteristics that are universal to all life forms include the organization of the DNA into (80)

__________________ and the processes of information flow starting with the (81) ___________________ of DNA into RNA and the (82) _________________ of proteins from an RNA template. The atmosphere of early Earth was reducing due to the lack of molecular (83) _______________. Evidence for rising oxygen levels comes from a type of sedimentary rock called (84) ___________________, which started to form around (85) _____________ billion years ago. The source of this gas was the (86) _______________, which evolved the ability to oxidize water and perform (87) ________________ photosynthesis. This provided a tremendous advantage as water is far more (88) ________________ than the other photosynthetic electron donors, allowing these organisms to become the dominant life forms on Earth. The rise in atmospheric oxygen was critical to the subsequent evolution of (89) __________________ cells.

Matching

Match each of the following terms with its correct descriptor or relevance to the origins of life on Earth.

90.	____	Stromatolites	A.	The structure that separates the external environment from a cell's cytoplasm
91.	____	Prokaryotes	B.	The theory that life on Earth had an extraterrestrial origin
92.	____	Plasma membrane	C.	A process of energy generation that oxidizes water for electrons
93.	____	Nucleoid	D.	Bacteria and Archaea
94.	____	Oxygenic photosynthesis	E.	Layered rock structures created by cyanobacteria
95.	____	Panspermia	F.	The central region in prokaryotic cells containing the cell's DNA

Short Answer

96. Discuss how scientists can determine whether carbon deposits are of biological origin.

____

____

97. Compare and contrast the stromatolites of Western Australia's Shark Bay with those from primordial Earth.

____

____

True/False

Mark if the statement is true or false. If the statement is false, justify your answer in the line below each statement.

98. ____________Photosynthetic bacteria that used water as an electron donor were likely the first life forms.

____

____

99. ____________Oxygen was present in Earth's atmosphere from the time the solar system and Earth formed.

____

____

100. ____________ Modern bacterial spores can survive exposure to radiation, desiccation, freezing, and extreme heat and could theoretically survive space travel.

____

____

101. ____________ Although much smaller and structurally simpler than eukaryotes, prokaryotes possess far greater metabolic flexibility and are far more numerous on Earth.

____

____

Labelling

Identify each numbered part of the following illustration.

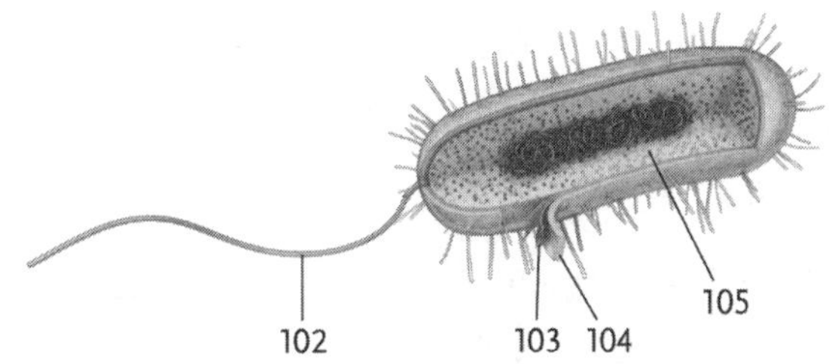

102. ____________

103. ____________

104. ____________

105. ____________

2.5 Eukaryotic Cells [pp. 35–43]

Section Review (Fill-in-the-Blanks)

Although prokaryotes and eukaryotes share certain universal features, eukaryotic cells are distinguished by the fact that the DNA is surrounded by a (106) ________________ envelope, the presence in the cytoplasm of (107) ____________________-bound compartments called (108) __________________, as well as highly specialized (109) __________________ proteins for movement. The endomembrane system is believed to have evolved by (110) __________________ of the (111) __________________ membrane. Components of this system include the (112) __________________ membrane, the endoplasmic (113) __________________, and the (114) _____________ complex. Among the organelles characteristic of eukaryotic cells are the energy-transducing (115) ________________ and (116) ______________________. According to the theory of (117) ___________________, these two organelles arose when free-living (118) ___________________ were engulfed by larger (119) ________________ cells, forming an initial mutually (120) _______________ relationship called a symbiosis. It is believed that slowly, over time, endosymbiotic cells became inseparable from the host cells, with the result that those endosymbionts that derived their energy from (121) ___________________ respiration became the (122) ________________ while the (123) ________________ photosynthetic endosymbionts became the (124)

__________________. Because virtually all eukaryotic cells possess (125) ____________________ but only the algae and green plants possess (126) ______________________, it is believed that the process of endosymbiosis occurred in (127) _____________. Another characteristic eukaryotic structure is the (128)___________________, which, at least in part, is responsible for the characteristic shape and internal organization of these cells. This is an interconnected system of protein (129) _________________ and (130) _________________. Cytoskeleton components are also involved in (131) _________________ both within the cell and of the cell itself. After the evolution of the eukaryotic cell, the evolution of (132) _________________ organisms is arguably the most profound transition in the evolution of life on Earth and one for which there is very little (133) ________________ evidence. However this happened, some researchers argue that the evolution of life on Earth was inevitable given the (134) _________________ and (135) ________________ conditions established by Earth's origins.

Short Answer

136. What kind of evidence is used to support the endosymbiont theory?

____

____

137. Which group of organisms is hypothesized as the ancestor of eukaryotic cells?

____

____

138. Compare and contrast eukaryotic cilia and flagella.

____

____

139. Discuss how motor proteins work.

____

____

True/False

Mark if the statement is true or false. If the statement is false, justify your answer in the line below each statement.

140.____________ The ER, Golgi, and nuclear membranes developed by endosymbiosis.

141.____________ All multicellular eukaryotes developed from the same unicellular organism.

142.____________ Mitochondria and chloroplasts contain ribosomes that resemble those of prokaryotes.

143. ____________ The flagella of eukaryotes evolved from the prokaryotic flagellum.

144. ____________ Not all eukaryotic ribosomes are found on the surface of the rough ER.

Matching

Match each of the following structures with its correct definition.

145.	____	Organelles	A.	Small, organized structures that carry out specialized functions inside the cell
146.	____	Smooth ER	B.	Internal membranous sacs that divide the cell into structural and functional compartments
147.	____	Ribosome	C.	Membrane-enclosed region of DNA
148.	____	Nucleus	D.	Non–membrane-bound organelles that assemble amino acids into proteins
149.	____	Plasma membrane	E.	Site of lipid synthesis
150.	____	Flagella	F.	Synthesis and transport of proteins
151.	____	Cilia	G.	Small membrane-bound compartments for transfer of material within the endomembrane system
152.	____	Endomembrane	H.	Lipid bilayer that serves as boundary of cell system
153.	____	Vesicles	I.	Modifies and distributes proteins
154.	____	Rough ER	J.	Long, hairlike structures on the cell that function in movement
155.	____	Golgi complex	K.	Motile structures that can occur in large numbers on cells and move by beating

Labelling

Identify each numbered part of the following illustration.

156. ______________

157. ______________

158. ______________

159. ______________

160. ______________

161. ______________

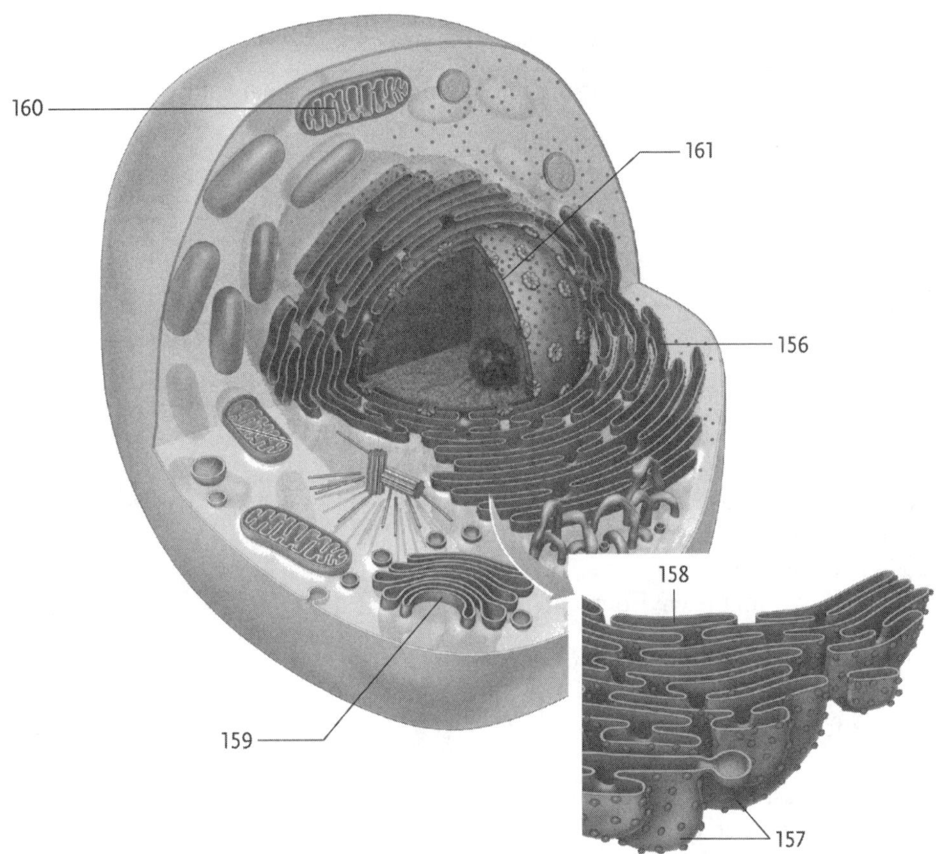

Complete the Table

162. Complete the following table.

Structure or process	*Function or purpose*
Exocytosis	A.
B.	The process of bringing material into the cell by invaginating and pinching off small portions of the plasma membrane
Secretory vesicle	C.
Microfilaments	D.
E.	Responsible for separating chromosomes during cell division and for the movement of flagella and cilia
F.	Part of the cytoskeleton of animal cells but not plant cells

SELF-TEST

1. Which of the following is not a characteristic of all living things? [pp. 24–25]

    a. They are able to reproduce
    b. They evolve
    c. They contain nucleic acids
    d. They extract energy from the environment

2. Which of the following is not a component of the cell theory? [p. 25]

    a. New cells arise from preexisting cells
    b. Cells contain genetic material
    c. The cell is the smallest unit of life
    d. Living things are composed of cells

3. Earth is estimated to have formed ______ billion years ago. [p. 26]

    a. 4.6
    b. 3.5
    c. 2.2
    d. 1.0

4. According to the Oparin–Haldane hypothesis, the abiotic synthesis of organic chemicals required which of the following atmospheric gases? [p. 27]

   a. H_2O, H_2, CH_4, and NH_3
   b. O_2, H_2, and H_2O
   c. O_2, N_2, H_2, and H_2O
   d. CH_4, NH_3, and O_2

5. The absence of which of the following in the primitive atmosphere is critical to the abiotic synthesis of organic chemicals? [p. 27]

   a. O_2
   b. N_2
   c. H_2O
   d. H_2

6. Which of the following is true for the experiments done by Miller and Urey? [p. 27]

   a. Their experiments did not require water
   b. They were able to use ozone to create organic chemicals
   c. They were able to make organic chemicals in the presence of oxygen
   d. They were able to make cells in their apparatus
   e. They were able to make organic chemicals using a reducing environment and electricity

7. Which of the following supports the origin of organic chemicals near the hydrothermal vents of the ocean floor? [p. 28]

   a. They support a wide range of life
   b. They release significant amounts of geothermally heated water, oxygen, and nitrogen gas
   c. They are characterized by high temperatures and reducing conditions
   d. They release significant amounts of geothermally heated water, methane, and ammonia

8. On primitive Earth, which of the following is believed to have been responsible for storage of information and support of chemical reactions? [pp. 29–30]

   a. DNA
   b. RNA
   c. Proteins
   d. Ribosomes

9. The oldest fossil of a prokaryote in stromatolite structures is estimated to be ________ billion years old. [p. 32]

   a. 4.6
   b. 3.5
   c. 2.2
   d. 1.0

10. Which of the following statements about prokaryotic and eukaryotic cells is true? [p. 33]

a. Prokaryotic cells have a plasma membrane, unlike eukaryotic cells
b. Eukaryotic cells have DNA, unlike prokaryotic cells
c. Eukaryotic cells have a nucleus, but prokaryotic cells do not
d. Prokaryotic cells have ribosomes, unlike eukaryotic cells

11. Which of the following statements regarding aerobic respiration is false? [p. 35]

a. It involves the oxidation of food molecules for energy generation
b. It involves the use of oxygen as a final electron acceptor
c. It occurs in the absence of oxygen
d. It was only possible after the evolution of oxygenic photosynthesis

12. Which of the following structures may have originated by in-folding of the plasma membrane? [pp. 35–36]

a. The mitochondrial membrane and ER
b. The ER and nuclear membrane
c. The mitochondria and chloroplasts
d. The Golgi complex and ribosomes

13. All of the following are part of the endomembrane system except [p. 36]

a. The endoplasmic reticulum
b. The nuclear envelope
c. The plasma membrane
d. The Golgi complex

14. The Golgi complex ________. [p. 36]

a. Modifies proteins
b. Produces lipids
c. Transports proteins to the nucleus
d. Assembles proteins using the mRNA template

15. Which of the following structures may have originated by an endosymbiotic event? [pp. 36–37]

a. The mitochondrial membrane and ER
b. The ER and nuclear membrane
c. The mitochondria and chloroplasts
d. The Golgi complex and ribosomes

16. Which of the following statements about mitochondria and chloroplasts is false? [pp. 36–38]

   a. They are both found in plant cells
   b. They are both found in animal cells
   c. They both contain DNA
   d. They both originated from ancient prokaryotes

17. Which of the following is not a key characteristic that supports the endosymbiotic origin of mitochondria and chloroplasts from prokaryotic ancestors? [p. 38]

   a. The presence of DNA
   b. The similarities of their ribosomes
   c. The use of electron transport chains to generate ATP
   d. Division by binary fission
   e. None of the above

18. The cytoskeleton __________. [pp. 38–39]

   a. Gives the cell shape
   b. Is composed of protein
   c. Includes microtubules
   d. Moves with the help of motor proteins
   e. All of the above

19. Cells are small because, in part, as size increases, the surface area to volume ratio _____. [pp. 40–42]

   a. Increases
   b. Doubles
   c. Decreases
   d. Remains constant

20. The oldest fossil of a eukaryote is estimated to be ________ billion years old. [p. 42]

   a. 4.6
   b. 3.5
   c. 2.2
   d. 1.0

INTEGRATING AND APPLYING KEY CONCEPTS

1. Discuss why the evolution of oxygenic photosynthetic bacteria was a critical and essential transition leading ultimately to the evolution of eukaryotic cells.
2. Identify the inherent problem cells face as they increase in size and describe the structural and functional differences in eukaryotic cells that permit their much larger size compared with the average size of their prokaryotic ancestors.
3. Try putting together your own idea map of the contents of this chapter.

OPTIONAL ACTIVITIES

1. Astrobiology is a recently developed multidisciplinary field of study that evolved out of the space race between the U.S. and the U.S.S.R. in the 1950s and 1960s. Research how the space race began and eventually led to this now international field of research.

2. In general, cells are small because of the surface area :volume changes with increasing size; however, a few notable exceptions include the "giant bacterium" *Thiomargarita namibiensis* and the eggs of birds. Research what is known about how these cells overcome the inherent problem of having such a relatively small surface area to volume ratio.

3. While it is commonly believed that all bacteria are unicellular, the *Myxobacteria* are a group of bacteria that have a multicellular and differentiated stage of growth. Research the role of these bacteria in nature and as research models.

3 Selection, Biodiversity, and Biosphere

CHAPTER HIGHLIGHTS

- The diversity of life is tremendous and is the product of selection leading to evolution.
  - Darwin's theory of evolution states that all life evolved from a common ancestor and the diversity is the result of variation in heritable traits combined with selective pressures that affect survival.
  - Those traits that provide a selective advantage allow organisms with those traits to survive and reproduce, thus altering the incidence of those traits from one generation to another.
- Selective pressures include abiotic factors in the biosphere: that is, the physical and chemical conditions anywhere where life is found, from the ocean floor to Earth's atmosphere.
  - The abiotic factors are ultimately the result of three things: Earth's spherical shape and the resulting latitudinal variation in the intensity of solar radiation, Earth's rotation on its axis, and Earth's rotation around the sun.
  - These all combine to create latitudinal differences in temperature, seasonal variations in the intensity and duration of sunlight, global wind patterns, ocean currents, and regional variations in precipitation.
- Selective pressures also include biotic interactions, such as symbiotic interactions: these can manifest as mutualism or parasitism.
- The history of life on Earth is characterized by certain evolutionary breakthroughs that resulted in adaptive radiations. Arguably, the two most dramatic examples are the evolution of oxygenic photosynthesis and terrestrial colonization.

STUDY STRATEGIES AND LEARNING OUTCOMES

- This chapter is not particularly complicated but has a lot of nonbiological detail. Make sure you learn the basic messages, paying attention to the section headings. Be careful not to get bogged down with the physical and meteorological details.
- If you lose track of a point, refer back to the idea map below.
- Take one section at a time and then work through the companion section(s) in the study guide. The study guide questions focus on the details that are pertinent to biological systems and processes.

By the end of this chapter, you should be able to

- Describe the modes of nutrition among living organisms
- Explain the role of selection in the generation of diversity of life on Earth, providing examples
- Distinguish between the "diversity" and the "unity" of life on Earth and relate both to the theory of evolution
- Describe Darwin's theory of evolution
- Explain how the evolution of oxygenic photosynthesis made the diversity of life on Earth possible
- Identify the abiotic factors that affect the diversity of life at local, regional, and global levels and explain how Earth's shape, axis, and rotation all impact major abiotic factors

TOPIC MAP

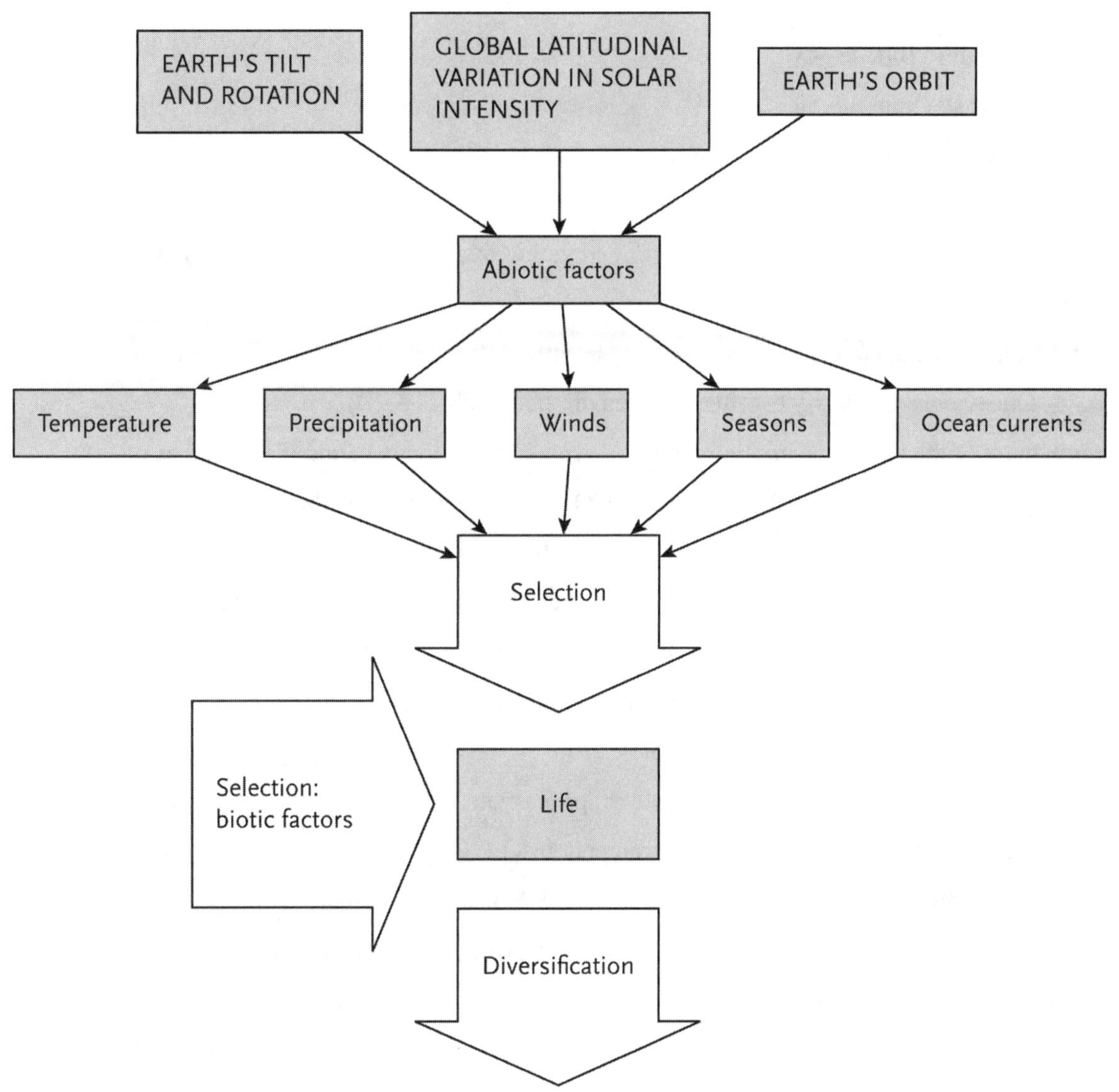

EARTH'S TILT AND ROTATION
GLOBAL LATITUDINAL VARIATION IN SOLAR INTENSITY
EARTH'S ORBIT
Abiotic factors
Temperature
Precipitation
Winds
Seasons
Ocean currents
Selection
Selection: biotic factors
Life
Diversification

INTERACTIVE EXERCISES

Why It Matters [pp. 47–48]

3.1 Biodiversity [pp. 48–50]

3.2 Selection [pp. 50–52]

Section Review (Fill-in-the-Blanks)

The biodiversity of life on Earth is vast, reflecting the diversity of environments where life exists, extending from the ocean (1) __________________ to the (2) __________________. Diversity, that is, the (3) __________________ types of organisms, can be measured many ways. One way is to classify organisms based on their sources of (4) __________________, a component of the backbone of all organic molecules within the cell, and their sources of (5) __________________. Organisms that use CO_2, an inorganic molecule, as their carbon source are called (6) __________________, while those that use organic carbon sources are called (7) __________________; organisms that derive their energy from the sun are called (8) __________________, whereas those that oxidize chemicals for energy are called (9) __________________. These are combined into four nutritional categories, two of which are only exhibited by the prokaryotes: (10) __________________ and (11) __________________. There are, however, prokaryotes that fall into the remaining two categories, and this is a reflection of the tremendous (12) __________________ diversity exhibited by this group of organisms. The driving factor for the incredible diversity of life on Earth is (13) __________________ of one form over another by a (14) __________________ factor. This factor is called the (15) __________________ pressure, and it affects some organisms and not others due to (16) __________________ variation within the population. A modern example of this phenomenon often occurs in hospital or health care settings, with (17) __________________ bacteria. Other examples include the evolution of the causative agent of the relatively new human disease (18) __________________ from an older, less serious, and non–sexually transmitted disease; the evolution of entirely new types of animals, as in the case of the evolution of the (19) __________________ from an extinct hippolike ancestor; or the evolution of the incredible diversity among the (20) __________________ plants.

Matching

Match each of the following terms with its correct descriptor or relevance to the origins of life on Earth.

21.	____	Cells	A.	Organisms that use CO_2 for carbon and oxidize chemicals for energy
22.	____	Populations	B.	Organisms that use CO_2 for carbon and harness the sun's energy
23.	____	Communities	C.	Members of the same species living in the same space
24.	____	Ecosystems	D.	Organisms that use organic molecules for carbon and harness the sun's energy
25.	____	Biosphere	E.	The smallest unit of life
26.	____	Chemoheterotroph	F.	Groups of communities interacting with their shared physical environment
27.	____	Chemoautotroph	G.	A group of unrelated organisms living in the same place
28.	____	Photoautotroph	H.	Organisms that use organics for carbon and energy
29.	____	Photoheterotroph	I.	All regions of Earth, including the atmosphere, that support life

Complete the Table

30. Complete the following table.

Nutritional category	*Carbon source*	*Energy source*	*Example life forms*
Photoheterotroph	Organics	Sunlight	A.
Photoautotroph	B.	Sunlight	Plants, algae, cyanobacteria
Chemoheterotroph	Organics	C.	Animals, some prokaryotes, fungi, protozoa, and some plants
D.	CO_2	Chemical	Only some prokaryotes

Sequence

31. Arrange the following hierarchical levels of life from simplest to most complex.

A. Cell

B. Community

C. Biosphere

D. Molecule

E. Organism

F. Population

G. Ecosystem

H. Organelles

____ ____ ____ ____ ____ ____ ____ ____

3.3 Evolution [pp. 52–55]

Section Review (Fill-in-the-Blanks)

According to Darwin's theory of (32) ____________________, the result of selection is the gradual change in characteristics of a population of organisms over time. This explains both the (33) ______________________ of life and, as discussed in the previous section, the (34) ______________________ of life on Earth. The former refers to the idea that all life on Earth is descended from a (35) ____________________ ancestor and therefore shares critical features such as the use of (36) ________________ for energy, (37) __________________ for information storage, and plasma membranes made of lipid (38) ____________________. The theory is based on the idea that individuals within a population exhibit genetic (39) __________________, and those traits that provide an advantage in a particular environment will allow those individuals not just to survive but to (40) ____________________, giving rise to more (41) _________________ with the same traits and, ultimately, to changes in the (42) _________________ of those traits within the population. Extraordinary diversification of life results from evolutionary breakthroughs called (43) ___________________ radiations. One dramatic example is the evolution of (44) ________________________ photosynthesis, a change that had a tremendous impact on Earth's atmosphere. As discussed in the previous chapter, the resulting increase in atmospheric oxygen levels in turn resulted in the evolution of (45) _____________________ respiration and the development of the (46) _________________ layer, which began to filter the sun's UV radiation. The results were the evolution of the eukaryotes, as previously discussed, and the colonization of (47) __________________ environments.

True/False

Mark if the statement is true or false. If the statement is false, justify your answer in the line below each statement.

48. ____________ Ammonia is not excreted by land animals because it is toxic, especially when not diluted by water.

____

____

49. _____________ Terrestrial animals expend more of their total energy budget acquiring oxygen than aquatic animals.

____

____

50. _____________ Molecular genetics is a useful tool for determining evolutionary relationships.

____

____

Short Answer

51. Discuss the changes in body structure necessitated by the movement of plants and animals to land.

52. Summarize the reasons why the evolution of oxygenic photosynthesis is considered an adaptive radiation.

Labelling

53. Label the following parts of the tree of life.

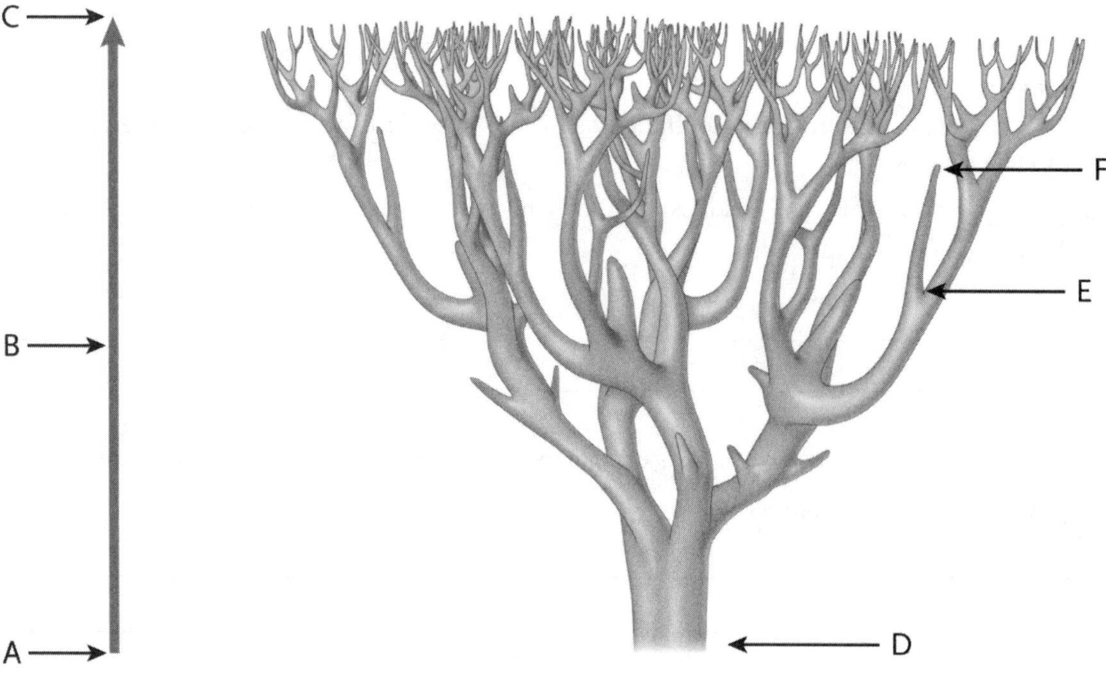

3.4 The Biosphere [pp. 56–61]

Section Review (Fill-in-the-Blanks)

The different physical environments in the biosphere vary in their (54) ______________ factors, which, in turn, combine to contribute to the region's (55) _________________, and this influences the evolution and (56) ________________ of organisms. The three main factors that ultimately affect all of the above derive from (i) Earth's spherical shape and a resultant (57) ___________________ variation in incoming (58) _________________ radiation, (ii) Earth's rotation on its (59) __________________, and (iii) its orbit around the (60) ________________. Because of the planet's shape, sunlight hits equatorial latitudes at (61) __________ degrees and is therefore much stronger, whereas further north and south, it travels a greater distance through the (62) __________________ and strikes at an (63) ________________ angle, therefore shining over a larger (64) _______________ and being intense. In addition, because Earth has a fixed (65) ________________ relative to its orbit around the (66) _________________, certain regions experience latitudinal variations in (67) ___________________ and (68) ___________________ of incoming solar radiation. The result is that regions in the (69) ____________________ hemisphere receive more sunlight with longer and warmer days in June, while regions in the (70) ________________ hemisphere experience warmer, longer days in December. These variations are most dramatic at the (71) _______________ and are least pronounced at the (72) ____________________. Regional differences in solar radiation give rise to uneven (73) ____________ of air masses, which gives rise to regional differences in (74) ____________ circulation, that is, (75) ___________. Another source of regional circulation differences is the (76) _____________ variation in the speed of Earth's (77) ________________ on its axis, which causes a resultant (78) ______________________ of the prevailing north–south path of air movements to a predominantly (79) ________________ direction in equatorial regions. Ultimately, all of these things, including regional wind patterns, have an effect on (80) ________________.

Short Answer

81. List the abiotic factors that affect evolution and diversity.

____

____

82. What other factors, in addition to latitudinal variations in solar radiation and regional differences in wind patterns, affect rainfall?

____

____

83. Define the phenomenon "adiabatic cooling" and explain the conditions that cause it.

____

____

True/False

Mark if the statement is true or false. If the statement is false, justify your answer in the line below each statement.

84. ____________ The tropical trade winds are a direct result of the Coriolis effect.

____

____

85. ____________ The continental climate is modified by ocean currents.

____

____

86. ____________ The rainy season experienced by tropical/equatorial regions is explained by the Coriolis effect.

____

____

3.5 Biotic Factors [pp. 61–63]

3.6 Cumulative Impact on Biotic and Abiotic Factors [pp. 63–64]

Matching

Match each of the following terms with its correct descriptor or example life form.

87.	____	Producers	A.	Organisms that eat other organisms
88.	____	Photoautotrophs	B.	Organisms that use organics for energy and carbon
89.	____	Consumers	C.	Harmful organisms that live inside a host organism
90.	____	Heterotrophs	D.	A form of symbiosis in which both partners benefit
91.	____	Decomposer	E.	Usually photoautotrophs
92.	____	Endoparasitism	F.	Harmful organisms that grow on the surface of a host organism
93.	____	Ectoparasitism	G.	Most plants except those that are achlorophyllous
94.	____	Mutualism	H.	Fungi

True/False

Mark if the statement is true or false. If the statement is false, justify your answer in the line below each statement.

95. The Barcode of Life project uses the DNA sequence of a gene associated with the chloroplast to identify species and help document the diversity of life on Earth.

____

____

96. ____________ Ammonia is both toxic for life and essential for life.

____

____

SELF-TEST

1. Which of the following taxonomic groups has the highest number of characterized species? [p. 50]

    a. Insects
    b. Fungi
    c. Algae
    d. Higher plants

2. Which group of organisms would exhibit a photoheterotrophic mode of nutrition? [p. 50]

    a. Most plants
    b. Fungi
    c. Some bacteria
    d. Animals

3. Which of the following statements is true? [p. 50]

    a. Autotophs use sunlight for energy
    b. Phototrophs use CO_2 for their carbon source
    c. Heterotrophs use CO_2 for their energy source
    d. Autotrophs use CO_2 for their carbon source

4. Which of the following statements is true? [pp. 51–52]
    a. Whales evolved from hippos
    b. Whales evolved from *Treponema pallidum*
    c. The ancestor of whales likely had hoofs
    d. The ancestor of whales was a recently discovered fish

5. Which of the following is not part of Darwin's theory of evolution? [pp. 52–53]

    a. Individuals vary in heritable traits
    b. There are generally more individuals in an area than the natural resources will support
    c. Those organisms that have a selective advantage are more likely to survive and reproduce
    d. Different life forms evolved from different ancestors

6. Which of the following would not be classified as an adaptive radiation? [pp. 51–53]

    a. The evolution of the syphilis bacterium from the bacterium that caused yaws
    b. The evolution of flowering plants
    c. The evolution of oxygenic photosynthesis
    d. The movement of plants onto land

7. Which of the following did not result from the evolution of oxygenic photosynthesis? [p. 53]

    a. The accumulation of oxygen in the atmosphere
    b. The increase in UV radiation on Earth's surface
    c. The evolution of aerobic respiration
    d. Terrestrial colonization

8. Which of the following is not an adaptation that accompanied terrestrial colonization? [p. 54]

    a. The excretion of urea waste products
    b. The excretion of ammonia waste products
    c. The development of skeletons
    d. The development of leaves on plants

9. The root of the tree of life represents ______________. [p. 55]

    a. The common ancestor for life
    b. Present time
    c. An extinct lineage
    d. Time

10. Which of the following is not an abiotic factor? [p. 56]

    a. Sunlight
    b. Parasitism
    c. Temperature
    d. Precipitation

11. Which of the following does not influence the global pattern of environmental diversity? [p. 56]

    a. Latitudinal variation in the intensity of solar radiation
    b. Earth's rotation around its axis
    c. Symbiotic interactions
    d. Earth's orbit around the sun

12. At the equator, the following conditions prevail: [pp. 56–58]

    a. Predominantly east–west winds
    b. Significant seasonal variation in daylight and temperature
    c. Predominantly north–south winds
    d. The angle of the sun's rays is oblique

13. The trade winds are the direct result of ____________. [pp. 57–58]

    a. Adiabatic cooling
    b. Latitudinal variation in the speed of Earth's rotation
    c. Seasonal variations in the duration and intensity of sunlight
    d. All of the above

14. Which of the following is not involved in the latitudinal variations in precipitation? [pp. 58–60]

    a. Latitudinal variation in the intensity of solar radiation
    b. Global wind patterns
    c. Adiabatic cooling
    d. All of the above

15. Adiabatic cooling is defined as ______________. [p. 58]

    a. A loss of heat without the loss of heat energy
    b. The rising and subsequent cooling of warm air masses
    c. The cooling of water at greater ocean depths
    d. The moderation of temperatures by coastal ocean currents

16. Which of the following does not play a role in the generation of ocean currents? [p. 59]

    a. Prevailing winds
    b. Positions of land masses
    c. Adiabatic cooling
    d. Latitudinal variation in the intensity of solar radiation

17. Which of the following has the greatest effect on survival and reproduction? [p. 60]

    a. Climate
    b. Microclimate
    c. Prevailing winds
    d. Regional precipitation patterns

18. Which of the following statements about sea slugs is true? [p. 61]

    a. They use chloroplasts for energy
    b. They are animals
    c. They are heterotrophic
    d. All of the above

19. Which of the following statements about plants is true? [p. 62]

    a. Some may be heterotrophic
    b. Some may be parasitic
    c. They are generally photoautotrophs
    d. All of the above

20. Which of the following best describes ectoparasitism? [p. 62]

    a. Harmful organisms that live on the surface of others
    b. Harmful organisms that live inside a host
    c. A symbiotic relationship where both partners are harmed
    d. Typical of the organisms in the rumen of ruminant animals such as cows

INTEGRATING AND APPLYING KEY CONCEPTS

1. Summarize how Earth's rotation around its axis, tilt, rotation around the sun, and spherical shape all combine to influence global temperature, wind and precipitation patterns, seasons, and ocean currents.

2. Define the term "adaptive radiation." Given that all life is believed to have derived from a common ancestor, discuss the kinds of adaptive radiations that must have been required in plants and animals to allow colonization in (a) the Arctic, (b) deserts, and (c) rain forests.

OPTIONAL ACTIVITIES

1. Human activities have given rise to documented increases in atmospheric levels of greenhouse gases such as CO_2, N_2O, and CH_4, which are implicated in global climate change. Research the most important sources of these gases and the kinds of approaches that are being used or proposed to mitigate these emissions.

2. Mycorrhizas are mutualistic fungi that grow in association with the roots/rhizosphere of certain plants. Some of these produce reproductive organs, which are considered edible delicacies. Research the applications of plant mycorrhizas in the food industry and for plant growth promotion.

4 Energy and Enzymes

CHAPTER HIGHLIGHTS

- Energy exists in two interchangeable forms: kinetic and potential energy.
    - The molecules in living cells have both forms of energy.
- The laws of thermodynamics explain energy flow.
- Metabolic reactions can be divided into
    - Catabolic reactions that break down molecules to release energy (exergonic reactions)
    - Anabolic reactions that use energy (endergonic reactions) to create larger molecules
- Energy released during exergonic reactions is stored as ATP.
- Living things use ATP as an immediate source of energy to support cellular functions.
- Enzymes
    - Are biological catalysts that speed up chemical reactions
    - Reduce the activation energy required to start a reaction
    - Catalyze a specific reaction and have a specific substrate and specific product(s)
    - Are not changed during a reaction and so are used repeatedly and made in very small numbers
    - May need cofactors and coenzymes for their function
    - Are affected by substrate concentration, temperature, and pH
    - Are often regulated by activators or inhibitors

STUDY STRATEGIES AND LEARNING OUTCOMES

- Take one section at a time and then work through the companion section(s) in the study guide.

By the end of this chapter, you should be able to

- Understand how energy and enzymes are involved in the chemical reactions of living systems
- Identify the two different classes of energy and the sources of energy for living organisms and explain into which class of energy these sources fall
- Explain the key concepts of thermodynamics—energy release and conversions—and describe the factors that determine whether an energy transformation will release energy or require energy
- Explain which types of metabolic reactions generally release energy and which normally require an input of energy
- Understand the thermodynamic basis for why living systems need a continuous supply of energy
- Describe how hydrolysis of ATP releases energy and supports cellular reactions
- Understand enzymes—their chemical nature, how they catalyze chemical reactions, their regulation, and why they are necessary components of any living system

TOPIC MAP

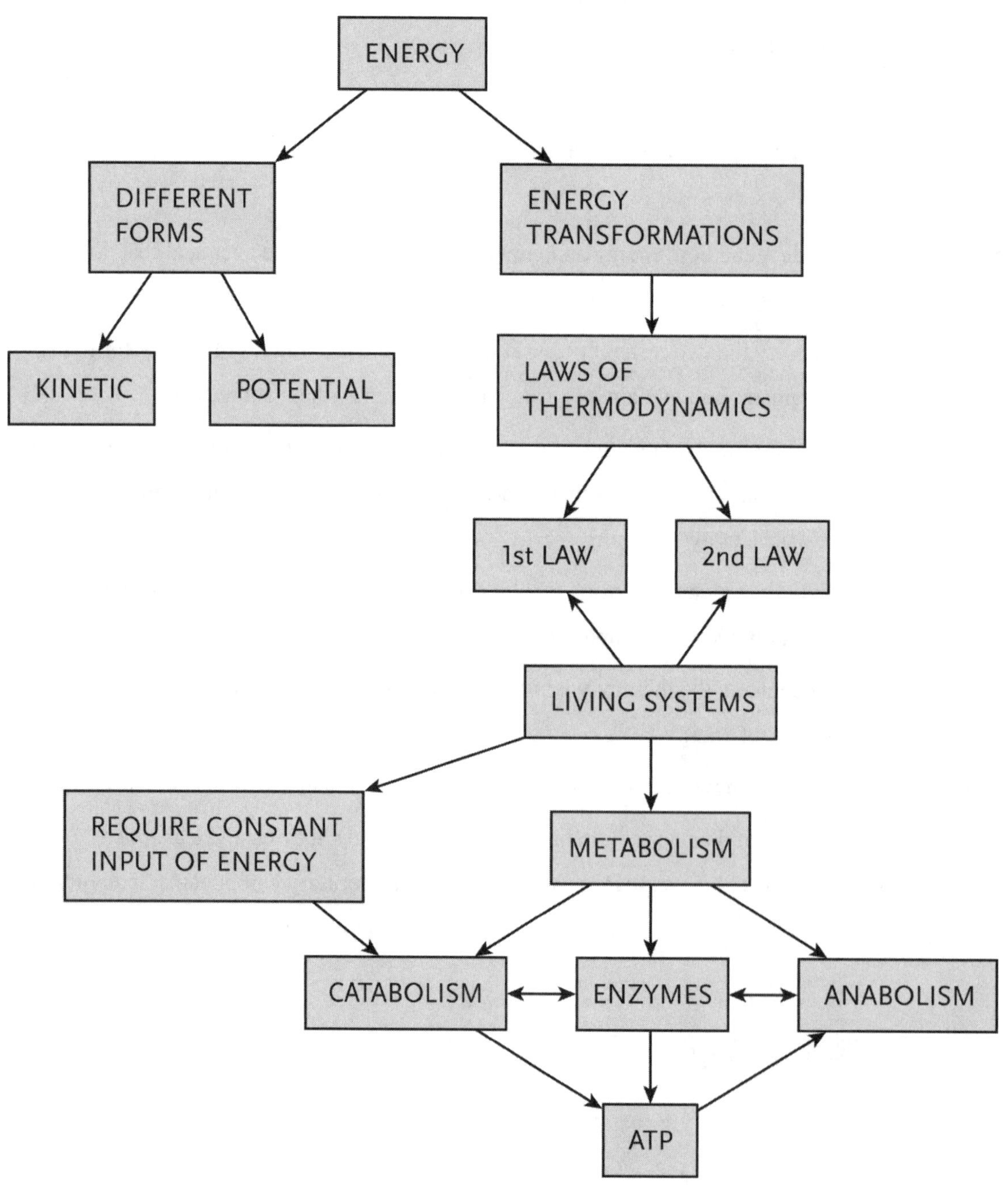
ENERGY
DIFFERENT FORMS
ENERGY TRANSFORMATIONS
KINETIC
POTENTIAL
LAWS OF THERMODYNAMICS
1st LAW
2nd LAW
LIVING SYSTEMS
REQUIRE CONSTANT INPUT OF ENERGY
METABOLISM
CATABOLISM
ENZYMES
ANABOLISM
ATP

INTERACTIVE EXERCISES

Why It Matters [pp. 71–72]

Section Review (Fill-in-the-Blanks)

As an alternative to using lethal temperatures to speed up chemical reaction rates, life forms use catalysts called (1) _______________. These molecules are so efficient that they have been found to speed up reactions many (2) ________________ of times. Without enzymes to speed up otherwise slow reactions, (3) _________________ could not exist.

4.1 Energy and the Laws of Thermodynamics [pp. 72–75]

Section Review (Fill-in-the-Blanks)

(4) ____________ is the capacity to do work and is classified as being in one of two states; (5) _______________ energy is the energy of motion, whereas (6) _____________ energy is the stored energy resulting from position or chemical structure. (7) ___________________ is the study of energy and its transformations. In these studies, the object being studied is called the (8) ____________ and everything else is called the (9) ______________. According to the first law of thermodynamics, also called the principle of the (10) _______________________ of energy, energy can be transformed or transferred, but neither (11) ______________ nor (12) _______________. According to the second law of thermodynamics, each time energy is transformed, there is an increase in disorder or (13) _______________. The measure of this disorder is called (14) ____________________, and it is denoted by the letter *S*. The second law therefore states that the total disorder, that is, (15) ___________________, of a system *and* its (16) _________________ always increase. Thus, while cells must maintain a high degree of order, the tendency is for them to physically (17) ___________________, and this can only be prevented with an input of (18) ________________. Meanwhile, by maintaining order and therefore low entropy at the cellular level, heat and chemical by-products are lost to the environment, thus increasing the (19) ___________________ of the universe or surroundings.

Matching

Match each of the following terms with its correct definition or descriptor.

20.	____	Energy	A.	A measure of randomness
21.	____	Thermodynamics	B.	Refers to stored energy
22.	____	Earth	C.	Open system(s)
23.	____	Heat	D.	The ability to do work
24.	____	Kinetic energy	E.	Refers to the energy due to the motion of an object
25.	____	Entropy	F.	Closed system(s)
26.	____	Potential energy	G.	The study of energy and its transformations
27.	____	Organisms	H.	The energy of random molecular motion

Short Answer

Explain how each of the following represents a source of potential energy.

28. A hamburger:

____

____

29. A skier at the top of a ski run:

____

____

30. Water at the top of a dam or waterfall:

____

____

31. Name the different types of kinetic energy:

____

____

For each of the following, provide a definition and an example.

32. An isolated system:

____

____

33. An open system:

____

____

34. A closed system:

____

____

Identification

For each of the following forms of kinetic energy, identify what it is that is moving.

35. Electricity:

____

____

36. Heat:

____

____

37. Light:

____

____

38. Ultraviolet radiation:

____

____

4.2 Free Energy and Spontaneous Reactions [pp. 75–78]

Section Review (Fill-in-the-Blanks)

Whether a reaction will happen spontaneously or not is influenced by whether the products of the reaction have less (39) ____________________ energy than the reactants as well as whether the reaction generates products that are more (40) ___________________ than the reactants. The thermodynamic term for chemical or physical reactions that take place without an input of energy is (41) ________________. The term for reactions that instead require an input of energy to proceed is (42) _______________. According to the second law of thermodynamics, not all of a system's potential energy or (43) ___________ (i.e., H) can be extracted to do work; some of the energy will be lost to an increase in (44) ____________________ (i.e., S). The (45) __________ energy (i.e., G) is that proportion that remains and is available to do work. To predict whether a reaction will result in a release of energy, one calculates the ΔG using the formula (46) _________________. If the ΔG is (47) ____________ than zero, then the reaction will proceed spontaneously and is (48) ___________________. Unstable systems have higher (49) _______________________ than stable ones, so spontaneous reactions naturally progress from an unstable state to a stable one until a state of (50) ____________________ is reached. This applies only to a(n) (51) _______________ system where there is no addition of reactants and products are not removed. In living systems, however, reactants are constantly being replenished and products used up, so the ΔG is always (52) _________________.

Matching

Match each of the following terms with its correct definition.

53.	____	Free energy	A.	Reactions that break down larger molecules into smaller chemicals
54.	____	Metabolism	B.	$\Delta G = 0$
55.	____	Endergonic	C.	Reactions that build larger molecules
56.	____	Catabolism	D.	Potential energy of a system
57.	____	ΔG	E.	Refers to reactions that release energy
58.	____	Anabolism	F.	Reactions that take place without outside help
59.	____	Exergonic	G.	Refers to all chemical reactions that take place in living things
60.	____	Spontaneous reactions	H.	Refers to reactions that require energy
61.	____	Equilibrium	I.	Refers to the energy available to do work
62.	____	Enthalpy	J.	Change in free energy

True/False

Mark if the statement is true or false. If the statement is false, provide the correct term and an explanation in the line below each statement.

63. _________ Cellular respiration is an anabolic pathway.

____

____

64. _________ Photosynthesis is an exergonic process.

____

____

65. _________ The melting of ice at room temperature is an endergonic process.

____

____

66. _________ Diffusion is a process with a negative ΔG.

____

____

Short Answer

67. Using thermodynamic terms, explain how one would calculate whether a chemical reaction is going to be spontaneous.

____

____

68. Explain why catabolic pathways are exergonic and anabolic pathways are endergonic.

____

____

69. Explain why equilibrium is attained for chemical reactions in closed systems but not in living organisms.

____

____

4.3 The Energy Currency of the Cell: ATP [pp. 79–81]

Section Review (Fill-in-the-Blanks)

The assembly of complex molecules from simpler ones has a (70) _____________ ΔG and requires an input of energy. The cell supplies the energy to drive these reactions through the (71) ___________________ of ATP. When ATP is hydrolyzed to ADP, the terminal phosphate is released, relieving the (72) ________________ of adjacent negative charges. The result is the release of large amounts of (73) __________. The reaction therefore has a negative ΔG and is (74) ______________________. In order to conserve the energy from the hydrolyzed phosphate bond, enzymes transfer the terminal (75) ___________________________ to a reactant molecule of an endergonic reaction. This phosphorylation causes the molecule to become less (76) _________________ and therefore more reactive. The coupling of the (77) _________________ of ATP to an endergonic biosynthetic reaction results in an overall (78) ___________________ reaction. Regeneration of ATP from ADP is a(n) (79) ______________ reaction and does not occur spontaneously. This reaction is coupled with the breakdown of complex molecules such as (80) ______________ by the cell. The continued breakdown and synthesis of ATP is called the (81) ___________________________.

Matching

Match each of the following terms with its correct definition or descriptor.

82.	____	Phosphorylation	A.	Exergonic reaction
83.	____	ATP	B.	Where an exergonic reaction is connected to an endergonic reaction
84.	____	Energy coupling	C.	Anabolic reaction
85.	____	ATP hydrolysis	D.	Addition of a phosphate group to a molecule
86.	____	Synthesis glutamine	E.	Primary coupling agent in all living things

Short Answer

87. What are coupled reactions?

____

____

88. Why are coupled reactions so important for life?

____

____

89. Give a short description of ATP structure and its role in coupled reactions.

____

____

90. Explain how shivering helps maintain body heat.

____

____

4.4 The Role of Enzymes in Biological Reactions [pp. 81–85]

Section Review (Fill-in-the-Blanks)

Although exergonic chemical reactions release free energy, they may not start spontaneously because a bond must be (91) ______________ and therefore less stable before it can be (92) ____________________. The energy required to create this less stable state is called the (93) _____________ energy. A reactant that has acquired this energy is more reactive and is said to be in a(n) (94) ____________ state. (95) _____________ increase the speed of a reaction by (96) _______________ the activation energy of the reaction. They do not take part in the reactions themselves and therefore are referred to as (97) ___________. They do not alter the (98) _______________________ of the reactants or products and therefore do not alter the (99) _________________ of the reaction. Enzymes are specific for a single type of (100) __________________, and this specificity is currently explained by the (101) __________________ hypothesis based on the finding that enzymes undergo (102) ________________ changes to best fit their substrates. Enzymes help attain the transition state by using their (103) __________ sites to bring the (104) ________________ closer together; providing (105) ___________ environments that promote catalysis; and changing the (106) ____________________ of the reactant(s) to help distort them toward the (107) ______________ state.

Matching

Match each of the following terms with its correct definition or descriptor.

108.	____	Cofactor	A.	Organic chemicals that help enzymes
109.	____	Catalyst	B.	Substances that speed up chemical reactions without changing themselves
110.	____	Activation energy	C.	A temporary state of enzyme binding with the substrate
111.	____	Active site	D.	Catalyzing only a specific chemical reaction
112.	____	Coenzyme	E.	These are often metals and are required by some enzymes
113.	____	Enzyme specificity	F.	Required to start a reaction
114.	____	Transition state	G.	The location on an enzyme where the substrate binds

Complete the Table

115. There are three mechanisms that contribute to the formation of the transition state. Complete the following table with specific information about these mechanisms.

A.
B.
C.

True/False

Mark if the statement is true or false. If the statement is false, provide the correct term and an explanation in the line below each statement.

116. _________ A spark is used to get propane to burn because the ΔG of its combustion is positive.

____

____

117. __________ The Michaelis–Menten equation is used to measure the rates of enzyme reactions.

____

____

118. __________ The activation energy of chemical reactions can be provided by heat.

____

____

4.5 Conditions and Factors that Affect Enzyme Activity [pp. 85–89]

Section Review (Fill-in-the-Blanks)

The activity of enzymes can be affected by a variety of factors. Increasing the concentration of substrate affects the rate of catalysis by increasing the rate of (119) _____________; however, the rate reaches a maximum when the active sites are (120) _______________. Molecules that slow or stop enzymatic reactions are called (121) ________________, and their effects are either (122) _________________ or (123) ________________, depending upon the strength of binding. (124) ____________________ inhibitors act by (125) _____________ the natural substrate and therefore (126) ___________ with the natural substrate for the active site. (127) _____________ inhibitors are molecules that bind away from the active site and generally act by altering the (128) __________________ of the enzyme. In (129) ________________ regulation of metabolic pathways, certain (130) __________________ may bind the allosteric site of the enzyme and, depending upon the (131) _______________________ changes that result from binding, may cause inhibition or (132) ________________________. In (133) _______________ inhibition, the end product of the pathway is the allosteric regulator. Enzyme activity is also affected by temperature and pH, with (134) ___________ activity being restricted to a narrow range for each.

Identification

Identify how each of the following types of inhibition or regulation is involved in the functioning of metabolic pathways.

135. Competitive inhibitor:

____

____

136. Noncompetitive inhibitor:

____

____

137. Allosteric regulation:

____

____

138. Feedback inhibition:

____

____

Identify how each of the following types of physical or chemical factors affects enzyme function.

139. Increasing temperature:

____

____

140. pH:

____

____

True/False

Mark if the statement is true or false. If the statement is false, justify your answer in the line below each statement.

141._____________ Cyanide binds irreversibly to an enzyme that is essential for cellular respiration.

____

____

142._____________ Enzymatic inhibitors bind the active site.

____

____

143. _____________ An allosteric inhibitor converts an allosteric enzyme to the high-affinity conformation.

____

____

SELF-TEST

1. Enzymes are _____________ molecules. [p. 83]

   a. Carbohydrate
   b. Protein
   c. Lipid
   d. RNA

2. Which of the following is not a component of ATP? [p. 79]

   a. Ribose
   b. Adenine
   c. P_i
   d. Phosphate groups

3. In which form is the energy that is stored in glucose? [p. 72]

   a. Potential energy
   b. Radiation energy
   c. Kinetic energy
   d. Electrical energy

4. Which of the following refers to the disorder in a system? [p. 74]

   a. Enthalpy
   b. Anabolism
   c. Free energy
   d. Entropy

5. All of the following would be true for a chemical reaction that has a negative ΔG except [p. 76]

   a. The reaction will be spontaneous
   b. The reaction releases free energy
   c. The reaction will be endergonic
   d. In an isolated system, the reaction would run to equilibrium

6. A place on the enzyme where a substrate binds is called the [p. 83]

   a. Regulatory site
   b. Allosteric site
   c. Active site
   d. Substrate site

7. All of the following are true for an enzyme except [p. 83]

   a. They catalyze a specific reaction
   b. They are unchanged by the reaction
   c. They temporarily combine with the substrate
   d. They make an endergonic reaction proceed spontaneously

8. Which of the following has a shape similar to an enzyme's substrate? [p. 86]

   a. Competitive inhibitor
   b. Allosteric inhibitor
   c. Noncompetitive inhibitor
   d. Coenzyme

9. Which of the following is an inorganic enzyme helper? [p. 84]

   a. Coenzyme
   b. Activator
   c. Cofactor
   d. Substrate

10. Which of the following is an organic enzyme helper? [p. 84]

    a. Coenzyme
    b. Inhibitor
    c. Cofactor
    d. Allosteric activator

11. Which of the following is true regarding spontaneous reactions? [p. 76]

   a. They have a positive ΔG
   b. They tend to happen when the products are more disordered than the reactants
   c. They tend to happen when the reactants have less potential energy than the products
   d. They are endergonic

12. Which of the following correctly describes a catalyst? [p. 83]

   a. It increases the activation energy of a reaction
   b. It undergoes a chemical change during a reaction
   c. It speeds up a chemical reaction
   d. It is always an inorganic molecule

13. All of the following are factors that affect enzyme activity except [p. 85]

   a. Temperature
   b. pH
   c. Substrate concentration
   d. Free energy

14. All of the following are true for a coupled reaction except [p. 80]

   a. They link exergonic and endergonic reactions
   b. They have an overall positive ΔG
   c. They are spontaneous
   d. They release free energy

15. All of the following statements about feedback inhibition are false except [p. 88]

   a. The inhibitor induces a conformational change from the low- to the high-affinity state
   b. It involves allosteric binding of pathway end products to early enzymes of a pathway
   c. It occurs when the binding site of an enzyme is saturated
   d. It results from competition for the active site

16. What is the approximate rate of ATP cycling in a typical cell? [p. 81]

   a. 10 million ATP/sec
   b. 10 billion ATP/sec
   c. 1000 ATP/sec
   d. 10 ATP/sec

17. Which of the following statements describes the enzyme cycle? [pp. 83–84]

   a. Enzymes are specific for a single type of molecule
   b. After catalysis, enzymes are released unchanged to bind another substrate
   c. Enzymes are continuously synthesized
   d. The active site cycles between high affinity and low affinity

18. The Michaelis–Menten equation is used to [p. 82]

    a. Determine the ΔG of an enzymatic reaction
    b. Determine the enthalpy of a substrate
    c. Calculate the rate of an enzymatic reaction
    d. Determine the entropy of a substrate

19. All of the following statements about activation energy are false except [p. 81]

    a. It is a kinetic barrier to spontaneous reactions
    b. It is provided by enzymes
    c. It is the energy required to hydrolyze ATP
    d. It is energy associated with the active site

20. Many cellular poisons, including cyanide, are toxic to cells because [p. 87]

    a. They bind reversibly to enzyme active sites
    b. They are competitive inhibitors
    c. They are allosteric inhibitors of essential enzymes
    d. They bind essential enzymes irreversibly

INTEGRATING AND APPLYING KEY CONCEPTS

1. Penicillin was the first antibiotic. Discuss how this drug acts and relates to the topics in this chapter. Using penicillin as an example, consider how antibiotics are selectively toxic to the infectious agent and not the host's cells.

2. Explain why ATP is called the primary coupling agent. Consider why ADP might not provide as much energy as a coupling agent compared to ATP.

3. Explain how heat can be used to speed up a chemical reaction and why life must rely on enzymes rather than heat to accomplish this.

OPTIONAL ACTIVITIES

1. Humans rely on many different types of energy for running their equipment, homes, offices, industries, and cars. Some of these sources of energy are renewable and some are not. Identify a number of examples of each and compare and contrast the generation of energy using nonrenewable resources with those that are renewable. Why are we currently unable to rely entirely on renewable resources?

2. The term "inborn error of metabolism" was coined in the early twentieth century to describe genetic diseases affecting the enzymes of metabolic pathways. One such disease is hereditary hemochromatosis, the most common genetic disorder in Canada, and is fatal if undetected. Investigate the basis for this disease, the symptoms, prevalence, and treatment(s).

5 Membranes and Transport

CHAPTER HIGHLIGHTS

- The plasma membrane is a selectively permeable boundary separating the inside of a cell from the outside.
- The composition and organization of membranes underlie their functions.
    - The structure is described by the fluid mosaic model.
    - Lipid composition and the presence of sterols influence the fluidity.
    - Proteins determine the particular characteristics and functions of the membrane.
- The plasma membrane controls the movement of substances into and out of the cell.
    - Material may move into or out of the cell by various energy-independent and energy-requiring processes.
    - Large substrates move into eukaryotic cells by endocytosis or, if particularly big, by phagocytosis.
    - Large substrates move out of eukaryotic cells by exocytosis.

STUDY STRATEGIES AND LEARNING OUTCOMES

- Review the topic map below to get an overview of the chapter.
- Do not try to work through all of this material in one setting. Do one section at a time, paying attention to new terms and processes, and then work through the corresponding section of the study guide.
- After completing the sections in the study guide, without looking at the topic map below, try to draw your own topic map, depicting the relationships between and among key concepts.
- Make sure you are able to clearly distinguish between all of the various forms of transport.

By the end of this chapter, you should be able to

- Describe the fluid mosaic model and how it relates to membrane transport and membrane structure
- Explain what makes the plasma membrane so important and how its structure can be altered in response to changes in temperature
- Clearly distinguish between all of the various forms of passive and active transport
- Explain the differences between endocytosis, exocytosis, and phagocytosis

TOPIC MAP

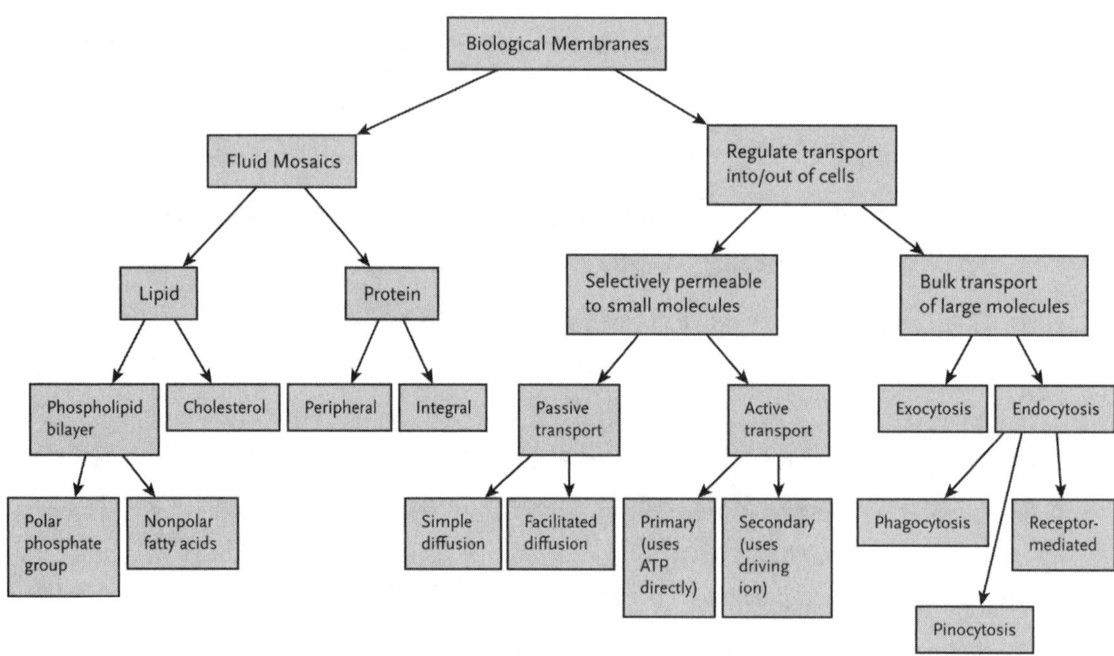

INTERACTIVE EXERCISES

Why It Matters [pp. 93–94]

5.1 An Overview of the Structure of Membranes [pp. 94–95]

Section Review (Fill-in-the-Blanks)

Cystic fibrosis is caused by (1) ________________ to a gene for a (2) ______________ transport protein. This protein resides in the plasma membrane and normally transports (3) ______________ out of lung and intestinal mucosal epithelial cells. When mutated, the normal (4) ______________ gradient established by these cells is destroyed. The reason such a balance can be established is the fact that the plasma membrane is a (5) ________________ permeable barrier. Its structure and that of other cellular membranes are described by the (6) ______________ ________________ model because molecules within the membrane can (7) ______________ and because membranes contain a wide assortment of (8) ______________, each with a specific function. These functions include (9) ________________, as described for the CF protein, attachment, as well as (10) ______________ for such processes as electron (11) ______________. These and other components of membranes are different between the two halves of the lipid (12) __________________, an organization that is referred to as membrane (13) __________________.

Short Answer

14. Describe the structure of membranes.

____

____

15. Describe the Frye–Edidin experiment.

____

____

True/False

Mark if the statement is true or false. If the statement is false, justify your answer in the line below each statement.

16. ________ Experimental demonstration that membranes are fluid was provided by the freeze-fracture technique in combination with electron microscopy.

____

____

5.2 The Lipid Fabric of a Membrane [pp. 96–99]

Section Review (Fill-in-the-Blanks)

The dominant lipids in membranes are (17) __________________, which consist of two (18) ______________ _______________ tails linked to one of several types of (19) __________________ or amino acids by a (20) __________________ group. The former makes this region very (21) _____________________ or nonpolar, whereas the latter is charged and therefore (22) _________________ or polar. Having both in the same molecule means that these molecules are (23) _____________________, a feature that is critical to the structure and function of biological membranes. In aqueous solution, this feature also means that the (24) _______________ structure of membranes forms spontaneously. Both the temperature and the lipid (25) __________________ determine the fluidity of the membrane. If the temperature drops below a certain point, the phospholipids become (26) ________________ packed and the membrane loses its fluid state, becoming a highly viscous semisolid (27) _______________. If the temperature is too high, membranes may become too (28) _______________, resulting in membrane (29) __________________, which may be lethal to the cell. The fluidity of membranes is primarily a function of the degree of (30) _______________ of the lipid fatty acids. The more (31) _________________ the membrane, the lower the gelling temperature. The explanation for this is that each (32) ___________ ____________ introduces bends or kinks in the fatty acid chain, creating more (33) ______________ between neighbouring lipids. Enzymes called (34) _______________ are responsible for controlling the fluidity of the membrane. They introduce (35) ________________ bonds between certain (36) _____________ atoms of the fatty acid tails. Because membrane fluidity is so critical for cell function and viability, many organisms can adjust the fatty acid (37) ________________ of their membranes to remain fluid over a broad range of (38) ______________. With decreasing temperatures, these organisms (39) _______________ the expression of the genes for these enzymes. Membrane fluidity is also affected by molecules called sterols, which act as membrane (40) _______________, helping to prevent membrane (41) _______________ at low temperatures and reducing (42) ___________________ at high temperatures.

Matching

Match each of the following terms with its correct descriptor or definition.

43.	____	Unsaturated fatty acid	A.	An organization that develops spontaneously and gives one of the lowest energy states possible
44.	____	Hydrophilic	B.	Possessing both polar and nonpolar regions in the same molecule
45.	____	Hydrophobic	C.	Contains linear carbon chain with one or more carbon–carbon double bonds
46.	____	Amphipathic	D.	Molecules found in the membranes of animals but not plants
47.	____	Desaturases	E.	Polar
48.	____	Lipid bilayer	F.	Nonpolar
49.	____	Cholesterol	G.	Enzymes that create double bonds in fatty acids

True/False

Mark if the statement is true or false. If the statement is false, justify your answer in the line below each statement.

50. ____________ Membranes with higher amounts of saturated fatty acids will remain fluid and functional at higher temperatures.

____

____

51. _____________ Desaturases remove hydrogen atoms from the phosphate groups of phospholipids.

____

____

52. ____________ At low temperatures, sterols help restrain the movement of membrane lipids.

____

____

Short Answer

53. Explain how the presence of double bonds in the fatty acids of phospholipids influences membrane fluidity.

____

____

54. Explain why detergents are able to remove oil stains from clothes.

____

____

5.3 Membrane Proteins [pp. 99–101]

Section Review (Fill-in-the-Blanks)

It is the nature of the membrane (55) ______________ that determines its function and makes each membrane unique. (56) ________________ proteins are responsible for controlling the movement of molecules into and out of the cell. Those with (57) ___________________ activity are critical for such processes as respiratory and photosynthetic (58) ___________________ transport. Receptor proteins are responsible for (59) _______________ _______________, which, upon binding of specific molecules such as hormones on the outside, trigger changes inside the cell. Finally, proteins on either side of the membrane can act as (60) _________________ points for cytoskeleton proteins or play a role in cell–cell (61) ________________. Proteins embedded in the lipid bilayer are called (62) _______________ membrane proteins, while those on either surface of the membrane are called (63) __________________ membrane proteins.

Complete the Table

64. Complete the following table.

Membrane protein	*Structural characteristics*
Peripheral	A.
Integral	B.

Short Answer

65. List the major functions of proteins associated with the membrane.

____

____

Labelling

66. Identify each labelled part in the following illustration of an animal cell plasma membrane.

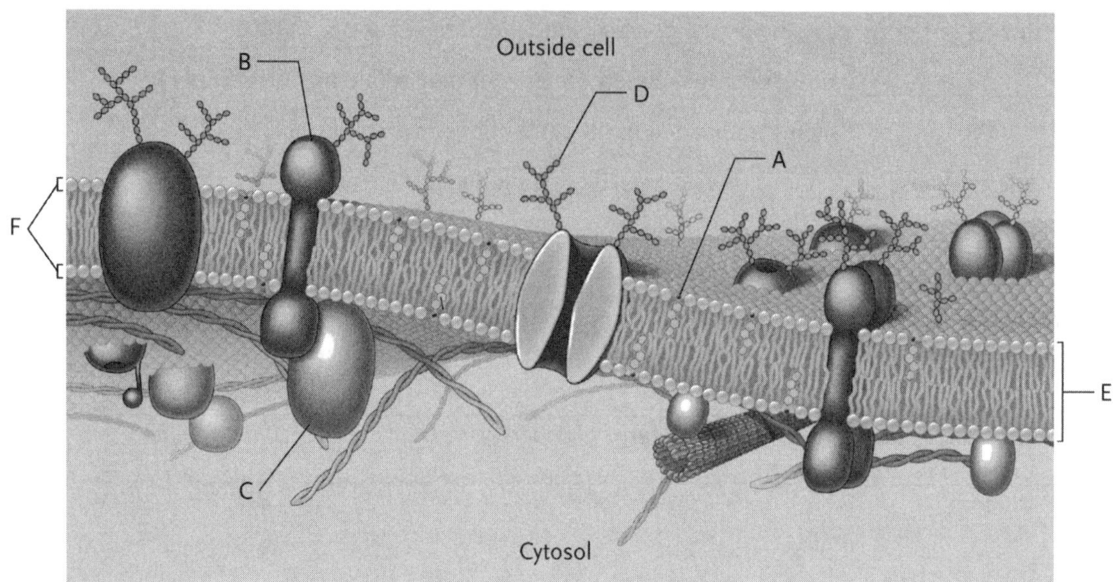

A. ____________

B. ____________

C. ____________

D. ____________

E. ____________

F. ____________

5.4 Passive Membrane Transport [pp. 101–104]

Section Review (Fill-in-the-Blanks)

Membranes control what enters and exits the cell because of their (67) ______________________ nature. While gases such as oxygen and CO_2, as well as certain other molecules, (68) __________________ quickly across membranes, most of the molecules and substances essential for life depend on (69) ____________________ proteins in order to cross the membrane. (70) ____________ is defined as the process whereby molecules move from a region of higher concentration to one of lower concentration, without the expenditure of (71) ________________. This type of process is driven by an increase in (72) ________________, that is, an increase in disorder. The rate of diffusion is determined by the (73) ________________ and (74) ______________ of a molecule. Simple diffusion occurs with molecules that are small and either (75) ___________________ or polar but uncharged so that they can move through the (76) ______________ interior of the membrane. This is not possible with ions because of their (77) _______________ shell and (78) _______________. The process of (79) ____________________ diffusion is faster than simple diffusion because of the involvement of transport proteins. Two types of integral membrane protein carry out this process: (80) ____________________ proteins and (81) ________________ proteins. The former create hydrophilic (82) ___________ in the membrane through which water and (83) ____________ pass, whereas the latter (84) ______________ to solute molecules such as sugar molecules and then undergo (85) ________________ changes that result in the movement of the (86) _______________-binding site from one side of the membrane to the other. Most ion channels in eukaryotes switch between open, closed, and intermediate stages and are therefore referred to as (87) _______________. Because these carry a single molecule in a single direction, they are called (88) _________________ proteins. Like enzymes, these proteins display a high degree of substrate (89) ______________ and can be saturated when all of the transporters are (90) ________________. Diffusion of water molecules across a membrane is specifically referred to as (91) _____________. This process occurs constantly in living cells and can occur by passive diffusion or through the help of channel proteins called (92) ______________.

Matching

Match each of the following terms with its correct descriptor or definition.

93.	____	Hypotonic	A.	Having an osmotic pressure or solute concentration equal to that of a reference solution
94.	____	Hypertonic	B.	A protein channel that switches between open, closed, or intermediate states
95.	____	Isotonic	C.	A membrane protein that forms a hydrophilic channel through which water and ions can pass
96.	____	Gated channel	D.	Having an osmotic pressure or solute concentration greater than a reference solution
97.	____	Channel protein	E.	Channel protein that facilitates the diffusion of water
98.	____	Carrier proteins	F.	A transport protein that binds to a single solute particle to move it across the plasma membrane
99.	____	Aquaporin	G.	Having an osmotic pressure or solute concentration less than that of a reference solution

True/False

Mark if the statement is true or false. If the statement is false, justify your answer in the line below each statement.

100.____________ During diffusion or osmosis, minimum entropy is reached at equilibrium.

____

____

101. ____________ Membranes are relatively impermeable to glycerol because it is a polar molecule and cannot pass through the nonpolar interior of the bilayer.

Short Answer

102. Compare and contrast simple diffusion with facilitated diffusion.

103. Discuss the basis for the specificity of aquaporins for water molecules.

104. What would happen if isolated animal cells were placed in a hypotonic solution?

Labelling

105. Identify each labelled part in the following figure.

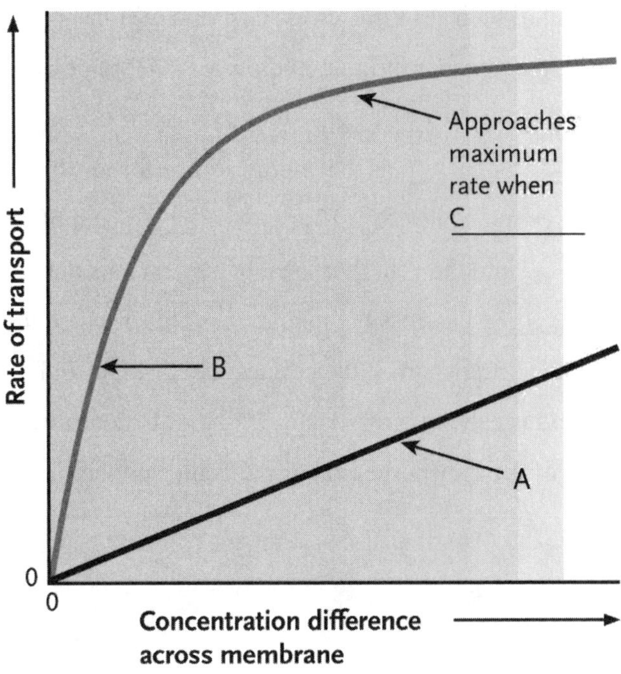

106. Label the following figures of cells undergoing osmosis.

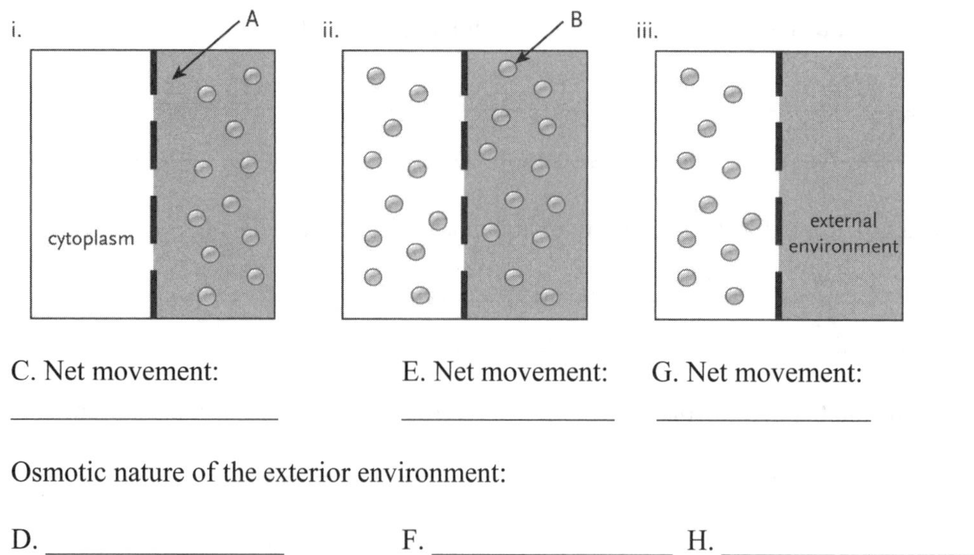

C. Net movement: ____________________

E. Net movement: ________________

G. Net movement: ________________

Osmotic nature of the exterior environment:

D. __________________ F. __________________ H. ____________________

5.5 Active Membrane Transport [pp. 104–107]

Section Review (Fill-in-the-Blanks)

Active transport uses (107) ___________ directly or indirectly, as well as carrier proteins, in order to transport ions or molecules against their concentration gradient. It is estimated that cells use approximately (108) ______________ percent of their ATP stores for this kind of transport. In (109) ___________________ active transport, the transport protein hydrolyzes ATP to power the transport of the bound solute. In contrast, (110) ________________ active transport uses a favourable concentration gradient of (111) ________________ as the energy source for transport. The Na^+/K^+ transporter simultaneously pumps three molecules of (112) _____________ out of the cell and two molecules of (113) _____________ into the cell, thus establishing an unequal distribution of ions that creates a (114) _______________ across the membrane. This is called the membrane (115) __________________. In secondary transport, if the cotransported solute and driving ion move in the same direction, the process is called (116) _____________. In contrast, (117) _______________ involves the movement of the transported ion or molecule and the driving ion in opposite directions.

Short Answer

118. Discuss the similarities of active transport and facilitated transport.

119. Explain why the membrane potential is referred to as an electrochemical gradient.

120. Discuss what kind of transporter and transport process would be dramatically higher in muscle cells than in many other types of tissue cells.

Complete the Table

121. Complete the following table by filling in the appropriate description of a listed transport process or by naming the transport process based on the listed description.

Process	*Description/Function*
A.	Uses energy to move ions or molecules against their concentration gradient
B.	Cotransported solute moves in same direction as driving ion
C.	Cotransported solute moves in opposite direction of driving ion

5.6 Exocytosis and Endocytosis [pp. 107–109]

Section Review (Fill-in-the-Blanks)

Molecules larger than amino acids or monosaccharides are transported into and out of eukaryotic cells by (122) ________________ and (123) __________________, respectively. In order for a cell to release waste products or secretory products, the cell creates (124) ________________ vesicles, which travel to, and fuse with, the (125) ________________ __________________. In (126) ___________, materials outside the cell become enclosed in a segment of plasma membrane, which bulges inward and pinches off to form a vesicle inside the cell. There are two distinct pathways by which this happens: (127) ____________________ endocytosis, sometimes called pinocytosis, and (128) _______________-mediated endocytosis. In the latter type, an (129) _______________ membrane protein binds the target molecule and forms a (130) ___________________-coated pit before pinching off and forming the (131) ___________________ vesicle. After losing the (132) _______________ coat, these vesicles generally fuse with (133) _____________, which digest the target molecule to small molecules, which then enter the cytoplasm through the action of (134) ___________________ proteins. Some cells, such as white blood cells and certain protists, take in much larger substrates, including whole cells, in a process called (135) __________________. This process also begins with receptor binding but is followed by the extension of (136) ________________ lobes around the substrate until it is engulfed. A large (137) ________________ then pinches off, and the contents are processed as with normal receptor-mediated endocytosis.

Matching

Match each of the following substrates with its correct method of entry or exit from the cell.

138.	____	Small solutes	A.	Exocytosis
139.	____	Protist prey	B.	Receptor-mediated endocytosis
140.	____	Peptide hormones	C.	Pinocytosis
141.	____	Milk proteins	D.	Phagocytosis
142.	____	Low-density lipoprotein		
143.	____	Mucus		
144.	____	Plant cell wall carbohydrates		
145.	____	Bacterial pathogens in the bloodstream of an infected person		

Short Answer

146. Explain why pinocytosis is also called "bulk-phase" endocytosis.

____

____

147. Explain how cells are able to extract valuable material from endocytic vesicles.

____

____

SELF-TEST

1. Phospholipids assemble into a bilayer because of their _____. [p. 96]

   a. Ability to dissolve in water
   b. Amphipathic nature
   c. Inability to associate with other phospholipids
   d. Lack of fatty acids

2. The bilayer arrangement of membranes represents a very ____________ energy state. [p. 96]

   a. Stable
   b. Low
   c. High
   d. Unstable

3. The interior of a phospholipid bilayer is _____. [pp. 96–97]

   a. Hydrophobic
   b. Amphipathic
   c. Hydrophilic
   d. Hypertonic

4. The gelling temperature of membranes is reduced with ____________. [pp. 96–99]

   a. Increased amounts of saturated lipids
   b. Increased amounts of unsaturated fatty acids
   c. Sterols
   d. (b) and (c)

5. Which of the following statements about desaturases is incorrect? [pp. 97–99]
   a. Their expression is increased at higher temperatures
   b. They remove two hydrogen atoms from adjacent carbon atoms in fatty acids
   c. They create double bonds
   d. None of the above

6. Which of the following is not a function of membrane proteins? [pp. 99–100]

   a. Transport
   b. Recognition/adhesion
   c. Receptor
   d. None of the above

7. The passive movement of a substance down its concentration gradient is called _____. [p. 101]

   a. Active transport
   b. Exocytosis
   c. Diffusion
   d. Symport

8. Which of the following statements about diffusion is true? [p. 101]

   a. Minimum entropy is reached at equilibrium
   b. The movement of solutes stops when equilibrium is reached
   c. The process is driven by entropy
   d. There is an indirect expenditure of energy

9. Which of the following is least likely to cross a plasma membrane by simple diffusion? [p. 102]

   a. O_2
   b. H_2O
   c. CO_2
   d. PO_4^{3-}

10. Aquaporins function in transport across the membrane by _____. [p. 104]

    a. Binding to a molecule of water and transporting it after undergoing a conformational change
    b. Using Na^+ as a driving ion to transport water in the same direction
    c. Forming a channel to facilitate the rapid diffusion of water
    d. Inhibiting the activity of the Na^+/K^+ pump

11. The energy-requiring transport of solutes against their concentration gradient is called _____. [p. 104]

    a. Osmosis
    b. Active transport
    c. Receptor-mediated endocytosis
    d. Facilitated diffusion

In a two-compartment system separated by a selectively permeable membrane, compartment A is filled with a 0.5 M solution of NaCl and compartment B is filled with a 1 M solution of NaCl. Assuming that NaCl dissociates completely in an aqueous solution and that the membrane is freely permeable to Na^+, Cl^-, and H_2O, characterize the following aspects of the system in questions 12 to 14. [pp. 101–104]

12. Compartment B is _____ compared with compartment A.

    a. Hypotonic
    b. Isotonic
    c. Hypertonic

13. Initially, there is _____.

   a. No net movement of solutes (Na^+ and Cl^-)
   b. Net movement of solutes (Na^+ and Cl^-) from A to B
   c. Net movement of solutes (Na^+ and Cl^-) from B to A

14. Initially, there is _____.

   a. No net movement of water
   b. Net movement of water from A to B
   c. Net movement of water from B to A

15. Primary active transport would be expected to transport which of the following? [p. 105]

   a. H^+
   b. Ca^{2+}
   c. K^+
   d. All of the above

16. Which of the following is not true of secondary active transport? [pp. 105–106]

   a. The transporter hydrolyzes ATP as an energy source to drive transport
   b. The transporter relies on ion gradients to drive transport
   c. Symporters are secondary active transporters
   d. Antiporters are secondary active transporters

17. Which of the following statements about Ca^{2+} pumps is false? [p. 105]

   a. They are primary active transporters
   b. They are rare in animal cells
   c. They are used to pump calcium out of the cytoplasm
   d. They are used to pump calcium into the lumen of the ER

18. Receptor-mediated endocytosis does not involve _____. [p. 107]

   a. Binding to a specific cell surface receptor
   b. Fusion of the endocytic vesicle with a lysosome
   c. Symport with a driver ion
   d. Formation of a clathrin-coated pit

19. Which of the following statements about clathrin is false? [p. 107]

   a. It is a type of protein
   b. It forms a network on the cytoplasmic side of the membrane
   c. It is a temporary component of some endocytic vesicles
   d. It forms a network on the exterior face of the membrane

20. Lap-Chee Tsui is famous for _________. [p. 110]

a. The discovery of the CF gene
b. The discovery of aquaporins
c. The discovery of receptor-mediated endocytosis
d. Developing carbon nanotubes for delivery of anticancer drugs.

INTEGRATING AND APPLYING KEY CONCEPTS

1. Although people are susceptible to hypothermia when exposed to freezing temperatures for any length of time without the proper clothing, it is estimated that the majority of biomass in the Southern Ocean of Antarctica are archaea. These prokaryotes thrive at subzero temperatures. Discuss one or more of the likely differences between humans and archaea that allow only the latter to thrive.

2. Cholesterol is transported in the blood in large complexes known as lipoproteins; a major cholesterol-containing lipoprotein is low-density lipoprotein (LDL). Some individuals have a genetic defect that results in chronic elevated levels of cholesterol in their blood. What defect with the plasma membranes of cells would explain this condition? How might cholesterol-lowering drugs such as Lipitor correct this? What are the dangers of having high blood cholesterol?

3. In this chapter, you learned that plants secrete carbohydrates for their cell walls. How is this accomplished, and what critical function would such cell walls serve? Speculate on how animals survive without these structures.

OPTIONAL ACTIVITY

1. There are a variety of food- and water-borne diarrheal diseases of humans that can sometimes be quite deadly. Probably the one that is most important worldwide is cholera. The primary symptom of this disease is the result of a toxin that interferes with ion transport in the intestinal epithelial cells. Research what this transporter does and how the toxin interferes with its function. What is the worldwide mortality rate for cholera?

6 Cellular Respiration

CHAPTER HIGHLIGHTS

- The majority of living things metabolize high-energy chemicals (glucose, fats, and proteins) to produce energy.
- This is accomplished using cellular respiration.
- This pathway is subdivided into three stages. Each stage
  - Has a series of chemical reactions catalyzed by specific enzymes
  - Has a beginning chemical and end product(s)
  - Produces energy that is stored in chemicals (ATP and NADP) or released as heat
  - Takes place at a specific location inside the cell
  - Is regulated by the presence of oxygen or by its end product(s)
- The last stage is electron transport through a membrane-bound electron transport chain to generate proton-motive force.
- The energy of this force is exploited by the cell through the process of chemiosmosis.
- ATP synthesis in cellular respiration occurs via substrate-level phosphorylation and oxidative phosphorylation.
- In the absence of oxygen, depending upon the organism or type of cell, fermentation or anaerobic respiration may allow the continuing generation of ATP.

STUDY STRATEGIES AND LEARNING OUTCOMES

- This chapter has extensive chemical pathways. Do not try to memorize all of the chemical reactions but focus instead on understanding how the process works to generate energy for the cell.
- Remember that the goal of this chapter is to understand how living things produce energy: this means you need to understand how the processes differ between prokaryotes and eukaryotes.
- Concentrate on the key concepts: how does each stage relate to the other two, what are the substrates and products of each.

- Make sure you understand any regulatory or environmental conditions that affect the flow of electrons through these pathways.
- Work through one section at a time and then the corresponding section of the study guide.

By the end of this chapter, you should be able to

- Identify how chemicals can be sources of energy
- Describe the purpose and basic components of cellular respiration, identifying for each the cellular location
- Describe what an electron transport chain is, where the respiratory chain is found, how it functions, and how its function is coupled to the production of energy
- Define chemiosmosis, proton-motive force, and oxidative phosphorylation and distinguish between the catabolic and anabolic roles of these pathways and products
- Describe the role of oxygen in cellular respiration and the alternatives to the use of oxygen

TOPIC MAP

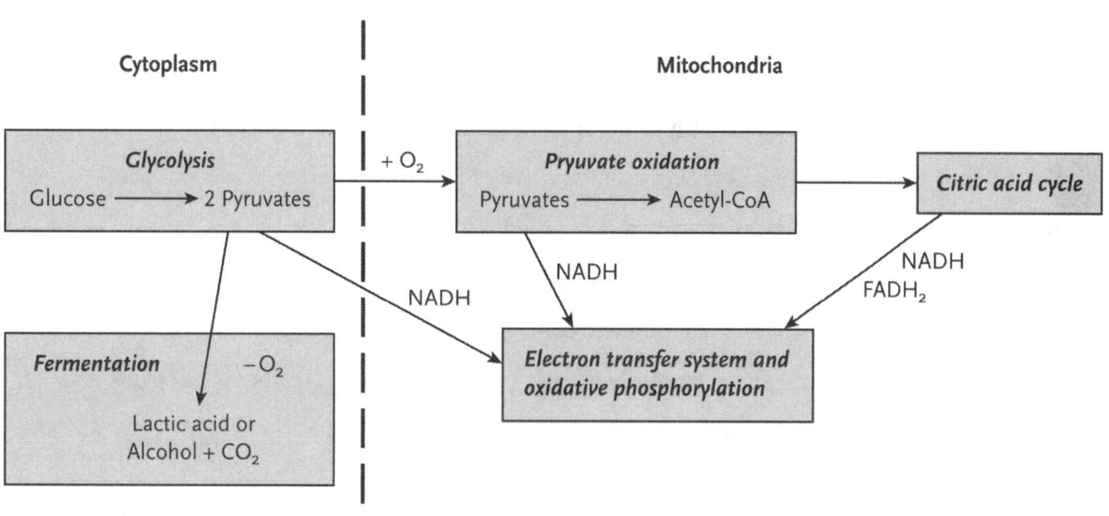

INTERACTIVE EXERCISES

Why It Matters [pp. 115–116]

6.1 The Chemical Basis of Cellular Respiration [pp. 116–119]

Section Review (Fill-in-the-Blanks)

The eukaryotic mitochondrion is the site of energy generation by cellular (1) ___________, (2) __________ transport, and (3) __________ synthesis. Evidence of the critical nature of this organelle comes from the fact that mitochondrial defects underlie a number of human (4) ____________, including age-related ones such as type I (5) ______________. The vast majority of energy in the biosphere enters through the trapping of solar energy in the process of (6) _____________. The trapped energy is then used to convert (7) _____________ and water to high-energy (8) _____________ molecules. Cellular (9) _______________ is essentially the opposite in that the chemical energy in fuel molecules is released through the process of (10) ________________; the liberated energy is converted to (11) _____________, which can be used for many energy-requiring cellular reactions. The complete oxidation of organic molecules results in the release of (12) ________________, allowing the cycle to start over again. The amount of energy that can be extracted from organic molecules increases with the number of (13) ____________ bonds. Because the bonding electrons are (14) _______________ between the two atoms, they have considerable (15) ____________. In contrast, because oxygen is very (16) _________________, molecules with more oxygen have less energy. The (17) ____________ of chemicals to obtain energy is a process that involves the (18) _________________ of electrons. Because these reactions are always coupled with (19) ______________ reactions, where electrons are transferred to another molecule, the term (20) _______________ is used. To maximize energy recovery for subsequent use, cellular respiration oxidizes molecules in a series of (21) ___________ steps. Enzymes called (22) _______________ remove electrons and protons and transfer them to a shuttle molecule, most commonly (23) _____________.

Matching

Match each of the following terms with its correct definition.

24.	____	Oxidation	A.	A chemical from which electrons have been removed
25.	____	Oxidized	B.	A group of enzymes that remove protons and electrons from molecules being oxidized
26.	____	Reduction	C.	A chemical to which electrons have been added
27.	____	Reduced	D.	A coenzyme that carries electrons and protons
28.	____	Redox reactions	E.	Coupled reactions where electrons released from one chemical are added to another chemical
29.	____	Dehydrogenases	F.	Process of removing electrons from a chemical
30.	____	NAD^+	G.	Process of adding electrons to a chemical

Complete the Table

31. Electron flow connects two processes: photosynthesis and cellular respiration. Complete the following table by selecting your answer from the parentheses, giving specific information about each process.

Events	*Photosynthesis*	*Cellular respiration*
A. Sugars (makes/breaks)		
B. O_2 (uses/releases)		
C. CO_2 (uses/releases)		
D. Net energy (stores/releases)		

True/False

Mark if the statement is true or false. If the statement is false, justify your answer in the line below each statement.

32. _____________ CO_2 is completely oxidized, so no further energy can be extracted from it.

____

____

33. _____________ Electrons that move closer to the atom's nucleus gain energy.

____

____

34. ____________ Oxidation reactions involve the transfer of electrons to oxygen.

____

____

Short Answer

35. For the combustion of methane, $CH_4 + O_2 \rightarrow CO_2 + H_2O$, identify which molecule(s) is(are) oxidized and reduced and why, without the obvious transfer of electrons, this is still a redox reaction.

____

____

6.2 Cellular Respiration: An Overview [pp. 119–120]

6.3 Glycolysis [pp. 120–122]

6.4 Pyruvate Oxidation and the Citric Acid Cycle [pp. 122–123]

Section Review (Fill-in-the-Blanks)

Cellular respiration consists of (36) _____________ stages, and each generates the high-energy molecule (37) _______________, the overarching goal of this process. In (38) ____________, which is the first stage, the six-carbon substrate (39) ______________ is oxidized to two molecules of the three-carbon product, (40) _______________. The first half of this pathway requires an input of two molecules of (41) ____________ per molecule of glucose, but the second part yields (42) __________ ATP molecules along with two molecules of (43) __________. The ATP molecules are generated by the enzymatic transfer of high-energy (44) ____________ groups from other molecules to (45) ____________, in a process referred to as (46) ________________-______________ phosphorylation. The first stage of cellular respiration is virtually universal among life forms, and in all of them, it is localized to the (47) ______________. In contrast, the localization of the second stage of respiration differs among life forms, taking place in the (48) ______________of prokaryotes but in the mitochondrial (49) ________________ of eukaryotes. To enter, pyruvate diffuses through (50) ________________ in the outer mitochondrial membrane and is transported across the inner membrane by specific (51) _________________ proteins. Pyruvate is then oxidized in a reaction that generates NADH, CO_2, and (52) _______________. The latter is transferred to coenzyme A, generating (53) ____________-____________, the high-energy substrate of the (54) ____________ ___________ cycle. Each turn of this cycle yields three molecules of (55) ______________, one molecule each of (56) ____________ and (57) ____________, and two molecules of (58) _______________. This last product tells us that the cycle has completed the (59) ___________ of glucose.

Matching

Match each of the following terms with its correct definition or descriptor.

60.	____	Substrate-level phosphorylation	A.	Removal of a COO^- group phosphorylation
61.	____	Glycolysis	B.	A cycle of reactions that generates ATP, NADH, and $FADH_2$
62.	____	Pyruvate oxidation	C.	Transfer of phosphate from a substrate to ADP to make ATP
63.	____	Citric acid cycle	D.	Process whereby pyruvate is converted to acetyl-CoA
64.	____	Decarboxylation	E.	Process whereby glucose is partially oxidized in the cytoplasm to form pyruvate

True/False

Mark if the statement is true or false. If the statement is false, justify your answer in the line below each statement.

65. ____________ Glycolysis is a typically eukaryotic process.

____

____

66. ___________ In respiratory cells, glycolysis and the citric acid cycle both occur in the cytoplasm.

____

____

Complete the Table

67. The following represents an energy "balance sheet" for the first two stages of cellular respiration.

Stage	*Pathway or reactions*	*Substrate*	*Net products*
First stage	A.	B.	C.
Second stage	D.	E.	F.

Short Answer

68. What is the evidence that glycolysis is one of the most ancient of all metabolic pathways?

____

____

69. Explain why the second stage of respiration is called a cycle.

____

____

70. Comment on the number of carbon atoms in the intermediates of the pathways of the glycolysis versus the citric acid cycle.

____

____

6.5 Electron Transport and Chemiosmosis [pp. 123–128]

Section Review (Fill-in-the-Blanks)

In the third stage of cellular respiration, electrons from the reduced coenzymes (71) ____________ and (72) ____________ are transferred through an electron transport chain to the final electron (73)_____________, which in aerobes is O_2. The chain consists of three major multiprotein (74) __________ and one single (75) ____________ membrane protein. The chain is located within the mitochondrial (76) ___________ ___________. In the (77) ______________, it is located in the plasma membrane. Electrons are shuttled between the protein components of complexes (78) ___________, (79) ____________, and (80) __________ by protein-associated (81) __________ groups. One such group is heme, which contains a redox-active iron that alternates between (82) _________ and (83) __________. Electrons enter the chain when complex I is reduced by (84) ___________ and complex II is reduced by (85) ___________. From these two complexes, the lipid-soluble molecule (86) ___________ shuttles electrons to complex III and the mobile carrier (87) _____________ shuttles electrons to (88) ___________. Each electron transfer is (89) _______________, meaning that free energy is released, so movement through the chain is (90) ______________. The free energy that is released is used to generate a (91) _____________ gradient across the membrane, which can then be used to do (92) ___________. The gradient is the result of the fact that the membrane is (93) ______________ to protons and that complexes (94) ________ and (95) __________ as well as the carrier (96) _____________ translocate protons to the (97) _______________ side of the membrane. Because protons are charged, the gradient has both a chemical and an (98) _______________ quality. This particular type of gradient is called the proton- (99) ___________ force. Through the process of (100) ____________, the force can be used to drive nutrient (101) ____________ or to generate ATP through the action of the membrane-bound ATP (102) ____________. This is a multienzyme complex that couples the flow of protons (103) ___________ their gradient with the synthesis of ATP from (104) _____________ and (105) _____________.

Matching

Match each of the following terms with its correct definition.

106.	____	ATP synthase	A.	Prosthetic group that accepts and donates electrons
107.	____	Fe/S group	B.	The use of a H^+ gradient to do work
108.	____	Oxidative phosphorylation	C.	Complex II of the electron respiratory transport chain
109.	____	Proton-motive force	D.	A proton and voltage gradient that is a source of potential energy
110.	____	Chemiosmosis	E.	An uncoupler of oxidative phosphorylation
111.	____	Succinate dehydrogenase	F.	Process whereby a H^+ gradient is used to add phosphate to ADP and make ATP
112.	____	Ionophore	G.	An enzyme that uses a H^+ gradient to generate ATP

Short Answer

113 . Differentiate between oxidative phosphorylation and substrate-level phosphorylation.

____

____

114. Explain the role of reduced coenzymes in generating ATP.

____

____

115. Compare the location of ATP synthase in the mitochondrion and prokaryotes.

____

____

116. Explain why the ATP synthase is described as a molecular motor.

____

____

Labelling

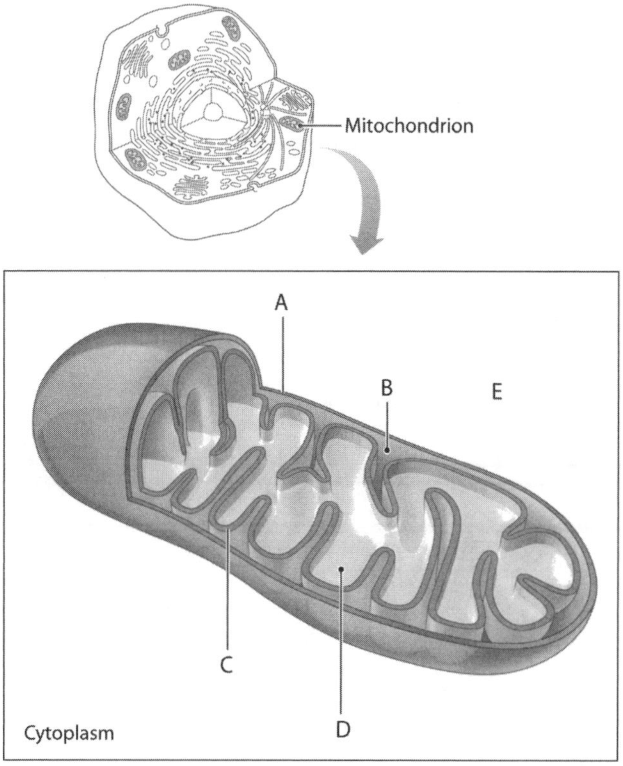

117. Identify each part of the mitochondrion.

A. ______________________

B. ______________________

C. ______________________

D. ______________________

Mitochondrion

A

B

E

C

D

Cytoplasm

Locate the following structures or processes in the cell:

118. ____ Glycolysis

119. ____ Pyruvate oxidation

120. ____ Citric acid cycle

121. ____ Electron transport chain

122. ____ ATP synthase

Sequence

123. Arrange the following steps of cellular respiration in the correct sequence.

A. Citric acid cycle

B. Electron transport chain

C. Glycolysis

D. Pyruvate oxidation

____ ____ ____ ____

6.6 The Efficiency and Regulation of Cellular Respiration [pp. 128–131]

Section Review (Fill-in-the-Blanks)

Biochemists have calculated that (124) ____________ protons are pumped across the mitochondrial inner membrane for each molecule of NADH oxidized and that one ATP is synthesized for every (125) _____________ that flows back into the matrix. Therefore, the oxidation of one NADH yields (126) _____________ ATP. This is in contrast to the coenzyme

(127) ____________, which only yields two ATP. One can therefore calculate the net yield of ATP from the complete oxidation of glucose to (128) ______________ ATP. Considering the caloric content of glucose, this represents an efficiency of approximately (129) ____________ percent. The pathways of cellular respiration are tightly (130) _________________ so that energy production closely (131) ______________ the cell's energy needs. The glycolytic pathway is controlled through the activation or inhibition of the enzyme (132) ___________________. This enzyme acts early in the pathway and is inhibited by (133) _______________ and two products of the citric acid cycle: (134) _________________ and (135) _____________. It is activated by (136) _________________. The citric acid cycle is inhibited at several steps by (137) ____________. Other organic molecules can also be oxidized by the pathways of cellular respiration. Molecules such as (138) _____________, (139) _______________, and (140) _________________ are processed first and then enter the pathways at several (141) ________________. As further evidence of the incredible metabolic (142) _____________ of cells, cellular respiration pathways are also used for (143) _____________.

True/False

Mark if the statement is true or false. If the statement is false, justify your answer in the line below each statement.

144. ____________ In cellular respiration, all of the ATP is produced in the mitochondrion.

____

____

145. ____________ Cellular respiration converts all of the energy in glucose to ATP.

____

____

Complete the Table

146. Complete the following table with specific information on each step of cellular respiration in eukaryotes.

Steps of cellular respiration	*Location in the cell*	*Number of ATP produced*	*Number of NADH and $FADH_2$ produced*	*O_2 used*	*Number of CO_2 produced*
A. Glycolysis					
B. Pyruvate oxidation					
C. Citric acid cycle					
D. Electron transfer system					

Short Answer

147. Explain how glycolysis and the citric acid cycle are regulated.

___

___

148. Explain how proteins and fats enter the cellular respiration pathways.

___

___

149. Explain how glycolysis and the citric acid are also able to provide intermediates for biosynthesis.

___

___

6.7 Oxygen and Cellular Respiration [pp. 131–136]

Section Review (Fill-in-the-Blanks)

In the absence of oxygen, some prokaryotes can perform (150) ___________ respiration, using molecules such as (151) __________, (152) ___________, and (153) ____________ as terminal electron acceptors. Some organisms and tissues respond to the absence of oxygen by (154) ____________. In muscle cells, this process is triggered when (155) ______________ starts to build up in the cytosol and the result is the reduction of pyruvate to (156) ______________. This reaction is reversible so that when oxygen levels increase, the fermentation product is (157) ____________ back to pyruvate and (158) ______________. In the (159) _____________ fermentation pathway of yeast, pyruvate undergoes a (160) _______________ reaction as well as a reduction. While of commercial importance in some cases, the metabolic role of these pathways is to recycle (161) ___________ to allow ATP synthesis by the (162) __________ pathway. The "paradox of aerobic life" refers to the fact that aerobic (163) ____________, while necessary for most life, generates (164) ___________ oxygen species or ROS. Aerobic organisms possess enzymes and molecules for detoxifying these molecules, but strict (165) ________________ do not and cannot survive in an aerobic atmosphere. Organisms that are incapable of fermentation or anaerobic respiration are called strict (166) _____________. Certain prokaryotes as well as vertebrate (167) _________ cells fall under this category.

Matching

Match each of the following terms with its correct definition.

168.	____	Fermentation	A.	Organisms that must have oxygen to survive
169.	____	Lactate fermentation	B.	Organisms that require an oxygen-free environment to survive
170.	____	Alcohol fermentation	C.	Process that converts pyruvate into lactate or alcohol
171.	____	Strict anaerobes	D.	Process that produces ethanol and CO_2
172.	____	Facultative anaerobes	E.	Fermentation process in muscle cells
173.	____	Strict aerobes	F.	Organisms that can switch between aerobic or anaerobic respiration/fermentation

Short Answer

174. Explain in general how ROS are detoxified.

____

____

175. Explain how the reduction of oxygen in aerobic respiration is controlled to prevent ROS generation.

____

____

Complete the Table

176. Complete the following table with specific information on each step.

Steps of anaerobic respiration	*Location in the cell*	*Number of ATP produced*	*Number of NADH and $FADH_2$ produced*	*O_2 used*	*CO_2 produced*
A. Glycolysis					
B. Fermentation					

True/False

Mark if the statement is true or false. If the statement is false, justify your answer in the line below each statement.

177. __________ Fermentation produces less energy than anaerobic respiration, which, in turn, produces less than in aerobic respiration.

____

____

178. ____________ Fermentation takes place in the mitochondrion.

____

____

179. ___________ Organisms that perform anaerobic respiration still perform glycolysis.

____

____

SELF-TEST

1. How many ATP molecules are invested in glycolysis? [p. 120]

   a. 2
   b. 4
   c. 32
   d. 38

2. Which of the following is the end product of glycolysis? [p. 120]

   a. Glucose
   b. Pyruvate
   c. Citric acid
   d. Acetyl-CoA

3. Where in the cell does glycolysis take place? [p. 120]

   a. Cytoplasm
   b. Mitochondrial matrix
   c. Intermembrane compartment
   d. Inner membrane of mitochondria

4. Before processing fat through cellular respiration, it must be broken down into ________. [p. 130]

   a. Amino acids
   b. Glycerol and fatty acids
   c. Nucleotides
   d. Monosaccharides

5. All of the following steps of cellular respiration give rise to ATP except ________. [p. 122]

   a. The citric acid cycle
   b. Pyruvate oxidation
   c. Glycolysis
   d. The electron transfer system

6. Only about 32% of the energy from glucose is stored in ATP. Why is this value not more efficient? [p. 129]

   a. It is stored in NADH
   b. Some of the energy is lost to entropy
   c. It is converted to fat
   d. It is wasted as light energy

7. During exercise, when oxygen is limited, additional ATP molecules are produced in the muscles by ________. [pp. 131–132]

   a. Aerobic respiration
   b. Glycolysis
   c. The citric acid cycle
   d. The electron transfer system

8. Where in the cell does the fermentation step take place? [p. 132]

   a. Cytoplasm
   b. Mitochondrial matrix
   c. Intermembrane compartment
   d. Inner membrane of mitochondria

9. Yogurt is a bacterial culture. As this culture gets older, it becomes increasingly sour due to the accumulation of lactic acid. Which of the following steps in bacteria is responsible for producing this acid? [p. 132]

   a. Glycolysis
   b. Citric acid cycle
   c. Electron transfer system
   d. Fermentation

10. While making the dough for bread, addition of yeast causes the dough to rise. Which of the following chemical(s) released by yeast makes this happen? [p. 132]

    a. Lactic acid
    b. Alcohol and carbon dioxide
    c. ATP and NADH
    d. Pyruvate and acetyl-CoA

11. Per glucose molecule, which of the following organisms would be more efficient and make more ATP molecules? [pp. 117, 132]

    a. Strict aerobes
    b. Strict anaerobes

12. Which of the following is(are) associated with defects in mitochondrial function? [p. 116]

   a. Alzheimer's disease
   b. Atherosclerosis
   c. Parkinson's disease
   d. All of the above

13. Which would contain more energy, a monosaccharide such as glucose ($C_6H_{12}O_6$) or a fatty acid? [pp. 116–117]

   a. They would both possess the same amount of energy
   b. Glucose has more energy
   c. The fatty acid would have more energy

14. Which of the following is not true? [p. 118]

   a. The combustion and the respiratory oxidation of glucose both result in the complete oxidation of the molecule
   b. Heat provides the activation energy for both the combustion and the oxidation of glucose
   c. The combustion and the respiratory oxidation of glucose are both energy-liberating processes
   d. CO_2 is produced by the combustion of glucose and by the respiratory oxidation of glucose

15. Peter Mitchell is famous for _______________. [pp. 126–127]

   a. Proposing the chemiosmotic theory
   b. Being the first to determine that certain diseases are caused by defects in mitochondrial function
   c. Describing the process of fermentation
   d. None of the above

16. Which of the following statements about prokaryotes is true? [pp. 127–128]

   a. The first stage of cellular respiration is different from that of eukaryotes
   b. Their electron transport chains translocate protons to the outside of the cell
   c. They only produce ATP through substrate-level phosphorylation
   d. All of the above are true

17. Which of the following statements about oxidative phosphorylation is true? [pp. 126–128]

   a. The substrates are ADP and a high-energy phosphorylated molecule
   b. Occurs in association with both respiratory and photosynthetic electron transport chains
   c. Only happens in eukaryotic cells
   d. None of the above

18. Which of the following statements about the ATP synthase is false? [pp. 126–128]

   a. It is not present in prokaryotic cells
   b. It is a molecular motor
   c. Its movement is related to the catalytic activity
   d. It is fully reversible

19. Which of the following statements about uncouplers is false? [pp. 128–129]

   a. They destroy the proton gradient across the membrane
   b. They are synthesized by certain cells in order to prevent electron transport chains from giving rise to a proton-motive force
   c. They inhibit the movement of electrons through electron transport chains
   d. They can form channels in the plasma membrane of prokaryotes

20. Which of the following statements about complex IV of the respiratory electron transport chain is false? [p. 134]

   a. It has four prosthetic groups that must be reduced before it will transfer them to oxygen
   b. It is a universal component of aerobic organisms
   c. It may be present in anaerobic organisms or cells
   d. None of the above

INTEGRATING AND APPLYING KEY CONCEPTS

1. Draw a diagram of an eukaryotic cell with a mitochondrion and, next to it, a diagram of an aerobic bacterial cell. Using these two diagrams, compare the location and direction of electron flow of the three stages of respiration. Now consider if the process would differ in the (a) bacteria that perform anaerobic respiration and (b) the bacteria used to make yogurt.
2. Provide an explanation for the localization of the mitochondrial electron transport chain based on the proposed endosymbiotic origin of the organelle.
3. What is the explanation for the burning feeling of overexercised muscles? Given this explanation, why is asphyxiation so deadly to vertebrate brain cells?

OPTIONAL ACTIVITY

1. Certain types of anaerobic bacteria are able to respire metals; that is, they can use a diverse array of metals as terminal electron acceptors. The discovery of such processes has given rise to the fledgling field of *geomicrobiology*, a field with applications to mining, bioremediation, geology and the evolution of life on Earth. Find out more about this field and some of its specific applications.

7 Photosynthesis

CHAPTER HIGHLIGHTS

- Autotrophs use chemical or light energy to convert simple inorganic chemical into complex organic chemicals.
- Photoautotrophs are the primary producers of the biosphere, trapping the energy from the sun to support their own growth as well as the growth, either directly or indirectly, of all other life forms.
- Photosynthesis has two steps:
  - The light-dependent reaction—uses light energy to make ATP and NADPH
  - The light-independent reaction or Calvin cycle—"fixes" carbon dioxide using ATP and NADPH to make organic molecules
- Each step
  - Consists of a series of events, chemical or electron transfer reactions
  - Produces or uses energy chemicals (ATP and NADPH)
  - Takes place at a specific location inside the eukaryotic chloroplast
- The enzyme for the Calvin cycle is the most abundant enzyme on Earth, yet it is remarkably inefficient.
  - Oxygen acts as a competitive inhibitor.
  - Because of the inefficiency and the concomitant loss of energy and carbon because of this inefficiency, photosynthetic organisms have evolved various ways to compensate. After millennia, the enzyme itself has not undergone any adaptations to improve the efficiency of the enzyme and reduce its affinity for oxygen.

STUDY STRATEGIES AND LEARNING OUTCOMES

- This chapter has a substantial amount of detail on electron transport and enzymatic pathways. Do not get lost in the details of each pathway and the names of any but the more central names.
- Remember that the goal of this chapter is to understand how phototrophs trap light energy in the form of ATP and NADPH and how these molecules are used to make the organic chemicals required by the cell.
- Concentrate on the key concepts and focus on how the pathways fit with each other.
- Pay attention to the figures: they will help you visualize the processes and see the bigger picture.
- Work through one section at a time and then the corresponding section of the study guide.

By the end of this chapter, you should be able to

- Describe what happens when photons are absorbed by pigments and distinguish between the molecular results of light absorption by an accessory pigment, an antenna chlorophyll, and a reaction centre chlorophyll
- Explain the function of the molecules in the antennae
- Distinguish between linear and cyclic electron transport
- Explain how photosynthetic electron transport is used to make ATP
- Describe the basic process of the Calvin cycle and where it happens and provide the name of the critical enzyme
- Understand what photorespiration is, why it happens, why it is a problem, and the various ways that have evolved to compensate

TOPIC MAP

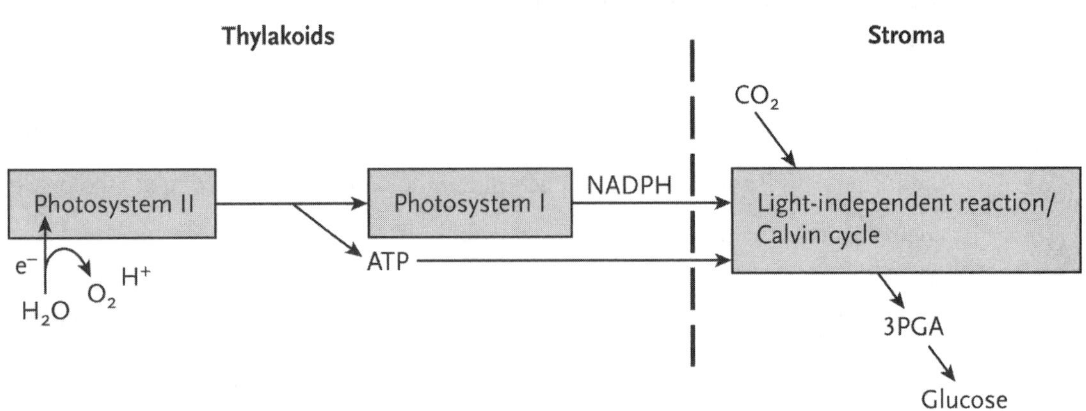

INTERACTIVE EXERCISES

Why It Matters [pp. 139–140]

Section Review (Fill-in-the-Blanks)

Life on Earth is entirely dependent on (1) _____________ organisms in both terrestrial and (2) _____________ habitats. These organisms use the energy they derive to convert inorganic carbon, (3) ___________, to complex organic molecules. Approximately (4) _______________ kg of carbon are fixed per year by these organisms. Aquatic (5) ___________ are responsible for approximately (6) ____________ of this. Due to their nutrient-poor waters, these organisms are less abundant in (7) _________ waters and are instead most abundant in the (8) ______________.

7.1 Photosynthesis: An Overview [pp. 140–142]

Section Review (Fill-in-the-Blanks)

Photoautotrophs are the major sources of (9) ___________ molecules for other organisms. They are therefore considered the primary (10) _____________ of this planet. Because animals eat plants and other animals, they are called (11) ____________. Eventually, the bodies of both die providing chemical energy for the (12) _____________. Heterotrophs such as certain bacteria and (13) ________ fall into this last class. Photosynthesis has two distinct phases: the first is called the (14) ___________-____________ reaction and the second the (15) _____________ cycle. In the first phase, two high-energy molecules are generated, (16) _____________ by electron transport and (17) ____________ by chemiosmosis. These are then used in the second phase to convert CO_2 into (18) ____________, a process referred to as CO_2– (19) ___________. Photoautotrophs can then use the immediate product of the second stage to make all of the other (20) ____________ molecules they require for growth. In eukaryotic organisms, both photosynthetic stages occur in (21) ______________, whereas in photosynthetic prokaryotes, they take place in the cytosol and (22) _____________ membrane.

Labelling

23. Identify each part of the chloroplast.

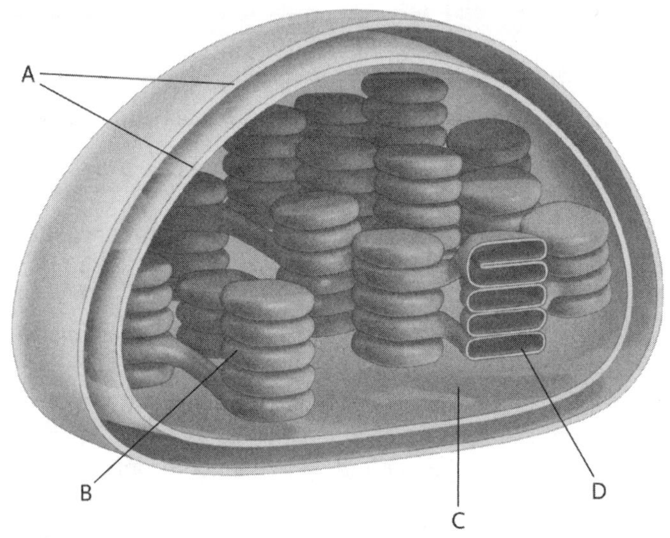

A. ________________________

B. ______________________

C. ______________________

D. ________________________

Locate the following reactions in the chloroplast:

24. Light-dependent reactions _____

25. Light-independent reactions/
Calvin cycle _____

True/False

Mark if the statement is true or false. If the statement is false, justify your answer in the line below each statement.

26. ____________ ATP and NADPH used in the Calvin cycle come from cellular respiration in the mitochondria.

____

27. ____________ Photosynthesis happens only in plants and algae.

____

Matching

Match each of the following terms with its correct definition.

28. ___	Photosynthesis	A. Conversion of CO_2 to carbohydrates
29. ___	Primary producers	B. Organelles in eukaryotic cells where photosynthesis takes place
30. ___	Consumers	C. Process by which certain organisms harness light energy to do work
31. ___	Decomposers	D. A stack of membrane sacs inside the chloroplast
32. ___	CO_2 fixation	E. Generic designation for carbohydrate units
33. ___	Chloroplasts	F. Fluid inside the chloroplast
34. ___	Thylakoids	G. Plants and other photosynthetic organisms that produce organic chemicals
35. ___	Stroma	H. Organisms that break down dead plants and animals to obtain organic chemicals
36. ___	$(CH_2O)n$	I. Organisms that live by eating plants and animals

Complete the Table

37. Electron flow connects two processes: photosynthesis and cellular respiration. Complete the following table by selecting your answer from the parentheses, giving specific information about each process.

Events	*Photosynthesis*	*Cellular respiration*
A. Sugars (makes/breaks)		
B. O_2 (uses/releases)		
C. CO_2 (uses/releases)		
D. Net energy (stores/releases)		

Short Answer

38. Provide the balanced equation for both stages of photosynthesis and identify where the two stages occur in the chloroplast.

____

____

7.2 The Photosynthetic Apparatus [pp. 142–145]

Section Review (Fill-in-the-Blanks)

(39) _______________ molecules absorb light of specific wavelengths. Photons of light are absorbed by (40) ____________, and this causes them to move from their (41) ___________ state to a(n) (42) ____________ state. Depending on the environment surrounding the pigment, the (43) ___________ state electron may be transferred to another molecule, which, in the case of photosynthesis, is called the (44) ____________ acceptor. The major photosynthetic pigment in plants, green algae, and cyanobacteria is (45) _____________. There are two dominant forms of this pigment: (46) _______ and (47) __________. The former is the only form that becomes (48) ___________. The latter and another type of pigment, (49) _____________, are referred to as (50) ___________ pigments because rather than being oxidized, their excited electrons transfer that energy to a neighbouring pigment, through a process called (51) __________ ________________. Photosynthetic pigments and proteins are organized as clusters called (52) _______________, which, in the chloroplast, are embedded in the (53) _____________ membranes. In plants, green algae, and cyanobacteria, there are two (54) ____________, PSI and II. Each contains a (55) _________ centre surrounded by an aggregate of accessory pigments and proteins called the (56) ____________ complex. This complex traps light energy and transmits it to the (57) ____________ centre, P700 in PSI and P680 in PSII, whereupon it becomes oxidized, donating its electrons to the (58) __________ acceptor.

Matching

Match each of the following terms with its correct definition.

59. ___	Chlorophylls	A.	Electrons with relatively low energy level
60. ___	Carotenoids	B.	Process by which the energy from an excited state electron is transferred to a neighbouring molecule
61. ___	Ground state electrons	C.	Chemical that accepts electron from reaction centre molecule
62. ___	Excited state electrons	D.	Chlorophyll *a* (P700 and P680) molecules that receive energized electrons from pigments of photosystems
63. ___	Fluorescence	E.	The amount of energy absorbed at different wavelengths of light
64. ___	Primary acceptor	F.	Green photosynthetic pigments present in plants, algae, and cyanobacteria
65. ___	Absorption spectrum	G.	Accessory photosynthetic pigments present in plants, algae, and cyanobacteria
66. ___	Action spectrum	H.	Light energy released by excited electrons returning to ground state
67. ___	Photosystem	I.	The effectiveness of each wavelength at driving photosynthesis
68. ___	Antenna complex	J.	Photosynthetic pigment found in photosynthetic prokaryotes other than cyanobacteria
69. ___	Reaction centre	K.	Electrons that have absorbed light energy and have a relatively high energy level
70. ___	Inductive resonance	L.	Cluster of light-absorbing pigments and proteins located in the thylakoid membranes
71. ___	Bacteriochlorophyll	M.	An aggregate of chlorophyll and carotenoid molecules in a photosystem

Short Answer

72. Explain the major role of the light-dependent reaction in photosynthesis.

____

____

____

73. Explain the role of accessory pigments in photosynthesis.

____

____

____

74. Explain the relevance of the P700 designation.

____

____

____

True/False

Mark if the statement is true or false. If the statement is false, justify your answer in the line below each statement.

75. ___________ _ One can determine the absorption spectrum for chlorophyll by measuring oxygen production as a function of wavelength.

____

76. ___________ P700 and P680 are different types of chlorophyll.

____

7.3 Photosynthetic Electron Transport [pp. 145–150]

Section Review (Fill-in-the-Blanks)

The evolution of (77) _____________ photosynthesis was a critical transition in the evolution of life on Earth because of the resulting liberation of oxygen through the splitting of (78) ____________. This oxygen is released into the (79) ____________; the resulting change in Earth's atmosphere and life was possible due to the evolution of photosystem (80) ___________. When PSII absorbs light energy, the energy is funnelled to the reaction centre chlorophyll, (81) ____________, which then transfers electrons to the (82) ___________ acceptor, and the electrons continue through a series of associated (83) ___________ transfer molecules. Having lost its electrons, the reaction centre pigment becomes a very powerful (84) ___________, and this enables it to split (85) ________. Similar to what happens in cellular (86) ___________, the energy from the electrons flowing through the chain is used to create a(n) (87) ___________ gradient, which triggers (88) __________ synthesis by a process called (89) ______________. In noncyclic transport, after losing their energy, the electrons from PSII are passed to $P700^+$, which has undergone (90) ____________ as a result of light absorption. Absorbing light, the electrons in this molecule are excited again, and a similar process of oxidation and (91) ___________ transfer ensues before the electrons are finally transferred to $NADP^+$ by the enzyme $NADP^+$ (92) _________. PSI can also work independently of PSII in a process of (93) ______________ electron transport. This process makes additional (94) __________ but no (95) ___________.

Matching

Match each of the following terms with its correct definition.

96. ___ ATP synthase
97. ___ Proton-motive force
98. ___ Chemiosmosis
99. ___ Noncyclic electron flow
100. ___ Cyclic electron flow
101. ___ $NADP^+$ reductase
102. ___ Cytochromes
103. ___ Plastoquinone
104. ___ P680*
105. ___ $P680^+$

A. A one-way flow of electrons in the light-dependent phase of photosynthesis
B. The use of the potential energy of proton-motive force to do work
C. The reaction centre pigment of PSII that forms when the molecule absorbs a photon of light
D. The reaction centre pigment of PSII after being oxidized by the primary acceptor
E. Membrane-bound enzyme that generates ATP
F. A mobile electron carrier that reduces P700
G. The proton gradient generated by electron transport chains
H. A flow of electrons that only involves PSI and generates ATP but no NADPH
I. The enzyme associated with noncyclic electron flow that reduces $NADP^+$, forming NADPH
J. A complex of electron carriers in electron transport chains with bound iron atoms that alternate between the Fe^{3+} and Fe^{2+} oxidation states

Sequence

106. Arrange the following steps of noncyclic electron flow in their correct sequence. One or more of these terms may be used twice.

A. Photosystem I

B. Photosystem II

C. NADPH

D. Electron transport chain

_____ _____ _____ _____

True/False

Mark if the statement is true or false. If the statement is false, justify your answer in the line below each statement.

107. ___________ _ The purpose of cyclic electron flow is to make extra NADPH.

____

108. ___________ ATP synthesis in photosynthetic processes happens by the chemiosmotic flow of electrons through the ATP synthase.

____

109. ___________ Photosynthetic electron flow results in a higher concentration of protons in the stroma.

____

Complete the Table

110. Complete the following table.

Process/structure	*Role in light-dependent reaction*
A. Noncyclic electron flow	
B. P680*	
C. Thylakoid membrane	
D. $P680^+$	

Short Answer

111. Explain the reason for and the process by which water is split in the light-dependent phase of photosynthesis.

____

____

____

112. Explain how cyclic photosynthesis relates to the subsequent needs to the Calvin cycle.

____

____

____

113. How many photons must be absorbed to generate one molecule of O_2? Explain how this is determined.

____

____

____

7.4 The Calvin Cycle [pp. 150–152]

7.5 Photorespiration and CO_2-Concentrating Mechanisms [pp. 152–157]

Section Review (Fill-in-the-Blanks)

In the Calvin cycle, ATP and NADPH are used to reduce CO_2 to (114) ____________. This cycle occurs in the (115) ___________ of the chloroplast and the (116) ____________ of photosynthetic bacteria. There are (117) ___________ stages in this cycle. In the first stage, CO_2 combines with (118) ____________, a five-carbon molecule, in an enzymatic reaction catalyzed by (119) ________________. The product is two molecules of (120) ___________. This is further reduced in the second stage, using two molecules of (121) _____________ and two molecules of (122) ___________. (123) ____________ is regenerated in the third stage of the cycle through a series of (124) _____________ reactions. Because of the need to regenerate this 5-C intermediate, it takes (125) ____________ cycles to produce one surplus (126) __________, which is used as a building block for the subsequent synthesis of many other (127) ____________ molecules. Although arguably the most important enzyme in the biosphere, (128) ___________ is surprisingly inefficient due to the fact that it can also act as a(n) (129) ____________, catalyzing the combination of O_2 with (130) _____________. The products of this reaction are (131) ______________ to the cell, and elimination of these molecules results in the release of (132) ______________. Because this process uses O_2 and produces CO_2, it is referred to as (133) __________________. The problem is compounded in hot environments because the increased (134) ____________ dramatically reduces the solubility of (135) __________ in the aqueous environment of the stroma. To prevent this depletion of energy, some plants have developed another carboxylation enzyme, (136) ______________ __________________. This enzyme does not react with oxygen and combines CO_2 with (137) ________________ to form the four-carbon molecule (138) _______________; plants that make this enzyme are therefore called (139) ____________ plants. Because this pathway has a higher (140) ___________ demand than the Calvin cycle, it is only used to supplement the activity of (141) ____________. This supplemental enzyme is either used in different leaf cells than rubisco or, in the case of the (142) _____________ plants, at a different time of day. Both strategies efficiently reduce exposure of rubisco to (143) ___________. In both strategies, the storage product of the C_4 reaction, malate, is ultimately (144) ____________ to release (145) ___________ for reaction with rubisco.

Matching

Match each of the following terms with its correct definition.

146. ___	Photorespiration	A.	Regulated openings on leaves for the diffusion of gases and water
147. ___	Oxaloacetate	B.	Plants that make oxaloacetate and release CO_2 for Calvin cycle
148. ___	Malate	C.	Plants that fix CO_2 to make G3P and cellular organic structures
149. ___	C_3 plants	D.	Plants that make malate at night and release CO_2 in chloroplasts during the day
150. ___	C_4 plants	E.	Light-stimulated use of oxygen and release of CO_2 in chloroplasts
151. ___	CAM plants	F.	Enzyme used by aquatic photoautotrophs for CO_2 concentration
152. ___	Oxygenase	G.	Immediate product of CO_2 fixation in C_4 cycle
153. ___	Stomata	H.	Four-carbon chemical used by some plants for CO_2 storage
154. ___	PEP carboxylase	I.	Enzymatic activity of rubisco that results in loss of carbon and energy from the cell
155. ___	ATP-dependent bicarbonate pump	J.	Enzyme used by C_4 and CAM plants for CO_2 fixation

Complete the Table

156. Complete the following table.

Chemicals/structure	*Role in Calvin cycle*
A. RuBP	
B. G3P	
C. 3PGA	
D. Rubisco	
E. Stroma	

157. Complete the following table with specific information on each step of photosynthesis.

Stage	*Location in chloroplast*	*ATP used/ produced*	*NADPH used/ produced*	*O_2 used/ produced/ n.a.*	*CO_2 used/ produced n.a.*	*Glucose produced*
A. Light dependent						
B. Calvin cycle						

True/False

Mark if the statement is true or false. If the statement is false, justify your answer in the line below each statement.

158. ____________ The Calvin cycle produces one molecule of glucose per cycle.

____

159. ____________ In sugar cane, all of the glucose produced during the Calvin cycle is stored as sucrose.

____

160. ____________ Oxygen is a competitive inhibitor of rubisco.

____

161. ____________ C_4 plants tend to predominate in warmer climates.

Short Answer

162. Discuss the hypothesis that is used to explain why rubisco is so widespread among photoautotrophic life forms despite its obvious inefficiency.

163. Discuss the three different strategies that have evolved among photoautotrophs for compensating for the inefficiency of rubisco.

164. What determines whether a plant is likely to use the physical separation of the C_4 and C_3 cycles or the separation in time, as is used in the CAM plants?

SELF-TEST

1. All of the following groups include photosynthetic organisms except [p. 143]
   a. Bacteria
   b. Algae
   c. Plants
   d. Fungi

2. Which of the following is not produced by the light-dependent reactions? [p. 140]
   a. NADPH
   b. ATP
   c. O_2
   d. H_2O

3. In eukaryotes, which of the following acts as the reaction centre molecule? [p. 143]
   a. Carotenoids
   b. Chlorophyll *a*
   c. Chlorophyll *b*

d. All of the above

4. Which of the following are present in photosystems? [pp. 143–144]

   a. Carotenoids
   b. Chlorophyll *a*
   c. Chlorophyll *b*
   d. All of the above

5. In cyanobacteria, which of the following provides electrons to light-dependent reactions? [p. 140]

   a. Water
   b. Chlorophyll
   c. ATP
   d. NADPH

6. Where are photosynthetic pigments present in plant cells? [p. 142]

   a. Cytoplasm
   b. Stroma of chloroplast
   c. Double membranes surrounding chloroplast
   d. Thylakoid membranes

7. Which of the following produces ATP and NADPH? [pp. 146–148]

   a. Calvin cycle
   b. Cyclic electron flow
   c. Noncyclic electron flow

8. All of the following are part of the Calvin cycle except [pp. 140, 150–151]

   a. Chlorophyll
   b. Rubisco
   c. RuBP
   d. G3P

9. How many times must the Calvin cycle go around before one molecule of G3P can be directed toward the synthesis of glucose and other organic molecules? [p. 151]

   a. 1
   b. 2
   c. 3
   d. 6

10. How many molecules of 3PGA are made during each Calvin cycle? [p. 151]

    a. 1
    b. 2
    c. 3
    d. 6

11. Most of the G3P formed during the Calvin cycle is used to regenerate ____________. [p. 151]

a. Rubisco
b. Glucose
c. RuBP
d. CO_2

12. Plants and other photosynthetic organisms that produce organic chemicals from CO_2 are called ___________. [pp. 140–142]

a. Photoautotrophs
b. Heterotrophs
c. Consumers
d. Decomposers

13. Structures through which plants exchange oxygen and CO_2 are called [p. 154]

a. Mesophyll cells
b. Bundle sheath cells
c. Cuticle
d. Stomata

14. Which of the following types of plants undergo photorespiration during hot days? [pp. 154–157]

a. C_3 plants
b. C_4 plants
c. CAM plants

15. C_4 and CAM store CO_2 as [pp. 155–157]

a. Rubisco
b. RuBP
c. G3P
d. Malate

16. Which of the following statements about cyclic photosynthesis is true? [pp. 149–150]

a. It produces ATP and NADPH
b. It produces ATP but no NADPH
c. It produces oxygen
d. It fixes CO_2

17. Which of the following statements is false? [p. 151]

a. The product of the Calvin cycle is glucose
b. The product of the Calvin cycle is the building block for all other cellular organic molecules
c. The product of the Calvin cycle is the building block for energy storage molecules such as starch
d. Most of the table sugar we use is the result of the Calvin cycle in sugar cane and sugar beets

18. Which of the following statements about rubisco is false? [pp. 152–153]

a. It is a multiprotein complex
b. Genes from the nucleus and the chloroplast are required for its synthesis
c. It has a higher affinity for oxygen than for CO_2
d. Under normal atmospheric conditions and moderate temperatures, it is only about 75% efficient

19. Which of the following statements about phytoplankton is true? [pp. 153–154]

a. Many produce PEP carboxylase in order to overcome the problem of photorespiration
b. Many have an active transport protein in their plasma membranes for pumping dissolved oxygen out of the cytosol
c. Many have an active transport protein in their plasma membranes for pumping bicarbonate into the cytosol
d. Many use the C_4 cycle to overcome the problem of photorespiration

20. Which of the following statements is true? [p. 156]

a. C_4 plants are as prevalent as C_3 plants in temperate regions such as Canada
b. CAM plants are found predominantly in regions with hot days and cool nights
c. C_4 plants use the C_4 pathway in bundle sheath cells and the Calvin cycle in mesophyll cells
d. CAM plants make and store malate in vacuoles during the day and use the Calvin cycle at night

INTEGRATING AND APPLYING KEY CONCEPTS

1. Discuss the problem of photorespiration and why certain environments are more problematic for this than others. Then discuss how photosynthetic organisms from different environmental niches have evolved to adapt to their conditions and mitigate the problems inherent in their environment that would lead to photorespiration.

2. Draw a single thylakoid within a larger "stroma" zone. Now draw the components of the light reaction, in their correct orientation (do not worry about the names of the electron transporters), including the ATP synthase. Show with arrows the absorption of photons, the movement of electrons and protons, the synthesis of any products, and where some may go for the Calvin cycle reactions.

3. Provide the equation for the conversion of CO_2 by the Calvin cycle to glucose, identifying the total numbers of ATP and NADPH required. Explain how this energy requirement is met by the light-dependent phase of photosynthesis.

OPTIONAL ACTIVITY

1. As discussed in this chapter, there are photosynthetic prokaryotes other than cyanobacteria. These other phototrophs possess bacteriochlorophyll rather than chlorophyll and do not split water during photosynthesis, so they are collectively referred to as the anoxygenic phototrophs. These organisms are probably more ancient than cyanobacteria and are important today in certain ecological niches. In addition, among the prokaryotes, there are organisms that have a chemoautotrophic metabolism, where they use inorganic chemicals such as H_2S for energy and CO_2 for their carbon source. In fact, recent evidence suggests that this latter group of prokaryotes may be the dominant primary producers in the world's oceans. Find out a little more about these various organisms.

8 Cell Communication

CHAPTER HIGHLIGHTS

- Cells communicate with one another by means of chemical messengers; such communication coordinates the activity of cells.
- The source cell produces the message, and the target cell receives the message.
- Receptors for chemical messengers can be located on the plasma membrane (cell surface) or in the interior of the target cell.
- Some cell surface receptors are also enzymes (e.g., tyrosine kinases), while others associate with G proteins to initiate the formation of second messengers inside the cell (e.g., cAMP, IP_3).
- Internal receptors regulate gene transcription.

STUDY STRATEGIES AND LEARNING OUTCOMES

- The study of cell communication is important for understanding how the activities of cells are controlled and coordinated. Many of these concepts will be important for understanding whole organism function later (growth, development, reproduction, osmoregulation, etc.).
- This chapter contains a lot of material that may be unfamiliar. Do not try to go through it in one sitting. Take one section at a time. Start by looking at the figures and legends and then read through the text, making note of boldface terms.
- Reexamining the textbook diagrams after going through a major section should help bring you back to the general process so that you are not overwhelmed by the various terms and names of the component parts. Remember that for many of these pathways, the common denominator is the establishment of a phosphorylation cascade.

- Remember, turning off a pathway is as important as turning it on when considering mechanisms of control. Therefore, carefully review both the activation and the inactivation of a particular signalling pathway.
- Work on the companion section of the study guide to cement the critical terms and concepts.
- Draw a big picture of a cell and the different signal transduction systems. Be sure to label the parts of each system. Repeated drawing will ensure that you are familiar with all components of each system.

By the end of this chapter, you should be able to

- Identify the three stages of signal transduction and explain what determines the specificity of a signal
- Identify the nature and role of protein kinases and explain how they accomplish signal amplification
- Identify the nature and role of receptor tyrosine kinases and G protein–coupled receptors, being able to explain for each how they transduce a signal, how they differ from each other, and how they accomplish signal amplification
- Identify the differences between, and examples of, effectors, first messengers, and second messengers
- Describe the role and function of steroid hormone nuclear receptors and explain how they differ from and how they are similar to the surface-receptor signal transduction pathways

TOPIC MAP

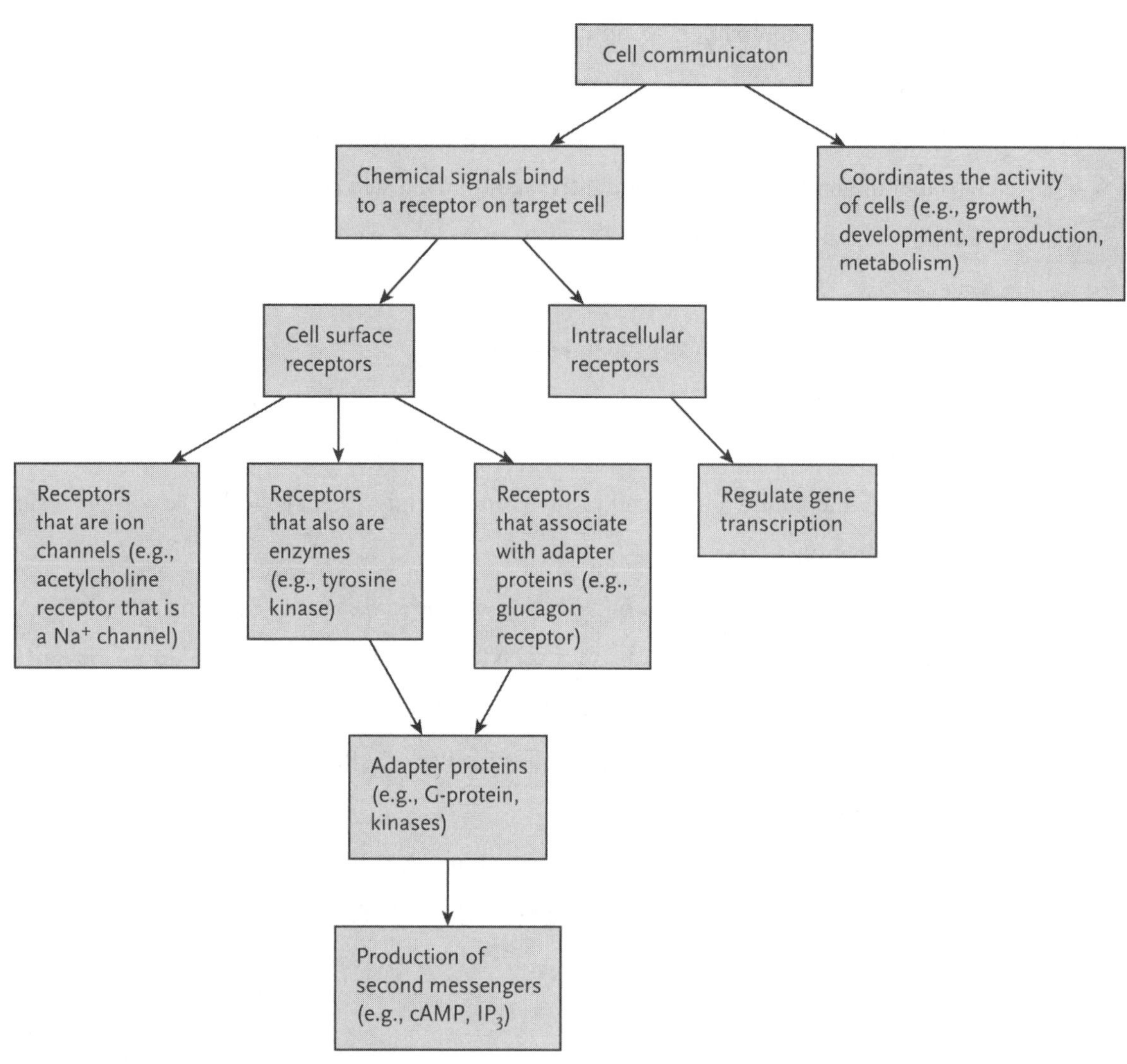

INTERACTIVE EXERCISES

Why It Matters [pp. 161–162]

8.1 Cell Communication: An Overview [pp. 162–164]

Section Review (Fill-in-the-Blanks)

Cell communication is critical for the function and survival of (1) __________ organisms, including animals but also (2) __________, protists, and fungi. Cells communicate with each other in three ways: through direct (3) _________ of communication, specific (4) __________ between cells, and (5) __________ chemical messengers. With the first type, small molecules and (6) ___________ pass through structures such as (7) ___________ _____________ in animal cells and (8) ______________ in plant cells. In the second type, surface molecules such as (9) ___________ allow direct contact with other cells. The most common method of cell communication is (10) ____________ communication, where a controlling cell releases a (11) __________ molecule that causes a response in target cells. Recognition of the signal by the target cells is the result of the specific (12) ___________ of the signal molecule by surface (13) _____________. This type of cell communication has three sequential steps: (14) ________, (15) ____________, and (16) __________. Quite often the second step may not be a single reaction but a series of reactions in what is referred to as a (17) ___________ cascade. The series of events through all three steps is referred to as signal (18) ____________.

Matching

Match each of the following terms with its correct definition.

19. ___	Gap junctions	A.	A form of bacterial intercellular communication that is based on population density
20. ___	Plasmodesmata	B.	A sequential series of reactions between different intracellular molecules leading to a cellular response
21. ___	Cell adhesion molecules	C.	The process of converting the signal into the form that triggers the cellular response
22. ___	Quorum sensing	D.	Communication channels between animal cells
23. ___	Signalling cascade	E.	Communication channels between plant cells
24. ___	Signal transduction	F.	Receptor-like glycoproteins in plasma membranes

Labelling

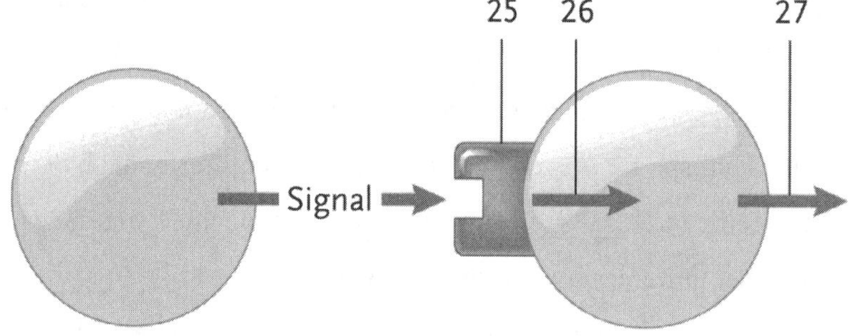

25. ____________

26. ____________

27. ____________

True/False

Mark if the statement is true or false. If the statement is false, justify your answer in the line below each statement.

28. ____________ Gap junctions are critical for communication between cardiac cells.

____

29. ____________ Cell adhesion molecules are particularly important in the regulation of embryonic development.

____

30. ____________ Cell surface signal receptors are peripheral membrane proteins.

____

Short Answer

31. What contribution(s) to science led to the award of the Nobel Prize to Earl Southerland? ________

____

____

8.2 Characteristics of Cell Communication Systems with Surface Receptors [pp. 165–167]

8.3 Surface Receptors with Built-in Protein Kinase Activity: Receptor Tyrosine Kinases [pp. 167–168]

8.4 G Protein–Coupled Receptors [pp. 168–172]

8.5 Pathways Triggered by Internal Receptors: Steroid Hormone Nuclear Receptors [pp. 172–174]

8.6 Integration of Cell Communication Pathways [pp. 174–175]

Section Review (Fill-in-the-Blanks)

In mammals and other (32) ____________, cell communication systems using cell surface receptors involve two major types of extracellular signal molecules: (33) ___________ and (34) ____________. There are three components to systems that involve surface receptors: (35) ___________ signal molecules, the surface receptors, and the internal response (36) ___________, which lead to the cellular response. Binding of the signal molecule to a receptor glycoprotein induces a (37) ____________ change in the receptor, which results in the (38) ____________ of the cytoplasmic domain of the receptor. Transduction inside the cell involves a common enzymatic modification, that is, (39) ______________. The enzymes that perform this reaction are called protein (40) ______________. In some cases, these enzymes act in a (41) ____________ or series of reactions. The response is reversed by enzymes called (42) ____________ and by removal of the receptor molecule from the membrane surface by (43) ____________. The stepwise magnification of events in a transduction pathway is called (44) ___________ and is the result of the involvement of (45) ___________ in each step of the pathway. Receptor tyrosine kinases add (46) _____________ groups to the (47) ____________ amino acids on the receptor molecules themselves and then to the (48) __________ proteins. Upon binding a signal molecule, these receptor proteins must move together to form (49) ____________ in the plasma membrane in order to activate their (50) ___________ activity. Target proteins are phosphorylated upon recognizing and (51) _____________ the (52) ____________ sites on the receptor. In signal transduction mediated by G protein–coupled receptors, binding of the (53) ___________ messenger, that is, the signal molecule, also triggers a (54) ___________ change in the receptor protein; however, instead of autophosphorylating, the activated domain triggers (55) ____________ binding by an associated G protein. This binding leads to signal transduction via a membrane-associated (56) ___________ enzyme that generates a (57) ___________ messenger, an internal, nonprotein signal molecule. In one of the two major pathways of this type, this enzyme is adenylyl cyclase, an enzyme that converts (58) ___________ to (59) ____________. This type of pathway results in the activation of specific protein (60) _____________. The second major G protein–dependent pathway uses (61)

________________ C as the effector. There are two second messengers in this pathway: (62) ____________ and (63) ______________; the former activates transport proteins in the ER to release stored (64) ________. These can act alone or with the second messenger to activate specific protein (65) _______________. When this happens in multiple pathways of this type, a phenomenon called cross- (66) ___________ may occur. This involves the participation of the activated protein (67) ___________ in their own as well as other signal transduction pathways, generating a complex (68) ____________ of interactions that gives rise to an integrated cellular response. Some pathways link receptor tyrosine kinases with the G protein (69) __________ to regulate gene expression. The activated G protein again turns on a (70) ___________ cascade. Another form of cellular communication involves signal molecules that penetrate through the (71) ___________ ___________ to bind internal receptors. These types of signals are primarily, but not exclusively, (72) _____________ ____________. The internal receptors are referred to as (73) _____________ receptors because upon binding their signal molecule, the complex enters the (74) _____________ where the activated (75) __________-binding domain of the receptor directly induces gene expression through its binding of a (76) ___________ sequence of a gene.

Matching

Match each of the following terms with its correct definition or descriptor.

77. ____	Autophosphorylation	A. A type of G protein that links receptor tyrosine kinases with regulation of gene expression
78. ____	Phosphatases	B. A class of peptide hormones that affect things such as cell differentiation and division
79. ____	Hormone	C. The magnification of events in signal transduction pathways due to the catalytic activity of enzymes in the pathway
80. ____	MAP	D. Addition of PO_4^{3-} group by a protein to another part of the same protein
81. ____	Growth factors	E. Enzymes that remove PO_4^{3-} groups from proteins
82. ____	Signal amplification	F. The enzyme that makes the second-messenger molecules IP_3 and DAG
83. ____	Ras protein	G. Usually steroid and peptide molecules released by specialized gland cells

84.	____	Receptor tyrosine kinases	H. Protein kinases that are activated by the Ras protein
85.	____	G protein–coupled receptors	I. The enzymes responsible for synthesis of cAMP
86.	____	G proteins	J. The interaction of signalling pathways
87.	____	Adenylyl cyclases	K. Surface receptor proteins that have autokinase activity
88.	____	Phospholipase C	L. Cytoplasmic molecule switch proteins activated by binding of GTP to give signal transduction
89.	____	Cross-talk	M. Surface receptor proteins characterized by seven transmembrane domains and a cytoplasmic domain

True/False

Mark if the statement is true or false. If the statement is false, justify your answer in the line below each statement.

90. ____________ Neurotransmitters may act as hormones.

____

91. ____________ Once released, signal molecules are very stable, persisting for long periods in the body.

____

92. ____________ Receptor tyrosine kinases are unique to animals.

____

93. ____________ In signal transduction involving G protein–coupled receptor proteins, the G protein generates the second messenger.

____

94. ____________ Both products of the effector protein phospholipase C activate protein kinases.

____

95. ____________ Steroid hormone receptors that bind DNA are hydrophilic proteins.

---

96. ____________ Multiple second-messenger pathways can be integrated to modify a cellular response.

Sequence

97. Arrange the following steps of a G protein–coupled response in the correct sequence.

A. Effector

B. Cellular response

C. Protein kinases

D. First messenger

E. Target proteins

F. G proteins

G. Second messenger

H. Receptor

_____ _____ _____ _____ _____ _____ _____ _____

Short Answer

98. Explain the basis of the specificity of hormones and other signal molecules with their target cells.

---

99. Compare and contrast receptor tyrosine kinases from G protein–coupled receptors.

---

100. Explain how a G protein is involved in regulation of cell division.

Complete the Table

101.

Messenger	*Function*
cAMP	A.
B.	Stimulates calcium release from the ER lumen into the cytoplasm
DAG	C.
Receptor	*Mode of action*
D.	Binds to DNA and activates gene expression
Receptor tyrosine kinase	E.
F.	Activates G proteins, which, in turn, stimulates effector enzymes
G.	Binds steroid hormone to enter the nucleus and activate gene expression

102.

Signal pathway	*Method in inactivation*
Receptor tyrosine kinase pathway	A.
cAMP-dependent G protein pathway	B.
IP_3–DAG pathways	C.
Ras-dependent	D.
Steroid nuclear receptor pathway	E.

SELF-TEST

1. Which of the following is not involved in signal transduction? [p. 164]
   a. Reception
   b. Transduction
   c. Response
   d. Synthesis of signal molecule

2. Which of the following is not a signal molecule? [pp. 163–165]
   a. Neurotransmitters
   b. Insulin
   c. Epinephrine
   d. Membrane glycoproteins

3. Which of the following does not activate protein kinases? [pp. 170–174]
   a. Steroid hormone nuclear receptors
   b. DAG
   c. Ras
   d. cAMP

4. Which of the following is not located on the surface of cells? [pp. 162, 167–168, 172–173]
   a. Steroid nuclear hormone receptors
   b. G protein–coupled receptors
   c. Receptor tyrosine kinase
   d. Gap junctions

5. Which of the following makes cAMP? [p. 170]
   a. Protein kinase
   b. Phosphodiesterase
   c. Phospholipase C
   d. Adenylyl synthase

6. The following are activated by a G protein: [pp. 149–153]
   a. Phospholipase C
   b. MAP kinases
   c. Adenylyl cyclase
   d. All of the above

7. Steroid nuclear hormone receptors evoke cellular responses by ______. [p. 173]
   a. Activating G proteins
   b. Binding to DNA
   c. Promoting Ca^{2+} release from ER lumen
   d. Activating protein kinases

8. Which of the following is not an activity associated with receptor tyrosine kinases? [pp.

167–168, 171–172]

a. Bind to DNA
b. Undergo autophosphorylation
c. Activate Ras
d. Phosphorylate other cellular proteins

9. Which of the following is involved in turning off signal transduction pathways? [pp. 166, 169–170, 175]

a. Phosphodiesterases
b. Protein phosphatases
c. Cross-talk
d. Endocytosis
e. All of the above

10. Which of the following is not transmitted by gap junctions to facilitate intracellular communication? [p. 175]

a. Ca^{2+}
b. cAMP
c. IP_3
d. G proteins

11. ______ is a cell adhesion molecule that can elicit response in adjacent cells. [p. 175]

a. Phospholipase C
b. Insulin
c. Epinephrine
d. Integrin

12. Which of the following statements is false? [pp. 162, 164, 172–173]

a. Bacteria exhibit intracellular communication
b. Plants exhibit intracellular communication
c. Steroid hormones only bind internal receptors
d. Certain Ras proteins are involved in cancer

13. Which of the following cell communication molecules would generate an amplification of the cellular response? [p. 166]

a. Receptor tyrosine kinases
b. Second messengers
c. IP_3
d. cAMP

14. Which of the following statements is true? [pp. 169–170, 173]

a. Some G protein–coupled receptor pathways use two types of high-energy phosphate molecules
b. The Ras protein is activated by binding GTP

c. G proteins are so named because they are activated by GDP
d. Steroid hormones must be actively transported across the plasma membrane

15. Which of the following statements about G protein–coupled receptor pathways is incorrect? [pp. 168, 170, 174–175]

a. They are involved in the perception of odour
b. The number of types of receptors in this system is far less than the number of different tyrosine kinase receptor proteins
c. They exhibit cross-talk
d. They can alter membrane transport

16. Which of the following bacterial infections involve a toxin that targets G proteins? [p. 169]

a. Whooping cough
b. Traveller's diarrhea
c. Cholera
d. All of the above

17. Which of the following is not a characteristic of steroid hormones? [pp. 165, 172–174]

a. They are small lipid-soluble molecules
b. They circulate in the bloodstream
c. They are produced by neurons
d. They may interact with internal or surface receptors

18. Which of the following statements about the ethylene signal molecule is true? [p. 174]

a. It is a gas at physiological temperature
b. It is a plant hormone
c. It is soluble in water and lipids
d. Its receptor is found within the hydrophobic interior of an internal membrane
e. All of the above

19. Which of the following statements about second messengers is true? [p. 169]

a. They are the G proteins
b. They travel through the bloodstream
c. They bind specific surface receptors
d. They bind specific internal receptors
e. All of the above

20. Which of the following proteins are not integral membrane proteins? [pp. 167–168, 173–174]

a. Tyrosine kinase receptor proteins
b. G protein–coupled receptor proteins
c. Steroid hormone nuclear receptor proteins

INTEGRATING AND APPLYING KEY CONCEPTS

1. Hormones are not unique to animals: plant growth and development are also controlled by hormones. Based on what you learned in this chapter, discuss how certain structural features and signalling pathways could combine to allow the hormone to propagate a response at a site distant from the source of the hormone signal molecule.
2. Individuals with insulin-independent diabetes mellitus can produce insulin, but it is not effective in stimulating glucose uptake. Explain potential defects in insulin signalling that would explain this insulin.
3. Discuss how a steroid hormone could trigger very different responses in the same cell.

OPTIONAL ACTIVITY

1. The science of genomics, which involves sequencing whole genomes of cells or organisms, has revolutionized how we study biological systems, medicine, and biotechnology. The leader in this area is J. Craig Venter, the person who spearheaded the human genome sequencing project and who now leads a sequencing institute. This institute has, in the last few years, reported novel findings about tyrosine kinases. In the Global Ocean Survey (GOS), bacterial DNA was isolated straight from ocean waters spread over an ~8000 km sampling route and then sequenced. The data revealed eukaryotic-like tyrosine kinase enzymes that outnumbered the known paradigm eukaryotic tyrosine kinases. In another report, sequencing the genomes of certain human brain tumour cells identified three associated mutations in tyrosine kinase receptor genes. Use the Internet to learn more about these studies. One place to start would be the home page for the J. Craig Venter Institute (http://www.jcvi.org/cms/home/).

9 Cell Cycles

CHAPTER HIGHLIGHTS

- Multicellular organisms grow through the process of mitosis.
- Mitosis results in the generation of two genetically identical daughter cells from a parent cell.
- Mitosis does not happen in certain mature cells where growth is not required.
  - The process of undergoing cell growth and division is subdivided into multiple steps, the total process representing the cell cycle.
  - In healthy organisms, this process is stringently regulated by internal and external factors.
  - Eukaryotic cells require special structures such as microtubules and mitotic spindles to ensure the proper division of cellular material to the daughter cells.
- Cancerous cells are characterized by their loss of control of the cell cycle.
- Senescence is the gradual loss of the ability of "older" cells to grow and divide.
- Cell division in prokaryotes involves binary fission instead of mitosis.

STUDY STRATEGIES AND LEARNING OUTCOMES

- As you are reading, pay attention to the main ideas first: what is the purpose of a step and the main features? Then worry about the various names of the molecules and structures.
- Make sure you understand the components and role of the mitotic spindle and how they as well as cytokinesis differ between plant and animal cells.
- Go through each section of the textbook and then work through the companion section of the study guide.
- After reading through all of the chapter sections of the textbook, try drawing a flow chart of the cell cycle, starting with a single diploid parental cell. Identify points at which there may be some control and work these control mechanisms or processes into your flow chart. Finally, indicate with the daughter cell the various possible alternative fates to normal cell cycling.

By the end of this chapter, you should be able to

- Describe the stages of the cell cycle and of mitosis and identify how they are regulated
- Understand the components and role of the mitotic spindle and how they, as well as cytokinesis, differ between plant and animal cells
- Define "Hayflick factors," tumour, and metastasis
- Explain how cell growth and division happen in prokaryotes

TOPIC MAP

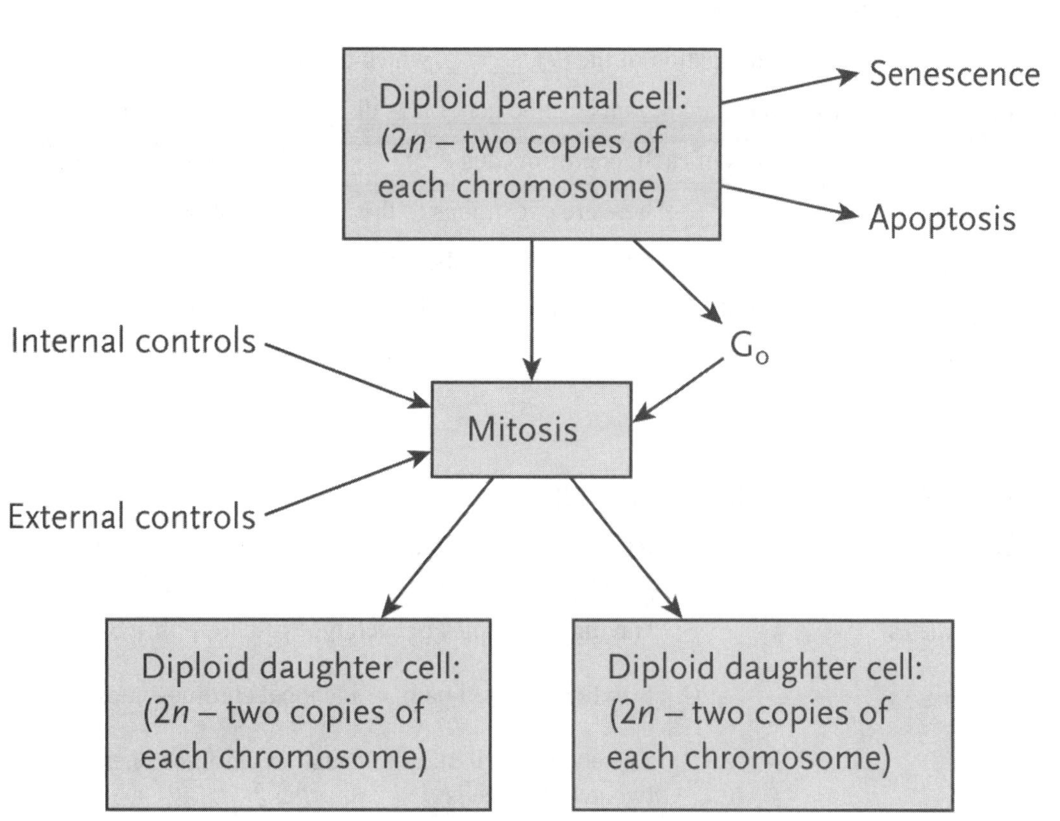

INTERACTIVE EXERCISES

Why It Matters [pp. 179–180]

9.1 The Cycle of Cell Growth and Division: An Overview [pp. 180–181]

In eukaryotes, the cell cycle consists of (1) ______ interrelated systems. These systems are growth, ending with (2)________ ____________, followed by the equal segregation of replicated (3) ______ and, finally, division of the cytoplasm or (4) ______. The result of this process is two (5) ______ cells that are identical to the (6) ______ cell. In eukaryotic cells, organs, and tissues, where (7) ______ identical copies of the parental cell are required, the process of replication that is used is (8) ______. The central aspect to this process is the replication and segregation of the (9) ______, which are linear DNA molecules with associated (10)______. In most eukaryotic organisms, the genome is distributed among several (11) _______, and most possess two copies of each, one from each (12)_______; their genome is therefore described as being $2n$, that is, (13) ______. There are exceptions to this, however: yeast can switch between $2n$ and $1n$, that is, (14)______. Regardless of the number of sets or the so-called (15) ______ of the organism, replication of each individual chromosome gives two identical copies, and these are referred to as sister (16) ______.

Matching

Match each of the following terms with its correct definition.

17. ____	Cell cycle	A.	Referring to the number of copies of each type of chromosome
18. ____	Cytokinesis	B.	Having two copies of each type of chromosome
19. ____	Diploid	C.	Newly replicated pairs of identical chromosomes
20. ____	Haploid	D.	The equal distribution of daughter chromosomes to each of two daughter cells
21. ____	Ploidy	E.	Growth followed by nuclear division and cytokinesis
22. ____	Sister chromatids	F.	Division of the cytoplasm
23. ____	Segregation	G.	Having one copy of each type of chromosome

True/False

Mark if the statement is true or false. If the statement is false, justify your answer in the line below each statement.

24. _____________ Some plant species may be polyploid.

____

25. ____________ In a multicellular organism, cells of different tissues contain different genetic information.

____

9.2 The Mitotic Cell Cycle [pp. 182–187]

There are two major phases of the cell cycle: (26) ______, which is the longest, and (27) ______. There are three specific stages in the former: (28)______, the period during which growth occurs and the one that varies most in duration; (29) ______, when the DNA is replicated; and (30)______, when the cell prepares to enter mitosis. Mitosis consists of five stages, beginning with (31) ______. In this phase, the chromosomes (32) ______, becoming visible in the microscope as thin threads. Also during this phase, the duplicated (33) ______ start to separate and generate the (34) ______. In the next stage, (35) ______, the (36) ______ membrane breaks down, the (37) ______ enters the former nuclear area, and the microtubules of the (38) ______ attach to each sister chromatid at a structure called a (39)______. In (40) _______ or stage 3, the (41) _______ is complete and has moved the chromosomes so that they are aligned at the (42) ______ plate. Then in (43) ______, the spindle separates the (44) _________ _______ and moves them to opposite (45) ______ poles, completing chromosome (46) ______. The chromosomes are now referred to as (47) ______ chromosomes. In (48) ______, the last phase, the chromosomes (49) ______, returning to their interphase state, the (50) ______ disassembles, and a new (51) ______ envelope forms around the chromosomes. This phase ends with the division of the cytoplasm in (52) _______. This process differs in animal and plant cells, involving (53) _______ in animal cells and cell (54) ______ formation in plant cells.

Matching

Match each of the following phases of the cell cycle with its correct definition.

55. ____	G_1 phase	A.	State of cellular arrest
56. ____	S phase	B.	Sister chromatids are apparent for the first time; spindle begins to develop
57. ____	G_2 phase	C.	Part of cytokinesis in plants; involves formation of a new cell wall from vesicles at the former spindle midpoint
58. ____	G_0 phase	D.	Stage during which DNA is replicated
59. ____	Prophase	E.	Stage preceding mitosis
60. ____	Prometaphase	F.	Spindle midpoint in metaphase
61. ____	Metaphase	G.	After nuclear division, cytoplasm divides and two daughter cells result
62. ____	Anaphase	H.	Stage during which spindles enter the former nuclear area
63. ____	Telophase	I.	Phase of cell growth and DNA replication

64. ____ Cytokinesis

65. ____ Interphase

66. ____ Mitosis

67. ____ Metaphase plate

68. ____ Cell plate

K. Sister chromatids move to opposite poles

L. Spindle disassembles, nuclear envelope reappears, chromosomes become less condensed

M. Sister chromatids are aligned at the midpoint of the spindle

N. Phase of cell growth after cytokinesis

O. Nuclear division; consists of 4–5 phases

True/False

Mark if the statement is true or false. If the statement is false, justify your answer in the line below each statement.

69. ____________ During G_1, cells are synthesizing RNA, proteins, and DNA.

____

70. ____________ In multicellular organisms, if cells or certain tissues are mature and do not need to increase in numbers or turn over, they enter G_0.

____

71. ____________ During G_2, cells are metabolically inactive.

____

Identification

72. For each image, identify the phase (top line) and the amount of DNA using the ploidy number (bottom line). For the first cell, A = G_1 phase and B = $2n$ DNA.

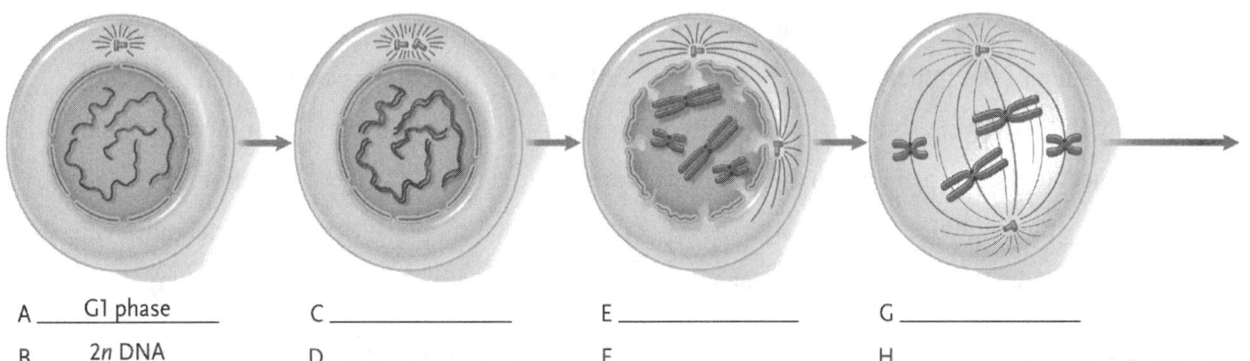

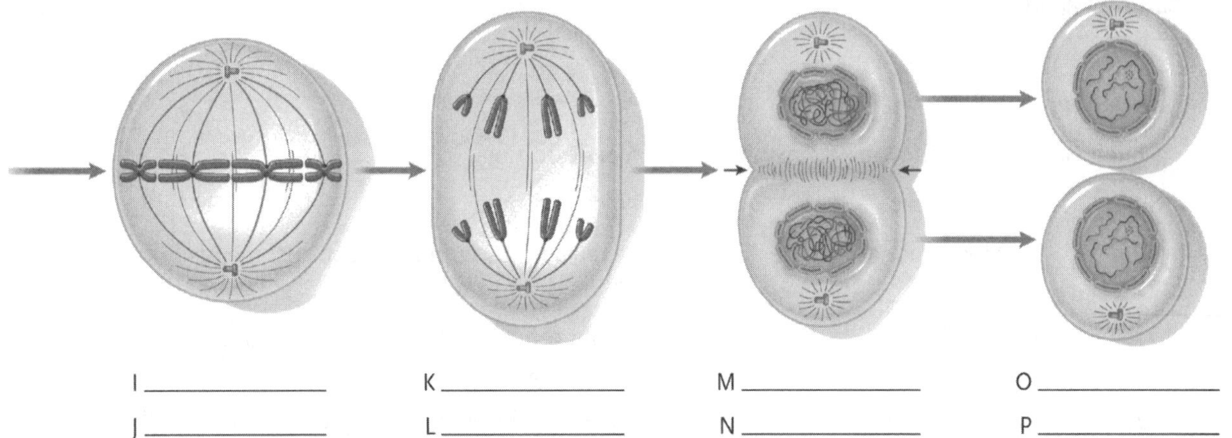

Short Answer

73. Explain what it is about the G_1/S transition that interests cancer researchers. ____________________

____

____

74. Explain the difference between chromosomes and sister chromatids. ____________________

____

____

75. Compare and contrast cell furrowing with cell plate formation. ____________________

____

____

9.3 Formation and Action of the Mitotic Spindle [pp. 187–189]

Mitosis and cytokinesis are both dependent on the formation of the (76)______. Spindle formation may develop differently in plants compared with in animals, due to the presence of a (77) ______ in animal cells but not in certain types of plants. This structure is considered the cell's microtubule (78) ______ centre, or MTOC. During mitosis, this structure divides and the two parts move apart, with a concomitant formation of the spindle (79)______. In plant cells, the microtubules assemble around the (80) ______. Microtubules in the spindle connect to the sister chromotids via (81) ______. Some microtubules do not connect to these structures, so these two types are referred to as (82) ______ microtubules and (83) ______ microtubules, respectively. The latter extend between opposite poles, overlapping with each other at the spindle (84) ______. During anaphase, chromosomes are believed to "walk" along their stationary microtubules, using motor proteins in the (85) ______. The microtubule (86) ______ as this structure passes over it. Meanwhile, the (87) ______ microtubules slide over each other, pushing the (88) ______ further apart.

Matching

Match each of the following terms with its correct definition.

89. ____	Centrosome	A. The centrosome
90. ____	Mitotic splindle	B. Two structures within the centrosome, usually arranged at right angles to each other
91. ____	MTOC	C. The microtubule structures involved in mitosis and cytokinesis
92. ____	Centrioles	D. The microtubule-organizing centre of the cell

True/False

Mark if the statement is true or false. If the statement is false, justify your answer in the line below each statement.

93. ____________ The centrioles are critical to mitosis and cytokinesis.

____

94. ____________ Certain types of plants possess multiple MTOCs.

____

9.4 Cell Cycle Regulation [pp. 190–194]

The cell cycle has internal (95) ______, which regulate the division process and prevent the cell from progressing to the next stage of the cycle unless the (96) ______ of the previous stage are complete. Internal regulation occurs primarily through the activity of (97) ______ complexes; the concentrations of these rise and fall with the cell (98) ______, hence the name. When in high enough concentrations, these complexes activate (99) ______ protein kinases, or CDKs, which, in turn, activate target proteins through (100) ______. This causes the cell to progress to the next stage of the cell (101) ______. After this transition, the (102) ______ are enzymatically degraded. The cell cycle is also regulated externally, primarily through (103) ______ ________ that recognize and bind signals such as (104) ______ hormones and (105) ______ factors or through (106) ______ with other cells. These trigger intracellular reactions that can include the addition of inhibitory or stimulating (107) ______ groups to cyclin/CDK:complexes. Healthy cells in organs and tissues are normally in contact with other cells and exhibit (108)______ ______; that is, they are in G_0 and are prevented from (109) ______. This control, as well as other cell cycle controls, is absent in cancer cells, which form masses of abnormal cells called (110) ______. Through a process called (111) ______, cells from the masses break free and spread to other sites within the (112) ______. In addition to regulating cell division, multicellular organisms regulate cell (113) ______ through an apparently ancient process called (114) ______. This is often used in development, when cells are (115)______, but it is also an important means of eliminating damaged or (116) ______ cells. A key player in this process is a protease from a family of normally inactive proteases called (117) _______.

Matching

Match each of the following terms/concepts with its correct definition.

118. ____ Contact inhibition	A.	Cyclin-dependent protein kinase
119. ____ Growth factors	B.	Cells are prevented from dividing due to the presence of adjacent cells
120. ____ Oncogene	C.	The gradual loss of proliferative ability in cells
121. ____ Tumour	D.	The cellular protein critical for regulation of cell cycle transition from G_1 to S
122. ____ Cellular senescence	E.	Gene that produces uncontrolled cell division
123. ____ Cyclin E	F.	External controls of cyclin:CDK activity
124. ____ CDK	G.	Mass of cell as a result of unregulated cell division

True/False

Mark if the statement is true or false. If the statement is false, justify your answer in the line below each statement.

125. ____________ Regulation of transitions between the different stages of the cell cycle is accomplished by varying the levels of CDKs.

____

126. ____________ Growth factors can cause a cell to enter G_1 from G_0.

____

Short Answer

127. Explain the term "cellular senescence" and why it happens.

____

____

____

128. Identify some of the cellular changes that can result in cancer.

____

____

____

9.5 Cell Division in Prokaryotes [pp. 194–195]

Prokaryotes do not undergo mitosis but divide by (129) ______ ______. In most prokaryotes, there is a single circular DNA (130) ______. Under optimum growth conditions, when cells are growing quickly, (131) ______ ______ occupies most of the time between cytokinesis events. Replication begins at a chromosomal sequence called the (132) ______ of replication and is catalyzed by enzymes located in the centre of the cell. Once this process is complete, the two (133) ______ migrate to the opposite poles. (134) ______ follows with the infolding of the plasma membrane and synthesis of a new cell (135) ______ between daughter cells.

True/False

Mark if the statement is true or false. If the statement is false, justify your answer in the line below each statement.

136. _____________ The newly replicated bacterial chromosomes migrate to the opposite poles of the cell through the action of bacterial spindles.

____

Short Answer

137. Define the term "binary fission."

____

____

____

SELF-TEST

1. Assuming that no errors occur in DNA replication, mitosis produces two daughter cells that are _________. [p. 181]

   a. Double in chromosomal content
   b. Genetically identical to the parental cell
   c. Able to grow indefinitely
   d. Slightly different from the parental cell

2. Which of the following phases is the most variable with respect to length? [p. 183]

   a. G_1 of interphase
   b. G_2 of interphase
   c. S phase of interphase
   d. Mitosis

3. When does the nuclear envelope break down? [p. 184]

   a. Prophase
   b. G_2 of interphase
   c. Telophase
   d. Prometaphase

4. Mature red blood cells lack a nucleus. In which phase of the cell cycle would you expect to find this cell? [p. 183]

   a. G_1 of interphase
   b. G_2 of interphase
   c. G_0 of interphase
   d. S phase of interphase

5. Which of the following is false with respect to a pair of sister chromatids? [pp. 184–185]

   a. They are genetically identical
   b. They will end up in different daughter cells
   c. They are held together by a single kinetochore
   d. Once separated, they are called daughter chromosomes

6. Plants differ from animal cells in that many plants ________. [p. 188]

   a. Lack a karyotype
   b. Lack microtubules
   c. Lack a centrosome
   d. Lack a microtubule organizing centre

7. Which of the following would prevent a cell from going from progressing to the M phase? [pp. 190–191]

   a. A decrease in the amount of cyclin B
   b. An increase in the amount of phosphate
   c. An increase in the amount of CDK2
   d. A decrease in phosphatase activity

8. What is the likely result of blocking contact inhibition? [pp. 191–192]

   a. Wounds would never heal
   b. A tumour might develop
   c. The cell would go into G_0 phase
   d. Both (b) and (c) are possible

9. A primary difference between mitosis in eukaryotic cells and binary fission in prokaryotes is that __________. [p. 194]

   a. Binary fission lacks cytokinesis
   b. Binary fission is not regulated
   c. Binary fission does not use centrioles
   d. DNA replication only occurs in mitosis

10. Which of the following statements about cells that undergo mitosis is false? [pp. 183–185]

    a. A cell in G_2 has twice as much DNA as in G_1
    b. DNA replication starts at a specific region, the origin of replication
    c. Chromatids move to opposite poles of the cell
    d. Checkpoints ensure that one phase is complete before the next phase is initiated

11. Which of the following statements about apoptosis is incorrect? [p. 193]

    a. It is the explanation for cellular senescence
    b. It is highly regulated
    c. It is required for the normal development of multicellular organisms
    d. It is an ancient mechanism

12. What is a "Haylick factor"? [p. 193]

    a. It is a molecule that helps regulate the cell cycle
    b. It is the MTOC of prokaryotic cells
    c. It is the critical enzyme that regulates apoptosis
    d. It is the term used for the as yet unknown cause(s) of cellular senescence

13. Which of the following statements about cell cultures is true? [p. 182]

    a. They are often used to test the toxicity or carcinogenicity (ability to cause cancer) of various chemicals
    b. They allow scientists to grow clones of an original cell
    c. They are possible for prokaryotic cells as well as eukaryotic cells
    d. All of the above

14. Which of the following statements about karyotypes is false? [p. 185]

    a. They can allow the identification of a species
    b. They are made using telophase chromosomes
    c. They allow the visual characterization of an organism based on the shape and size of sister chromatids
    d. None of the above

15. Cancer may be associated with _________. [p. 192]

    a. A mutation in a plasma membrane nutrient transport protein
    b. Yeast cells
    c. Loss of contact inhibition

d. (b) and (c)

16. External controls of the cell cycle may include _______. [pp. 190–191]

a. CDK2
b. Peptide hormones
c. Cyclins
d. Cyclases

17. Which of the following statements about mitosis is false? [p. 195]

a. It is believed to have evolved from binary fission
b. Because the eukaryotic yeast cells have a mitotic system that more closely resembles the binary fission process of prokaryotes, their process of mitosis is believed to be the ancestor of the more common process in animals and higher plants
c. Its evolution was necessary as the genomes of eukaryotic cells became larger and more complex
d. Some aspects of this process vary in certain eukaryotic organisms

18. Cell plates are _________. [pp. 185–186]

a. Formed by vesicles from the ER and Golgi complex at the site of the spindle midpoint of plant cells
b. Formed by endocytic vesicles at the site of the spindle midpoint of plant cells
c. Formed by microfibres at the site of the spindle midpoint of plant cells
d. Required for segregation of chromosomes in certain types of plant cells

19. The enzyme telomerase is ________. [pp. 192–193]

a. An excellent prospect for an anticancer treatment
b. An enzyme that controls a cell's entry into telophase from anaphase
c. Involved in DNA replication
d. Involved in triggering apoptosis

20. HeLa cells are ___________. [p. 182]

a. The first cancer cell line developed for tissue culture
b. Able to grow indefinitely in tissue culture
c. A type of cell that has been used in research for over 50 years
d. All of the above

INTEGRATING AND APPLYING KEY CONCEPTS

1. Explain why prokaryotes can successfully divide every 20 minutes and eukaryotic division is significantly longer.

2. In the "Why It Matters" section of this chapter, the process of regeneration of the zebrafish's dorsal fin is described. Having read through the rest of this chapter, go back and discuss the various processes that would be involved in this regeneration. Once regenerated, what would happen to the cell cycle of the new fin?

3. Compare and contrast G_0 with the processes of senescence and apoptosis.

OPTIONAL ACTIVITY

1. As described in the textbook, once mature, nerve cells generally stop dividing and enter G_0. Reversing this process in a controlled manner, thereby getting cells to reenter G_1, is a highly active field of spinal chord research. Try to find out what the current challenges and prospects are for developing a cure for these injuries.

10 Genetic Recombination

CHAPTER HIGHLIGHTS

- Genetic diversity is a critical component of evolution and arises with mutation of DNA sequences.
    - These happen at a low frequency.
    - Genetic recombination allows amplification of the diversity arising from mutations.
    - In eukaryotic organisms, further amplification is accomplished through sexual reproduction.
- Genetic recombination between homologous sequences takes place in bacteria by three mechanisms:
    - Conjugation, which is thought of as bacterial sexual reproduction and involves direct transfer of DNA from one cell to another through a "cytoplasmic bridge"
    - Transformation, which is the absorption of free DNA from dead and ruptured bacterial cells. Only certain species have this ability, but it can be "forced" in the lab.
    - Transduction, either specialized or generalized, which is the transfer of DNA from one cell to another via lysogenic or lytic viruses, respectively
- Genetic recombination among homologous sequences in eukaryotes occurs during the meiosis.
    - This is a specialized process of cell division in which four haploid daughter cells are generated from a single diploid parental cell. These are the gametes of sexual reproduction.
    - Homologous pairs of chromosomes combine in tetrad complexes of tightly associated DNA and protein. Recombination occurs within these tetrads via crossover events between alleles of homologous chromosomes.
    - Further diversity arises with the random segregation of homologous chromosomes and, later in the process, random segregation of the recombinant chromatids.
- Nonhomologous recombinations occur in both prokaryotes and eukaryotes and involve DNA sequences called transposable elements that can move from one site in the genome to another.
    - In prokaryotes, transposition from the chromosome to a conjugative plasmid allows for vertical transmission of genetic material within and between species.

- Transposable elements cause changes in the genome and affect the expression of the genes. In humans and other complex animals, these changes have been linked to cancer.

STUDY STRATEGIES AND LEARNING OUTCOMES

- This is a long chapter with a lot of material, so it is important to go through one section at a time.
- Remember while you are reading that the topic is recombination mechanisms as a means of increasing genetic variability: there are really only two methods of recombination: recombination of nonhomologous sequences, mediated by transposable elements, and recombination of homologous sequences, which is everything else covered in the chapter.
- Focus on the main message, which is generally laid out in the heading for each subsection. Pay attention to the figures; they will make it easier to picture and remember the mechanism of recombination.
- In reading the section on meiosis, focus on the stages of meiosis I and II, paying attention to whether the cells are $2n$ and $1n$ and looking for the three places that amplify genetic diversity, and review the similarities and differences with mitosis.
- Finally, as you complete a section in the text, work on the accompanying section in the study guide before moving on.

By the end of this chapter, you should be able to

- Define and distinguish between conjugation, transformation, and transduction and explain how each participates in recombination in prokaryotes
- Identify the stages of meiosis and explain how they differ from mitosis, including how the end products differ
- Understand how recombination occurs during meiosis and how sexual reproduction then adds to genetic diversity
- Understand what mobile elements are and identify the differences between those of prokaryotes and those of eukaryotes

TOPIC MAP

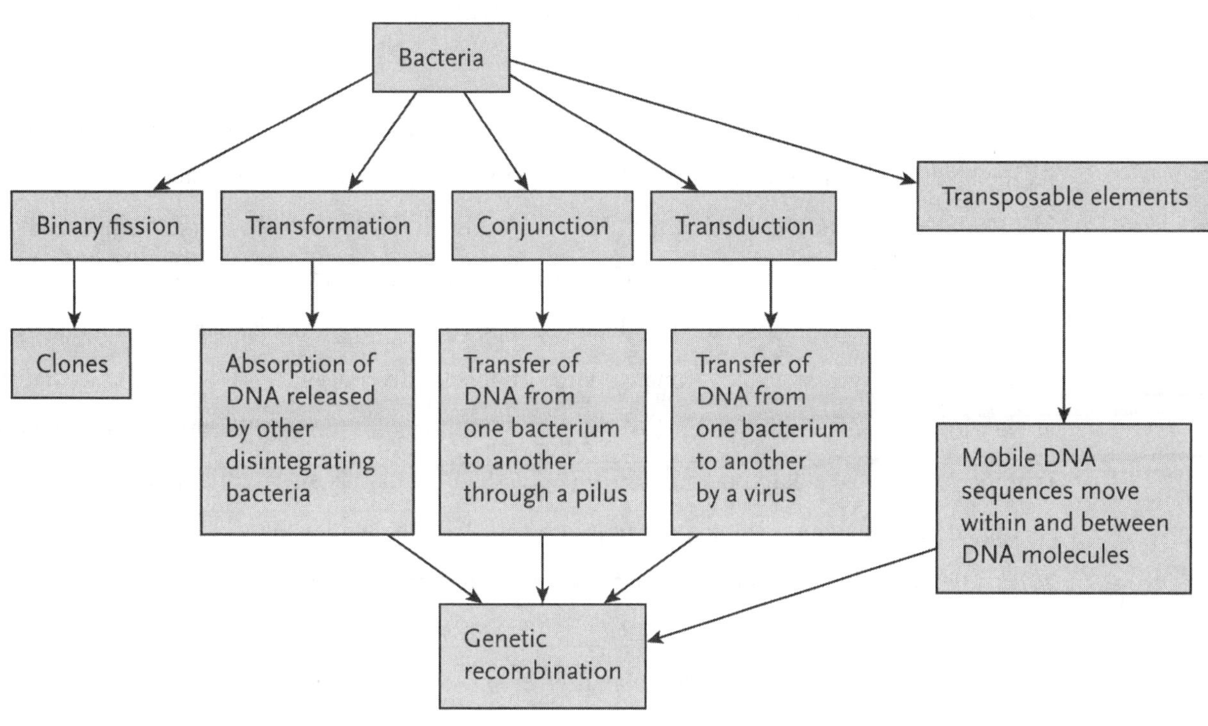

Bacteria
Binary fission
Transformation
Conjunction
Transduction
Transposable elements
Clones
Absorption of DNA released by other disintegrating bacteria
Transfer of DNA from one bacterium to another through a pilus
Transfer of DNA from one bacterium to another by a virus
Mobile DNA sequences move within and between DNA molecules
Genetic recombination

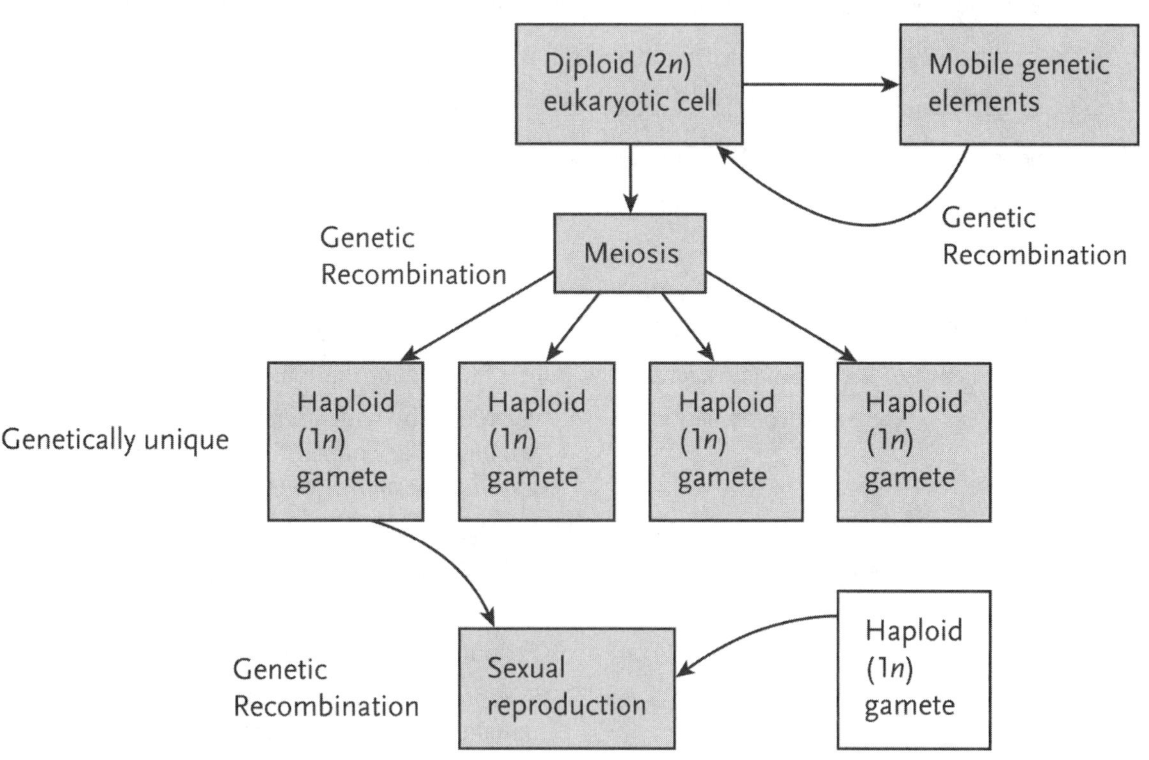

Diploid (2n) eukaryotic cell
Mobile genetic elements
Genetic Recombination
Meiosis
Genetic Recombination
Genetically unique
Haploid (1n) gamete
Haploid (1n) gamete
Haploid (1n) gamete
Haploid (1n) gamete
Genetic Recombination
Sexual reproduction
Haploid (1n) gamete

Why It Matters [pp. 199–200]

10.1 Mechanism of Genetic Recombination [pp. 200–201]

Section Review (Fill-in-the-Blanks)

There is a tension in biology between (1) _______, as generated by mitosis in eukaryotes, and difference. At the level of a multicellular organism, it is important that all of the individual cells are (2) _______ identical. However, at the level of the (3) _______, evolutionary changes arise from natural (4) _______ of genetically (5) _______ members of a particular species. In order for selective advantages to be passed on, differences between individuals must be (6) _______, that is, passed from one generation to the next. In all organisms, the ultimate source of genetic diversity is (7) _______; however, these occur at very low frequencies; the process of (8) _______ allows for the amplification of genetic diversity and is at the heart of the eukaryotic process of (9) _______, which is an essential component of sexual reproduction. Genetic recombination requires two DNA molecules that (10) _______ from each other in at least two places, a mechanism for bringing the two molecules into close (11) _______ and enzymes to cut the covalent bonds of the DNA (12) _______, (13) _______ the ends, and (14) _______ the DNA molecules back together. Most recombination events occur between (15) _______ regions of DNA, that is, regions that are very similar. This similarity allows the two molecules to line up and (16) _______ precisely. Enzymes cut each (17) _______ of both molecules, the free ends are (18) _______, and enzymes then attach the free ends by making a total of (19) _______ new covalent bonds. Despite the fact that (20) _______ bonds are cut and reformed, this process leads to a (21) _______ recombination event. If two (22) _______ DNA molecules were involved in a recombination event, as is the case between various bacterial DNA molecules, the result would be the (23) _______ into a single larger molecule.

Matching

Match each of the following scientists with their major contribution.

24.	____ Genetic recombination	A.	A process by which populations undergo change through the natural selection of certain individuals in the population based on heritable traits
25.	____ Asexual reproduction	B.	Identical
26.	____ Evolution	C.	Development of new combinations of genetic information to generate genetic variability
27.	____ Homologous	D.	A reproductive process that generates two identical daughter cells from a parental cell

Short Answer

28. Explain how the product of a single recombination event between two linear molecules would result in a different number of products than if it happened between two circular molecules. __________

____

____

10.2 Genetic Recombination in Bacteria [pp. 201–208]

Section Review (Fill-in-the-Blanks)

Although prokaryotes do not reproduce sexually by meiosis, they do undergo genetic (29) _______. Through three main processes, these organisms bring DNA from different (30) _______ into their cell, recombine homologous sequences, and produce (31) _______ that are different from either parent. Studies with the laboratory workhouse, the bacterial species (32) _______, laid the foundations for our knowledge of prokaryotic recombination. The first method of acquiring DNA, (33) _______, is considered a type of sexual reproduction in bacteria. It involves the passage of DNA from a (34) _______ cell to a recipient cell through a structure called a (35) _______. In the best characterized example, donor strains are distinguished by the presence of the *E. coli* (36) _______, an extrachromosomal DNA molecule. These strains are referred to as being (37) _______; recipients are designated F^{-}. In some donor strains, this molecule becomes (38) _______ into the bacterial chromosome, and these strains are designated (39) _______ strains. When these particular strains participate in a "mating," (40) _______ genes of the donor can be transferred to the recipients, along with a portion of the F (41) _______. By using mating partners that differ in their (42) _______ requirements, that is, their auxotrophies, growth media designed to monitor recombination of these traits, and determining the time of (43) _______ into recipient cells, researchers were able to use Hfr donors to (44) _______ genes on the *E. coli* chromosome. In another type of bacterial recombination, certain species directly absorb DNA released by (45) _______ cells. This process is called (46) _______, and studies using one type of bacterium that is capable of being (47) _______, *Streptococcus*, allowed researchers to determine that DNA carried the (48) _______ material for pathogenic traits. The third method of acquiring DNA for recombination, that of (49) _______, involves bacterial viruses, that is, (50) _______. This transfer is the result of a mistake these agents make when they leave a previous (51) _______ cell, inserting bacterial DNA instead of, or in addition to, its own in the phage head. Bacteriophage such as P22 of *Salmonella*, which undergo a (52) _______ life cycle and enzymatically break down the host's chromosome during infection, will accidentally package (53) _______ fragments of host DNA: this is referred to as (54) _______ transduction. In contrast, (55) _______ transduction results when a bacterial prophage transfers host sequences lying next to the point of insertion. The best characterized agent of this type of transduction is bacteriophage (56) _______, and the type of life cycle that allows these phages to integrate into the host chromosome is referred to as (57) _______.

Matching

Match each of the following scientists with their major contribution.

58.	____	Theodore Escherich	A.	Scientists who used interrupted-mating experiments to map genes on the *E. coli* chromosome
59.	____	Joshua Lederberg and Edward Tatum	B.	Demonstrated transformation in bacteria
60.	____	François Jacob and Elie Wollman	C.	Discovered conjugation in bacteria
61.	____	Oswald Avery	D.	Demonstrated transduction in *Salmonella typhimurium*
62.	____	Fred Griffith	E.	Discoverer of the bacterium that helped lay the foundations for our understanding of recombination
63.	____	Joshua Lederberg and Norton Zinder	F.	Proved that DNA is the transforming factor

Match each of the following terms with its correct definition.

64.	____	Virulent phages	A.	An extrachromosomal circular DNA molecule in *E. coli* that is transferred by conjugation
65.	____	Temperate phages	B.	Where bacterial cells make a physical connection to pass part of their genetic material from one to another
66.	____	Clone	C.	Bacterial cells that have the F plasmid incorporated into their chromosome
67.	____	Sex pilus	D.	Bacteriophage that follows the lysogenic cycle
68.	____	Prophage	E.	Genetically identical cells
69.	____	Conjugation	F.	Bacterial cells that have an F plasmid
70.	____	Hfr cells	G.	A tubelike structure that bacteria develop to connect one cell to another

71.	____	F^+ cells	H.	Mutants that require additional nutrients added to the minimal medium
72.	____	F^- cells	I.	Refers to determining the location of the genes on a chromosome
73.	____	Genetic map	J.	Bacteriophages that follow a lytic life cycle
74.	____	Transformation	K.	The form of a temperate phage when it is incorporated into the host chromosome
75.	____	F plasmid	L.	The state of bacteria after receiving DNA from another cell but before undergoing genetic recombination
76.	____	Transduction	M.	Where a bacterium absorbs DNA released by other dead bacteria
77.	____	Auxotroph	N.	Where a virus transfers DNA from one bacterium to another
78.	____	Partial diploid	O.	Cells that lack the F plasmid and act as recipients

Complete the Table

79. Bacterial cells divide asexually to produce clones; however, they can achieve genetic recombination through the following three mechanisms.

Mechanism of recombination	*Brief description*
A. Transformation	
B. Conjugation	
C. Transduction	

80. Complete the following table to describe the mechanisms of multiplication in bacteriophages.

Bacteriophage type	*Type of multiplication*	*Brief description*
A. Virulent		
B. Temperate		

True/False

Mark if the statement is true or false. If the statement is false, justify your answer in the line below each statement.

81. ____________ When bacteria divide, they produce clones.

____

82. ____________ All varieties of bacteria are capable of undergoing transformation.

____

83. ____________ During conjugation, DNA is transferred from the donor to the F^- recipient. Eventually, all bacteria in a culture will become F^+.

____

84. ____________ Temperate viruses follow both the lysogenic and lytic life cycles.

____

Short Answer

85. Compare and contrast the type of transduction process of bacteriophage P22 with that of bacteriophage λ.

____

____

86. Differentiate between a phage and a prophage.

____

____

10.3 Genetic Recombination in Eukaryotes: Meiosis [pp. 209–218]

Section Review (Fill-in-the-Blanks)

Sexual reproduction in eukaryotes requires a specialized division process called (87) _______. This process has two major results: it generates daughter cells with (88) _______ of the number of chromosomes present in the G_1 nucleus of the species and, just as importantly, it generates diversity through genetic (89) _______ between homologous sequences. The time and place of this type of cell division follow one of (90) _______ major pathways in the life cycles of eukaryotes, and these are distinguished by the proportion of the life cycle spent in the (91) _______, that is, $1n$ and (92) _______ or $2n$ phases and when the asexual or (93) _______ divisions occur. In animals, the (94) _______ phase dominates and meiosis is not followed by mitosis but leads directly to the formation of (95) _______. In contrast, plants and most fungi alternate between both phases, either of which can dominate, and (96) _______ can happen during each phase. The third type of pathway is seen in some fungi where the diploid phase is limited to a single cell that is produced by (97) _______ and that is immediately followed by (98) _______. Although there are certain similarities between the phases and events of meiosis and mitosis, there are essential differences, as evidenced by the final products: (99) _______ daughter cells instead of two, each of which is (100) _______ if the parent cell was diploid, and each of which is genetically (101) _______ from the parent cell. The process begins with DNA replication during the premeiotic (102) _______ and is followed by meiosis (103) _______. This begins with (104) _______, during which, as with mitosis, the sister (105) _______ condense and the spindle starts to form. A critical difference is that the homologous chromosomes then come together in a process called (106) _______, to form a tightly associated complex in which (107) _______ occurs through the exchange of segments of homologous chromosomes. This phase is followed by the breakdown of the nuclear envelope in (108) _______. Also during this stage, and unlike prometaphase in mitosis, the kinetochores of each sister chromatid join to a single (109) _______ microtubule, so that when separated, the homologous pairs move to (110) _______ poles of the cell. During metaphase (111) _______, as in mitosis, the chromosomes line up at the (112) _______ plate, but unlike mitosis, the aligned chromosomes are tetrads, and in the next phase, (113) _______, the pairs of homologous chromosomes are separated so that each pole has a (114) _______ number of chromosomes. Meiosis I ends with telophase I and interkinesis, during which there may be (115) _______ of the spindle, may or may not be formation of two new nuclear envelopes, and may or may not be limited (116) _______ or unfolding of the chromosomes. The two haploid cells from meiosis I then progress to meiosis II, the phases of which are quite similar to those of mitosis: in (117) _______, the spindles reform and chromosomes condense; in (118) _______, the spindles enter the former nuclear area and a kinetochore microtubule attaches to (119) _______ sister chromatid kinetochore; during (120) _______, the chromosomes line up at the metaphase plate; sister chromatids separate during (121) _______; and in (122) _______, the spindles disassemble and new nuclear envelopes form. In animals, all of this happens in the (123) _______ tissue and the cells that are generated, the (124) _______, remain in G_0 until sexual reproduction happens, giving rise to a diploid (125) _______. Ultimately, the combined processes of meiosis and fertilization give rise to genetic variability through four mechanisms: the first is genetic recombination, an inevitable product of prophase (126) _______; the second is the random segregation of chromosomes during anaphase (127) _______, giving cells that have a random assortment of (128) _______ and (129) _______ chromosomes; the third is the random segregation of (130) _______ chromatids during anaphase II; and, finally, diversity arises with the process of (131) _______, when different paternal and maternal gametes fuse to give a diploid zygote.

Matching

Match each of the following events with the most appropriate division of meiosis.

A. Meiosis I B. Meiosis II

132. ____ DNA replication prior to this division

133. ____ Recombination of alleles occurs

134. ____ The product is four daughter cells

135. ____ Homologous pairs line up in synaptonemal complexes

136. ____ Sister chromatids separate

137. ____ Homologous pairs separate

Match each of the following terms or concepts with its correct definition.

138. ____	Chiasmata	A.	A protein–DNA structure that is the substrate for crossover events
139. ____	Alleles	B.	X and Y in humans
140. ____	Synaptonemal complex	C.	The microscopically visible product of crossover events between nonsister chromatids
141. ____	Nondisjunction	D.	Genetically distinct versions of the same gene
142. ____	Sex chromosome	E.	Occurs when chromosome segregation fails in meiosis I or II

Sequence

143. Arrange the following events of meiosis in the correct hierarchical order and for each identify the phase of meiosis in which it occurs.

A. Homologous pairs undergo synapsis

B. Four haploid daughter cells result

C. Sister chromatids separate

D. Homologous pairs separate

E. Crossing over

F. Two haploid daughter cells result

_____ _____ _____ _____ _____ _____

Short Answer

144. Explain the basis of Down syndrome in humans. ____________________

____

____

____

145. What is interkinesis? ____________________

____

____

146. What is the relationship between synapsis and tetrads?

____

____

True/False

Mark if the statement is true or false. If the statement is false, justify your answer in the line below each statement.

147. ____________ Nondisjunction is a common problem in meiosis.

____

148. ____________ During gamete formation in humans, the X and Y chromosomes behave as homologues.

____

Labelling

149. For each of the labelled sections, respond to the following:

A. Number of homologous chromosome pairs: _______. This cell is _____. (haploid or diploid)

B. Assuming this cell is in prophase I, DNA replication has _____. (occurred or not occurred)

C. During this phase, genetic variability is increased by _____.

D. During this division, the _____ _____ separate. This is done in a _____ fashion. Does DNA replication occur again? _____ (Yes or No)

E. During this division, the _____ _____ separate. Daughter cells are _____. (haploid or diploid)

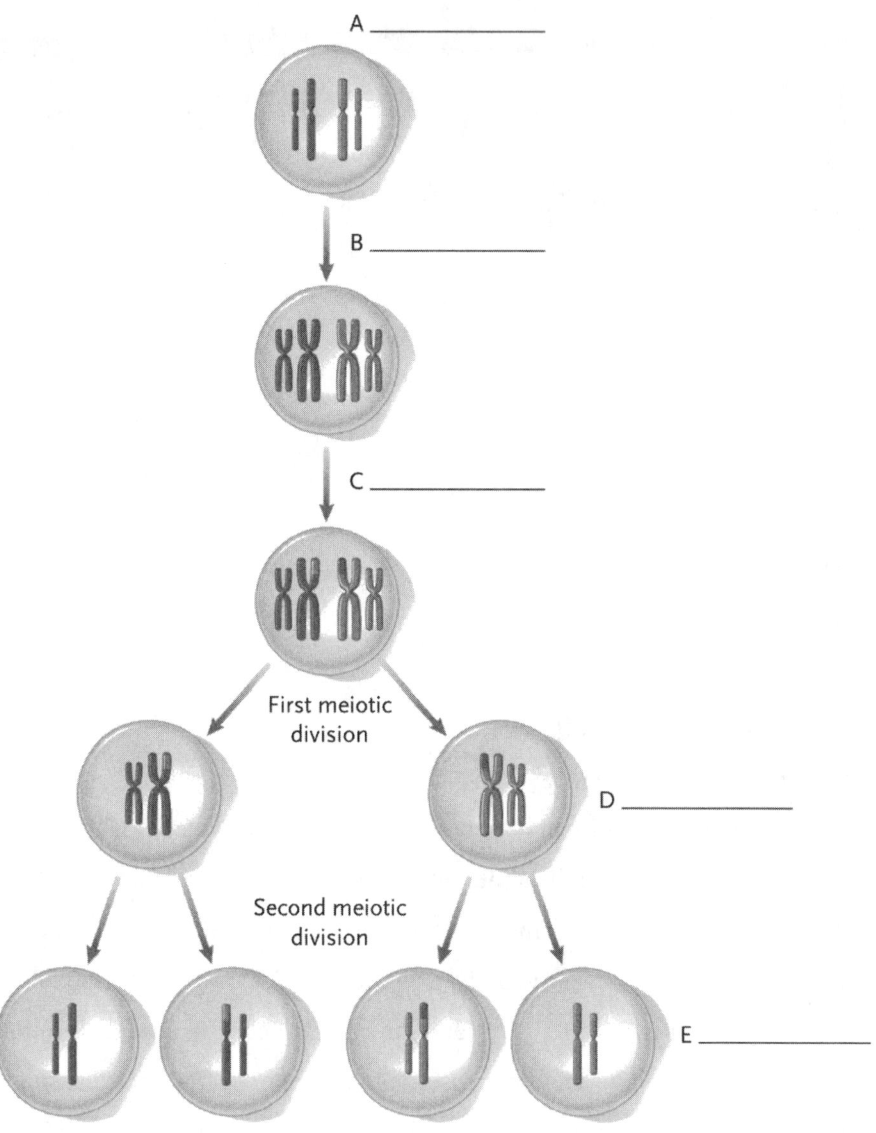

150. For each of the following descriptions, choose the most appropriate label(s) from the diagram.

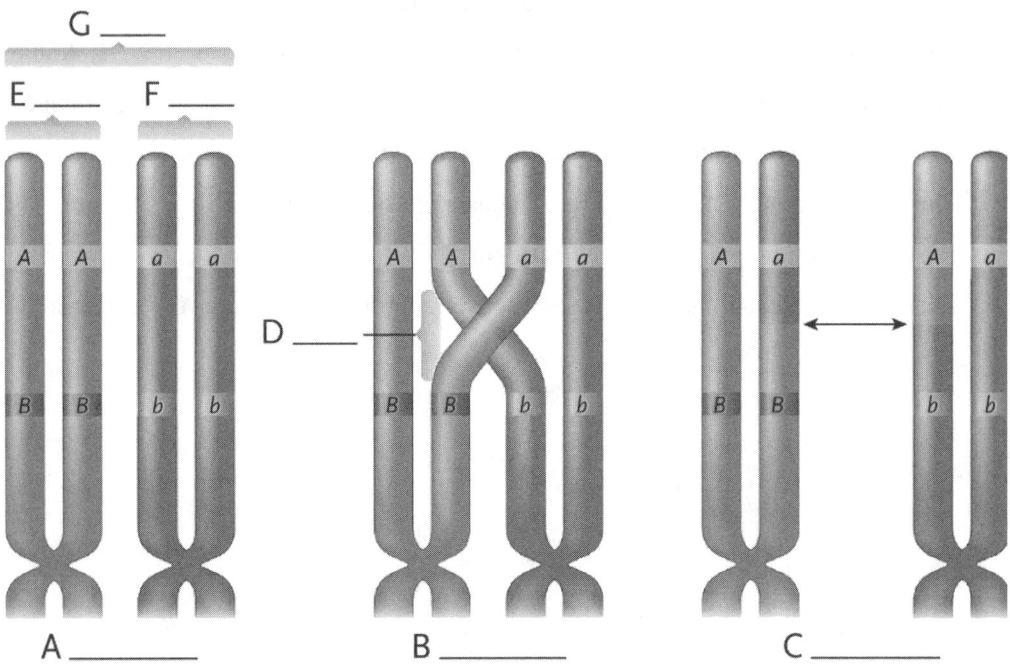

i. Sister chromatids of paternal/maternal source _____

ii. Crossover/chiasma _____

iii. Recombinant chromatids _____

iv. Occurs in anaphase I _____

v. Homologous pair ____

vi. Occurs in prophase I _____

10.4 Mobile Elements [pp. 219–224]

Section Review (Fill-in-the-Blanks)

Another source of genetic variability in both prokaryotes and eukaryotes is DNA sequences or (151) ______ _______ (TEs), which have the ability to move from place to place within DNA molecules. In contrast to the recombination mechanisms above, this movement involves nonhomologous sequences and is referred to as (152) ______. Through the action of their (153) ______ enzyme, TEs either move by cutting and (154) ______ or by (155) ______ and inserting the copy at another location. TEs may insert into a gene and (156) ______ it, or they may insert into regulatory sequences, causing increased or decreased gene (157) ______. The latter has been implicated in certain types of (158) ______ in humans and other complex animals. Because of their effects on gene expression, TEs are biological (159) ______ as well as sources of genetic variability. There are two major types of bacterial TEs: small (160) ______ ______ (IS) and longer (161) ______. The latter differ from the IS elements by encoding, in addition to the (162) ______ enzyme, extra genes such as those for antibiotic (163) ______. In *E. coli*, creation of Hfr strains results from recombination between homologous IS elements in the (164) ______ and the chromosome. In addition, transposition between the chromosome and (165) ______ plasmids allows for the spread of genes within and between species. The TEs in eukaryotes exist as transposons or (166) ______; the latter are similar to a family of viruses called (167) ______. Both move by transcribing an (168) ______ copy, replicating it, or reverse (169) ______ it into a DNA copy and inserting it into a new location while leaving the "parent" copy at the original site. In eukaryotes, TE-mediated change becomes heritable if it occurs in cells that go on to become (170) ______.

Matching

Match each of the following terms/concepts with its correct definition.

171. ____ Barbara McClintock
172. ____ Target site
173. ____ Transposable elements
174. ____ Transposition
175. ____ Retrovirus
176. ____ Transposase
177. ____ Retrotransposons
178. ____ Reverse transcriptase
179. ____ Provirus

A. Eukaryotic viruses that encode reverse transcriptase
B. TE-encoded enzyme that catalyzes cut-and-paste transposition
C. The location where a TE inserts
D. Eukaryotic TEs that move to different locations via an RNA intermediate
E. Segments of DNA that can move from one place to another
F. Process of moving segments of DNA from one place to another
G. Enzyme that copies RNA to make complementary DNA
H. State of virus where its DNA is inserted into eukaryotic genome
I. Discoverer of transposable elements

True/False

Mark if the statement is true or false. If the statement is false, justify your answer in the line below each statement.

180. ____________ Bacterial transposons are small sequences that generally contain a single gene for the transposase enzyme.

____

181. ____________ Approximately 40% of the human genome consists of retrotransposons and retroviruses.

____

Short Answer

182. Provide an explanation for the multicoloured nature of maize. ______________________________

____

____

183. Discuss the opposing concepts of genome plasticity and stability. __________________________

____

____

SELF-TEST

1. All of the following cause genetic variation in bacterial populations except _________. [p. 208]

    a. Binary fission
    b. Transduction
    c. Conjugation
    d. Transformation

2. Which of the following would describe a type of bacterial variant that may result from exposure to X-rays or UV light? [p. 201]

    a. Parental
    b. Temperate
    c. Virulent
    d. Auxotroph

3. The structure that connects two bacterial cells during transfer of DNA is called _______. [p. 203]

    a. A plasmid
    b. A nucleoid
    c. A pilus
    d. A flagellum

4. Which of the following bacteria would have the F plasmid integrated into their chromosome? [p. 205]

    a. F^-
    b. Hfr
    c. F+

5. During conjugation between an Hfr and an F^- recipient, ________. [pp. 204–205]

    a. Some of the genes from the Hfr are combined with the DNA of the recipient
    b. The recipient becomes an Hfr
    c. The recipient becomes F^+
    d. Genes from the F^- are transferred to the Hfr

6. Which of the following refers to a virus transferring DNA from one bacterium to another? [p. 206]

    a. Binary fission
    b. Transduction
    c. Conjugation
    d. Transformation

7. All varieties of bacteria are able to undergo transformation in nature. [p. 206]

    a. True
    b. False

8. During which cycle does a bacteriophage DNA become incorporated inside the host genome? [pp. 207–208]

   a. Lytic cycle
   b. Lysogenic cycle

9. During which cycle does a virus rupture the host cell in order to be released? [pp. 207–208]

   a. Lytic cycle
   b. Lysogenic cycle

10. A prophage or provirus is _________. [pp. 206, 224]

   a. A virus that contains DNA
   b. A virus that has an RNA intermediate
   c. A virus whose DNA is incorporated into the host genome
   d. A virus that is assembled and ready to be released

11. Bacterial transposable elements can move DNA segments _________. [pp. 220–221]

   a. From one location in the chromosome to another location in the chromosome
   b. From a plasmid to the main chromosome
   c. From one plasmid to another plasmid
   d. Any of the above

12. Transposable elements in bacteria may be ________. [p. 224]

   a. Insertion sequences
   b. Transposons
   c. Retrotransposons
   d. (a) and (b)
   e. (a), (b), and (c)

13. If a diploid organism normally has 16 chromosomes in somatic cells, then _________ are of maternal origin and _________ are of paternal origin. [p. 211]

   a. 32; 32
   b. 8; 8
   c. 4; 4
   d. 16; 16

14. On a pair of homologous chromosomes, the alleles ______. [p. 211]

   a. Are always the same
   b. Are always different
   c. May or may not be the same
   d. None of these

15. In the process of meiosis, how many times does DNA replication occur? [p. 211]

   a. Once
   b. Twice
   c. Replication happens during the interphase that precedes meiosis
   d. None of the above

16. Homologous pairs separate during ________. [p. 213]

   a. Anaphase I
   b. Anaphase II
   c. Telophase I
   d. Prophase II

17. If the diploid number for an organism is six and one homologous pair underwent nondisjunction during meiosis I, how many chromosomes would be present in the daughter cells at the end of meiosis I? [p. 213]

   a. Both daughter cells would have three
   b. One would have two and the other three
   c. One would have one and the other five
   d. One would have two and the other four

18. With respect to the sex chromosomes in humans, females and males will produce gametes with which possible sex chromosomes? [p. 213]

   a. Females only X; males only Y
   b. Females only X; males both X and Y
   c. Females both X and Y; males only Y
   d. Both females and males can produce both types of gametes—X and Y

19. If the diploid number for an organism is X and a crossover occurred during premetaphase I, the amount of DNA at the end of meiosis I will be ______. [p. 212]

   a. 0.5 X
   b. 0.75 X
   c. X
   d. 2 X

20. Which of the following life cycle patterns would likely give rise to the most genetic variability? [p. 209]

   a. Diploid dominant
   b. Haploid dominant
   c. Diploid–haploid alternate
   d. All of these life cycle patterns produce genetic variability

21. Regardless of the life cycle pattern, gametes that are produced by meiosis are genetically ________. [p. 211]

   a. Different from each other
   b. Identical to each other but different from the parental cell
   c. Different from each other and from the parental cell
   d. None of the above

22. Mobile elements in eukaryotes are called _________ . [p. 221]

   a. Transposons
   b. Retrotransposons
   c. Insertion sequences
   d. (a) and (b)
   e. (a), (b), and (c)

23. Which of the following statements about retroviruses is false? [pp. 223–224]

   a. They may have been the evolutionary predecessor to retrotransposons
   b. They synthesize DNA from an RNA sequence
   c. They are found in bacterial and eukaryotic cells
   d. They can exist in up to 100 different provirus insertions in the average human chromosome

INTEGRATING AND APPLYING KEY CONCEPTS

1. Discuss the effects of a mutation that prevented synapsis.
2. Discuss the advantages and disadvantages of gametes that are genetically identical versus gametes that are genetically unique.
3. Discuss the effects of a hypothetical process that would allow recombination between nonhomologous chromosomes rather than just homologous pairs.
4. You want to do some chromosomal mapping using an interrupted mating experiment and you have, in your lab, the following strains:
   a. *E. coli* Hfr with the traits his^+ met^- phe^- arg^+
   b. *E. coli* Hfr with the traits his^- met^+ phe^+ arg^-
   c. *E. coli* F^- with the traits his^+ met^- phe^- arg^+
   d. *E. coli* F^- with the traits his^- met^+ phe^+ arg^-

   where each of the four gene names represents a gene for the biosynthesis of an essential amino acid and the minus sign denotes a mutation, while the plus sign indicates the normally functioning gene. Identify the following:

   i. Which mating pairs would you need to use (in two separate experiments) to map each of the four genes?
   ii. What kind of growth media would you need for each of the two experiments?
   iii. What kind of unit would you be likely to use for the "map" distance between two genes?
   iv. What kind of control(s) would you need?

OPTIONAL ACTIVITY

1. Although the textbook indicates that cells that suffer nondisjunction rarely lead to a live birth, there are more examples of nondisjunction than Down syndrome. Nondisjunction in humans leads to some gametes and children suffering monosomy (lacking one of the two homologous pairs) or trisomy (having an additional copy of the homologous pair, e.g., Down syndrome). Some examples of nondisjunction disorders include Turner syndrome, Klinefelter syndrome, and trisomy X. Do some research on these (or other) nondisjunction disorders, their symptoms, life expectancy, and possible treatments.

11 Mendel, Genes, and Inheritance

CHAPTER HIGHLIGHTS

- Gregor Mendel determined the basic principles of inheritance patterns and genetics.
  - Genes are present in individuals in pairs or alleles. This was later shown to be a reflection of the diploid nature of animal and plant cells.
  - The genetic nature of a particular trait is the genotype; the visible result is the phenotype.
  - Alleles segregate independently when gametes are formed. This was later correlated with the events of meiosis.
  - When looking at more than one character or gene, the separation of both during formation of the gametes is independent. This was also shown later to correlate with the events of meiois and reflect the fact that genes are present on chromosomes.
- The principles of segregation and independent assortment are discussed using the Punnett square method of predicting the genotype and phenotype of the offspring and their ratios in crosses involving a single gene and two genes.
- Some traits do not obey Mendelian principles, where a single gene may exist in only two forms: dominant or recessive.
  - Some traits are the product of multiple genes (polygenic traits).
  - Some genes are incompletely dominant or display codominance.
  - Some genes give rise to a wide variety of effects; that is, they are pleiotropic.
  - Epistasis occurs when one gene has an effect on the expression of another trait.

STUDY STRATEGIES AND LEARNING OUTCOMES

- First, focus on the terminology and Mendel's experiments: look at the figures as they make things easier to picture and interpret.
- Next, using testcrosses and Punnett square analysis, predict the probability of gene expression in offspring. There are plenty of problems to work on in the self-test questions at the end of the chapter and in the companion study guide chapter.
- Be able to relate allele and chromosome activities in meiosis to gene expression.

- Finally, focus on the complexity of multiple alleles, gene interactions, and the enormous potential for variation.
- As always, work through the companion section of the study guide after finishing each section of the textbook.

By the end of this chapter, you should be able to

- Describe Mendel's three hypotheses and distinguish between an allele and a locus, a phenotype, and a genotype and relate these to the chromosome theory of inheritance
- Use a Punnett square and define a testcross and a dihybrid cross
- Distinguish between complete dominance, incomplete dominance, codominance, epistasis, pleiotropy, and polygenic inheritance, giving examples of each

TOPIC MAP

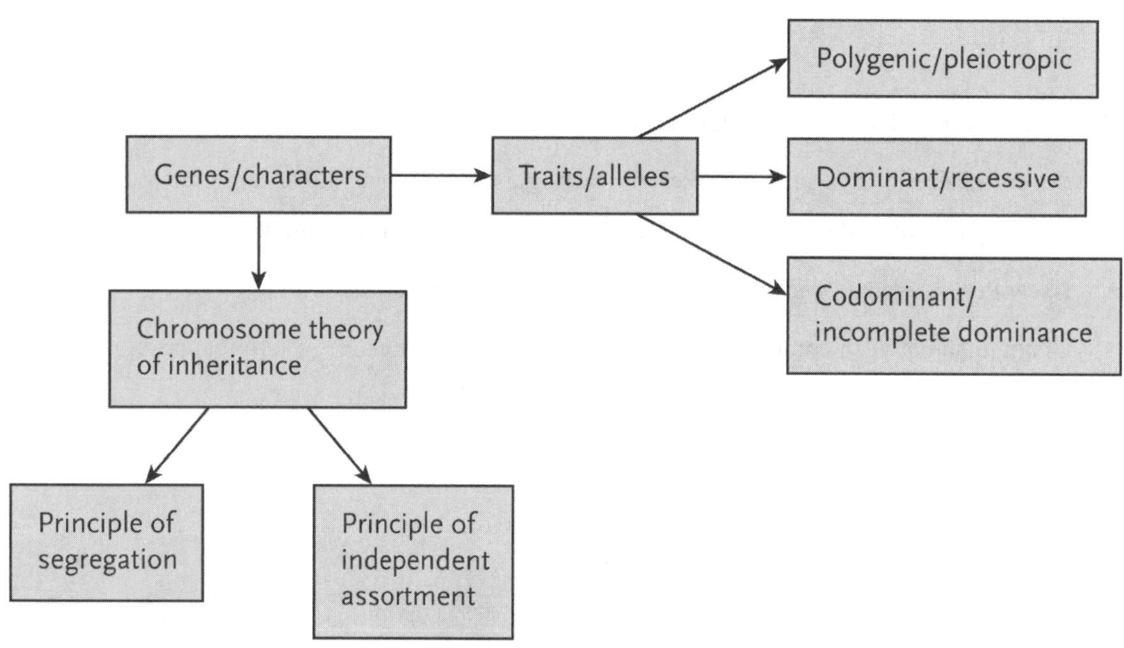

Why It Matters [pp. 229–230]

11.1 The Beginnings of Genetics: Mendel's Garden Peas [pp. 230–239]

Section Review (Fill-in-the-Blanks)

In the 1860s, the monk (1) ______ used careful scientific experimentation combined with quantitative analyses of the data to disprove the (2) ______ theory of inheritance, proving instead that traits were passed from one (3) ______ to the next in predictable ratios and combinations. His choice of the (4) ______ as an experimental organism was a good one for three reasons: they were easy to cultivate, had easily observed, (5) ______-breeding characters, and allowed him to control the (6) ______ or sexual reproduction of the plants. This second trait meant that when (7) ______, parental traits were passed unchanged to the next generation. This allowed him to evaluate the effects of mating pea plants with varying traits, the term he used for what we now call (8) ______. In his crosses between two parental plants, that is, the (9) ______ generation, he designated the products as the filial or (10) ______ generation; when the F_1 plants were allowed to self-pollinate or (11) ______, the offspring from that cross were designated the (12) ______ generation. After many crosses, he was able to make conclusions about alleles and chromosome movement during the process of (13) ______, long before genes (what he called (14) ______), chromosomes, and cell division for gamete formation were understood. He also provided an explanation for why organisms may look the same but be genetically different: that is, the relationship between (15) ______ and (16) ______, respectively. Based on his work with (17) ______ or single-character crosses, Mendel showed that genes that govern traits in individuals occur in (18) ______; that if different alleles are present in an individual, one allele will be (19) ______ over the other; and that the two alleles of a gene segregate (20) ______ and occur in single copy in the gametes, a behaviour known as the principle of (21) ______. To confirm his findings and interpretations, Mendel used (22) ______: he crossed an F_1 heterozygote and a homozygous recessive or (23) ______ breeding parent. Using this approach, one can determine easily the (24) ______ of an individual who has a dominant phenotype, that is, homozygous dominant or heterozygous. The use of (25) ______ crosses led Mendel to a fourth hypothesis: the alleles of two separate characters segregate (26) ______. This is now known as the principle of random (27) ______. The findings and predictions made by Mendel are the underpinnings of today's science of genetics and chromosome theory of (28) ______.

Matching

Match each of the following genetics terms and concepts with its correct definition or explanation.

29.	____	Dominance	A.	Testcross to evaluate one gene (character)
30.	____	Dominant trait	B.	The location on a chromosome of a particular gene
31.	____	Recessive trait	C.	The appearance of an organism
32.	____	Homozygote	D.	Trait that is masked, unless in the homozygous state
33.	____	Heterozygote	E.	The term used by Mendel for a heritable characteristic; now known as a gene
34.	____	Monohybrid cross	F.	Masking of a allele (trait)
35.	____	Dihybrid cross	G.	Alleles (traits) for a given gene (character) are different
36.	____	Locus	H.	Both alleles (traits) are the same
37.	____	Character	I.	The genetic constitution of an organism
38.	____	Self	J.	A cross between an individual with a dominant trait with one that is homozygous recessive, in order to determine if the individual with the dominant trait is homozygous or heterozygous
39.	____	Genotype	K.	Trait that is expressed, when in the heterozygous state
40.	____	Phenotype	L.	Testcross to evaluate two genes (characters)
41.	____	Testcross	M.	In genetics terms, this refers to the process of allowing the F_1 generation to self-pollinate to generate the F_2 generation and

thereby determine the genotype of the first generation of offspring

Matching

For each of the following statements, choose the most appropriate genetic term.

A. Genotype B. Phenotype

42. ____ Expressed alleles

43. ____ When you look in the mirror

44. ____ Purple flowers

45. ____ Plant that is heterozygous for flower colour

46. ____ Alleles that are present

Complete the Table

47. In the following table of garden pea individuals, "T" (Tall) is dominant over "t" (short) for plant height and "R" (red) is dominant over "r" (white) for flower colour. For the genotypes given, identify the number and type of all possible gametes (even if the gametes are identical)—see the example below:

	Number of potential gametes	*Types of all possible gametes*			
Tt	2	T		t	
TT	2	T		T	
Rr	A. _____	B. _____		C._____	
TTRr	4	D. _____	E. _____	F. _____	G. _____
TtRr	H. _____	I. _____	J. _____	K. _____	L. _____

48. Given the table above, fill in the blanks in the summary below:

The genotype of the Rr individual is (A) _____, while the flower colour is (B) _____. The flower colour is not the genotype but the (C) ___________. The phenotype of the TTRr individual is (D) ____ in height and (E) _____ in colour. Another way to indicate the genotype of the TTRr individual is (F) _____ _____ for the height and (G) _____ _____ for the flower colour. In the case of flower colour, the only way to have a white flower is to have a genotype of (H) _____ ____. The only way to have a short plant is to have a genotype of (I) _____ _____. In both cases, the explanation for the phenotype of the heterozygote is that the T and R alleles are (J) _________, whereas the t and r alleles are (K) _________. In order to determine whether a red flower was due to a homozygous or heterozygous genotype, one would do a (L) ___________-cross and examine the phenotypes of the F_1 generation.

Short Answer

49. Define the principle of segregation. ____

____

____

50. Define the principle of independent assortment. ______________________________

____

____

51. What is a testcross, and why is it used? ______ ______________________________

____

____

52. Describe the chromosome theory of inheritance. ______________________________

____

____

Problems

53. In garden pea plants, tall (T) is dominant over dwarf (t). Use the Punnett square method for the monohybrid cross Tt × tt in order to answer the following:

    A. The genotypes of the possible gametes from both parents

    B. The genotype probabilities of the offspring

    C. The phenotype probabilities of the offspring

    D. Would this be a monohybrid cross or a testcross?

	______	______
______		
______		

54. The parental or P generation is a cross between two true-breeding plants. One plant is homozygous dominant for both plant height (tall) and stem strength (strong), and the other plant is homozygous recessive for both traits: TTSS × ttss.

    A. What is the genotype of the possible gametes for the two P generation plants? ______________________________

    B. What are the genotype probabilities of the F_1 offspring? ______________________________

    C. What are the phenotype probabilities of the F_1 offspring? ______________________________

    Using the Punnett square method for the dihybrid cross in which the F_1 offspring were selfed, answer the following questions:

	____	____	____	____
____				
____				
____				
____				

    D. The genotype probabilities of the F_2 offspring ______________________________

____

E. The phenotype probabilities of the F_2 offspring ______________________________

____

11.2 Later Modifications and Additions to Mendel's Hypotheses [pp. 239–246]

Section Review (Fill-in-the-Blanks)

Subsequent research has led to some modifications and additions to Mendel's hypotheses. It was determined than some alleles are neither fully (55) ______ nor fully (56) ______ so that the phenotype of the heterozygote is not like that of either homozygote. There are two situations in which this can arise. In (57) ______ dominance, the activity of one allele is insufficient to compensate for the inactivity of another, such as in the case of red, white, and (58) ______ snapdragons or diseases such as the genetic disorder that affects red blood cells, (59) __________ __________ ___________. In contrast, in (60) __________ dominance, both alleles in the heterozygote are equally active, as in the example of the (61) __________ blood group. Unlike flower colour in garden peas, many genes within a population have more than one form or (62) __________. The implications of this can have far-reaching effects, as in the situation of trying to find matches for (63) __________ transplants in humans. The coat colour phenotype of Labrador retrievers provides an example of (64) __________, where one gene product has an effect on the expression of genes at other loci. In this breed, the combined effects of two genes determine coat colour: one for production of melanin, a brownish-black pigment, and one for the (65) __________ of the pigment. A homozygous recessive individual for the latter will have virtually no pigment (66) __________, even if the pigment is produced at high levels in a homozygous dominant dog. A (67) __________ trait is one that arises from the combined effects of several genes so that one sees a (68) __________ variation of the trait within a population. Plotting the distribution of height of adult humans is a good example of this variation: with (69) __________ traits, such plots generally yield a (70) __________ curve. In almost the opposite phenomenon, (71) __________ describes the situation where one gene affects more than one character of an organism.

Short Answer

72. Discuss the three types of dominance for the heterozygote Pp.______________________________

____

____

73. Compare and contrast pleiotropic and polygenic characters.______________________________

____

____

SELF-TEST

1. Which of the following of Mendel's hypotheses agrees with the events of meiosis? [pp. 233, 237–238]

   a. The principle of segregation
   b. The principle of independent assortment
   c. Genes that govern genetic characters occur in pairs in individuals
   d. All of the above

2. In comparing two matings, TT × tt and TT × Tt, assuming T is dominant over t, which would exhibit the greatest phenotypic diversity in the F_1 generation? [p. 233]

   a. TT × tt
   b. TT × Tt
   c. They would be the same

3. Bob has blood type A, and his children all have blood type O. Bob thinks the mailman might be the father because his wife is also blood type A. You come to the rescue and explain to Bob that he is getting the _____ confused with the ________. [p. 242]

   a. Phenotype; genotype
   b. Dominance; recessive
   c. Monohybrid; dihybrid
   d. Homozygous; heterozygous

4. An individual genotype is AaBB. What is(are) the type(s) and number of different gamete combinations? [p. 233]

   a. Aa and BB, two
   b. AaB, one
   c. AB and aB, two
   d. A and a and B, three

5. For the cross AA × aa, if the F_1 generation was a different phenotype than either of the parental individuals, which of the following is the best and simplest explanation? [pp. 239–242]

   a. Codominance
   b. Multiple alleles
   c. Pleiotropy
   d. Incomplete dominance
   e. It could be either codominance or incomplete dominance

6. Blue flower colour is dominant over white. Predict the phenotypic ratio if a heterozygous blue flower is crossed with a white flower. [pp. 234–235]

   a. 100% blue flowers
   b. 75% blue flowers and 25% white (3:1)
   c. 50% blue flowers and 50% white (1:1)
   d. 25% blue flowers and 75% white (1:3)

7. Refer to question 6. Predict the genotypic ratio. [pp. 234–235]

   a. All Bb

b. 75% Bb and 25% bb
c. 50% Bb and 50% bb
d. 25% BB and 75% bb

8. Albert has type A blood, and Barbara has type B blood. Not knowing their genotypes, identify all the possible blood types of their children. [p. 242]

a. Type A and B
b. Type A, B, AB, or O
c. Type A, B, or AB
d. Not able to determine

9. The genetics of the ABO blood types demonstrate ______. [p. 242]

a. The phenomenon of codominance
b. The phenomenon of incomplete dominance
c. The phenomenon of epistasis
d. The phenomenon of pleiotropy

10. You have a new puppy, a yellow Lab. The parents of your new puppy were both black Labs. The only way this could have occurred is that both parents must have been _____. [p. 243]

a. *BbEe*
b. *BBEe*
c. *BBee*
d. (a) or (b)

11. The genetics of the coat colour for Labrador retrievers demonstrate ______. [p. 243]

a. The phenomenon of codominance
b. The phenomenon of incomplete dominance
c. That the epistasis gene has an effect on the expression of pigment, the product of a different gene
d. The phenomenon of pleiotropy

12. Harry knew that he was adopted. He was 1.88 m, and both parents were under 1.67 m. In fact, Harry checked his pedigree for three generations and no one was over 1.72 m. How would you counsel Harry? [pp. 244–245]

a. Harry is correct; he must be adopted
b. Height is determined by polygenic inheritance
c. Height is a result of the product rule
d. Height is pleiotropic

13. An example of pleiotropy would be _______. [pp. 244–245]

a. Human blood types
b. Adult human height
c. Familial hypocholesterolemia
d. Sickle cell anemia

14. The distinction between codominance and incomplete dominance is _______. [pp. 239–242]

a. In codominance, both alleles are equally expressed, and in incomplete dominance, one is completely expressed, but the other is not expressed at all

b. In codominance, one allele is completely expressed, but the other is not expressed at all, and in incomplete dominance, both alleles are equally expressed
c. In codominance, both alleles are equally expressed, and in incomplete dominance, one allele is only partially masked by the other
d. In codominance, one allele is only partially masked by the other, and in incomplete dominance, both alleles are equally expressed

15. In comparing two matings, TT × tt and TT × Tt, assuming T and t are codominant, which would exhibit the greatest phenotypic diversity in the F_1 generation? [pp. 233, 242]

a. TT × tt
b. TT × Tt
c. They would be the same

16. In comparing two matings, TT × tt and TT × Tt, assuming T is incompletely dominant, which would exhibit the greatest phenotypic diversity in the F_1 generation? [pp. 233, 242]

a. TT × tt
b. TT × Tt
c. They would be the same

17. Given the prevailing idea about inheritance at the time Mendel began his research, if he had chosen to work with snapdragons instead of garden peas, what might the results have been?

a. His results might have provided evidence for the blending theory of inheritance
b. He would have still come up with the same hypotheses and principles
c. Impossible to say in the absence of more information about snapdragons

18. Which of the following is unlikely to be a polygenic trait? [pp. 239, 244–245]

a. Skin colour
b. Body weight in humans
c. Cystic fibrosis
d. Eye colour

19. Which of the following human traits does not display inheritance patterns that follow Mendelian principles? [pp. 239, 240–241]

a. Sickle cell disease
b. Webbed fingers
c. Albinism
d. None of the above follow Mendelian principles

INTEGRATING AND APPLYING KEY CONCEPTS

1. Sometimes recessive alleles are defective or produce an unwanted phenotype and in homozygous individuals can cause health problems. Starting with two healthy, heterozygous mice, design an experiment that would eliminate the recessive allele.
2. After being in a serious traffic accident, Suzie arrives at the hospital and undergoes emergency surgery. She has to have multiple blood transfusions, but the medical staff are worried that they may not be able to

get enough from the blood bank, which is running low. If Suzie's blood type is A, her father's is type B, and her mother's is type O, would either of her parents be able to donate blood to Suzie? Who or why not?

3. Joe is accused of fathering a baby with type AB blood. Joe has type O blood, and the mother has type B blood. Does Joe have a case? Defend your response.

OPTIONAL ACTIVITY

1. The textbook describes the horrible effects of sickle cell disease in homozygous individuals and discusses the fact that heterozygous individuals are also unhealthy, although with less severe symptoms: an example of incomplete dominance. Given that the disease is the product of a single genetic mutation, it is perhaps surprising that natural selection has not worked over millennia to eliminate this allele from the human population. The explanation for this, and for the prevalence of the allele specifically among people of African background, is the fact that heterozygous individuals have a survival advantage when infected with the mosquito-borne parasite that causes malaria. This is a disease that is endemic in Africa. Do some research on this fascinating relationship and the disease.

12 Genes, Chromosomes, and Human Genetics

CHAPTER HIGHLIGHTS

- Thomas H. Morgan developed the fruit fly or *Drosophila melanogaster* as a model organism for genetics and discovered that genes are genetically linked on chromosomes.
    - Those genes may behave as single genes.
    - If recombination between those two genes occurs, genetic changes arise, so that in the progeny, the genotypes and their ratios are not what would be predicted through Mendelian inheritance.
    - The frequency of recombination between two linked genes is generally proportional to the physical distance separating them, and this frequency can be used to give a relative measure of this distance. This is the map unit or centimorgan.
- In humans and fruit flies, one pair of chromosomes is different between males and females: females are XX and males are XY.
    - One of the X chromosomes in the female is inactivated to ensure the same gene dosage of X-linked genes as in males. This is seen microscopically as a condensed chromosome called the Barr body.
- The inheritance pattern of sex-linked genes differs from the inheritance pattern of genes on the autosomal or nonsex chromosomes.
- Abnormalities in recombination events during meiosis, or exposure to certain harmful agents, can cause chromosomal alterations that have genetic consequences.
- A number of examples of common dominant, recessive, and sex-linked genetic disorders are identified and explained, describing how their mode of inheritance can be used to determine if the trait is carried on autosomal or sex chromosomes and if it is recessive or dominant.

- There are also a few nontraditional modes of inheritance:
  - Cytoplasmic inheritance is based on inheriting the mitochondria from the mother.
  - Certain genes in the gametes are imprinted, so that the gene is permanently turned off by chemical modification of the control sequences of the gene. This can alter inheritance patterns and give rise to disease if fertilization produces a zygote with one imprinted allele and one mutant allele.

STUDY STRATEGIES AND LEARNING OUTCOMES

- First, focus on the concepts of linkage groups and genetic recombination events. Go through the genetic tests involving the fruit flies, paying particular attention to the method of determining chromosomal maps of specific genes and to the use of reciprocal crosses.
- Be able to relate sex-linked disorders to sex-linked genes; use a Punnett square to go through the various possibilities of X-linked versus autosomal linkage and inheritance patterns.
- Given either genotypes or phenotypes of parents with genetic disorders, predict the probabilities of these genetic disorders in offspring.
- Carefully examine the figures and tables as well as the various subject headings to keep the ideas clearly organized in your mind.
- As always, go through one section of the textbook at a time and then work through the companion section of the study guide.

By the end of this chapter, you should be able to

- Understand the idea of linkage and sex-linked genes and explain how one would prove one or the other
- Understand how recombination frequencies can be used to map alleles
- Comprehend the differences in the sex chromosomes of mammals and *Drosophila* and explain what determines female versus male and what is done to control the gene dosage of females

- Identify the different kinds of chromosomal alterations and the basic process that generates each (e.g., unequal recombination, nondisjunction)
- Understand the genetics of human diseases, autosomal recessive, autosomal dominant, and X-linked recessive, and know how the inheritance patterns differ between these. Finally, you should understand how they would be identified by a genetic counsellor or physician.
- Define cytoplasmic inheritance and genomic imprinting

TOPIC MAP

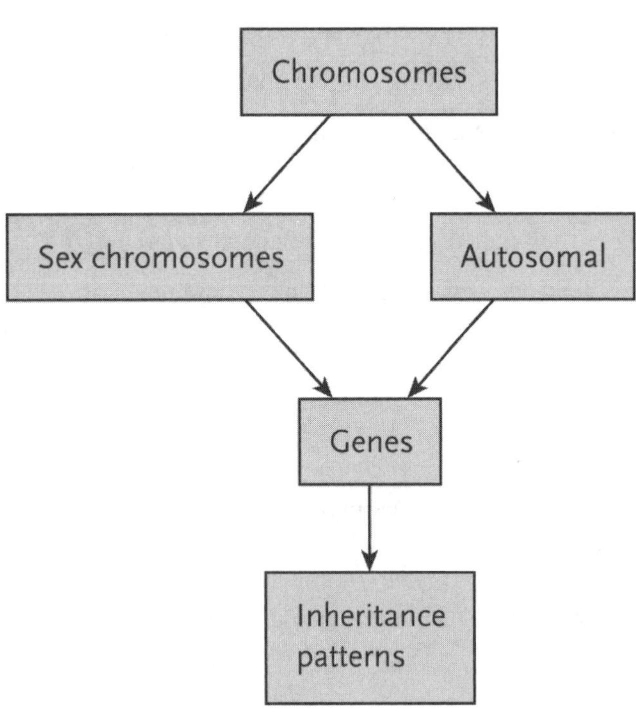

INTERACTIVE EXERCISES

Why It Matters [pp. 249–250]

12.1 Genetic Linkage and Recombination [pp. 250–254]

Section Review (Fill-in-the-Blanks)

Mendel's principle of independent assortment was based on the concept that different genes were not (1) _____, that is, located on the same chromosome. We now know that organisms have far more genes than (2) _____, so that genes that are not located on the same chromosome assort (3) _____ and display the expected Mendelian inheritance pattern, while (4) _____ genes assort together, behaving as a single gene during meiosis. However, because prophase I involves (5) _____ between homologous chromosomes, the inheritance patterns of linked genes generally fails to match those expected for single genes; the conclusion is that they have undergone a (6) _____ event during prophase I. In most instances, the (7) _______ event leads to the equal exchange of an allele on one chromosome with its corresponding gene on the (8) _____ chromosome. The amount of recombination between any two alleles is (9) _______ to the amount of intervening DNA; that is, as linkage distance between two genes increases, so does the recombination (10) _____. This is the basis for determining the (11) _______ distance between two alleles whereby genes are assigned relative locations with respect to (12) ______ ______ and the distance separating them is described in (13) _____ units. Because recombination (14) _______ vary to some degree from one position to another on chromosomes, this unit of measurement is a (15) _____ one and does not correlate with the actual bases of intervening DNA. In order to detect linkage and map the distance between two genes A/a and B/b, dihybrid testcrosses are done: by definition (16) _____ crossed with (17) _____. If all progeny classes are equally frequent, then the genes are determined to be (18) _____. However, in cases where map distances are extremely long and (19) _____ crossover events occur, the calculated frequency of recombination would suggest that the linked genes are, in fact, (20) _____ _____.

Matching

Match each of the following genetics terms and concepts with its correct definition or explanation.

21.	____	Locus	A.	The term applied to a genotype or phenotype that is arbitrarily designated as "normal"
22.	____	Linkage	B.	Alternative name for map unit
23.	____	Thomas H. Morgan	C.	Measures distances between adjacent genes using map units
24.	____	Wild-type	D.	The physical location of a gene on a chromosome
25.	____	Linkage map	E.	The phenomenon explained by the inheritance behaviour of genes located on the same chromosome
26.	____	Centimorgan	G.	Discovered the principles of linkage through studies of *Drosophila melanogaster*

27. Explain how the map unit distance between two genes is calculated.

____

____

28. Provide one type of evidence for linkage groups.

____

____

True/False

Mark if the statement is true or false. If the statement is false, justify your answer in the line below each statement.

29. ____________ Mendel was fortunate in his choice of garden pea characters as they were all unlinked, allowing him to determine his principle of independent assortment.

____

____

30. ____________ A linkage group would be a collection of genes that are linked on a chromosome.

____

____

12.2 Sex-Linked Genes [pp. 254–260]

Section Review (Fill-in-the-Blanks)

In many organisms, males and females have one or more (31) ______ of chromosomes that differ. These are called the (32) _____ ______; the remaining identical pairs of chromosomes are called the (33) _____ _____. In humans and (34) _______, females have two copies of the (35) ______ chromosome, whereas males only have one. In males, the chromosome that pairs with this during meiosis is the male-specific (36) _____ chromosome. These sex-based differences are reflected in certain non-Mendelian patterns of (37) _____, and these are best identified using (38) ______ crosses with true-breeding females. When looking at the expression of recessive (39) _____ traits, if males inherit the gene, it will be expressed, but expression in females requires inheritance from both (40) _____. The determination of a human's (41) _______ is determined by the *SRY* gene located on the Y chromosome. In mammals, the dosage of (42) _______ genes is equalized between males and females by (43) _____ of one of the X chromosomes during early embryonic development. This can be seen microscopically since the method of inactivation is chromosomal (44) _____; the inactivated chromosome is referred to as the (45) _____.

Matching

46.	____	Reciprocal cross	A.	Describes the non-Mendelian pattern of inheritance of traits carried by the sex chromosomes
47.	____	Autosomal	B.	A chart depicting all of the relatives of an individual over many generations, including the sex and presence of a particular trait
48.	____	Sex-linked inheritance	C.	The condensed, inactivated X chromosome visible in the nuclei of female cells
49.	____	Sex-linked trait	D.	The effective number of genes in a cell; for X-linked genes, this is equalized between males and females by inactivation
50.	____	Pedigree	E.	Performing a mating in which the phenotypes of the parental organisms are switched
51.	____	Barr body	F.	Chromosomes that are the same between males and females
52.	____	Gene dosage	G.	A gene that is linked to a chromosome that is inherited differently in males and females

True/False

Mark if the statement is true or false. If the statement is false, justify your answer in the line below each statement.

53. ____________The gene for hemophilia is located on the Y chromosome.

____

____

54. ____________ The sex of a child is determined by a gene on the Y chromosome.

____

____

55. ____________ In a female with different alleles of an X-linked gene, the same chromosome and allele will be inactivated in all cells.

____

____

Short Answer

56. Explain how a female can have a sex-linked genetic disorder.

____

____

57. An XY individual has a defective *SRY* gene. Explain the phenotypic sex of the individual.

____

____

58. Explain how the two sex chromosomes would be accurately segregated during meiosis.

____

____

59. Explain how an X-linked trait "skips" a generation.

____

____

Problem

60. Suzie's dad has an X-linked recessive disorder, SQ. Sam, Suzie's husband, has no known genetic disorders in his family. Suzie and Sam want to have children. As their genetic counsellor, you need to determine the answers to the following questions. Hint, the Punnett square method will help.

A. The genotype and phenotype of Suzie

____

____

B. The probability of this couple having a child with the genotype and phenotype of SQ

____

____

C. The possible phenotypes and genotypes of this couple's male children

____

____

D. The possible phenotypes and genotypes of this couple's female children

____

____

12.3 Chromosomal Alterations That Affect Inheritance [pp. 260–263]

12.4 Human Genetics and Genetic Counselling [pp. 264–266]

Section Review (Fill-in-the-Blanks)

Changes to the chromosome structure will affect (61)______ and may affect gene expression. Such alterations can result from exposure to (62)_______, certain chemicals, or the enzymes of some (63)_______. The immediate result is breakage of the DNA backbone leading to four common alterations: (64) _______, ________, ________, and (65)_______. The latter involves the movement of a chromosomal segment to a non– (66)_____ chromosome. Duplications have been suggested to play an important role in (67)_____ as the duplicated gene is free to undergo changes in the DNA sequence without adversely affecting the individual. However, in order for this to happen, the genetic changes must occur in (68)_____ line cells. Abnormalities in the number of chromosomes can also arise, most often from (69)_____ during meiosis. This can happen during meiosis I when (70)_______ chromosomes fail to separate or during meiosis II when (71) _____ ______ fail to separate. In rare cases, a cell will have a whole extra set of chromosomes; these cells are called (72)_______ and generally arise when the spindle fails to function during (73)_____ of germ-line cells. There are (74)_____ major types of inheritance patterns in humans, and determining which governs inheritance of abnormal genes and genetic disorders can be of critical importance. In (75)_______ recessive inheritance, both male and female heterozygous individuals are (76) _____, while homozygous recessive individuals exhibit the trait. An indication of this form of inheritance is the birth of an affected child to (77) ______ parents. In contrast, with autosomal (78) _______ inheritance, heterozygous and homozygous dominant individuals will both exhibit the trait, but homozygous (79)_____ individuals will not. In (80)_____ recessive inheritance, male individuals and homozygous recessive females will exhibit the trait and (81)_____ females are carriers. Examples of each mode of inheritance are associated with various genetic disorders, and it often falls to genetic (82)_____ to characterize the mode of inheritance based on a family (83) _____. Having this information gives prospective parents an idea of the inherent risks to potential (84) _______.

Matching

Match each of the following terms/concepts with its correct definition.

85.	____	Duplication	A.	Segment of a chromosome that is attached to a nonhomologous chromosome
86.	____	Euploids	B.	Individual with a 2n that is either –1 or +1 chromosome
87.	____	Translocation	C.	The other name for Down syndrome, a disorder resulting from having three copies of chromosome 21
88.	____	Nondisjunction	D.	Individual is 2n with no abnormalities
89.	____	Aneuploids	E.	Extra set of chromosomes
90.	____	Deletion	F.	Segment of a chromosome is present more than once
91.	____	Polyploids	G.	Segment of a chromosome that is present, but in the reverse order
92.	____	Inversion	H.	Segment of a chromosome is missing
93.	____	Trisomy 21	I.	Failure of homologous pairs to separate during meiosis I

True/False

Mark if the statement is true or false. If the statement is false, justify your answer in the line below each statement.

94. _____________ Aneuploidy of sex chromosomes is less likely to have lethal effects than autosomal aneuploidy.

____

____

95. _____________ Polyploidy in eukaryotes is generally fatal during the embryonic stage of development.

____

____

Short Answer

96. For a genetic disorder, distinguish between autosomal recessive and autosomal dominant inheritance patterns.

____

____

12.5 Nontraditional Patterns of Inheritance [pp. 266–268]

Section Review (Fill-in-the-Blanks)

(97) _______ inheritance and genomic (98) ______ are both modes of inheritance that do not follow the traditional patterns. In the first type, the inherited trait is encoded by the (99) ______ chromosome, and since these are found in the cytoplasm and the female gamete has far more cytoplasm in the egg than the sperm, these traits generally exhibit (100) _____ inheritance. Genetic disorders with this type of inheritance generally result in problems with cellular (101) _____ _______. In the second type of nontraditional inheritance, the allele inherited from one parent is (102) _______, while the other is not. This is a form of gene expression control called (103) _________ and depends on (104) _______ of certain bases within the control region of the gene. This reaction happens in the (105) _______ and so precedes fertilization. Genetic disorders associated with (106) _______ alleles arise when the homologous allele has suffered a deletion. Certain cancers have also been associated with these genes; however, in these cases, the problem is from failure to (107) _______ a gene, a phenomenon called loss of (108) _____, which gives rise to an increased (109) _______ dosage effect.

True/False

Mark if the statement is true or false. If the statement is false, justify your answer in the line below each statement.

110. _____________ Health problems associated with loss of imprinting are due to a methylation of an essential gene.

____

____

Short Answer

111. Define the term "uniparental inheritance."

____

____

112. If your genetic counsellor told you that a particular trait is associated with the "egg parent," what would you conclude about this trait?

____

____

SELF-TEST

1. All individuals of a group of cats have rounded ears and short whiskers. The simplest explanation of this pattern of inheritance is ________. [p. 250]

   a. The two traits are sex-linked
   b. The two traits are linked
   c. Pleiotropic
   d. Independent

2. Two moths with short wings have the following offspring: 100 short-winged males, 97 long-winged males, and 205 short-winged females. The gene that determines wing length is _______. [pp. 256–257]

   a. Sex-linked and recessive
   b. Sex-linked and dominant
   c. Autosomal recessive
   d. Autosomal dominant

3. Siobhan and James just had a baby. The doctor comes in and tells the happy couple the child is a carrier for red–green colour blindness. The sex of the child is ________. [p. 258]

   a. Male
   b. Female
   c. Impossible to tell

4. You are analyzing karyotypes of two patients, a male and a female. The phone rings, and you are distracted for a few minutes. When you return, you know you are looking at the female karyotype because you clearly see ______. [p. 265]

   a. Mitochondria
   b. Centrioles
   c. A Barr body
   d. Only one Y chromosome

5. A normal chromosome sequence is ABCDEFG. You discover the sequence in your patient to be ABGFEDGFEDC. Which of the following genetic alteration(s) most likely occurred? [pp. 260–261]

   a. Deletion and duplication
   b. Duplication and translocation
   c. Inversion
   d. Inversion and duplication

6. After six months of experimental crosses with a plant species that has 50 chromosomes, you discover offspring with 150 chromosomes. What must have happened? [p. 263]

   a. Nondisjunction
   b. Euploidy
   c. Tetraploidy
   d. Triploidy

7. You need to determine if a patient has trisomy 21 or Down syndrome. The best way to make an accurate diagnosis is to evaluate a ________. [p. 262]

   a. Pedigree
   b. Karyotype
   c. Maternal inheritance
   d. Mitochondrion

8. Referring to question 7, this condition most likely occurred because of ________. [pp. 261–262]

   a. Nondisjunction
   b. Polyploidy
   c. Translocations
   d. Inversions

9. If 75% of the offspring have a genetic condition, it can be concluded the condition is ____________. [p. 264]

   a. Autosomal recessive
   b. Sex-linked autosomal recessive
   c. Autosomal dominant
   d. Sex-linked autosomal dominant

10. Assume you are expecting a child. Your doctor wants to test for PKU. Which approach would he or she take? [p. 266]

   a. Chorionic villus sampling
   b. Screening test done after birth
   c. Amniocentesis
   d. Karyotyping

11. Which of the following is not an X-linked trait? [pp. 255, 258]

   a. Hemophilia
   b. *SRY*
   c. Red–green colour blindness
   d. All of the above are X-linked

12. Because she has had two previous miscarriages, Julie, who is 8 weeks pregnant, has a sonogram to confirm that there are no problems with her baby. She and her husband tell the doctor they would like to know the sex of the baby. If you were the doctor, what would you tell them? [p. 255]

   a. Everything is too small at that point to allow accurate identification of the sex
   b. They will only be able to tell if the baby is a boy
   c. They cannot tell at that point because development of the male features does not even begin until after 6 to 8 weeks
   d. No problem!

13. Which of the following statements is false? [p. 259]

   a. Calico cats are all female
   b. The father of calico cats will either be black or orange
   c. The calico colour in cats is due to imprinting
   d. The calico coat of some cats is the result of epistasis

14. Which of the following statements about gene duplications is false? [p. 260]

   a. Duplications may have beneficial effects
   b. Gene duplications are the source of evolutionary changes
   c. Some of the human genes for hemoglobin appear to be the result of gene duplication
   d. Some of the shark genes for hemoglobin appear to be the result of gene duplication

15. Which of the following statements about aneuploidy is false? [p. 262]

   a. In humans, aneuploidy of autosomal chromosomes is quite often lethal
   b. In humans, aneuploidy of autosomal chromosomes quite often has no obvious adverse effects
   c. ~70% of human miscarriages are the result of aneuploidy
   d. Aneuploidy results from nondisjunction

16. Which of the following is autosomal recessive?[p. 264]

   a. Cystic fibrosis
   b. Trisomy 21
   c. Red–green colour blindness
   d. Achondroplasia

17. Which of the following would be used for prenatal diagnosis of Down syndrome? [p. 266]

    a. Genetic screening
    b. Amniocentesis combined with karyotyping
    c. Chorionic villus sampling combined with karyotyping
    d. (b) and (c)

18. How would you determine whether a trait was cytoplasmically inherited or X-linked recessive? [pp. 258, 267]

    a. If only the males exhibited the trait, then it would be cytoplasmically inherited
    b. If only the males exhibited the trait, then it would be X-linked
    c. If females had to be homozygous for the trait but males did not, then it would by X-linked
    d. If males and females showed the trait only if their mother also showed it, it would be cytoplasmically inherited
    e. (c) and (d)

19. In the cross AABB × aabb, if you allowed the F_1 generation to breed and you got the ratio expected for unlinked genes, then you would conclude that _________. [pp. 252–254]

    a. A and B are not linked
    b. A and B are linked but widely separated
    c. A and B are either linked but widely separated or unlinked

20. If in a dihybrid cross for two widely separated but linked genes you wanted to determine the map unit distance between them, you would do which of the following? [pp. 253–254]

    a. Do a reciprocal cross
    b. Determine the distance to genes close to one or the other
    c. Determine the distance from each gene to one in between
    d. There is no way to determine this distance

INTEGRATING AND APPLYING KEY CONCEPTS

1. After completing the chapter in the textbook and the accompanying sections in the study guide, devise your own topic map for this material.
2. Why might polyploidy be advantageous in plants but not in animals?
3. Given the information in your text, predict which is better tolerated, aneuploidy in autosomes or sex chromosomes. Defend your response.
4. Explain how a reciprocal cross can demonstrate that a trait is X-linked

OPTIONAL ACTIVITY

1. Epstein-Barr virus (EBV) causes the common human disease infectious mononucleosis or "mono." Most people are exposed to this virus at some point in their lives but do not even fall ill. The vast majority of people who do develop mono make a full recovery. Like the chickenpox virus, EBV remains in the body, dormant, for the rest of their lives. However, outside of the Americas, the infection has been associated with the cancer you read about in this chapter, Burkitt lymphoma. Find out more about this association and the genetic changes that give rise to this cancer.

13 DNA Structure, Replication, and Organization

CHAPTER HIGHLIGHTS

- Many researchers over the course of almost 100 years worked to elucidate the nature of the hereditary material.
    - DNA was established as the hereditary material in 1952.
    - Shortly after, Watson and Crick resolved the three-dimensional structure of the DNA molecule.
    - Nucleotides are the major structural component of DNA, which has a helical, double-stranded, antiparallel structure.
- DNA is replicated in a semiconservative fashion with each strand of the parent molecule serving as a template for synthesis of a complementary daughter strand.
    - Many enzymes participate in the process of replication; DNA polymerase is the major replicative enzyme.
    - DNA polymerase has proofreading ability, so if it incorporates the wrong base, it can recognize this, remove the base, back up, and recommence replication.
- DNA replication of the circular prokaryotic chromosome uses the same mechanism as in eukaryotes.
- Telomeres are highly repeated noncoding sequences found on the ends of the eukaryotic linear chromosomes. These get shorter with increasing rounds of replication, and this shortening may be involved in the aging process.
- In combination with post-replication repair, the error rate during DNA replication is extremely low.
    - Errors can lead to mutations, which are required for evolution.

- Chromosomes are organized into condensed structures so that they can fit into very much smaller cells.
    - In eukaryotic chromosomes, various proteins participate in the condensation of the chromosomes of euchromatin or further into inactive heterochromatin or further still into the structures seen during mitosis and meiosis.
    - Condensation may play a role in controlling gene expression.
    - Prokaryotic cells also organize the circular chromosome so that it fits into the cell. The organization of the condensed structure is different from the linear eukaryotic chromosomes, as are the proteins.

STUDY STRATEGIES AND LEARNING OUTCOMES

- First, focus on the structure of DNA—the three basic components of a nucleotide, including the two types of nitrogenous bases, and the underlying structural reasons for complementary base pairing.
- If given a series of DNA nucleotides, be able to provide the complementary bases. Make sure you understand what antiparallel means and what 5′ and 3′ ends mean and how they are related to each other.
- DNA replication has multiple steps; focus on one step at a time.
- Carefully examine the figures and tables as well as the various subject headings to keep the ideas clearly organized in your mind.
- As always, go through one section of the textbook at a time and then work through the companion section of the study guide.

By the end of this chapter, you should be able to

- Identify the key people and experiments that gradually revealed the nature, structure, and replication of the genetic material
- Describe the structure of DNA and know which bases are purines, which are pyrimidines, and which ones form base pairs
- Identify the step-by-step process of DNA replication in eukaryotes, identifying the key enzymes and activities

- Explain the purpose of telomeres, the importance of telomerase, and knowing how prokaryotic chromosomes are replicated and why these are both unnecessary in prokaryotes
- Understand how errors in replication are corrected
- Describe the increasing levels of organization of eukaryotic chromosomes, identifying which is present at the various stages of cell growth and what maintains each level of organization

TOPIC MAP

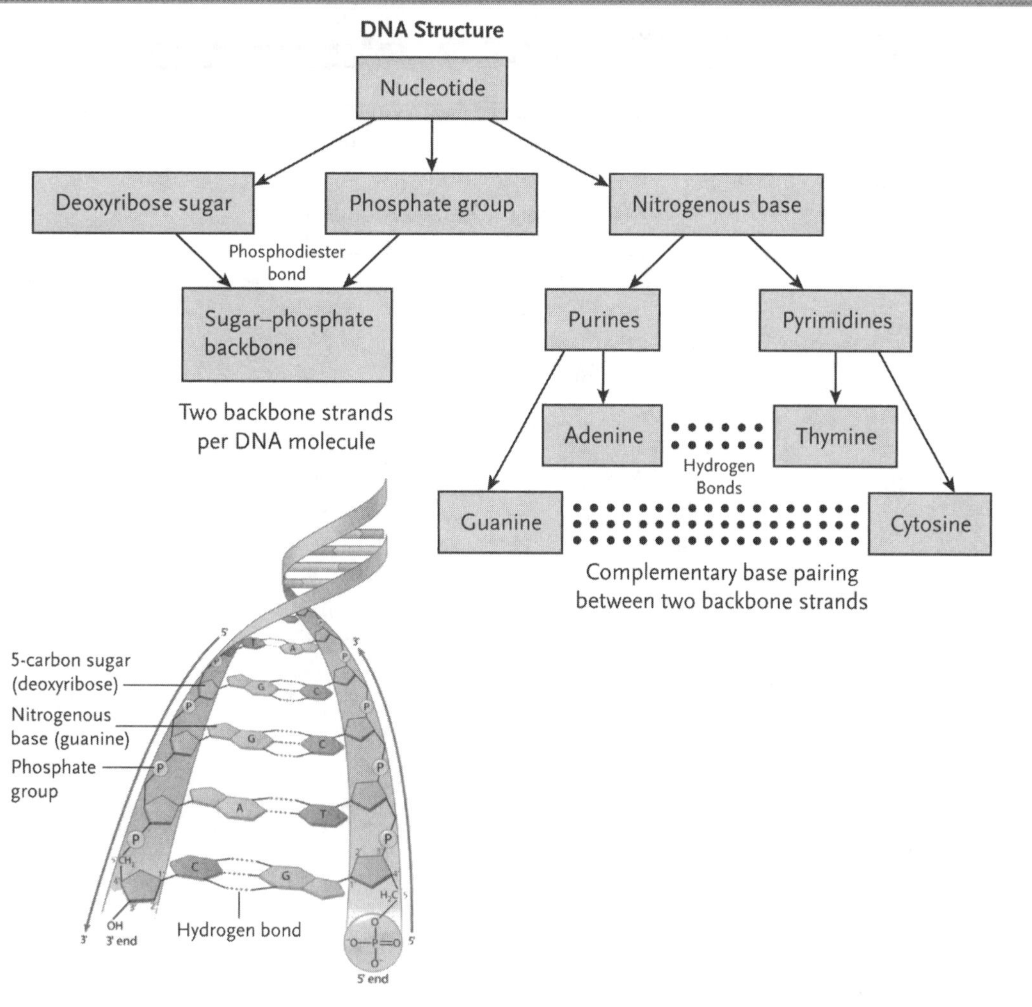

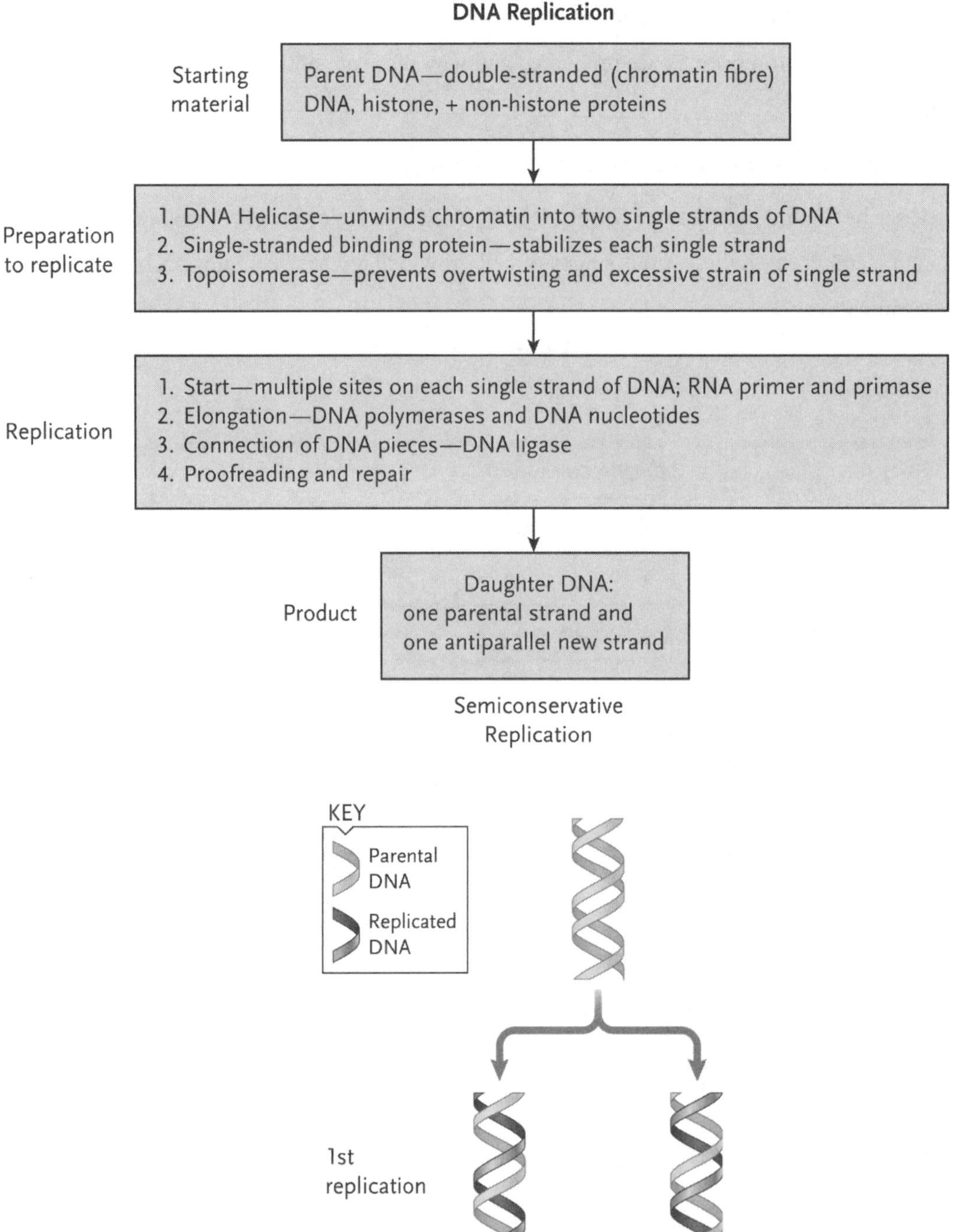

INTERACTIVE EXERCISES

Why It Matters [pp. 271–272]

13.1 Establishing DNA as the Hereditary Molecule [pp. 272–275]

Section Review (Fill-in-the-Blanks)

As a result of his studies of the cell nucleus, Johann Friedrich Miescher became the first person to isolate and begin the chemical characterization of (1) _____. He called the material he extracted nuclein, which later became known as (2) _____. The material he extracted was found to be (3) _____ and have a high (4) _____ content. Although Mendel had published his research only a few years before, it was not until the mid- (5) _____ century that scientists were able to further characterize the molecular structure and identify DNA as the (6) _____ material. Scientists knew that this material would need to contain a vast amount of information, and many felt that (7) _____ provided the necessary variety due to the higher number of different (8) _____ _____ available to make the polymer. Griffith's experiments with pathogenic (*S* strain) and nonpathogenic (*R* strain) bacteria showed that *R*-type bacteria could be altered due to the process of (9) _____. Something in the *S* strain was passed to the *R* strain, even when the *S* strain was (10) _____. He called this material the (11) _____ principle. The nature of this material was identified by (12) _____ and colleagues after they used *S*-strain bacteria that had been (13) _____-_____, purified the (14) _____ candidate macromolecules, and showed that after (15) _____ degradation of the proteins and RNA, the remaining DNA was able to (16) _____ the *R*-strain bacteria to *S*-strain bacteria. Because some people still did not believe that DNA was the hereditary material, Hershey and Chase provided further evidence when they infected (17) _____ with radioactively labelled (18) _____. They demonstrated that when only the protein coat of the infecting virus was labelled, the viral progeny were (19) _____, but when the viral (20) _____ was labelled, the viral progeny resulting from the infection were (21) _____.

Matching

Match each of the following names or terms with the correct experiment, experimental conclusions, or definition.

22.	____	Watson and Crick	A.	Using mice, found that nonpathogenic bacteria could be transformed into pathogenic bacteria
23.	____	Griffith	B.	A type of virus that infects bacteria
24.	____	Hershey and Chase	C.	The name given to the mysterious molecule that transformed his nonvirulent strain to virulence
25.	____	Avery and coworkers	D.	Using bacteriophages and radioactive isotopes of sulphur and phosphate, concluded that DNA is the genetic material rather than proteins

26.	____	Transforming principle	E.	Analyzed chemical and physical data to determine the three-dimensional structure of DNA
27.	____	Bacteriophage	F.	Found that enzymatically treated material purified from heat-killed bacteria could still transform nonpathogenic bacteria into pathogenic bacteria

13.2 DNA Structure [pp. 276–278]

Section Review (Fill-in-the-Blanks)

Prior to the studies of Watson, Crick, and Franklin, DNA was known to contain four nitrogenous (28) _____. These were the purines (29) _____ and (30) _____, both characterized by two-ring structures in contrast to the single-ring structures of the (31) _____, cytosine and thymine. The studies of organic chemist Erwin Chargaff established that the number of adenines was always equal to the number of (32) _____ and the number of cytosines was always equal to the number of (33) _____; this became known as (34) _____ rules. The molecule was known to consist of chains of (35) _____ with a backbone of sugars linked to phosphates by (36) _____ bonds. The backbone had a polarity so that on one end of a chain, there was a phosphate group attached to the (37) _____ carbon of a deoxyribose (38) _____ and at the other was a (39) _____ group linked to the 3′ carbon of the sugar. These ends of the chain are called the (40) _____ and (41) _____ ends, respectively. With this information, combined with the X-ray diffraction studies of Wilkins and Franklin, (42) _____ were able to deduce that the DNA molecule contains (43) _____ sugar–phosphate chains that run (44) _____ to each other, and in between are two (45) _____, one associated with each sugar–phosphate chain. (46) _____ base pairing binds one (47) _____ with one pyrimidine. Based on Chargaff's rules, adenosine was determined to pair with (48) _____ and cytosine with (49) _____. These pairings were determined to be stabilized by (50) _____ bonds; these also held the two chains together, and the orientation of the two chains was determined to be (51) _____ with the 5′ end of one strand opposite the 3′ end of the other. Finally, they determined that the information that was stored in this (52) _____ molecule was coded by the particular (53) _____ of the bases.

Matching

Match each of the following names or terms with the correct experiment, experimental conclusions, or definition.

54	____	Watson and Crick	A.	Used X-ray diffraction studies to analyze the structure of DNA
55.	____	Chargaff's rules	B.	The 5′ end of one DNA strand is opposite the 3′ end of the other strand in the double helix
56.	____	Antiparallel	C.	The twisting shape of a DNA molecule
57.	____	Wilkins and Franklin	D.	The association of purines and pyrimidines
58.	____	Complementary base pairing	E.	Analyzed chemical and physical data to determine the three-dimensional structure of DNA
59.	____	Double helix	F.	In a DNA molecule, the number of adenines is equal to the number of thymines and the number of cytosines is equal to the number of guanines

True/False

Mark if the statement is true or false. If the statement is false, justify your answer in the line below each statement.

60. ____________ A-T base pairs are held together by three covalent bonds.

____

____

Short Answer

61. Describe the basic structural features of the DNA molecule.

____

____

Labelling

62. Given the backbone of DNA, label the directional ends of the backbone and number the carbons of deoxyribose.

A. __________

B. __________

C. __________

D. __________

E. __________

F. __________

G. __________

H. __________

I. __________

J. __________

13.3 DNA Replication [pp. 278–287]

Section Review (Fill-in-the-Blanks)

DNA is duplicated in a (63) _____ fashion with each strand acting as a (64) _____ for the synthesis of a (65) _____ copy. When replication begins, the enzyme (66) _____ unwinds the double helix, producing a site called a (67) _____ _____. Replication is primed with short pieces of a(n) (68) _____ _____ that are produced by an enzyme, (69) _____. The nucleotides are then incorporated into the growing chain by the enzyme (70) _____ _____. This enzyme moves down the template in the (71) _____ → _____ direction and assembles the growing chain in the (72) _____ → _____ direction: that is, during assembly, the last nucleotide added to the growing chain has a(n) (73) _____ group exposed. The reason for this direction of polymerization is the specificity of the DNA polymerase: it can only add a nucleotide at the (74) _____ end of a nucleotide chain. Replication progresses in opposite directions from the (75) _____ of replication. Because of the antiparallel orientation of the two strands and the movement of the DNA (76) _____, only one strand can be synthesized continuously as it keeps pace with the advancing replication (77) _____: this is the (78) _____ strand. The second DNA strand is synthesized in short lengths in the opposite direction of the fork and is called the (79) _____ _____. These short lengths of DNA are called (80) _____ _____. Another DNA polymerase removes the (81) _____ primers and fills in any gaps in the nucleotide strands, and then the enzyme (82) _____ _____ closes all of the remaining single-chain nicks. Because eukaryotic chromosomes are linear and DNA polymerase only moves in the (83) _____ → _____ direction, removal of the RNA primer from the (84) _____ end results in chromosomes that become (85) _____ in length with each round of DNA (86) _____. This could prove lethal if this resulted in deletion of (87) _____. In most chromosomes, this is prevented by a buffer of noncoding DNA called (88) _____, which consists of short repeats that occur hundreds to (89) _____ of times. With each replication cycle, these sequences become (90) _____; however, the enzyme (91) _____ can maintain the buffer by adding more repeats. This enzyme becomes inactive after a number of (92) _____ so that eventually the cell stops dividing and (93) _____. The (94) _____ process and the development of (95) _____ have both been associated with (96) _____ length and (97) _____ activity.

Matching

For each of the following statements, choose the most appropriate enzyme that catalyzes the reaction in DNA replication.

A. DNA polymerase　B. DNA helicase　C. DNA ligase

D. Telomerase　E. Primase

98. ____ Catalyzes the unwinding of the DNA double helix
99. ____ Produces an RNA strand that acts as a starting point for DNA replication
100. ____ Connects Okazaki fragments in the lagging strand
101. ____ Adds repeating units to the end of chromosomes
102. ____ May be responsible for cancer cell development
103. ____ Adds nucleotides only at the 3′ end of an existing nucleotide chain

True/False

Mark if the statement is true or false. If the statement is false, justify your answer in the line below each statement.

104. ____________The semiconservative nature of DNA replication was determined by Watson and Crick.

____

____

105. ____________ The aging process can be prevented by reactivating telomerase.

____

____

Short Answer

106. Given the following DNA molecule and assuming the arrow denotes the direction of the helicase, identify which strand is the leading strand and which is the lagging strand:

A. ____ 3′ AATCCGTACGGT 5′

| | | | | | | | | | | | →

B. ____ 5′ TTAGGCATGCCA 3′

107. Refer to the previous question and identify how the leading and lagging strands correlate with the continuity and discontinuity of replication and why this occurs.

____

____

108. What is meant by the term "semiconservative replication"?

____

____

13.4 Mechanisms that Correct Replication Errors [pp. 287–289]

Section Review (Fill-in-the-Blanks)

DNA polymerases make very few errors during replication, and most of the errors that are made are base-pair (109) _____. These can be corrected by the (110) _____ mechanism of DNA polymerase. This involves the reversal of the enzyme on the template, the removal of the (111) _____ nucleotide, and the resumption of synthesis in the (112) _____ direction. If a base pair mistake remains after replication, this will result in a (113) _____ of the double helix, and this is detected by enzymes responsible for (114) _____ repair. The enzymes involved in this process remove a portion of the new chain, including the (115) _____ nucleotides; the gap is filled in by a (116) _____ _____; and the nicks are repaired by DNA (117) _____. In combination, DNA proofreading and repair mechanisms ensure a high level of (118) _____ during replication; errors that do remain are a primary source of (119) _____, and these are extremely important for the (120) _____ process.

Short Answer

121. Discuss, using an example, why DNA repair mechanisms are so critical.

____

____

13.5 DNA Organization in Eukaryotes and Prokaryotes [pp. 289–292]

Section Review (Fill-in-the-Blanks)

Eukaryotic chromosomes on their own are far too big to fit into the (122) _____ of the cell. The DNA is therefore compacted by a group of proteins called the (123) _____. These proteins are (124) _____ charged and associated with the oppositely charged (125) _____ groups of the DNA. At regular intervals, the DNA is wrapped around a complex of proteins made of up (126)____, _____, _____, and _____; this complex is called the (127) _____. The DNA is further organized by the (128) _____ protein, which binds the DNA in the linker region between these complexes and generates a coiled structure called the (129) _____. This organization is necessary to allow the DNA to fit into the nucleus but probably also protects it from mechanical and (130) _____ damage. The term (131) _____ is used to refer to the complex of DNA and its associated proteins. In interphase cells, the DNA contains regions that are loosely packed, specifically referred to as (132) _____, and regions that are tightly packed, called the (133) _____. The latter is believed to represent a mechanism by which expression of large blocks of genes is (134) _____. This is supported by the fact that inactivation of one of the X chromosomes in females occurs through condensation and the resulting structure, the (135) _____ body, is composed entirely of (136) _____. Another group of DNA-associated proteins, the (137) _____ proteins, are involved in the regulation of individual (138) _____. Some exert their effect by modifying the interaction of the (139) _____ proteins with the DNA so that they are more or less tightly associated with the DNA, thereby modifying the level of (140) _____ of the chromatin. Although, in general, prokaryotes have only one (141) _____ chromosome, it is still far bigger than the cell. The chromosome is organized into (142) _____ by a variety of (143) _____-charged proteins. The resulting structure resides in the (144) _____ of the cell. Prokaryotes may also have smaller DNA molecules called (145) _____. These are also generally (146) _____ in shape; both these and the chromosomes replicate through the same basic process as seen in eukaryotes; however, replication is initiated at a single (147) ______ _____ ______.

Matching

Match each of the following terms/concepts with its correct definition.

148.	____	Euchromatin	A.	The highly condensed and inactivated X chromosome seen in the nucleus of human females
149.	____	Heterochromatin	B.	The regularly occurring "beads" of eukaryotic DNA consisting of the chromosomal DNA wrapped around a complex of histone proteins
150.	____	H1	C.	A small extrachromosmal DNA molecule found in many prokaryotes
151.	____	Nucleosome	D.	The method of DNA replication exhibited by plasmids undergoing conjugation
152.	____	Barr body	E.	The loosely packed regions of the eukaryotic chromosome
153.	____	Plasmid	F.	The histone protein that binds in the linker regions of the eukaryotic chromosome
154.	____	Rolling circle replication	G.	The tightly packed regions of the eukaryotic chromosome

Labelling

155. Label the various levels of organization of eukaryotic chromatin and chromosomes.

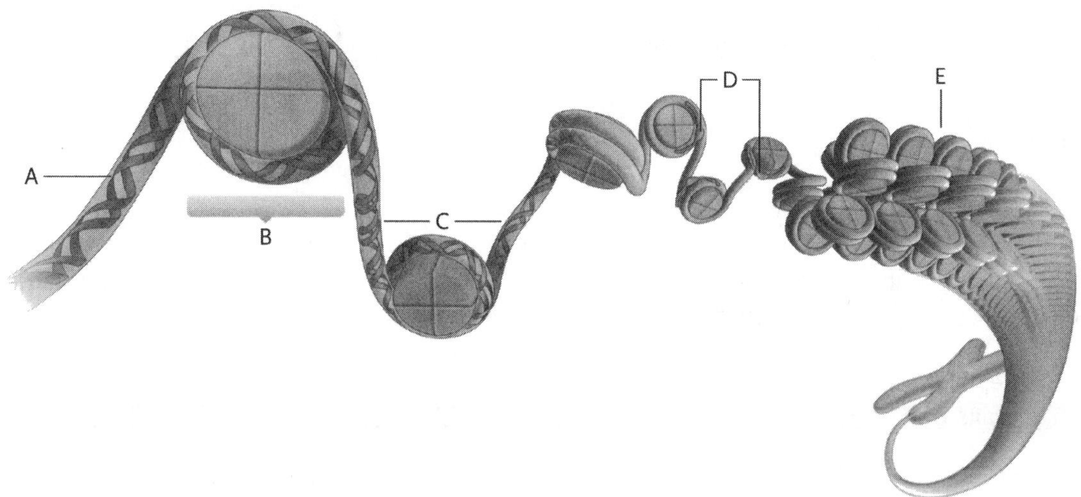

A. ______________

B. ______________

C. ______________

D. ______________

E. ______________

Short Answer

156. Compare and contrast euchromatin and heterochromatin.

____

____

157. Discuss the various structures of DNA.

____

____

SELF-TEST

1. In the early 1900s, the hereditary material was thought to be ________. [p. 272]

   a. RNA
   b. DNA
   c. Carbohydrates
   d. Proteins

2. Griffith's experiments with bacteria identified that an agent from *S* strain (pathogenic) bacteria could ________ an *R* strain (nonpathogenic) to pathogenic. [p. 273]

   a. Replicate
   b. Transform
   c. Transduce
   d. Exchange

3. Hershey and Chase used radioactive labelling experiments to identify the hereditary material. Radioactive isotopes of both sulphur and phosphate were used. Sulphur is found in ______ but not in ________. [pp. 274–275]

   a. DNA; proteins
   b. Proteins; DNA
   c. Carbohydrates; DNA
   d. DNA; lipids

4. Nucleotides consist of deoxyribose, a phosphate group, and ________. [p. 276]

   a. A purine
   b. A pyrimidine
   c. Either a purine or pyrimidine
   d. A phosphodiester bond

5. Which of the following techniques led to the conclusion that the DNA molecule was a double helix? [p. 277]

   a. Complementary base pairing
   b. Transformation
   c. X-ray diffraction
   d. Replication

6. Experiments of DNA replication identified that the daughter DNA consisted of one strand of parent DNA and one strand of new DNA. This type of replication is ________. [p. 279]

   a. Semiconservative
   b. Conservative
   c. Dispersive
   d. Semidisruptive

7. When DNA is replicating, a nucleotide can only be added to the ________ end of an existing nucleotide chain. [p. 280]

   a. 3′
   b. 5′
   c. Either 3′ or 5′
   d. Phosphodiester end

8. Activity in which enzyme would indicate that DNA was about to replicate? [p. 282]

   a. DNA ligase
   b. DNA polymerase
   c. Primase
   d. DNA helicase

9. Due to the directionality of DNA polymerase, the lagging strand of DNA is replicated in a _______ fashion. [p. 283]

   a. Discontinuous
   b. Continuous
   c. Okazaki
   d. Antiparallel

10. If the replication fork were blocked, the activity of ________ would be affected first. [p. 282]

    a. Helicase
    b. DNA polymerase
    c. Primase
    d. DNA ligase

11. If DNA proofreading fails, _________ may occur. [pp. 287–289]

    a. Transformations
    b. Mutations
    c. Telomerization
    d. Mismatch base pairing

12. If you wanted to study the genes involved in eukaryotic DNA replication, the most likely portion of the chromosome to look for genes being actively expressed would be the ________. [p. 290]

    a. Heterochromatin
    b. Nucleoid
    c. Nucleosome
    d. Euchromatin

13. Avery and coworkers did which of the following? [p. 274]

   a. Proved the semiconservative replication of DNA
   b. Correctly identified the helical nature of DNA based on X-ray diffraction data
   c. Purified DNA from test tube–grown, heat-killed S-*Streptococcus* and used it to transform R-*Streptococcus*
   d. Determined that the number of adenine is equal to the number of thymines and the number of cytosines is equal to the number of guanines

14. In performing their experiments with *E. coli* and bacteriophage T2, what would Hershey and Chase have observed? [pp. 274–275]

   a. The ^{35}S-labelled bacteriophage proteins were isolated from infected cells as well as from the phage progeny
   b. The ^{32}P-labelled bacteriophage proteins were isolated from infected cells as well as from the phage progeny
   c. The ^{35}S-labelled bacteriophage DNA were isolated from infected cells as well as from the phage progeny
   d. The ^{32}P-labelled bacteriophage DNA were isolated from infected cells as well as from the phage progeny

15. Which of the following is not true about purines? [p. 277]

   a. They do not base pair with each other because they are too big and would distort the DNA double-helical structure
   b. They possess two ring structures
   c. They are the adenines and guanines
   d. They are the thymines and cytosines

16. If DNA replication were conservative rather than semiconservative, what would Meselson and Stahl have observed in their CsCl density gradients after the first round of replication following transfer from the ^{15}N-containing media to the ^{14}N-containing media? [pp. 279–281]

   a. A single band in the density gradient with a density corresponding to a uniform distribution of ^{15}N
   b. Two bands in the density gradient, one of a density corresponding to a uniform distribution of ^{15}N and one of a density corresponding to a uniform distribution of ^{14}N
   c. A single band in the density gradient with a density corresponding to a uniform distribution of ^{14}N
   d. Two bands in the density gradient, one of a density corresponding to a uniform distribution of ^{15}N and one of a density corresponding to DNA that was half ^{14}N and half ^{15}N

17. Energy for the polymerization of DNA during replication comes from [p. 280]

    a. Hydrolysis of the last phosphate group in the nucleoside triphosphate
    b. Hydrolysis of the bond between the inner phosphate group and the terminal pyrophosphate of the nucleoside triphosphate
    c. DNA polymerase
    d. There is no energy requirement for the process

18. Which of the following statements about primase is false? [pp. 283–285]

    a. It synthesizes primers of approximately 10 bases in length
    b. It synthesizes primers made of RNA
    c. It synthesizes primers made of DNA
    d. It does not require a 3′-hydroxyl to begin replication

19. Why do bacterial cells not require a telomerase enzyme? [pp. 286, 291]

    a. The bacterial chromosome is circular, so no gaps are left after the RNA primers are removed
    b. The DNA polymerase in bacteria does not need a 3′ hydroxyl of an existing strand to begin replication
    c. The mode of DNA replication in bacteria is totally different from that in eukaryotic cells
    d. Bacteria do not live long enough for telomere shortening to become lethal

20. Which of the following statements is true about rolling circle replication? [p. 291]

    a. It is the method of replication used for replication of the bacterial chromosome
    b. It is the method of replication used during conjugative transfer of plasmids such as the F plasmid in bacteria
    c. It is the method used to replicated the telomeres
    d. It is the type of replication used by telomerase enzymes

INTEGRATING AND APPLYING KEY CONCEPTS

1. Draw a picture of a bacterial chromosome in the process of being replicated. Assuming bidirectional progression of the replication forks, figure out which newly replicated DNA will be the leading and lagging strands. How do these strands relate to each other when comparing one replication fork with the other?

2. Predict the effects of mutations in histone and nonhistone proteins.

3. If telomerase activity could be conclusively linked to cancer cell development, predict some of the possible types of treatment or agents to inhibit or stop cancer growth and spread to other tissues.

OPTIONAL ACTIVITY

1. Although eukaryotic cells appear to have evolved from prokaryotic predecessors, linear chromosomes are extremely rare in bacteria and archaea but are the only form of DNA found in eukaryotic cells. In addition, it is believed that viruses have been around almost as long as life has existed on Earth and they all have linear chromosomes. This has led some researchers to propose that viruses may have been the predecessor to the linear chromosomes of eukaryotic cells. Do some research through the Internet on the various proposals for the appearance of linear chromosomes in the eukaryotes.

14 Gene Structure and Expression

CHAPTER HIGHLIGHTS

- The ability to make the required cellular proteins is determined by the genome of an organism.
- DNA molecules contain units called genes or transcription units. These are copied into various types of RNA.
- There are two major steps required to make a protein from a gene: transcription and translation.
- Transcription:
    - The base sequence on DNA is read.
    - RNA polymerase is used to make a corresponding mRNA sequence.
- Translation:
    - Triplet base sequence on mRNA are "codons" that provide the code for specific amino acids.
    - The genetic code is universal among life forms.
    - Ribosomes read the codes on mRNA.
    - tRNA brings the coded amino acid to the ribosome.
    - The ribosome is a ribozyme that catalyzes the formation of peptide bonds between amino acids to form a polypeptide.
- Although the basic process of making proteins is similar in prokaryotic and eukaryotic cells, there are a few differences, including the speed and location of the two processes as well as the need for processing of transcribed RNA in eukaryotes before the mature mRNA is available.

- The flow of information from DNA to RNA to protein is called the central dogma as it is universal among life forms.
- Mutations in a transcriptional unit can have significant effects on the final protein product of the gene.

STUDY STRATEGIES AND LEARNING OUTCOMES

- Remember that the goal of this chapter is to understand how cells make proteins.
- There are step-by-step processes that must be understood and followed.
- Concentrate on the key concepts of transcription and translation.
- Once the key ideas are understood, you can then learn the steps within each pathway.
- To have a complete appreciation of the various steps, pay attention to the differences between prokaryotes and eukaryotes.
- Carefully examine the figures and tables as well as the various subject headings to keep the ideas clearly organized in your mind.
- As always, go through one section of the textbook at a time and then work through the companion section of the study guide.

By the end of this chapter, you should be able to

- Understand the difference between the one gene–one enzyme hypothesis and the one gene–one polypeptide hypothesis and be able to relate the latter to the central dogma of biology
- Describe the structure of mRNA, the genetic code, and its redundancy and know the initiation codon and amino acid and the three stop codons
- Understand the process of transcription and be able to compare the transcriptional process of prokaryotes with that of eukaryotes
- Explain the molecular process of mRNA splicing
- Understand the process of translation and be able to compare translation in prokaryotes with that of eukaryotes
- Describe the structure of ribosomes, including the nature of the catalytic component
- Understand how proteins are targeted to different cellular locations in eukaryotes

TOPIC MAP

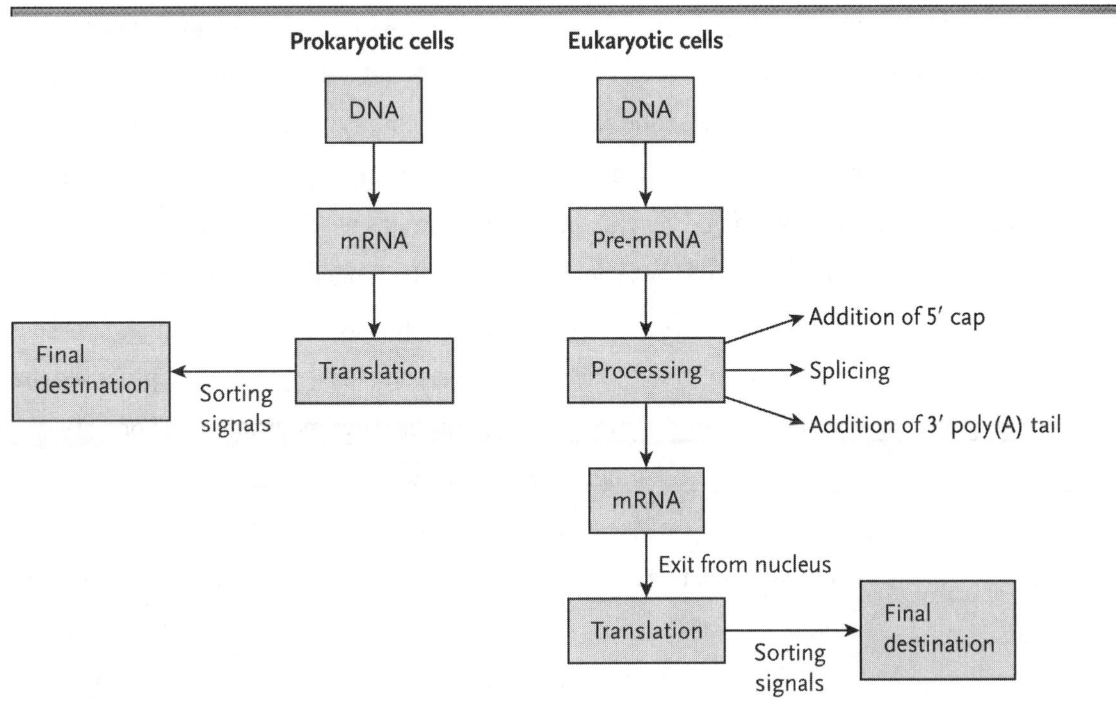

INTERACTIVE EXERCISES

Why It Matters [pp. 295–296]

Section Review (Fill-in-the-Blanks)

The marine mussel *Mytilus* produces fibres made of (1) _______ that, together with other proteins, make an exceptionally strong underwater adhesive. In order to produce large amounts of this strong adhesive, genetic engineers are inserting the (2) ______ from the mussel to yeast cells. Yeast cells are then able to rapidly translate the mussel genes into (3) ______, which, it is hoped, will give rise to the fibres of this adhesive. This chapter is about reading the code in (4) ____ to make corresponding copies of messenger (5) _____. These are then read by (6) _______ to make (7) ______.

14.1 The Connection between DNA, RNA, and Protein [pp. 296–300]

Section Review (Fill-in-the-Blanks)

The studies performed by (8) ________ and (9) _________ using the haploid fungus *Neurospora crassa* revealed a direct correlation between gene mutations and alterations of the (10) ___________ required to catalyze metabolic reactions. They demonstrated that wild-type *Neurospora* can grow on (11) __________ medium containing salts, sugar, and vitamins but that after exposing the fungus to mutagenic X rays, the resulting strains needed additional (12) _________ added to the medium. These new variants of the fungus are referred to as (13) ________. The scientists hypothesized that each mutant had a defect in a (14) ________________ coding for an enzyme needed to synthesize a (15) ______. They proposed the one (16) ___________–one (17) ________________ hypothesis, later modified to the one (18) ___________–one (19) ________________ hypothesis. The path from genes to (20) ________ involves two steps: (21) ____________ of the gene sequence in the DNA to a complementary (22) ___________ sequence, and then this sequence is (23) _________ to the amino acid code of the (24) __________. In 1956, Francis (25) __________ named this flow of information the (26) ___________ ___________. While this holds true for genes that encode proteins, some genes do not code for proteins but code for other types of RNA molecule, such as (27) ________, ________, and __________. In the first step toward synthesizing a protein from a gene, the specific sequence in one of the two DNA strands, that is, the (28) ____________ strand, is copied to make a complementary messenger (29) _________. The enzyme involved in this process is called (30) _______ ___________. The code in the mRNA is then read by the (31) ____________, which joins (32) ________ ________ into the appropriate sequence to synthesize the (33) _________. Although the two steps of protein synthesis are similar in (34) __________________ and eukaryotic cells, they differ in that the two processes happen (35) _________ in prokaryotic cells, while in eukaryotic cells, the (36) _______ is synthesized and processed in the (37) ___________ before it is translated in the (38) _________ of the cell. The genetic code is a (39) ___________-base code, and because it is almost identical among all life forms, it is called the (40) __________ _________. The code was deciphered by synthesizing artificial (41) __________ of known nucleotide sequences and then determining the amino acids used by the ribosomes to make a(n) (42) ____________. Using the four bases in mRNA, there are (43) ______ possible combinations, more than what is needed to code (44) ______ amino acids. The phenomenon of multiple codes for an amino acid is referred to as (45) ______ or redundancy. Among all of the possible triplet combinations of nucleotides, that is, (46) ______, there are three that do not code for amino acids, and these are called the (47) __________ or (48)

__________ codons; these act as periods in the polypeptide-encoding sequence. Another codon, AUG, not only codes for the amino acid (49) ________________ but is also the (50) _________ codon.

Matching

Match each of the following terms with its correct definition.

51.	____	Auxotroph	A.	Organelles that read mRNA to assemble a polypeptide
52.	____	Polypeptide	B.	AUG—that codes for methionine and starts the translation process
53.	____	Transcription	C.	Mutants that require additional nutrients added to the minimal medium
54.	____	Translation	D.	UAA, UGA, UAG—that code for the termination of the translation process
55.	____	Central dogma	E.	A chain of amino acids
56.	____	RNA polymerase	F.	Named the flow of information in cells the central dogma
57.	____	Ribosomes	G.	The process of copying DNA sequence to make mRNA
58.	____	Codons	H.	Proposed the one gene–one enzyme hypothesis
59.	____	Francis Crick	I.	Refers to the genetic code being the same in all living things
60.	____	Start codon	J.	The process of reading codes on mRNA to assemble amino acids
61.	____	Stop codon	K.	The enzyme that helps in making mRNA
62.	____	Universal code	L.	A set of three nucleotides on mRNA that codes for a specific amino acid
63.	____	Beadle and Tatum	M.	The flow of information from DNA to RNA to proteins

Labelling

64. Label the major steps of protein synthesis in eukaryotic cells.

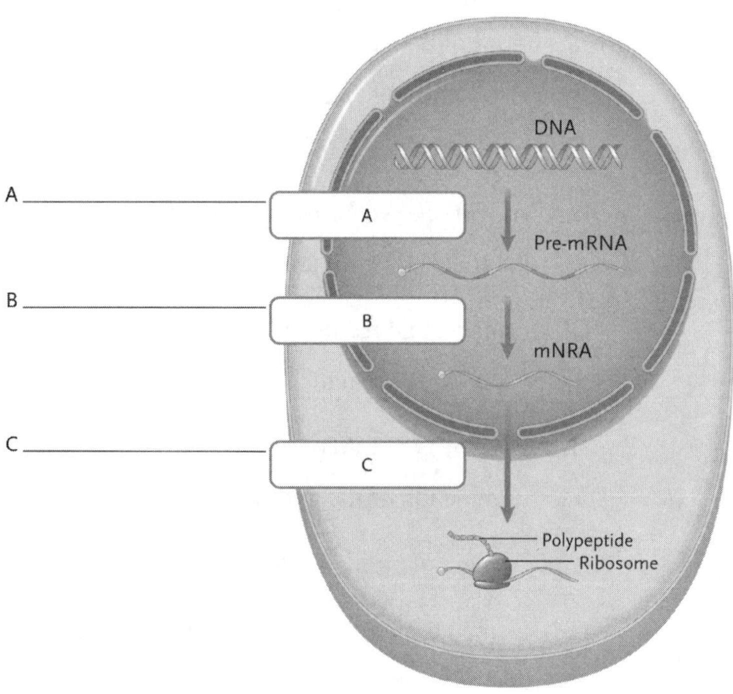

True/False

Mark if the statement is true or false. If the statement is false, justify your answer in the line below each statement.

65. ____________ The terms "proteins" and "polypeptides" are synonyms.

____

____

Short Answer

66. Explain what is meant by the term "inborn error of metabolism."

____

____

67. What is uracil, and where is it found?

____

____

14.2 Transcription: DNA-Directed RNA Synthesis [pp. 301–304]

Section Review (Fill-in-the-Blanks)

Copying DNA sequence to make a complementary RNA copy is called (68) _________. This process is similar to making copies of DNA, with the exceptions that adenine bases in the template DNA are base-paired with (69) _________ in the RNA; only one strand of the DNA (70) __________ is copied into RNA; the product is (71) ________-stranded rather than double-stranded; a different polymerase, the (72) _________ polymerase, is used; and unlike DNA polymerase, this does not require a (73) ________ hydroxyl from a previously synthesized (74) ________. The replication process begins when the enzyme binds to the (75) __________ sequence within the DNA. The enzyme unwinds the DNA strands and, like DNA polymerase, moves from (76) _____ → (77) _____. The complementary (78) _________ is polymerized in the (79) _____ → (80) _____ direction. The new mRNA temporarily binds to the (81) _______ template. At the end of the transcribed sequence, both the enzyme and the (82) __________ release from the (83) _______. Although the major steps of transcription are similar for prokaryotic and eukaryotic cells, prokaryotic cells use a single type of RNA polymerase to transcribe (84) ____ types of genes, whereas in eukaryotes, there are (85) ________ different RNA polymerases. (86) _____ transcribes protein-coding genes, while RNA polymerases (87) ____ and (88) ______ transcribe non–protein-coding genes, such as those for (89) ___________ and (90) ___________. In prokaryotes, the RNA polymerase on its own can recognize and bind to the (91) ____________ region; however, in eukaryotes, proteins called (92) ___________ factors recruit the polymerase after binding to a specific sequence within the promoter. This sequence is called the (93) __________ box. A final difference between prokaryotes and eukaryotes is in the method of transcription (94) ___________. The former have two different types of sequence at the end of the transcribed region called the (95) ___________; there is no equivalent in eukaryotes. Instead, the 3′ end of the (96) __________ itself determines the end of transcription.

Matching

Match each of the following terms with its correct definition.

		Term		Definition
97.	____	DNA polymerase	A.	DNA sequence that encodes a mRNA
98.	____	RNA polymerase	B.	Enzyme(s) for transcription of protein-coding genes in eukaryotic cells
99.	____	RNA polymerase II	C.	Enzyme(s) for DNA replication
100.	____	RNA polymerases I and III	D.	Sequence located before the transcription unit, where RNA polymerase binds
101.	____	Transcription unit	E.	Sequence located at the end of the transcription unit in prokaryotes that signals the end of transcription
102.	____	Protein-coding genes	F.	Enzyme(s) responsible for transcription of protein-coding and non–protein-coding genes in prokaryotes
103.	____	Non–protein-coding genes	G.	Region of DNA that has the specific sequence for a complementary RNA
104.	____	Promoter region	H.	Enzyme(s) for transcription of non–protein-coding genes in eukaryotic cells
105.	____	RNA primer	I.	A sequence in the promoter region of eukaryotic cells that is recognized by transcription factors before RNA polymerase binds
106.	____	TATA box	J.	Genes that code for RNA molecules such as tRNA
107.	____	Transcription factors	K.	A short RNA sequence that is needed before enzymes can replicate DNA
108.	____	Terminator region	L.	Proteins that must bind to the promoter region before RNA polymerase binds in eukaryotic cells

Sequence

109. Arrange the following steps of transcription in the correct sequence.

A. The new mRNA temporarily binds to the DNA template.

B. RNA polymerase binds to the promoter region.

C. The enzyme reaches the end of the transcription unit.

D. The enzyme moves along the transcription unit.

E. The enzyme unwinds the DNA double helix.

F. mRNA and the enzyme are released from DNA.

G. Complementary nucleotides are added.

_____ _____ _____ _____ _____ _____ _____

Complete the Table

110. Complete the following table for the differences between DNA replication and transcription.

Points of comparison	*DNA replication*	*Transcription*
A. Types of base pairs		
B. Number of DNA strands copied		
C. Number of new strands formed		
D. Enzyme involved		
E. Primer formed		

111. Complete the following table for differences in transcription of prokaryotic and eukaryotic cells.

Points of comparison	*Prokaryotic cells*	*Eukaryotic cells*
A. Transcription enzyme(s)		
B. Binding of RNA polymerase to promoter region		
C. Termination of transcription		

Short Answer

112. Compare the types of RNA polymerase(s) found in prokaryotic and eukaryotic cells and how they differ in their functions.

____

____

113. What is a TATA box, and what is its role in eukaryotic transcription?

____

____

114. As one RNA polymerase passes over the sequence of the transcription unit, what is happening on the template strand immediately behind the enzyme?

____

____

True/False

Mark if the statement is true or false. If the statement is false, justify your answer in the line below each statement.

115. _________ Transcription takes place in the cytoplasm of prokaryotic cells and in the nucleus of eukaryotic cells.

____

____

14.3 Processing of mRNAs in Eukaryotes [pp. 304–307]

Section Review (Fill-in-the-Blanks)

Prokaryotic mRNA contains a coding region that is exactly the size needed to code for the (116) ___________; however, this is not the case for the mRNA transcripts in eukaryotic cells. In eukaryotes, mRNAs require (117) _______ to give translatable mRNAs prior to leaving the (118) __________. During transcription, a(n) (119) __________ triphosphate is added to the 5′ end so that it is joined by three phosphates and the 3′ –OH is facing out. This structure is called the (120) ____-______, and it serves as the binding site for the (121) _________ at the beginning of translation. This addition is added by a(n) (122) __________ and is not based on complementary base pairing. Unlike prokaryotic cells, eukaryotic cells do not have a (123) _____________ sequence at the end of protein-coding genes. Instead, a (124) ________ signal at the 3′ end of the gene is transcribed, cleaved, and extended with a chain of adenosines by the enzyme (125) __________ ____________. This is called the (126) ________-_________ tail, and its function is to protect the transcript from degradation by enzymes while it is moving from the (127) ____________ to the (128) ______________. One of the most remarkable differences between prokaryotic and transcripts is that the pre-mRNA in eukaryotes contains (129) _____________, protein-coding segments, which are interspersed with (130) ___________, non–protein-coding sequences. The latter are removed through a process called (131) _______ _________ to give a mature mRNA. In this process, (132) ___________ bind the (133) _________, loop them out of the pre-mRNA, and then clip them at each (134) __________ boundary and finally join the adjacent (135) ____________ together. While highly accurate, this splicing process allows for the splicing of (136) ________ in different combinations to create a greater number and variety of proteins without requiring a concomitant increase in the number of (137) ______ encoded by the genome. This processing of pre-mRNAs may also provide a mechanism for the evolution of new proteins by (138) _________ ___________.

Matching

Match each of the following terms with its correct definition.

139.	____	Untranslated regions	A.	Sequence of adenines added to pre-mRNA to protect it from degradative enzymes
140.	____	5′ cap	B.	The enzymatically added RNA sequences at the 5′ and 3′ ends of eukaryotic mRNAs
141.	____	3′ poly(A) tail	C.	Sequence in prokaryotic mRNA that regulates protein synthesis
142.	____	Pre-mRNA	D.	Process where introns are removed and exons are joined together
143.	____	Introns	E.	Small ribonucleoprotein particles used for splicing
144.	____	Exons	F.	RNA found in the nucleus that combines with snRNPs to form spliceosomes
145.	____	mRNA splicing	G.	The form of an intron after being released by splicing
146.	____	Spliceosome	H.	Multiple ways to edit pre-mRNA to form different types of mRNA
147.	____	Promoter	I.	A process where different protein regions or domains in the gene are mixed in new combinations to create new proteins
148.	____	snRNPs	J.	Protein-coding sequence in eukaryotic pre-mRNA
149.	____	Alternative splicing	K.	Larger form of mRNA made by eukaryotes
150.	____	Exon shuffling	L.	A complex formed by ribonucleoproteins attached to pre-mRNA during editing process
151.	____	Lariat	M.	5′ end of pre-mRNA that serves as the binding site for the ribosomes

True/False

Mark if the statement is true or false. If the statement is false, justify your answer in the line below each statement.

152. _____________ Introns are believed to have evolved from the prokaryotic predecessors of the eukaryotes.

____

____

153. _____________ The activity of splicesomes is the result of ribozyme activity.

____

____

Short Answer

154. Why is the presence of and removal of introns in eukaryotic pre-mRNA not a waste of cellular resources?

____

____

Complete the Table

155. Complete the following table with specific information on eukaryotic transcription.

Eukaryotic transcription	*Role in transcription*
A. Pre-mRNA	
B. 5′ GTP cap	
C. 3′ poly(A) tail	
D. mRNA splicing	

14.4 Translation: mRNA-Directed Polypeptide Synthesis [pp. 307–317]

Section Review (Fill-in-the-Blanks)

Translation is performed by (156) ___________, organelles that comprise rRNA and (157) _______. These translate the mRNA sequence or code into the (158) _________ ________ sequence of the polypeptide. In prokaryotes, this process takes place in the (159) __________ and begins as soon as the mRNA emerges from the (160) ___________ _______________. In eukaryotic cells, the process generally takes place in the (161) __________ as well; however, some proteins are made in the mitochondrion and (162) ________. In eukaryotes, the pre-mRNA is processed in the (163) _________ and then exits to join with the (164) _________. Translation also requires (165) __________ RNA or tRNA molecules to carry the (166) _________ ________. tRNA is made of only 75 to 90 nucleotides and has (167) _______ regions of internal base pairing to give rise to a (168) _______ ______ shape. At one end of the folded structure is a three-base sequence, the (169) ______________, which matches a specific codon sequence on the (170) ___________. The other end of the folded tRNAs carries the (171) _________ _________; these are linked to their appropriate tRNA by the (172) ___________-____________ synthetases. The accuracy of (173) __________ ultimately stems from the accuracy of these enzymes. Ribosomes are made of a (174) ___________ and (175) __________ subunit. While similar in structure and (176) _________, it is the differences in molecular makeup between prokaryotic and eukaryotic ribosomes that allow doctors to use certain (177) __________ to treat infections. Ribosomes contain (178) ______ binding sites: the (179) _______ site where aminoacyl-tRNA binds; the (180) ____ site, where tRNA shifts after its amino acid has joined the growing peptide; and the (181) ____ site, where the uncharged tRNA leaves the ribosome. Translation has three major steps: in the first, (182) ____________, the ribosome binds to the start codon, (183) _________, on the (184) ___ end of mRNA, and the first tRNA, the initiator (185) _____________ tRNA, binds the (186) ___ site of the ribosome. During the (187) ____________ stage, amino acids linked to (188) _________ are added one at a time to the growing (189) __________ chain. The (190) __________ transferase activity of a (191) ___________ within the ribosome catalyzes the formation of the bonds between the amino acids. The ribosome hydrolyzes (192) _______ as it advances from one codon to the next: in so doing, it shifts the newly uncharged tRNA to the (193) ___ site, whereupon the molecule diffuses out of the organelle. As the ribosome advances, the tRNA with the growing peptide chain shifts to the (194) ___ site; opening up the A site is for the next (195) ___-tRNA. The ribosome continues to move toward the (196) ___ end of the mRNA until it reaches the last codon, that is, the (197) _______ codon. In the (198) _______ stage of translation, a (199) __________ factor binds to the stop codon within the (200) ___ site and triggers release of the polypeptide, separation of the (201) ______ from the ribosome, and separation of the ribosomal subunits. Further (202) _____ of the polypeptide may involve the removal of (203) ________ _________ or addition of organic molecules such as (204) ____________ or ___________. Finally, the protein is folded, and this usually requires the participation of helper proteins called (205) ______________, which help them achieve the proper three-dimensional shape. Proteins that are to remain in the cytoplasm are made by (206) ________ ribosomes; however, proteins to be transported to other locations have amino acid sorting signals called (207) ___________ sequences that direct the proteins to their final destination. For many cellular destinations, the ribosome–mRNA complex combines with a (208) _________ recognition particle (SRP) and then attaches to the (209) _______ receptors on rough ER for further processing and sorting through the various ER compartments. Mutations or changes in the (210) ______________ sequence can have a variety of effects on the structure and (211) ________ of proteins; the effect is dependent on the nature of the mutation. Base-pair (212) ________ result in a change to the (213) _______, and if this also causes a change in the amino acid at that position, the mutation is called a (214) ___________ mutation. If the base change gives rise to a stop codon, the mutation is called a (215) _______ mutation, and the result is premature (216) ________ of the polypeptide. Frameshift mutations result from the deletion or insertion of a(n) (217) ________ ________ giving rise to a change in the (218) _________ frame beyond the site of the mutation. The result is a change in the amino acid (219) _______ from that point on, generally also rendering the protein nonfunctional.

Matching

Match each of the following terms with its correct definition.

220.	____	Translation	A.	RNA that carries a specific amino acid to ribosomes
221.	____	Ribosomes	B.	A set of three nucleotides on tRNA that matches a specific codon on mRNA
222.	____	tRNA	C.	Process of reading the codes on mRNA to put amino acids together in a polypeptide
223.	____	Codon	D.	Set of 20 enzymes that join a specific amino acid to a tRNA
224.	____	Anticodon	E.	Exit site on the ribosome where tRNA binds after its peptide transfers to the new tRNA
225.	____	Aminoacylation	F.	An enzyme that is made of RNA instead of a protein
226.	____	Aminoacyl-tRNA synthetases	G.	Organelles that assist translation
227.	____	rRNA	H.	Site where charged tRNA first binds
228.	____	E site	I.	Site where tRNA moves after formation of the peptide bond attaches the growing peptide to its amino acid
229.	____	Ribozyme	J.	A protein made by cells lining the stomach; requires processing by removal of a segment of amino acids to give the active, shorter enzyme
230.	____	A site	K.	A set of three nucleotides on mRNA that codes for a specific amino acid
231.	____	P site	L.	Type of RNA that makes up ribosomes
232.	____	Polysome	M.	Proteins that help fold certain polypeptides to achieve a specific three-dimensional shape
233.	____	Pepsinogen	N.	A series of ribosomes bound to an mRNA
234.	____	Chapterones	O.	Process of attaching specific amino acid to a tRNA

True/False

Mark if the statement is true or false. If the statement is false, justify your answer in the line below each statement.

235. _______ Regardless of the protein or the organism, methionine is always the first amino acid in a protein.

____

____

236. _______ The signal sequences in prokaryotes are very different from those of eukaryotes.

____

____

Complete the Table

237. Fill in the appropriate mutation or definition

Type of mutation	*Definition*
Base substitution	A.
B.	Mutation where the change in DNA sequence causes a change in the amino acid sequence and alters the function of the protein
C.	Mutation that results in abrupt termination of the polypeptide, leaving the protein totally dysfunctional.
Frameshift mutation	D.

Short Answer

238. Prokaryotic cells can transcribe and translate simultaneously. Explain why that is not possible in eukaryotic cells.

____

____

239. Explain why a substitution mutation may not cause a change in protein sequence.

____

____

Labelling

240. The following depicts the general process of translation. Label the diagram to identify the appropriate structures and components.

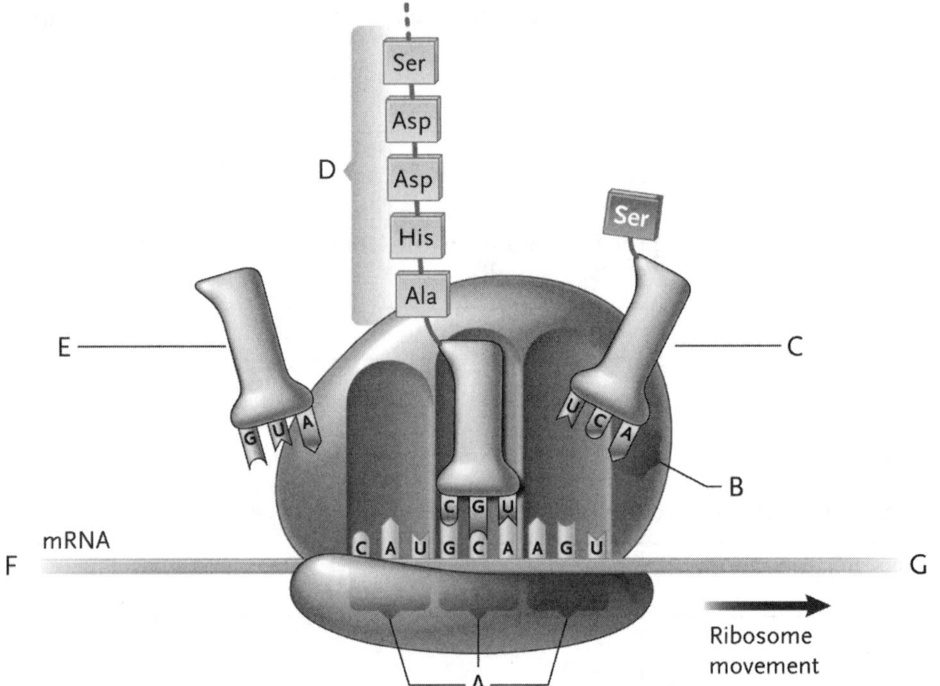

A. ______________

B. ______________

C. ______________

D. ______________

E. ______________

F. ______________

G. ______________

SELF-TEST

1. In which step of protein synthesis is RNA polymerase involved? [p. 297]

   a. Transcription
   b. Translation
   c. Both steps
   d. Neither of the two steps

2. Codons are read in [p. 299]

   a. Transcription
   b. Translation
   c. Both steps
   d. Neither of the two steps

3. Which of the following is the start codon? [p. 300]

   a. TAC
   b. UAA
   c. AUG
   d. UAC

4. Which of the following is the amino acid coded by the start codon? [p. 300]

   a. Glutamine
   b. Methionine
   c. Lysine
   d. Phenylalanine

5. How many codons are possible using the four types of nitrogenous bases? [p. 299]

   a. 20
   b. 40
   c. 46
   d. 64

6. Where does protein synthesis take place in a prokaryotic cell? [p. 297]

   a. In the cytoplasm
   b. In the nucleus
   c. Part in the nucleus and part in the cytoplasm
   d. In mitochondria

7. Which of the following is formed when a gene is transcribed in a eukaryotic cell? [p. 304]

   a. Protein
   b. mRNA
   c. Pre-mRNA
   d. Pepsinogen

8. In mature eukaryotic mRNA, all of the following are present except [p. 304]

   a. Introns and exons
   b. 5′ GTP cap
   c. Methionine
   d. 3′ poly(A) tail

9. Which of the following helps in splicing pre-mRNA in eukaryotic cells? [p. 305]

   a. snRNP
   b. Ribosome
   c. RNA polymerase
   d. tRNA

10. Which of the following has the anticodon? [p. 308]

   a. mRNA
   b. DNA
   c. tRNA
   d. rRNA

11. Which of the following sites in the ribosome allows aminoacyl-tRNA to bind? [p. 308]

   a. A and P
   b. P
   c. E
   d. A

12. At which site in ribosome does the polypeptide grow? [pp. 308, 310]

   a. A
   b. P
   c. E
   d. D

13. The enzyme peptidyl transferase is [p. 311]

   a. A protein
   b. rRNA
   c. A ribozyme
   d. (b) and (c)

14. Which of the following types of mutation is usually considered most devastating? [p. 317]

   a. Base substitution
   b. Silent
   c. Frameshift
   d Missense

15. Only one ribosome can translate an mRNA at one time. [p. 313]

   a. True
   b. False

16. Which of the following provides an explanation for the observation from the Human Genome Project that there are approximately 25 000 genes in the human genome but a much greater number of proteins synthesized? [p. 306]

   a. The wobble hypothesis
   b. Alternate splicing
   c. Exon shuffling
   d. (b) and (c)

17. Which of the following statements about mRNA processing is true? [pp. 304–306]

   a. It is a rare occurrence in eukaryotes
   b. It occurs in both prokaryotes and eukaryotes
   c. It is a process that allows eukaryotes to increase their protein-coding capacity without having to increase the number of genes in the genome
   d. It happens in approximately 75% of human mRNAs

18. The wobble hypothesis explains which of the following? [p. 308]

   a. The fact that the universal code is redundant, but there are only 20 different tRNAs
   b. The fact that tRNAs can base-pair with more than a single unique codon on the mRNA
   c. The fact that the introns in eukaryotic pre-mRNA may be spliced at alternate positions
   d. (a) and (b)

19. Which of the following statements is false? [pp. 296–297, 308]

   a. Francis Crick proposed the one gene–one enzyme hypothesis
   b. Francis Crick coined the term "central dogma"
   c. Francis Crick proposed the wobble hypothesis
   d. Beadle and Tatum fist provided experimental evidence for the relationship between genes and proteins

20. Under optimum conditions, the bacterium *E. coli* reproduces itself every 20 minutes; however, humans cells require a much longer time to do this. Which of the following would help explain the difference in generation time? [pp. 297, 304, 312–313]

    a. The elongation reactions in prokaryotes happen much faster than in eukaryotes
    b. Prokaryotes do not have to export their mRNA from a nucleus to the site of translation
    c. Prokaryotes do not have to process their mRNA before it is in a translatable form
    d. Prokaryotes generally do not need to process their polypeptides after translation
    e. All of the above

INTEGRATING AND APPLYING KEY CONCEPTS

1. Design an experiment to prove (a) the following proposed eukaryotic biosynthetic pathway and (b) that each enzyme is encoded by a single DNA sequence and therefore is heritable.

Precursor A Enzyme B Intermediate B Enzyme C Intermediate C Enzyme X Amino acid X

2. Discuss the impact of a splicing error where the spliceosome moved the splice site by four bases relative to the intron–exon boundary.

3. Explain the relationship between gene and protein as referred to by each of the three terms/descriptions:
    (a) One gene–one protein
    (b) One gene–one polypeptide
    (c) One gene–one particular polypeptide under particular conditions

4. Compare the process of translation of a protein destined for the rough ER with that of a nuclear protein.

OPTIONAL ACTIVITY

1. The genetic code is redundant, meaning that there are generally a number of synonymous codons for a given amino acid. You have learned that there are 61 distinct codons coding for a total of 20 amino acids and that there are far less than 61 different tRNAs. However, despite the relaxed base pairing exhibited by tRNA molecules, the redundancy of the genetic code is accompanied by a phenomenon called "codon bias" among different organisms and viruses, which can have a tremendous impact on genetic engineering (i.e., the ability to express the gene from one organism in a different organism). This phenomenon is demonstrated by the fact that while an organism may have one tRNA for each of the 20 amino acids, the sequence of a tRNA for a given amino acid, specifically in the anticodon region, differs from that of another organism. The factor that seems to impact codon bias the most is the relative amount of guanine and cytosine bases in the genome, that is, the mol % G+C. This varies tremendously among prokaryotes, with a range from ~25% to 75%; however, the phenomenon is not restricted to the prokaryotes. Find out more about the fascinating genetic and evolutionary events and pressures underlying this phenomenon.

15 Control of Gene Expression

CHAPTER HIGHLIGHTS

- Prokaryotes are simple, single-celled organisms that adapt quickly to changes in their environment.
- Adaptation is mainly accomplished through regulation of gene expression at the transcriptional level.
    - Responses are mainly short-term, rapid, and reversible alterations in their metabolic pathways. The genes for these pathways are generally organized as operons, so they are expressed as a single mRNA under the control of a single promoter.
    - Regulation of transcription occurs through the binding of repressor or activator proteins to the operator or activator binding sites near the promoter.
- Eukaryotes are more complex unicellular or multicellular organisms.
- Gene expression is regulated at transcriptional, posttranscriptional, translational, and posttranslational levels.
    - Differentiation is based on long-term changes in gene expression.
    - Adaptation to changes in their environment is through short-term changes in gene expression.
- Cancer is a disease characterized by the loss of the normal controls for expression of genes involved in regulating cell division.

STUDY STRATEGIES AND LEARNING OUTCOMES

- This chapter has extensive gene control pathways. Trying to memorize all of the steps is the most common mistake.
- Remember that the goal of this chapter is to understand how genes are controlled.
- Concentrate on the key steps for protein synthesis: transcription and translation.
- Focus on what each step is about and where changes can be made.

- Once the key steps are understood, you can then learn the regulatory pathways.
- Carefully examine the figures and the various subject headings to keep the ideas clearly organized in your mind.

By the end of this chapter, you should be able to

- Describe negative regulation in prokaryotes: that is, the *lac* and *trp* operons
- Describe positive expression in prokaryotes: that is, the CAP–cAMP control of the *lac* operon
- Describe how transcriptional control in eukaryotes differs from that of prokaryotes and be able to explain the various levels of transcriptional control in eukaryotes
- Describe how gene expression is controlled posttranscriptionally, translationally, and posttranslationally and relate the breakdown of regulation of gene expression to cancer

TOPIC MAP

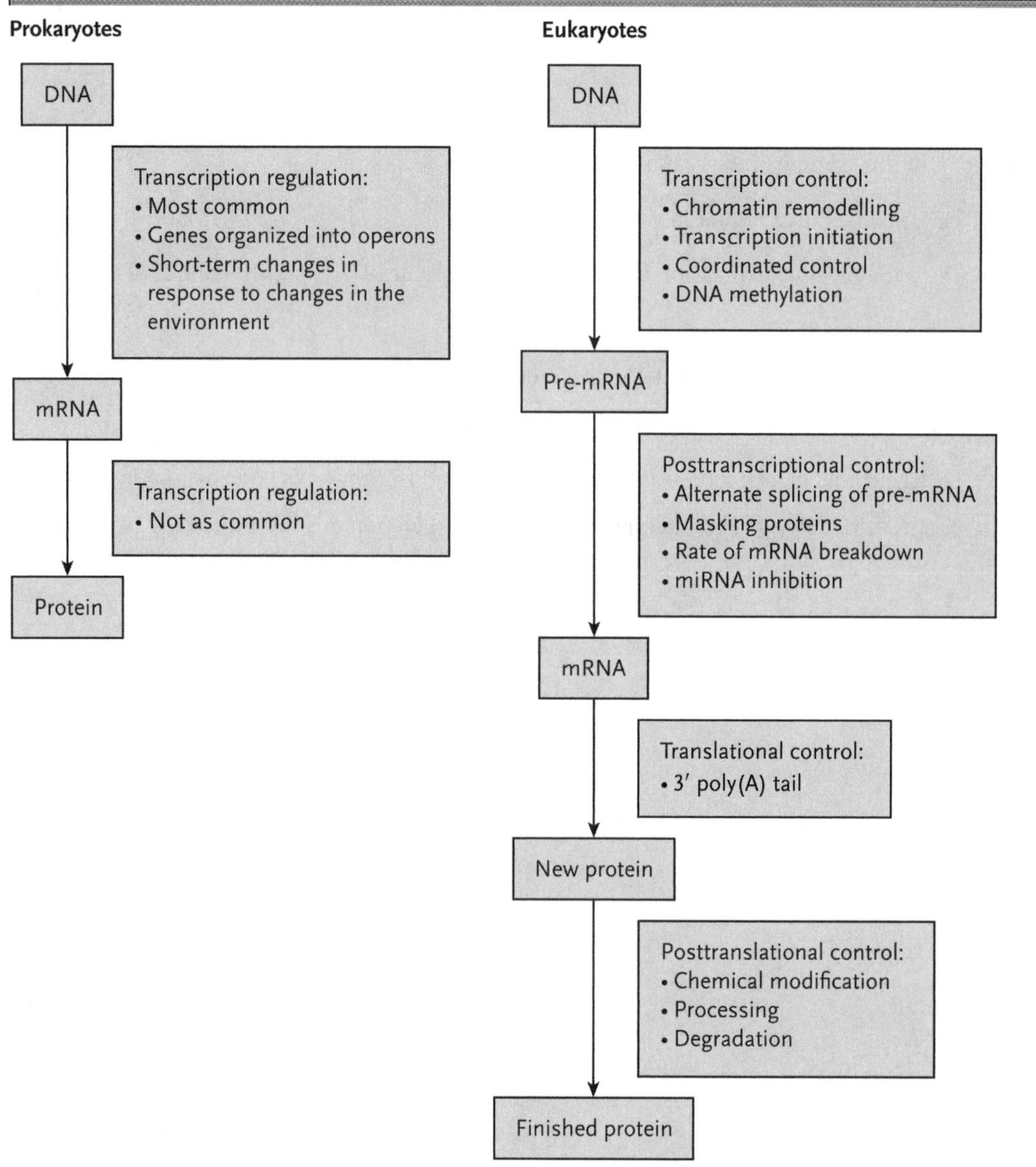

Prokaryotes
DNA
Transcription regulation:
• Most common
• Genes organized into operons
• Short-term changes in response to changes in the environment
mRNA
Transcription regulation:
• Not as common
Protein
Eukaryotes
DNA
Transcription control:
• Chromatin remodelling
• Transcription initiation
• Coordinated control
• DNA methylation
Pre-mRNA
Posttranscriptional control:
• Alternate splicing of pre-mRNA
• Masking proteins
• Rate of mRNA breakdown
• miRNA inhibition
mRNA
Translational control:
• 3′ poly(A) tail
New protein
Posttranslational control:
• Chemical modification
• Processing
• Degradation
Finished protein

INTERACTIVE EXERCISES

Why It Matters [pp. 321–322]

Section Review (Fill-in-the-Blanks)

At the time of release, a human egg is metabolically (1) ______________. As soon as the (2) ______________ fertilizes the egg, the egg cell becomes (3) ___________ and starts dividing. Initially, all the embryonic cells are structurally and functionally similar, but they soon start to (4) _______________. That is, even though they have the same set of (5) ___________, some are turned on and others are turned (6) ______. Prokaryotic organisms make rapid, (7) __________-term changes in their transcriptional step to respond to changes in their environment. Eukaryotic organisms make (8) ________-term changes to respond to changes in their environment but (9) __________-term changes for differentiation of their cells.

15.1 Regulation of Gene Expression in Prokaryotes [pp. 322–327]

Section Review (Fill-in-the-Blanks)

Prokaryotic cells are relatively simple, single-celled organisms that can rapidly and (10) _____ alter gene expression to adapt to the changes in their (11) ____________. Generally, these changes involve turning (12) _____ genes for metabolic processes that are not required, for example, pathways for the breakdown of nutrients no longer (13) _____ and turning (14) _____ genes for processes such as the biosynthesis of amino acids no longer (15) _____ from the environment. In prokaryotic cells, genes involved in a common pathway are generally part of a gene cluster called an (16)_______, a term coined by (17) ____________ and (18) __________, the scientists who proposed the model. These clusters are transcribed on a (19) _______ mRNA and are subject to (20) ________ control through the promoter, the (21) _______, and other sequences for the binding of regulatory proteins. The result of this binding can be (22) _______ or activation of the genes in the cluster. Expression of the three genes in the *lac* (23) _______ is subject to both. Regulatory proteins involved in negative regulation, that is, repressors, may be synthesized inactive and require the participation of a (24) ______ or may be synthesized fully (25)______, as is the case with the *lac* repressor. The genes in this cluster encode proteins required to (26) ______ the disaccharide lactose. When lactose is (27) ______, the *lac* repressor binds the (28) ______ sequence, located so its binding inhibits the function of the RNA polymerase. When lactose is available, a lactose metabolite called allolactose combines with the repressor to relieve (29)______, allowing the (30) _____ ______ to bind the promoter and read through the operon. This type of operon is a(n) (31) ______ operon, and in this case, allolactose is the (32) ______. For biosynthetic operons, such as the one for tryptophan biosynthesis, expression is normally on unless the end product is (33) ______ from the environment. Repressors for these operons are synthesized in a(n) (34) ______ form but combine with the end product to give the (35) ______ repressor. The end product is therefore considered a (36) ______, and the operon is referred to as a(n) (37) ______ operon. Genes and operons may also be subject to positive regulation where they bind (38) _______ proteins. These are normally synthesized in a(n) (39) ______ form. The positive regulation of the *lac* operon provides the mechanism whereby, given the simultaneous availability of lactose and the preferred carbon source, (40) ______, *E. coli* will not express the (41) _____ ______ until the glucose is used up. The presence of glucose results in the inactivation of the enzyme (42) _________ cyclase and therefore its production of the small molecule (43) _______. The latter, when present, binds the (44) ______ _______ protein or CAP to give a functional (45) ______. Once glucose disappears, repression by the *lac* repressor is relieved, as is repression of the (46) _____ cyclase, so (47) _________ accumulates and combines with the CAP protein, giving rise to the functional activator; the complex binds the activator sequence, that is, the (48) _____ site, and recruits RNA polymerase to transcribe the operon.

Matching

Match each of the following terms with its correct definition.

		Term		Definition
49.	____	*lac* operon model	A.	Encodes a repressor or activator protein
50.	____	Operon	B.	An operon where the metabolite molecule represses or decreases the expression of the cluster of genes
51.	____	Promoter sequence	C.	Proposed by Jacob and Monod to explain gene expression in *E. coli* for lactose metabolism
52.	____	Regulatory gene	D.	Regulation mechanism where the active repressor turns off the gene expression
53.	____	Repressor protein	E.	Regulation mechanism where an activator protein turns on gene expression
54.	____	Operator sequence	F.	Refers to a cluster of genes expressed as a single transcript
55.	____	Inducer	G.	DNA sequence where RNA polymerase binds
56.	____	Inducible operon	H.	A chemical that turns on expression of genes in an operon
57.	____	Repressible operon	I.	A metabolite molecule combines with the repressor protein to shut off the operon
58.	____	Corepressor	J.	Protein that is made by the regulatory gene in active or inactive form and binds to the operator to block the transcription of the cluster of genes
59.	____	Negative gene regulation	K.	A sequence where the repressor protein binds and is present between the promoter and the cluster of genes
60.	____	Positive gene regulation	L.	An operon where the metabolite molecule enhances or increases the expression of the cluster of genes
61.	____	CAP	M.	Binding sequence for the catabolite activator protein–cAMP complex
62.	____	CAP site	N.	A specific activator protein that binds to its recognition sequence and helps in the binding of RNA polymerase to the promoter

Complete the Table

63. Prokaryotic cells have different types of operons and gene regulation mechanisms. Provide descriptions of each in the following table.

A. Inducible operon	*B. Repressible operon*
C. Negative gene regulation	*D. Positive gene regulation*

Short Answer

64. What would be the phenotype of a $LacI^-$ *E. coli* (i.e., Lac repressor mutant) strain growing on minimal medium with lactose as the only carbon source?

____

____

65. What would be the phenotype of an *E. coli* strain that had a mutation in its *CAP* gene (i.e., Cap^-)?

____

____

66. What would be the phenotype of an *E. coli* strain that had a mutation in its *Trp* repressor gene (i.e., Trp^-)?

____

____

Labelling

67. Label the following diagram of the *E. coli lac* operon.

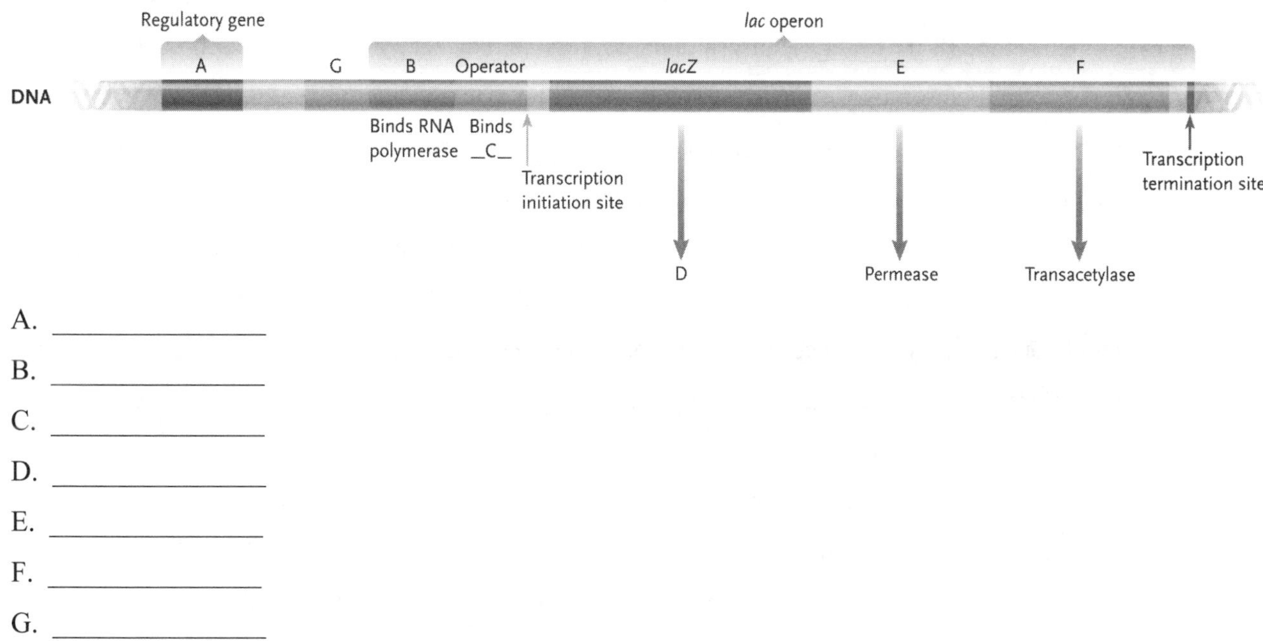

A. _______________

B. _______________

C. _______________

D. _______________

E. _______________

F. _______________

G. _______________

15.2 Regulation of Transcription in Eukaryotes [pp. 327–334]

Section Review (Fill-in-the-Blanks)

Unlike prokaryotes, the eukaryotic genes for related functions are scattered throughout the genome rather than being organized in (68) ______________; they are, however, (69) _____ in a coordinated manner. There are two major categories of regulation in eukaryotes: (70) ________-term regulation to respond to frequent changes in the environment and (71) ________-term regulation for development and differentiation in the organism. Unlike prokaryotes, eukaryotic cells not only regulate transcription but also regulate gene expression (72) ______-transcription, translation, and (73) _____-translation.

Genes that are transcriptionally active have a looser (74) ______ structure than inactive genes. The structural change that must precede transcriptional activation involves (75) _____________ remodelling. This process begins when an (76) ______ protein binds a regulatory sequence upstream of the promoter and either recruits the (77) ______ complex to displace a nucleosome or recruits an (78) ______ enzyme to chemically modify the (79) _______ proteins, causing them to become less tightly associated with the DNA; both mechanisms result in the promoter becoming (80) _______. Transcriptional control is the most important method of regulation: very low levels of expression arise from the binding of (81) __________ factors to the (82)

________ box of the promoter to recruit RNA polymerase II. Expression is increased further by the binding of (83) ______ proteins to the promoter (84) ______ elements. Maximal expression occurs when, in addition to these elements, other activators bind to the (85) ______, further upstream from the promoter. These interact with a (86) ______, which simultaneously binds the proteins at the (87) ______. Transcriptional repressors may act by competing with the (88) ______ for their binding sites, or they may instead bind nearby and prevent the activator from binding the (89) ______. A third mechanism involves their recruiting (90) ________ deactylation enzymes so that the (91) _______ is compacted and the gene becomes inaccessible for transcription. Coordination of gene expression depends on the nature of the activator and (92) _____ sequences and the relative (93) ______ of binding of the different proteins. Specific regulatory proteins have specific recognition sequences, and by (94) ______ regulatory sequences for different regulators, eukaryotes control transcription of all their protein-coding genes using relatively (95) _____ regulator proteins. This phenomenon is called (96) ______ gene regulation. The mechanism by which the organism (97) ______ expression of genes for related functions and coordinately regulates different genes in different (98) _____ is demonstrated clearly with steroid hormone regulation. These lipid-soluble regulators enter the cytoplasm by diffusion through the (99) ______ ______ and combine with specific steroid hormone (100) ______; these may differ between different (101) ______. The binding gives rise to an (102) ______ transcriptional regulator, which, upon entry into the nucleus, binds the steroid hormone (103) _____ _______. Eukaryotes may also effect transcriptional regulation through (104) _______ of DNA sequences. When this is used to inactivate part or all of a chromosome, the effect is referred to as (105) ______. In genomic (106) ______, this type of modification is used for a specific (107) ______.

Matching

Match each of the following terms with its correct definition or descriptor.

108.	____	Histones	A.	Part of the eukaryotic promoter sequence that is recognized by the transcription factors necessary for RNA polymerase binding
109.	____	Chromatin remodelling	B.	Proteins that help in the binding of RNA polymerase to the promoter
110.	____	Nucleosome	C.	Refers to proteins around which DNA is wrapped to form the nucleosome
111.	____	TATA box	D.	The complex of DNA and 8 units of histones
112.	____	Transcription factors	E.	Refers to activation of genes by loosening up chromatin
113.	____	Transcription initiation complex	F.	Proteins that bind to the promoter proximal elements and enhance transcription
114.	____	Promoter proximal elements	G.	The process of chemically modifying cytosines in DNA to turn off the genes
115.	____	Activators	H.	A sequence located before the promoter sequence for the binding of activators that enhances transcription
116.	____	Coactivator	I.	Occurs when an individual inherits either a maternal or paternal allele that is permanently inactivated through DNA methylation
117.	____	Hormone-receptor complex	J.	A large multiprotein complex that loops DNA by joining activators
118.	____	Methylation	K.	Formed by binding of transcription factors and RNA polymerase II at the promoter sequence
119.	____	Silencing	L.	The enzymatic modification of cytosine bases in DNA
120.	____	Genomic imprinting	M.	The functional transcriptional activator formed by steroid hormones

Complete the Table

121. Eukaryotic cells have different types of gene regulation mechanisms at the transcriptional level. Complete the following table giving a short description of each type.

Mechanism	*Description*
A. Chromatin control	
B. Transcription initiation control	
C. Coordinated control	
D. DNA methylation control	

True/False

Mark if the statement is true or false. If the statement is false, justify your answer in the line below each statement.

122. ____________ Long-term regulation of gene expression occurs in multicellular eukaryotes but not in unicellular eukaryotes.

____

____

123. ____________ Acetylation is the chemical basis for chromatin remodelling, and methylation is the chemical basis for silencing.

____

____

Short Answer

124. Compare the effect(s) of a deletion of the steroid X hormone response element in the zygote versus a differentiating cell in the embryo.

____

____

15.3 Posttranscriptional, Translational, and Posttranslational Regulation [pp. 334–337]

Section Review (Fill-in-the-Blanks)

Posttranscriptional, translational, and posttranslational controls allow cells to fine-tune gene expression through the regulation of the (125) _____ of proteins. Posttranscriptional controls regulate (126) _______ processing, mRNA availability for (127) _______, and the rate of mRNA (128) ______. Transcription of eukaryotic genes generally yields a longer transcript called the (129) ___________, which must undergo (130) _____________ to give a translatable mRNA. In this process, different combinations of (131) __________ can be fused together to form different mRNAs, which, in turn, yield different types of proteins. Another method of posttranscriptional regulation involves the use of a (132) ______ protein, which blocks translation. This method is used in (133) ______ animal eggs and is relieved through removal of the proteins once the egg is (134) ______. Yet another mechanism of posttranscriptional control is through the modulation of mRNA (135) ______. In mammary glands, the peptide hormone prolactin is used to extend the (136) _____-______ of the casein mRNA. The result is (137) _______ levels of casein production. The recently discovered (138) _____-______ or mi-RNAs are involved in the regulation of genes involved in (139) ______, growth, and behaviour. These regulatory RNA molecules are transcribed as (140) ______ miRNAs, which, due to internal regions of complementarity, form a (141) _____-______ structure. This structure is cleaved by a protein called (142) ______ to give a short double-stranded RNA molecule; further degradation by a protein complex gives the single-stranded (143) _____-_____, which, remaining bound to that protein complex, base-pairs with target mRNA molecules and either causes their (144) ______ or prevents their (145) ______. Translational control occurs in virtually all cell types and species and is commonly accomplished by the enzymatic increase or decrease in the length of the (146) _______ ______ of mRNA: the result is an (147) ______ or ______ in the rate of translation, respectively. Once a protein is made, its function can be regulated by chemical (148) _______, processing of an inactive (149) _______ protein, or controlling the rate of (150) _______ of the protein. Short-lived proteins are targeted for degradation by attachment of small proteins called (151) ___________; this results in the recognition and digestion of the protein by a (152) ______. Examples of the first type of regulation include (153) _______ of histones in chromatin remodelling as well as the phosphorylation of proteins in both signal (154) _______ pathways and regulation of the cell (155) ______. The digestive enzyme pepsin and the peptide hormone (156) _______ are both made as inactive (157) _____________ proteins that are activated by removal of a segment of amino acids.

Matching

Match each of the following terms with its correct definition.

158.	____	Alternative splicing	A.	The molecule that is involved in the phenomenon of RNA interference
159.	____	Masking proteins	B.	Enzyme that breaks down proteins
160.	____	miRNA	C.	The time required for the concentration of something to reduce to half the initial amount
161.	____	Poly(A) tail	D.	Proteins that bind to mRNA to block translation in an unfertilized egg
162.	____	Ubiquitin	E.	A short chain of adenines attached to the 3′ end of mRNA whose length affects the translation process
163.	____	Proteasome	F.	A regulatory mechanism that involves complementary base pairing of small, single-stranded RNA molecules
164.	____	RNAi	G.	Small protein that attaches to some of the short-lived proteins of the cell, encouraging their quick breakdown
165.	____	Half-life		Process of removing different combinations of exons and introns in a pre-mRNA to form different mRNAs and different proteins

Complete the Table

166. Eukaryotic cells have different types of gene regulation mechanisms in addition to the transcriptional level. Complete the following table giving a short description of each type.

Mechanism	*Description*
A. Posttranscriptional control	
B. Translational control	
C. Posttranslational control	

True/False

Mark if the statement is true or false. If the statement is false, justify your answer in the line below each statement.

167. ____________ Viruses may use RNAi to inhibit translation of host proteins.

____

____

168. ____________ Although the 3′ nontranslated region of eukaryotic mRNA molecules may play a role in regulation of gene expression, this is not true of the 5′ untranslated region.

____

____

Short Answer

169. Explain how the phenomenon of RNAi can be exploited in the laboratory.

____

____

15.4 The Loss of Regulatory Controls in Cancer [pp. 337–339]

Section Review (Fill-in-the-Blanks)

Cell division is regulated by specific (170) __________, which, if mutated, may allow the cells to continue dividing and form (171) ____________. The cells revert from being specialized to embryonic form, a process that is called (172) _________________. If the dividing cells remain as a single mass, they are referred to as (173) _____________. If the cells separate and spread to other tissues and organs, the tumour is called (174) _____________. This is also referred to as (175) _____________. The spreading of malignant tumours is called (176) _______________. In normal cells, (177) _______-______________ code for proteins that stimulate cell division. In cancer cells, these genes are converted to (178) ____________ by mutation, translocation, or viral infection. In contrast, (179) _______-___________ genes in their normal form encode proteins that inhibit cell division; when mutated, they lose their (180) ___________ activity. The best known of these genes is *TP53*, which encodes the (181) ____________-___________ protein p53 and which is defective in many different types of cancer. Although mutations in these genes are associated with cancer, most cancers result from a (182) ______________ progression of changes in multiple genes.

Matching

Match each of the following terms with its correct definition.

183.	____	Mutation	A.	Tumour-suppressor gene that is altered in many different types of cancer cells
184.	____	Benign tumour	B.	The tumour whose cells separate and spread to other tissue and organs
185.	____	Malignant tumour	C.	A change in base sequence in DNA
186.	____	Dedifferentiation	D.	The tumour that remains at its original site
187.	____	Metastasis	E.	Modified proto-oncogenes that stimulate the cell to become cancerous
188.	____	Proto-oncogenes	F.	Genes that encode proteins that inhibit cell division in normal cells
189.	____	Oncogenes	G.	The process by which specialized cells revert to the embryonic state
190.	____	*TP53*	H.	Genes that encode proteins that stimulate cell division in normal cells
191.	____	Tumour-suppressor genes	I.	Spreading of malignant tumors

Complete the Table

192. Complete the following table to compare terms connected to tumours.

A. Benign tumours	*B. Malignant tumours*
C. Proto-oncogenes	*D. Oncogenes*
E. Differentiation	*F. Dedifferentiation*
G. Tumour-suppressor genes	*H. p53*

Short Answer

193. Discuss the various types of mutation that can give rise to an oncogene.

SELF-TEST

1. Which of the following terms refers to a cluster of genes and the sequences involved in their regulation in prokaryotes? [p. 322]

   a. Promoter
   b. Operator
   c. Operon
   d. Repressor

2. To which region does the enzyme RNA polymerase bind? [pp. 322–323]

   a. Operator
   b. Promoter
   c. Regulatory
   d. Inducer

3. In the *lac* operon, lactose or an isomer acts as a(an) [p. 323]

   a. Repressor
   b. Corepressor
   c. Activator
   d. Inducer

4. In the *trp* operon, tryptophan acts as a(an) [p. 324]

   a. Repressor
   b. Corepressor
   c. Activator
   d. Inducer

5. In eukaryotes, RNA polymerase binds directly to the promoter sequence. [p. 329]

   a. False
   b. True

6. In both prokaryotic and eukaryotic cells, gene regulation takes place mostly at the level of transcription. [p. 327]

   a. False
   b. True

7. Which of the following are universally involved in packing DNA to form chromatin in eukaryotes? [p. 328]

   a. Repressors
   b. Activators
   c. Histones
   d. Inducers

8. Steroid hormones are able to exert their effect on several genes due to the presence of the same ________________ sequence before all of those genes. [p. 333]

   a. Regulatory
   b. Promoter
   c. Operator
   d. Repressor

9. Which of the following is added to some of the bases of DNA in order to inactivate certain genes? [p. 333]

   a. Methyl
   b. Acetyl
   c. Phosphate
   d. Carbonyl

10. TATA box is part of the eukaryotic ______________ sequence that is recognized by the transcription factors before RNA polymerase can bind. [p. 329]

   a. Operator
   b. Promoter
   c. Regulatory
   d. Inducer

11. An unfertilized egg stores a lot of mRNA whose translation is blocked until after the sperm has fused with it. The proteins that block mRNA are called the _______________ proteins. [p. 335]

   a. Repressor
   b. Masking
   c. Activator
   d. Enzyme

12. Which of the following is involved in translational control of eukaryotes? [p. 336]

   a. 3′ poly(A)
   b. Ubiquitin
   c. Acetylation of histones
   d. Proteasome

13. Which of the following tumours spreads from one tissue to another tissue or organ? [p. 338]

   a. Benign
   b. Malignant

14. Which of the following genes stimulates a normal cell to become cancerous? [p. 338]

   a. Proto-oncogenes
   b. Oncogenes

15. Which of the following refers to spreading of the cancer cells? [p. 338]

a. Dedifferentiation
b. Benign
c. Metastasis
d. Mutation
e. Oncogenes

16. Which of the following genes stimulates a normal cell to become cancerous? [p. 338]

a. *TP53*
b. *TP53*$^{-}$

17. Which of the following statements about eukaryotic gene expression is false? [pp. 322, 327]

a. Regulation of gene expression in eukaryotes is more complex than in prokaryotes because of the presence of a nucleus
b. Regulation of gene expression in eukaryotes is more complex than in prokaryotes because many undergo cellular differentiation
c. Regulation of gene expression in eukaryotes is more complex than in prokaryotes because of the need to process mRNA before it can be translated
d. Regulation of gene expression in eukaryotes is more complex than in prokaryotes because of the more rapid and extreme changes in their environmental conditions

18. Which of the following is involved in the phenomenon of gene silencing? [pp. 328, 333–334]

a. Acetylation of the cytosines in all or part of a chromosome
b. Acetylation of the cytosines for a specific gene, resulting in its inactivation
c. Methylation of all or part of a chromosome
d. Phosphorylation of all or part of a chromosome

19. miRNA molecules are _________. [p. 335]

a. Single-stranded RNAs for posttranscriptional regulation
b. Single-stranded RNAs for posttranslational regulation
c. Eukaryotic mRNA prior to processing
d. Microbial RNA

20. Which of the following allows prokaryotic cells to maintain their rapid generation times and abilities to respond rapidly to changing environmental conditions? [p. 322–327]

a. Very short mRNA half-life
b. No nuclear membrane
c. No requirement for mRNA processing
d. All of the above

INTEGRATING AND APPLYING KEY CONCEPTS

1. This chapter in the textbook has presented the details of control systems for gene expression in prokaryotes and eukaryotes and revealed the relative simplicity of the systems used in the relatively simple prokaryotic cells versus those in the larger, more complex, and often multicellular eukaryotes. One of the terms that is used to describe the CAP–cAMP system of control in prokaryotes is a regulon (p. 326). Compare and contrast the prokaryotic regulon with the steroid hormone–regulated systems/mechanism of regulation in eukaryotes.

2. All body cells have the same set of genes. Explain how red blood cells are able to make hemoglobin while the other body cells do not.

3. Researchers think that there are approximately 120 genes for miRNA in worms but approximately 250 in humans. (A) Speculate on why there would be such differing levels of these genes in these two organisms. (B) Explain how miRNA exerts control over gene expression and how the viral equivalent, siRNA, can play a role in viral infections.

OPTIONAL ACTIVITY

1. The possible applications of miRNA and siRNA extend beyond the study of gene expression and its regulation to possible therapies for various diseases, including cancer. This type of therapy would be characterized as "gene therapy." Find out more about what that term means, the hope for this type of approach, and some of the research that is being done with respect to the correlation between gene regulation and cancer and potential miRNA/siRNA treatments for various types of cancer.

16 DNA Technologies and Genomics

CHAPTER HIGHLIGHTS

- DNA is cloned/copied in a bacterium with the help of
    - Restriction endonucleases that make cuts in DNA
    - Bacterial plasmids that allow DNA segments to be inserted into bacteria for multiplication
- DNA can also be cloned in vitro using the polymerase chain reaction (PCR).
- DNA from eukaryotes can be cloned by reverse transcription and replication of mRNA.
- Applications for DNA technologies include
    - Testing for genetic diseases
    - Resolving forensic and paternity cases
    - Distinguishing individuals of one species from another
    - Determining ancestry
    - Altering genes in a cell or an organism to correct genetic disorders
    - Inserting genes in plants and animals for agriculture
    - Inserting genes in plants and animals for large-scale production of pharmaceuticals
- Developments and the use of DNA technologies are monitored by national and world agencies.
- The genomes of certain species were completed by the Human Genome Project (HGP).
    - Many more organisms have since had their genome sequenced.
- DNA sequencing and analysis are important for understanding the
    - Functions of genes
    - Role of other parts of the genome in regulating the genes
    - Phylogenetic relationships between species

- DNA chips and protein chips analyze gene expression and protein expression of all genes and proteins in specific types of cells or organisms.
- Metagenomic sequencing sequences DNA purified from the environment and does not depend upon the ability to culture or grow the organism in the laboratory.
    - Metagenomic studies are revealing unimagined diversity in the microbial world and in previously unknown protein-coding sequences.

STUDY STRATEGIES AND LEARNING OUTCOMES

- In order to understand this chapter, you must review DNA structure, its multiplication, and its organization in prokaryotic and eukaryotic cells.
- This chapter has extensive terminologies. Trying to memorize rather than understand the terms is the most common mistake.
- Carefully examine the figures and the various subject headings to keep the ideas clearly organized in your mind.

By the end of this chapter, you should be able to

- Explain what restrictions enzymes are and how they can be used for cloning or examining RFLPs
- Describe the two ways you can amplify a gene: by cloning and PCR
- Understand what cDNA is and why one would clone cDNA rather than the gene itself
- Explain what a gene library is, a cDNA library
- Describe what DNA fingerprinting means and how and why it is used
- Describe some of the applications of genetic engineering
- Explain how genome sequencing is done and, for humans, what the data revealed
- Explain what a DNA array is and describe what is meant by the processes of transcriptomics and proteomics

TOPIC MAP

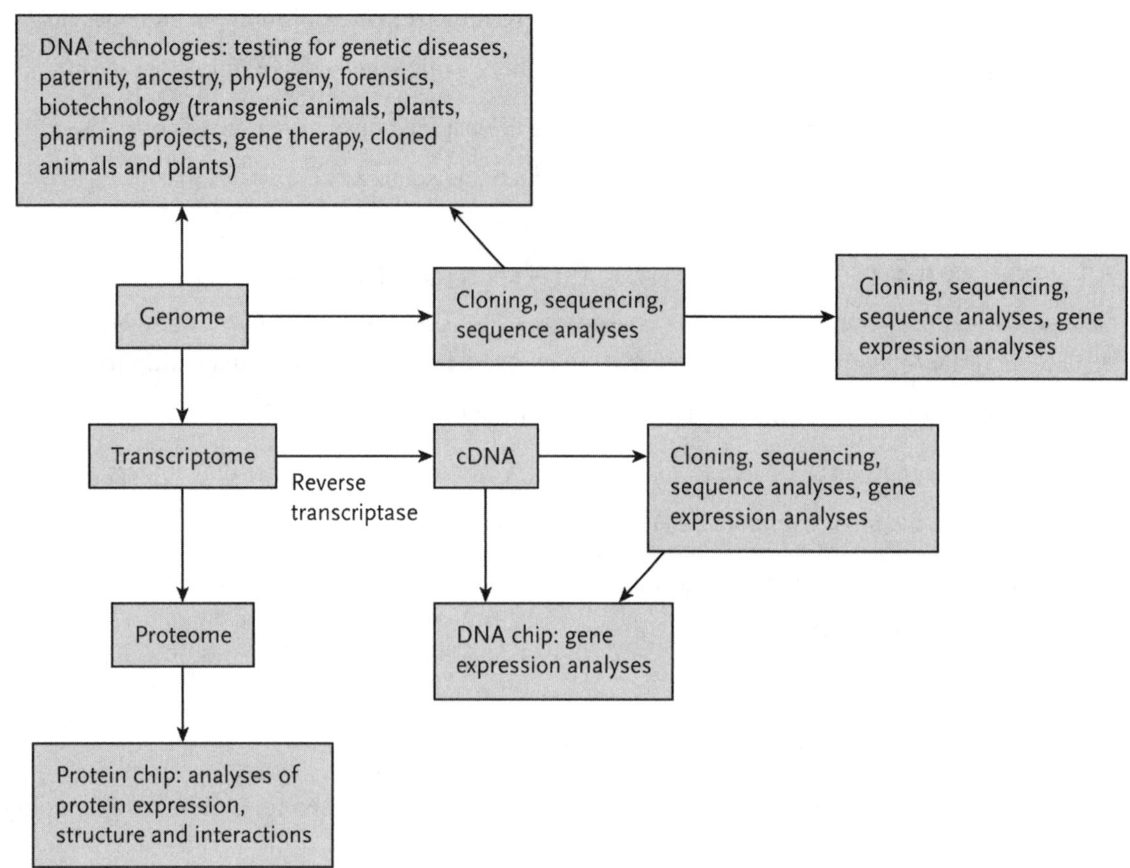
DNA technologies: testing for genetic diseases, paternity, ancestry, phylogeny, forensics, biotechnology (transgenic animals, plants, pharming projects, gene therapy, cloned animals and plants)
Genome
Cloning, sequencing, sequence analyses
Cloning, sequencing, sequence analyses, gene expression analyses
Transcriptome
cDNA
Reverse transcriptase
Cloning, sequencing, sequence analyses, gene expression analyses
Proteome
DNA chip: gene expression analyses
Protein chip: analyses of protein expression, structure and interactions

INTERACTIVE EXERCISES

Why It Matters [pp. 343–344]

Section Review (Fill-in-the-Blanks)

(1) ________________ refers to any technique that is used to make changes or improvements in living organisms or their products. DNA (2) ________________ is the use of techniques to isolate, analyze, and manipulate DNA, while genetic (3) ________________ is the use of genes that benefit us. The recent technique of (4) _______ allows scientists to sequence DNA straight from an environmental sample so that we are no longer limited by the ability to (5) _______ a species in the laboratory. This approach is revealing vast numbers of novel (6) _______, which may be exploited in biotechnology.

16.1 DNA Cloning [pp. 344–352]

Section Review (Fill-in-the-Blanks)

DNA (7) _____________ is a technique that generates multiple copies of genes. These multiple copies can then be used to understand the structure and (8) ______________ of genes and how they are regulated, or they can be manipulated by, for example, introducing (9) _______ in the gene of interest to then learn about its function or regulation. They can be inserted into a bacterium, which will transcribe and translate the foreign genes to make large amounts of (10) _______________. To clone DNA, both the genetic material containing the (11) __________ of interest and a cloning vector such as a bacterial plasmid are cut with the same (12) _________________ enzyme. These enzymes cleave the DNA backbone at specific (13) _______ and often generate single-stranded (14) _______ ends. The DNA fragments are joined together using the enzyme DNA (15) ______ to produce a (16) _______ plasmid. These are introduced by (17) ________ into a living cell such as a bacterium, and as the bacterium grows and divides, the plasmid is (18) ______. These cloning vectors often have two features that simplify the process of identifying bacteria containing (19) _______ plasmids: they have genes that confer resistance to (20) _______, and they have restriction enzyme cleavage sites within the gene for β-galactosidase. The former trait allows for the selective growth of previously sensitive host bacteria that have been (21) ______ by the plasmid, while the latter allows for identification of recombinant clones based on the (22) ________ of the ability to make the β-galactosidase enzyme and is visualized by the presence of (23) _________-coloured colonies. Bacteria containing the (24) _________________ plasmid with the specific gene are identified using a complementary nucleic acid probe, a technique called DNA (25) ________________. Once the colony of interest is identified, the bacteria in that clone are grown to large numbers and the specific gene or (26) _____________ product can be isolated. A genomic (27) ____________ is a collection of plasmids containing a copy of every DNA sequence in the genome. A cDNA library is one generated using the (28) ______ purified from a cell. This approach has two advantages: if the source of the genetic material is eukaryotic, then the transcript has already been processed to remove the noncoding sequences, that is, the (29) ___________. The second advantage is it allows one to specifically clone genes that are being (30) ______ by that cell or under specific growth conditions. The technique requires the use of the retrovirus enzyme (31) ___________ _______________ to make a single-stranded DNA that is complementary to the (32) _______. The enzyme DNA (33) ________________ is used to make the second DNA strand to give rise to a double-stranded (34) ___________________ DNA, or cDNA, which can then be cloned using restriction enzymes. Sometimes one can obtain multiple copies of a gene of interest without having to clone; the technique uses in vitro DNA replication and is called the (35) __________ _________ _______, or PCR. This allows you to replicate a specific sequence, as defined by a pair of (36) ___________. The quantity of the initial target sequence is increased by successive rounds of (37) _______ of the template DNA using high temperatures, (38) ______ the primers by reducing the temperature, and (39) _______ the primers to give, with each round, (40) _______ the amount of DNA than in the previous cycle. A sample of the amplified DNA can then be analyzed by agarose (41) _____ ___________.

Matching

Match each of the following terms with its correct definition.

42.	____	Biotechnology	A.	To make multiple copies of a DNA segment
43.	____	DNA technology	B.	Plasmid in which genes of interest are inserted
44.	____	Genetic engineering	C.	Refers to any technique used to make improvements in living organisms or their products
45.	____	DNA cloning	D.	Special enzymes isolated from bacteria to cut DNA at specific sites
46.	____	Bacterial plasmids	E.	Short, single-stranded ends of a DNA fragment created by special restriction enzymes
47.	____	Restriction enzymes	F.	Complementary DNA that is formed by copying mRNA by reverse transcriptase
48.	____	DNA ligase	G.	Small, circular molecules of DNA found in certain bacteria in addition to the main chromosomes
49.	____	Recombinant plasmid	H.	A technique used to separate DNA, RNA, or protein mixtures in a gel that is subjected to electrical field
50.	____	Gel electrophoresis	I.	Polymerase chain reaction technique used to make multiple copies of small amounts of DNA
51.	____	Sticky ends	J.	A collection of recombinant plasmids that together contain each DNA sequence from the genome of an organism
52.	____	Transformation	K.	Using techniques to isolate, analyze, and manipulate DNA
53.	____	Genomic library	L.	To use technologies to manipulate genes that benefit us
54.	____	cDNA	M.	Enzymes that repair the nicks in double-stranded DNA molecules
55.	____	PCR	N.	A short single strand of DNA that is often used in DNA technologies to start DNA synthesis
56.	____	DNA primer	O.	Absorption of DNA by bacteria to create genetic recombination

Short Answer

57. Many different restriction enzymes have been identified in bacteria. How do they differ from each other?

58. What is reverse transcriptase? What role does it play in DNA cloning?

59. What is DNA hybridization? What role does it play in DNA cloning?

60. Explain why PCR is only an option under certain circumstances.

True/False

Mark if the statement is true or false. If the statement is false, justify your answer in the line below each statement

61. ______________ Metagenomics requires the purification of DNA from a pure source.

62. ______________ Reverse transcriptase is a viral enzyme.

63. ____________ Paul Berg and Stanly Cohen received a Nobel Prize for the development of the PCR technique to amplify DNA.

16.2 Applications of DNA Technologies [pp. 352–362]

Section Review (Fill-in-the-Blanks)

Recombinant (64) _________ and PCR techniques are now used in research in all areas of biology and have tremendous practical applications. One important application is their use for the diagnosis of (65) ______ diseases. These tests are used to determine if a patient has a normal or a (66) ___________ allele. Testing often involves treating the genomic DNA with a specific (67) ______________ enzyme, which cuts within the gene of interest. Depending upon the specific sequence, DNA fragments of varying lengths, called (68) ______________ ____________ ___________ ______________ or RFLPs, are seen upon comparison of a normal allele with an altered allele. The fragments are separated by agarose gel (69) ______, and the bands are compared by (70) ________________ __________ using a labelled nucleic acid probe. DNA technology can also be used to (71) ______ individuals for forensics or paternity cases or to trace ancestry. Such studies rely on (72) ______ fingerprints produced by analysis of a number of loci within the genome; these loci are called (73) ___________ _______________ _______________ or STRs, and they vary in length between individuals, except between (74) ______ ______. Each of these loci is amplified from genomic DNA using (75) ________, and the various DNA products from the reaction are separated into bands using (76) __________ ______________. The (77) ______ of each band identifies the allele the individual has for that repeated sequence. Genetic (78) _______ is used to introduce new genes or genetic information from humans and other animals, (79) ______, and microorganisms. The primary uses have been the improvement of (80) _______ animals and (81) ______ plants, to provide large amounts of (82) _______ for research and medicine, as well as being central to the relatively new area of gene therapy, where normal alleles are introduced into mutant (83) _______ or germ-line cells of organisms to correct (84) __________ defects. Although (85) ________ ________ has enormous potential for research, medicine, agriculture, and industry, there is also cause for concern. The public is particularly concerned about the potential unintended effects of the use of (86) ________ ________ ________ or GMOs. On a global level, the use of such organisms is governed by the Cartagena Protocol on (87) ________.

Matching

Match each of the following terms or names with the appropriate definition or contribution.

88.	____	RFLP	A.	Where normal genes are inserted into a fertilized egg to correct a genetic disorder
89.	____	Southern blot analysis	B.	The tumour-inducing DNA that is produced by certain bacteria to cause crown gall in plants
90.	____	Short tandem repeats	C.	Refers to restriction fragment length variations after restriction enzyme digestion of DNA
91.	____	Somatic gene therapy	D.	A technique whereby DNA is digested with restriction enzyme and fragments are separated by gel electrophoresis, transferred to a filter paper, and analyzed using labelled probes
92.	____	Germ-line gene therapy	E.	Sequences in noncoding regions of DNA that are used to identify individuals through DNA fingerprinting
93.	____	Ti plasmid	F.	Developed a technique whereby DNA is digested with restriction enzyme and fragments are separated by gel electrophoresis, transferred to a filter paper, and analyzed using labelled probes
94.	____	Edward Southern	G.	Where normal genes are inserted into diploid cells of the body to correct a genetic disorder

Short Answer

95. Discuss how DNA technology can be used to determine if someone has sickle cell trait.

____

____

96. Provide an explanation for the term "transgenic."

____

____

97. Describe how an animal might be used as a source of foreign protein.

____

____

True/False

Mark if the statement is true or false. If the statement is false, justify your answer in the line below each statement.

98. _____________ Normal genes introduced into somatic cells with a defective gene will ensure that the next generation does not inherit the mutant allele.

____

____

99. _____________ A tobacco plant with a firefly gene is called a transgenic plant.

____

____

100. _____________ Germ-line therapy can be used to correct certain human genetic diseases.

____

____

16.3 Genome Analysis [pp. 362–369]

Section Review (Fill-in-the-Blanks)

Genome analysis can be divided into two parts: (101) _______________ genomics is the actual sequence of the genome, while (102) _________________ genomics is the location and function of the recognized genes. (103) _______________ combines biology with computers and mathematics to analyze DNA and study its evolutionary relationships. Sequencing is most commonly done using a whole-genome 104) ______ method where the entire genome of an organism is broken up into random, overlapping (105) ________, which are then (106) ______ and sequenced. The traditional method for sequencing on this scale is a chain (107) _______ technique called dideoxy sequencing. Each reaction consists of the enzyme DNA (108) ________, the four deoxyribonucleotides, the template, and a primer, as well as a mixture of fluorescently labelled (109) __________. Incorporation of these modified substrates will cause the (110) ________ of that particular molecule to terminate. The lengths of the terminated chains combined with the particular (111) ________ on each provide the sequence for that region of the DNA. Computer (112) _______ then assemble the overlapping sequences into the (113) ________ sequence. Once a DNA sequence is obtained, a computer can again be used to identify the sequence of (114)_______-coding genes, looking for the ATG start (115) ________ and then advancing (116) _______ bases at a time until a stop codon is reached. The goal of the (117) ________ ________ ________ was to sequence the DNA of humans and other selected organisms. It showed that the human genome is made of 3.2 billion (118) ________ __________. There are 20 000 to 25 000 (119) _____________-_____________ genes, far less than what would be expected to encode the approximately (120) ________ proteins produced in humans. Comparison to the genome sequences of other eukaryotes has revealed that 25% to 50% of eukaryotic genomes contain (121) _______ sequences, mostly in the form of (122) ________ sequences with no known function. As part of genome research, investigators are interested in comparing which genes are (123) _______ in different cell types and different organisms. The tool that is used to accomplish this task is the DNA (124) ________ or DNA chip. This technique exploits complementary base pairing of (125) ________ labelled cDNA purified from a cell type or organism with synthetic oligonucleotide probes representing every (126) ______-_______ gene in the genome. The data reveal what genes are being expressed as well as the (127) _______ of expression. This same concept is now being used to analyze the complete set of (128) _______ in an organism or cell type. The name used to describe these studies is (129) _______.

True/False

Mark if the statement is true or false. If the statement is false, justify your answer in the line below each statement.

130. _____________ J. Craig Venter and Celera Genomics are famous for being the first to sequence the entire genome of a bacterium using the shotgun method.

____

____

131. _____________ The human genome has about 100 000 protein-coding genes.

____

____

132. ____________ Germ-line therapy can be used to correct certain human genetic diseases.

____

____

SELF-TEST

1. Which of the following would be considered transgenic? [p. 356]

   a. A bacterial strain with an antibiotic resistance plasmid
   b. A bacterial chromosome with the *lac* operon
   c. A human cell encoding a growth hormone
   d. A bacterial strain containing the insulin gene

2. cDNA is made by copying [p. 349]

   a. mRNA
   b. DNA
   c. Proteins
   d. Plasmids

3. The primer for PCR is ___________. [pp. 349–350]

   a. RNA
   b. DNA
   c. Proteins
   d. Plasmids

4. Which of the following enzymes is used in PCR? [pp. 349–350]

   a. RNA polymerase
   b. DNA polymerase
   c. Reverse transcriptase
   d. DNA ligase

5. Gel electrophoresis is used to separate _________. [p. 351]

   a. DNA
   b. RNA
   c. Protein
   d. Any of the above

6. Which of the following is associated with "sticky ends"? [p. 345]

   a. Recombinant bacterial plasmids with the insulin gene
   b. mRNA isolated from human cells
   c. DNA fragments with single-stranded ends
   d. Proteins that stick to DNA

7. Over 50% of the human genome consists of repeated sequences that have no apparent function. [p. 366]

   a. True
   b. False

8. Only about 2% of the human genome contains protein-coding sequences. [p. 366]

   a. True
   b. False

9. In cloning Dolly the sheep, the scientists used the nucleus from [p. 359]

   a. A somatic cell
   b. An egg
   c. A zygote
   d. A sperm

10. Reverse transcriptase uses which of the following as a template? [p. 349]

    a. RNA
    b. DNA
    c. Proteins
    d. Plasmids

11. Reverse transcriptase is obtained from __________. [p. 349]

    a. *Rhizobium radiobacter*
    b. Retroviruses
    c. Humans
    d. Plants

12. Which of the following scientists developed PCR? [p. 349]

    a. Kary Mullis
    b. Edward Southern
    c. Paul Berg
    d. Frederick Sanger

13. Gene therapy has been successful in animals other than humans. [pp. 357–358]

    a. True
    b. False

14. Which of the following would not be a step in the generation of a recombinant bacterium that produces large amounts of a human protein such as insulin? [pp. 349, 356]

   a. Isolate and clone the gene from human genomic DNA
   b. Isolate the mRNA from human cells
   c. Make cDNA with restriction sites on the ends
   d. Clone restriction-digested DNA containing the insulin gene into the similarly digested cloning vector

15. Which of the following statements is false regarding gene therapy? [pp. 357–358]

   a. Normal genes are introduced into somatic cells that have a defective gene
   b. Normal genes are introduced into fertilized eggs that have a defective gene
   c. Germ-line gene therapy has been successfully used with humans
   d. Somatic gene therapy has been successfully used with humans

16. Which of the following statements about transgenic animals is true? [p. 358]

   a. They can be used as pharmaceutical protein factories
   b. A transgenic sheep was made to secrete a human blood-clotting factor in its milk
   c. They are mostly used in pharming projects
   d. All of the above

17. Which of the following statements about the cloning of animals is false? [p. 359]

   a. This was first performed with a sheep
   b. The procedure involves removing the nucleus from an unfertilized egg and fusing it with a diploid somatic cell
   c. This procedure was very inefficient and has not been duplicated
   d. One can now purchase cloned cows for dairy farming

18. Which of the following statements about transgenic plants is false? [pp. 359–361]

   a. Transgenic plants have been created to produce vitamins
   b. Transgenic plants are relatively easy to make
   c. Transgenic plants are relatively widely accepted GMOs
   d. One common method of generating a transgenic plant is to create a tumour on it

19. The term “pharming projects” refers to ________. [p. 358]

   a. The cloning of animals
   b. Gene therapy
   c. The use of transgenic animals to produce proteins of medicinal value
   d. The use of antibiotic resistance genes in bacterial cloning vectors

20. J. Craig Venter is famous for ___________. [pp. 343–344, 363]

a. Doing metagenomic studies of DNA from the ocean to identify previously unknown viruses, microbes, and proteins
b. Having his genome sequenced
c. His involvement in sequencing the first genome of an organism
d. All of the above

INTEGRATING AND APPLYING KEY CONCEPTS

1. Explain why somatic gene therapy is being experimented in humans but germ-line gene therapy is not permitted in humans.

2. Genetically engineered food products are slowly coming onto the market. What is the rationale for the skepticism in our community to buy these products?

3. The Human Genome Project identified a much smaller number of protein-coding genes in the human genome than was expected. How did they come up with the predicted value of 100 000 protein-coding genes, and how do they explain the discrepancy between that predicted value and the observed value of 20 000 to 25 000 protein-coding sequences? Explain how they would have come up with the lower figure of 20 000 to 25 000.

OPTIONAL ACTIVITY

1. As DNA sequencing technologies improve, becoming faster and cheaper, the prospect of medical genome sequencing becomes more of a possibility in the near future. However, before a patient has his or her genome sequenced, many issues would need to be resolved, both by the patient and by an ethical review panel. Consider what specific issues would need to be addressed and what kinds of information, as a patient, you would want to be given and what kinds you would prefer withheld. Then see what you can find out about how these are being addressed by the Human Medical Genome Sequencing Project (National Institutes of Health).

17 Microevolution: Genetic Changes within Populations

CHAPTER HIGHLIGHTS

- Individuals in populations vary, and the variations may be caused by differences in their genetic makeup, environmental factors, or the interaction of the two.
- All populations have a genetic structure that can be described, and mathematical models can be used to test hypotheses relating to the evolution of those populations.
- The Hardy–Weinberg principle describes how populations will achieve genetic equilibrium if certain assumptions are met.
- Three modes of natural selection (directional selection, stabilizing selection, and disruptive selection) have differing effects on phenotypic variation within populations.
- Sexual selection may lead to the evolution of showy structures or display behaviour and sexual dimorphism.
- A variety of factors maintain genetic and phenotypic variation in populations despite the effects of stabilizing selection and genetic drift.
- Natural selection leads to the evolution of adaptive traits, although there are constraints on this evolution.

STUDY STRATEGIES AND LEARNING OUTCOMES

- This chapter has a lot of new terms and concepts. Make sure you pay attention to the examples that demonstrate each one; remembering these will make it easier to distinguish one from the other.
- Make sure you understand the Hardy–Weinberg principle. As you read through the chapter, keep relating the various agents of microevolution back to the requirements of this principle and understand which aspect of the principle is being violated or how each contributes to genetic variation.
- Carefully examine the figures, examples, and various subject headings to keep the concepts clearly organized in your mind.

- As always, go through one section of the textbook at a time and then work through the companion section of the study guide.

By the end of this chapter, you should be able to

- Understand the sources of variation in populations and devise experiments that can identify the basis of those variations
- Describe the Hardy–Weinberg principle and know the conditions that must exist for a population to be in genetic equilibrium
- Explain the three modes of natural selection and their effect on the phenotypic variation of a population

TOPIC MAP

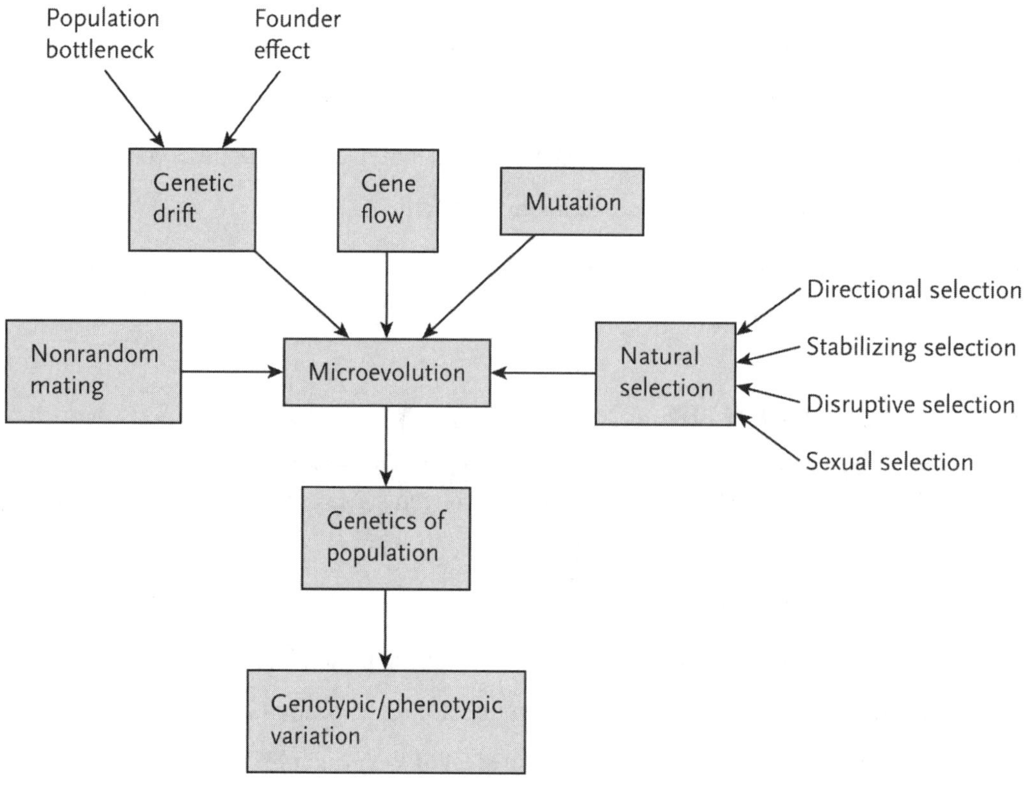

INTERACTIVE EXERCISES

Why It Matters [pp. 373–374]

Section Review (Fill-in-the-Blanks)

(1) _____ is a drug used to fight bacteria that enter through the skin and saved many lives after the Cocoanut Grove nightclub fire in 1942. Its overuse led to the evolution of (2) _____ strains of *Staphylococcus*. The evolution of these strains is an example of (3) _________, that is, a heritable change in the genetics of a population. A population of organisms includes all the individuals of the same (4) _____ that live together in the same place and time.

17.1 Variation in Natural Populations [pp. 374–377]

Section Review (Fill-in-the-Blanks)

The individuals in most populations differ with regard to external and internal structure and function; that is, they show (5) ________ variation. If those variations are generally small and incremental, they are said to be (6)___________; if variations occur in two or more discrete states, for example, blood type in humans, they are said to be (7) ________. The existence of discrete variants of a character is called a(n) (8) ________. Data relating to the former are generally depicted on a (9) ________ graph if it is a small population or as a (10) ________ if the population is large. The width of the plot reflects the amount of (11) _________ among the individuals, and the mean describes the average value of the character. For traits that display discrete variants, these are described quantitatively as percentages or (12) ________ of each trait.

Short Answer

For each of the following statements, provide an example that supports the statement.

13. Phenotypic variation in individuals may have an environmental as well as a genetic basis.

____

____

14. The behaviour of animals may be influenced by genetic differences between individuals.

Provide an explanation for each of the following statements.

15. Only variation based on genetic differences is subject to evolutionary change.

16. Differences in the DNA of organisms do not always produce phenotypic differences.

17. Height in humans is a trait that varies quantitatively.

Complete the Table

Complete the following table.

Term/Process	*Description/Function*
Polymorphism	18.
Microevolution	19.
20.	A technique that can identify different forms of a protein and that is useful in analyzing genetic differences between individuals
Population	21.

17.2 Population Genetics [pp. 377–379]

Section Review (Fill-in-the-Blanks)

To predict how certain factors may influence genetic (22) ________, population geneticists first describe the structure of the population. The sum of all alleles at all gene loci in all individuals in a population is referred to as its (23) ________. The proportion of individuals in a population possessing a specific genotype is that population's (24) ______________, while the relative abundance of each allele is described by the (25) _________. For a gene locus with two alleles, scientists specify one as (26) _______ and the other as (27) ________. The (28) __________– _____________ principle describes the conditions under which a population achieves genetic (29) ____________, a state where the genetic structure of a population is not evolving. Since natural populations are often difficult to manipulate experimentally, much of what we know about them comes from (30) ___________ data. Study of these populations is often based on (31) ___________ models, which predict what would happen if a specific factor had no effect.

Short Answer

List the five conditions that must be met for a population to be in Hardy–Weinberg equilibrium.

32.

____

____

33.

____

____

34.

____

____

35.

____

____

36.

____

____

17.3 The Agents of Microevolution [pp. 379–388]

Section Review (Fill-in-the-Blanks)

If any of the conditions necessary for a population to be in Hardy–Weinberg equilibrium are not met, (37) ________ (a change in allele frequencies over a number of generations) will occur. A (38) _________ is a heritable change in a gene or chromosome and may impact allele frequencies. Likewise, (39) _________ caused by immigrants entering a population or emigrants leaving it will change the genetic structure of the population. Changes in allele frequencies may also occur due to chance events, an effect referred to as (40) ___________; this is minimized in very large (or infinite) populations. When populations are dramatically reduced in size, the result may be the loss of alleles or a reduction in genetic (41) ________. Two general circumstances give rise to this situation: population (42) _______ and (43) ________ effects. The former results when disease, drought, or starvation kills many individuals in a population and eliminates some (44) __________ in the process; this is what is being seen with (45) __________ species. The latter results when a small number of individuals start a new (46) ____________; this will inevitably result in a limited number of alleles compared to "normal," as is seen with captive (47) __________ programs. The phenomenon of genetic (48) ________ therefore has serious implications for conservation. Also involved in (49) ___________ is natural selection. This is the process where certain individuals with specific, (50) ___________ traits produce more offspring than individuals lacking these traits. This difference in reproductive success is reflected in an individual's (51) ___________, a measure that reflects the number of surviving offspring an individual produces compared with other members of the population. Three modes of natural selection can cause changes in the phenotypic variation of a population. One is (52) __________ selection, in which individuals possessing traits at the extremes of the phenotypic continuum are selected against and phenotypic variation is reduced. Alternatively, in (53) __________ selection, only one end of the distribution of a trait is selected against, and the mean value of that trait changes; this is the phenomenon involved in breeding programs. A third mode, (54) ____________ selection, involves selection against the most common phenotype, favouring both extremes of the distribution instead. The process of (55) ________ selection may result in the evolution of showy structures and complex courtship behaviours, all of which would be produced as a result of (56) _________ selection, that is, resulting from interactions between males and females. (57) __________ selection would give rise to large body size or structures used in competition with other members of the same sex. Both may also lead to sexual (58) _____________, a condition wherein males and females differ in size or appearance.

Matching

Match each of the processes on the left with the most appropriate example on the right.

59.	____	Directional selection	A.	Babies with very low or very high birth weights suffer higher than average mortality.
60.	____	Disruptive selection	B.	A horse breeder mates the fastest males with the fastest females.
61.	____	Intersexual selection	C.	During droughts, cactus finches with very long or very deep bills enjoy higher survival than those with average-sized bills.
62.	____	Intrasexual selection	D.	Female bowerbirds prefer males with extremely long tails.
63.	____	Stabilizing selection	E.	Very large elephant seals win battles with smaller males. The winner of one of these encounters attains dominant status and gets to mate with females.

17.4 Maintaining Genetic and Phenotypic Variation [pp. 388–391]

17.5 Adaptation and Evolutionary Constraints [pp. 391–392]

Section Review (Fill-in-the-Blanks)

Despite the combined effects of stabilizing selection and genetic drift, most natural populations exhibit a tremendous amount of (64) _________ and (65) ___________ variation. One reason for this is that the (66) ________ state characteristic of most eukaryotes can hide recessive alleles from the effects of natural selection. By definition, a recessive trait is masked in (67) _________ individuals, meaning that the trait, even if undesirable, will not be eliminated due to its ability to hide in these individuals. Another reason for such high levels of variation is the phenomenon of balanced (68) __________. This occurs when circumstances allow for two or more phenotypes to be maintained in fairly stable proportions in a population; it is commonly seen in situations where heterozygotes enjoy a higher relative (69) ___________ than either homozygote. The name for this effect is (70) _________ _________, and an example is the situation with people who have sickle cell trait and live in areas where (71) ____________ is endemic. Genetic variability can also be maintained by (72) __________ environments or when rare phenotypes have higher (73) __________ compared with more common phenotypes. The latter phenomenon is referred to as (74) __________-____________ selection. Some biologists believe in the (75) __________

___________ hypothesis, which states that much of the variation in populations has no measurable effect on fitness. These variations are said to be (76) _________ _________. This hypothesis explains why large populations and ones that have not experienced a recent population (77) ___________ exhibit the highest levels of genetic variation. While the scientists who support this hypothesis do not believe that every genetic variant has been maintained by (78) ___________ ___________, they do not question that (79) _________ traits, that is, those that increase the relative fitness of an individual, are products of natural selection. The accumulation of these traits over time, that is, (80)___________, is constrained by several factors: most adaptive traits represent compromises among conflicting needs; most environments are constantly (81) __________, and natural selection can only work on (82) _________ traits; it cannot create new (better) traits.

Matching

Match each of the processes on the left with the most appropriate example on the right.

83.	____	Adaptation	A.	A mutation changes base sequence in DNA, but the protein it codes for is unchanged.
84.	____	Frequency-dependent selection	B.	A change in plumage resulting from natural selection that makes a mouse more difficult to be seen by predators.
85.	____	Heterozygote advantage	C.	Different shell patterns in snails are favoured in different environments.
86.	____	Selectively neutral	D.	Homozygous individuals have lower fitness than heterozygotes.
87.	____	Selection in varying environments	E.	Rare or unusual phenotypes in a population enjoy higher than average reproductive success.

SELF-TEST

1. How does diploidy maintain genetic variability in a population? [p. 388]

   a. It doubles the number of gametes produced by individuals
   b. It hides recessive alleles from natural selection
   c. It increases mutation rate
   d. It promotes genetic drift

2. An individual's phenotype is the result of ________. [pp. 375–376]

   a. Environmental factors
   b. Genetic factors
   c. Quantitative factors
   d. (a) and (b)

3. Only ________ variation in phenotypes is subject to evolutionary change. [p. 376]

   a. Acquired
   b. Anatomic
   c. Genetic
   d. Environmental

4. ________ create new alleles, while ________ shuffles existing alleles into new combinations, producing novel genotypes and phenotypes. [pp. 376, 379]

   a. Artificial selection; stabilizing selection
   b. Founder effect; nonrandom mating
   c. Gene flow; genetic drift
   d. Mutations; sexual reproduction

5. Population bottlenecks are a cause of ________. [p. 382]

   a. Gene flow
   b. Genetic drift
   c. Mutation
   d. Polymorphism

6. The number of surviving offspring an individual produces compared with other members of a population is that individual's ________. [p. 383]

   a. Allele frequency
   b. Gene flow
   c. Heterozygote advantage
   d. Relative fitness

7. Which of the following will not alter allele frequencies in future generations? [pp. 383, 385]

   a. Genetic drift
   b. Mutation
   c. Natural selection
   d. Nonrandom mating

8. Which of the following will reduce genetic variability in a population? [p. 385]

   a. Directional selection
   b. Disruptive selection
   c. Stabilizing selection
   d. Sexual selection

9. The evolution of showy structures and behaviours in males of many species is the result of ________. [p. 387]

   a. Directional selection
   b. Disruptive selection
   c. Sexual selection
   d. Stabilizing selection

10. Which of the following statements is true? [pp. 390, 392]

    a. No organism is perfectly adapted; evolutionary compromises must always be made
    b. In the absence of human intervention, natural selection produces organisms perfectly adapted to all conditions
    c. Biologists are in agreement that all genetic change has fitness implications
    d. Traits evolved for a specific function cannot be co-opted for other purposes

11. Which of the following statements is false? [pp. 376, 388]

    a. Sexual reproduction contributes to phenotypic diversity
    b. In humans, 10^{600} combinations of alleles are estimated to be possible in human gametes
    c. Sexual reproduction can create new alleles
    d. Inbreeding is a form of nonrandom mating that increases the frequency of homozygous individuals

12. Which of the following is true about the evolution of antibiotic resistance? [pp. 373–374]

    a. It is the result of balanced polymorphism
    b. It is an example of directional selection
    c. It is an example of frequency-dependent selection
    d. Antibiotics cause the accumulation of mutations

13. Which of the following statements is true about gel electrophoresis? [pp. 376–377]

   a. If the protein products of two alleles from the same locus differed in size, this would be identified by electrophoresis
   b. If the sequence of two alleles from the same locus differed in one or two codons, this would be identified by electrophoresis of the proteins
   c. If the protein products of two alleles from the same locus differed in shape, this would be identified by electrophoresis
   d. If the protein products of two alleles from the same locus differed in net charge, this would be identified by electrophoresis

14. Genetic variation is not caused by ________. [pp. 374, 376]

   a. The overuse of antibiotics
   b. Rearrangement of existing alleles
   c. Fusion of gametes
   d. Mutation

15. The definition of the gene pool is ___________. [p. 378]

   a. All alleles at all gene loci in all individuals in a population
   b. All alleles of all genes loci in a given individual
   c. All of the various alleles of a particular gene/locus
   d. The degree of variation at a given locus within a population

16. Balanced polymorphism arises when ___________. [p. 388]

   a. A small group of individuals moves into a new environment
   b. Heterozygote individuals have increased fitness
   c. Males develop exaggerated traits to attract mates
   d. There is inbreeding

17. Which of the following is a qualitative variation? [pp. 378, 389]

   a. Birth size
   b. People with dwarfism
   c. People with sickle cell disease
   d. People with sickle cell trait

18. Which of the following would be measured in terms of frequencies? [p. 378]

   a. Birth size
   b. Height of people
   c. People with dwarfism
   d. Athletic ability

19. Which of the following explains why lethal alleles that manifest late in life are not eliminated from the population? [p. 383]

    a. The heterozygous individuals are unaffected and hide the allele from natural selection
    b. The allele is manifested after the individuals mate
    c. Only homozygous recessive individuals are affected
    d. (a) and (b)

20. If a particular virus were introduced into a small population and proved lethal for the very young and the elderly, geneticists would identify this as __________. [pp. 382, 385]

    a. A population bottleneck
    b. Stabilizing selection
    c. A founder effect
    d. (a) and (b)

INTEGRATING AND APPLYING KEY CONCEPTS

1. Investigators studying genetic variation in humpback whales, with an average life span of 30 to 40 years, were surprised to learn that two of the three populations that survived until whale hunting was outlawed retained relatively high levels of variability. Would you expect to obtain similar results if you were studying a bird species with a comparatively short generation time? Explain why or why not.

2. If a population geneticist finds that a population is not in Hardy–Weinberg equilibrium, he or she will probably want to determine which of the required assumptions is violated. What sorts of things might he or she do to determine whether each condition is met?

3. Genetic studies of individuals who are resistant to HIV infection have identified, in at least some, a homozygous mutation in an immune cell surface receptor gene, *ccr5*. Heterozygous individuals appear to have higher levels of resistance compared to homozygous wild-type (normal) individuals. Discuss how these observations may have been determined and how researchers would ethically prove or disprove this association.

OPTIONAL ACTIVITY

1. In one of the previous chapters, the idea of animal cloning was discussed, and, in addition to the famous cloned sheep Dolly, it was pointed out that a single, prize-winning dairy cow had been cloned and many of those clones sold. Discuss the population genetics aspects of these cows being used for breeding stock.

18 Species

CHAPTER HIGHLIGHTS

- There are four ways that biologists commonly think of the concept of "species." They are the morphological, ecological, biological, and phylogenetic species concepts.
- Pre- and postzygotic reproductive isolating mechanisms maintain species integrity.
- Most biologists recognize three ways in which new species arise: sympatric, parapatric, and allopatric speciation.
- The accumulation of genetic variation between allopatric populations can lead to speciation.
- Polyploidy is a mechanism by which sympatric speciation may occur in plants.
- Speciation may result from chromosomal alterations.

STUDY STRATEGIES AND LEARNING OUTCOMES

- This chapter has a lot of new terms and concepts. Make sure you pay attention to the subject headings and examples that demonstrate each concept or process. Between both, this should help you keep things straight.
- Make sure you are clear on the meaning of the prefixes allo-, para-, sym-, and auto- as well as the suffix -ploidy. Remember the difference between mitosis and meiosis and the change in chromosome number resulting from the latter.
- As always, go through one section of the textbook at a time and then work through the companion section of the study guide.

By the end of this chapter, you should be able to

- Discuss the four species concepts used by biologists and explain why biologists cannot agree on a single concept and why, specifically, there are limitations with the Biological Species Concept
- Understand the importance of reproductive isolation between species in maintaining species integrity and understand the various pre- and postzygotic isolating mechanisms

- Identify the three generally recognized models of speciation and understand how, with each, reproductive isolation can evolve
- Describe the genetic mechanisms of speciation and explain how they also give rise to reproductive isolating mechanisms

TOPIC MAP

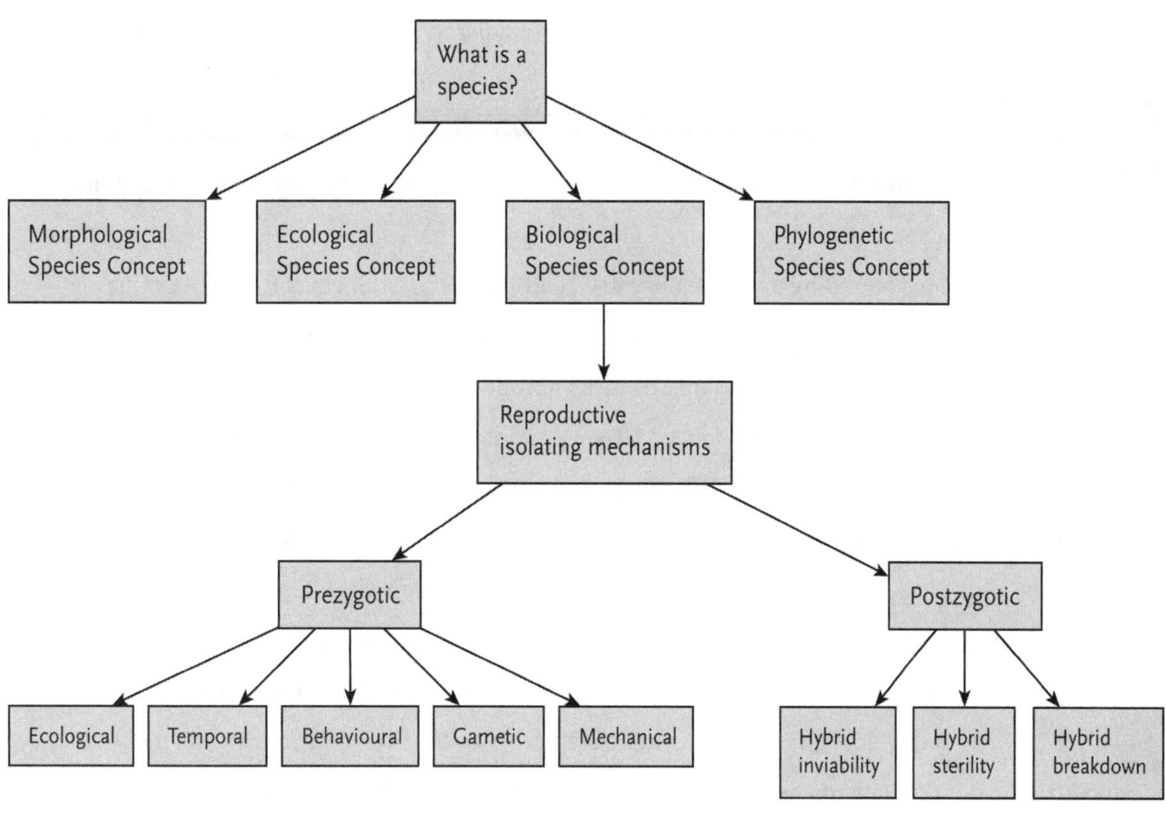

INTERACTIVE EXERCISES

Why It Matters [pp. 373–374]

18.1 What's in a Name? [p. 396]

18.2 Definition of "Species" [pp. 396–397]

Section Review (Fill-in-the-Blanks)

Names contribute to the precision of (1) _______ and in biology can convey a great deal of (2) ______. Biologists use Latinized names to identify (3) ______, subspecies, or varieties. Each described species has a (4) _______ name. Other animals rely on their own method of naming, using (5) ______ to refer to different organisms. Species are the fundamental taxonomic units of biological (6) ______. There are many ways that biologists define the term "species." The (7) ______ Species Concept is based on the ability of different groups of organisms to breed in nature and produce fully fertile offspring. The (8) ______ Species Concept defines a species as a group of organisms bound by a unique ancestry. The (9) _______ Species Concept defines a species as a group of organisms that share a distinct ecological niche. There are many others. There are several problems with the first of these definitions: it does not include organisms, such as any of the prokaryotes, that reproduce (10) ______; it also does not apply to organisms that hybridize and produce (11) ______ offspring. Androdioecous species also pose a problem: they do not have (12) _______ but are composed of males and (13) _______ that produce both sperm and eggs. (14) _______ species consist only of females, and their eggs begin development and lead to reproduction after being mechanically stimulated by the male sperm of another species. (15) ________ is a principal advantage of sexual reproduction, and this advantage helps explain its prevalence in the biological world; genetic studies of water fleas reproducing sexually versus those that were reproducing asexually provided support for an advantage as the asexually reproducing fleas experienced higher levels of (16) _______ in their mitochondrial protein-encoding genes.

Matching

For each of the following concepts, choose the most appropriate definition from the list below.

17.	____	Biological Species Concept	A.	A group of organisms that share a distinct ecological niche
18.	____	Ecological Species Concept	B.	A group of organisms that share a unique evolutionary ancestry
19.	____	Phylogenetic Species Concept	C.	A group of organisms that interbreed to produce fully fertile offspring

True/False

Mark if the statement is true or false. If it is false, justify your answer in the line below each statement.

20. ___________ Bacteria reproduce asexually and so are not given species names.

____

____

21. ___________ Gynogenetic organisms do not reproduce sexually.

____

____

22. ___________ Zebras and horses cannot interbreed.

____

____

Short Answer

23. Explain what is meant by the term "hybridization."

____

____

24. Describe what is meant by hybrid vigour.

____

____

18.3 One Size Does Not Fit All [pp. 397–399]

18.4 Gene Flow: Four Examples [pp. 399–401]

Section Review (Fill-in-the-Blanks)

No single definition of species satisfies all biologists or applies to all (25)______; this is not surprising as organisms are the product of (26) ______, a very dynamic process. The Biological Species Concept assumes a genetic (27) _______ of species, and gene (28) _______ is thought to be the "glue" that holds members of a species together. This concept also assumes that members of one species are genetically (29) _______ from members of another species so that they cannot interbreed, and, in fact, the process of (30) _______ is often defined as the evolution of reproductive isolation between different organisms. Biologists often use (31) _______ similarities as markers of genetic isolation and to describe new species. This approach is based on the (32) _______ Species Concept of the 1800s, which identifies organisms of the same species based on shared (33) _______ traits. This approach is still used by paleontologists to associate (34) ______ with particular species and by naturalists in the field. Although gene (35) ______ is assumed to explain the genetic cohesiveness of members of the same species, this cannot explain the similarities among geographically isolated members of the same species. Distribution influences (36)_____ ______, and widely but sparsely distributed species may experience less (37) _______ _______ than those with continuous distribution; in contrast, increases in habitat continuity can result in an increase in (38) _____ ______, causing a reduction in biodiversity of the species. This has been demonstrated using (39) _______ technology to monitor the impact of increased habitat continuity for the Eurasian red squirrels of northern England and Scotland. Instead of relying on continuity of habitat, some species may depend on others for their (40) _______, as has been demonstrated for the *Balea perversa* snails. These snails are distributed from Tristan de Cunha in the South Atlantic to Iceland in the North Atlantic as a result of their ability to (41) _______ on migrating birds. In such cases, the (42) _____ cohesiveness of these widely separated populations would have been inexplicable without the modern techniques of (43) _______ sequencing and (44) _______ sequence databases. In a strikingly different situation, (45) ______ behaviour can limit gene flow between organisms that share a continuous distribution. An example of this comes from studies of house mice, which indicated that their social structure prevented males from (46) ______ with females of other populations of the same species. Using genetic markers to follow (47)_____ ______, scientists showed that rather than being completely reproductively isolated, the females of different populations are more easily assimilated and will (48) ______.

Short Answer

49. Explain how distribution can affect gene flow of a species.

____

____

50. Explain how social behaviour can influence gene flow of a species.

____

____

51. Explain how the gene flow of one species may be dependent upon another.

____

____

18.5 A Dynamic Situation [p. 402]

18.6 Geographic Variation [pp. 402–404]

Section Review (Fill-in-the-Blanks)

Members of a species can vary in time and in space, showing morphological or other changes over the passage of time or as they move into a new (52) ______. When there is genetic or (53) ______ variation, the different populations are referred to as (54) ______ or, sometimes, races, breeds, or, in the case of plants, cultivars or variants. Members of different subspecies may interbreed when their distributions overlap, and the offspring generally have a (55) ______ that is intermediate between the parents. With certain geographic distributions, there may be (56) _____ ______ between adjacent populations but not directly between distant populations. These populations are called (57) _______ species. Similarly, species that occupy a wide geographic range often exhibit some genetic and phenotypic variation as they adapt to local conditions. This variation is often gradual and continuous, and the traits are referred to as a (58) _____.

Short Answer

59. Explain how a particular type of geography gives rise to a ring species.

____

____

60. Explain how a particular type of geography gives rise to clinal variation.

____

____

True/False

Mark if the statement is true or false. If it is false, justify your answer in the line below each statement.

61. _______ Rat snakes exhibit clinal variation.

____

____

18.7 Reproductive Isolation [pp. 404–407]

Section Review (Fill-in-the-Blanks)

(62) _______ ______ mechanisms are biological characteristics that prevent the gene pools of two sympatric species from mixing. This can be accomplished in two basic ways: (63) _______ isolating mechanisms prevent the fusion of gametes and the production of a zygote, whereas (64) ______ isolating mechanisms operate after fertilization has occurred. There are at least five mechanisms that can prevent fertilization: one is (65) _______ isolation, where individuals occupy different habitats in the same area. (66) _______ isolation occurs between species occupying the same area but reproducing at different times. One of the most common mechanisms is (67) _______ isolation, wherein various courtship displays indicate species identity to potential (68) ______ but are not recognized by individuals of other species. In some instances, the reproductive structures of males and females are only compatible with other members of the same species, a situation called (69) _______ isolation. Finally, (70) ______ isolation occurs when the sperm of one species are incompatible with the eggs of another species. When the sperm of one species does fertilize an egg of another species, the offspring of such a mating, that is, the (71) ______ hybrids, may have reduced fitness compared with (72) ______ offspring, resulting in postzygotic isolation. One way in which this arises is through hybrid (73) ______; this is due to the incompatibility of embryonic developmental programs of the two parent species. Sometimes development proceeds and produces otherwise healthy offspring that suffer hybrid (74) ______ due to differences in the number or structures of their chromosomes and a resulting failure of meiosis. Hybrid (75) ______ occurs when interspecific hybrids are fertile and can breed with other hybrids and either parental species, but their offspring exhibit reduced

survival or fertility. This is a result of harmful types of genetic (76) ______ or chromosomal abnormalities.

True/False

Mark if the statement is true or false. If it is false, justify your answer in the line below each statement.

---

77. _______ Song birds of different species are reproductively isolated by behavioural isolation.

___

___

78. _______ Sympatric species are the product of interspecific matings.

___

___

Short Answer

79. Provide an explanation for the evolution of species-specific courtship displays.

___

___

___

80. Provide a molecular explanation of how gametic isolation occurs.

___

___

___

81. Provide an explanation of how mechanical isolation would be accomplished by plants.

___

___

___

Complete the Table

82. Fill in the blanks with the proper term or description.

Prezygotic mechanism	*Description*
A.	Individuals occupy different habitats.
Temporal isolation	B.
Behavioural isolation	C.
Mechanical isolation	D.
E.	Incompatiblity between gametes of different species

83. Fill in the blanks with the proper term or description.

Postzygotic mechanism	*Description*
A.	Abnormal embryonic development
B.	Hybrid offspring do not make functional gametes
Hybrid breakdown	C.

18.8 Geography of Speciation [pp. 407–412]

18.9 Genetic Mechanisms of Speciation [pp. 412–416]

18.10 Back to the Species Concept in Biology [p. 416]

Section Review (Fill-in-the-Blanks)

If a population is subdivided by a physical barrier to gene flow, speciation may occur: this is called (84) ______ speciation. Sometimes the barrier may subsequently be removed, allowing (85) ______ contact. If the evolution of prezygotic isolating mechanisms is not complete, individuals from the two populations may interbreed in (86) _______ zones, producing fertile offspring. If the hybrid offspring have lowered fitness, selection will favour individuals that mate only with members of their own species, a phenomenon known as (87) ________. (88) _______

speciation occurs when a species is distributed across a discontinuity in environmental conditions so that natural selection favours different alleles and (89) ________ on either side of the discontinuity. (90) _______ speciation occurs among individuals of a continuously distributed population. Extensive studies with the apple maggot, a possible example of this type of speciation, point to genetic changes yielding a switch in preferred (91) _______ plant species and, thus, reproductive isolation. When this kind of isolation results in the development of a subspecies, the subspecies is called a(n) (92) ______ race. Geographic isolation results in the loss of gene (93) ______, which, in turn, allows allopatric populations to accumulate (94) ________ differences; these may result in postzygotic reproductive isolation upon secondary (95) _______. The genetic basis for postzygotic isolating mechanisms may be as simple as the two alterations that cause hybrid (96) _______ through the development of lethal tumours in the hybrid offspring of swordtails and platys aquarium fish; in other cases, many more genes are involved. When hybrid offspring instead have reduced fitness, natural selection may then (97) ______ the evolution of prezygotic isolating mechanisms. As with postzygotic mechanisms, these mechanisms may also have a simple genetic basis, as in the case of the single mutation that causes the (98) ________ isolation of oppositely coiled snails. Reproductive isolation among sympatric populations can result from changes in the numbers of sets of chromosomes, a phenomenon that is more common in (99) _______ than in animals and that is referred to as (100) ______. Individuals that are > $2n$ may result from an error in mitosis (in the germ line) or (101) ______ so that gametes are unreduced in the parent; this is called (102) ______. Unreduced gametes may fuse in a self-fertilizing plant or between two plants of the same species to give tetraploid offspring, which can either reproduce by self-pollination or by breeding with other (103) ______. Speciation may result because tetraploids are (104) _______ isolated from the original diploid population as a result of differences in chromosome numbers; that is, such hybrids would be (105) _______. Speciation resulting from (106) ______ occurs when two closely related species hybridize. Although the two sets of parental chromosomes are unable to pair in meiosis, nondisjunction during mitosis of germ-line cells will (107) ______ the chromosome number, giving $4n$ germ cells, which can be reduced during meiosis to give viable (108) _____ or $2n$ gametes. These can fuse by self-pollination or with another diploid gamete; either process gives rise to (109) _______ or $4n$ offspring and can yield a new (110) _______ within a single generation, without geographic isolation.

Building Vocabulary

Using the definitions of the prefixes below, assign the appropriate prefix to the appropriate suffix for the following:

Prefixes	Meaning
allo-	different
auto-	self
para-	beside
sym-	together

Prefix	Suffix	Definition
_____	-patric	111. A type of speciation that results from a sudden discontinuity in environmental condition
_____	-polyploidy	112. The production of unreduced gametes; it may lead to speciation
_____	-polyploidy	113. Hybrid offspring become polyploid; it may lead to speciation
_____	-patric	114. A type of speciation that occurs within one continuously distributed population
_____	-patric	115. Speciation that requires geographic isolation of subpopulations

True/False

Mark if the statement is true or false. If it is false, justify your answer in the line below each statement.

116. _______ Natural selection selects for prezygotic reproductive isolation mechanisms in allopatric populations.

____

____

117. _______ In the case of behavioural isolation, individuals can often successfully interbreed in captivity, producing viable and fertile offspring.

____

____

Short Answer

118. Provide an explanation of how structural alterations of chromosomes can give rise to a new species.

____

____

119. Define a species cluster.

____

____

120. Explain why an organism that is $4n$ is reproductively isolated from its $2n$ parent species.

____

____

SELF-TEST

1. Two populations that can and do interbreed in nature and produce fully fertile offspring would exemplify the ________. [p. 396]

   a. Biological Species Concept
   b. Morphological Species Concept
   c. Phylogenetic Species Concept
   d. All of the above

2. Which of the following species concepts relies mainly on visible anatomical characters to define species? [p. 399]

   a. Biological Species Concept
   b. Morphological Species Concept
   c. Phylogenetic Species Concept
   d. All of the above

3. Local variants of species that interbreed where their distributions overlap are called ________. [p. 403]

   a. Races
   b. Clines
   c. Subspecies
   d. (a) and (c)

4. Traits that vary along an environmental gradient represent a ________. [p. 404]

   a. Cline
   b. Hybrid zone
   c. Reproductive isolating mechanism
   d. Subspecies

5. The elaborate courtship displays of many animal species have evolved as ________. [p. 405]

   a. Mechanical isolating mechanisms
   b. Ecological isolating mechanisms
   c. Behavioural isolating mechanisms
   d. Temporal isolating mechanisms

6. In some cases, the genitalia of closely related insect species do not fit and, thus, prevent successful mating. This is an example of ________. [p. 406]

   a. Behavioural isolation
   b. Ecological isolation
   c. Hybrid inviability
   d. Mechanical isolation

7. Mules, the hybrid offspring of horses and donkeys, are sterile. This is the result of ________. [p. 406]

   a. A postzygotic isolating mechanism
   b. A prezygotic isolating mechanism
   c. Allopolyploidy
   d. Clinal variation

8. An earthquake causes the course of a river to change and separates a subpopulation of mammals from the main body of the population. Over time, this isolated population accumulates enough genetic variation to be considered a new species. This is an example of ________. [pp. 407–408]

   a. Allopatric speciation
   b. Autopolyploidy
   c. Parapatric speciation
   d. Sympatric speciation

9. Polyploidy is a mechanism that can allow ________ speciation to occur. [pp. 413–414]

   a. Allopatric
   b. Nongenetic
   c. Parapatric
   d. Sympatric

10. Which of the following statements is false? [p. 413]

    a. Autopolyploidy is common in animals
    b. Chromosomal rearrangements may foster speciation
    c. Speciation may occur between adjacent populations
    d. Sympatric speciation may occur within a continuously distributed population

11. Which of the following can affect gene flow? [pp. 399–401]

    a. The ecology of a habitat
    b. Habitat availability
    c. Behaviour
    d. All of the above

12. Which of the following is not a synonym for subspecies? [p. 404]

    a. Race
    b. Cline
    c. Variant
    d. Cultivar

13. Which of the following describes a species that spreads through an ecologically varying region surrounding an uninhabitable zone with the accumulation of habitat-specific traits among the different populations but where the overlapping populations can breed and give fertile, fully viable hybrid offspring? [pp. 403–404]

    a. Cline
    b. Race
    c. Ring species
    d. Species cluster

14. Which of the following describes a trait that exhibits a gradual and continuous variation over a large, environmentally diverse region? [p. 404]

    a. Cline
    b. Race
    c. Ring species
    d. Species cluster

15. Which of the following describes a species that spreads into isolated pockets, which allows the evolution of behavioural differences that prevent the mating upon secondary contact even though the different individuals can breed and give fertile, fully viable hybrid offspring in captivity? [p. 413]

    a. Different species
    b. Race
    c. Ring species
    d. Species cluster

16. Which of the following statements is true? [pp. 413–414]

    a. Polyploidy is common in some animals and plants
    b. Plant breeders use certain chemicals to create polyploidy plant species
    c. Polyploidy is a temporary and harmful condition
    d. Alloploidy can result in speciation only if the polyploidy individuals are geographically isolated from the parent species

17. Humans and chimps may have evolved from their recent common ancestors as a result of _________. [pp. 414–416]

    a. Geographic isolation
    b. Temporary polyploidy
    c. Chromosomal alterations
    d. Mutations

18. Which of the following is true? [pp. 415–416]

    a. Chromosomal alterations can be visualized with certain dyes
    b. Chromosomal alterations can only be identified by DNA sequence analysis
    c. Unless chromosomal inversions include the centromeric region, they will not have an identifiable effect on the organism.
    d. The degree of genetic divergence between two species is directly related to the phenotypic differences between those species

19. Which of the following is true? [pp. 414–415]

    a. Sterile autopolyploids can be more robust compared to their diploid parents
    b. Sterile allopolyploids can have a higher relative fitness than their diploid parents
    c. Polyploid crop plants are rare
    d. The wheat strain commonly used for bread making was created by hybridization and allopolyploidy possibly thousands of years ago

20. Which of the following does not accurately describe the process of speciation? [p. 416]

    a. It is a highly dynamic process
    b. It requires geographic isolation and adaptation to differing environments
    c. It can happen in a single generation
    d. There is more than one way to define a species

INTEGRATING AND APPLYING KEY CONCEPTS

1. Most of the small number of "speciation genes" that have been identified so far affect postzygotic reproductive isolating mechanisms. Speculate on why this might be so and why it may be more difficult to identify genes that contribute to the evolution of prezygotic reproductive isolating mechanisms.

2. Studies of fruit flies in Hawaii indicate that flies from older islands in the Hawaiian archipelago colonized the young island of Hawaii at least 19 different times. Describe the mechanisms of speciation that would have been involved and speculate on how sexual selection could have hastened the evolution of prezygotic isolating mechanism and, hence, speciation.

3. If you observed organisms in a region surrounding an uninhabitable zone and they seemed to be morphologically similar but not identical, how would you determine if these were all subspecies of a ring species or had evolved to be separate species?

OPTIONAL ACTIVITY

1. The beginning of this chapter discusses the lack of agreement between biologists on the definition of "species." Although four definitions were presented here, there are at least eight other prominent species concepts. Do some research on the other species concepts and find out whether, with all of these, reproductive isolation is an essential or coincidental component of the concept.

19 Evolution and Classification

CHAPTER HIGHLIGHTS

- The study of systematics aims to classify species and reconstruct their evolutionary history.
- Organisms are organized into a hierarchical classification system.
- Morphological characters provide clues about evolutionary relationships.
- Cladistics uses shared derived characters to trace evolutionary history.
- Molecular characteristics have clarified many evolutionary relationships.

STUDY STRATEGIES AND LEARNING OUTCOMES

- This chapter has a lot of new terms and concepts. Make sure you pay attention to the subject headings and examples that demonstrate each concept or process. Between both, this should help you keep things straight.
- Pay particular attention to the various phylogenetic trees that are depicted, noting the specific information that each provides.
- As always, go through one section of the textbook at a time and then work through the companion section of the study guide.

By the end of this chapter, you should be able to

- Understand the problem with determining phylogeny and relationships based strictly on morphology or observed similarities
- Understand what is meant by the terms "Linnaean classification," "systematics," "phylogeny," and "taxonomy"
- Understand how phylogenetic trees are derived, using both the traditional evolutionary systematic approach and the cladistic approach; you should be able to explain why cladistics is now more widely used than the traditional approach and be able to derive a simple phylogenic tree given some minimal information
- Explain the process of molecular phylogeny and why, giving examples, it often sheds light on relationships that previously could not be determined.

TOPIC MAP

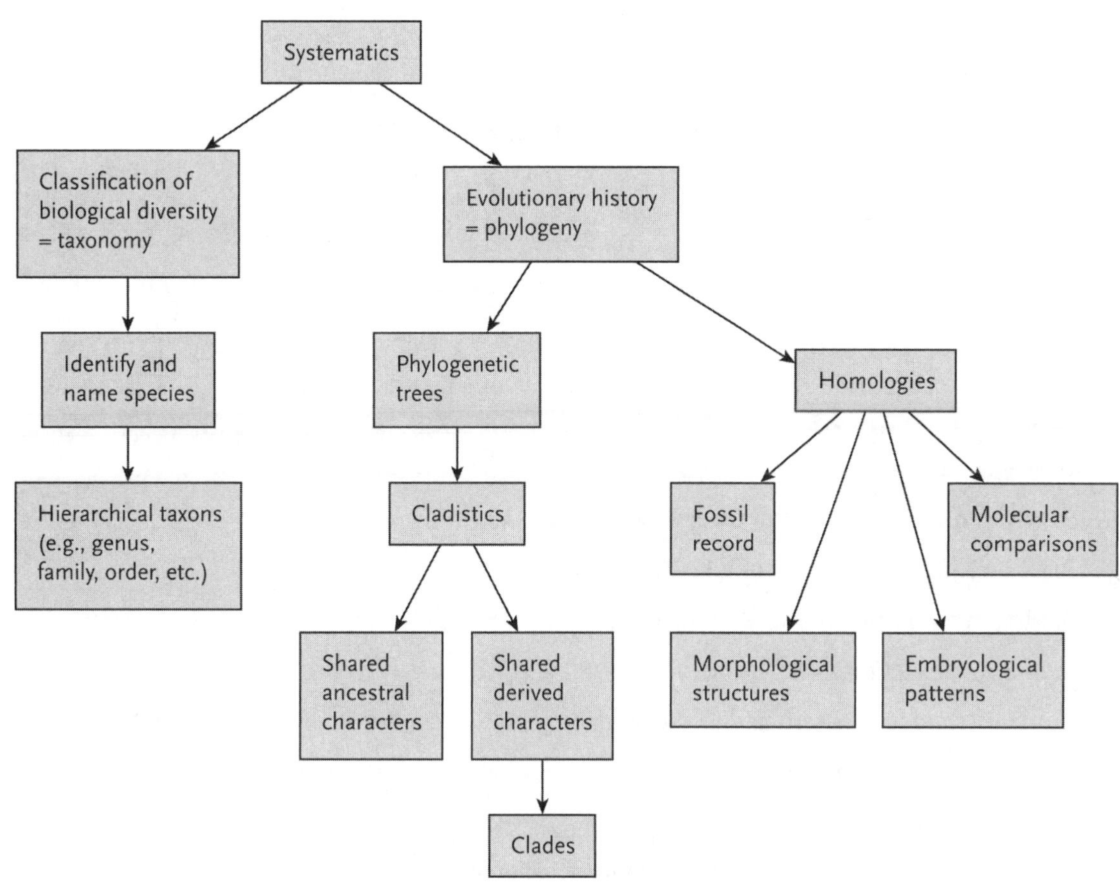

Systematics
Classification of biological diversity = taxonomy
Evolutionary history = phylogeny
Identify and name species
Phylogenetic trees
Homologies
Hierarchical taxons (e.g., genus, family, order, etc.)
Cladistics
Fossil record
Molecular comparisons
Shared ancestral characters
Shared derived characters
Morphological structures
Embryological patterns
Clades

INTERACTIVE EXERCISES

Why It Matters [pp. 419–421]

19.1 The Significance of Similarities and Differences [p. 422]

Section Review (Fill-in-the-Blanks)

Organisms living under the same (1) ______ tend to evolve the same body forms: this is called (2) ______ evolution if the two organisms are closely related and (3) _______ evolution if they are more distantly related. One cannot assume, therefore, that organisms that look alike are, in fact, closely related: the similarities between edible true morels and the (4) _____ false morels provide an example.

19.2 Systematic Biology: An Overview [pp. 422–424]

19.3 The Linnaean System of Classification [p. 424]

19.4 From Classification to Phylogeny [p. 424]

Section Review (Fill-in-the-Blanks)

(5) ______ biology, classification, and taxonomy help organize the living world. One goal of systematics is to reconstruct the (6) ______, or evolutionary history of organisms; such relationships are illustrated using (7) _____ _______; these can be used to distinguish similarities inherited from a common ancestor from those that evolved (8) _______ (e.g., wings). The second goal is the identification and naming of species, which is called (9) ______. A(n) (10) _______ is an arrangement of organisms into hierarchical groups that reflect the relatedness of the organisms; these generally reflect the (11) ______ history of the group of organisms in question. The (12) _______ system of classification was developed in the 1700s by Carl von Linné and is still used today. Organisms are grouped into increasingly inclusive categories, so this system demonstrates (13) _______ hierarchy. According to his system, the most specific (14) _____ is the species and the most inclusive is the (15) ______. While he based his classifications on organismal traits, mainly (16) _______, more recently, other criteria have been incorporated into classification schemes: these include chromosomal anatomy, physiology, morphology of (17) ______ structures, cells and (18) _______ structures, gene (19) _______, and even patterns of behaviour.

Sequence

Arrange each of the following taxa in the correct sequence, from the most inclusive group to the least inclusive.

20. ____	A. Family
21. ____	B. Phylum
22. ____	C. Genus
23. ____	D. Order
24. ____	E. Domain
25 ____	F. Species
26. ____	G. Kingdom
27. ____	H. Class

Matching

Match each of the following with its correct definition.

28.	____	Species	A.	An illustration that identifies likely evolutionary relationships among species
29.	____	Taxon	B.	Similar traits evolved by distantly related organisms living under the same conditions
30.	____	Carnivorous flowers	C.	The science devoted to naming and organizing organisms
31.	____	Phylogeny	D.	A poisonous mushroom that resembles the edible chanterelles
32.	____	Taxonomy	E.	The science of determining evolutionary histories of organisms and naming and classifying those organisms
33.	____	Taxonomic hierarchy	F.	Similar traits evolved by closely related organisms living under the same conditions
34.	____	Phylogenetic tree	G.	An arrangement of organisms into ever more inclusive groups
35.	____	Jack-o'-lantern	H.	The group of organisms at a given taxonomic level
36.	____	Convergent evolution	I.	The least inclusive of Linnaean taxa
37.	____	Parallel evolution	J.	A group of organisms that trap insects as a source of nitrogen and that do not all share the same recent common ancestor
38.	____	Systematics	K.	The science of determining evolutionary histories of organisms

True/False

Mark if the statement is true or false. If it is false, justify your answer in the line below each statement.

39. __________The evolution of flight is convergent between vertebrates and parallel between insects and vertebrates.

____

____

40. __________ All life on Earth can be classified into three kingdoms.

____

____

Short Answer

41. Explain why phylogenetic trees are considered to be scientific hypotheses.

____

____

19.5 Evaluating Systematic Characters [pp. 424–427]

19.6 Phylogenetic Inference and Classification [pp. 427–429]

Section Review (Fill-in-the-Blanks)

In order to prepare a phylogeny, two important criteria must be met: the phenotypic variation must reflect (42) ______ differences rather than environmental ones and the traits must be genetically (43) _______, that is, reflecting different parts of the organism's genome. (44) _______ characters are those inherited from a common ancestor and therefore reflect underlying (45) _______ similarities. Phenotypic similarities that evolved independently in different lineages are called (46) ______ or homoplasious characters. Because the latter do not have a common (47) _______ basis, they are not used when determining common ancestry and relatedness. The term (48) ______ evolution refers to the fact that some characters evolve slowly while others evolve rapidly. As a result, every species displays old forms of traits, that is, (49) ______ characters, and new forms of traits, that is, (50) _______ characters that are inherited by all subsequent descendents, and is therefore a marker for that (51) _____. In (52) ______ comparison, characters identified in the group under study are compared to distantly related species not included in the analysis. A (53) ______ taxon includes a single ancestral species and all of its descendants. In contrast, a (54) _______ taxon includes species from separate evolutionary lineages and a (55)

______ taxon includes an ancestor and some but not all of its descendants. The (56) _______ ________ _______ holds that because a particular evolutionary change is rare, it is unlikely that the change evolved more than once in the same lineage. Traditional evolutionary systematics groups together species that share both ancestral and (57) _______ characters, and classifications therefore reflect both (58) _______ branching and (59) ________ divergence. (60) ______ is a classification approach based solely on evolutionary relationships. In this approach, species that share derived characters and form a monophyletic lineage are grouped together in a (61) ______. Trees that illustrate hypothesized evolutionary branching are called (62) ______ and are built using the principle of (63) _______. The advantage with this system is the clarity arising from basing the classification on a single phenomenon, that is, (64) _______ ________.

Matching

Match each of the following terms with its correct definition.

65.	____	Cladogram	A.	A phenomenon that describes how some characters evolve slowly and others evolve rapidly
66.	____	Outgroup comparison	B.	Traits that have remained unchanged in a species
67.	____	Homoplasies	C.	A group that possesses a unique set of derived characters
68.	____	Mosaic evolution	D.	The assignment to a taxon based on presumed evolution from a single ancestral species
69.	____	Clade	E.	Taxon consisting of several evolutionary lines and not including a common ancestor
70.	____	PhlyoCode	F.	The idea that the same change evolved more than once in a lineage is extremely unlikely
71.	____	Principle of monophyly	G.	A tree that illustrates a hypothesized sequence of evolutionary branchings
72.	____	Polyphyletic	H.	Similarities that result from shared ancestry
73.	____	Paraphyletic	I.	Traits not present in ancestors
74.	____	Homologies	J.	An alternative system for identification and naming based on clades rather than familiar taxonomic levels
75.	____	Ancestral characteristics	K.	A technique that compares a group under study to a distantly related species
76.	____	Derived traits	L.	Taxon consisting of a common ancestor and some but not all of its descendants
77.	____	Principle of parsimony	M.	Similarities that evolved independently in different lineages

True/False

Mark if the statement is true or false. If it is false, justify your answer in the line below each statement.

78. ___________ Cladistics is based on derived and ancestral features.

____

____

79. ___________ Homologous characters can differ markedly in form and function.

____

____

80. ___________ Depending upon whether you are comparing birds to insects or birds to bats, their wings could be considered homologous or analogous.

____

____

81. ___________ The PhyloCode system divides organisms into domains, kingdoms, phyla, classes, and orders but not families, genera, and species.

____

____

Short Answer

82. Explain how embryonic developmental patterns are related to homologous characters.

____

____

83. Explain why derived traits provide more useful information about evolutionary relatedness than ancestral traits.

____

____

Complete the Table

84. A particular character may be considered ancestral or derived, depending upon what organisms are being considered. For the following pairwise comparisons of organisms, determine if the vertebral column is derived or ancestral.

Paired comparison	*Presence/absence of vertebral column*	*Ancestral or derived trait*
Birds: insects	Yes/no	A.
Humans: chimps	Yes/yes	B.
Dogs: cats	Yes/yes	C.
Crocodiles: wolves	Yes/yes	D.

E. For the purposes of deriving a cladogram of all of the eight organisms in the above table, the outgroup would be ________.

19.7 Add Molecular Data [pp. 429–434]

19.8 Clarifications from Molecular Phylogenetics [pp. 434–435]

19.9 Putting It Together [pp. 435–436]

Section Review (Fill-in-the-Blanks)

Because DNA is inherited, differences in DNA, RNA, or (85) _____ sequences of two species can serve as independent (86) ______ for determining relatedness. The advantages of using molecular sequence data include the facts that there are many (87) _____ genes even among organisms that share no morphological similarities and the sequences are not affected by developmental or (88) ______ factors. Some sequences evolve more rapidly than others: for example, mitochondrial DNA evolves relatively (89) ______, whereas chloroplast DNA evolves relatively (90) _______, a demonstration of the phenomenon of (91) ______ evolution. Regardless of whether a sequence evolves rapidly or slowly, as long as it evolves at a (92) ______ rate, one can use the degree of change as an indicator for the time of (93) _____. Large differences would therefore indicate divergence in the (94) ______ past, and minor differences would be indicative of divergence in the (95) _____ past. Such sequences therefore serve as molecular (96) ______, and if they are calibrated with the (97) _______ record, they can give precise information on the timing of evolutionary (98) ______. Computer programs are used to precisely (99) _____ sequence data, taking into account insertions and deletions. Similarities and differences are then used to construct (100) _______ ________, often using a statistical approach called the (101) _______ ________ methods in order to determine the likelihood of a particular sequence change being a shared (102) ________ trait (i.e., inherited from a common ancestor) rather than having arisen independently. Such independently arising changes are more likely with DNA sequences than protein sequences as there are only (103) ______ possible bases for any position in the sequence. Molecular phylogenetics has revolutionized our view of the (104) _______ _________ _________ because it has provided resolution of the relationships between the prokaryotes, the protists, and other extant life forms. This revolution is the result of the work of a scientist named (105) _____ ______ who used rRNA sequence comparisons to derive a phylogenetic tree that divides living organisms into three (106) ______: Bacteria, Archaea, and Eukarya. Within the Eukarya are found the familiar animals, plants, and fungi as well as many lineages formerly included in the group (107) _______.

True/False

Mark if the statement is true or false. If it is false, justify your answer in the line below each statement.

108. ___________ Sequence data from mitochondrial DNA are more useful for dating evolutionary divergence from the distant past.

____

____

109. ___________ Prior to the development of molecular phylogenetics, the eukaryotes were subdivided into four monophyletic kingdoms, but the prokaryotes were assigned to a single polyphyletic kingdom.

____

____

Short Answer

110. Explain how the analysis of rRNA was able to provide a far more accurate view of the tree of life than had previously been possible.

____

____

111. Describe what is meant by the term "molecular clock."

____

____

SELF-TEST

1. The taxon that comprises related species is _________. [p. 424]

   a. Class
   b. Family
   c. Genus
   d. Order

2. Systematics is the study of ____. [pp. 422–423]

   a. The evolutionary history, identification, and naming of organisms
   b. The evolutionary history of organisms
   c. The identification and naming or organisms
   d. The development of organisms

3. Phenotypic similarities that evolved independently in different lineages are called ________. [p. 425]

   a. Homologous characters
   b. Evolutionary mosaics
   c. Feathers
   d. Analogous characters

4. Evidence indicates that these are derived characters. [pp. 426–427]

   a. Vertebral column
   b. Insects with six walking legs
   c. Insects with four walking legs
   d. (a) and (c)

5. Cladistics would consider which of the following when classifying organisms? [p. 429]

   a. Shared ancestral characters
   b. Shared derived characters
   c. A combination of ancestral and derived characters
   d. Morphology

6. Molecular systematists conduct phylogentic analyses using which kinds of molecules? [p. 429]

   a. Protein
   b. DNA
   c. RNA
   d. All of the above

7. Why are changes in the amino acid sequence of a protein of more value in determining evolutionary divergence than DNA base changes? [pp. 429–431]

   a. Protein sequences evolve at a more constant rate compared to DNA sequences
   b. Because there are only four bases but 20 amino acids, there is a higher likelihood that similarities in DNA sequence arose independently
   c. Because there are only four bases but 20 amino acids, there is a higher likelihood that similarities in protein sequence arose independently
   d. Translation is more accurate than DNA replication

8. Molecular phylogentics has identified which three major lineages? [pp. 434–435]

   a. Prokarya, Eukarya, Protista
   b. Eukarya, Fungi, Bacteria
   c. Bacteria, Archaea, Eukarya
   d. Protista, Monera, Plantae
   e. Archaea, Animalia, Plantae

9. The hyomandibula, a bone that braced the lower jaw in early jawed fishes, is homologous to the ____ of tetrapod vertebrates. [p. 425]

   a. Inner ear
   b. Mandible
   c. Stapes
   d. Femur

For questions 10 to 14, refer to the following figure.

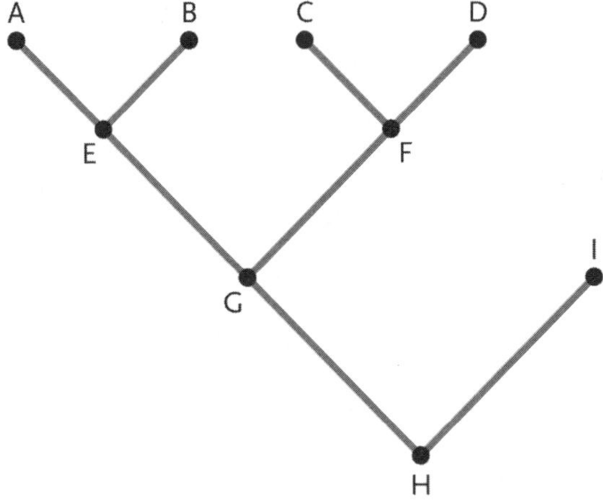

10. Species C and D are part of a monophyletic taxon. Which is the ancestor for this taxon? [p. 427]

    a. Species E
    b. Species F
    c. Species A
    d. Species B

11. Species B and C are part of a taxon that is best described as ____. [p. 427]

    a. Monophyletic
    b. Polyphyletic
    c. Paraphyletic

12. What is the common ancestor for species A and B? [p. 427]

    a. Species E
    b. Species G
    c. Species H

13. In order to determine which shared characteristics are derived versus ancestral when comparing organisms A, B, and C, which would be a likely outgroup? [p. 427]

    a. Species E
    b. Species G
    c. Species H
    d. Species I

14. If this tree were derived using cladistics, the following would be true: [p. 429]

    a. The various branches would be monophyletic
    b. The phylogenetic tree would be based on evolutionary and morphological divergence
    c. The various species (A–I) could be PhyloCode designations rather than species, genera, families, etc.
    d. (a) and (c)

15. Which of the following statements about carnivorous plants is false? [pp. 422, 435–436]

    a. They are monophyletic
    b. They are examples of parallel evolution
    c. They are distinguished by molecular phylogeny
    d. They use different types of insect traps

16. Tunas, sharks, and dolphins are examples of _______. [p. 419]

    a. Convergent evolution
    b. Divergent evolution
    c. Parallel evolution
    d. (a) or (c) depending upon the evolutionary relatedness

17. Which of the following statements about the animal *Castor canadensis* is not true? [pp. 422, 424]

    a. It is the classification of the Canadian beaver
    b. It is closely related to the "Mesozoic beaver" identified by fossil evidence from China
    c. *Canadensis* is the species name
    d. *Castor* is the genus name

18. Which of the following statements is true? [p. 424]

    a. The class taxon is more inclusive than the phylum taxon
    b. The family taxon is more specific than the genus taxon
    c. Pine trees all belong to the genus *Pinus*
    d. All prokaryotes belong to the same domain

19. Homologous characters are _______. [pp. 425–426]

    a. Indicators of common ancestry
    b. Indicators of genetic relatedness
    c. Affected by environmental conditions
    d. The product of comparable embryonic structures
    e. All of the above except (c)

20. Comparing a phylogenetic tree for four species, A to D, if both agreed with the evidence but the first tree had less branches than the second, one would say that ________. [p. 427]
    a. The first was most likely to be corrrect
    b. The second was most likely to be correct
    c. The one based on molecular data would be more likely to be correct

INTEGRATING AND APPLYING KEY CONCEPTS

1. Given that bacteria were first discovered more than 300 years ago, why is it only with the advent of molecular phylogeny that the polyphyletic nature of the Monera has come to light?

2. Remembering what you learned about transcription and translation, why is rRNA such a good molecular clock and mRNA is not?

3. Construct a cladogram based on the information below and identify (a) which organism would be the outgroup and (b) for the most right-hand branch, the derived trait.

Species	*Vertebral column*	*Swim bladder or lungs*	*Mammary glands*	*Extraembryonic membrane*
Lancelets	–	–	–	–
Amphibians	+	+	–	–
Birds	+	+	–	+
Lizards	+	+	–	+
Mammals	+	+	+	+
Sharks	+	–	–	–

OPTIONAL ACTIVITY

1. You may have noted in the chapters you have read mentions of the DNA Barcode Initiative, which began in 2003 as a research program in the laboratory of Paul Hébert at the University of Guelph, in Guelph, Ontario. It has since evolved into the International Barcode of Life Initiative (iBOL). Canada is still a major participant, providing DNA sequencing facilities at the Canadian Centre for DNA Barcoding (CCDB), located in the Biodiversity Institute of Ontario on the University of Guelph campus. This institute is also the headquarters for the researchers in the Canadian Barcode of Life Network. Use the Internet to find out more about the goals of this project, what exactly a DNA barcode is, the CCDB, and the international scope of this phylogenetic undertaking.

20 Darwin, Fossils, and Developmental Biology

CHAPTER HIGHLIGHTS

- Darwin's theory of evolution by natural selection was preceded by the studies and publications of many others, beginning with the classifications of the natural world by the Greek philosopher Aristotle. These all set the stage for the general acceptance of his theory.
- Macroevolution is simply microevolution occurring over vast periods of time resulting in large-scale change in morphology and species diversity.
- Much of the evidence for macroevolution comes from the analysis of fossils by paleobiologists.
- Earth's history includes many changes in its physical environment, often the result of plate textonics, but, as a result, Earth's history also includes many changes in the distribution and diversity of organisms.
- Convergent evolution results in distantly related organisms adapting in similar ways to similar selection pressures.
- The rate of macroevolution may be gradual, as one species slowly undergoes change in response to changing environmental conditions, or may be rapid, with one species diverging into two of more descendant species.
- Increases in biodiversity (speciation) and decreases (extinction) have occurred frequently in evolutionary history.
- In recent years, the study of macroevolution has emphasized evolutionary developmental biology (evo-devo).

STUDY STRATEGIES AND LEARNING OUTCOMES

- Remember that macroevolution is essentially the same as microevolution but over much longer time periods with large-scale changes. Keep checking the topic map to help you maintain this perspective.
- Make sure you pay attention to the subject headings and especially the examples that demonstrate each concept or process.
- As always, go through one section of the textbook at a time and then work through the companion section of the study guide.

By the end of this chapter, you should be able to

- Provide the names and contributions of the various people who contributed to the world-view of evolutionary change
- Understand that Earth is a dynamic planet and that changes in its physical properties (e.g., climatic change, movement of continents) result in changes in species composition and diversity. New species evolve and existing species go extinct.
- Understand the role of paleobiology in our knowledge of evolutionary theory, including why the fossil record paints an incomplete picture
- Understand why convergent evolution results in superficial similarities among distantly related species
- Identify and be familiar with both types of evolutionary change, anagenesis and cladogenesis, as well as their underpinnings: the gradualist hypothesis and the punctuated equilibrium hypothesis
- Identify some of the macroevolutionary trends in morphology
- Understand how changes in the timing of various developmental events can lead to evolutionary change

TOPIC MAP

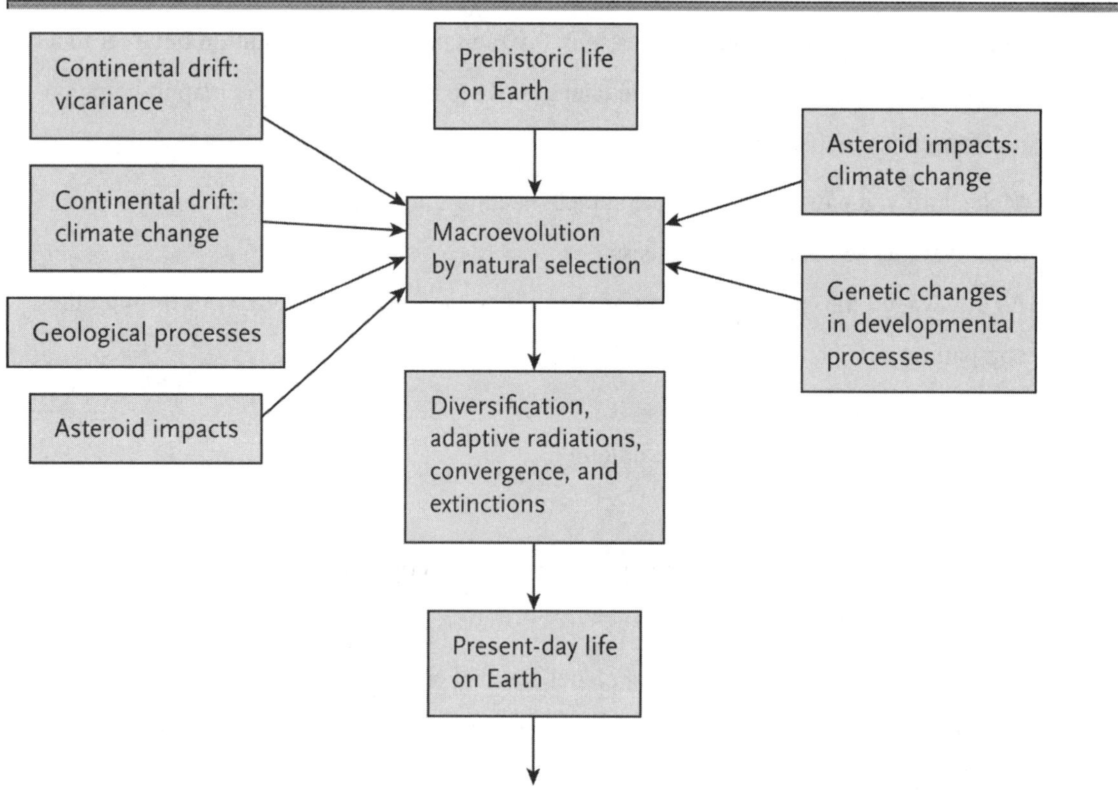

INTERACTIVE EXERCISES

Why It Matters [pp. 439–441]

20.1 Recognition of Evolutionary Change [pp. 441–442]

20.2 Changes in Earth [p. 442]

20.3 Charles Darwin [pp. 442–445]

Section Review (Fill-in-the-Blanks)

The Greek philosopher (1) ______ was the first to create a ladder-like classification of nature, from simplest to most complex; this ladder was the precursor to the classification of Linnaeus. Centuries later, Sir Francis Bacon established the importance of the three critical aspects of scientific enquiry: (2)______, (3) _______, and (4) ______ reasoning. Subsequently, three new scientific disciplines arose that reflected a growing awareness of change in the natural world: (5) ______, comparative morphology, and (6) _____. In the 1700s, French scientist (7) ___________ became the first to question the idea that organisms were "created" in perfect form and never changed, proposing that the useless (8) _______ structures present in some animals had some function in ancestral species. In 1796, Georges Cuvier added to this idea of change, suggesting that layers of (9) _______ represented species that had gone extinct and that abrupt changes between geological strata resulted from dramatic shifts in ancient (10) _______, giving rise to the theory of "catastrophism." At the same time, Jean Baptiste de Lamarck proposed two mechanisms for evolutionary change: the principle of (11) ______ and ______, whereby body parts grow or shrink based on how much they are used and that these changes are inherited by the offspring through his proposed "inheritance of (12) ________ ________." Although subsequently shown to be invalid, Lamarck's contributions beyond these two mechanisms included four important ideas: all species (13) ______ through time; these are inherited by the next generation; organisms change in response to their (14) ______; and there are specific (15) _______ that underlie these evolutionary changes. During the same time period, Scottish geologist James Hutton proposed the theory of (16) ______, that is, that Earth changed slowly over its history. This theory was expanded by English geologist (17) _________, who proposed the concept of (18) _________, which stated that the geological processes that sculpted Earth's surface over geological time scales are the same as modern-day processes: volcanism, earthquakes, erosion, and the formation and movement of (19) _______. Subsequently, many years after his 5-year voyage on the *HMS Beagle*, (20) __________ reported his observations of many variations among living organisms and the theory that variations were natural components of (21) _______ and that only those individuals with the more successful variations in a given environment survive and (22) _______, thus giving rise to a change in the population's characteristics in subsequent (23) _______. His theory of evolution through natural selection paralleled a proposal by (24) _____________, and both theories were presented to the Linnaean Society of London on July 1, 1858. A few months later, Darwin published his book, (25) _______________________.

Complete the Table

26. Complete the following table with the scientist's name or contribution to the recognition of evolutionary change.

Name	*Contribution*
A.	British economist whose essay on the inevitable deaths from starvation as a result of a higher rate of population growth versus agricultural growth in England provided a model for Darwin's theory of natural selection
Charles Lyell	B.
Alfred Russell Wallace	C.
D.	Proposed the mechanisms of "use and disuse" and the "inheritance of acquired characteristics"
James Hutton	E.
F.	Proposed that "organisms were conceived by Nature and produced by Time"
G.	Developed the "Great Chain of Being"
Georges Cuvier	H.

Short Answer

27. Briefly explain how Darwin's readings and observations led him to propose a mechanism of selective breeding in nature.

____

____

28. Explain what is meant by Darwin's idea of "descent with modification."

____

____

20.4 The Fossil Record [pp. 445–447]

Section Review (Fill-in-the-Blanks)

(29) ______ is the study of fossils and is our primary source of data about the evolutionary history of organisms. Fossils can be formed in a number of ways, but most commonly they are the result of (30) _______ of parts of organisms. Ideal conditions for fossil formation include low (31) ______ levels and high (32) _____. Hard structures such as the (33) _______, bones, and shells of animals and the (34) _______ and wood of plants are more likely to form fossils through the slow replacement of the original material by dissolved (35) ______. Some fossils are formed as casts or (36) _______. In environments where oxygen is scarce, (37) _______ does not occur, so even soft-bodied organisms may be preserved. Unfortunately, only certain types of environment give rise to fossils, so the fossil record is extremely (38) ______. There are a number of methods used to estimate the relative age of fossils, but one technique, (39) _______ _______, allows paleobiologists to precisely determine the age of fossils by measuring the amount of an unstable isotope in a fossil and knowing the (40) _____-_____ of that isotope. Fossils that still contain organic material can be dated by determining the ratio of (41) ______:______. Relative dating is done by comparing (42) ________ geological strata to (43) ________ strata.

True/False

Mark if the statement is true or false. If it is false, justify your answer in the line below each statement.

44. _______ Fossils form when organisms are buried under sediment or are preserved in mountain forests.

____

____

45. ________ The fossil record provides a complete record of life throughout evolutionary history.

____

____

46. ________ The half-life of ^{14}C is approximately 5600 years. A piece of wood with one-quarter of the ^{14}C found in living specimens would therefore be roughly 11 200 years old.

____

____

Short Answer

47. Explain why soft-bodied animals are preserved in peat bogs, tar pits, and glacial ice.

____

____

48. What is the relevance of amber to paleontology?

____

____

20.5 Earth History, Biogeography, and Convergent Evolution [pp. 447–450]

Section Review (Fill-in-the-Blanks)

Living organisms are affected by prevailing climates and (49) _______ ________. These, in turn, are affected by shifts in geography, the most dramatic of which result from plate (50) _______, that is, the movement of pieces of Earth's crust on its semisolid (51) _______. This movement underlies the phenomenon known as (52) ______ ________ and explains the previous gradual changes in Earth's climate; both of these have substantially impacted the evolution of the diversity of life. In addition, catastrophic events such as massive volcanic eruptions or (53) _______ _______ can result in massive extinctions over relatively short geological time scales. (54) ________ is the study of the distribution of living and ancient organisms over the face of Earth, and (55) ________ biogeographers attempt to determine how the biogeography of organisms has developed over evolutionary time spans. Species that live in all suitable habitats throughout a geographic area are said to have a (56) ______ distribution and require no historical explanation. On the other hand, the phenomenon of (57) ______ distribution refers to organisms that occupy widely separated areas and is explained by two historical processes: (58) ______, or the movement of organisms away from their place of origin, and (59)_______, the fragmentation of a continuous distribution caused by external factors. An example of such a fragmentation is the breakup of the supercontinent, (60)______, resulting in the isolation of the continents as we know them today. Each of the modern-day continents has a different mix of organisms, and their isolation and environmental conditions have led to the evolution of unique (61) ______ on each. Alfred Russell Wallace used these to define the six biogeographic (62) ______ that we still recognize today. Some of these biogeographic regions, Australia in particular, contain many (63) ______ species that occur nowhere else on Earth. However, sometimes distantly related species that are biogeographically isolated may evolve similar (64) _______ features as a result of similar environmental pressures and lifestyles. These similarities are the result of (65) _______ evolution.

True/False

If the statement is true, write a "T" in the blank. If the statement is false, make it correct by changing the underlined word(s) and writing the correct word(s) in the answer blank.

66. _____________ Historical biogeography attempts to explain the geographic distribution of organisms.

67. _____________ The distribution of magnolia trees is explained by vicariance.

68. _____________ The theory of plate tectonics helps explain continental drift and historical biogeography.

69. _____________ The redfin darter (a small fish) is found only in the rivers of the Ozarks; thus, it is endemic to the region.

70. _____________ The ring-billed gull is found in all coastal regions of the continental U.S. Its distribution is disjunct.

Short Answer

71. Provide the explanation for the morphological similarities between many fish and aquatic mammals.

____

____

20.6 Interpreting Evolutionary Lineages [pp. 451–454]

20.7 Macroevolutionary Trends in Morphology [pp. 455–457]

Section Review (Fill-in-the-Blanks)

The evolution and adaptive radiation of horses reveal that the evolution of a species is not (72) ______: many previous species have arisen and then subsequently become (73) ______. Possession of (74) ______ traits may not be evidence of direct ancestry. The gradual evolutionary transformation of an existing species into a new one is called (75) ______ and does not increase the number of existing species. In contrast, (76) ______ is the evolution of two or more descendant species from a single ancestral species, giving rise to an increase in the number of existing species. The rate of change in the former reflects the (77) ______ hypothesis; that is, speciation results from the slow, continuous accumulation of small changes at a steady rate. According to the (78) ______ ______ hypothesis, macroevolutionary change occurs rapidly when a species undergoes (79) ______. There are examples of both in the (80) ______ record. Some evolutionary lineages exhibit a trend toward greater size and (81) _______, while others exhibit the development of novel (82)_______. The nineteenth-century paleobiologist (83) _______________ noticed that vertebrates generally exhibit the former trend: this is now known as *Cope's Rule*. Novel morphological structures often appear suddenly in the fossil record, and one explanation for this is (84) _______; a trait that serves one function is later co-opted for another. Another mechanism that may contribute to speciation is called (85) ________ growth and is based on differential growth rates of specific anatomical structures. Yet another mechanism leading to differences in species is a change in the timing of developmental events, or (86) _______. At least two phenomena can be involved: (87) ______ and (88) ______. The former occurs when larvae acquire the ability to (89) ______, a trait normally associated with adults, not the juvenile larval stage. The latter occurs when adults retain (90) _______ _______ and may arise from a simple mutation.

Building Vocabulary

Using the definitions of the prefixes below, assign the appropriate prefix to the appropriate suffix for the following:

Prefixes	Meaning
allo-	different
clado-	a branch
hetero-	different
pedo-	child

Prefix	Suffix	Definition
	-morphosis	91. Retention of juvenile characteristics into adulthood
	-chrony	92. Changes in timing of developmental events
	-metric	93. Different relative sizes of structures in related species
	-genesis	94. Evolution of two or more descendant species from a common ancestor

Matching

Match each of the following names or terms with the correct concept or example.

95.	____	Preadaptation	A.	An evolutionary trend toward greater size in a lineage
96.	____	Allometric growth	B.	Punctuated equilibrium
97.	____	Phylogeny of horse species	C.	Some species of amphibians retain gills, a juvenile trait, into adulthood, possibly due to a mutation in a gene for metamorphosis.
98.	____	Edward Drinker Cope	D.	Feathers and winglike forelimbs may have originally evolved among some dinosaurs as a means of capturing prey and conserving body heat.
99.	____	Pedomorphosis	E.	The skulls of newborn chimps resemble those of newborn humans; however, by the time skull development is complete, significant changes are apparent.

True/False

Mark if the statement is true or false. If it is false, justify your answer in the line below each statement.

100. __________ Heterochrony is believed to be the explanation for the difference in flower morphology and pollination species in the more recently evolved species *Delphinium nudicaule* compared with the ancestral species *D. decorum*.

____

____

20.8 Evolutionary Developmental Biology [pp. 457–462]

20.9 Evolutionary Biology since Darwin [p. 462]

Section Review (Fill-in-the Blanks)

(101) ________ ________ ________ or "evo-devo" is the study of changes in regulatory genes that affect development and can create morphological changes. These studies show that alterations in the timing or (102) _______ of various developmental processes can lead to new forms. If the changes are due to changes in genes that regulate development, (103) _______ innovations may result. (104) _______ genes encode transcription factors that bind to regulatory sequences associated with genes that contribute to the organism's (105) ______. Several hundred of these genes are found in most animals and are referred to as the (106) "_______ _______-______" because, collectively, they govern the basic design of the body plan. The (107) _______

family of genes controls the overall body plan of animals. The genes in this family are characterized by a 180-nucleotide sequence, the (108)_______, which encodes the (109) _______ that confers the transcription factor function of the protein products. Different *Hox* genes are expressed at different positions along the (110) _____-to-_____ axis of the developing embryo. *Pax-6* is another highly conserved (111) _____-_____ gene: it controls the development of (112) _______-_______ organs and, based on the presence of a homeobox, encodes a transcription factor. Although highly conserved among organisms, tool-kit genes may have differing (113) ______ of activity at different (114) ______ in the embryonic development of different species. This may cause different sequences of (115) ________ and underlie differing morphologies. The regulatory sequences that these tool-kit gene products bind have been equated to genetic (116) ______; based on this concept, (117) _______ _______ could arise from evolutionary changes in developmental switches that cause certain body parts to grow larger or more quickly than others. Other changes in developmental switches would also provide an explanation for (118) _______ so that the development of adult characteristics is either delayed or the development of reproductive maturity is sped up.

True/False

Mark if the statement is true or false. If it is false, justify your answer in the line below each statement.

119. __________ *Pitx1* regulates genes that control the development of light-sensing organs in a wide variety of animals.

____

____

120. ___________ Evolutionary developmental biology was made possible by the evolution of the science of genomics.

____

____

121. ___________ Homeotic genes are present in multicellular organisms as well as unicellular organisms.

____

____

122.__________ The morphological differences between mice and fruit flies are the result of the evolution of different tool-kit genes.

____

____

Short Answer

123. How does evo-devo help shed light on macroevolutionary trends and adaptive radiations?

____

____

124. Explain how a single mutation combined with natural selection could provide an explanation for the differing morphologies of stickleback fish.

____

____

125. What is the evidence that the *Pax-6* gene has been conserved over the course of hundreds of millions of years?

____

____

SELF-TEST

1. Natural theology refers to ______. [p. 441]

   a. The scientific pursuit of classifying all of God's creations prevalent in fourteenth-century Europe
   b. The study of the form and variety of organisms in their natural environments
   c. The story of creation
   d. The classification of all animals on Earth as being below the gods of the spiritual realm

2. Le Comte de Buffon explained the existence of two toes on some mammals that never touch the ground by proposing _______. [p. 441]

   a. The hypothesis of adaptive radiation
   b. The theory of gradualism
   c. The theory that some animals must have changed since their creation
   d. The theory of cladogenesis

3. The theory of catastrophism _______. [p. 441]

   a. Refers to the presence of fossils in abruptly changing geological strata
   b. Refers to the extinction of certain lineages of an organism
   c. Was proposed by Charles Lyell
   d. Is used to determine the relative ages of fossils

4. Which of the following were specific proposals made by Jean Baptiste de Lamarck that ultimately contributed to an evolutionary world-view? [pp. 441–442]

   a. Vestigial structures must have served a purpose in an ancestral species
   b. Organisms change in response to their environments
   c. Slow and continual geological processes shape and continue to shape Earth's geology
   d. All life forms are evolved from the same common ancestor

5. Darwin's idea of selection of the fittest was based on the writings of ________. [p. 444]

   a. Charles Lyell
   b. Alfred Russell Wallace
   c. Thomas Malthus
   d. Jean Baptiste de Lamarck

6. Fossils are most likely to form when _________. [p. 445]

   a. Aquatic invertebrates are exposed to the air and are preserved by dehydration
   b. Hard parts of organisms are buried under sediments in anaerobic conditions
   c. Soft-bodied organisms are buried under sediments in aerobic conditions
   d. Volcanoes erupt

7. Radiometric dating is based on the ________ of various radioactive ________. [pp. 446–447]

   a. Decay rate; isotopes
   b. Position; strata
   c. Variation; isotopes
   d. Particle size; strata

8. Continental drift is responsible for the _________ of some populations. [pp. 447–448]

   a. Continuous distribution
   b. Disjunct distribution
   c. Dispersal
   d. Preadaptation

9. Which of the following is not one of Wallace's six biogeographical realms? [p. 449]

   a. Australian
   b. Ethiopian
   c. Neotropical
   d. Arctic

10. The evolution of the horse lineage is an example of ________. [pp. 451–452]

   a. Heterochrony
   b. Pedomorphosis
   c. Cladogenesis
   d. Tectonics

11. Periods of stability followed by periods of rapid evolutionary change are cornerstones of ________. [p. 451]

   a. Convergence
   b. Gradualism
   c. Heterochrony
   d. Punctuated equilibrium

12. Cope's Rule states that there is a trend toward increasing ________ within evolutionary lineages. [p. 455]

   a. Allometric growth
   b. Pedomorphosis
   c. Size
   d. Vicariance

13. The adaptive radiation of a species into a new adaptive zone may result from ________. [p. 455]

   a. Preadaptation
   b. Convergence
   c. Evo-devo
   d. Vicariance

14. The change in relative size of various anatomical features of humans through development is an example of ________. [p. 456]

   a. Adaptive radiation
   b. Allometric growth
   c. Pedomorphosis
   d. Preadaptation

15. Homeotic genes serve as a "tool kit" that ________. [p. 457]

   a. Links regulatory genes on a single chromosome
   b. Is species specific
   c. Regulates the timing and sequence of developmental processes
   d. Encodes different morphological structures

16. Which of the following was an observation made by Darwin that underpinned his theory of evolution by natural selection? [p. 444]

    a. Individuals within a population compete for limited resources
    b. Certain hereditary characteristics allow some individuals to survive longer and reproduce more than others
    c. Many variations in an individual have a genetic basis and so are inherited by subsequent generations
    d. A population's characteristics will change over generations

17. Continental drift _______. [p. 447–449]

    a. Is explained by plate tectonics
    b. Is the result of currents in Earth's mantle
    c. Explains disjunct populations
    d. All of the above

18. Endemic species are ones that _________. [p. 449]

    a. Exhibit a continuous distribution
    b. Exhibit a disjunct distribution
    c. Are unique to a single location or region
    d. Are from lineages that became extinct

19. North American cacti and African spurges are almost identical because ______. [p. 450]

    a. They share the same recent common ancestor
    b. They are the product of convergent evolution
    c. They are different species of the same genus
    d. They are the product of a recent vicariance

20. Anagenesis refers to ________. [p. 451]

    a. The gradual change of individuals in a lineage as they adapt to changing environments
    b. The rapid diversification of a population into two or more descendant species
    c. The differential growth rates of different parts of the body
    d. Changing in the timing of developmental events

INTEGRATING AND APPLYING KEY CONCEPTS

1. Based on what you have learned about homeotic genes, why is it not surprising that organisms with serially repeating segments are most likely to show extreme organ displacement, for example, substitution of legs for antennae in Drosophila? Arthropods, which exhibit this pattern of segmentation, are the most diverse group in the animal kingdom. How might these two facts relate to each other?

2. The evolution of placental mammals resulted in the extinction of most marsupials in most parts of the world. Having learned about the impact of continental drift on the distribution of flora and fauna, what would be the explanation for the fact that marsupials are so common in Australia? What do you think would be the fate of Australian marsupials if placental animals from other parts of the world were introduced without regulation?

3. The research described on page 460, "From Fins to Fingers," indicates that tetrapod digits are a morphological novelty. Can you suggest any further experiments or observations that could determine whether fish have the genetic capacity to form digits?

OPTIONAL ACTIVITY

1. Nick Lane is a biochemist, an honourary reader at University College in London, England, and a novelist. His most recent book is entitled *Life Ascending: The Ten Great Inventions of Evolution*. In this publication, he presents what he feels are the 10 key inventions of evolution that have given rise to the great diversity we see on present-day Earth. These include DNA, photosynthesis, sex, sight, and even death, that is, adaptive radiations. Try to come up with your "top 10" adaptive radiations and then find out more about Lane's book and his proposals.

21 Prokaryotes

CHAPTER HIGHLIGHTS

- Prokaryotes
  - Have a very simple structure
  - Are metabolically more diverse than eukaryotic life forms
  - Generally reproduce asexually but have far higher genetic variability than the eukaryotes
  - Have approximately 99% of their world yet to be isolated and characterized
- The domain Bacteria
  - Is classified by molecular techniques into more than 12 evolutionary lineages
  - Includes photosynthetic bacteria that do not produce oxygen and some that are photoheterotrophs, in addition to the ancestor of the eukaryotic chloroplast
  - Includes many human, animal, and plant pathogens but many more nonpathogenic species
  - Includes bacteria that are chemoautotrophs
- The domain Archaea
  - Is classified by molecular techniques into three evolutionary lineages, one of which has only been identified through environmental isolation of DNA
  - Contains many extremophilic species: halophiles, extreme thermophiles, and psychrophiles
  - Also contains nonextremophilic species that may be symbionts of animals, insects, or humans and mesophilic species that form a major component of the marine plankton
  - Shares characteristics with common bacteria as well as eukaryotes

STUDY STRATEGIES AND LEARNING OUTCOMES

- This chapter is fairly simple: pay particular attention to the various examples of the diversity of these organisms. There are some new terms to remember; going through the companion section of the study guide will help you learn these terms.
- Since the organisms discussed here are extremely small, understanding requires some imagination.
- As always, go through one section of the textbook at a time and then work through the companion section of the study guide.

By the end of this chapter, you should be able to

- Describe the structure of prokaryotic cells and explain how Archaea and Bacteria are differentiated
- Identify the four types of metabolism in Bacteria and Archaea and the different types of oxygen requirements
- Understand why prokaryotes are critical to biogeochemical cycling, using the nitrogen cycle as an example
- Explain how bacteria can cause illness, including through formation of biofilms, and how they may resist the effects of antibiotics
- Provide examples of the different types of lineages of the Bacteria and Archaea and why that particular lineage is important in nature

TOPIC MAP

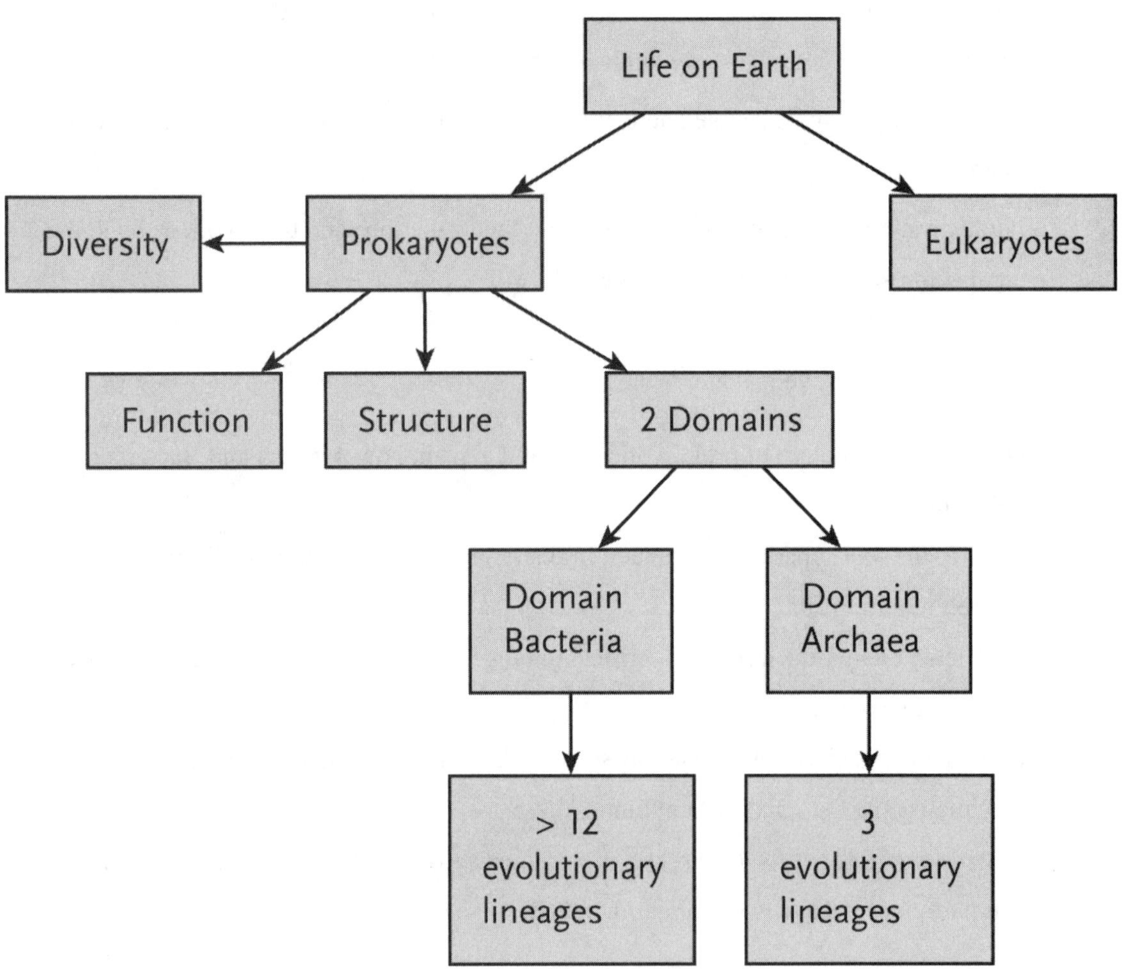
Life on Earth
Prokaryotes
Eukaryotes
Diversity
Function
Structure
2 Domains
Domain Bacteria
Domain Archaea
> 12 evolutionary lineages
3 evolutionary lineages

INTERACTIVE EXERCISES

Why It Matters [pp. 464–465]

21.1 The Full Extent of Prokaryote Diversity Is Unknown [p. 466]

21.2 Prokaryote Structure and Function [pp. 466–475]

Section Review (Fill-in-the-Blanks)

Escherichia coli is an essential component of the human large intestine and belongs to the group of organisms called (1) ______. The organism also resides in the gut of cows, and one particular strain, *E. coli* O157:H7, can cause severe human (2) ______. It may also be resistant to the (3) ______ used to treat infections. This single species demonstrates the diversity typical of this Bacteria and Archaea, collectively called the (4) _____. The Bacteria and Archaea represent two of the three (5) _______ of life. Although the most abundant and metabolically (6) ______ life forms on Earth, the true extent of their (7) ______ is unknown because of our inability to culture them in the lab. Prokaryotes are small and lack all of the membrane-bound (8) ______ seen in eukaryotic cells. They generally have a single, (9) _____ chromosome, packaged into a (10) _____, and they may also possess a smaller, circular DNA molecule called a (11) _____, which replicates independently of the chromosome and may be spread to other cells by (12) ______ gene transfer. Their ribosomes are (13) ______ than eukaryotic ribosomes, and in nonresistant bacteria, (14) _____ may act by interfering with the activity of the ribosomes. Outside the plasma membrane is the prokaryotic (15) ______ ______. A staining procedure allows bacteria to be classified into two major groups, depending upon the nature of the (16) _____ ______. The common component in both is a (17) _______ layer, which provides mechanical strength and which may be targeted by antibiotics such as penicillin. Gram- (18) ______ bacteria have a single thick (19) ______ layer, whereas Gram- (20) ______ bacteria have a thin layer of (21) ______ surrounded by an outer, lipopolysaccharide-containing (22) ______. Archaeal cell walls are much more variable and so do not stain reliably with the (23) ______ stain. Many Bacteria and Archaea are surrounded by a protective polysaccharide layer called the (24) ______; this may have a critical role in (25) ________, *Streptococcus pneumoniae* being one example. Many prokaryotes have (26) _______, which allow them to swim. Some prokaryotes also have rigid shafts of proteins on their surfaces called (27) _____ that allow them to attach to other prokaryotic cells, host cells, or help in transfer of bacterial (28) ______ between cells. In the case of *Geobacter*, they may be used to conduct (29) ______. Prokaryotes are metabolically very diverse and can be divided into four types based on their source of (30) _____ and carbon. Two of these types are also seen among the eukaryotes: (31) _______, as is typical of animals, and (32) ______, as is typical of plants. Two other types are unique to the prokaryotes: (33) _______ obtain energy from the oxidation of inorganic substances and use CO_2 as their carbon source, and (34) _______ obtain energy through photosynthesis but use organic chemicals as their carbon source. Prokaryotes are also diverse in their (35) _______ or tolerance for oxygen; in addition to the (36) ______ that, like the eukaryotes, require oxygen for energy generation, there are (37) ______ that do not require oxygen and the (38) _____ anaerobes that die upon exposure to oxygen. The (39) _______ ________ switch between the two modes of growth, depending upon oxygen availability. Because of the metabolic diversity, prokaryotes are critical for (40) _______ cycling, transforming elements from one form to another in reactions that sometimes cannot happen any other way. For example, some prokaryotes are unique among life forms in being able to reduce atmospheric nitrogen to ammonia, a process called nitrogen (41) _______. Others can perform (42) _______, that is, the conversion of ammonium to nitrate; the combination of the two reactions provides nitrogen in a usable form for the synthesis of (43) ______ and nucleic acids in all other organisms. Prokaryotes normally reproduce asexually and extremely rapidly, by (44) ______ fission, producing two genetically identical daughter cells from the parental cell. However, they are still capable of generating vast amounts of genetic variation, primarily through having a much higher (45) ______ rate than eukaryotes, but also through mechanisms of (46) _____ gene transfer, including bacterial conjugation using the sex pilus. It was through horizontal gene transfer that the pathogenic strain *E. coli* O157:H7 acquired the (47) _____ gene that makes it so deadly. The product of this gene is an exotoxin, one of many that are associated with bacterial disease; the other major type of toxin is specific to the Gram-negative bacteria: (48) ______. This molecule is a component of the outer membrane (49) _______.

Matching

Match each of the following terms related to prokaryotic cell structure and function with its correct definition.

50.	____	Biofilm	A.	Additional small DNA molecules that have supplemental genes in some prokaryotes
51.	____	Plasmid	B.	A slimy, polysaccharide substance that is secreted for biofilm formation
52.	____	Nucleoid	C.	An antibiotic that kills bacteria by interfering with peptidoglycan synthesis
53.	____	Pilus	D.	The region in the cytoplasm where the chromosome is organized
54.	____	Flagella	E.	A rigid protein structure that allows certain bacteria to attach to their host or transfer some of their DNA to another bacterial cell
55.	____	Capsule	F.	The conversion of atmospheric nitrogen to ammonia
56.	____	EPS	G.	Helical protein structures that move by rotation, providing motility to the prokaryotic cell
57.	____	Nitrogen fixation	H.	A complex aggregation of microorganisms attached to a solid surface
58.	____	Quorum sensing	I.	A polysaccharide layer that may surround prokaryotic cells
59.	____	Coccus	J.	Spherical prokaryote
60.	____	Gram	K.	A cell wall component of Gram-negative bacteria that occurs outside the peptidoglycan layer
61.	____	Penicillin	L.	Bacterial types that stain purple with the Gram staining technique and have a thick wall of peptidoglycan
62.	____	Peptidoglycan	M.	The toxic component of the Gram-negative bacterial cell wall
63.	____	Outer membrane	N.	A method of intercellular communication among bacteria that is important for regulation of gene expression and biofilm formation
64.	____	Endotoxin	O.	The rigid cell wall layer of bacteria

Complete the Table

65. Complete the following table with the appropriate name or description for processes that generate genetic variability in prokaryotes.

Process	*Description*
A.	A change in DNA sequence due to exposure to mutagens
Transformation	B.
Conjugation	C.
D.	When a virus transfers DNA from one bacterium to another to cause genetic recombination

True/False

If the statement is true, write a "T" in the blank. If the statement is false, make it correct by changing the underlined word(s) and writing the correct word(s) in the answer blank.

66. _____________ Plasmids are the main DNA molecule in all prokaryotes that contain the genes required for most metabolic activities of the cell.

67. ______________ Prokaryotic genes for such things as antibiotic resistance gene and conjugation are found on the chromosome.

68. ______________ Gram-negative bacteria stain red with the Gram staining technique and have a thin wall of peptidoglycan and an outer membrane.

69. ______________ It is estimated that we have identified only 10% of the prokaryotes that exist in nature.

Short Answer

70. Why is it not surprising that Bacteria and Archaea are so much more abundant and metabolically diverse than the eukaryotes?

____

____

71. What function does the Gram-negative outer membrane serve?

____

____

72. Identify some of the beneficial as well as detrimental roles of biofilms.

____

____

73. Identify some of the mechanisms by which bacteria may become resistant to the effects of antibiotics.

____

____

Complete the Table

74. Complete the following table with the appropriate name or description of the metabolic process in prokaryotes.

Metabolic process	*Description*
A.	Obtain energy by oxidizing inorganic substances and use CO_2 as their carbon source
Chemoheterotroph	B.
Photoautotrophs	C.
D.	Use light as their energy source and organic chemicals as their carbon source
E.	Use oxygen when present; otherwise produce energy anaerobically
F.	Do not require oxygen and cannot tolerate its presence
Obligate aerobes	G.

21.3 The Domain Bacteria [pp. 475–478]

21.4 The Domain Archaea [pp. 478–480]

Section Review (Fill-in-the-Blanks)

Based on molecular phylogeny, the bacteria are currently divided into more than a dozen (75) ______ branches. The highly diverse Gram-negative bacteria belong to the group (76) _______, a group that also includes the presumed ancestor of this lineage, the photosynthetic (77) ______ bacteria. These latter bacteria may be photoautotrophs or (78) ______; they do not produce oxygen, and they owe their colour to a unique type of (79) ______. Other members of this branch, such as *E. coli*, are chemoorganotrophs, and many are human or plant (80) _____. The mitochondrion is believed to be derived from one of the chemoheterotrophic (81) ______. The other photosynthetic bacteria are found in two other lineages: the (82) _____ _____, which may also be photoautotrophs or (83) ______ and which also do not produce (84) _____, and the (85) ______, believed to be the ancestor to the eukaryotic (86) ______ and which, like the eukaryotic organelle, are photoautotrophic and produce (87) _____. These bacteria are important to the biosphere because of their ability to fix (88) ______ gas. Another branch, that of the (89) _____ _____, includes bacteria that may be pathogenic, such as *Streptococcus* species, as well as those that are beneficial, such as the fermentative (90) _______ species. The chlamydias, another branch of the bacteria, are unusual in that they lack (91) ______ in their cell wall and are all obligate (92) ______ parasites of animals. The other prokaryotic domain, the Archaea, were originally thought to be associated only with extreme environments and were therefore often referred to as (93) _____; they have since been identified in less extreme environments. Prokaryotes in this domain have some characteristics in common with bacteria, some that are more like the (94) _____, while others, such as the (95) _____ of the plasma membrane, are unique. Based on molecular phylogeny, the Archaea are classified into three evolutionary lineages, including one that has not yet been cultured in the lab: the (96) _____. Of the other two major groups, the (97) ______ include organisms that are (98) _____, that is, they require high salt concentrations; (99) ______, which live in low oxygen environments, including as symbionts of cows and termites; and some extreme (100) ______, that is, organisms that live at very high temperatures. The third major group of Archaea, the (101) ______, vary in their optimum growth temperature: some are cold-loving (102) ______ and some are extreme (103) ______, with those that grow around deep sea vents having the highest growth temperatures of any living organism. In contrast, (104) ______ Archaea from this group constitute a major portion of marine (105) _______.

Matching

Match each of the following terms related to Archaea with its correct definition.

106.	____	Extremophile	A.	Archaea that live in high-temperature environments
107.	____	Mesophile	B.	Archaea that serve as food source to other marine organisms would be an example
108.	____	Methanogens	C.	Psychrophiles, thermophiles, and halophiles would all be examples
109.	____	Halophiles	D.	Archaea that live in high-salt environments
110.	____	Thermophiles	E.	Archaea that live in cold environments
111.	____	Psychrophiles	F.	Archaea associated with low oxygen environments such as the rumen of cows and sheep and the large intestine of dogs and humans

True/False

Mark if the statement is true or false. If it is false, justify your answer in the line below each statement.

112. ____________ Green bacteria perform the same type of photosynthesis as green plants and are believed to be the ancestor of the chloroplast.

____

____

113. ____________ *Treponema pallidum* is a coccus-shaped, Gram-positive bacterium that causes strep throat.

____

____

114. ____________ The ability of certain Archaea to survive boiling in harsh detergents is at least partly due to the unique structures of their membrane lipids.

____

____

Short Answer

115. Compare the ribosomes of Archaea and Bacteria.

____

____

116. What was one biotechnological development that is attributed to the discovery of thermophilic Archaea?

____

____

SELF-TEST

1. The reason we have only isolated and identified an estimated 1% of the prokaryotes is _______. [p. 466]

   a. Most of them are too dangerous to try to isolate
   b. Most of them grow in inaccessible places
   c. Most of them cannot yet be cultured in the lab
   d. (b) and (c)
   e. All of the above

2. Which of the following statements is false? [p. 466]

   a. The total biomass of the prokaryotes exceeds that of the animals
   b. An average human contains more prokaryotic cells than human cells
   c. Prokaryotes appeared on Earth shortly before the eukaryotes
   d. The average prokaryotic cell measures only 1 to 2 microns

3. Prokaryotic cells do not contain ________. [p. 467]

   a. A plasma membrane
   b. A cell wall
   c. Ribosomes
   d. A nucleus

4. All of the following are true of prokaryotic DNA except [pp. 467–468]

   a. They have a single DNA chromosome
   b. Their chromosome is generally circular
   c. They often have multiple chromosomes and a single plasmid
   d. Their DNA is located inside a nucleoid

5. All of the following are true of Gram-negative bacteria except [p. 468–469]

   a. They have a thin peptidoglycan layer
   b. They have an outer membrane
   c. They often cause diseases
   d. They stain purple with the Gram staining technique

6. Which of the following structures gives additional protection to some bacteria? [p. 469]

   a. The plasma membrane
   b. The nucleoid
   c. The cell wall
   d. The capsule

7. Eukaryotic and prokaryotic flagella are made of microtubules and move by a whipping motion. [pp. 469–470]

   a. True
   b. False

8. Pili may function in _________. [p. 470]

   a. Attachment to other prokaryotes
   b. Bacterial sex
   c. Electricity generation
   d. All of the above

9. Which of the following type of bacteria are able to produce energy with and without oxygen? [p. 471]

   a. Facultative anaerobes
   b. Obligate aerobes
   c. Obligate anaerobes

10. Nitrogen fixation _______. [p. 471]

    a. Can be done by certain bacteria and certain eukaryotes
    b. Is the conversion of ammonium to nitrate
    c. Is an essential part of biogeochemical cycling
    d. Is also called nitrification

11. All of the following underlie the tremendous genetic variability in bacteria except _______. [p. 472]

    a. Binary fission
    b. Mutation
    c. Horizontal gene transfer
    d. Their much greater numbers in the biosphere

12. Which of the following are associated with bacterial pathogenesis? [p. 472]

    a. The plasma membrane
    b. The nucleoid
    c. Lipopolysaccharide
    d. Antibiotics

13. Which of the following types of bacteria use inorganic chemicals as their energy source and CO_2 as their carbon source? [pp. 470–471]

    a. Chemoautotrophs
    b. Chemoheterotrophs
    c. Photoautotrophs
    d. Photoheterotrophs

14. Which of the following types of bacteria obtain energy from sunlight and use organic chemicals as their carbon source? [pp. 470–471]

    a. Chemoautotrophs
    b. Chemoheterotrophs
    c. Photoautotrophs
    d. Photoheterotrophs

15. The following may be associated with antibiotic resistance: [p. 473]

    a. Horizontal transfer of plasmids
    b. Mutation
    c. Enzymatic inactivation
    d. Overuse of antibiotics
    e. All of the above

16. Which of the following statements about biofilms is false? [pp. 474–475]

    a. Outside of the natural environment, they are detrimental
    b. They require the production of an extracellular polymeric substance
    c. They are the predominant form of prokaryotic growth in nature
    d. They generally form from a complex assortment of different microbes

17. Which of the following statements about spirochetes is false? [p. 478]

    a. They are associated with human disease
    b. They are often nonpathogenic components of the human mouth
    c. They may be symbiotic partners with fungi in lichens
    d. They are spiral shaped

18. Which of the following have been identified solely on the basis of DNA isolated from the environment? [pp. 479–480]

    a. Korarchaeota
    b. Crenarchaeota
    c. Green bacteria
    d. Euryarchaeota

19. Which of the following types of Archaea are typically found in high-salt environments? [p. 480]

    a. Psychrophiles
    b. Halophiles
    c. Thermophiles
    d. Mesophiles

20. Which of the following types of Archaea are typically found in freezing environments? [p. 480]

    a. Psychrophiles
    b. Halophiles
    c. Thermophiles
    d. Mesophiles

INTEGRATING AND APPLYING KEY CONCEPTS

1. As you learned, the chlamydias are all obligate intracellular parasites of animals. They also have the smallest genomes of any other known life form. Can you draw a connection between these two facts?
2. Since the purple and green photosynthetic bacteria do not produce oxygen as a by-product, unlike the cyanobacteria, speculate on where they might be isolated compared with the cyanobacteria.
3. Why would penicillins be able to kill bacteria such as *Streptococcus pneumoniae* but not kill the host cells? Why might organisms in nature, such as the fungus *Penicillium*, produce such chemicals?

OPTIONAL ACTIVITY

1. *Geobacter* is a type of bacterium that "breathes" metals. Another type of metal "breathing" bacterium is *Shewanellai* species. Both of these organisms form part of research projects aiming to harness their metabolic capacity of bioremediate areas contaminated with metals, such as areas around the Colorado River that are contaminated with the uranium isotope U^{6+}. One of the major players in this research is Derek Lovely, at the University of Massachusetts. Find out more about the study and use of metal-breathing bacteria.

22 Viruses, Viroids, and Prions: Infectious Biological Particles

CHAPTER HIGHLIGHTS

- Viruses
  - Lack cellular structure
  - Infect all groups of organisms
  - Depend upon the host for all metabolic and reproductive activities
- Viroids and prions
  - Have a structure that is simpler than that of viruses
  - Are rare, but viroids infect animals and prions infect plants

STUDY STRATEGIES AND LEARNING OUTCOMES

- This chapter is fairly simple: pay particular attention to the similarities and differences between the viruses of the different host types. There are some new terms to remember, particularly those associated with the different types of viral life cycle.
- Pay attention to the figures: they are quite helpful in depicting the various viral morphologies and life cycle processes.
- Compare the viroids and prions to the viruses.
- As always, go through one section of the textbook at a time and then work through the companion section of the study guide.

By the end of this chapter, you should be able to

- Describe the structure of viruses and their life cycles
- Describe the structure and infectious processes of viroids and prions
- Understand the challenges of treating viral infections

TOPIC MAP

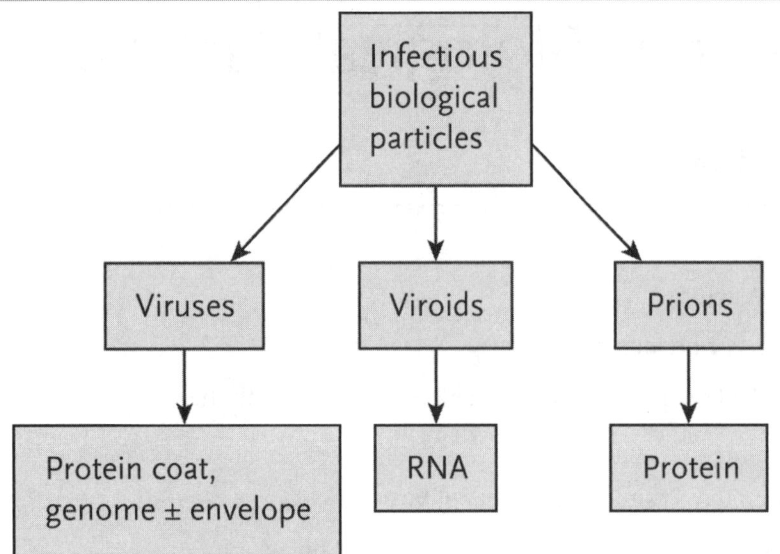

INTERACTIVE EXERCISES

Why It Matters [pp. 483–484]

22.1 What Is a Virus? Characteristics of Viruses [pp. 484–486]

22.2 Viruses Infect Bacterial, Animal, and Plant Cells by Similar Pathways [pp. 486–491]

22.3 Viral Infections Are Typically Difficult to Treat [pp. 491–492]

22.4 Viruses May Have Evolved from Fragments of Cellular DNA or RNA [p. 492]

Section Review (Fill-in-the-Blanks)

Viruses consist of one or more (1) _____ _____ molecules, surrounded by a protein coat called a (2) _____. Some viruses are surrounded by a(n) (3) ______ that is derived from the host cell's membrane. Viruses are not cells; they cannot generate (4) ______ or reproduce on their own, so most biologists do not consider them (5) _____. Their genome may be either (6) _____ or (7) _____, (8) _____-stranded or (9) _____-stranded. Regardless of the nature of the genome, they minimally encode the genes for their coat proteins, chromosome replication, and (10) _____ proteins that allow them to recognize, bind, and infect host cells. Most viruses take one of two forms: (11) ____ viruses have their coat proteins spirally arranged around the genome and (12) ______ viruses have triangular protein units joined to form a geodesic sphere. The former shape is typical of (13) _____ viruses, while the latter is typical of (14) _____ viruses. Both forms may additionally be surrounded by a(n) (15) _____. Viruses are classified into (16) _____, families,

(17) _____, and species based on a variety of criteria, including the nature of their genome. Although more than 4000 (18) ____ have been characterized, like the prokaryotes, this represents a very small fraction of the viral (19) _____ in nature, and many more remain to be identified. Viruses cause disease in humans as well as wild and domestic (20) _____, including insect pests such as the spruce budworm, and in (21) _____, where they can cause significant crop losses. They also infect (22) _____ and in this respect are an essential component of the ecosystem, preventing it from being overrun by these rapidly multiplying organisms. Viruses that infect bacteria are called (23) ______ or phages. They often have a complex morphology, with a head that contains the genome and a helical (24) ______ that injects the viral genome into the host. Bacteriophages follow two types of life cycle: (25) _____ phages that kill their host during each cycle of infection and (26) ______ phages that enter an inactive phase within the host cell and can be passed through several generations before becoming active and killing the host. T-even phages such as the T4 phage of *E. coli* are an example of (27) ______ phages. After injection of the viral chromosome, they use the host's machinery to make progeny viral particles, which, after assembly in the cytoplasm, are released by (28) _____. This type of life cycle is therefore referred to as (29) _____. Bacteriophage (30) ______ is a well-characterized temperate phage of *E. coli*. After injection of its chromosome, it generally progresses through a (31) _____ life cycle, where the viral chromosome gets inserted into the host's chromosome. It remains there as a (32) _____ until certain environmental cues such as UV irradiation trigger it to excise and commence a (33) _____ life cycle, where progeny viral particles are released by lysis. Animal viruses may also undergo lytic or temperate life cycles, the latter being referred to as (34) _____ viruses; herpesvirus is one example. The method of entry in animal cells is different from that with bacteria: enveloped viruses enter by (35) _____ of their membrane with the host's membrane, and nonenveloped viruses are taken up by (36) _____. While the latter are released by (37) ______, enveloped viruses are released by passing through the host's (38) _____ _______. While many viral infections are asymptomatic, symptoms of viral disease can be due to destruction of host cells by lysis, a rapid rate of (39) ______, as is seen with HIV, or may result from altered gene (40) _____ when viruses insert into the host's DNA. Plant viruses are commonly rod shaped or polyhedral but do not have a(n) (41) _____. Due to the presence of a cell wall, they enter the host cell through mechanical (42) _____ to leaves and stems or can be spread by pollination, by certain biting (43) _____, or from one generation to the next through the (44) _____. Once inside the plant, they spread through the (45) ______ that interconnect plant cells. Viruses cannot be treated with antibiotics that were developed to treat (46) _____ infections, and most viral infections are just left to run their course. There are limited numbers of antiviral drugs for some

of the serious infections, and some, such as the anti-influenza drug zanamavir, may target viral (47) ______ from the host cell. Some anti-HIV drugs target the viral enzyme (48) ______ _______; however, the virus has become resistant to these. This resistance, and one of the reasons that the influenza vaccine is not effective for long, is that viruses have a high rate of (49) ______.

Matching

Match each of the following terms with its correct definition.

50.	____	Capsid	A.	Viruses that have a membrane acquired from the host
51.	____	Helical viruses	B.	Part of a complex virus such as T4 or lambda that contains the viral chromosome
52.	____	Polyhedral viruses	C.	A cycle followed by bacteriophage where the viral genome is incorporated into the bacterial chromosome and multiplies with it
53.	____	Enveloped viruses	D.	The retrovirus that evolved from viruses of chimps and gorillas to cause AIDS in humans
54.	____	T-even phage	E.	Protein coat of a virus
55.	____	Virus head	F.	A virus made of triangular units joined together to form a geodesic shape
56.	____	Virus tail	G.	An animal virus that remains incorporated in the host's genome for the life of the host
57.	____	Bacteriophage	H.	Cytoplasmic connections between plant cell
58.	____	Prophage	I.	A cycle followed by a variety of viruses where the progeny viruses are released by rupturing of the bacterial cell
59.	____	Lytic cycle	J.	Part of a complex virus such as T4 or lambda that helps to inject its genome into the host cell
60.	____	Lysogenic cycle	K	Bacteriophage that are made of a head containing the genome and a tail to inject the viral DNA
61.	____	Plasmodesmata	L.	A virus that attacks bacteria
62.	____	Herpesvirus	M.	Tobacco mosaic virus is an example
63.	____	HIV	N.	State of a bacteriophage when its genome is incorporated into the bacterial chromosome

True/False

Mark if the statement is true or false. If it is false, justify your answer in the line below each statement.

64. ____________ Viruses are not living but can grow and replicate.

____

____

65. ____________ Viruses can multiply inside or outside a living host.

____

____

66. ____________ Antibiotics such as penicillin may be helpful in treating some viral infections.

____

____

Short Answer

67. Compare generalized transduction with specialized transduction.

____

____

68. Explain why a new "flu" vaccine is developed every year.

____

____

69. Explain some of the ideas for the evolution of viruses.

____

____

22.5 Viroids and Prions Are Infective Agents Even Simpler in Structure than Viruses [pp. 492–494]

Section Review (Fill-in-the-Blanks)

Viroids are simpler than viruses and are small infectious pieces of single-stranded (70) _____. They infect (71) ____ _____ and cause diseases through an unknown mechanism. Recent evidence suggests that they may act by interfering with processing of the host's (72) _____. Prions are infectious (73) ______ that infect animals and cause a group of diseases called (74)

_______ _______, so named because of the spongy appearance of the brains of affected animals. Examples of prion diseases include (75) ______, also known as mad cow disease; (76) _____ of sheep; (77) ______, a disease seen in cannibals of New Guinea; and (78) ________-_______ _______ or CJD. Prions are misfolded proteins of a normal protein of the brain; they are believed to cause disease by triggering the (79) _______ of the normal protein causing neural damage.

Complete the Table

80. Complete the following table with the chemical composition of virus, viroids, and prions.

Infectious particle	*Genetic material*	*Protein*	*Envelope*
Virus			
Viroid			
Prion			

Short Answer

81. Explain the connection between scrapie, BSE, and variant CJD.

____

____

SELF-TEST

1. All of the following are considered infectious particles except [p. 484]

   a. Bacteria
   b. Viruses
   c. Viroids
   d. Prions

2. The Spanish flu was _________. [p. 484]

   a. Responsible for a pandemic
   b. Responsible for killing 1 in 20 people worldwide
   c. A virus that infected almost half the world's population in 1918
   d. All of the above

3. Which of the following would not be associated with viruses? [p. 484]

   a. Ribosomes
   b. RNA chromosome
   c. Double-stranded RNA chromosome
   d. Multiple chromosomes

4. Which of the following statements is true? [p. 485]

   a. Most viruses have a broad host range
   b. Every living organism is likely permanently infected with one or more viruses
   c. The chromosome of an enveloped virus includes genes for the synthesis of the viral envelope
   d. Viruses are believed to be an essential part of the world's oceans
   e. (b) and (d)

5. The T-even bacteriophages _________. [p. 487]

   a. Are temperate bacteriophages
   b. Are virulent bacteriophages
   c. May infect plants
   d. Are enveloped

6. Generalized transduction ________. [pp. 487–488]

   a. May occur with some lytic viruses
   b. Occurs with lysogenic viruses
   c. Transfers genes from bacteria to animal cells
   d. Is done by bacteriophage lambda

7. Bacteriophage lambda ___________. [pp. 488–489]

   a. Infects plants
   b. Is polyhedral
   c. May perform specialized transduction
   d. Is an enveloped virus

8. Which of the following statements about herpesvirus is false? [p. 489]

   a. It is an enveloped virus
   b. It is a latent virus
   c. It causes symptoms when it causes host cells to lyse
   d. It is released by lysis

9. Which of the following statements about HIV is false? [pp. 489–490]

   a. It has eight chromosomes
   b. It appears to have evolved from chimp and gorilla viruses
   c. It has RNA chromosomes that are converted to DNA and inserted into the host's chromosome
   d. It has a very high rate of mutation

10. All of the following explain why influenza virus is hard to treat except [p. 491]

    a. It has a high mutation rate and so can evolve into new strains
    b. It has multiple chromosomes, which, in mixed infections, can randomly assort into new viral particles
    c. Scientists have been unable to develop any drugs that will target the virus
    d. It can change its protein coat

11. Which of the following is a possible explanation for the evolution of viruses? [p. 492]

    a. They started as escaped fragments of eukaryotic and prokaryotic DNA and RNA
    b. They may have evolved from the primordial gene pool before life evolved
    c. Viruses may have formed when nuclei escaped from their host cells
    d. (a) and (b)

12. Viroids have the following characteristics: [p. 492]

    a. They may be DNA or RNA molecules
    b. They are double-stranded molecules
    c. They are single-stranded RNA molecules
    d. They cause host proteins to misfold

13. Which of the following is involved in prion infection? [pp. 492–493]

    a. The misfolded protein enters by consumption of contaminated meat
    b. The misfolded protein enters through contaminated corneas in corneal transplants
    c. The misfolded protein enters through contaminated surgical instruments
    d. All of the above

14. Which of the following statements about prion diseases is true? [pp. 492–493]

    a. Scrapie is a disease seen in cannibal populations of New Guinea
    b. BSE seems to have resulted from changes in rendering processes of sheep and cattle meal and may have come from contaminated sheep carcasses
    c. CJD is a slowly progressing disease resulting from the consumption of beef contaminated with the BSE prion
    d. Stanley Prusiner is famous for developing the first vaccine against prion diseases

15. Which of the following are made of only protein? [p. 492]

    a. Viruses
    b. Viroids
    c. Prions
    d. Bacteria

INTEGRATING AND APPLYING KEY CONCEPTS

1. Provide an explanation for the fact that people who get cold sores will get them for the rest of their lives but only have occasional symptoms.

2. Why do common antibiotics not work against viruses?

3. Why does the World Health Organization (WHO) determine the composition of a new flu vaccine each year, and why are the WHO and the world's governments worried about a flu pandemic?

OPTIONAL ACTIVITY

1. As you have learned, HIV infections can be treated with antiviral drugs; however, this does not provide a cure. Because of the high rate of mutation and the resulting probability of resistance, the current strategy is a type of combination therapy that involves the simultaneous use of multiple drugs that target multiple viral structures and processes to reduce this possibility. This treatment is called HAART: highly active antiretroviral therapy. Find out more about this therapy and the problems with it, in terms of accessibility, long-term effects, and patient noncompliance, and about the current life span of HIV-positive individuals.

23 Protists

CHAPTER HIGHLIGHTS

- Protists are eukaryotes that mostly live in aquatic or moist environments.
  - Some are responsible for major human diseases.
  - Some are responsible for significant crop losses.
  - Some are major components of the phytoplankton in the world's marine and freshwaters.
- Protists may be photoautotrophic and/or heterotrophic.
- Protists may be unicellular, colonial, or multicellular.
- Protists may be microscopic or some of the world's largest organisms.
- Protists reproduce sexually by meiosis and asexually by mitosis or, in some cases, by binary fission.
- Because of their diverse structure, metabolism, and reproduction, they are difficult to classify; however, it is clear that they do not all share a common ancestor.
- Major groups:
  - Excavates—single celled, flagellated, lack mitochondria, produce ATP anaerobically, heterotrophic, parasitic
  - Discicristates—single celled, flagellated, photoautotrophic or heterotrophic, disc-shaped mitochondrial cristae
  - Alveolates—single celled, flagellated or nonmotile, photoautotrophic or heterotrophic, specialized structures such as membrane-bound vesicles called the alveoli
  - Heterokonts—funguslike, photoautotrophic or heterotrophic, flagellated gametes
  - Cercozoa—single celled, photoautotrophic or heterotrophic, move or ingest food by producing stiff projections called the axopods that project through hard outer shell called tests
  - Amoebozoa—irregular cell shape, use flexible pseudopodia to move and engulf food particles, heterotrophic

- Archaeplastida—multicellular, higher photoautotrophic, closest relatives to to land plants
- Opisthokonts—unicellular or colonial, heterotrophic, flagellated, closest relatives to fungi and animals

- The endosymbiont theory explains the origin of chloroplast in protists:
  - Some of the photosynthetic protists are the product of primary endosymbiosis and some are the product of secondary endosymbiosis.

STUDY STRATEGIES AND LEARNING OUTCOMES

- There is a huge amount of new information and new names for organisms and structures. Do not worry too much about remembering all the details; the goal of this chapter is to understand what protists are, why they are important, and how they differ in their structure, metabolism, and reproduction.
- Keep track of the headings and subheadings as you go through the various protist groups. Look for figures or examples (e.g., of pathogens) that will help you remember something about that particular group.
- As always, go through one section of the textbook at a time and then work through the companion section of the study guide.
- Use the "Study Break" sections of the textbook and the "Review" section of the textbook as well as this study guide section to help guide you on the level of detail.

By the end of this chapter, you should be able to

- Understand how the mitochondria and chloroplasts evolved; it began with these organisms
- Identify what features distinguish the protists from the prokaryotes, fungi, animals, and plants and what features resemble these other groups
- Explain why the classification of the protists is so problematic
- Identify some of the protists that cause disease or blights
- Identify some of the other important groups: the brown algae with the kelp "forests," the green algae and their relationship with land plants, and the opisthokonts and their relationship with the animals and fungi
- Identify the very basic characteristics for the different groups

TOPIC MAP

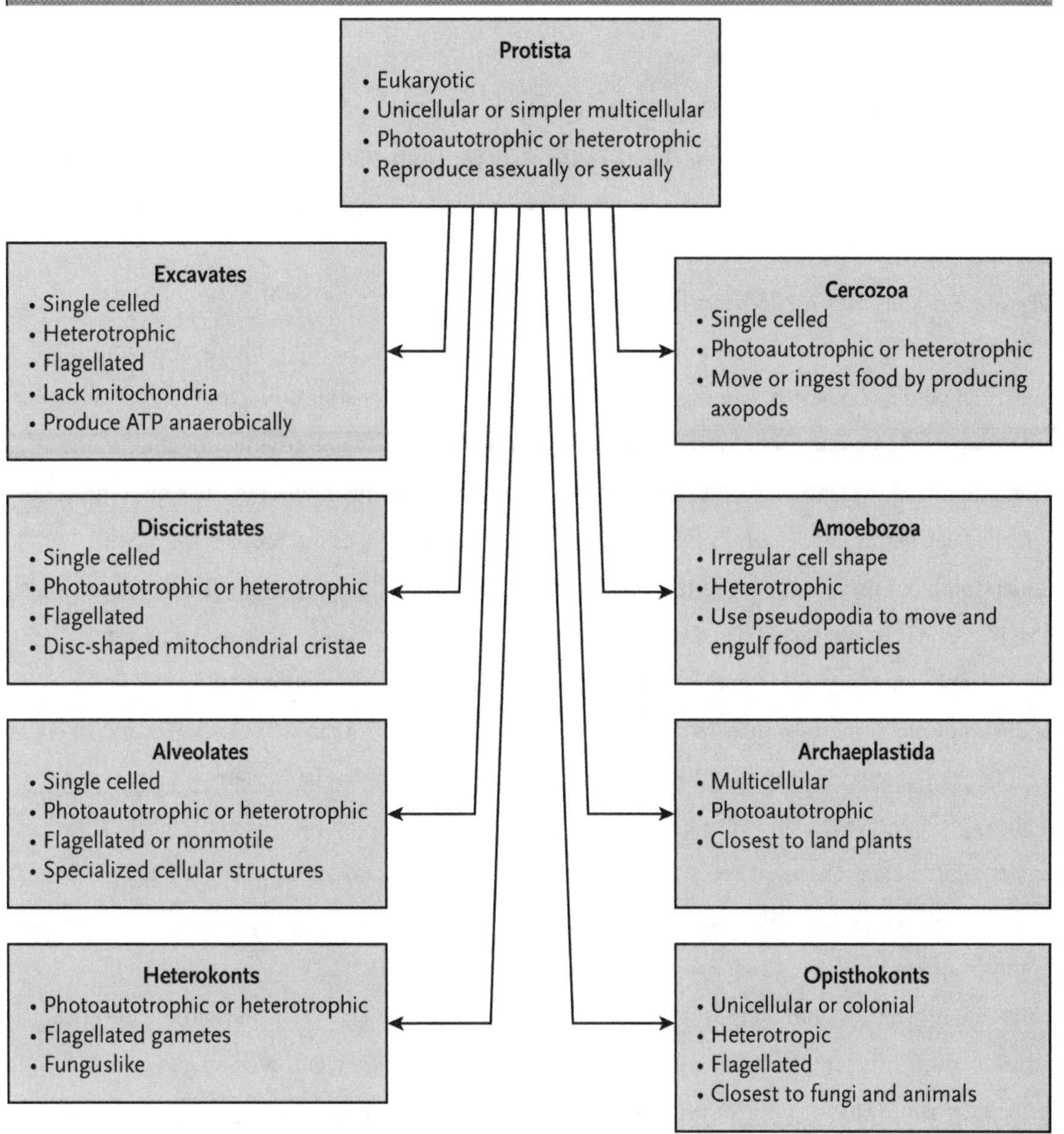

Protista
• Eukaryotic
• Unicellular or simpler multicellular
• Photoautotrophic or heterotrophic
• Reproduce asexually or sexually
Excavates
• Single celled
• Heterotrophic
• Flagellated
• Lack mitochondria
• Produce ATP anaerobically
Cercozoa
• Single celled
• Photoautotrophic or heterotrophic
• Move or ingest food by producing axopods
Discicristates
• Single celled
• Photoautotrophic or heterotrophic
• Flagellated
• Disc-shaped mitochondrial cristae
Amoebozoa
• Irregular cell shape
• Heterotrophic
• Use pseudopodia to move and engulf food particles
Alveolates
• Single celled
• Photoautotrophic or heterotrophic
• Flagellated or nonmotile
• Specialized cellular structures
Archaeplastida
• Multicellular
• Photoautotrophic
• Closest to land plants
Heterokonts
• Photoautotrophic or heterotrophic
• Flagellated gametes
• Funguslike
Opisthokonts
• Unicellular or colonial
• Heterotropic
• Flagellated
• Closest to fungi and animals

INTERACTIVE EXERCISES

Why It Matters [pp. 497–498]

23.1 Evolution of Protists Involved Endosymbiosis [pp. 498–499]

23.2 What Is a Protist? Characteristics of Protists [pp. 499–500]

23.3 Protists' Diversity Is Reflected in Their Metabolism, Reproduction, Structure, and Habitat [pp. 500–502]

Section Review (Fill-in-the-Blanks)

"Beaver fever" is a water-borne infection caused by the protist (1) _____, the most common intestinal parasite in North America. Protists are a very heterogeneous collection of 200 000 eukaryotes that are not very closely (2) _____. Traditionally, they are grouped together in the kingdom (3) ______; however, this is a catch-all group that is convenient but does not represent lineages derived from a single common (4) _____. The evolution of the protists is not well understood, but they are believed to have first appeared (5) _______ years ago by endosymbiosis. The (6) ______ arose through a single endosymbiotic event involving a free-living aerobic, heterotrophic prokaryote. In contrast, the (7) _____ evolved through more than one endosymbiotic event in protists that already had a (8) ______. Protists are eukaryotes and so share a number of characteristics with the other eukaryotic kingdoms, that is, the (9) _______ _______ and _______; however, they also have some striking differences. Unlike the (10) _____, they have a motile stage and their cell walls contain (11) ______ but not chitin. Unlike plants, they lack the differentiation of the body into (12) ______, _______, ________, and ______ and do not retain embryos in parental tissue; unlike animals, they are usually (13) _______ and lack collagen, nerve cells, and a digestive tract. Protists are highly (14) ______ in their metabolism, reproduction, structure, and (15) ______. Metabolically, protists are aerobic and can live as heterotrophs or (16) _______ or use a combination of the two. Protists live in (17) ______ or moist terrestrial habitats or within (18) _____ ______. They may be single celled, (19) ______, or multicellular. The cellular structure of the single-celled species is extremely complex, with a wide variety of (20) ______ structures such as (21) ______ vacuoles that collect and expel excess water driven into cells in hypotonic environments. (22) ______ and cilia are hairlike structures that help some of the protists move; others move by extending their cytoplasm or cell, forming (23) ______. Reproduction may be asexual or (24) _______ and the fusion of gametes.

Matching

Match each of the following terms with its correct definition.

25.	____	Pellicle	A.	Small photosynthetic protists that collectively live in ponds, lakes, and oceans
26.	____	Contractile vacuole	B.	Heterotrophic protists, bacteria, and small animals that live in ponds, lakes, and oceans and depend on protists for organic nutrients and oxygen
27.	____	Phytoplankton	C.	Catch-all term for all photosynthetic protists
28.	____	Zooplankton	D.	Most common intestinal parasite in North America
29.	____	Pseudopodia	E.	A membrane-bound organelle that helps a protist get rid of excess water
30.	____	*Giardia lamblia*	F.	A layer of protein fibres under the plasma membrane
31.	____	Algae	G.	Cytoplasmic and cellular extensions of protists that help them move

Complete the Table

32. Complete the following table for the characteristics of protists.

Major characteristic	*Description*
Habitat	A.
Cell structure	B.
Metabolism	C.
Reproduction	D.

True/False

Mark if the statement is true or false. If it is false, justify your answer in the line below each statement.

33. ______________ The kingdom Protista defines the lineages that descended from a common ancestor.

____

____

34. ______________ Protists are generally unicellular and microscopic.

____

____

Short Answer

35. How would you distinguish sea weeds from aquatic plants?

____

____

23.4 The Protist Groups [p. 502–518]

Section Review (Fill-in-the-Blanks)

Molecular data have aided in the phylogenetic classification of the protists; however, the only thing that biologists agree on is that the protists do not all share a (36) _____ _____. Based on a consensus of molecular and (37) _______ data, they can be assigned to the following groups:

A. Excavates—flagellated, single cells that lack typical (38) _______ but often have organelles derived from these. They are heterotrophic but limited to using (39) ______ for generating ATP. They are named for the "excavated" ventral (40) ______ groove found in most members and are exemplified by the Diplomonadida, which includes the pathogen (41) _______ _______, and Parabasala, which includes the pathogen (42) _______ _______.

B. Discicristates—flagellated, single cells that may be photosynthetic and/or (43) ______. They are often called protozoa, are named for their disc-shaped mitochondrial (44) ______, and are exemplified by the Euglenoids, including (45) _____, which is photosynthetic, and the Kinetoplastids, including the pathogenic (46) _______ species.

C. Alveolates—single cells that may be (47) _______ or heterotrophic. They are named for their small, membrane-bound vesicles called (48) ______; these are located in a layer under the plasma membrane. They comprise three major groups, the first two being motile and primarily free living and the last, the (49) ______, being nonmotile and parasitic. One of the two motile groups is the (50) ______, or ciliates, which swim using cilia, for example, *Paramecium*, and have complex cytoplasmic structures such as contractile vacuoles. The second group is the (51) _______, a largely photosynthetic group of species that have a shell of (52) ______ plates and two flagella that make them move in a whirling or spinning motion. The Apicomplexans are named for their (53) ______ _______, associated with their ability to attach to and invade host cells; the pathogenic genus *Plasmodium* is an example.

D. Heterokonts—These are named for their two different (54) _______, one that is smooth and one that is hairy; these are normally only found on their reproductive cells. The heterokonts include four major groups: the funguslike (55) _______, which includes the water moulds, the Bacillariophyta or (56) ________, the (57) _______ or golden algae, and the (58) ______ or brown algae. Although molecular data indicate that the oomycetes are not related to (59) ______, they are similar in their secretion of degradative enzymes and growth by the formation of (60) ______ and mycelia. The most notorious member is *Phytophthora infestans*, the causative agent of the Irish (61) ______ of the 1800s. Diatoms have a glassy shell made of (62) ______ and are photoautotrophs; golden algae are photoautotrophs like plants but have an additional pigment called (63) _______, which masks the green of the chloroplyll; brown algae are multicellular photoautotrophs and differentiate into leaflike (64) ______, stalklike (65) ______, and rootlike (66) ______. The biggest of the brown algae are the (67) ______.

E. Cercozoa—These are amoebas with filamentous (68) ______, and many produce hard outer shells called (69) _____. Radiolarians have microtubule-supported extensions called (70) ______; Foraminifera or Forams are named for the (71) _______ in their shells through which extend cytoplasmic projections. Both engulf their prey using these (72) ______. Chloroarachniophyta are photosynthetic and heterotrophic amoeba, using (73) _____ to capture their food.

F. Amoebozoa—This group is heterotrophic and contains most of the amoebas as well as the cellular and (74) ______ slime moulds; all use (75) ______ to move and to engulf food particles. Amoebas in this group are single celled and only reproduce asexually and use (76) _______ _______ to do so; (77) _______ is a slime mould that is extensively used in research because it differentiates to form a fruiting body under unfavourable conditions. Plasmodial slime moulds exist as a (78) ______ plasmodium in which individual nuclei are suspended in a common cytoplasm, surrounded by a single (79) ______ _______.

G. Archaeplastida—This group includes the (80) _____ or red algae, the (81) ______ or green algae, and the land plants of the kingdom (82) ______. Red algae have additional pigments called (83) _______; green algae have the same pigments as land plants, and one group, the (84) _______, may have been the ancestor of land plants.

H. Opisthokonts—This group is named for the single, posterior (85) ______ and includes the Choanoflagellates, which may have been the ancestor to the (86) _______.

Matching

Match each of the following terms with its correct definition.

87.	____	Bioluminescent	A.	Refers to leaflike structures of algae
88.	____	Fucoxanthin	B.	Refers to rootlike structures of algae
89.	____	Blades	C.	Structures found in ciliates such as Paramecium for eliminating excess water
90.	____	Stipes	D.	Light released by dinoflagellates when they are disturbed
91.	____	Holdfast	E.	Ground fossilized protist shells used in toothpaste and other commercial products
92.	____	Axopods	F.	The mouthlike region of *Paramecium* where food is taken in
93.	____	Phycobilins	G.	Slender strands of cytoplasmic extensions that are typically present in Cerozoa
94.	____	*Plasmodium*	H.	Named for the small sunbeam they resemble
95.	____	Pseudopodium	I.	Additional pigment present in red algae
96.	____	*Euglena*	J.	Brown algae that includes the giant kelp
97.	____	Contractile vacuole	K.	Literally "false foot"
98.	____	Gullet	L.	Refers to stemlike structures of algae
99.	____	Diatomaceous earth	M.	A multinucleate form of slime mould
100.	____	Phaeophyta	N.	Photosynthetic organisms that possess a light-sensing eyespot
101.	____	Radiolarians	O.	Additional pigment present in golden algae

Complete the Table

102. Fill in the following table with the name of the organism or the disease/disease features associated with the organism.

Organism	*Disease*
A.	Sleeping sickness
B.	Beaver fever
Trichomonas vaginalis	C.
Plasmodium species	D.
E.	Paralytic shellfish poisoning
F.	Disease contracted by inhalation of cysts from cat feces that may cause brain damage or death of an unborn fetus
G.	Caused the Irish potato famine of the 1800s and has emerged as a serious disease of potato crops in Canada and the United States
H.	Amoebic dysentery

Building Vocabulary

Using the definitions of the prefixes below, assign the appropriate prefix to the appropriate suffix for the following:

Prefix	Meaning
hetero-	different
pseudo-	false
chloro-	green
rhodo-	rose
diplo-	double
proto-	first
choano-	collar
dino-	spinning
opistho-	posterior
foramin-	little hole

Prefix	Suffix	Protist group	
	-monads	103.	Organisms with two equal and functional nuclei; include *Giardia*
	-zoa	104.	Alternate name for the discicristates based on the fact that they ingest food and move by themselves, similar to ancient animals
	-flagellates	105.	Protists with a shell formed from two cellular plates and flagella arranged in between so the organisms move by spinning
	-konts	106.	Possess two different flagella: one smooth and one covered with bristles
	-ifera	107.	Protists with pseudopodia projecting through small perforations in their shells
	-phyta	108.	Red algae used to wrap sushi and as a source of agar and carageenan and important in corals
	-phyta	109.	Green algae that include the likely ancestor to land plants
	-konts	110.	Protist group named for the single flagellum located at the end of the cells from some stages of their life cycle
	-flagellata	111.	Protist group named for the collar surrounding the flagellum; includes what is thought to be the ancestor of animals

23.5 Some Protist Groups Arose from Primary Endosymbiosis and Others from Secondary Endosymbiosis [pp. 518–520]

Section Review (Fill-in-the-Blanks)

Several groups of protists and the land plants contain chloroplasts that arose through primary endosymbiosis where a eukaryotic cell engulfed a (112) ______, which then became the chloroplast. Evolutionary divergence from this ancestral phototroph gave rise to the (113) ______ and _____ algae and the land plants. Other photosynthetic protists, such as the (114) ______ (the "eyespot organisms"), (115) ______ (the "spinning" protists), (116) ______ (with their smooth and hairy flagella), and chlorarachniophytes, were produced by secondary endosymbiosis in which a nonphotosynthetic protist engulfed a photosynthetic (117) ______. Organisms that formed from secondary endosymbiosis have chloroplasts surrounded by (118) ______ membranes derived from the original host cell.

SELF-TEST

1. Protists are [p. 498]

   a. Prokaryotic
   b. Eukaryotic

2. The Protista _________. [pp. 498–500]

   a. All share a common ancestor
   b. Are all photosynthetic
   c. Are a catch-all group that simply designates eukaryotes that are not animals, fungi, or plants
   d. All have mitochondria

3. Which of the following statements about the differences between the "animal-like" protists and animals is true? [p. 500]

   a. The "animal-like" protists may be multicellular but not always
   b. The "animal-like" protists lack differentiated structures such as limbs
   c. The "animal-like" protists may possess collagen in their cell walls but not always
   d. All of the above are true

4. Which of the following statements about the differences between the photosynthetic protists and plants is true? [p. 500]

   a. The photosynthetic protists form leaves and stems but not roots
   b. The photosynthetic protists retain parental tissue in their embryos
   c. The photosynthetic protists may also live as heterotrophs
   d. All of the above are true

5. Which of the following structures is involved in getting rid of excess water? [p. 501]

   a. Food vacuole
   b. Contractile vacuole
   c. Pseudopodia
   d. Axopods

6. The heterotrophic protists produce ATP using mitochondria. [p. 502]

   a. True
   b. False

7. Which of the following protists have an eyespot? [p. 504]

   a. *Euglena*
   b. *Paramecium*
   c. *Giardia lamblia*
   d. Kelp

8. Which of the following statements about the ciliates (Ciliophora) is true? [pp. 505–506]

   a. They possess two functionally and structurally different nuclei
   b. They use cilia rather than flagella to swim
   c. They possess contractile vacuoles
   d. All of the above are true

9. Which of the following statements about dinoflagellates is false? [pp. 506–507]

   a. They are part of the zooplankton
   b. They cause red tides
   c. They may cause food-borne intoxication that results in paralysis
   d. They glow in the dark

10. Which of the following statements about the apicomplexans is false? [p. 507]

   a. They have an apical complex involved in attaching to and invading hosts
   b. They are parasites of animals
   c. They may be motile or nonmotile
   d. They include the genus *Plasmodium*

11. Which of the following statements is false? [pp. 508–509]

  a. The Oomycota include the water moulds
  b. The Oomycota were responsible for the Irish potato famine
  c. The Oomycota are fungi that used to be considered protists
  d. The Oomycota are key decomposers in freshwater environments

12. Which of the following statements is false? [pp. 509–510]

  a. The diatoms may be unicellular or multicellular
  b. The diatom shells represent a major component of the sediments of lakes and seas
  c. The diatoms have silica cell walls
  d. The diatom shells are ground up and put in toothpaste

13. Which of the following pigments gives brown and golden algae their colour? [p. 510]

  a. Phycobilins
  b. Fucoxanthin
  c. Chlorophylls
  d. Carotenoids

14. Which of the following is(are) not associated with brown algae? [p. 510]

  a. Gas bladders
  b. Agar
  c. Blades
  d. Holdfasts

15. The Cercozoa do not ___________. [pp. 511–512]

  a. Contain organisms that move through cytoplasmic extensions called pseudopods
  b. Contain the "shape-shifting" amoeba
  c. Contribute to the formation of the sedimentary rock
  d. Have anything to do with the distinctive appearance of the White Cliffs of Dover

16. Which of the following are not part of the Amoebozoa? [pp. 513–515]

  a. The water moulds
  b. The slime moulds
  c. The plasmodial slime moulds
  d. The amoeba

17. The protists that have been studied very closely to understand the process of differentiation belong to the group ______. [p. 564]

  a. Opisthokonts
  b. Cercozoa
  c. Alveolates
  d. Amoebozoa

18. Which of the following pigments gives red algae their colour? [p. 517]

a. Phycobilins
b. Fucoxanthin
c. Chlorophylls
d. Carotenoids

19. Which of the following lack mitochondria? [p. 504]

a. Opisthokonts
b. Archaeplastida
c. Excavates
d. Amoebozoa
e. Alveolates

20. Which of the following groups includes the land plants? [p. 515]

a. Opisthokonts
b. Archaeplastida
c. Excavates
d. Amoebozoa
e. Alveolates

21 Which of the following may have been the ancestor to the animals? [p. 517]

a. Opisthokonts
b. Archaeplastida
c. Excavates
d. Amoebozoa
e. Alveolates

22. In primary endosymbiosis, a heterotrophic eukaryotic cell engulfed a ________________ that was not digested but became a permanent resident, ultimately transforming into a chloroplast, as in red and green algae and land plants. [p. 518]

a. Nonphotosynthetic prokaryotic cell
b. Photosynthetic prokaryotic cell
c. Nonphotosynthetic eukaryotic cell
d. Photosynthetic eukaryotic cell

23. In secondary endosymbiosis, a heterotrophic eukaryotic cell engulfed a ________________ that became a permanent resident and transformed into a chloroplast with multiple membrane layers. [pp. 518–520]

a. Nonphotosynthetic prokaryotic cell
b. Photosynthetic prokaryotic cell
c. Nonphotosynthetic eukaryotic cell
d. Photosynthetic eukaryotic cell

24. Protista includes organisms that reproduce asexually by binary fission. [p. 513]

   a. True
   b. False

INTEGRATING AND APPLYING KEY CONCEPTS

1. What are the distinguishing characteristics of protists that separate them from organisms in other kingdoms?

2. Discuss why it is difficult to classify protists.

3. Using one example for each, explain the process of primary and secondary endosymbiosis, the evidence, and why proving such events would have been impossible prior to the advent of genomics.

OPTIONAL ACTIVITY

1. The scientist who proposed the endosymbiont theory to explain the evolution of the chloroplasts and mitochondria is Lynn Margulis, now a distinguished university professor at the University of Massachusetts, Department of Geosciences. She has had an extremely active career and written countless scientific articles as well as numerous books, including many coauthored with her son, Dorion Sagan. She is also credited for her contributions to the Gaia Theory of James Lovelock, a well-known atmospheric chemist. The Gaia Theory proposes that the living and nonliving components of Earth and its atmosphere have evolved to function as a vast, self-regulated system. Find out more about this fascinating woman and her work.

24 Fungi

CHAPTER HIGHLIGHTS

- Fungi are heterotrophic organisms that obtain their nutrients by extracellular digestion and absorption. They are critical components of the biosphere for their saprotrophic abilities.
- Fungi may also cause disease in animals and plants, be of commercial importance, or cause food spoilage.
- Fungi reproduce by asexually and sexually with spores.
- Kingdom Fungi is divided into five major groups based mostly on the structures produced for sexual reproduction:
    - Chytridiomycota are mostly aquatic fungi that produce flagellated spores.
    - Zygomycota have aseptate hyphae and reproduce sexually by producing zygospores.
    - Glomeromycota form arbuscular mycorrhizae with many types of plant roots and produce asexual spores at the tips of specialized hyphae.
    - Ascomycota produce ascocarp fruiting bodies that contain sacs called asci, which contain four to eight haploid ascospores.
    - Basidiomycota produce basidiocarp fruiting bodies such as mushrooms. These contain saclike basidia with four haploid basidiospores.
- Deuteromycota or Fungi Imperfecti is a temporary group for those species in which sexual reproduction has not yet been observed.
- Lichens are compound organisms of a mycobiont (fungal partner) and a photobiont (algal or cyanobacterial partner). These are symbiotic relationships, which may be mutualistic but, in many cases, appear to be more of a parasitic relationship.
- Mycorrhizas are mutualistic relationships formed between the glomeromycetes, ascomycetes, and some basidiomycetes with plant roots.
- Endosymbionts are poorly characterized relationships between fungi and above-ground plant growth. It remains to be determined if these relationships are parasitic or mutualistic.

STUDY STRATEGIES AND LEARNING OUTCOMES

- This chapter contains a fair amount of new information and new names for organisms, phyla, and structures. Do not worry too much about remembering all the details; the goal of this chapter is to understand what fungi are, why they are important, and how they differ in their reproduction, reproductive structures, and lifestyles.
- Keep track of the headings and subheadings as you go through the six fungal groups. Look for figures or examples (e.g., of pathogens, commercially important fungi) that will help you remember something about that particular group.
- As always, go through one section of the textbook at a time and then work through the companion section of the study guide.
- Use the "Study Break" sections of the textbook and the "Review" section of the textbook as well as this study guide section to help guide you on the level of detail.

By the end of this chapter, you should be able to

- Identify what features distinguish each of the five evolutionary lineages of the fungi
- Understand why the classification of the Deutromycetes and fungi in general is likely to change with time and increasing molecular data
- Understand the various important roles of the fungi as saprotrophs and symbionts
- Identify some of the fungi that cause disease or crop losses
- Explain what lichens, mycorrhiza, and endophytes are

TOPIC MAP

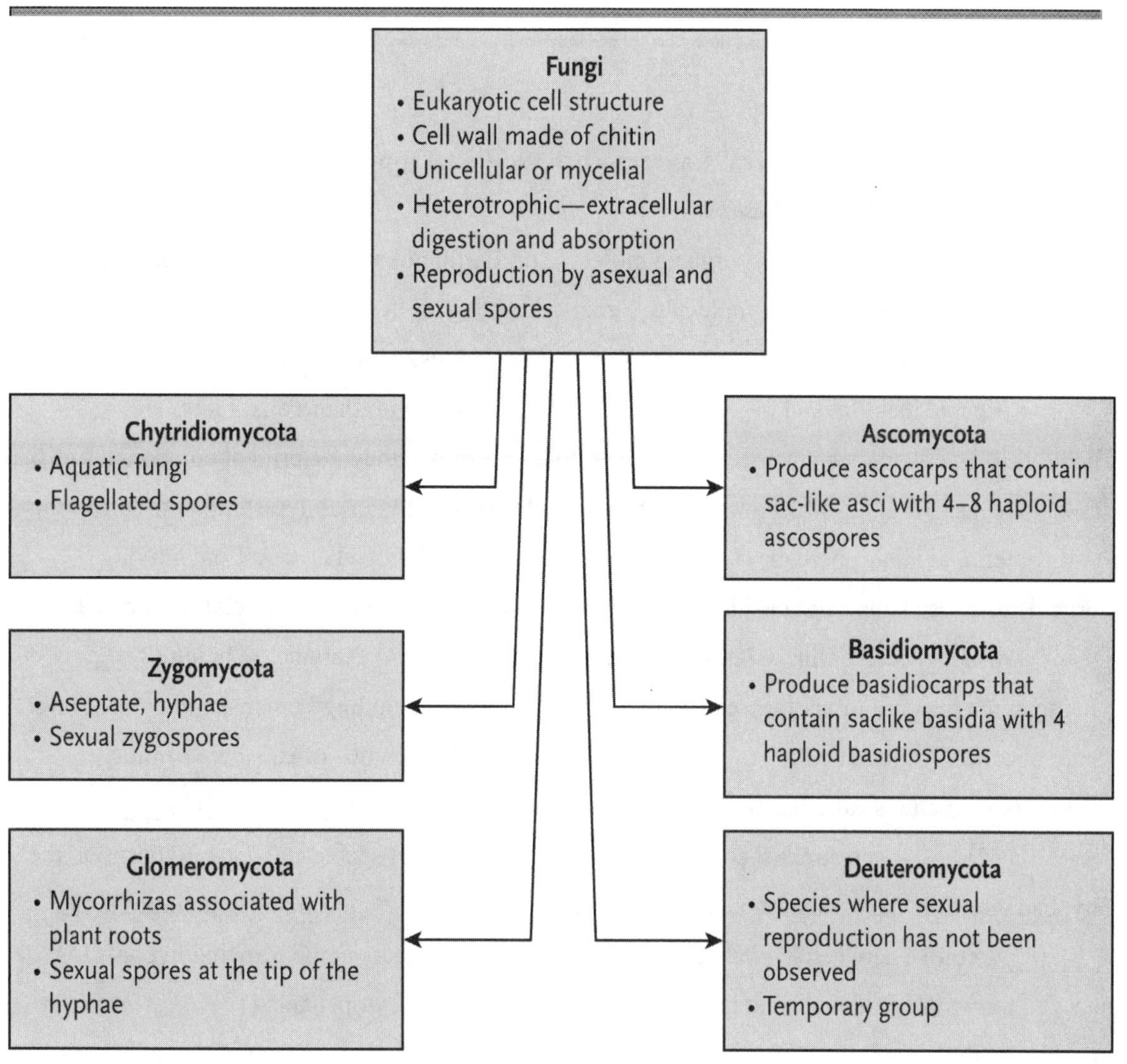

Fungi
• Eukaryotic cell structure
• Cell wall made of chitin
• Unicellular or mycelial
• Heterotrophic—extracellular digestion and absorption
• Reproduction by asexual and sexual spores
Chytridiomycota
• Aquatic fungi
• Flagellated spores
Zygomycota
• Aseptate, hyphae
• Sexual zygospores
Glomeromycota
• Mycorrhizas associated with plant roots
• Sexual spores at the tip of the hyphae
Ascomycota
• Produce ascocarps that contain sac-like asci with 4–8 haploid ascospores
Basidiomycota
• Produce basidiocarps that contain saclike basidia with 4 haploid basidiospores
Deuteromycota
• Species where sexual reproduction has not been observed
• Temporary group

INTERACTIVE EXERCISES

Why It Matters [pp. 523–524]

24.1 What Is a Fungus? General Characteristics of Fungi [pp. 525–527]

Section Review (Fill-in-the-Blanks)

(1) ______ is the most abundant organic molecule on Earth, but most organisms lack the (2) _____ to degrade this complex molecule. Fungi are among the few organisms that do have this ability as they produce and secrete a wide variety of enzymes, which allows them to (3) _____ ______ organic material and (4) _____ the smaller molecules into their cells. Fungi are heterotrophic eukaryotes, and because of their unique features, they are critical components of the biosphere, as (5) _____ or decomposers, breaking down dead material, returning nutrients to the soil, water, and atmosphere, or (6) _____, obtaining their carbon and energy from a living organism. These latter relationships span the continuum from the (7) _____ relationships with many plants and even animals, leaf-cutter ants being an excellent example, to being (8) _____, living at the expense of another, causing disease of animals, including humans, and crops. Fungi may occur as single-celled (9) _____ or as multicellular, filamentous organisms, forming threadlike structures called (10) _____, which give rise to a mass referred to as (11) _____. A cell wall of (12) _____ surrounds the plasma membrane and in most species (13) _____ partition the hyphae into cell-like compartments. Pores in the septa permit cytoplasm and sometimes (14) _______ to move between hyphal cells. Substrate digestion occurs at the growing hyphal (15) _____; nutrients are absorbed at these sites and then spread by cytoplasmic (16) _____ through the rest of the organism. All fungi reproduce asexually by producing large numbers of haploid (17) _____ but also sometimes through budding or the (18) _____ of the parent body. Most also reproduce sexually, usually involving three stages: (19) _____ or the fusion of the cytoplasm of two haploid hyphae or gametes, followed at some point by (20) ______ or fusion of the two parental nuclei to form a (21) _____ zygote ($2n$); this is then followed by meiosis to give (22) _____ spores ($1n$). The second stage may be considerably delayed in some species, resulting in a prolonged (23) _____ stage, designated as ($n + n$).

Matching

Match each of the following terms with its correct definition.

24.	____	Hypha	A.	Involves secretion of degradative enzymes to release small molecules that are easily transported into the cell
25.	____	Mycelium	B.	Fusion of the two nuclei in a dikaryon
26.	____	Saprobes	C.	Organisms that live on other living things and harm them
27.	____	Parasites	D.	Involves the hyphal tips only and considerable mechanical force
28.	____	Mutualism	E.	A cell with two haploid nuclei ($n + n$)
29.	____	Spores	F.	Threadlike structure that make up most fungi
30.	____	Plasmogamy	G.	Where two organisms live together benefiting each other
31.	____	Dikaryon	H.	The movement of nutrients to nonabsorptive regions of the organism
32.	____	Karyogamy	I.	Organisms that live on dead organisms to get nutrients
33.	____	Absorptive nutrition	J.	Single cells that are produced by fungi as part of reproduction
34.	____	Apical growth	K.	A mass of hyphae
35.	____	Cytoplasmic streaming	L.	Fusion of the cytoplasm of two hyphae or two gametes

Complete the Table

36. Complete the following table by giving definitions of the terms used to describe different ways fungi get their nutrients.

Terms	*Definitions*
A. Saprobes	
B. Parasites	
C. Mutualism	

37. Complete the following table by giving the major characteristics of fungi.

Characteristic	*Description*
A. Cell type	
B. Cell wall	
C. Method of growth, structures	
D. Mode of nutrition, nutrient acquisition	
E. Reproduction, unusual aspects	

True/False

Mark if the statement is true or false. If it is false, justify your answer in the line below each statement.

38. ______________ Leaf-cutter ants produce antibiotics to control the growth of their fungal "garden."

____

____

39. ______________ Fungi only grow and obtain nutrients at their fungal tips.

____

____

Short Answer

40. Discuss some of the commercial applications of the fungi.

____

____

41. Discuss the relevance to the evolution of the fungi and the colonization of land by plants.

____

____

42. Define the term "secondary metabolite."

____

____

24.2 Evolution and Diversity of Fungi [pp. 527–538]

Section Review (Fill-in-the-Blanks)

Traditionally, scientists have classified fungi based on distinctive structures associated with (43) ______. More recently, molecular techniques have provided the data to classify the fungi into (44) _____ phyla. This will assuredly be revised as more data become available given that two of these phyla, the chytridiomycota and the zygomycota, are (45) _____. In addition, a sixth temporary group called (46) ______ (otherwise known as Fungi Imperfecti) is used as a "holding pen" for fungi for which sexual reproduction has not been observed.

Chytridiomycota: The chytrids are the only fungi to produce motile (47) ______ spores, and this allows them to live in (48) _____ or moist habitats. Chytrids are generally (49) ______ that break down dead organic matter. Others are symbionts: some living as (50) ______, breaking down cellulose in the gut of cows and other herbivores, and others as (51)_____, causing diseases that include chytridomycosis, which is responsible, at least in part, for the worldwide decline in (52) ______. While generally unicellular, cells may also form chains and (53) ______ that anchor the fungus to a substrate. Reproduction is generally asexual, with the haploid spores forming in a sac called a (54) _____.

Zygomycota: These fungi are important soil (55) ______ but also include the (56) ______ that form on bread, fruits, vegetables, and grains. Some have (57) _____ applications, steroid production being one. They generally form (58) _______ hyphae, that is, hyphae that lack cell partitions. During sexual reproduction, specialized (59) _____ of the + and – mating types grow toward each other, producing sex organs called (60) _____ at their tips. These fuse and form a thick-walled (61) _____, which may remain (62) _____ for years. Under favourable conditions, the zygospore germinates to produce a (63) _____ and the diploid nuclei undergo (64) _____ to generate haploid spores. Under favourable conditions, zygomycetes reproduce (65) ______ through the formation of mycelia and sporangia and the production of haploid spores by (66) ______.

Glomeromycota: These fungi are an important component of soil and form (67) ______ mycorrhizas in the (68) _______ of many land plants, including crops such as (69) ______. They reproduce asexually by walling off a section of (70) _____ to generate a haploid (71) _____.

Ascomycota: Some members of this group have been important models for (72) ______ research, and others, such as the bread and brewing yeast (73) _______ _____, are commercially important. In nature, they may be saprotrophs or symbionts, forming mycorrhizas and (74) ______, or they may be pathogenic to plants and animals. (75) _______ _______, which causes thrush or yeast infections in humans, is a well-known example. Although some grow in this yeast form, reproducing asexually by (76) ______ or by binary fission, most grow as haploid, septate (77) ______. In these fungi, asexual reproduction does not involve a sporangium: instead, modified hyphae produce numerous exposed spores called (78) ______. Sexual reproduction involves fusion of the + and – hyphae and (79) ______ or fusion of the cytoplasms. (80) _______, or fusion of the two nuclei, is somewhat delayed, giving rise to short-lived (81) ______ or ($n + n$) hyphae. Sacs called (82) ______ form at the tips of these hyphae, inside of which the two nuclei fuse and then undergo meiosis and usually also mitosis, to form (83) ______ haploid ascospores. These sacs are often found in a multicellular fruiting body called a(n) (84) ______.

Basidiomycota: These fungi have the unusual and essential ability to digest (85) ______ from woody plants; because of structural similarities with lignin, they are also able to degrade pollutants such as (86) ______ and _______. Many of these wood-decay fungi obtain their nitrogen by trapping and consuming (87) _______. Some of the basidiomycetes form mycorrhizas, while the (88) ______ and ______ cause serious diseases of crops and other plants. (89) _______ are found in this group, some of which are edible, but others, such as *Amanita phalloides*, are (90) _______. While some basidiomycetes may grow as unicellular (91) ______, most are mycelial. Sexual reproduction occurs through the fusion of two haploid mycelia, after which (92) ______ is considerably delayed; as a result, mycelia of these fungi are quite often (93) _______, with each compartment of the growing mycelium containing one nucleus of each type. When conditions are (94) ______, a tight cluster of hyphae produces a large, short-lived fruiting body called a (95) ______; mushrooms are familiar examples. This structure consists of a stalk with a cap and develops numerous sacs called (96) ______ that form on the "gills" of the cap. Each sac contains a diploid nucleus that undergoes (97) ______ to eventually form four haploid, exposed (98) _______. The asexual stage of these fungi involves budding of yeasts and formation of (99) ______ in mycelia organisms.

Deuteromycota: This temporary group includes some fungi that are well known, including the genus (100) ______, which includes species that produce antibiotics and others used in the production of Camembert and Roquefort cheeses.

Matching

Match each of the following terms with its correct definition.

101.	____	Basidiospores	A.	The sexual spores of the ascomycetes
102.	____	Mycorrhizas	B.	The multicellular fruiting body of the Ascomycota that bears sexual spores
103.	____	Arbuscule	C.	Include subgroups that are not all descended from a single ancestor
104.	____	Sporangium	D.	Mushrooms are an example
105.	____	Conidia	E.	A symbiotic relationship between fungi and plant roots
106.	____	Asci	F.	A multicellular structure that may anchor chytrids to a substrate
107.	____	Ascocarp	G.	The mycorrhizal form of Glomeromycota where a cluster of hyphae penetrate the root cells
108.	____	Basidiocarp	H.	A terminal sac that is filled with asexual spores in Chytridiomycota and Zygomycota
109.	____	Chemotaxis	I.	A thick-walled spore that may remain dormant for years, it will germinate to form a sporangium
110.	____	Basidia	J.	A temporary group of fungi for which no sexual stage has been identified
111.	____	Paraphyletic	K.	A sac that contains 4–8 sexual spores of Ascomycota
112.	____	Ascospores	L.	Swimming in response to chemicals, a characteristic of the flagellated chytrid spores
113.	____	Phizoid	M.	The asexual spores formed by basidiomycetes and ascomycetes
114.	____	Zygospore	N.	Associated with the "gills" of mushrooms, these sacs contain the basidiospores
115.	____	Fungi Imperfecta	O.	The sexual spores associated with the mushrooms

Complete the Table

116. Complete the following table by giving the major characteristics of each phylum of the kingdom Fungi.

Phyla/Group	*Major characteristics*
A. Ascomycota	
B. Zygomycota	
C. Basidiomycota	
D. Chytridiomycota	
E. Glomeromycota	

117. For each of the phyla below, identify an associated pathogen, disease, or benefit that will help you remember this group.

Phyla/Group	*Major characteristics*
A. Ascomycota	
B. Zygomycota	
C. Basidiomycota	
D. Chytridiomycota	
E. Glomeromycota	

True/False

Mark if the statement is true or false. If it is false, justify your answer in the line below each statement.

118. _____________ It is estimated that there are approximately 1.6 million undiscovered species of fungi.

____

____

119. _____________ Because fungi are so different from animals, fungal infections are difficult to treat.

____

____

120. _____________ The products of meiosis in the fungi are spores rather than gametes.

____

____

121. _____________ Favourable environmental conditions trigger sexual reproduction in the fungi.

____

____

Short Answer

122. Why is the fungal group Deuteromycota not considered a phylum?

____

____

123. Describe some of the current evidence for the evolutionary history and origin of the fungi.

____

____

24.3 Fungal Lifestyles [pp. 538–542]

Section Review (Fill-in-the-Blanks)

All fungi are heterotrophs but can obtain their carbon from dead matter as (124) ______ or from living hosts as (125) ______. As saprotrophs, the fungi are critical components of the ecosystem, degrading (126) ________ and returning key nutrients such as phosphorus and nitrogen to the soil and recycling organic carbon to (127) ______. A lichen is a (128) ______ organism formed by a fungal species, generally an ascomycete or sometimes a basidiomycete, and either a species of (129) ______ _______ or (130) _______; the fungal partner is called the (131) _______, and the photosynthetic partner is called the (132) ______. The fungus makes up most of the body or (133) ______ of the lichen, with the photosynthetic partner usually confined to a thin layer inside of this. Despite being a compound organism, lichens are classified as species based on the (134) ______. Lichens reproduce asexual when specialized fragments called (135) _______ break off from the lichen and are dispersed by the wind, water, or passing animals; each of these structures consists of (136) ______ cells wrapped in (137) _______. The nature of the symbiotic relationship is not clear, but given that the (138) ______ ______ may penetrate the phycobiont cells and that the latter are the sole source of (139) ______ for the mycobiont, most lichens seem to represent a parasitic symbiosis. Some, however, are (140) ______ as the green algal phytobionts benefit in being protected from (141) ______ and damaging UV rays. Fungi in the (142) _______, Ascomycota, and Basidiomycota form mycorrhizal relationships with plant roots. This is a (143) _______ relationship as the plant provides organic carbon to the fungal partner, while the fungus, because it proliferates in the soil beyond the roots and has a huge (144) ______ ______, increases the plant's access to (145) ______ _______ and sometimes organic nutrients. The (146) _______ mycorrhizas formed by Glomeromycota are the oldest and most (147) _______ type of mycorrhiza, and their occurrence in the fossil record suggests that they may have been critical for the (148) _______ of ______ by plants. Ectomycorrhizas are less widespread and evolved more recently and are formed by (149) ________ and some (150) _______. They are particularly important components of Canada's (151) _______ forest and coastal (152) ______ _______. In addition to forming symbiotic relationships with plant roots, (153) ______ fungi form symbiotic relationships with above-ground plant growth. While these relationships are very poorly characterized, at least in some cases, they seem to be (154) ______ associations, with the fungal partner producing toxins that deter (155) _______.

Matching

Match each of the following terms with its correct definition.

156.	____	Thallus	A.	A cluster of fungal and algal cells used for asexual reproduction
157.	____	Soredia	B.	Edible fruiting bodies formed by ectomycorrhizal ascomycetes and oak trees
158.	____	Mycobiont	C.	Symbiotic relationship between Ascomycetes and some basidiomycetes and plant roots
159.	____	Photobiont	D.	Symbiotic relationship between fungi and above-ground plant growth
160.	____	Arbuscular mycorrhiza	E.	A chemical isolated from certain endophytic fungi
161.	____	Ectomycorrhiza	F.	The photosynthetic partner of a lichen
162.	____	Endophyte	G.	The fungal partner of a lichen
163.	____	Truffles	H.	The main body of the lichen
164.	____	Taxol	I.	Symbiotic relationship between Glomeromycota and plant roots

True/False

Mark if the statement is true or false. If it is false, justify your answer in the line below each statement.

165.______________ Lichens may be used to monitor air quality.

____

____

Short Answer

166. In addition to providing increased access to nutrients, what other potential benefit might the plant partner of a mycorrhizal relationship derive?

____

____

SELF-TEST

1. Fungi are _________. [p. 525]

   a. Prokaryotic
   b. Eukaryotic

2. Fungi are _________. [p. 525]

   a. Autotrophic
   b. Heterotrophic

3. All of the fungi produce asexual as well as sexual spores. [p. 527]

   a. True
   b. False

4. All fungi are multicellular. [p. 526]

   a. True
   b. False

5. Which of the following is the mode of nutrition in fungi? [p. 525]

   a. They are autotrophic
   b. They ingest their food
   c. They secrete enzymes to digest the nutrients in their substrate and then absorb them

6. Which of the following phyla includes species that produce flagellated spores? [p.529]

   a. Chytridiomycota
   b. Ascomycota
   c. Basidiomycota
   d. Zygomycota
   e. Glomeromycota

7. Which of the following phyla includes species that produce mushrooms? [p. 534]

   a. Chytridiomycota
   b. Ascomycota
   c. Basidiomycota
   d. Zygomycota
   e. Glomeromycota

8. Which of the following phyla includes moulds? [p. 529]

   a. Chytridiomycota
   b. Ascomycota
   c. Basidiomycota
   d. Zygomycota
   e. Glomeromycota

9. Which of the following phyla includes species that produce four to eight haploid sexual spores in a sac? [p. 531]

   a. Chytridiomycota
   b. Ascomycota
   c. Basidiomycota
   d. Zygomycota
   e. Glomeromycota

10. Which of the following phyla includes species that form arbuscular mycorrhizas? [p. 531]

    a. Chytridiomycota
    b. Ascomycota
    c. Basidiomycota
    d. Zygomycota
    e. Glomeromycota

11. Which of the following terms would not be an accurate description of the relationship between the mycobiont and the phytobiont of lichens? [pp. 539–540]

    a. Parasitic
    b. Decomposers
    c. Symbiotic
    d. Mutualistic

12. Which of the following is not a true evolutionary lineage of the kingdom Fungi? [p. 537]

    a. Chytridiomycota
    b. Ascomycota
    c. Deuteromycota
    d. Zygomycota
    e. Glomeromycota

13. Which of the following are both paraphyletic phyla? [p. 528]

    a. Chytridiomycota and Ascomycota
    b. Ascomycota and Basidiomycota
    c. Basidiomycota and Zygomycota
    d. Zygomycota and Chytridiomycota
    e. Glomeromycota and Deuteromycota

14. Fungi Imperfecti are given that name because ________. [p. 537]

   a. They do not have a definite shape
   b. They are parasitic fungi
   c. They do not reproduce
   d. Their sexual reproduction has not been observed

15. The fungal partner in a lichen is referred to as the [p. 539]

   a. Mycobiont
   b. Photobiont
   c. Mycobiont and photobiont

16. Soredia are [p. 540]

   a. Clusters of fungal spores
   b. Clusters of algal spores
   c. Clusters of the mycobiont and phytobiont

17. *Claviceps purpurea* is ______. [pp. 533–534]

   a. An ascomycete
   b. An edible fungus
   c. A spoilage fungus that may have caused ergotism in Salem, Massachusetts, which resulted in the Salem witch trials
   d. A deadly mushroom sometimes mistaken for an edible one

18. Which of the following has a prolonged dikaryon stage? [p. 536]

   a. Chytridiomycota
   b. Ascomycota
   c. Zygomycota
   d. Basidiomycota
   e. Glomeromycota

19. In terms of obtaining nutrients, fungi _________. [pp. 525–526, 532, 535]

   a. Obtain their carbon from organic material of dead or living organisms
   b. May obtain nutrients from invertebrates and insects
   c. May obtain nutrients from pollutants such as DDT and PCBs
   d. All of the above

20. Endophytes are _______. [p. 542]

   a. Of interest to biotechnology and pharmaceutical companies
   b. Poorly understood
   c. Possibly mutualistic with plants
   d. All of the above

INTEGRATING AND APPLYING KEY CONCEPTS

1. Millions of plants and animals die each year. Discuss how fungi play an important role in the decay and recycling processes.

2. Fungi are considered opportunistic organisms by virtue of their reproductive diversity. Explain how their reproduction helps them adapt to their environment.

3. Fungi are now believed to be closer to animals than plants. Discuss the evidence.

OPTIONAL ACTIVITY

1. In this chapter, you read about and saw pictures of edible delicacies called morels and truffles. These are both fruiting bodies of mycorrhizal fungi. They are also extremely expensive. Like many commercial goods, rare gems, and precious metals, the price is a reflection of the labour involved in obtaining these "crops." Why can we not just farm them like we do button mushrooms? Find out more about these fascinating foods, how they are obtained (including the traditional "truffle-hunting" pigs in France), where they grow, and the difficulties of trying to grow them rather than finding them in the wild.

25 Plants

CHAPTER HIGHLIGHTS

- Plants are thought to have evolved from green algae 425 to 490 million years ago.
- Adaptations displayed by early land plants include a waxy cuticle, lignified tissues, and internal chambers to protect the developing gamete.
- Later adaptations to land included development of root and shoot systems, vascular tissue, a shift from long-lived haploid (gametophyte) generations to a larger, long-lived diploid (sporophyte) generation, and a shift from homospory to heterospory with separate male and female gametophytes.
- Living nonvascular plants (bryophytes) include liverworts (thought to be the first land plants), hornworts, and mosses.
- Living seedless vascular plants include lycophytes (club mosses) and pterophytes (ferns, whisks, and horsetails).
- Gymnosperms were the first seed plants. A seed forms when an ovule matures following fertilization and functions to protect and help disperse the embryonic sporophyte.
- Angiosperms are flowering seed plants. Ovules are protected by ovaries, which mature into fruit that nourishes the embryo and aids in seed dispersal.

STUDY STRATEGIES AND LEARNING OUTCOMES

- This chapter makes use of the concepts of evolution, cell structure, mitosis, meiosis, and sexual reproduction developed earlier. You should briefly review these concepts to make sure that you understand them.
- Pay attention to the differences between the various groups and the evolutionary adaptations as you move from the least evolved bryophytes to the most evolved angiosperms.

- This chapter contains a lot of new information, names, and concepts. Do not worry about trying to memorize everything; the goal of this chapter is to understand the challenges faced by the movement to land and how the various adaptations to life on land help survive the conditions of this environment, attract a mate, reproduce, and disseminate offspring.
- Keep track of the headings and subheadings as you go through the different plant groups. Look for figures or examples (e.g., of a familiar type of plant) that will help you remember something about that particular group. As you progress through the chapter, keep referring back to Tables 25.1 and 25.2 as well as Figures 25.8 and 25.10; these should help you keep track of the main ideas and concepts.
- As always, go through one section of the textbook at a time and then work through the companion section of the study guide.
- Use the Study Break sections of the textbook and the Review section of the textbook, as well as this study guide section, to help guide you on the level of detail.

By the end of this chapter, you should be able to

- Identify the defining features of land plants
- Understand the alternation of generations and be able to follow the process through the various major groups of plants: the bryophytes, lycophytes, ferns, gymnosperms, and angiosperms
- Understand the distinction between sporophytes and gametophytes, spores and gametes: understand that spores and gametes are both haploid and that the spores, rather than the gametes, are the product of meiosis
- Identify the shared features (ancestral traits) as well as the distinctions (evolved traits) between the charophyte algae and the least evolved land plants, the liverworts
- Explain the role of mycorrhizas in the success of the early land plants
- Identify the key adaptations that allowed the adaptive radiations of the plants: by learning these, you will also be able to distinguish between the bryophytes, seedless vascular plants, gymnosperms, and angiosperms

TOPIC MAP

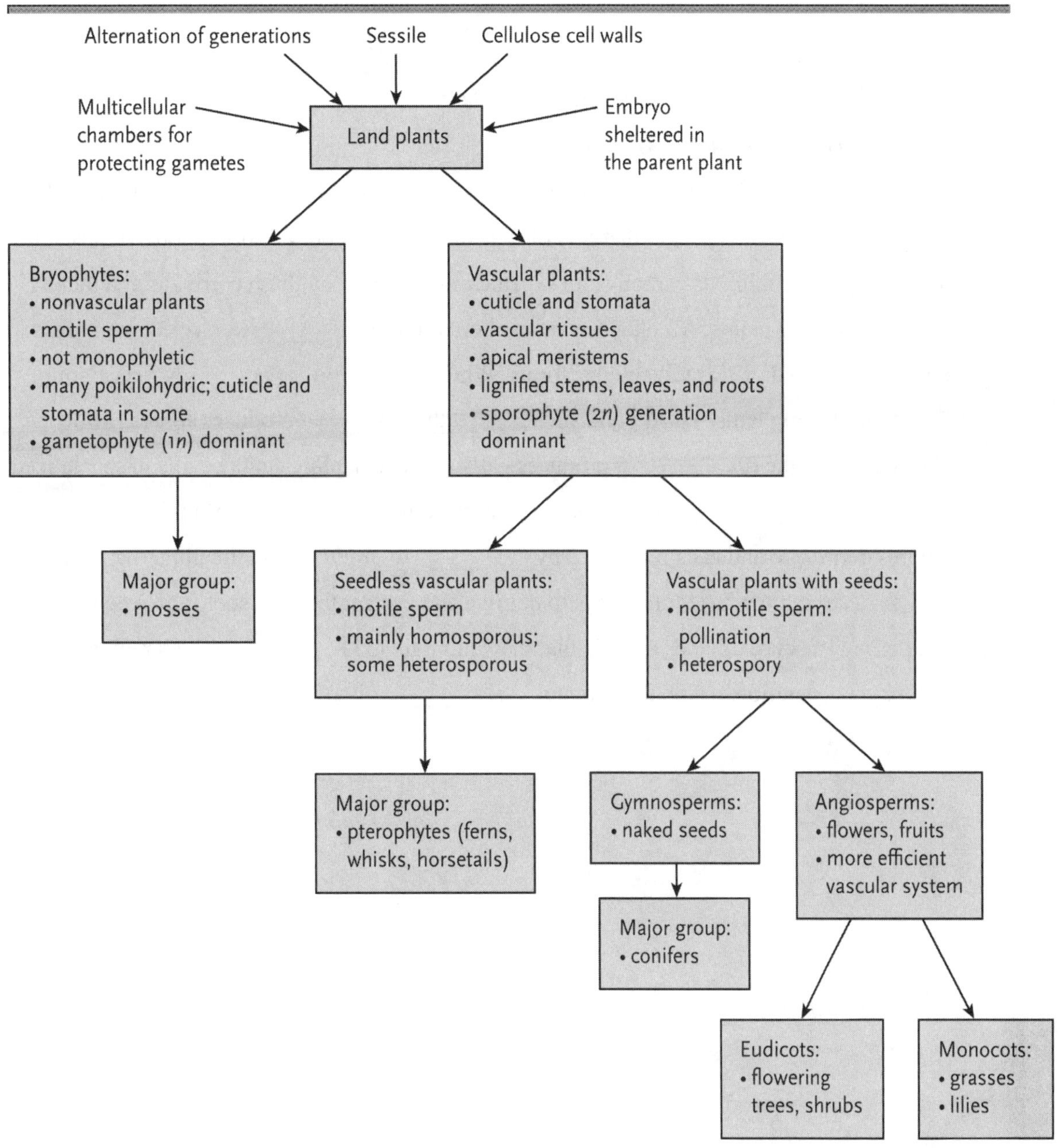
Alternation of generations
Sessile
Cellulose cell walls
Multicellular chambers for protecting gametes
Land plants
Embryo sheltered in the parent plant
Bryophytes:
• nonvascular plants
• motile sperm
• not monophyletic
• many poikilohydric; cuticle and stomata in some
• gametophyte (1n) dominant
Vascular plants:
• cuticle and stomata
• vascular tissues
• apical meristems
• lignified stems, leaves, and roots
• sporophyte (2n) generation dominant
Major group:
• mosses
Seedless vascular plants:
• motile sperm
• mainly homosporous; some heterosporous
Vascular plants with seeds:
• nonmotile sperm: pollination
• heterospory
Major group:
• pterophytes (ferns, whisks, horsetails)
Gymnosperms:
• naked seeds
Angiosperms:
• flowers, fruits
• more efficient vascular system
Major group:
• conifers
Eudicots:
• flowering trees, shrubs
Monocots:
• grasses
• lilies

INTERACTIVE EXERCISES

Why It Matters [pp. 547-548]

25.1 Defining Characteristics of Land Plants [p. 548]

Section Review (Fill-in-the-Blanks)

Land plants are multicellular eukaryotes with (1) ________ cell walls. Most, but not all, are (2) ________. Indian pipe is unusual in that it is (3) ______ and does not perform photosynthesis; it is a (4) ________, obtaining its carbon from organic molecules through mycorrhizas and the (5) _____ of neighbouring plants. All plants have an alternation of (6) _____ life cycles, going from diploid or $2n$ to haploid or $1n$ multicellular forms. Which generation is (7) _____ varies among the different groups of plants. The diploid stage, called the (8) _____, produces spores, while the haploid stage, called the (9) ________, produces gametes. The haploid stage begins in specialized cells of the $2n$ (10) ________, producing haploid spores through the process of (11) ________. These spores germinate and cells reproduce by (12) _____ to produce a multicellular haploid gametophyte: a structure whose function is to nourish and protect the next sporophyte generation. The final defining feature of land plants is that they all retain (13) ________, which represent the new sporophyte generation, inside the parental (14) ________ tissue.

Complete the Table

15. Complete the following table, providing definitions of the terms relating to land plant structures and processes.

Term	*Definition*
A. Haploid	
B. Diploid	
C. Sporophyte	
D. Gametophyte	
E. Fertilization	
F. Spores	
G. Gametes	
H. Zygote	

True/False

Mark if the statement is true or false. If it is false, justify your answer in the line below each statement.

16. ______________ Plants have cell walls and are sessile.

____

____

Short Answer

17. Explain what is meant by the term "alternation of generations."

____

____

25.2 The Transition to Life on Land [pp. 548–556]

Section Review (Fill-in-the-Blanks)

Plants are thought to have evolved from (18) ________ green algae between 425 and 490 million years ago. The earliest groups were the nonvascular (19) ________: these lack internal transport vessels and are largely (20) ________, meaning that they have little control over their internal water content. They grow close to the ground and obtain water by (21) ________; mosses are modern-day examples. Later plants evolved a waxy (22) ________ on their surface to reduce water loss and small openings or (23) ________ that allowed for control of gas exchange and water evaporation. The second group of land plants is the (24) ________ plants, named for their internal transport vessels; ferns, conifers, and flowering plants are modern-day examples. The vascular tissue consists of two types of vessel: (25) ________, which distributes water and ions, and (26) ________, which distribute sugars made during photosynthesis. This system evolved to allow plants to maximize access to (27) ______ by growing taller; growing upright was also made possible by the production of (28) ________, a polymer that strengthens cell walls, allowing plants to grow vertically and horizontally, against the force of gravity. Vascular plants also developed a(n) (29) ______ _______, a region of unspecialized constantly dividing cells near the tips of shoots and roots. (30) ________ anchor the plant to the substrate and absorb water and nutrients; for the vast majority of plants, roots are associated with fungal (31) ________, a system that vastly increases the ability of the plant's root system to acquire (32) ______. Because fungi are believed to have colonized land first, and the soil quality would have initially been poor, these symbiotic associations are believed to have been critical to the evolution of (33) ______ ______. As plants became more evolved, the diploid (34) ______ generation became dominant. These eventually produce capsules called (35) ______, which give rise to spores by (36) ________. Some plants make only one type of spore and are (37) _____, in some, producing a "bisexual" (38) ________; other plants are (39) _____ and produce two types of spores: a smaller type, a microspore, that develops into a (40) ________ gametophyte, and a larger type, a megaspore, that develops into a (41) ________ gametophyte, which became increasingly protected inside (42) ________ tissues.

Building Vocabulary

Using the definitions of the prefixes below, assign the appropriate prefix to the appropriate suffix for the following:

Prefixes	Meaning
gamet(o)-	sex cells (egg, sperm)
spor(o)-	spore
homo-	same
hetero-	different

Suffix	Meaning
-angi(o)(um)	vessel, container
-phyte	plant
-spor(e)(ous)	of or relating to a spore

Prefix	Suffix	Definition	
	-monads	43.	structure in which gametes are formed
		44.	structure in which spores are produced
		45.	the gamete-producing phase in the life cycle of a plant
		46.	the spore-producing phase in the life cycle of a plant
		47.	production of one type of spore
		48.	production of two different types of spores

Matching

Match each of the following structures with its correct definition.

49.	____	Stomata	A.	A region of unspecialized dividing cells near shoot and root tips
50.	____	Lignin	B.	Plants lacking internal transport vessels
51.	____	Spore	C.	Vascular tissue that distributes sugars made during photosynthesis
52.	____	Apical meristem	D.	Vascular tissue that distributes water and ions
53.	____	Cuticle	E.	The haploid generation of a plant
54.	____	Bryophytes	F.	The diploid generation of a plant
55.	____	Xylem	G.	A plant that makes two different kinds of spores
56.	____	Phloem	H.	A horizontal modified stem that can penetrate a substrate and provide support only
57.	____	Sporophyte	I.	Structure in which spores develop
58.	____	Gametophyte	J.	Formed through meiosis
59.	____	Sporangium	K.	Tissue specialized for the uptake of water and nutrients
60.	____	Heterosporous	L.	Tissue specialized for the absorption of light and the uptake of CO_2 from air
61.	____	Root system	M.	The waxy layer on the surface of a plant
62.	____	Shoot system	N.	Openings in the surface of the plant for the control gas exchange and H_2O evaporation
63.	____	Rhizome	O.	A polymer that strengthens cell walls

25.3 Barophytes: Nonvascular Land Plants [pp. 556–559]

25.4 Seedless Vascular Plants [pp. 559–564]

Section Review (Fill-in-the-Blanks)

Because they lack a system for conducting water, (64) ________ commonly grow in moist places. They are generally poikilohydric, lacking a vascular system and (65) ________, relying instead on rhizoids, stems, and leaves; they do, however, have leaflike and stemlike parts. The group includes (66) ________, thought to be the first land plants, (67) ______, and the familiar (68) ________ of the phylum Bryophyta. The 1*n* or (69) ______ stage is dominant. The gametes develop in a protective layer of cells called a (70) ________; for the female eggs, this structure is flask-shaped and referred to as the (71) _____, while the sperm develop in a rounded structure called an (72) ______. The sperm are flagellated so (73) ______ of the egg requires a layer of water. In mosses, the life cycle starts when a 1*n* spore lands on moist soil and (74) ______, elongating and branching into a thin web of tissue called a (75) ________. After several weeks, these give rise to leafy, green (76) ______ anchored by (77) ________. The tips of the male gametophytes develop the (78) ______, while the tips of the female gametophytes develop the (79) ________. Once released, the sperm swim down a channel in the neck of the (80) _______ attracted by a chemical attractant secreted by each (81) ______. The fertilized egg develops into the next (82) ______ generation, inside the gametophyte. The mature sporophyte consists of a stalk and a (83) ________. Some moss gametophytes possess rudimentary vessels for conducting (84) ________, and in some mosses, these vessels may be surrounded by (85) ________-conducting tissue. These structures did not, however, give rise to the xylem and (86) ________ of vascular plants. The first vascular plants did not produce (87) _____ and were the dominant plant life during the Carboniferous period. Existing members include the (88) ______ or club mosses, and the pterophytes, which includes the whisk ferns, horsetails, and (89) ______. Like bryophytes, these plants release spores and have (90) ______ sperm; however, unlike the bryophytes, they have well-developed (91) ______ tissues, roots, stems, and (92) ________. Another difference is seen in the life cycle: the diploid or (93) ______ generation is dominant and independent of the gametophyte. Spores develop in (94) ______, which, in ferns, develop on the underside or margins of the leaves. Spores are released and germinate into (95) _______, which produce two different gametangia, sometimes on the same gametophyte: the (96)________, in which the egg forms, and the (97) _______, in which the sperm form. Fertilization of the egg within its protective gametangium produces a zygote that develops into a (98) _______, which in ferns remains attached to the gametophyte until it becomes (99) ________ independent, at which point, the parent gametophyte degenerates and dies. Some seedless vascular plants are (100)

________, producing two different types of spores in two different sporangia: microspores give rise to the (101) ______ gametophyte, and (102) ______ give rise to the female gametophyte.

Matching

Match each of the following types of junction with its correct definition.

103.	____	Gametangium	A.	Cup-shaped gametophyte that produces the female egg
104.	____	Protonema	B.	The phylum that contains the ferns
105.	____	Poikilohydric	C.	Spores that give rise to the female gametophytes
106.	____	Liverworts	D.	The phylum that contains the club mosses
107.	____	Antheridium	E.	Spores that give rise to the male gametophytes
108.	____	Archegonium	F.	Structure in which gametes form
109.	____	Lycophytes	G.	Thought to be the first land plants
110.	____	Pterophytes	H.	Filamentous web of tissue
111.	____	Microspores	I.	The gametophyte that produces the sperm
112.	____	Megaspores	J.	Having little control over internal water

Complete the Table

113. Complete the table with the appropriate characteristics.

Type of plant	*Major characteristics*
Bryophytes	A.
Seedless vascular plants	B.
Life cycle	*Major features*
Mosses	C.
Ferns	D.

True/False

Mark if the statement is true or false. If it is false, justify your answer in the line below each statement.

114. ______________ The pterophyta, containing the ferns, whisks, and horsetails, are extremely diverse, growing in tropical climates, the Arctic, semi-arid deserts, and mangrove swamps.

____

____

115. ______________ In the dry arctic tundra, the bryophytes are a critical component of the food web.

____

____

116. ______________ Molecular studies have recently confirmed that bryophytes are all descended from a common ancestor.

____

____

Short Answer

117. In the fern life cycle, explain the advantage of switching from a single bisexual gametophyte to many male gametophytes.

____

____

118. Explain why bryophyte "stems" and "leaves" are not homologous structures to the stems and leaves of vascular plants.

____

____

25.5 Gymnosperms: The First Seed Plants [pp. 564–568]

25.6 Angiosperms: Flowering Plants [pp. 568–574]

Section Review (Fill-in-the-Blanks)

(119) ________, that is, the conifers and their relatives and (120) ______, the flowering plants, are the seed-bearing vascular plants. Collectively, their evolution involved important reproductive adaptations: pollen and (121) ______, the ovule and the (122) ______. The name "gymnosperm" comes from Greek and means (123) ______ seed, a description that refers to the fact that the seeds are not protected inside a fruit. The predominant gymnosperms are the conifers: in these, the male sporophyte gives rise to microspores and megaspores, which in pine trees are contained within, respectively, (124) ________ cones, generally on the lower branches, and (125) ________ cones, on the upper branches. The microspores undergo mitosis to become (126) ________ grains, immature male gametophytes. The megaspores are produced inside a protective ovule: four result from meiosis, but only one survives to become a (127) ______. When pollen grains land on a female cone, it germinates, producing a(n) (128) _____ ______, and sperm form while the (129) _____ _____ invades the ovule. Fertilization gives rise to a (130) ________, and this develops into the embryo, which remains inside the gametophyte tissue and the seed coat. In the case of pine trees, the seed is the (131) ________ ________. Three other members of the gymnosperms are very restricted in numbers and habitat: for example, the (132) ________ are restricted to warmer climates. The other two types are both remnants of ancient lineages: the (133) ________, which grows in temperate forests in China, and the (134) ______. The most successful group of plants is the (135) ________; currently, they are the most diverse plant group. In addition to evolving a more efficient (136) ______ system, defining and key reproductive structures of this group are a protective (137) ______ surrounding the ovule, a(n) (138) ______ that results from double fertilization, flowers that attract (139) _____, and (140) ______ that protect and disperse the seeds. Angiosperms are assigned to the phylum (141) ________, most of which are classified as (142) ________ or (143) ________, grasses and lilies being examples of the former and most of the flowering shrubs and trees being examples of the latter. The fruit that develops in these plants results from the fact that two sperm are delivered to the female (144) _______, with one fertilizing the egg and the other fertilizing the (145) ________-producing cell. Eventually, the ovary develops into the (146) ______, which protects the seeds but also aids in (147) ________ of the seeds through the feces of fruit-eating animals.

Matching

Match each of the following terms with its correct definition.

148.	____	Gymnosperms	A.	Sporophyte structure in which female gametophyte develops
149.	____	Angiosperms	B.	A structure that nourishes the angiosperm embryo and aids in seed dispersal
150.	____	Microspores	C.	Embryos generally possess two leaves
151.	____	Megaspores	D.	The interdependent evolution of two or more species
152.	____	Monocots	E.	An immature male gametophyte that gives rise to sperm
153.	____	Eudicots	F.	Produced in male cones and will form a pollen grain
154.	____	Ovule	G.	Plants with "naked" seeds
155.	____	Ovary	H.	The process of transferring pollen to female reproductive structures
156.	____	Pollen grain	I.	The defining feature of angiosperms
157.	____	Fruit	J.	Seeds develop in a carpel
158.	____	Coevolution	K.	Produced in female cones and will form a gametophyte
159.	____	Seed	L.	The process by which the seed of flowering plants gives rise to an embryo and to nutritive tissue
160.	____	Flowers	M.	Embryos possess a single leaf
161.	____	Double fertilization	N.	A structure formed after fertilization that includes the embryo sporophyte, nutritive tissue, and a protective cover
162.	____	Pollination	O.	Part of the carpel; matures into a fruit following fertilization

Complete the Table

163. Complete the table with the appropriate characteristics

Type of plant	*Major characteristic(s)*
Cycads	A.
B.	Only one living species; found in China and used for herbal remedies
Conifers	C.
D.	Contains only 3 genera with a total of approximately 70 species; ancient lineage
Type of plant	*Representative(s)*
Monocots	E.
Eudicots	F.

True/False

Mark if the statement is true or false. If it is false, justify your answer in the line below each statement.

164. ______________ Coconuts are the seeds of palm trees like pinenuts are the seeds of pine trees.

____

____

165. ______________ Pine resin is the sap that flows through the phloem of these trees.

____

____

166. ______________ Despite their distinctive structures, the fossil record of the evolution of the angiosperms is very poor.

____

____

Short Answer

167. Explain why plants that evolved to produce ovules had a selective advantage.

____

____

168. Explain how the evolution of pollen represented an adaptive advantage.

____

____

169. Explain the phenomenon of coevolution as it relates to the angiosperms.

____

____

SELF-TEST

1. All of the following are factors that contributed to the adaptive success of angiosperms except ____. [pp. 571–572]

   a. More efficient transport of water and nutrients
   b. Enhanced nutrition and physical protection of embryos
   c. Enhanced dispersal of seeds
   d. Requirement of water for sperm to fertilize egg

2. Which of the following is not associated with pine trees? [p. 561]

   a. Double fertilization
   b. Megasporangia on female cones
   c. Separate male and female parts on the same tree
   d. Two sizes of spores in separate cones
   e. Female cones that are larger than male cones

3. In gymnosperms, the pollen develops from ____. [pp. 565–566]

   a. The gametophyte generation
   b. Microspore cells
   c. The female gametophyte
   d. Meiosis of cells in microspore
   e. (a) and (b)

4. A strobilus is not ____. [p. 561]

    a. On a diploid plant
    b. On a vascular plant
    c. Found on lycophytes (club mosses)
    d. On a haploid plant

5. Which of the following statements about the sporophyte generation of a plant is incorrect? [p. 548]

    a. It is the haploid generation
    b. It is the diploid generation
    c. It produces spores by meiosis
    d. It is multicellular

6. The sporangia on ferns is ____. [pp. 561–562]

    a. Formed by the haploid generation
    b. Usually on the fiddleheads
    c. Often arranged in a sorus
    d. A precursor to the fiddlehead

7. Land plants are thought to have evolved from ____. [p. 549]

    a. Charophytes
    b. Mycorrhizas
    c. Bryophytes
    d. Mosses
    e. Cyanobacteria

8. The gametophyte generation of a plant ________. [p. 548]

    a. Is diploid
    b. Is haploid
    c. Produces haploid spores
    d. Produces haploid gametes by meiosis
    e. (b) and (d)

9. Plant sperm form in _________. [p. 557]

    a. Diploid gametophyte plants
    b. Archegonia
    c. Haploid sporophyte plants
    d. Antheridia

10. Which of the following is not true about spores? [p. 548]

    a. They grow into gametophytes
    b. They grow into sporophyte plants
    c. They are haploid
    d. They form from meiosis

11. Which of the following adaptations facilitated an increase in vertical growth as plants colonized land? [pp. 550–551]

    a. Lignified stems and leaves
    b. Root systems
    c. Phloem
    d. Xylem
    e. All of the above

12. The following adaptation is not involved in increased protection of the gametes of embryo: [pp. 564–565, 572]

    a. Fruit
    b. Ovary
    c. Pollen
    d. Seeds

13. Which of the following is not an adaptation that facilitated dissemination of offspring on land? [pp. 565, 571–572]

    a. Archegonia
    b. Pollen
    c. Carpels
    d. Fruit
    e. Seeds

14. Which of the following is not evidence to support the liverworts as the most closely related of the land plants to charophytes? [pp. 548, 556–558]

    a. They both have haploid and diploid multicellular phases
    b. They may have pores but do not have stomata
    c. They share more features with the charophytes than any other land plant or any of the other bryophytes
    d. Molecular data

15. The fossil record of angiosperms ________. [pp. 568–569]

    a. Is incomplete
    b. In nonexistent
    c. Lacks any transition stages
    d. Is quite extensive
    e. (a) and (c)

16. The following is not a gymnosperm: [pp. 565–569]

   a. Pine trees
   b. Coconut trees
   c. Fir trees
   d. Balsam trees

17. Rhizospheres and roots ________. [pp. 552]

   a. Are homologous structures
   b. Provide an anchor for land plants
   c. Are each associated with mycorrhizas
   d. Are part of a vascular system

18. Apical meristems ________. [p. 551]

   a. Are differentiated cells
   b. Give rise to sporangia
   c. Are the growth sites of roots and shoots
   d. Are present in all branching bryophytes

19. Mycorrhizas __________. [pp. 550–551]

   a. Are fungal symbionts
   b. Are associated with most land plants
   c. Are essential for plant nutrition
   d. Were likely critical to colonization of land by early plants
   e. All of the above

20. Double fertilization is required for _______. [pp. 571–572]

   a. Development of an endosperm in angiosperm plants
   b. Development of a zygote in heterosporous plants
   c. Development of the fruit in angiosperms
   d. Pollination in gymnosperms

21. Pollination of plants by animals involved _______. [pp. 572–573]

   a. Coevolution of flower shape and beak shape
   b. Coevolution of flower colour and visual abilities
   c. Coevolution of odour and sense of smell
   d. All of the above

INTEGRATING AND APPLYING KEY CONCEPTS

1. What is the adaptive significance of flowers? What has made angiosperms so successful? Are they more successful than the gymnosperms? If so, why?

2. What are the problems with tracing the evolution of the land plants? How was the current classification of land plants, as depicted in Figure 25.10, derived? Is it likely to change? If so, why? If not, why not?

3. Without looking at the topic map at the beginning of this study guide chapter, try to derive your own topic map, focusing on the major evolutionary steps in the structures and reproductive strategies of land plants, starting with the presumed first ones (or most ancient), the liverworts, and progressing through to the angiosperms.

OPTIONAL ACTIVITY

1. There are two very different types of plants briefly discussed in this chapter: achlorophyllous plants and the carnivorous Venus flytrap. How exactly do these plants differ from the other land plants, and why might they have evolved these unique traits? Is there an advantage? Find out more about these unique traits in plants: how widespread are they, and what is the speculated origin of these traits—mutations, or are they adaptive traits?

26 Protostomes

CHAPTER HIGHLIGHTS

- An animal is defined with an overview of animal phylogeny and classification.
- Adaptations associated with animals are explained.
- Characteristics and species examples of metazoans are described.
- The protostomes, both Lophotrochozoan and Ecdysozoan lineages, are described.

STUDY STRATEGIES AND LEARNING OUTCOMES

- Become familiar with an overview of animal phylogeny, comparing and contrasting characteristics that produced the groupings.
- Examine the morphological innovations, developmental patterns, and molecular analyses associated with animals.
- Make comparisons within and between the various animal classification groups, focusing on diversity and success of animal groups in their environments.

By the end of this chapter, you should be able to

- Explain the present animal phylogenetic tree in terms of the overall classification down to phylum within the protostomes
- Describe morphological innovations and developmental patterns that are the basis for animal classification
- Define terms relating to structural, functional, and developmental differences between groups of animals

TOPIC MAP

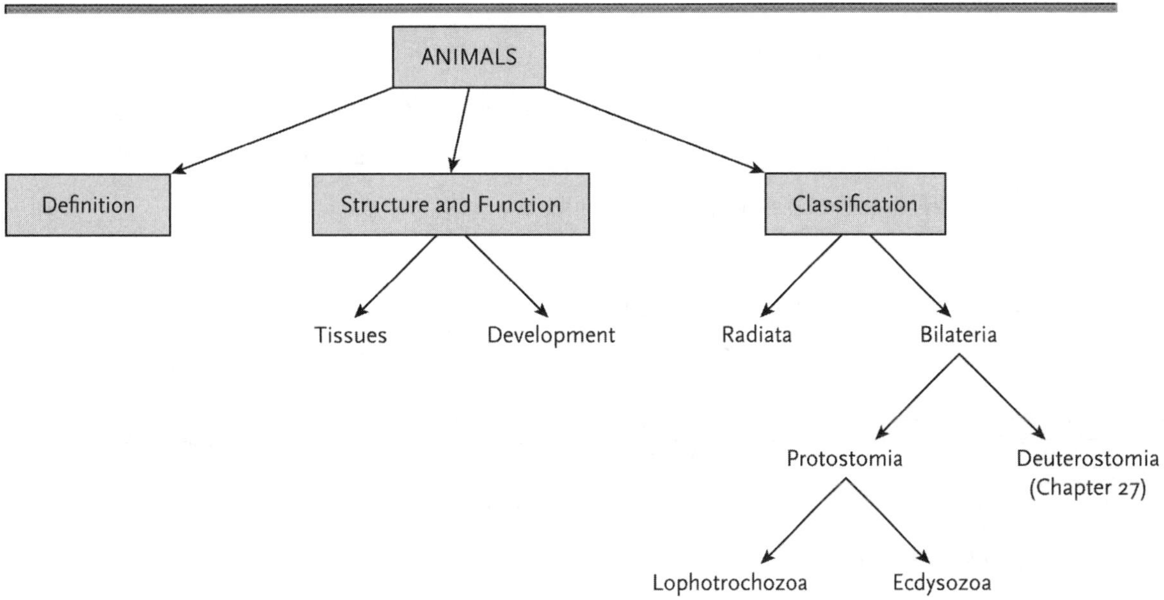
ANIMALS
Definition
Structure and Function
Classification
Tissues
Development
Radiata
Bilateria
Protostomia
Deuterostomia
(Chapter 27)
Lophotrochozoa
Ecdysozoa

INTERACTIVE EXERCISES

Why It Matters [pp. 577–578]

Section Review (Fill-in-the-Blanks)

During the (1) "_______ Explosion," approximately 540 million years ago (mya), a large number of new (2) _______ evolved. Due to a (3) _______ of the Cathedral Escarpment about 509 mya, fossilized animals in the mud were preserved and have now been discovered in (4) _______, an area in British Columbia. Although some of these fossils appear familiar, such as (5) _______ and trilobites, other species appear strange (such as *Opabinia* with its five eyes) and can only be seen as fossils, and they are now (6) _______.

26.1 What Is an Animal? [pp. 578–579]

26.2 Key Innovations in Animal Evolution [pp. 579–583]

Section Review (Fill-in-the-Blanks)

Animals are a (7) _____ group, a kingdom that has arisen from a common ancestor. Like plants, animals are multicellular organisms composed of (8) _____ (nucleus-containing) cells; however, unlike plants, animal cells (9) _______ cell walls. Animals acquire (10) _____ for their processes and activities by eating plants and other (11) _______; thus, they are (12) _______. Typically, animals are (13) _____, able to move away from any inhospitable environments, although some animals are (14) _____ at least in a portion of their life cycle. Reproduction is either (15) _______, requiring the production of egg or (16) _____, or (17) _____. Animals have a wide array of (18) _____ plans. In terms of development, more highly developed animals, the (19) _______, contain embryonic tissues. Embryos that are (20) _____ are composed of only an outer tissue, the (21) _______, and an inner tissue, the (22) _______, whereas those with three germ layers are (23) _____ and contain a third tissue in the middle called the (24) _____. Another body plan addresses animal shape, primarily in terms of the amount of symmetry present. Animals such as sponges are (25) _____, showing no symmetry; animals such as crayfish show (26) _______ symmetry since a mirror image is only observed between the right and left sides; and animals such as jellyfish show (27) _______ symmetry since mirror images can be observed in many planes along a (28) _______ axis. The body plan for bilateral animals can be (29) _______ if there is no body cavity present, (30) _____ if there is a body cavity but no (31) _______ lining the endoderm, or (32) _______ if there is a body cavity and the endoderm is lined with mesoderm tissue.

Matching

Match each of the following terms with its correct definition.

33.	____	Front	A.	True fluid-filled, lined body cavity
34.	____	Lower	B.	Cleavage pattern of protostomes
35.	____	Head end	C.	Solid body, no body cavity
36.	____	Upper	D.	Cavity from the splitting of mesoderm
37.	____	Back	E.	Blocks of repeating tissue
38.	____	Coelom	F.	Developmental fate determined at cleavage
39.	____	Peritoneum	G.	Ventral
40.	____	Spiral cleavage	H.	Posterior
41.	____	Determinate cleavage	I.	Cleavage pattern of deuterostomes
42.	____	Acoelomates	J.	Space pinched off from outpockets of archenteron
43.	____	Indeterminate cleavage	K.	Anterior
44.	____	Mesenteries	L.	Surrounds internal organs that are suspended in the coelom
45.	____	Schizocoelom	M.	Cephalization
46.	____	Enterocoelom	N.	Dorsal
47.	____	Segmentation	O.	Tissue that lines a coelom
48.	____	Radical cleavage	P.	Development fate determined after cleavage

26.3 An Overview of Animal Phylogeny and Classification [pp. 583–585]

Section Review (Fill-in-the-Blanks)

Relationships between animals have previously been based on (49) _____ (study of structure) and (50) _______ (study of prebirth development). With advances in technology, these relationships are now also based on (51) _______ analysis, usually resulting from studies of (52) _____ sequences from ribosomal RNA, (53) _______ DNA, or specific (54) _______. In some cases, the molecular analyses have supported the major (55) _______ already identified, such as the grouping of the (56) _______ and the Bilateria; however, in other cases, the molecular analyses have compelled re-evaluation of an original classification, such as the new evidence that indicated that segmentation evolved by convergent evolution independently in (57) _______ lineages rather than in two lineages. A(n) (58) _______ tree could be revised at any time as further molecular studies are completed.

26.4 Phylum Porifera [pp. 585–586]

26.5 Metazoans with Radial Symmetry [pp. 586–590]

Section Review (Fill-in-the-Blanks)

Originally believed to be Parazoan, the (59) _______ phylum is actually (60) _______, although it has no symmetry and no nerves. Sponges are (61) _______ feeders, taking in water through (62) _____ (perforations) located in the two layers of cells that compose the body wall. (63) _____ as adults, sponges are only motile as (64) _____. There are two phyla of metazoans with radial symmetry: the (65) _____, which include corals and jellyfish, and the (66) _______, the comb jellies. Organisms in both of these phyla have body walls composed of (67) _______ layers of cells that surround a (68) _______ cavity, which has a large opening, the (69) _______ at one end. Cnidarians, whether they have an upright (70) _______ form or a bell-shaped (71) _____ form, have an interesting way of capturing prey; specialized cells called (72) _______ release whipping threads, (73) _______ that sting or paralyze their prey. Ctenophores, which use (74) _______ filaments to capture their prey, are the largest animals to use (75) _______ for movement.

Choice

For each of the following terms, choose the most appropriate phylum (or phyla).

76.	____	Nematocysts	A.	Phylum Porifera
77.	____	Mesoglea	B.	Phylum Cnidaria
78.	____	Spongocoel	C.	Phylum Ctenophora
79.	____	Filter feeder	D.	Phyla Cnidaria and Ctenophora
80.	____	Gastrovascular cavity	E.	Phyla Porifera and Cnidaria
81.	____	Both polyp and medusa	F.	Phyla Porifera, Cnidaria, and Ctenophora
82.	____	Have sessile stages		
83.	____	Use cilia for locomotion		
84.	____	Anthozoans		
85.	____	Comb jellies		
86.	____	Sea anemone		
87.	____	Venus flower basket		
88.	____	Corals		
89.	____	Asymmetrical		
90.	____	Metazoan		

26.6 Lophotrochozoan Protostomes [pp. 590–603]

Section Review (Fill-in-the-Blanks)

The protostomes, all of which show (91) _______ symmetry, can be divided into (92) _____ major lineages: the (93) _______ , which consists of eight phyla, and the (94) _______, which consists of (95) _____ phyla. In the lophotrochozoan lineage, three of the phyla possess a(n) (96) _____, a fold with ciliated tentacles around the mouth that the organisms use to (97) _______ food as well as for (98) _______ exchange. Of these three lophophorate phyla, organisms belonging to the (99) _______ phylum live as colonies, organisms of the (100) _______ phylum are protected by two shells composed of (101) _____, and in the third phylum, the (102) _____ build tubes of chitin to live in. The animals of the phylum (103) _____ have flat bodies and so are commonly known as (104) _______. They have relatively complex systems, including a (105) _______ system, which can include a protruding pharynx; a (106) _______ system, which includes an anterior concentration of nerve tissue called a (107) _______; a (108) _______ system, which contains organs such as an ovary and testis; and a(n) (109) _______ system, which uses (110) _______ cells to remove wastes. Of the (111) _______ flatworm groups or classes, only one, class (112) _______, is free-living; the rest are (113) _______. An individual of the phylum (114) _____ has a wheel-like structure around its head that is covered in (115) _______, which it uses for locomotion and to obtain food. When environmental conditions are good, rotifers reproduce by (116) _____, producing female offspring from diploid (117) _______ eggs. Individuals in both Rotifera and Nemertea phyla have (118) _______ digestive tracts since food enters the (119) _____, passes through the stomach and intestines, and exits from the (120) _______. Organisms belonging to the phylum Nemertea are unique in that they have a chamber that houses a (121) _______, which they use to capture prey when (122) _______. The characteristics of the phylum Mollusca include a body that is divided into (123) _____ regions: the (124) _______, which contains many of the major organ systems; the (125) _____, which is primarily used for locomotion; and the (126) _______, which usually functions as a protective covering. This phylum is composed of (127) _______ classes, the largest being class (128) _____, snails and slugs, and the most intelligent of the invertebrates being class (129) _______, octopuses and related species. The last phylum of this lineage, phylum (130) _______, is composed of (131) _______ classes of organisms. All annelids have bodies that are divided into (132) _______ by partitions called (133) _______. Some organ systems appear as repeating units in each of the segments, such as the paired (134) ________ of the excretory system. Bristles called (135) _______ are used for traction in both class (136) _______, the bristle worms, and class (137) _______, the earthworms, but they are lacking in class (138) _______, the leeches.

Matching

Match each of the following with its correct definition.

139.	____	Platys	A.	Excretory system of flatworms
140.	____	Moll	B.	Soft
141.	____	Trochophore	C.	Few
142.	____	Metanephridia	D.	Wheel
143.	____	Helmis	E.	Folding door
144.	____	Rota	F.	Bristles
145.	____	Parthenogenesis	G.	Ring
146.	____	Gaster	H.	Excretory system of segmented worms
147.	____	Radula	I.	Cephalopods
148.	____	Pod	J.	Belly
149.	____	Oligos	K.	Unfertilized egg develops into a female
150.	____	Flame cells	L.	Food scraping device of molluscs
151.	____	Closed circulation	M.	Foot
152.	____	Valva	N.	Worm
153.	____	Anellus	O.	Larval form of molluscs
154.	____	Chaetae	P.	Flat

Complete the Table

Phylum	*Organism (common name)*
155.	Fluke
Nemertea	156.
157.	Earthworm
158.	Brachiopod
Rotifera	159.
160.	Clam

26.7 Ecdysozoan Protostomes [pp. 603–614]

Section Review (Fill-in-the-Blanks)

The adaptation of shedding a hard exterior cuticle, called a(n) (161) _______, is one of the major characteristics of the group of (162) _______ phyla that comprise the (163) _______ protostome lineage. Most of the animals in the phylum (164) _______, the round worms, are very successful (165) _____, infecting organisms in most other phyla globally. Velvet worms, belonging to the phylum (166) _____, have superficial segmentation and live primarily in the (167) _______ hemisphere. The phylum (168) _______ consists of a large and diverse group of animals, composing approximately (169) _______ of all living species. An arthropod sheds it exoskeleton by the process of (170) _______ and then replaces it as the animal (171) _______. Its body is composed of (172) _______ body segments, the head, thorax, and (173) _______, although in some species, the head and thorax fuse to form a (174) _______.

Short Answer

175. Identify and discuss one disadvantage of shedding (moulting) the external covering or exoskeleton.

____

____

176. Explain how parasitic nematodes are so successful.

____

____

Matching

Match each of the following with its correct definition.

177.	____	Crusta	A.	Claw
178.	____	Lobos	B.	Shell
179.	____	Arthros	C.	Horn
180.	____	Nemata	D.	Six
181.	____	Cheol	E.	Jointed
182.	____	Instar	F.	Thread
183.	____	Hex	G.	Lobe
184.	____	Cera	H.	Development stage

Choice

Given the animal characteristics, select the most appropriate phyla or subphyla.

185.	____	Cephalothorax	A.	Nematodes
186.	____	Antennae	B.	Arthropoda
187.	____	Book lungs	C.	Chelicerata
188.	____	Compound eyes	D.	Hexapode
189.	____	Malpighian tubules	E.	Onchophora
190.	____	Numberous unjointed legs		
191.	____	Metamorphosis		

SELF-TEST

1. From the following, what is the most unique to animals? [pp. 578–583]

   a. Sexual reproduction
   b. Eukaryotic
   c. Heterotrophic
   d. Multicellular

2. Select the pair that are not appropriately matched. [pp. 578–583]

   a. Protostomes—spiral cleavage
   b. Deuterostomes—radial cleavage
   c. Bilateral symmetry—cephalization
   d. Platyhelminthes—coelomate

3. Identify the characteristic that is different between protostomes and deuterostomes. [pp. 578–583]

   a. Segmentation
   b. Origin of mesoderm
   c. Bilateral symmetry
   d. Triploblastic

4. If an animal has determinate cleavage and the nervous system located on the ventral side of the body, it is most likely a ________. [pp. 578–583]

   a. Protostome
   b. Deuterostome
   c. Coelomate
   d. Asymmetrical animal

5. In protostomes, the blastopore develops __________. [pp. 578–583]

   a. Into a mouth
   b. Into an anus
   c. From the mesoderm
   d. From the schizocoleom

6. What embryonic tissue is the lining of the gut formed from? [pp. 579–583]

   a. Endoderm
   b. Mesoderm
   c. Ectoderm

7. Animals that have undergone cephalization in their development are __________. [pp. 579–583]

   a. Asymmetrical
   b. Radially symmetrical
   c. Bilaterally symmetrical

8. A phylogenetic tree resulting from the collection of morphological innovation, embryological patterning, and molecular analyses will not require future revision. [pp. 583–585]

   a. True
   b. False

9. If an animal has the ability to produce both eggs and sperm, it would be described as _____. [pp. 585–590]

   a. Able to undergo parthenogenesis
   b. Monoecious
   c. Able to undergo metamorphosis
   d. Parasitic

10. Some sponges produce a cluster of cells that germinate only under favourable conditions. What are these called? [pp. 585–590]

    a. Nematocyst
    b. Gemmule
    c. Polyp
    d. Porocyte

11. Which of the following does <u>not</u> undergo polymorphic development during its life? [pp. 585–590]

    a. *Obelia*
    b. Sponges
    c. Coral
    d. *Hydra*

12. Where would you expect to find animals that have a radula? [pp. 590–603]

    a. Bottom of water body or among rocks
    b. In trees
    c. Living in the respiratory system of other animals
    d. Living on the surface of a fish

13. Cephalopods have a closed circulatory system. This is thought to be due to _______. [pp. 590–603]

    a. Increased mobility
    b. High oxygen requirement for movement
    c. Slow movement along the bottom
    d. (a) and (b)

14. Select the pair that is <u>not</u> appropriately matched. [pp. 590–603]

    a. Flame cell excretory system—Platyhelminthes
    b. Setae—Annelida
    c. Corona and mastax—Nemertea
    c. Mantle—Mollusca

15. What tiny Lophotrochozoan makes up a large component of fresh water zooplankton? [pp. 590–603]

    a. Flatworm
    b. Mussel
    c. Chiton
    d. Rotifer

16. Insects differ from spiders in that spiders have ________. [pp. 603–614]

    a. The first pair of appendages modified into pedipalps
    b. Chelicerae
    c. Malpighian tubules
    d. Complete metamorphosis

17. If each segment has a pair of metanephridia and ganglia, the most likely animal would be __________. [pp. 600–614]

    a. A roundworm
    b. An annelid
    c. An insect
    d. A spider

18. Which group of arthropods consists of individuals with a separate head and thorax (not fused)?

    a. Millipedes
    b. Centipedes
    c. Insects
    d. Shrimps
    e. Spiders

19. Insects that undergo complete metamorphosis are successful because the larvae and the adults _____. [pp. 603–614]

    a. Are both able to fly
    b. Are both able to undergo pupation
    c. Occupy different habitats and eat different food
    d. Have the same body form

INTEGRATING AND APPLYING KEY CONCEPTS

1. Discuss the major characteristics that make an organism an animal.

2. Why are three "worms," tapeworms, round worms, and earthworms, not grouped in the same phylum?

3. Why is segmentation considered to be a result of convergent evolution?

OPTIONAL ACTIVITIES

1. Address several reason why nematodes and arthropods are so successful.

2. What metazoans are found in a hydrothermal vent community, and what characteristics do they have that allow them to live there?

27 Deuterostomes: Vertebrates and Their Closest Relatives

CHAPTER HIGHLIGHTS

- Characteristics of deuterostomes, from invertebrates through phylum Chordata, are presented.
- Vertebrate diversification, specialization, and adaptation to habitat are described.

STUDY STRATEGIES AND LEARNING OUTCOMES

- First, focus on major groupings of deuterostomes, identifying general characteristics associated with each.
- Next, make comparisons within and between groups, looking at the tremendous diversity of these animals.
- Examine how adaptations have enhanced success in a given habitat.

By the end of this chapter, you should be able to

- Differentiate between organisms within and between each of the major groups in terms of their characteristics and their habitats
- Explain the importance of interpreting information, such as molecular data, to classification
- Describe the two hypotheses identifying the location and starting time of the evolution of modern humans

TOPIC MAP

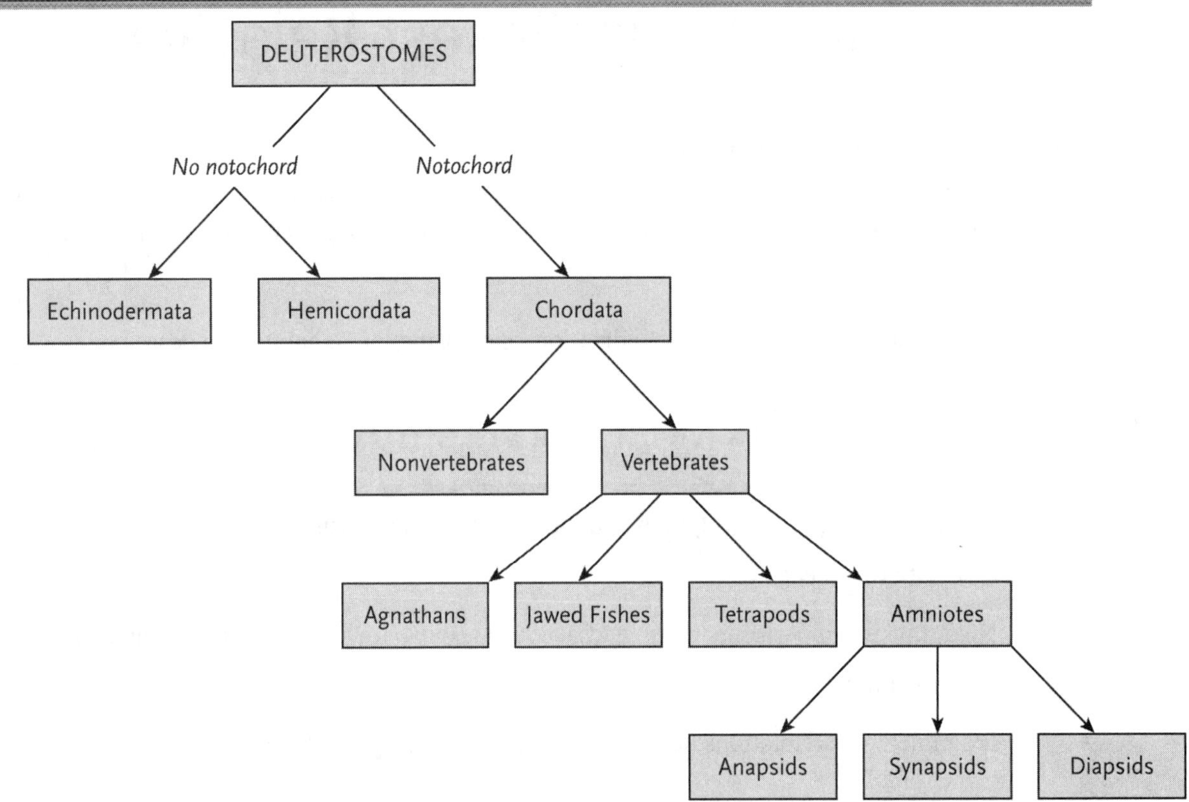
DEUTEROSTOMES
No notochord
Notochord
Echinodermata
Hemicordata
Chordata
Nonvertebrates
Vertebrates
Agnathans
Jawed Fishes
Tetrapods
Amniotes
Anapsids
Synapsids
Diapsids

INTERACTIVE EXERCISES

Why It Matters [pp. 617–619]

Section Review (Fill-in-the-Blanks)

The use of (1) _______, as well as morphology and development patterns, has provided information for (2) _______ of animals into (3) _______ trees. However, difficulties can occur when ambiguities result from analysis of information. An example is the classification of a group of ciliated marine worms called (4) _______. There was confusion as to whether this group of organisms is a protostome (belonging to the phylum (5) _______ or a (6) _______. It was determined that it is a (7) _____ when it was realized that some of the genetic and (8) _______ data belonged to mollusc prey that had been (9) _______ by the *Xenoturbella.*

27.1 Deuterostomes [p. 619]

27.2 Phylum Echinodermata [pp. 619–623]

27.3 Phylum Hemichordata [p. 623]

Section Review (Fill-in-the-Blanks)

One of the unifying characteristics of a deuterostome is that during development, the (10) _______ arises from a(n) (11) _______ opening and the (12) _______ from a second opening. In the nervous system of phylum Echinodermata, nerve cords surround the (13) _______ and radiate outwards into rays (or (14) _______), whereas in the phylum Hemichordata, a dorsal nerve (15) _______ is present but is not as developed as that found in the phylum Chordata. A(n) (16) ________ echinoderm develops from a(n) (17) _______ symmetrical larva, which, as it develops, forms a(n) (18) _______ symmetrical body, often with (19) _______ rays. This phylum uses a (20) _______ vascular system for movement, which involves the flow of fluid into and out of (21) _______ feet. The phylum (22) _______ is a small group of (23) _____ worms. These animals are organized with an anterior (24) _______ that tunnels into soil, as well as mouth and gill slits (connected by pharynx) through which water flows (25) _______ and (26) _______ the organism, respectively, trapping food particles.

Matching

Match each of the following roots with its correct definition.

27.	____	Chord	A.	Half
28.	____	Derm	B.	Opening
29.	____	Echino	C.	Skin
30.	____	Hemi	D.	Second
31.	____	Stomia	E.	String
32.	____	Deuteron	F.	Spiny

For each of the following characteristics or structures, match the term with its description from the list below.

33.	____	Proboscis	A.	Calcium-stiffened structures of endoskeleton
34.	____	Water vascular system	B.	Surface opposite to the mouth
35.	____	Aboral	C.	Sievelike plate through which water enters
36.	____	Ring canal	D.	Used to burrow into sand or mud
37.	____	Tube feet	E.	Location where water exits in hemichordates
38.	____	Madreporite	F.	Small pincers at base of spines to remove fallen debris
39.	____	Gill slits	G.	Fusion of ossicles; protection but reduced flexibility
40.	____	Test	H.	Protrusions for movement using internal flow of fluids
41.	____	Pedicellariae	I.	System of fluid-filled canals used for movement
42.	____	Ossicle	J.	Tube that surrounds esophagus

27.4 Phylum Chordata [pp. 623–625]

27.5 The Origin and Diversification of Vertebrates [pp. 625–628]

Section Review (Fill-in-the-Blanks)

All organisms in the phylum Chordata have four characteristics, two of which are a (43) _______ (a flexible rod of fluid-filled cells) and paired (44) _______ in the pharynx. Within this phylum, two of the (45) _______ subphyla do not have a (46) _______ column, so these organisms are (47) _______. The subphylum that has a(n) (48) _____, or spine, usually made of (49) _____, is the subphylum (50) _______. An organism in this group also has a protective covering for the brain, a (51) _______, and (52) _______ crest cells. The vertebrates are a diverse group of organisms, much of the diversity arising from the repeated (53) _____ of a homeotic gene called the (54) _______ gene. These genes are arranged in a particular order to form a (55) _______, such that each of the component genes controls the (56) _______ of a particular structure (such as eyes, wings, or legs) and therefore the (57) _______ shape of the organism. It appears that animal groups with a more (58) _______ structure, such as humans, have (59) _______ *Hox* genes than animal groups with a simpler structure, such as a hagfish.

There is one group of vertebrates that lacks a(n) (60) _______, the agnathans. All other vertebrates are considered to belong to the lineage (61) _______. Within this group, the change from an aquatic habitat to a more (62) _______ habitat was accomplished by (63) _______, who use (64) _______ limbs for (65) _______. One lineage of tetrapods, the (66) _______, produces specialized eggs that can survive on land, further increasing success in a(n) (67) _______ habitat.

Matching

Match each of the following roots with its correct definition.

68.	____	Noto	A.	Not
69.	____	Tetra	B.	Head
70.	____	Gnath	C.	Back
71.	____	Stoma	D.	Mouth
72.	____	Uro	E.	Tail
73.	____	A	F.	Foot
74.	____	Pod	G.	Jawed
75.	____	Cephalo	H.	Four

Choice

For each of the following characteristics or structures, select the most appropriate animal group from the list below.

76.	____	Dorsal hollow nerve chord	A.	Subphylum Vertebrata
77.	____	Atriopore	B.	Phylum Chordata
78.	____	Vertebral column	C.	Subphylum Urochordata
79.	____	Head-to-tail notochord	D.	Subphylum Cephalochordata
80.	____	Incurrent siphon		
81.	____	Bone		
82.	____	Segmented muscle in body wall and tail		
83.	____	Cranium		
84.	____	Neural crest		
85.	____	Gill slits in pharynx		
86.	____	Oral hood		

27.6 Agnathans: Hagfishes and Lampreys, Conodonts, and Ostracoderms [pp. 628–629]

27.7 Jawed Fishes: Jaws Expanded the Feeding Opportunities for Vertebrates [pp. 629–635]

Section Review (Fill-in-the-Blanks)

Only two groups of jawless fishes, the (87) _______, are living: the (88) _______, which feed on invertebrates and dead or dying fish, and the (89) _____, many of which feed by rasping a hole in a live fish with its (90) _______ disk and ingesting the body fluids. Jaws evolved from fish gill (91) _______, allowing fish to eat (92) _______ food pieces. One group of jawed fish, the class (93) _______, has derived a trait from the ancestral bony endoskeleton; these fish are entirely cartilaginous. Both skates and rays, which have a (94) _______ flattened shape, belong to this group, as well as (95) _______, a dominant predator in the ocean. (96) _______ fish have a lightweight (97) _______, consisting of a cranium, a vertebral column, and bones supporting fins, as well as a covering of smooth and lightweight (98) _______ that are coated with (99) _______. The majority of fish living are (100) _______ fish belonging to the subclass (101) _______.

These fish have a(n)) (102) _____ to protect each chamber containing the gills, as well as an improved sensory system, which includes a (103) _______ line system and (104) _______ nostrils. The diversity of body shape within this group is a result of the variation in (105) _______ (food ranges from plankton to other vertebrates) and (106) _______ (power and control of swimming movements).

Matching

Match each of the following roots, terms, or structures with its correct definition.

107.	____	Plac	A.	Oil in liver of sharks
108.	____	Chondr	B.	Sarcopterygii
109.	____	Lateral line system	C.	Ray
110.	____	Ichthy	D.	Increases surface area in digestive tract
111.	____	Sturgeons	E.	Hydrostatic organ
112.	____	Claspers	F.	Fish
113.	____	Acanth	G.	Plate
114.	____	Squalene	H.	Cartilage
115.	____	Acti	I.	Flap covering chamber containing gills
116.	____	Spiral valve	J.	Actinopteryglii
117.	____	Swim bladder	K.	Male reproductive specializations
118.	____	Lungfish	L.	Spine
119.	____	Ptery	M.	Detects vibrations in water
120.	____	Operculum	N.	Fin

27.8 Early Tetrapods and Modern Amphibians [pp. 635–637]

Section Review (Fill-in-the-Blanks)

Life on land requires several important characteristics, including a system for locomotion and adaptations for survival in a(n) (121) ______-breathing environment. For example, early tetrapods developed a (122) _______ system that included a more robust vertebral column and a (123) _______ girdle that was not fused with the cranium (both of which aided in movement), as well as a (124) _______ on either side of the head to detect (125) _______ waves. Many adult amphibians have a thin, scaleless (126) _______ through which gas exchange can occur, although it must be kept (127) _______. Amphibians at the (128) _______ stage, the developmental portion of their life cycle, often require a(n) (129) _______ habitat, but as adults, they live on land or in water or move between (130) _______ habitats. There are three groups of amphibians: the (131) _______, which consists of frogs and (132) _______; the (133) _______, which are the salamanders; and the (134) _______, which are the caecilians.

Matching

Match each of the following with its correct definition.

135.	____	Gymnos	A.	Bone homologous to support structure of jaws
136.	____	Bios	B.	Both
137.	____	Stapes	C.	Membrane that vibrates with sound waves
138.	____	Del	D.	Snakelike
139.	____	Amphi	E.	Naked
140.	____	Ophioneos	F.	Visible
141.	____	Tympanum	G.	Life

27.9 The Origin and Mesozoic Radiations of Amniotes [pp. 638–642]

Section Review (Fill-in-the-Blanks)

Amniotes are characterized by a fluid-filled (142) _______ that surrounds the (143) _______, called the (144) _______. These animals have several important characteristics that enhance their ability to survive in a(n) (145) _______ environment. First, their (146) _______ will not dehydrate in air because its cells are filled with (147) _____ and lipids, which makes it (148) _____. Second, the (149) _______ egg has a hard, leathery (150) _______ and four (151) _______ that makes it resistant to (152) _______ (drying up) but facilitates exchange of (153) _____, as well as water, between the egg and its environment. In addition, the egg (154) _______ provides energy and (155) _____ provides nutrients and water to the embryo. The third characteristic of amniotes is that they produce nontoxic (156) ______ (or urea) waste instead of (157) _____. There are three groups of amniotes: the (158) _______, which produce no temporal arches; the (159) _______, which produce one temporal arch; and the (160) _______, which produce two temporal arches in the temporal region of the (161) _______.

Matching

Match each of the following with its correct definition.

162.	____	Morph	A.	Lizard
163.	____	Di	B.	Scale
164.	____	Apsid	C.	With
165.	____	Sauro	D.	Two
166.	____	Syn	E.	Ruler
167.	____	Archo	F.	Form
168.	____	Lepi	G.	Connection

27.10 Subclass Testudines: Turtles and Tortoises [pp. 642–643]

27.11 Living Diapsids: Sphenodontids, Squamates, and Crocodilians [pp. 643–645]

27.12 Aves: Birds [pp. 645–649]

Section Review (Fill-in-the-Blanks)

Turtles and tortoises are amniotes that are (169) _______ since they have no temporal arches in their skulls. An animal in this group is characterized by its boxlike (170) _______, which has a (171) _____ fused to it. When under attack, it can retract its (172) _______ and all four (173) _______ into it. Another of the amniotic groups, the (174) _______, have (175) ______ temporal arches. Animals classified within this lineage include the lizardlike group (176) _______, the group composed of lizards and (177) _____, the group composed of crocodiles, (178) _______, and gavials, and the group composed of animals with wings, the (179) _______. There is extensive variation within the aves group due to their ecological specializations; for example, (180) _______ size and shape reflect diet and (181) _______ size and shape reflect swimming or flight requirements.

Choice

For each of the following characteristics or structures, select the most appropriate animal group from the list below.

182.	____	Third eye	A.	Turtles
183.	____	Keeled sternum	B.	Birds
184.	____	Lacking legs	C.	Crocodiles and alligators
185.	____	Furculum	D.	Tuatara
186.	____	Overlapping keratinized scales	E.	Lizards and snakes
187.	____	Muscle analogous to diaphragm in mammals		
188.	____	Boxlike shell		
189.	____	Uncinate processes on ribs		
190.	____	Girdles inside rib cage		

27.13 Mammalia: Monotremes, Marsupials, and Placentals [pp. 649–652]

27.14 Evolutionary Convergence and Mammalian Diversity: Tails to Teeth [pp. 652–657]

27.15 The Evolution of Humans [pp. 657–662]

Section Review (Fill-in-the-Blanks)

There is extensive variation among (191) _______, the amniotes that are (192) _______ because they have only one (193) _______ arch. Although there are no definitive characteristics, most mammals are (194) _______, a body covering that helps to maintain a stable and elevated temperature, since they are (195) _______. Other structures usually found in these organisms include a layer of muscle separating the chest cavity and viscera, the (196) _______, two (197) _______ condyles where the skull attaches to the neck, a secondary (198) _______ comprising the roof of the mouth, and a brain (199) _____ for processing information. Although some mammals have no teeth, many are (200) _____, having different teeth for different functions, as well as (201) _______, producing an earlier set of (202) _______ teeth and a later set of (203) _______ teeth. Many mammals feed their young (204) _______, which is produced in mammary glands. One of the major differences between groups of mammals is their mode of (205) ______.

One group, the (206) _______, lay eggs; another group, the (207) _____ mammals, produce offspring that complete their embryonic development in the uterus; and a third group, the (208) _____, produce offspring that begin their development in the uterus but finish their development attached to a (209) _______ located in a (210) _______. In mammals, it is common for both evolutionary (211) _____, such as the (212) _______ flattened tail found in different groups of mammals, and high (213) _______, such as the variation in mammalian teeth, to occur.

(214) _______, the lineage that includes humans, are characterized by (215) _______ locomotion and an (216) _______ stature. Hands are available to hold objects with a firm (217) _____ grip or a manipulative (218) _______ grip. There are two (219) _______ that identify the time and location of the start of modern human evolution: the (220) _______ Hypothesis, which suggests that humans evolved in many regions simultaneously less than 0.5 million years ago, and the (221) ______ ______ Hypothesis, which suggests that *Homo sapiens* arose in Africa less than (222) _______ years ago. Both hypotheses are still being debated.

Matching

Match each of the following derivatives and characteristics with its correct definition.

223.	____	Eu	A.	Wild beast
224.	____	Heterodont	B.	Hoof
225.	____	Trema	C.	Between
226.	____	Precocial	D.	Helpless at birth
227.	____	Marsupium	E.	Two generations of teeth
228.	____	Theri	F.	Perforation
229.	____	Proto	G.	Quickly mobile when born
230.	____	Atricial	H.	Good
231.	____	Meta	I.	Cheek teeth
232.	____	Ungula	J.	First
233.	____	Molar	K.	Purse
234.	____	Diphyodont	L.	Individual teeth specialized for different functions

SELF-TEST

1. Which characteristics are correct for Echinoderms? [pp. 619–623]

   a. Radially symmetrical larvae
   b. Bilaterally symmetrical adults
   c. Tube feet that attach to substrate when filled with water
   d. Internal sexual reproduction; male gametes entering through madreporite

2. If an animal has pedicellariae and a larval form with bilateral symmetry, it belongs to which group? [pp. 619–623]

   a. Ophinuroidea
   b. Asteroidea
   c. Echinoidea
   d. Urochordata

3. One of the major differences between animals of Echinodermata and Hemichordata is ______. [pp. 619–623]

   a. Anterior proboscis
   b. Gill slits
   c. A dorsal hollow nerve chord
   d. (a) and (b)

4. Sea cucumbers _________________. [pp. 619–623]

   a. Have radially symmetrical bodies as adults
   b. Have 10 rows of tube feet
   c. Have a symbiotic relationship with pearl fish
   d. Belong to the Astroidea group

5. What subphylum of chordates are free-living larvae and sessile adults? [pp. 623–628]

   a. Echinodermata
   b. Hemichordata
   c. Urochordata
   d. Cephalochordata
   e. Vertebrata

6. Which group would you expect to have the greatest number of *Hox* gene complexes? [pp. 623–628]

   a. Lamprey
   b. Hagfish
   c. Placoderm
   d. Seahorse

7. Which of these classes of fish do not contain any bone (totally cartilaginous)? [pp. 628–635]

   a. Actinopterygii
   b. Chondrichthyes
   c. Placodermi
   d. Sacropterygii

8. Which of the following are retained and used in the adult jawed fishes? [pp. 628–635]

   a. Gill arches
   b. Radial symmetry
   c. Segmented nervous system
   d. None of these

9. Which of the following belong to the Gnathostomata lineage? [pp. 628–635]

   a. Hagfishes
   b. Lampreys
   c. Conodonts
   d. Acanthodian

10. If the swim bladder of a bony fish were destroyed, what would be the effect? [pp. 628–635]

    a. There would be no effect
    b. The fish would not be able to breath underwater
    c. The fish would be closer to the surface of the water
    d. The fish would be farther away from the surface of the water

11. One of the major differences between life in an aquatic versus a terrestrial environment is that terrestrial animals could _____. [pp. 635–637]

    a. Become easily dehydrated
    b. Become disoriented due to a smaller brain
    c. Be less likely to find their mates
    d. Die from starvation due to lack of food

12. Why are blood vessels under the skin in adult amphibians important? [pp. 635–637]

    a. Keep animals warm
    b. Carry gases to and from the site of gas exchange
    c. Provide a method of sensing sound waves
    d. Provide support for this class of tetrapods

13. In addition to the amniotic egg, which of the following is advantageous in a terrestrial environment? [p. 638–642]

    a. Keratin and lipid in skin cells
    b. Ammonium as a waste product
    c. Perforated pharynx in the adult form
    d. All of these

14. Which of the diapsids have a four-chambered heart? [pp. 642–649]

    a. Birds
    b. Crocodiles
    c. Turtles
    d. (a) and (b)
    e. (b) and (c)

15. Of the following, which is the most advantageous for flight? [pp. 642–649]

    a. Amniotic egg
    b. Bipedal locomotion
    c. Hollow limb bones
    d. Dorsal hollow nerve chord

16. Which group of amniotes detects prey using a long tongue and the sensory receptors in the roof of its mouth? [pp. 642–649]

    a. Birds
    b. Alligators
    c. Turtles
    d. Tortoises
    e. Snakes

17. Which of the following has a prototheria reproductive mode? [pp. 649–662]

    a. Monotremes
    b. Mammals that lay eggs
    c. Duck-billed platypus
    d. All of these

18. Which of the following is an example of convergent evolution? [pp. 649–662]

    a. Dorsoventrally flattened tail
    b. Development of protective spines from hair
    c. Teeth used for different functions, such as crushing or cutting
    d. (a) and (b)

19. How many times can mammals replace teeth? [pp. 649–662]

    a. > 10 times
    b. 2–10 times
    c. One time
    d. Mammals cannot replace their teeth

20. Hominids are able to play the piano because of _______. [pp. 649–662]

    a. A power grip
    b. A precision grip
    c. Bipedal locomotion
    d. (a) and (b)

INTEGRATING AND APPLYING KEY CONCEPTS

1. Discuss how the adult echinoderm—a sea star—is significantly different from a human, yet both are considered to be deuterostomes.

2. Address the major characteristics or adaptation that enhanced or allowed movement from an aquatic to a terrestrial environment.

OPTIONAL ACTIVITIES

1. Jack Berrill's work on tunicates indicates that they are able to undergo regeneration. How do these animals do this, and how does this process fit in with morphogenesis?

2. How are geckos able to "unstick" to surfaces?

28 The Plant Body

CHAPTER HIGHLIGHTS

- Plants can survive on land by having a shoot system and a root system to provide necessary resources and mechanisms.
- Meristems are responsible for the development of all new tissue and therefore plant primary and secondary growth. There are fundamental differences between monocots and eudicots.
- Ground tissue, vascular tissue, and dermal tissue systems all have important adaptive functions in the life of plants.
- The primary shoot system and root system both function in support, growth, and transport of water and solutes and storage; only the primary shoot system functions in photosynthesis, and only the root system is responsible for acquisition of water and minerals.

STUDY STRATEGIES AND LEARNING OUTCOMES

- Concentrate on the key concepts of the overview of plant structure and growth before tackling the specifics.
- Focus on terminology; otherwise, understanding concepts will be difficult.

By the end of this chapter, you should be able to

- Describe differences between plant and animal structure and growth
- Differentiate between root system and shoot system structures, components, and their functions
- Identify and describe three different ways to classify angiosperms
- Describe plant primary and secondary growth for stems and roots

TOPIC MAP

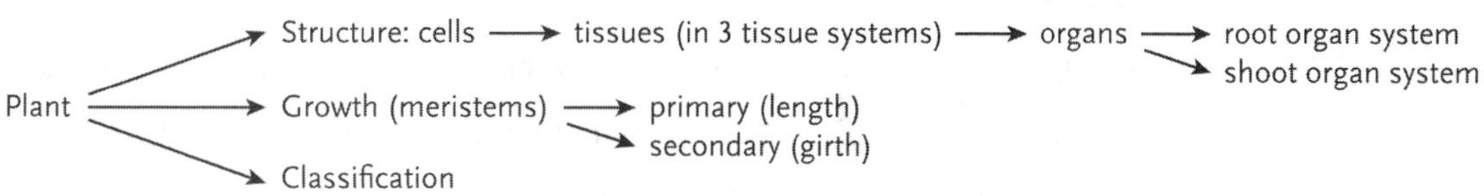

INTERACTIVE EXERCISES

Why It Matters [pp. 667–668]

Section Review (Fill-in-the-Blanks)

Most plants develop from seeds located in soil, producing an underground (1) _____ system that accesses (2) _____ and nutrients and an above-ground (3) _____ system that provides energy through the process of (4) _____. Banyan trees, however, produce seeds that germinate in the (5) _____ of a host tree, form (6) _____ downward around the host's trunk and into the ground, and then (7) _____ upward from which large leaves outcompete the host for light.

28.1 Plant Structure and Growth: An Overview [pp. 668–673]

Section Review (Fill-in-the-Blanks)

Both the shoot and root plant systems are composed of (8) _____, which are structures that are composed of several tissues that have a specific form and function. Stems, leaves, and, if present, flowers are organs that make up the (9) _____ of a plant. Within an organ, each tissue is composed of groups of (10) _____ that function together to perform specific tasks. Although animal and plant cells contain many of the same components, only plant cells contain (11) _____, which function in photosynthesis; (12) _____, which can provide storage and assist in cell elongation and cell rigidity; and (13) _____, which give cells strength and flexibility. Many plants have a secondary cell wall within their cells that is laid down inside the primary, cellulose cell wall. The secondary cell wall contains (14) _____, which is a waterproof substance that also strengthens the cell wall.

Animals usually grow to a certain size and then stop. This is known as (15) _____. Plants, on the other hand, can continue to grow throughout their lifetime, so they experience (16) _____. New tissues in plants arise from (17) _____, which is embryonic tissue. All vascular plants contain (18) _____ at the tip of the shoot and root, which give rise to the (19) _____ of a plant during primary growth and make up the (20) _____. Some plants, especially ones that produce woody tissues, produce (21) _____ from cylinders of embryonic tissue known as (22) _____, and this results in growth in diameter. This type of growth is known as (23) _____.

The two major lineages of flowering plants have different body plans. These two lineages are known as (24) _____ and (25) _____. Plants also vary in their life spans. Plants that can complete the entire life cycle in one growing season are known as (26) _____, those that need two full growing seasons and may produce limited amounts of secondary tissue are (27) _____, and plants that continue vegetative growth and reproduction throughout their lives are known as (28) _____.

True/False

If the statement is true, write a "T" in the blank. If the statement is false, make it correct by changing the underlined word(s) and writing the correct word(s) in the answer blank.

29. _____________ Plants undergoing totipotency are unable to dedifferentiate.

30. _____________ Solutes move between adjacent plant cells through plasmodesmata.

31. _____________ Increases in the height of a plant are due to cell divisions in the root apical meristem.

32. _____________ Lateral meristems give rise to primary tissues.

33. _____________ Herbaceous plants show little or no secondary growth.

Complete the Table

Complete the following table by filling in the appropriate description of monocots or eudicots for each of the traits identified.

Trait	*Monocots*	*Eudicots*
Flower parts	34.	Multiples of 4 or 5
Vascular bundles in stem	35.	Arranged in a ring
Leaf veins	Parallel	36.
Root systems	Fibrous	37.

28.2 The Three Plant Tissue Systems [pp. 673–678]

Section Review (Fill-in-the-Blanks)

Plants have three tissue systems that are the basis of plant organs. These tissue systems are the (38) _____, which makes up most of the plant body; the (39) _____, which makes up the plant's circulatory system; and (40) _____, which forms the "skin" of the plant.

There are three types of cells that make up ground tissue. Thin-walled and irregularly shaped (41) _____ cells comprise most of the primary body (roots, stems, leaves, flowers, and fruits), (42) _____ cells provide flexible support and tend to be elongated, and (43) _____ cells provide rigid support and protection. These latter cells typically have thick, lignified secondary cell walls that ultimately cause cell death.

The vascular tissue system is specialized for conduction of water and solutes through cells abutted (44) _____ forming pipeline-type structures. (45) _____ conduct water and minerals upward from the roots. It is composed of two types of cells: (46) _____, which are elongated cells with overlapping tapered ends, and/or (47) _____, which are shorter cells with perforations in their ends. The other type of vascular tissue is the (48) _____. It is primarily composed of cells called (49) _____, which conduct solutes, mostly sugars. In many plants, adjacent, specialized parenchyma cells called (50) ____ communicate with sieve tube members via plasmodesmata and help regulate their function.

Dermal tissue is composed of several types of cells. The (51) _____ covers the plant body in a continuous layer and is covered with a waxy, waterproof layer called the (52) _____. Specialized epidermal cells include paired and crescent-shaped (53) _____, which regulate gas exchange by controlling the size of the central spaces within (the (54) _____). Another type of specialized epidermal cell is a (55) _____, which may function to increase root surface area (maximize water and mineral uptake) or contain sugars or irritants.

Choice

For each of the following plant structures, choose the most appropriate cell type from the list below.

a. Collenchyma b. Parenchyma c. Sclerenchyma d. Tracheid

e. Sieve tube member f. Companion cell

56. ____ The hard outer layer of a cherry pit

57. ____ The plant stem; aids sieve tube members

58. _____ The "string" in a stalk of celery

59. ____ The plant stem; transports sugars

60. _____ The plant leaves; where photosynthesis occurs

61. _____ The plant stem; transports water

28.3 Primary Shoot Systems [pp. 678–682]

Section Review (Fill-in-the-Blanks)

A leaf stalk is a (62) _____ that attaches to a stem at a(n) (63) _____. The distance between two of them is a(n) (64) _____. Buds at the tip of the main shoot are (65) _____, while buds that produce branches are (66) _____. Terminal buds produce a hormone that inhibits the development of nearby lateral buds, a phenomenon known as (67) _____. When a cell in an apical meristem divides, one of the daughter cells gives rise to one of three types of (68) _____, unspecialized tissues. One of these tissues, the (69) _____, becomes the stem's epidermis; a second tissue, the (70) _____, ultimately forms the ground tissue of the plant; and the primary vascular tissues are derived from the third type of tissue, the (71) _____. These vascular tissues are usually arranged into chords of xylem and phloem known as (72) _____. In the roots and stems of most eudicots, these chords are arranged in a(n) (73) _____, separating the ground tissue such that cells toward the outside of the cell comprise the (74) _____, while those toward the centre of the cell form the (75) _____.

Along the sides of shoot apical meristems (76) _____ give rise to mature leaves. A mature leaf has an upper and a lower epidermis, between which is the (77) _____ region, which can be divided into a palisades layer and a spongy layer. The vascular bundles of leaves are arranged in various patterns visible on the surface and are known as (78) _____.

Matching

Match each of the following plant parts with its correct name.

79.	____	Onion "head"	A.	Bud
80.	____	Strawberry stem	B.	Bulb
81.	____	Potato "eye"	C.	Rhizome
82.	____	Cactus leaf	D.	Petiole
83.	____	Celery stalk	E.	Stolon
84.	____	Ginger "root"	F.	Spine

28.4 Root Systems [pp. 683–685]

28.5 Secondary Growth [pp. 685–691]

Section Review (Fill-in-the-Blanks)

The roots of most eudicots exhibit a(n) (85) _____ system, which consists of a main, large root with smaller branching roots called (86) _____. Most monocots have a highly branched root systems know as a(n) (87) _____ system. Roots that emanate from the stem or some other region of a plant, such as the prop roots of corn, are called (88) _____ roots. The root apical meristem is covered by a(n) (89) _____, which protects it as it grows through the soil. The more actively dividing cells behind it are in the (90) _____, where one daughter cell becomes a primary meristem (as discussed for primary shoots). Above this region is the (91) _____, where most cells increase in length, followed by the (92) _____, where cells complete their differentiation. The outer layer of root cortex cells may develop into a(n) (93) _____, which limits water loss and regulates ion uptake, whereas the innermost layer becomes the (94) _____, which regulates water and ion uptake into the (95) _____ tissue system. The cells immediately interior to the endodermis comprise the (96) _____, which, under the influence of chemical growth regulators, can give rise to (97) _____ roots. Secondary growth arises from two types of lateral meristems: (98) _____, which gives rise to secondary xylem and phloem, and (99) _____, which produces (100) _____ cells to replace the ruptured cortex and (101) _____ that were formed during primary growth. In stems and roots, (102) _____ cells form from the vascular cambium toward the centre and (103) _____ cells form outward from the vascular cambium. As secondary xylem accumulates and ages, it hardens into the tissue known as (104) _____. Older wood, near the centre of the stem, often accumulates material that clogs the vessels as it dries and hardens. These xylem cells make up the (105) _____, while the younger, moister xylem cells closer to the vascular cambium comprise the (106) _____.

All the living tissue between the vascular cambium and the surface of the stem comprises the (107) _____. It consists of secondary phloem, cork cambium, and cork.

True/False

If the statement is true, write a "T" in the blank. If the statement is false, make it correct by changing the underlined word(s) and writing the correct word(s) in the answer blank.

108.____________ Primary phloem cells are present in woody plants.

109.____________ Heartwood consists of xylem vessels that are clogged and no longer conduct water.

110.____________ Root hairs form in the zone of elongation.

111.____________ The three primary meristems are formed in the zone of cell division.

SELF-TEST

1. The height of a lily plant's stem is an example of its [p. 668]

   a. Anatomy
   b. Physiology
   c. Morphology
   d. Life span

2. The layer between the primary cell wall of adjacent plant cells is the ________. [p. 669]

   a. Middle lamella
   b. Protoplast
   c. Secondary cell wall
   d. Vascular cambium

3. Herbaceous plants do not usually contain which of the following? [p. 669]

   a. Primary tissues
   b. Secondary tissues
   c. Root systems
   d. Vegetative shoots

4. A ________ shoot produces flowers and, ultimately, fruits. [p. 670]

   a. Floral
   b. Nodal
   c. Reproductive
   d. Vegetative

5. The order of the major components of a plant cell in a woody plant from centre outward is [p. 670]

   a. Cell membrane, cytoplasm, primary cell wall, secondary cell wall
   b. Cytoplasm, cell membrane, secondary cell wall, primary cell wall
   c. Cell membrane, primary cell wall, cytoplasm, secondary cell wall
   d. Cytoplasm, secondary cell wall, cell membrane, primary cell wall

6. The self-perpetuating embryonic cells of plants is called ________ tissue. [p. 671]

   a. Dermal
   b. Ground
   c. Meristematic
   d. Vascular

7. The tissue responsible for growth in the diameter of a plant is ________. [p. 672]

   a. Lateral mersitem
   b. Root apical meristem
   c. Shoot apical meristem
   d. (b) and (c)

8. Which of the following characteristics would you expect to find in a lily? [p. 673]

   a. Two cotyledons
   b. Petals in multiples of 3
   c. Netlike leaf veins
   d. Taproot

9. Which type of ground tissue provides the greatest support as it produces cells with thick secondary walls? [p. 675]

   a. Collenchyma
   b. Sclerenchyma
   c. Parenchyma

10. Which cell type is most likely to be photosynthetic? [p. 675]

    a. Collenchyma
    b. Parenchyma
    c. Sclerenchyma
    d. Stone cell

11. What cells of the vascular tissue system are found in most angiosperms? [p. 676]

    a. Tracheids
    b. Vessel members
    c. Sieve tube members
    d. (a) and (b)
    e. (a), (b), and (c)

12. Which type of cell assists in the functioning of sieve tube cells? [p. 677]

a. Companion cells
b. Endodermal cells
c. Pith cells
d. Stone cells

13. The distance between two leaves on a stem is known as a(n) ________. [p. 678]

a. Axil
b. Internode
c. Leaf gap
d. Node

14. Which of the following is NOT a primary meristem? [p. 679]

a. Ground meristem
b. Procambium
c. Protoderm
d. Vascular cambium

15. In bamboo, the hollow centre of the stem forms from the breakdown of the [p. 679]

a. Cortex
b. Pith
c. Xylem
d. Phloem

16. Trichomes are specialized structures that can take the form of [p. 681]

a. Leaf protrusions containing a toxin
b. Salt bladder
c. Cuticle
d. (a) and (b)

17. Most of the photosynthesis in a leaf takes place in the ________. [p. 682]

a. Guard cells
b. Lower epidermis
c. Palisades mesophyll
d. Spongy mesophyll

18. Why is a root cap required to cover the root apical meristem? [p. 684]

a. Protection
b. Guide for root growth
c. Produces new cells
d. (a) and (b)

19. The annual rings that form in trees in temperate climates are formed by alternating layers of ________. [p. 687]

    a. Fusiform initials and ray initials
    b. Heartwood and sapwood
    c. Pericycle and endodermis
    d. Spring wood and summer wood

20. A cedar tree that has 50 very narrow tree rings [p. 687]

    a. Is 100 years old
    b. Has grown little each year
    c. Is growing in optimal environmental conditions

INTEGRATING AND APPLYING KEY CONCEPTS

1. Why is it important that both plant shoot systems and root systems are dendritic?

2. Within a vegetative bud, cells must differentiate to form each of the necessary plant tissue systems. Identify what tissue system that each of the primary meristems develops into and give their resulting functions.

3. You planted a red oak tree in your backyard three years ago. Explain how growth this summer will be different from the growth that occurred in the tree's first summer.

OPTIONAL ACTIVITIES

1. Why is there interest in using switchgrass as a biofuel, and what problems are associated with this?

2. *Welwitschia mirabilis* is an amazing species, having such a long life span in such a harsh environment. How is it able to survive through a single hot, dry day (i.e., adaptations, specialized structures, etc.)?

29 Transport in Plants

CHAPTER HIGHLIGHTS

- Passive and active transport moves materials into and out of plant cells, and these movements play a role in moving substances within a plant body.
- Water and minerals travel through roots toward the vascular tissue system via apoplast, symplast, and/or transmembrane routes.
- Physical properties of water assist in explaining how water and solutes move through the vascular tissue system within a plant.
- Solutes are translocated through the phloem by a pressure differential between different parts of the plant body.

STUDY STRATEGIES AND LEARNING OUTCOMES

- Concentrate on understanding the mechanisms functioning at different steps of the pathways of water and solute movement.
- Remember that water moves from areas with high water potential to areas of low water potential (similar to its movement from high concentration to low concentration).

By the end of this chapter, you should be able to

- Differentiate between the different mechanisms that move substances across a cell membrane and how they apply to water and solute movement in plants
- Define water potential and describe how each of its two components affects its value
- Describe the pathway and mechanisms of water and minerals movement into the plant, to the xylem, within the xylem, and out of the plant
- Describe the pathway and mechanisms of organic solutes movement from the source, through the phloem, and to the sink

TOPIC MAP

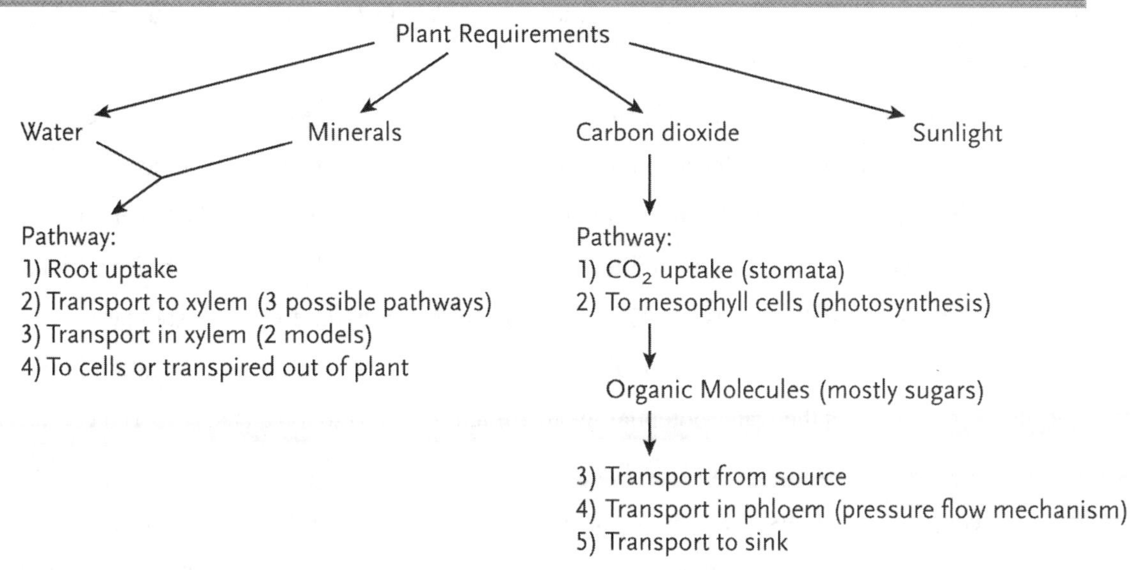

INTERACTIVE EXERCISES

Why It Matters [pp. 695–696]

Section Review (Fill-in-the-Blanks)

Plants must move (1) _____ and dissolved nutrients upward throughout the plant body without the use of a (2) ______; animals differ as they have hearts that function in this way. The cumulative effect of the rather weak forces such as (3) _____ and (4) _____ can be sufficient to move water in some plants over 100 m, counteracting the downward force of (5) _____.

29.1 Principles of Water and Solute Movement in Plants [pp. 696–700]

Section Review (Fill-in-the-Blanks)

Two types of transport move water and solutes through plants. If the process requires no energy and involves movement of material down a concentration gradient, it is (6) _____. If energy is involved, it is a(n) (7) _____ process. Sometimes proteins embedded in the cell membrane called (8) _____ may assist in movement of material across the membrane. In plant cells, the cytoplasm contains more negatively charged ions than the extracellular fluid. This charge difference is known as a(n) (9) _____, which refers to the fact that it represents a source of potential energy. Usually, more H^+ ions are outside a cell relative to inside, as a result of primary active transport. This allows for two kinds of secondary active transport of ions against a(n) (10) _____ gradient to take place. As the H^+ ions move passively back into the cell, (11) _____ the electrochemical gradient, the energy released may power the simultaneous (i) uptake of ions against this gradient, a process known as (12) _____, or (ii) movement of ions against the gradient in the opposite direction of the H^+ movement, a process known as (13) _____.

Within the vascular tissues of plants, water and solutes move by (14) _____, the movement of groups of molecules due to the pressure difference between two regions. In the xylem, the solution of water and dissolved materials is known as (15) _____. Between individual cells, water moves passively across individual cell membranes by (16) _____, which is driven by energy within the water known as its (17) _____. By convention, pure water at standard atmospheric pressure has a water potential value of (18) _____ megapascals. The addition of dissolved solutes alters the water potential by an amount referred to as (19) _____. If a cell has a lower water potential relative to outside the cell, water will move (20) _____ the cell, increasing the physical pressure against the cells' walls (called (21) _____). If this pressure increases such that it prevents further water from entering the cell, this (22) _____ offsets the solute potential, resulting in a water potential value of (23) _____, and the cell is distended or (24) _____.

Most plant cells contain a large, water-filled (25) _____ surrounded by a membrane called a(n) (26) _____, which maintains the turgor pressure of the cell. If the cytoplasm loses water, it is replaced from the central vacuole through channel proteins in the tonoplast membrane known as (27) _____. If a plant loses more water than it takes in over an extended period, (28) it will _____.

True/False

If the statement is true, write a "T" in the blank. If the statement is false, make it correct by changing the underlined word(s) and writing the correct word(s) in the answer blank.

29. ____________ In osmosis, water flows from an area of <u>higher water potential to an area of lower water potential</u>.

30. ____________ <u>Antiport</u> will move a substance across a cell membrane in the same direction as the flow of protons.

31. ____________ A slight charge difference on the two sides of a cell membrane is known as a <u>membrane potential</u>.

32. ____________ Facilitated diffusion is a type of <u>passive</u> transport that uses transport proteins.

33. ____________ Wilting is a result of loss of turgor pressure.

Choice

For each of the following situations, select the change that would result to a solution that has a water potential of zero.

34. ____	Addition of solutes	A. Increase in water potential
35. ____	Addition of pure water	B. Decrease in water potential
36. ____	Addition of tension	C. No change to water potential
37. ____	Addition of pressure	

29.2 Transport in Roots [pp. 701–702]

29.3 Transport of Water and Minerals in the Xylem [pp. 703–708]

Section Review (Fill-in-the-Blanks)

Water enters the plant root in one of three ways. The first is the (38) ____, in which water travels through intercellular spaces between cells and cell walls in the root epidermis and cortex. The second is the (39) _____, in which the water moves through the cytoplasm of adjacent cells using plasmodesmata. Lastly, water can pass from cell to cell by travelling across cell membranes (not using plasmodesmata), a route called the (40) _____. Water using any of these pathways will move through the root cortex until it encounters the last layer of cells called the (41) _____, each of which are wrapped transversely and (42) _____ in bands of waterproof suberin called the (43) _____. These bands force the soil solution to pass through the membranes of living endodermal cells in order to reach the stele, thereby regulating water and minerals entering the stele.

The model that describes the long distance movement of xylem sap is called the (44) _____. It is based on two important physical properties of water, (45) _____ and (46) _____, and the driving force for the movement is the process of (47) _____ from leaves. Since plant water loss is primarily through (48) _____ in the leaves, plants have various controls, such as regulating the size of the pores using surrounding (49) _____. Another model to explain movement of xylem sap in plants, especially over short distances in nonwoody plants, is the (50) ____ model, which involves the active transport of ions from the soil into the stele to create a water potential difference across the (51) _____. As water entering the xylem builds, so does a significant pressure, forcing water upward and potentially out of leaf pores, a phenomenon known as (52) _____. Some plants that live in hot, dry environments utilize (53) _____ to reduce plant water loss, which involves opening stomata and fixing CO_2 into an organic acid during the (54) _____ (when humidity is higher and temperatures are lower) and manufacturing sugar during the (55) _____ while stomata remain closed.

Matching

Match each of the following terms with its correct definition.

56.	____	Abscisic acid (ABA)	A.	Model that explains movement of water through xylem over short distances, resulting in guttation
57.	____	Adhesion	B.	Plant adapted to hot, dry conditions
58.	____	Root pressure	C.	Mechanism to move mineral ions directly into symplast pathway
59.	____	Xerophyte	D.	Active transport of K^+ following passive movement of H^+ into guard cells
60.	____	Symport	E.	Tendency of water molecules to stick to other types of molecules, including the cell walls of xylem
61.	____	Transpiration	F.	Evaporation of water from intercellular spaces to the atmosphere
62.	____	Active transport	G.	Plant hormone responsible for stomatal closure under stressful conditions

29.4 Transport of Organic Substances in the Phloem [pp. 709–711]

Section Review (Fill-in-the-Blanks)

The long-distance movement of sugars and many other substances (such as proteins, fats, and hormones) through the sieve tubes of phloem is called (63) _____, and the mix of water and dissolved substances that flows within them is called (64) _____. The point of origin of these materials is called the (65) _____, and their destination is the (66) _____. The force that drives this movement is known as the (67) _____ mechanism because solutes in the phloem move down (68) _____ gradients. The loading of solutes in the phloem (69) _____ the water potential of the phloem sap at the source, which causes water to follow using the process of (70) _____. The solutes (and water) travel by (71) _____ to lower pressure regions (at the sink since solutes and water exit from the sieve tube members). In some plants, companion cells are modified in order to actively transport large amounts of substances into the sieve tube members. These cells are called (72) _____.

True/False

If the statement is true, write a "T" in the blank. If the statement is false, make it correct by changing the underlined word(s) and writing the correct word(s) in the answer blank.

73. _____________ Fructose is the most common type of sugar translocated in phloem.

74. _____________ Phloem transports sap downward from the source.

75. _____________ Translocation originates at the source and ends at the sink.

76. _____________ Movement of solutes into companion cells is accomplished by active transport.

77. _____________ A root is a sink throughout the growing season.

SELF-TEST

1. Facilitated diffusion uses ________ to move ions through pores down an electrochemical gradient. [p. 696]

    a. Transport proteins
    b. Carrier proteins
    c. Channel proteins
    d. No proteins

2. Symport and antiport are examples of ________. [p. 697]

    a. Facilitated diffusion
    b. Primary active transport
    c. Pressure flow
    d. Secondary active transport

3. ________ is the effect dissolved materials have on water's tendency to move across a membrane. [p. 699]

   a. Membrane potential
   b. Pressure potential
   c. Solute potential
   d. Water potential

4. What would be the water potential in a cell in which the solute potential was –0.38 and the pressure potential was 0.25? [p. 699]

   a. –0.63
   b. –0.13
   c. 0
   d. 0.63

5. The membrane surrounding the central vacuole of a plant cell is the ________. [p. 699]

   a. Apoplast
   b. Protoplast
   c. Symplast
   d. Tonoplast

6. A plant cell that contains enough water to exert pressure on the cell wall is said to be ________. [p. 700]

   a. Dehydrated
   b. Flaccid
   c. Plasmolyzed
   d. Turgid

7. A molecule of soil water is taken up by a root hair and travels through plasmodesmata in the root cortex cells. This molecule is taking the ________ pathway. [p. 701]

   a. Apoplastic
   b. Transmembrane
   c. Symplastic

8. Water can enter a root and move toward the stele without entering the cytoplasm of a cell until it reaches the ________. [p. 701]

   a. Cortical parenchyma
   b. Endodermis
   c. Epidermis
   d. Pericycle

9. Which of the following cells are <u>not</u> alive at maturity? [pp. 702, 709]

   a. Tracheids
   b. Vessel elements
   c. Sieve tube elements
   d. (a) and (b)

10. Transpiration is the ________ of water from leaves. [p. 703]

   a. Cohesion
   b. Adhesion
   c. Evaporation
   d. Condensation

11. The "stretching" of hydrogen bonds is due to ________ on the xylem sap resulting from transpiration. [p. 704]

   a. Pressure
   b. Cohesion
   c. Tension
   d. Viscosity

12. The most important players in the "transpiration–photosynthesis compromise" are ________. [p. 706]

   a. Companion cells and sieve tube members
   b. Guard cells and stomata
   c. Spongy and palisades mesophyll
   d. Xylem and phloem

13. Which ions are most important in regulating stomata function? [p. 706]

   a. Ca^{++} and Mg^{++}
   b. H^{+} and Cl^{-}
   c. H^{+} and K^{+}
   d. K^{+} and Mg^{++}

14. When a pair of guard cells have a high water potential, the central stoma will ________. [p. 706]

   a. Be open
   b. Be closed
   c. Quickly alternate between opening and closing

15. Which of the following will cause stomata to close? [p. 707]

   a. Low concentration of CO_2 in leaf air spaces
   b. Daytime light levels
   c. Abscisic acid production in roots
   d. Moist soil

16. Which plant would most likely utilize CAM photosynthesis? [p. 708]

   a. Maple tree
   b. Moss
   c. Sedum
   d. Water lily

17. Phloem sap contains ________. [p. 709]

   a. Sugars
   b. Amino acids
   c. Fatty acids
   d. (a), (b), and (c)

18. The "honeydew" from the anus of an aphid provided evidence that phloem sap in sieve tubes __________. [p. 710]

   a. Has a high water potential
   b. Has a low solute potential
   c. Is under high pressure
   d. Is under no pressure

19. The sink for photosynthate ________. [p. 710]

   a. Can be anywhere on the plant
   b. Must be above the source
   c. Must be below the source
   d. Must be below the surface of the soil

20. In late summer, which of these organs is a source of organic substances? [p. 710]

   a. Leaves
   b. Roots
   c. Buds

INTEGRATING AND APPLYING KEY CONCEPTS

1. Explain how the process of causing a stoma to open is an example of a symport mechanism.

2. Why is xylem transport considered to be unidirectional whereas phloem transport is multidirectional?

3. Provide three examples (at least one for the root system and one for the shoot system) of adaptations used by plants to survive in drought-stressed environments. Explain how these adaptations increase water uptake or decrease water loss.

OPTIONAL ACTIVITIES

1. Suberin can be found in other locations in a plant, not just the Casparian strip. Where else may it be found, and how do its characteristics provide a beneficial function to the plant?

2. Current research indicates that plasmodesmata are not simple, static channels between adjacent cells but dynamic, changing structures. Can you think of any advantages and/or disadvantages associated with this flexibility?

30 Reproduction and Development in Flowering Plants

CHAPTER HIGHLIGHTS

- Angiosperms (flowering plants) reproduce sexually and asexually.
- At maturity, diploid sporophyte angiosperms produce flowers containing stamens and/or carpels for sexual reproduction.
- Sexual reproduction and development involves an alternation of generations life cycle; diploid plant flowers give rise to haploid male and female gametophytes, which produce male and female gametes, respectively, and which, after fertilization, complete the cycle by giving rise to a developing diploid sporophyte plant.
- Pollination, followed by double fertilization, results in the production of a diploid zygote (first cell of new sporophyte) and a triploid endosperm (nourishment), both of which are protected within a seed during embryo development.
- After seed germination, a seedling develops and eventually matures into a flowering plant.
- Asexual reproduction (or vegetative propagation) results in offspring that have DNA identical to the parent's DNA (clone).

STUDY STRATEGIES AND LEARNING OUTCOMES

- If you are already familiar with the reproduction process in animals, some of the processes in plants are similar.
- Concentrate on the key concepts and on the sequence of events in each process.

By the end of this chapter, you should be able to

- Define plant structure and process terminology and apply these terms to sexual and asexual plant reproduction processes
- Describe an imperfect flower, differentiating between dioecious and monoecious plants

- Describe the life cycle of a sexually reproducing angiosperm, incorporating the terms sporophyte, gametophyte, haploid, and diploid, as well as structures and processes involved
- Explain why honeybees are important pollinators
- Explain the importance of the *S* gene and describe two situations in which pollination does not lead to fertilization
- Identify two functions of fruit and differentiate between different types of fruit
- Describe how you could create a new plant from a plant cutting

TOPIC MAP

Using the following terms, identify each of the processes occurring in angiosperm reproduction: dedifferentiation, differentiation, double fertilization, germination, maturation, meiosis, mitosis, pollination. Note that terms can be used more than once.

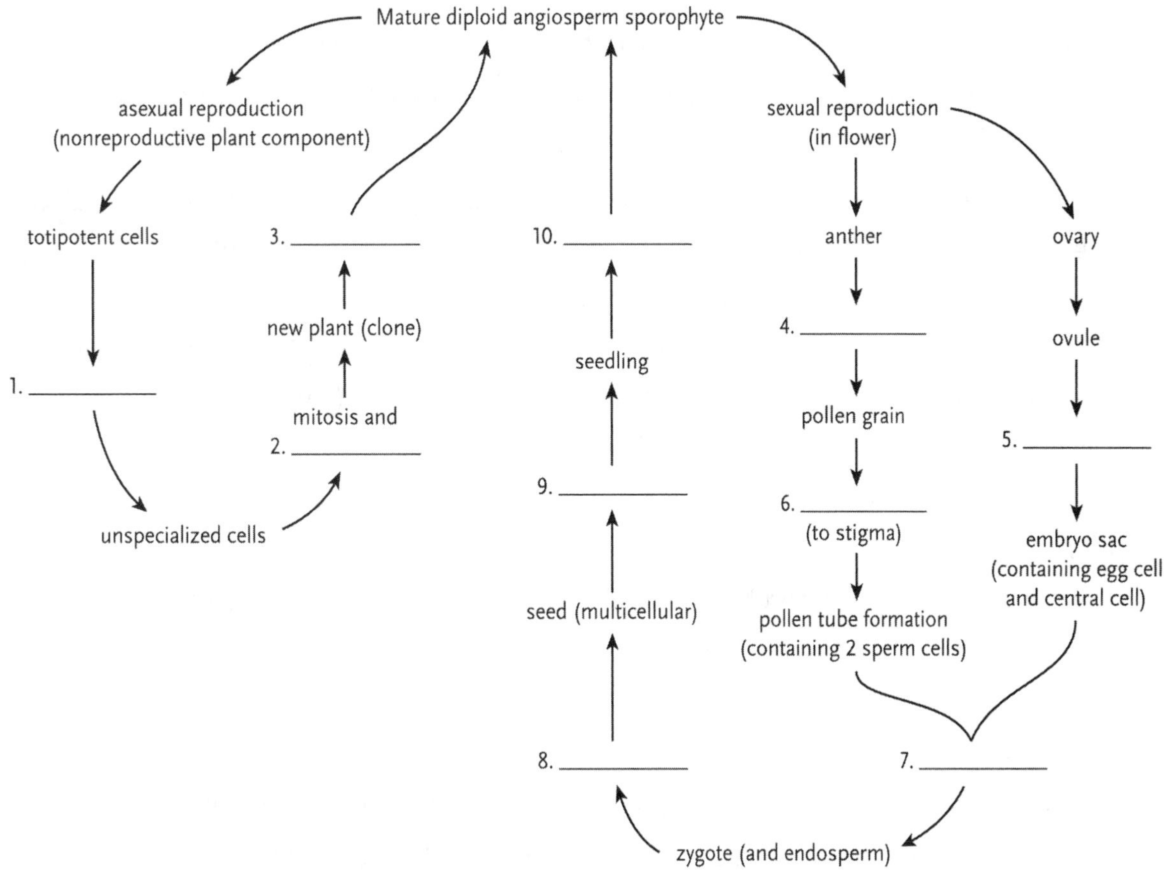

INTERACTIVE EXERCISES

Why It Matters [pp. 715–716]

Section Review (Fill-in-the-Blanks)

Angiosperms reproduce sexually using male and female reproductive structures found in (1) ____________. Male gametes in pollen grains are transferred to female gametes in other plants by the process called (2) ________________. Approximately one-third of North American plants are pollinated using (3) ________________. The reason for the decline in this pollinator's numbers, referred to as (4) __________ __________ __________ (or CCD), is unknown. If CCD spreads, angiosperm (5) ________, such as cucumbers, berries, and apples, could be at risk.

30.1 Overview of Flowering Plant Reproduction [pp. 717–718]

30.2 The Formation of Flowers and Gametes [pp. 718–721]

Section Review (Fill-in-the-Blanks)

A plant is a diploid (6) ____________. It bears flowers where meiosis leads to the formation of male and female (7) ____________ spores. After mitosis, each male spore forms a three-celled (8) ________ __________ and a female spore forms a seven-celled (9) __________ _______. The fusion of the haploid male and female (10) ________________ forms a (11) __________ zygote. The zygote divides by (12) __________ to eventually form a new sporophyte plant.

Once a vegetative shoot becomes a(n) (13) ____________ ______________, it bears a flower or a group of flowers called a(n) (14) ____________________. Each flower has four concentric whorls: A. The outermost, first whorl is made of leaflike (15)__________ to enclose the flower as a bud. B. The second whorl is made of colourful (16) __________ to attract animals for pollination. C. The third whorl has the (17) ____________, each of which is made of a thin (18) __________ with a sac called the (19) ____________ containing the (20) __________ grains. D. The innermost whorl has one or more (21) _________. The lower part is called the (22) __________ and contains the ovules. Its extension is called the (23) ____________ and ends in a flattened surface, the (24) ______________.

In an anther, each (25) ______________ (or microspore mother cell) undergoes meiosis, resulting in the formation of four (26) ______________. After mitosis and maturation, microspores will form (27) __________ ______, each containing two (28) __________ cells and a(n) (29) _______ _______ cell.

In a(n) (30) _____________, there may be one or more ovules formed. Inside each ovule, meiosis of a(n) (31) ____________ (or megaspore mother cell) results in the formation of (32) ______ mature megaspore since the three others degenerate. A megaspore undergoes mitosis divisions to form an eight-celled female (33) _____________ sac. As well as three antipodal cells, each sac has (34) _______ egg cell, (35) __________ synergids, and (36) ____ large central cell. All cells in the sac are haploid except for the (37) ___________ cell, which contains two nuclei. The egg cell and the synergids are located near the (38) ______________ of the ovule.

Matching

Match each of the following terms with its correct definition.

39.	____	Alternation of generations	A.	A flower that lacks one of the four whorls
40.	____	Sporophyte flower	B.	Where a plant bears the male and female flowers or flower parts on different plants
41.	____	Gametophyte	C.	The diploid plant that produces diploid microsporocytes and megasporocytes
42.	____	Inflorescence	D.	The haploid stage of the plant that bears the gametes
43.	____	Imperfect flower	E.	Where the life cycle alternates between a diploid sporophyte and a haploid gametophyte
44.	____	Monoecious	F.	Flowers that are borne in a group
45.	____	Dionecious	G.	Where a plant bears both the male and female flowers or flower parts on the same plant

Match each of the following structures found in flowers with its correct definition.

46.	____	Embryo sac	A.	The long tubelike portion of a carpel through which the pollen tube travels to reach the micropyle
47.	____	Sepal	B.	The third whorl of a flower that is composed of male reproductive structures
48.	____	Style	C.	The innermost whorl that is composed of female reproductive structures
49.	____	Petal	D.	The tip of the carpel where the pollen lands during pollination
50.	____	Stamen	E.	A leafy component of the outermost whorl
51.	____	Filament	F.	The male gametophyte of a plant
52.	____	Anther	G.	One of the brightly coloured parts of a flower that attracts pollinators
53.	____	Carpel	H.	The female gametophyte of a plant
54.	____	Ovary	I.	The thin stem of a stamen that bears an anther at the tip
55.	____	Stigma	J.	The lower part of a carpel that bears ovules
56.	____	Ovule	K.	A round structure formed inside an ovary where the female gametophyte develops
57.	____	Pollen grain	L	The sac located at the tip of a stamen that bears the pollen grains

Match each cell of the male and female gametophyte structures with its correct definition.

58. ____ Sperm cell

59. ____ Pollen tube cell

60. ____ Egg cell

61. ____ Synergids

62. ____ Antipodal cells

63. ____ Central cell

A. The female gamete of the plant

B. The male gamete of the plant

C. The cell of the male gametophyte that grows inside the style of the carpel and transports the sperm cells to the ovule

D. Three cells located in the far end of the embryo sac that eventually degenerate

E. The large diploid cell that fuses with one of the sperm cells and forms the triploid endosperm

F. The two cells that flank the egg cell in the embryo sac and degenerate to guide the sperm cells to their receptive cells in the sac

30.3 Pollination, Fertilization, and Germination [pp. 721–728]

Section Review (Fill-in-the-Blanks)

Air, water, birds, or other organisms aid in (64) ________________, transferring pollen grains from the (65) _____________ of one plant to the (66) ______________ of the carpel of another plant. Incompatibility between two plants of the same species or plants of differing species is identified using the multiple (67) _______________ system of the *S* gene. If the pollen and stigma have the same (68) _______________, pollen may not germinate or some other process will occur to prevent fertilization. If the pollen and stigma are compatible, after a pollen grain germinates, a(n) (69) ___________ tube grows through the (70) ___________ of the carpel to reach the micropyle end of an ovule. Two haploid (71) __________ cells enter the embryo sac and (72) __________ ___________ occurs, with one sperm cell fusing with the egg cell to form a diploid (73) ______________ and the other sperm cell fusing with the diploid central nucleus to form a(n) (74) _______________ cell. The first division of the zygote forms a(n) (75) _______________ cell, which will form the embryo, and a basal cell, which will form a(n) (76) _____________ and provide nutrients to the developing embryo from the parent plant. Eudicot seeds have two (77) _______________, which can be thick or thin and function to provide stored nutrients to the growing embryo. Monocots have only one (78) ________________ and a nutrient-rich (79) __________________. The embryo in both has a(n) (80) ________________ that develops into a root and a shoot apical meristem that becomes a(n) (81) __________________ (leafy shoot). A seed also develops a protective seed (82) _____________. In monocots, a sheath of cells covers the embryo, (83) ________________, which covers the root meristem, and (84) ________________, which covers the shoot meristem.

In most angiosperms, while a seed develops in a(n) (85) ___________, a fruit develops from the wall of the (86) __________ within the carpel. Its formation is stimulated by hormones released by the (87) _____________ grain and the fruit wall becomes a(n) (88) __________________. The fruit provides both (89) _______________ and aids in seed (90) ________________. Once a seed is formed, it usually goes into a period of (91) ________________ during which it becomes metabolically inactive. Upon receiving proper environmental conditions, the process of (92) _________________ begins. A seed absorbs water, a process called (93) _________________, which causes the seed to swell and the seed (94) ________ to split. The increased water and oxygen reaching the embryo cause (95) _____________ to be produced, which catalyze many reactions, including the release of nutrients. Cells rapidly produce new cells by (96) ___________. Germination has ended when the (97) ______________ emerges from inside the seed coat to become the primary (98) ____________, which is followed by the (99) _________________ growing to become the new foliage.

Complete the Table

100. Complete the following table by describing each process.

Events	*Description*
A. Pollination	
B. Fertilization	
C. Germination	

101. Complete the following table by describing each fruit type.

Fruits	*Description*
A. Simple fruits	
B. Aggregate fruits	
C. Multiple fruits	

Matching

Match each part of a seed with its correct definition.

102. ____ Cotyledon

103. ____ Endosperm

104. ____ Seed coat

105. ____ Radicle

106. ____ Plumule

107. ____ Coleorhiza

108. ____ Coleoptile

A. The fleshy structure that is formed by the triploid cells when the haploid sperm cell fuses with the diploid central cell

B. The protective covering of a seed

C. The part of the embryo in the seed that becomes the shoot of the seedling

D. A protective covering of the plumule in a monocot seed

E. An embryonic structure that often becomes one of the first leaves of a eudicot seedling

F. The part of the embryo in the seed that becomes the primary root of the seedling

G. A protective covering of the radicle in a monocot seed

Short Answer

109. Why is it advantageous for angiosperms to produce endosperm as a nutrient source rather than use gametophyte tissue?

____

____

110. Differentiate between the function of the cotyledons in a eudicot such as a sunflower seed and the cotyledon in a monocot such as a corn seed.

____

____

30.4 Asexual Reproduction of Flowering Plants [pp. 728–731]

30.5 Early Development of Plant Form and Function [p. 731]

Section Review (Fill-in-the-Blanks)

Producing new plants from nonreproductive parts of a parent plant is asexual reproduction, often referred to as (111) __________________ reproduction. This is possible because many plant cells can dedifferentiate and divide to form a complete plant, a property called (112) _______________. One type of asexual reproduction is (113) ________________, which occurs when wounded cells in a plant part (such as a cutting) dedifferentiate and produce other plant parts to form a new complete plant. A somatic embryo forms in a seed when a(n) (114) ______________ egg or a diploid cell around the embryo sac of an ovule undergoes cell division but not fertilization. This type of asexual reproduction is (115) ______________. Plant cells grown in culture often form an undifferentiated cell mass called the (116) ____________ that can be stimulated to redifferentiate by providing specific (117) __________________. The result of all types of asexual reproduction is a new plant that is genetically (118) ____________ to the original plant or plant cells.

As soon as a plant zygote divides, a(n) (119) _______________________ axis is established where the (120) ______________ cell will eventually form shoots and the (121) _____________ cell will form the root system. These two early cells contain different transcription-regulating proteins called (122) ____________ _____________. This results in different genes being expressed in these cells, which leads to (123) _____________ between these cells and all cells arising to form the specialized shoot and root systems.

Matching

Match each of the following terms with its correct definition.

124.	____	Totipotency	A.	Type of vegetative reproduction when a piece of a plant detaches and forms a new plant
125.	____	Dedifferentiation	B.	Proteins that regulate transcription of genes
126.	____	Fragmentation	C.	When a cell has the potential to become a new organism
127.	____	Callus	D.	When a specialized cell becomes undifferentiated
128.	____	Apomixis	E.	When an undifferentiated cell becomes a specialized cell
129.	____	Transcription factors	F.	Embryo development of epidermal, ground, and vascular tissues in their correct relative locations
130.	____	Differentiation	G.	A mass of cells in culture grown from cells taken from a plant
131.	____	Outside-to-inside	H.	When a plant develops from an unfertilized egg or a organization diploid cell of the embryo sac
132.	____	Root-shoot axis	I.	Result of early embryo development when it is decided which cell will form root versus shoot

Short Answer

133. Differentiate between totipotent, pluripotent, and multipotent.

____

____

SELF-TEST

1. A three-celled pollen grain is a [p. 719]

   a. Male sporophyte
   b. Male gametophyte
   c. Female sporophyte
   d. Female gametophyte

2. Which of the following represents the cells in a pollen grain? [p. 719]

   a. One sperm cell, two pollen tube cells
   b. Two sperm cells, one synergid, one central cell
   c. Three antipodal cells and one sperm cell
   d. One pollen tube cell and two sperm cells

3. Which specific part of the flower transforms into a seed after fertilization? [p. 718]

   a. Ovary
   b. Stamen
   c. Ovule
   d. Embryo sac

4. Which cell of the embryo sac fuses with the sperm cell to form triploid endosperm? [p. 724]

   a. Egg cell
   b. Synergids
   c. Antipodal cells
   d. Central cell

5. The type of fruit that is formed from ovaries from different flowers in an inflorescence, such as a pineapple produces, is a(n) [p. 724]

   a. Simple fruit
   b. Aggregate fruit
   c. Multiple fruit

6. All the cells in a callus are genetically alike. [p. 731]

   a. True
   b. False

7. A radicle forms [p. 726]

   a. Stems
   b. Leaves
   c. Flowers
   d. Roots

8. Which of the following terms refers to specialized cells becoming unspecialized? [p. 729]

   a. Differentiate
   b. Dedifferentiate
   c. Redifferentiate

9. The pathway the growing pollen tube travels is through [p. 724]

   a. Style, then ovule, then stigma
   b. Ovule, then stigma, then anther
   c. Stigma, then style, then micropyle
   d. Filament, then micropyle, then anther

10. The correct order of processes occurring during germination is [p. 727]

    a. Imbibition, seed swelling, seed coat rupturing, radicle growing
    b. Seed swelling, plumule growing, imbibition, seed coat rupturing
    c. Radicle growing, seed coat rupturing, seed swelling, imbibition
    d. Seed coat rupturing, imbibition, plumule, growing, seed swelling

11. Why does a developing eudicot embryo become heart-shaped? [p. 725]

    a. The seed coat is heart-shaped
    b. The suspensor cells divide unevenly
    c. Two cotyledons are forming
    d. The radicle and plumule are forming

12. Pollinators function to transfer [p. 715]

    a. Sperm cells to an ovule
    b. Pollen grains to a stigma on a different plant
    c. Egg cells to an anther on a different plant
    d. Female gametophytes to a filament

13. An advantage of reproducing asexually is that [p. 730]

    a. There is greater variation between the plants produced than those resulting from sexual reproduction
    b. There is a lower energy cost than for sexual reproduction
    c. There is a greater chance of survival if the environmental conditions change than through sexual reproduction

14. In monoecious species, [p. 719]

    a. All flowers contain both male and female reproductive structures
    b. All plants contain flowers with only male reproductive structures and flowers with only female reproductive structures
    c. Plants contain either flowers with male reproductive structures or flowers with female reproductive structures

15. Self-fertilization is reduced due to the incompatibility of a pollen grain and a stigma having the same *S* gene allele. [p. 722]

    a. True
    b. False

16. A somatic embryo can develop following [p. 730]

    a. Fertilization
    b. Pollination
    c. Differentiation of totipotent cells
    d. Maturation of seedlings
    e. Meiosis of microspores

17. For seeds requiring a dormancy period, dormancy ends when _______________ begins. [p. 727]

    a. Fertilization
    b. Pollination
    c. Microspore formation
    d. Megaspore formation
    e. Germination

18. How many nuclei are present in endosperm? [p. 724]

    a. One
    b. Two
    c. Three
    d. Four

19. Which of the following can promote germination? [p. 726]

    a. Increased day length
    b. Increased soil moisture
    c. Increased temperature
    d. All of the above

20. In a developing seed, the suspensor [p. 725]

    a. Is formed from the cotyledon(s)
    b. Transfers nutrients from the parent plant to the embryo
    c. Forms a protective seed coat
    d. Develops into root and shoot meristems

INTEGRATING AND APPLYING KEY CONCEPTS

1. Totipotency is demonstrated in plants. Explain the concept by giving an example.
2. Describe the *S* gene concept to explain incompatibility between male and female reproductive tissues.
3. What is CCD, and why is it a problem for farmers growing apple trees and/or berry plants?

OPTIONAL ACTIVITIES

1. Octenol is one example of an attractant to lure insects to flowers to aid in pollination. Identify two other methods used by angiosperm species to attract pollinators.
2. Bee pollination is one example of angiosperm pollination. Provide an example of angiosperm pollination by (a) an insect pollinator and (b) a bird pollinator, identifying the animal and flower species involved and how the process occurs.
3. Locate and describe a study that has provided information about development of the root-shoot axis using the model organism *Arabidopsis thaliana*.

31 Control of Plant Growth and Development

CHAPTER HIGHLIGHTS

- Plants produce hormones (auxins, gibberellins, cytokinins, ethylene, brassinosteroids, abscicic acid, and jasmonates and oligosaccharins) that regulate growth, development, and reproduction.
- Plants produce chemicals that protect them from infections and predators.
- Plants respond to light (phototropism), gravity (gravitropism), and physical stress (thigmotropism) and exhibit nondirectional movement (nastic movement).
- Plants maintain a 24-hour circadian rhythm that affects their movements, growth, flowering, seed dormancy, and germination.

STUDY STRATEGIES AND LEARNING OUTCOMES

- This chapter can be confusing because there are a number of overlapping functions of chemicals that plants produce.
- It helps to keep track of major functions first and then try to understand their interactions.

By the end of this chapter, you should be able to

- Differentiate between the different plant hormones, including where each is synthesized and its effect on the plant
- Compare and contrast different types of general and specific plant defence responses
- Describe how plants respond to signals, from hormones or the environment, at the cellular level
- Differentiate between the different types of tropism, describing what causes these changes to occur
- Explain how biological clocks can affect different plants

TOPIC MAP

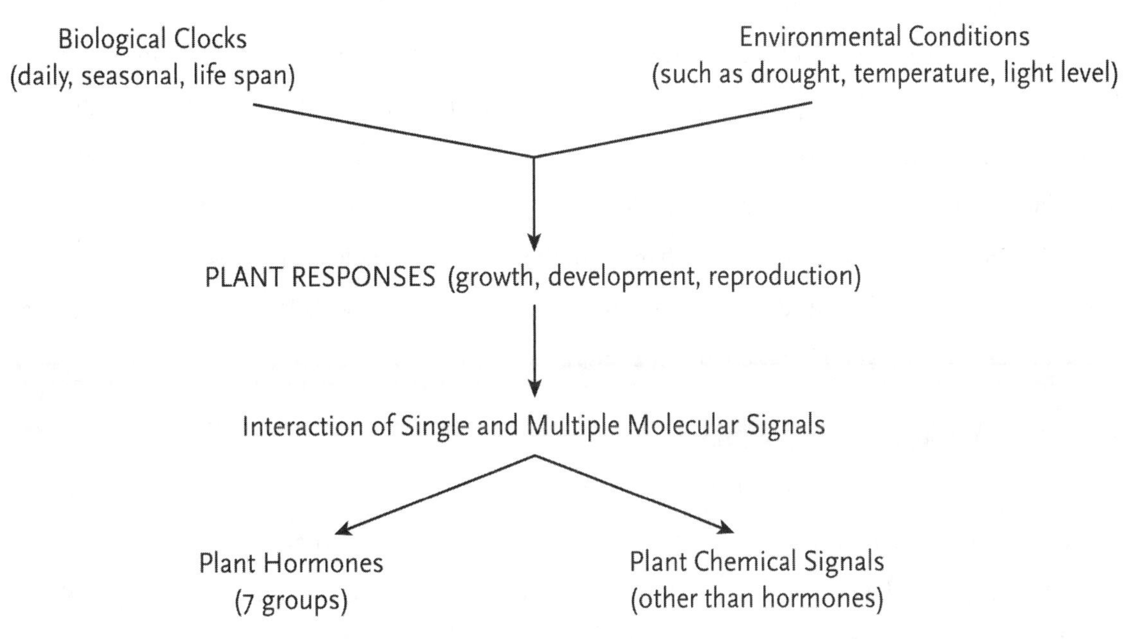
Biological Clocks
(daily, seasonal, life span)
Environmental Conditions
(such as drought, temperature, light level)
PLANT RESPONSES (growth, development, reproduction)
Interaction of Single and Multiple Molecular Signals
Plant Hormones
(7 groups)
Plant Chemical Signals
(other than hormones)

INTERACTIVE EXERCISES

Why It Matters [pp. 735–736]

Section Review (Fill-in-the-Blanks)

As (1) _____ organisms, plants cannot run from predators, so they have adapted by developing (2) _____ responses. Plants produce chemicals that can be transported to other cells in plants, which act as triggers for various metabolic events, such as seed (3) ____ (to begin growth of a new plant), shifting from vegetative phase to (4) _____ phase, or end of season (5) _____ of body parts. These responses are initiated by changes in the (6) _____, such as changes in day length, temperature, and soil moisture. Some plants produce (7) _____compounds, which, when released in the air, signal animal and neighbouring plant responses.

31.1 Plant Hormones [pp. 736–746]

Section Review (Fill-in-the-Blanks)

Plants respond to internal (8) _____ changes and external (9) ______ conditions by producing special chemicals called (10) ______. These chemicals are small and organic and are required in very (11) _____ amounts. They diffuse from one cell to another, or they may be transported by the (12) ______ tissues. Although hormones show large variability in their effects, they can be classified into seven major groups:

A. Auxins—They are made by apical (13) _____ and young leaves and stems. They (14) _____ growth of stems and lateral roots, primarily through cell (15) _____. Since auxins diffuse away from (16) _____, differential cell growth can occur that causes shoot tips or other plant parts to (17) _____ toward the light. This response is called (18) _____. Auxins also (19) _____ fruit development.

B. Gibberellins—They are made mostly by root and shoot (20) _____. They (21) ______ cell division and cell elongation, most notably in stems. In some species, gibberellins cause elongation of floral stalks in rosette plants, increasing the length of (22) _____, such that leaves no longer appear to arise from the same node. This process is called (23) _____. Gibberellins also help in breaking seed and bud (24) _____ at the beginning of the growing season.

C. Cytokinins—Produced mainly in root (25) ________, cytokinins stimulate cell (26) _____ rather than cell elongation. Unlike gibberellins, which are transported in phloem tissue, cytokinins are transported in (27) _____ tissue.

D. Ethylene—Produced by most parts of the plant, ethylene is a (28) _____ that assists in regulating many processes. For example, during seedling growth, it increases cell (29) _____ relative to cell elongation to increase stem girth, assisting in growth out of the soil. Ethylene promotes (30) _____ ripening, as well as stimulating the process of (31) _____ (aging and cell death). With aging, ethylene stimulates digestive enzymes that break down cell walls in (32) _____, located at the base of petioles, which results in the dropping of leaves, flowers, and fruits.

E. Brassinosteroids—These are (33) _____ hormones that were first discovered in the mustard family. Produced primarily in (34) _____ tips, seeds, and embryos, they stimulate cell division and (35) _____. In roots, ethylene (36) _____ root elongation.

F. Abscicic acid—This hormone is derived from (37) _____ in leaf chloroplasts. It (38) _____ growth and promotes dormancy, in response to (39) _____ conditions. It often accumulates in seed coats during seed development and requires a dormancy period to slowly break down before the seed can (40) ________.

G. Jasmonates and Oligosaccharins—Jasmonates are derived from (41) _____ acids, and oligosaccharins are derived from (42) _____. Although both of these hormones (43) ____ growth of one or more plant parts, they are best known for protecting plants from (44) _____.

Matching

Match each of the following terms with its correct definition.

45.	____	Bolting	A.	Aging process in plants that leads to breakdown and death of their cells
46.	____	Senescence	B.	Dropping of leaves, flowers, and fruits in response to environmental changes
47.	____	Phloem	C.	Tissue that cytokinins travel through
48.	____	Abscission	D.	Period when seeds and buds go into growth-inhibiting phase
49.	____	Dormancy	E.	Mechanism used by auxins to travel through a plant
50.	____	Xylem	F.	Sudden elongation of the floral stem in rosette plants
51.	____	Polar transport	G.	Tissue that gibberellins travel through

Match each of the following scientists with their major contribution.

52. ____ Eiichi Kurosawa

53. ____ Francis and Charles Darwin

54. ____ Frits Went

55. ____ Winslow Briggs

A. Performed the first experiments to explain phototropism in grasses

B. Demonstrated that the growth-promoting chemical travels from the shoot tip downward

C. Showed that auxins move laterally in response to light exposure

D. First one to discover the effect of gibberellins produced by a fungus on rice plants

Complete the Table

56. Complete the following table by describing the main actions of each hormone.

Hormone	*Action*
Auxins	
Gibberellins	
Cytokinins	
Ethylene	
Brassinosteroids	
Abscisic acid	
Jasmonates and oligosaccharins	

31.2 Signal Responses at the Cellular Level [pp. 747–748]

31.3 Plant Chemical Defences [748–752]

Section Review (Fill-in-the-Blanks)

In plants, a signal molecule, such as a hormone, initiates a (57) _____ pathway that involves converting the signal molecule into a (58) _____ that causes a change in a target cell's (59) _____ or metabolism. Generally, the binding of a signal molecule to a (60) _____ molecule, located on the plasma membrane, on the endoplasmic reticulum, or in the (61) _____, will cause a reaction to occur inside the cell. Possible reactions include activation of an enzyme or some other (62) _____, opening of a channel for ion flow, or translation of mRNA, all of which result in a physiological response. In some cases, the reaction between the (63) _____ molecule (the primary messenger) and the receptor stimulates the synthesis of an internal (64) _____ messenger. It is this intermediary molecule that acts as the main signal to alter how the (65) _____ cell functions.

Plants are attacked by (66) ______________, ______________, ______________, and ______________. If a plant provides a general defence to an insect attack, it will likely immediately initiate a complex signalling (67) _____that, with the involvement of ethylene, (68) _____, or some other hormone, will produce a chemical or (69) _____ defence at the wound site. In tomato plants, the synthesis of jasmonates stimulates the expression of a protease (70) _____ that prevents the insect predator from digesting plant (71) _____. If attached by bacteria or fungi, many plants are able to provide a (72) _____ response, which involves isolating the infection site by signalling nearby cells to release (73) _____ compounds, such as hydrogen peroxide, to kill these cells. The pathogen is unable to survive without the plant (74) _____ supply. Furthermore, the release of (75) ______ _____ (SA) triggers cells to produce (76) _____ proteins (PR proteins), which contain hydrolytic enzymes that harm the pathogen by breaking down its cell walls.

Plants also produce protective chemicals called the (77) ______ metabolites, such as (78) _____, which act like antibiotics, or caffeine, cocaine, strychnine, tannins, and terpenes that ward off feeding herbivores. Since terpenes are (79) _____, they are released into the air during the attack, attracting (80) _____ of the pathogen.

Gene-for-gene recognition is a (81) _____ defence response since it requires a dominant (82) _____ gene (*R* gene) in the plant cell that produces a plasma membrane (83) _____, which matches with a protein molecule produced from a dominant (84) _____ gene (*Avr* gene). When this occurs, a defence response, usually the hypersensitive response, will be triggered and PR proteins will be produced.

Plants that survive pathogen attacks produce (85) _____, which stimulates production of (86) _____. Both of these chemicals provide (87) _____ protection from further attacks.

Short Answer

88. How does a secondary messenger differ from a secondary metabolite?

____

____

89. PR proteins function in various plant defences. What do they do, and why is this important?

____

____

Matching

Match each of the following terms with its correct definition.

90.	____	Salicylic acid	A.	A chemical produced by plants to prevent herbivores from consuming them
91.	____	Phytoalexin	B.	A chemical produced by plants infected with bacteria or fungi that acts like an antibiotic
92.	____	Secondary metabolite	C.	A chemical similar to aspirin that is produced by some wounded plants to protect themselves from infections
93.	____	Hydrogen peroxide	D.	A structure on a cell's plasma membrane that a particular hormone can bind to
94.	____	Receptor	E.	A chemical that some plants release to kill cells surrounding a wound

31.4 Plant Responses to the Environment: Movements [pp. 752–756]

31.5 Plant Responses to the Environment: Biological Clocks [pp. 756–760]

31.6 Plant Responses to the Environment: Responses to Temperature Extremes [p. 760]

Section Review (Fill-in-the-Blanks)

Plants respond to certain environmental factors by growing away or toward the factor, a process referred to as (95) _____. One type is (96) _____, which is plant growth toward a unidirectional light. This occurs when a blue light–absorbing pigment, such as (97) ________, stimulates lateral movement of auxins that eventually leads to differential growth of cells on the shaded and nonshaded sides of a plant part. (98) _____ is growth in response to gravity where roots are stimulated to grow (99) _____ and shoots to grow (100) _____. It is hypothesized that amyloplasts in some cells sink to the bottom of cells, an indicator of direction to ensure that auxins stimulate cell (101) _____ to promote growth in the appropriate direction. (102) _____ is growth in response to physical contact with another object where auxins and ethylene stimulate bending of the plant. Specialized stems, such as (103) _____ on a pea plant, are able to grow around an object, thereby receiving necessary (104) _____ for their slender stems. (105) _____

movement is different from the tropisms above since it is reversible, a temporary response to a nondirectional stimulus. This type of movement is due to changes in cell (106) _____ pressure and results in movements such as folding of leaves at night, opening or closing of stomata, and shifting of flowers or leaves as they follow the movement of the sun.

Plants, like animals, have a (107) _____ that acts as an internal time-measuring mechanism. A 24-hour cycle, called the (108) _____ rhythm, ensures that necessary activities occur repeatedly at a set time within this cycle. Plant activities also tend to change with seasons, often due to changes in (109) _____, a rhythm based on the division of light and dark periods in a 24-hour period. Plants primarily use a blue-green pigment, (110) _____, which exists in two reversible forms (P_r and P_{fr}) to measure changes in photoperiodism. Plants may be classified as (111) _____-_____ plants, which tend to flower in the spring, or (112) _____-_____ plants, which flower in the fall. It is the length of (113) _____ in a 24-hour cycle that determines when a plant will initiate flowering. Some plants will not flower unless they go through a period of cold temperature, a process called (114) _____.

Plants that have seeds and buds that go through a phase of no growth (even if conditions are suitable) experience (115) _____. Environmental conditions that initiate this state include (116) _____ nights, cold temperature, dry soil, and nitrogen deficiency.

Matching

Match each of the following terms with its correct definition.

117.	____	Phototropism	A.	Amyloplasts that move in the plant root and shoot tips to help in responding to gravity
118.	____	Gravitropism	B.	A blue-green pigment that exists in two forms and signals the light switch
119.	____	Thigmotropism	C.	Refers to a low-temperature stimulation of flowering
120.	____	Nastic movement	D.	Movement or growth of a plant in response to contact with an object
121.	____	Circadian rhythm	E.	Movement or growth of a plant in response to unidirectional light
122.	____	Phytochrome	F.	Movement or growth of a plant in response to gravity
123.	____	Vernalization	G.	Reversible or temporary movement in response to unidirectional stimulus
124.	____	Statoliths	H.	A 24-hour cycle of response in plants

SELF-TEST

1. Which of the following was identified as the first plant hormone? [p. 737]

   a. Auxins
   b. Gibberellins
   c. Cytokinins
   d. Ethylene
   e. Abscisic acid

2. How did Went prevent light "contamination" when performing his experiment using auxins? [p. 738]

   a. He grew only seedlings sensitive to high light levels
   b. He kept seedlings in the dark
   c. He provided light for a limited period each day
   d. He provided light for the full 24-hour cycle each day

3. Which of the following hormones exhibit(s) polar transport and move(s) away from unidirectional light? [p. 740]

   a. Jasmonates
   b. Auxins
   c. Oligosaccharins
   d. Cytokinins
   e. Abscisic acid

4. When cell elongation occurs, how do auxins (and gibberellins) cause the cell walls to stretch (according to the acid-growth hypothesis)? [p. 741]

   a. Loosening the cell wall's cellulose microfibril mesh
   b. Breaking down the cell's plasma membrane
   c. Reducing the turgor pressure in the cell

5. Which of the following hormones is(are) involved in the breaking of seed and bud dormancy? [p. 742]

   a. Auxins
   b. Gibberellins
   c. Abscisic acid
   d. Ethylene
   e. Brassinosteroids

6. Which of the following hormones is(are) involved in increasing internode length in rosette plants? [p. 742]

   a. Gibberellins
   b. Jasmonates
   c. Auxins
   d. Ethylene
   e. Abscisic acid

7. Which of the following hormones coordinate(s) growth of roots and shoots in concert with the auxins? [p. 743]

   a. Abscisic acid
   b. Gibberellins
   c. Cytokinins
   d. Salicylic acid

8. Which of the following hormones is(are) involved in ripening of fruits? [p. 745]

   a. Auxins
   b. Salicyclic acid
   c. Systemin
   d. Ethylene
   e. Abscisic acid

9. You order tree seedlings for delivery in early spring. What hormone(s) will the tree nursery apply to your seedlings to promote dormancy and minimize damage during transport? [p. 746]

   a. Ethylene
   b. Brassinosteroids
   c. Abscisic acid
   d. Auxins

10. Which of the following terms is used for dropping of flowers, fruits, and leaves? [p. 744]

   a. Senescence
   b. Abscission
   c. Tropism
   d. Bolting
   e. Vernalization

11. Which of the following terms is used for extension of the floral stem in rosette plants? [p. 742]

   a. Senescence
   b. Abscission
   c. Tropism
   d. Bolting
   e. Vernalization

12. Which of the following describe(s) systemin? [p. 749]

   a. Functions as a defence response in tomato plants
   b. First peptide hormone discovered in plants
   c. Binds to a receptor in a target cell plasma membrane
   d. (a) and (b)
   e. (a), (b), and (c)

13. What is shoot growth a response to? [p. 752]

   a. Positive gravitropism
   b. Negative gravitropism
   c. Neutral gravitropism

14. Which of the following refers to movement or growth of a plant in response to contact with an object? [p. 754]

   a. Phototropism
   b. Gravitropism
   c. Thigmotropism
   d. Nastic movement
   e. Photoperiodism

15. Which of the following refers to a temporary, reversible response to a nondirectional stimulus? [p. 755]

   a. Phototropism
   b. Gravitropism
   c. Thigmotropism
   d. Nastic movement
   e. Photoperiodism

16. Which of the following refers to the response of a plant due to changes in the length of light and dark periods during each 24-hour period? [p. 756]

   a. Phototropism
   b. Gravitropism
   c. Thigmotropism
   d. Nastic movement
   e. Photoperiodism

17. Which of the following terms is used for low-temperature stimulation of flowering? [p. 758]

   a. Senescence
   b. Abscission
   c. Tropism
   d. Bolting
   e. Vernalization

18. Which of the following terms is used for aging in plants? [p. 744]

   a. Senescence
   b. Abscission
   c. Tropism
   d. Bolting
   e. Vernalization

19. In which of the following is phytochrome involved? [p. 756]

   a. Phototropism
   b. Gravitropism
   c. Thigmotropism
   d. Nastic movement
   e. Photoperiodism

20. Gibberellins are made by plants and fungi. [p. 742]

   a. True
   b. False

INTEGRATING AND APPLYING KEY CONCEPTS

1. Knowing that ethylene is involved in fruit ripening, discuss how this information is used by fruit growers.

2. How does a plant use its phytochrome pigments to identify changes in the photoperiod?

3. Describe how an action potential leads to a change in cell turgor pressure during a nastic movement.

OPTIONAL ACTIVITIES

1. Experiments by the Darwins and Went indicated that one or more hormones from a shoot tip travelled to the tissue below and caused it to bend toward the light. How was it determined that auxin was the hormone involved?

2. Salicylic acid (SA) can play multiple roles in plant defence, one of which involves interacting with jasmonates. It is believed that some pathogens have evolved strategies to suppress the SA-mediated response. Describe one of these strategies.

32 Introduction to Animal Organization and Physiology

CHAPTER HIGHLIGHTS

- The organization of the animal body is presented—both structure and the associated function.
- Homeostasis is a dynamic equilibrium to meet the changing demands of both the internal and external environments of animals.
- Explanation of feedback systems, the control mechanisms of homeostasis, is introduced.

STUDY STRATEGIES AND LEARNING OUTCOMES

- First, focus on major levels of organization of animals—cells, tissues, organs, and organ systems.
- Examine and learn the structure of the various tissues, organs, and organ systems and then associate them with their functions.
- Evaluate the control mechanisms of feedback systems and maintenance of a constant yet dynamic internal environment.

By the end of this chapter, you should be able to

- Differentiate between the four tissues found in animals in terms of their structure and function
- Explain how the major levels of organization in animals are related to each other
- Describe how animals are able to maintain homeostasis, despite changing external and internal environments

TOPIC MAP

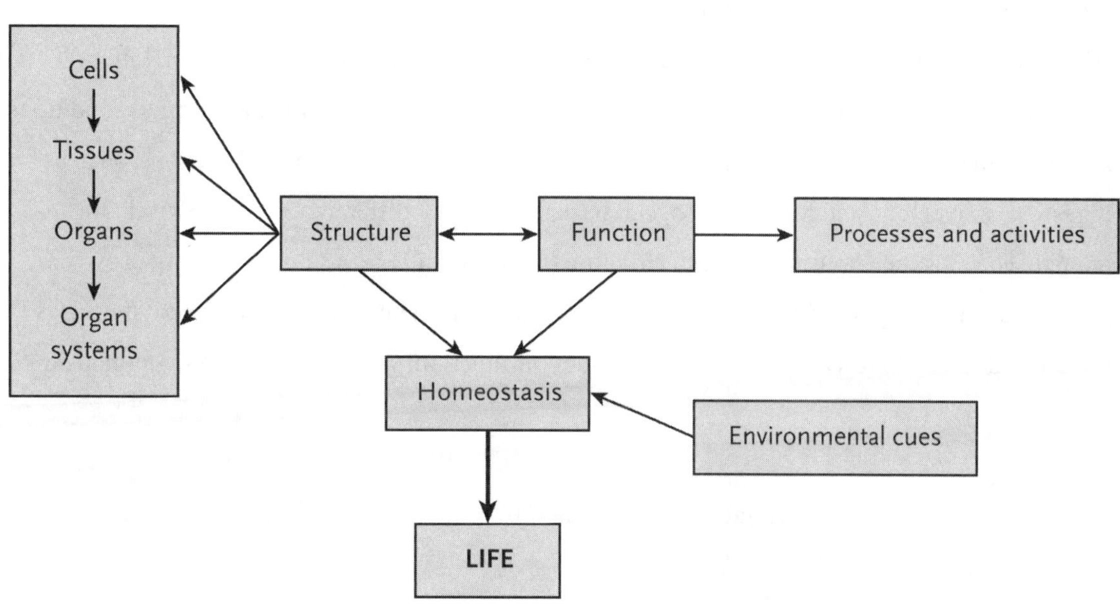
Cells
Tissues
Organs
Organ systems
Structure
Function
Processes and activities
Homeostasis
Environmental cues
LIFE

INTERACTIVE EXERCISES

Why It Matters [pp. 765–766]

Section Review (Fill-in-the-Blanks)

All living organisms are exposed to a variety of environmental conditions. An animal must maintain (1) _____, a stable yet dynamic (2) _____ environment, despite the external conditions and the demands within the organism. Under these conditions, various body functions, such as digestion, movement, and reproduction, can be carried out by different (3) _____ within the animal. The close relationship between structure (at the cell, (4) _____, organ, or organ system level) and (5) _____ explains how animals are able to coordinate their necessary activities. It is therefore difficult to research (6) _____, the study of function, without also studying organism structure, (7) _____.

32.1 Organization of the Animal Body [pp. 766–767]

Section Review (Fill-in-the-Blanks)

Organisms that are (8) _____ contain (9) _____, also known as extracellular fluid, between their cells, which assists cells in various ways, such as supplying required ions and (10) _____ to cells for cellular activities and accepting by-products, the (11) _____ molecules, produced from cellular activities. The concentration of ions and molecules in the extracellular fluid must be balanced with neighbouring (12) _____ to prevent excessive water movement by (13) _____. A group of similar cells, specialized to perform a specific (14) _____, forms a(n) (15) _____. A(n) (16) _____ is composed of two or more (17) _____ tissues that together perform a specific function. At the highest level within an organism, a(n) (18) _____ _____ coordinates the activities of two or more (19) _____ to carry out a major body activity, such as digestion or reproduction.

32.2 Animal Tissues [pp. 767–775]

Section Review (Fill-in-the-Blanks)

The structure and function of a tissue are primarily determined by the (20) _____ within the cells it is composed of, the (21) _____ matrix surrounding the cell, and the (22) _____, which holds the cells within the tissue together. Complex animals are composed of four major types of tissues. One type, (23) _____ tissue, is classified by the number of layers of (24) _____ above a basement membrane and the (25) _____ of the cells. Connective tissue structure and function are varied but are consistently composed of cells surrounded by a(n) (26) _____ matrix embedded with protein

(27) _____. The consistency of the matrix ranges from liquid to (28) _____ to porous, depending upon the type of connective tissue. There are (29) _____ types of (30) _____ tissue, cardiac, skeletal, and (31) _____; only the former two contain striations. The fourth type of tissue is (32) _____ tissue, which is composed of (33) _____ (also known as nerve cells) and supporting (34) _____ cells.

Matching

Match each of the following tissues with its correct function.

35.	____	Epithelial	A.	Supports, binds, provides structure
36.	____	Connective	B.	Responds to stimuli and communicates with other cells and tissues
37.	____	Nervous	C.	Contains actin and myosin—forcibly shortens
38.	____	Muscle	D.	Covers or lines body surfaces and cavities

Match each of the following tissues or cells with its correct description.

39.	____	Neurons	A.	Short, branched cells with intercalated disks
40.	____	Cardiac	B.	Moves bones
41.	____	Skeletal	C.	Walls of tubes and cavities, primarily autorhythmic
42.	____	Glial cells	D.	Respond to stimuli with axon and dendrites
43.	____	Smooth	E.	Support and electrical insulation

Choice

For each of the following statements, choose the most appropriate connective tissue components or type from the list below.

A. Collagen B. Fibrous connective tissue C. Loose connective tissue
D. Cartilage E. Bone F. Blood G. Adipose

44. _____ Primary cell type is fibroblast; forms mesenteries

45. _____ Lipid storage

46. _____ Fibrous glycoprotein

47. _____ Function of oxygen delivery to tissues as well as immunity

48. _____ Fibroblast cells; attaches muscles to bones

49. _____ Primary cell type is the osteocyte

50. _____ Chondrocytes in collagen and chondroitin sulphate matrix

51. _____ Fibroblast cells; attaches bones to bones

52. _____ Resilient like rubber, cushioning disks between vertebrae

53. _____ Calcium–phosphate mineral deposits in extracellular matrix

Complete the Table

54.

Tissue type	*Function*	*Location*
Epithelial	A.	B.
C.	Support, protection, attachment	Most body structures and systems
Muscle	D.	E.
Nervous	F.	Central and peripheral nervous systems

Short Answer

55. Identify the type of junction found between cells of epithelial tissue lining the urinary bladder and explain why it is found there.

____

____

56. How does the function of an osteoblast differ from that of an osteoclast?

____

____

57. What is the benefit of storing excess carbohydrates as fat?

____

____

32.3 Coordination of Tissues in Organs and Organ Systems [pp. 775–776]

Section Review (Fill-in-the-Blanks)

In multicellular organisms, the (58) _____ is the unit of life, yet it does not function in isolation. It must perform basic metabolic activities to (59) _____, and it must also perform at least one function for each of the tissue, organ, and (60) _____ levels to which it belongs. Most animals have (61) _____ major organ systems. These organ systems work in a coordinated fashion to accomplish basic functions, including acquiring, processing, and distributing (62) _____ and other substances throughout the body and disposing of (63) _____, synthesizing macromolecules, sensing and (64) _____ to environmental changes, protecting the body, and (65) _____, and then nourishing young.

Choice

For each of the following organs, choose the most appropriate organ system(s) it belongs to from the list below.

66. ____	Brain	A.	Endocrine
67. ____	Heart	B.	Muscular
68. ____	Skin	C.	Nervous
69. ____	Thymus	D.	Circulatory
70. ____	Bones	E.	Lymphatic
71. ____	Thyroid	F.	Integumentary
72. ____	Stomach	G.	Reproductive
73. ____	Lungs	H.	Excretory
74. ____	Cardiac and smooth	I.	Digestive
75. ____	Spleen	J.	Respiratory
76. ____	Testes	K.	Skeletal
77. ____	Hair and nails		
78. ____	Pancreas		
79. ____	Liver		
80. ____	Spinal cord		
81. ____	Uterus		
82. ____	Kidneys		

32.4 Homeostasis [pp. 776–779]

Short Answer

83. Explain the importance of homeostasis with respect to a changing external environment.

____

____

84. Compare and contrast a negative and a positive feedback mechanism when a change results in excess product.

____

____

Matching

Match each of the following components associated with a homeostatic mechanism with its correct definition.

85.	____	Response	A.	Control centre—compares change to set point
86.	____	Stimulus	B.	Detects change—temperature, pH, touch
87.	____	Effector	C.	Output
88.	____	Sensor	D.	Acceptable level for output
89.	____	Integrator	E.	Responds to stimulus; produces the output
90.	____	Set point	F.	Input—environmental change

SELF-TEST

1. The demands placed on a cell are both _____. [p. 766]

   a. Internal and external
   b. Physical and anatomical
   c. Functional and physiological
   d. None of these

2. What is a muscle, such as the biceps muscle of your arm, composed of? [p. 766]

   a. Similar tissues
   b. Different tissues
   c. Similar cells
   d. Different organs
   e. Similar organs

3. Select the structure in the highest level of organization (i.e., contains the other structures). [p. 766]

   a. Liver
   b. Epithelium
   c. Mitochondria
   d. Hepatic (liver) cell

4. Where is interstitial fluid found in an organism? [p. 766]

   a. In the cytoplasm
   b. In the cell membrane
   c. In the cell wall
   d. Between cells

5. What is the study of animal structure called? [p. 766]

   a. Homeostasis
   b. Osmosis
   c. Anatomy
   d. Physiology

6. Given that it lines a body cavity and has little or no extracellular matrix, select the correct tissue type. [pp. 767–775]

   a. Nervous
   b. Muscle
   c. Connective
   d. Epithelial

7. What type of junction allows ions and molecules to flow between cells by way of channels? [pp. 767–775]

   a. Tight junction
   b. Gap junction
   c. Anchoring junction

8. This tissue has cells involved in immunity as well as oxygen delivery to cells. [pp. 767–775]

   a. Nervous
   b. Muscle
   c. Connective
   d. Epithelial

9. This tissue is found in every organ or organ system. It is the most varied and is characterized by various cell types, fibres, and an extracellular matrix. [pp. 767–775]

   a. Nervous
   b. Muscle
   c. Connective
   d. Epithelial

10. A unit consisting of cells, concentric layers of calcium–phosphate minerals, a blood vessel, and nerve endings is a(n) __________. [pp. 767–775]

    a. Fibroblast
    b. Osteon
    c. Chondrocyte
    d. Osteoblast

11. What type of junction connects smooth muscle cells such that they can contract as a unit? [pp. 767–775]

   a. Tight junction
   b. Gap junction
   c. Anchoring junction

12. What type of tissue contains erythrocytes and leukocytes? [pp. 767–775]

   a. Epithelial
   b. Bone
   c. Blood
   d. Skeletal muscle
   e. Nervous

13. In what direction does an electrical signal move through a neuron? [pp. 767–775]

   a. Dendrite, cell body, then axon
   b. Axon, dendrite, then cell body
   c. Cell body, dendrite, then axon
   d. Axon, cell body, then dendrite

14. What type of a cell can a chemical signal be passed to after leaving a neuron? [pp. 767–775]

   a. Gland cell
   b. Muscle fibre
   c. Another neuron
   d. (a), (b), and (c)

15. Maintenance of osmotic balance, electrolytes, and pH is an important function of this organ system. [pp. 775–776]

   a. Circulatory
   b. Excretory
   c. Respiratory
   d. Endocrine

16. This organ system has sweat glands and provides protection to the organism. [pp. 775–776]

   a. Muscular
   b. Integumentary
   c. Skeletal
   d. Nervous

17. What organ system functions in acquiring nutrients, such as proteins and carbohydrates, and breaks them down? [pp. 775–776]

   a. Endocrine
   b. Circulatory
   c. Lymphatic
   d. Respiratory
   e. Digestive

18. The element of a homeostatic mechanism that compares a detected change with a set point is the __________. [pp. 776–779]

   a. Effector
   b. Response
   c. Integrator
   d. Sensor
   e. Stimulus

19. If the stimulus resulted in amplification of the response, this would be an example of ____. [pp. 776–779]

   a. Negative feedback
   b. Positive feedback
   c. Homeostasis
   d. Integration

20. If body temperature exceeded the set point of the hypothalamus, you would expect that the stimulus to activate sweat glands would _____. [pp. 776–779]

   a. Increase
   b. Decrease
   c. Not change
   d. Either (b) or (c)

INTEGRATING AND APPLYING KEY CONCEPTS

1. Where is sweat produced, and how does it reach the stem's surface?

2. Compare and contrast the bone, cartilage, and adipose tissue in terms of their structure and function.

3. Differentiate between structure and function of the three types of muscle tissue.

OPTIONAL ACTIVITIES

1. Compare and contrast the "division of labour" of a eukaryotic cell and the organ systems of a cat.

2. Why was the discovery of neural stem cells so important?

33 Information Flow: Nerves, Ganglia, and Brains

CHAPTER HIGHLIGHTS

- Neuronal circuits are made up of neurons that (1) sense environmental information, (2) integrate that information, and (3) effect an appropriate response to the information.
- Neurons communicate with other neurons in a circuit or with other targets (muscles, glands) via synapses. The vast majority of vertebrate neurons communicate by means of chemical synapses, where neurons release neurotransmitters that interact with receptors on the postsynaptic cell.
- Neurons are specialized cells that transmit information by electrical impulses. The impulses result from changes in ion distribution across the membrane of the neuron.
- Some neurotransmitters bind to ion channels (ligand-gated channels) to directly alter ion movement, whereas other neurotransmitters bind to G protein receptors that, in turn, produce second messengers and alter cellular activity, including ion movement.
- Neurons integrate stimulatory and inhibitory information to initiate electrical impulses.
- Nervous systems consist of networks of neurons.
- The complexity of nervous systems increases with the complexity of the animal group. The nervous systems of invertebrates are generally simple, whereas those of vertebrates are generally elaborate and display pronounced cephalization.
- In vertebrates, the central nervous system (CNS) consists of the brain and spinal cord. The peripheral nervous system consists of all the neurons and ganglia that connect the brain and spinal cord to the rest of the body.
- The peripheral nervous system of vertebrates has two main components: the somatic nervous system, which connects the CNS to the body wall (including the skeletal musculature), and the autonomic nervous system, which connects the CNS to the internal organs and blood vessels.
- The autonomic nervous system has two antagonistic components: the sympathetic division, which predominates in stressful or strenuous situations, and the parasympathetic division, which predominates during low-stress and feeding situations.

- Memory, learning, and consciousness involve modifications of neuron behaviour and connections between and among neurons in different parts of the brain.

STUDY STRATEGIES AND LEARNING OUTCOMES

- The study of neurons and neural circuits is important for understanding how animal function is controlled.
- This chapter contains a lot of new and unfamiliar material. Go slowly. Take one section at a time and then work through the companion section(s) in the study guide.
- Practise drawing neurons and labelling their component parts. Also practise drawing action potentials, being sure to label each phase and describing the bases for each phase.

By the end of this chapter, you should be able to

- Discuss the organization of nervous systems
- Draw the structure and identify the major parts of a neuron
- Explain how neurons communicate with each other
- Explain how resting membrane potential is established and how it changes during an action potential
- Provide a graphic representation of an action potential and discuss the reasons for the changes
- Explain how a nerve impulse travels down a neuron and how this nerve impulse is conducted to another neuron or muscle or gland
- List different types of neurotransmitters and receptors involved in chemical synapses
- Explain how neurons integrate incoming signals from other neurons
- Compare the organization of the nervous systems of protostomes
- Identify the major divisions and structures of the brain of vertebrates and discuss their roles
- Explain how the spinal cord relays signals between the peripheral nervous system and the brain and controls reflexes
- Compare the organization and roles of the sympathetic and parasympathetic nervous systems
- Discuss the organization of the somatic nervous system

- Discuss memory, learning, and consciousness

TOPIC MAP

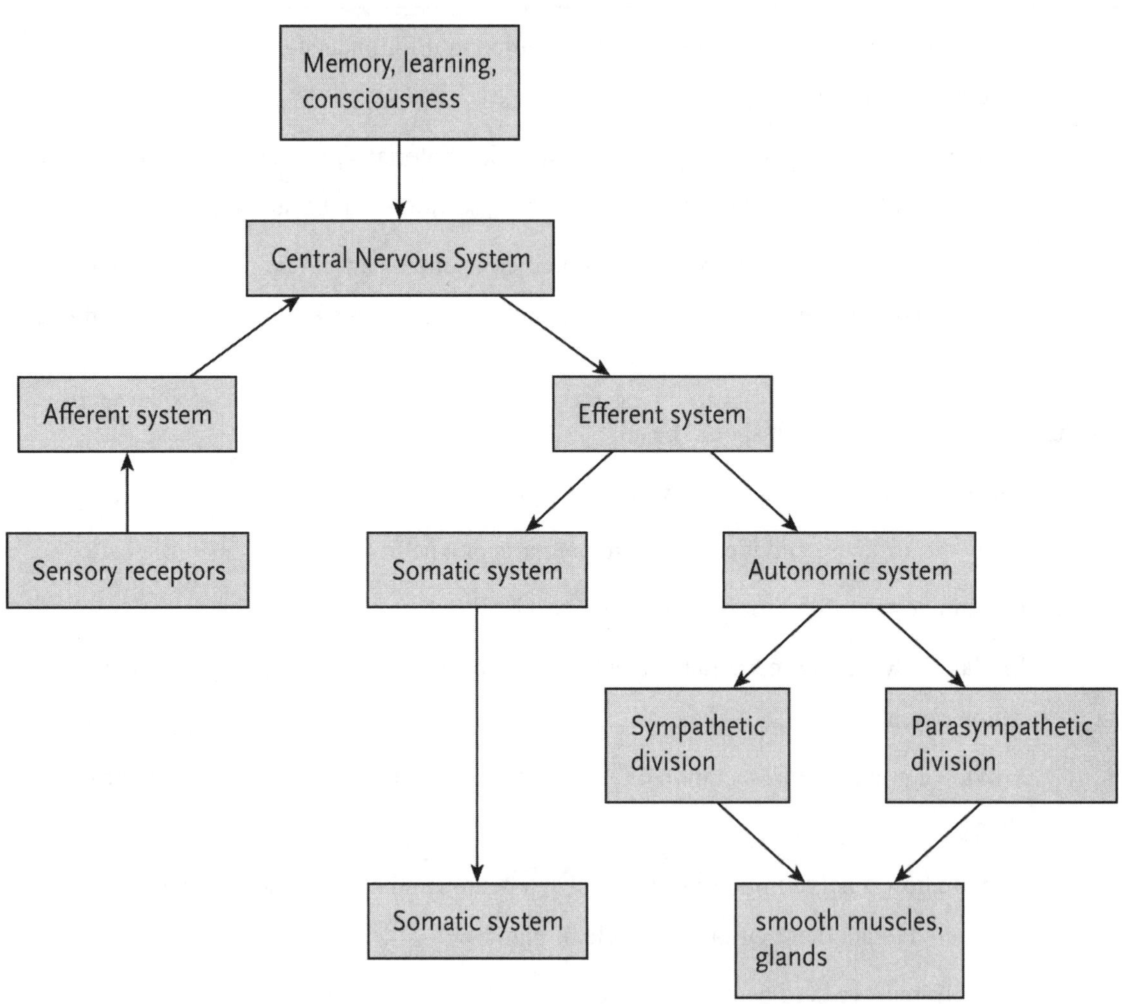

INTERACTIVE EXERCISES

Why It Matters [p. 783]

Section Review (Fill-in-the-Blanks)

The (1) ____ integrates a variety of sensory inputs and then makes compensating adjustments to the activities of the body.

33.1 Neurons and Their Organization in Nervous Systems: An Overview [pp. 784–787]

Section Review (Fill-in-the-Blanks)

(2) ____ is communication by (3) ____, which has four components. (4) ____ is the detection of a stimulus. (5) ____ is the sending of a message along a neuron and then relaying the message to another neuron or to a muscle or gland. (6) ____ is the sorting and interpretation of the message. (7) ____ is the output or action that results from the message. (8) ____, also known as (9) ____, transmit sensory stimuli to (10) ____, which integrate the information to formulate an appropriate response. (11) ____ carry response signals to (12) ____, such as muscles and glands. The (13) ____ of a neuron contains the nucleus. (14) ____ receive signals and transmit them toward the cell body, whereas (15) ____, which arise from the (16) ____, conduct impulses away from the cell body. Axons end in buttonlike swellings called (17) ____. The association of neurons (connected from the axon of one to the dendrite of another) forms a chain called a(n) (18) ____. (19) ____ provide nutrition and support to neurons. Star-shaped (20) ____ provide physical support and help maintain ion concentrations in the extracellular fluid. In vertebrates, (21) ____ in the central nervous system and (22) ____ in the peripheral nervous system wrap around axons to form sheaths. Gaps between Schwann cells along an axon are called (23) ____. A(n) (24) ____ is the junction between a neuron and another neuron or between a neuron and a muscle cell. On one side of the synapse is the (25) ____ and on the other side is the (26) ____. Communication across the synapse occurs in one of two ways: (a) by direct flow of electrical current, making a(n) (27) ____, or (b) by chemical transmission, making a(n) (28) ____. A(n) (29) ____ is the chemical released by a presynaptic cell at a chemical synapse, which crosses a narrow gap or (30) ____ and then binds to receptors on the postsynaptic cell.

Building Vocabulary

Using the definitions of the prefixes below, assign the appropriate prefix to the appropriate suffix for the following:

Prefix	Meaning
pre-	before, prior to
post-	after, following
inter-	between, among

Prefix	Meaning
-neuro(n)	of or relating to a nerve

Prefix	Suffix	Definition
____	-synaptic	31. the neuron that transmits a signal to a specific synapse
____	-synaptic	32. the neuron that receives the signal at a specific synapse
____	_____	33. neurons that integrate information

Choice

For each of the following characteristics, choose the most appropriate type of synapse.

a. Chemical synapse b. Electrical synapse

34. ____ Plasma membranes of presynaptic cells and postsynaptic cells are in direct contact, and the electrical response flows between the two cells through gap junctions.

35. ____ Plasma membranes of presynaptic cells and postsynaptic cells are separated by a cleft, and the electrical impulse is conveyed by neurotransmitters.

For each of the following descriptions, choose the most appropriate type of cell.

a. Presynaptic cell b. Postsynaptic cell

36. ____ Neuron that transmits signal

37. ____ Neuron that receives signal

For each of the following descriptions, choose the most appropriate type of cell.

a. Afferent neuron b. Interneuron c. Efferent neuron

38. ____ Transmits stimuli collected by sensory neurons

39. ____ Integrates sensory information to formulate appropriate response

40. ____ Carries impulses to effectors such as muscles and glands

Making Comparisons

41. Distinguish between motor neurons and efferent neurons.

____

____

____

Sequence

Arrange the following events in neural signalling in the correct order.

42. ____ a. Integration

43. ____ b. Transmission

44. ____ c. Response

45. ____ d. Reception

Complete the Table

Type of cell	*Description*
Glial cells	46.
47.	Star-shaped cells that provide physical support and help maintain ion concentration
Oligodendrocytes	48.
Schwann cells	49.
50.	Cells of the nervous system specialized to generate electrical impulses

Matching

Match each of the following with its correct definition.

51.	____	Synapse	A.	Communication by neurons
52.	____	Neural circuits	B.	Targets of efferent neurons such as muscles and glands
53.	____	Neurotransmitter	C.	A chain of neurons
54.	____	Node of Ranvier	D.	Gaps between Schwann cells that expose axons
55.	____	Effector	E.	A special junction between a neuron and another neuron or between a neuron and other effector cell
56.	____	Synaptic cleft	F.	Chemical released from an axon terminus into a synapse
57.	____	Neural signalling	G.	The narrow gap of a chemical synapse

Identify

Identify the various parts on the neuron.

58. ______________

59. ______________

60. ______________

61. ______________

62. ______________

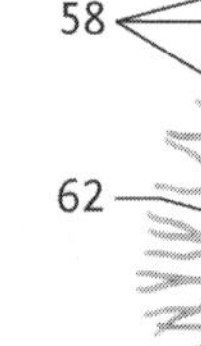

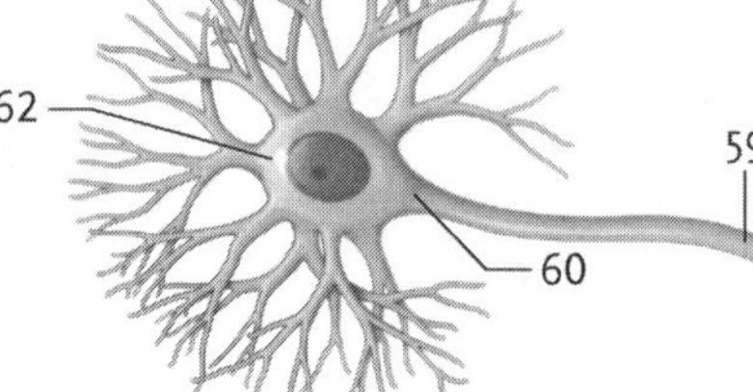

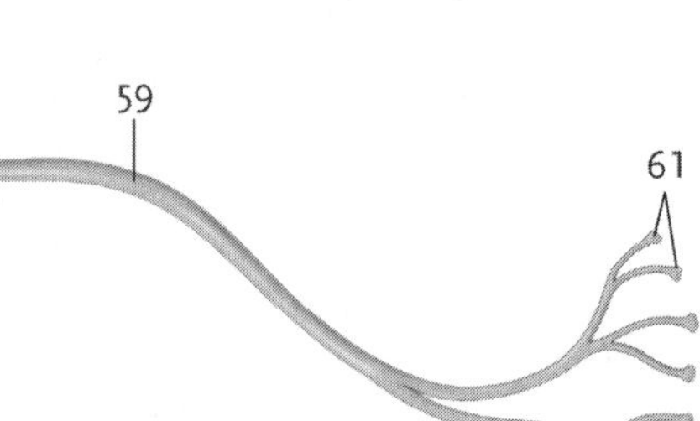

33.2 Signal Conduction by Neurons [pp. 787–794]

33.3 Conduction across Chemical Synapses [pp. 794–798]

33.4 Integration of Incoming Signals by Neurons [pp. 798–799]

Section Review (Fill-in-the-Blanks)

(63) ____ is the difference in charge across the membrane of all animal cells. The steady negative membrane potential of a neuron that is not conducting is called the (64) ____, which makes the cell (65) _____. When a neuron conducts an impulse, there is an abrupt transient change in membrane potential called the (66) ____. As the membrane becomes less negative, the membrane becomes (67) ____. Depolarization proceeds slowly at first until (68) ____ is reached, and then the potential changes rapidly, rising to as much as +30 mV inside with respect to the outside. The potential then falls, sometimes dropping below the resting level and resulting in the membrane becoming (69) ____. An action potential only results if the stimulus is strong enough to cause the membrane potential to reach threshold; once triggered, the action potential results regardless of stimulus strength. This phenomenon is known as the (70) ____. At the peak of the action potential, the membrane enters a(n) (71) ____, which lasts until the resting potential is reestablished. During the refractory period, the neuron cannot be stimulated again, which keeps the impulses travelling only in one direction. The action potential is produced by movements of Na^+ and K^+ through (72) ____. (73) ____ of the electrical impulse occurs along the axon as a wave of depolarization. Impulses are sped when action potentials hop along the axon in a process called (74) ____.

In chemical synapses, neurotransmitters are released from the (75) ____, diffuse across the cleft, and alter ion conduction by activation of (76) ____ on the (77) ____. Neurotransmitters are stored in presynaptic cells in (78) ____ and are released by the process of (79) ____. (80) ____ directly bind ligand-gated channels to alter the flow of ions. (81) ____ bind to G protein–coupled receptors that trigger the generation of a second messenger (e.g., cAMP); the second messenger then affects ion channels to alter ion movement. Many different types of chemicals serve as neurotransmitters, including amines, amino acids, and peptides.

Neurotransmitters may stimulate or inhibit the generation of action potentials in the postsynaptic cell. Stimulation opens a ligand-gated sodium channel, depolarizing the cell and driving closer to threshold. Such a potential change is called a(n) (82) ____. An inhibitory neurotransmitter opens a ligand-gated channel that allows Cl^- to flow into the cell and K^+ to leave, causing the cell to become hyperpolarized and taking the cell further from threshold; such a potential change is called a(n) (83) ____. In contrast to all-or-nothing action potentials, EPSPs and IPSPs are (84) ____. (85) ____ is the accumulation of EPSPs from a single presynaptic neuron over a short period of time. (86) ____ is the accumulation of EPSPs produced from several different presynaptic neurons.

Building Vocabulary

Using the definitions of the prefixes below, assign the appropriate prefix to the appropriate suffix for the following:

Prefix	Meaning
neuro-	of or relating to a neuron
hyper-	above, to a greater extent
de-	to undo

Prefix	Suffix	Definition
____	-transmitter	87. The chemical released at a chemical synapse
____	-polarized	88. A change in membrane potential that makes the cell less negative inside compared with outside
____	-polarized	89. A change in membrane potential that makes the cell more negative inside compared with outside

Choice

For each of the following characteristics, choose the most appropriate term from the list below.

a. Presynaptic membrane   b. Postsynaptic membrane

90. ____ Membrane from which exocytosis occurs to release neurotransmitters

91. ____ Membrane on which neurotransmitters bind

For each of the following characteristics, choose the most appropriate neurotransmitter from the list below.

a. Indirect neurotransmitter   b. Direct neurotransmitter

92. ____ Neurotransmitter that binds to a ligand-gated channel

93. ____ Neurotransmitter that binds to a receptor to trigger production of a second messenger

For each of the following, choose the most appropriate potential from the list below.

a. Excitatory postsynaptic potential b. Inhibitory postsynaptic potential

94. ____ Change in membrane potential of postsynaptic cells that brings it closer to threshold

95. ____ Change in membrane potential of postsynaptic cells that brings it further from threshold

For each of the following characteristics, choose the most appropriate type of summation from the list below.

a. Temporal summation b. Spatial summation

96. ____ Change in membrane potential of postsynaptic neuron produced by the firing of several presynaptic neurons

97. ____ Change in membrane potential of postsynaptic neuron produced by successive firing of a single presynaptic neuron over a short period of time

For each of the following characteristics, choose the most appropriate type of channel from the list below.

a. Ligand-gated channel b. Voltage-gated channel

98. ____ Membrane proteins that control ion flow by opening and closing as membrane potential changes

99. ____ Membrane proteins that control ion flow in response to binding a neurotransmitter

Making Comparisons

100. Distinguish between membrane potential and resting potential.

____

____

Matching

Match each of the following aspects of neural signalling with its correct description.

101.	____	All-or-nothing principle	A.	An abrupt, transient change in membrane potential
102.	____	Synaptic vesicles	B.	Secretory vesicles that contain neurotransmitters
103.	____	Action potential	C.	Rapid conduction of an impulse in which action potentials "hop" along an axon
104.	____	Propagation	D.	The progression of action potentials along an axon
105.	____	Salutatory conduction	E.	Once triggered, the change in membrane potential that takes place regardless of stimulus strength
106.	____	Graded potential	F.	Potential that moves up or down in response to stimulus without triggering an action potential; amplitude depends on the strength of the stimulus

Identify

Identify the various parts of an action potential.

107. __________

108. __________

109. __________

110. __________

111. __________

112. __________

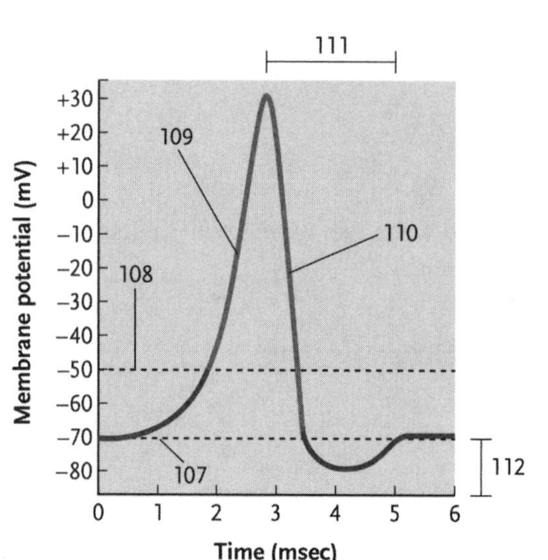

33.5 Integration in Protostomes: Network, Nerves, Ganglia, and Brains [pp. 799–802]

33.6 Vertebrates Have the Most Complex Nervous Systems [pp. 802–804]

Section Review (Fill-in-the-Blanks)

Radially symmetrical animals such as cniderians (jellyfish) and echinoderms (starfish, sea urchins) possess loose meshes of neurons called (113) ___. Groups of nerve cells with localized interconnections are called (114) ___. There is an evolutionary trend toward cephalization in which a distinct head region forms that contains a central ganglion or (115) ____ connected to one or more (116) ____ that extend to the rest of the body. The brain and nerve cord constitute the (117) ____, and the nerves that extend from the CNS to the rest of the body make up the (118) ____. During embryonic development, the nervous system of vertebrates arises from a hollow (119) ____, the anterior portion of which develops into the brain, and the rest gives rise to the (120) ____. The central cavity of the neural tube persists in adults as the (121) ____ of the brain and the (122) ____ of the spinal cord. The (123) ____, (124) ____, and (125) ____ are the three distinct regions of the brain early in development that give rise to the adult brain.

Building Vocabulary

Using the definitions of the prefixes below, assign the appropriate prefix to the appropriate suffix for the following:

Prefix	Meaning
hypo-	under, below
post-	behind, after
pre-	before, prior to

Prefix	Suffix	Definition
____	-thalamus	126. Part of the brain located below the thalamus and primary integration centre for regulation of the viscera
____	-ganglionic	127. Relates to a neuron leading away from a ganglion
____	-ganglionic	128. Relates to a neuron leading into a ganglion

Choice

For each of the following descriptions, choose the most appropriate term from the list below.

a. Ventricle   b. Central canal

129.____ Space within the brain

130.____ Space within the spinal cord

Complete the Table

Structure	*Feature/Description*
Nerve ring	131.
132.	Functional clusters of neurons
Nerve cord	133.

33.7 The Central Nervous System (CNS) and Its Functions [pp. 804–810]

33.8 The Peripheral Nervous System (PNS) [pp. 810–812]

Section Review (Fill-in-the-Blanks)

The (134) ____ control body movements that are primarily voluntary. The (135) ___ controls largely involuntary processes such as digestion, secretion by sweat glands, and circulation of blood. Situations involving adaptation to stress or strenuous activity are mediated by the (136) ____, whereas the (137) ____ predominates during digestion and periods of quiet and relaxation. The (138) ____ surrounds and protects the brain and spinal cord. (139) ____ circulates in the central canal and in the ventricles. The butterfly-shaped core of the spinal cord is (140) ____, which is surrounded by (141) ____. Interneurons in the grey matter are involved in (142) ____. The (143) ____ connects the forebrain with the spinal cord. The surface layer of the forebrain is the (144) ____. Tight junctions between cells of brain capillaries form a(n) (145) ____ and prevent the movement of many substances.

Incoming sensory information is filtered by the (146) ____ before going to other CNS centres. The (147) ____ helps fine-tune balance and body movements. The (148) ____ relays sensory information to regions of the cerebral cortex concerned with motor responses. The (149) ____ helps coordinate temperature and osmotic homeostasis. The (150) ____ surrounds the thalamus and moderates voluntary movements directed by motor centres in the cerebrum. The (151) _____ is made up of the thalamus, hypothalamus, and basal nuclei, as well as the (152) ____, (153) ____, and (154) ____. The two cerebral hemispheres are connected via the (155)

____. The (156) ____ in each hemisphere of the cerebral cortex registers information on touch, pain, temperature, and pressure. (157) ____ of the cerebral cortex integrate information from the sensory areas, formulate responses, and pass them on to the (158) ____. The localization of some brain functions in one of the two hemispheres is called (159) ____.

Choice

For each of the following descriptions, choose the most appropriate structure from the list below.

a. Spinal nerves b. Cranial nerves

160.____ Connect with the CNS at the level of the brain

161.____ Connect with the CNS at the level of the spinal cord

For each of the following descriptions, choose the most appropriate characteristic from the list below.

a. Efferent b. Afferent

162.____ Neuron that conducts information away from the CNS

163.____ Neuron that conducts information toward the CNS

For each of the following descriptions, choose the most appropriate characteristic from the list below.

a. Grey matter b. White matter

164.____ Consists of nerve cell bodies and dendrite

165.____ Consists of axons, many of which are surrounded my myelin sheaths

Complete the Tables

Component	*Subcomponent*	*Function*
Somatic		166.
167.		Control internal organs and blood vessels (most involuntary)
	168.	Predominates in situations involving stress, excitement, etc.
	169.	Predominates in low-stress situations

Structure	*Function*
Corpus callosum	170.
171.	Registers information about touch, pain, temperature, and pressure
172.	Causes movement of specific part of the body (each hemisphere controls the opposite side)
Association areas	173.

Matching

Match each of the following type of junction with its correct definition.

174.	____	Meninges	A.	Connective tissue layers that surround and protect brain and spinal cord
175.	____	Basal nuclei	B.	Fluid that circulates in central canal and the ventricles of the brain
176.	____	Amygdala	C.	Programmed movement that takes place without conscious effort
177.	____	Lateralization	D.	Results from tight junction between capillary cells in the brain
178.	____	Cerebral cortex	E.	Phenomenon in which some brain functions are localized to one hemisphere or the other
179.	____	Reflex	F.	Connects brain with spinal cord
180.	____	Blood–brain barrier	G.	Surface layer of cerebrum
181.	____	Thalamus	H.	Part of the brain stem that connects thalamus to spinal cord
182.	____	Hippocampus	I.	Receives sensory information and relays it to high CNS centres
183.	____	Hypothalamus	J.	Moderates voluntary movements directed by the cerebrum
184.	____	Reticular formation	K.	Sends information to frontal lobes
185.	____	Olfactory bulb	L.	Relays information about experience and emotions
186.	____	Cerebrospinal fluid	M.	Relays olfactory information to cerebral cortex
187.	____	Brain stem	N.	Regulates basis homeostatic functions

33.9 Memory, Learning, and Consciousness [pp. 812–814]

Section Review (Fill-in-the-Blanks)

(188) ____ is the storage of an experience. (189) ____ involves a change in the response to a stimulus based on information or experience stored in memory. (190) ____ is awareness of oneself and one's surroundings. (191) ____ stores information for up to an hour or so, whereas (192) ______ stores information for days, years, or even for life. (193) ____ is a lasting increase in the strength of synaptic connections in neural pathways. Changes in neural activity can be recorded by a(n) (194) ____. During (195) ____, an individual's heart rate and respiration decrease, their limbs twitch, and their eyes move rapidly behind closed eyelids.

Choice

For each of the following descriptions, choose the most appropriate structure from the list below.

a. Short-term memory b. Long-term memory

196.____ Lasts up to an hour or so

197.____ Lasts for days, years, or even life.

Matching

Match each of the following types of junction with its correct definition.

198.	____	Memory	A.	Storage/retrieval of a sensory or motor experience
199.	____	Long-term potentiation	B.	Changes in response to a stimulus
200.	____	Rapid eye movement (REM) sleep	C.	Awareness of self and surroundings
201.	____	Consciousness	D.	Long-lasting increase in strength of response
202.	____	Learning	E.	A phase of sleep in which brain waves similar to those in the waking state are observed

SELF-TEST

1. Neural signalling involves [p. 785]

   a. Differentiation
   b. Integration
   c. Response
   d. Transmission
   e. Reception

2. Efferent neurons [p. 785]

   a. Innervate muscles and glands
   b. Integrate information
   c. Transmit electrical impulses away from interneurons
   d. Transmit electrical impulses from sensory receptors

3. The nodes of Ranvier [p. 786]

   a. Speed up conduction of electrical impulses
   b. Make up a myelin sheath
   c. Expose the axon
   d. Are gaps between adjacent Schwann cells

4. Which of the following are glial cells? [p. 786]

   a. Astrocyte
   b. Interneuron
   c. Schwann cell
   d. Neurosecretory cell
   e. Oligodendrocyte

5. Electrical synapses [p. 787]

   a. Are plasma membranes of pre- and postsynaptic cells that are in direct contact
   b. Have narrow gaps called clefts between cells
   c. Are electrical impulses that flow between cells through gap junctions
   d. Involve release of neurotransmitters

6. The resting potential results from [pp. 788–789]

   a. Differential permeability of membranes that results in accumulation of proteins and other anions on the inside of a cell
   b. An opening of voltage-gated sodium channels
   c. An imbalance of Na^+ and K^+ inside and outside the cells created by the Na^+/K^+ pump
   d. An opening of ligand-gated sodium channels

7. The ion movement accomplished by the Na^+/K^+ pump. [pp. 788–789]

   a. Na^+ in, K^+ out
   b. K^+ in, Na^+ out
   c. Both Na^+ and K^+ in
   d. More K^+ out than Na^+ in
   e. More Na^+ out than K^+ in

8. The refractory period [p. 789]

   a. Is initiated with the closing of the inactivation gate of the Na^+ voltage-gated channel
   b. Is initiated with the opening of the activation gate of the Na^+ voltage-gated channel
   c. Ceases with the opening of the activation gate of the K^+ voltage-gated channel
   d. Lasts from the peak of the action potential to the reestablishment of resting potential

9. Nerve impulses [pp. 791–794]

   a. Always travel in both directions along an axon
   b. Sometimes travel in both directions along an axon
   c. Travel in only one direction along an axon
   d. Always travel with the same intensity along the length of an axon
   e. Diminish in intensity as they travel along axon

10. Neurotransmitters [pp. 794–795]

   a. Travel through gap junctions
   b. Are released from presynaptic cells through the process of exocytosis
   c. Can bind to ligand-gated ion channels
   d. Can bind to G protein–coupled receptors

11. Inhibitory postsynaptic potentials (IPSPs) [p. 798]

   a. Result from K^+ influx into a cell
   b. Result from hyperpolarization of a postsynaptic cell
   c. Are all or nothing
   d. Are graded

12. Spatial summation [p. 799]

   a. Is the change in membrane potential of a postsynaptic cell brought on by the firing of different presynaptic neurons
   b. Is the change in membrane potential of a postsynaptic cell brought on by the successive firing of a single presynaptic neuron over a short period of time
   c. Is the total of the membrane potentials that occur along an axon as the electrical impulse is transmitted

13. Which is the principal integration centre of homeostatic regulation and leads to the release of hormones? [p. 807]

    a. Cerebellum
    b. Cerebrum
    c. Thalamus
    d. Hypothalamus
    e. Association area

14. The regulation of blood pressure is primarily under the control of ___. [pp. 811–812]

    a. Somatic nervous system
    b. Autonomic nervous system
    c. Parasympathetic division
    d. Sympathetic division

15. The regulation of feeding and digestion is primarily under the control of ___. [pp. 811–812]

    a. Somatic nervous system
    b. Autonomic nervous system
    c. Parasympathetic division
    d. Sympathetic division

16. Cerebrospinal fluid [p. 804]

    a. Circulates in the central canal
    b. Circulates in the ventricles of the brain
    c. Is within layer of the meninges
    d. Mixes with blood
    e. Cushions brain and spinal cord from jarring movements

17. Which is the part of the mammalian brain that integrates information about posture and muscle tone? [pp. 806–807]

    a. Cerebrum
    b. Cerebellum
    c. Myencephalon
    d. Medulla oblongata

18. Which is the part of the brain that coordinates muscular activity? [pp. 806–807]

    a. Cerebrum
    b. Cerebellum
    c. Myencephalon
    d. Medulla oblongata

19. The brain and the spinal cord are wrapped in a connective tissue layer called the ____. [p. 804]

    a. Grey matter
    b. Meninges
    c. Dura mater
    d. Sclera

20. In non-REM sleep compared to REM sleep, there is(are) ____. [pp. 813–814]

    a. More delta waves
    b. Faster breathing
    c. Lower blood pressure
    d. More dream consciousness

21. If only one temporal lobe is damaged, you would expect ____. [p. 808]

    a. Blindness in one eye
    b. Total blindness
    c. Loss of hearing in one ear
    d. Partial loss of hearing in both ears

INTEGRATING AND APPLYING KEY CONCEPTS

1. The salivary glands of vertebrates are innervated with adrenergic neurons (neurons that release epinephrine from their axon termini). Epinephrine stimulates both fluid and amylase (a carbohydrate hydrolyzing enzyme) secretion from the salivary gland. The fluid secretion is calcium dependent (can be inhibited by calcium channel blockers), while amylase secretion is cAMP dependent. Explain.

2. Discuss the selective pressure for cephalization in animals. Would there be a difference between those with sessile or motile life histories?

OPTIONAL ACTIVITIES

1. In cases of diagnosed hypertension, medication can be prescribed as a means to lower blood pressure. Identify different types of medications used for these purposes and discuss how they lower blood pressure (explain in terms of your knowledge gained from your study of the material in this chapter).

2. Many types of local anesthetic compounds are used to reduce pain sensations during medical procedures. Identify different types of local anesthetics and explain how they work to reduce pain sensations.

34 Sensory Systems

CHAPTER HIGHLIGHTS

- Sensory receptors convey information about the external and internal environments to the nervous system.
- Mechanoreceptors detect pressure and vibration and provide information about movement and position.
- Photoreceptors detect radiant energy and are involved with perception of visual images.
- Chemoreceptors respond to different kinds of chemicals and are involved with the perception of taste and smell.
- Thermoreceptors detect heat and are involved with temperature regulation. Nociceptors detect pain and protect animals from dangerous stimuli.
- Electroreceptors detect electrical fields and are used for communication and to detect prey. Magnetoreceptors detect magnetic fields and are used for navigation.

STUDY STRATEGIES AND LEARNING OUTCOMES

- This chapter makes use of the concepts of cell structure, cell communication, and nerve function developed earlier. You should briefly review these concepts to make sure that you understand them.
- While many of the terms in this chapters may be familiar, others probably will not be, so do not be hasty in your study.
- This chapter is extremely long. DO NOT try to go through it all in one sitting. Take one section at a time and then work through the companion section(s) in the study guide.
- Draw pictures of the various sensory systems, being sure to label the various parts and noting their linkage to the nervous system.

By the end of this chapter, you should be able to

- Explain how sensory information is detected and how this message is transduced and decoded by the central nervous system
- Draw the general structure of sensory receptors

- List and define the different types of sensory receptors
- Describe the structure and function of the accessory organs and mechanoreceptors involved in the detection of touch and pressure and to maintain body balance and orientation
- Describe the structure and function of the accessory organs and mechanoreceptors involved in the detection of sound
- Describe the structure and function of the accessory organs and photoreceptors involved in the detection of light and formation of images
- Describe the structure and function of the accessory organs and chemoreceptors involved in the detection of chemicals
- Describe the structure and function of the accessory organs and thermoreceptors and nociceptors involved in the detection of heat and pain
- Describe the structure and function of the accessory organs and electroreceptors and magnetoreceptors involved in the detection of electric and magnetic fields
- Identify, label the parts of, and draw the different accessory organs involved in the detection of external and internal information

TOPIC MAP

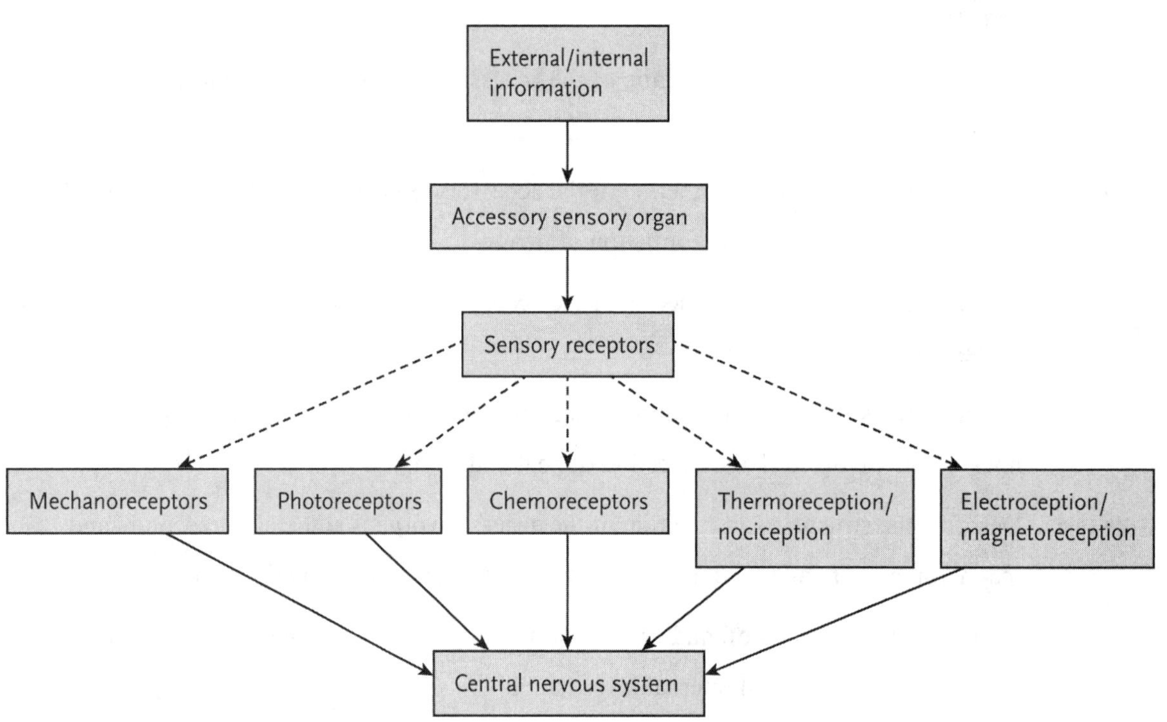
External/internal information
Accessory sensory organ
Sensory receptors
Mechanoreceptors
Photoreceptors
Chemoreceptors
Thermoreception/ nociception
Electroception/ magnetoreception
Central nervous system

INTERACTIVE EXERCISES

Why It Matters [pp. 819–820]

Section Review (Fill-in-the-Blanks)

Information about the internal and external environment is conveyed by the (1) ____ to the central nervous system.

34.1 Overview of Sensory Receptors and Pathways [pp. 820–822]

34.2 Mechanoreceptors and the Tactile and Spatial Senses [pp. 822–830]

34.3 Mechanoreceptors and Hearing [pp. 831–833]

Section Review (Fill-in-the-Blanks)

(2) ____ are formed by dendrites of a neuron or are specialized cells that synapse with afferent neurons. The conversion of a stimulus into a change in membrane potential is called (3) ____. The five types of sensory receptors are (4) ____, (5) ____, (6) ____, (7) ____, and (8) ____. The intensity and extent of a stimulus are registered by the (9) ____ and the (10) ____. (11) ____ is the effect of reducing the response to a stimulus at a constant level.

Some mechanoreceptors that detect touch and pressure are free nerve endings, while others, such as (12) ____, have structures that surround the nerve ending to help detect stimuli. (13) ____ detect stimuli that provide information about the position of limbs and are used to maintain balance. (14) ____ are fluid-filled chambers that contain (15) ____ and (16) ____ to detect position in some invertebrates. Fish and some amphibians detect vibrations and water current with a(n) (17) ____. Some fish also have dome-shaped (18) ____ that contain sensory hair cells covered with (19) ____ that extend into a gelatinous (20) ___ and are used to detect orientation and velocity.

The (21) ____ perceives position and motion of the head and consists of three (22) ____ and two chambers, the (23) ____ and the (24) ____, which contain small crystals called (25) ____. The two types of (26) ____ that detect the position and movement of limbs are called (27) ___ in muscles and (28) ____ in tendons.

Some invertebrates have auditory organs that consist of a thinned region of exoskeleton or (29) ____ stretched over a hollow chamber. The (30) ____ of the (31) ____ focuses sound waves into the auditory canal, where they strike the (32) ____. The (33) ____ is an air-filled cavity containing three interconnected bones, the (34) ____, (35) ____, and (36) ____, the latter of which is attached to an elastic (37) ____. The (38) ____ contains several fluid-filled compartments, (39) ____, (40) ____, and (41) ____, as well as a spiralled (42) ____. Within the cochlea is the (43) ____, which contains sensory hair cells that detect sound vibrations that dissipate when they reach the (44) ____. Many vertebrates locate prey and avoid obstacles by (45) ____, a process that involves generating sounds and listening for the echoes that bounce back.

Building Vocabulary

Using the definitions of the prefixes below, assign the appropriate prefix to the appropriate suffix for the following:

Prefix	Meaning
chemo-	chemical
oto-	ear
proprio-	one's own
thermo-	heat, warm
photo-	light
noci-	pain, injury

Suffix	Meaning
-lith	stone

Prefix	Suffix	Definition
____	-receptor	46. Specialized to detect chemical stimuli
____	-receptor	47. Specialized to detect heat/temperature
____	____	48. Calcium carbonate crystals in inner ear of vertebrates
____	-ceptor	49. Sensory receptors in muscles, tendons, and joints
____	-receptor	50. Specialized for the detection of radiant energy at particular wavelengths
____	-ceptor	51. Specialized for the detection of tissue damage/noxious chemicals

Complete the Table

Structure	*Description*
52.	Detects vibrations and currents in water
Statocysts	53.
Vestibular apparatus	54.
55.	Detects stretch of muscle
56.	Contains sensory hairs that detect sound waves
Golgi tendon organ	57.
58.	Provides information about orientation and velocity of fish

Matching

Match each of the following structure with its correct definition.

59.	____	Proprioceptors	A.	Type of mechanoreceptor that detects position and movement of body parts
60.	____	Echolocation	B.	Type of proprioceptor that detects position and movement of limbs
61.	____	Tympanum	C.	Calcium carbonate crystal inside vestibular apparatus
62.	____	Cupula	D.	Sensory cell with long hairlike projection of plasma membrane
63.	____	Otoliths	E.	Gelatinus matrix inside neuromasts
64.	____	Stereocilia	F.	Stonelike body inside statocysts
65.	____	Stretch receptor	G.	Microvilli on hair cells of neuromasts
66.	____	Statoliths	H.	Process of generating sound waves and detecting echoes
67.	____	Sensory hair cell	I.	Thinned region of exoskeleton on some invertebrates specialized for detecting vibrations

Labelling

Identify each numbered part of the following illustration.

68. ______________

69. ______________

70. ______________

71. ______________

72. ______________

73. ______________

74. ______________

75. ______________

76. ______________

77. ______________

78. ______________

79. ______________

80. ______________

81. ______________

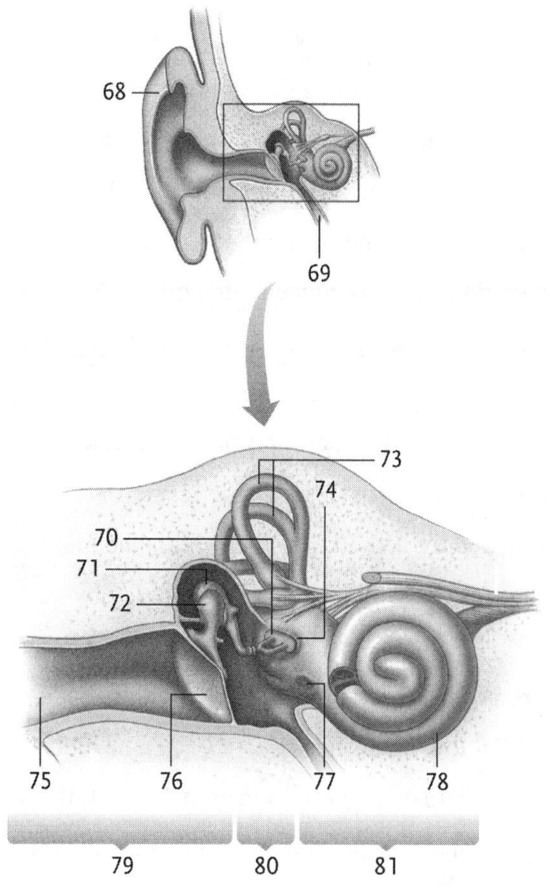

34.4 Photoreceptors and Vision [pp. 833–840]

34.5 Chemoreceptors [pp. 841–843]

34.6 Thermoreceptors and Nociceptors [pp. 844–846]

34.7 Electroreceptors and Magnetoreceptors [p. 847]

Section Review (Fill-in-the-Blanks)

The simplest eye in invertebrates, a(n) (82) ____, detects light but does not form an image. One type of image-forming eye in invertebrates is the (83) ____, which has hundreds to thousands of faceted visual units called (84) ____, which in insects are focused by a(n) (85) ____ onto a bundle of photoreceptive cells containing the (86) ____, rhodopsin. The other type of image-forming eye in invertebrates is the (87) ____, which resembles the eye of vertebrates in its camera-like operation. Light is concentrated by the (88) ____ onto the (89) ____, a layer of photoreceptors at the back of the eye. Muscle of the (90) ____ adjusts the size of the (91) ____ to regulate the amount of light entering the eye. (92) ____ is a process by which the lens changes to enable the eye to focus on objects at different distances. The structures of the vertebrate eye are similar to those of the invertebrate single-lens eye. (93) ____ fills the space between the cornea and the lens, and the jellylike (94) ____ fills the main chamber of the eye between the lens and the retina. The lens of many vertebrates is focused by changing its shape by contraction of muscles of the (95) ____ and adjusting the tension of the ligaments that anchor the lens to the muscles. The two types of photoreceptors in the vertebrate eye are (96) ____, which are specialized for detection of light at low intensities, and (97) ____, which are specialized for detection of different wavelengths (colours). In mammals and birds, cones are concentrated in and around the (98) ____. Images focused there can be seen distinctly, while the surround image is termed (99) ____. (100) ____ are composed of the light-absorbing molecule (101) ____, covalently bound to proteins called (102) ____. (103) ____ is the photopigment in rods. Photoreceptors are linked to a network of neurons that integrate and process initial visual information. (104) ____ synapse with rods and cones on one end and with (105) ____ on their other end. In addition, (106) ____ connect photoreceptor cells and bipolar cells. (107) ____ connect bipolar with ganglion cells. In (108) ____, horizontal cells inhibit bipolar cells that are outside a spot of light, striking the retina; this visual processing sharpens the edges of the image and enhances contrast. Many animals have colour vision, which depends on the number and types of cones. Humans and other primates have three types of cones, based on the form of (109) ____ that they possess. Just behind the eye, optic nerves converge and a portion of each optic nerve crosses over to the opposite side of the brain, forming the (110) ____. Axons enter the (111) ____ of the thalamus, where they synapse with neurons leading to the visual cortex.

Insects have taste receptors inside hollow sensory bristles called (112) ____. (113) ____ of vertebrates have distinct receptors that respond to sweet, sour, salty, bitter, and umami (savoury). Olfactory receptor cells possess (114) ____; the density of these receptors determines olfactory sensitivity.

The (115) ____-gated calcium channel family acts as heat receptors; different channels have different temperature thresholds. Pain receptors do not exhibit adaptation.

Electroreceptors are specialized for detecting electric fields. (116) ____ detect Earth's magnetic field and help provide directional information important for navigation.

Distinguish between Members of the Following Sets of Terms

117. Compound eye and single-lens eye

____

____

118. Opsins, rhodopsins, and photopsins

____

____

Describe

119. What is accommodation?

____

____

120. Describe the difference between electroreceptors and magnetoreceptors.

____

____

Complete the Tables

Cell type	*Function*
121.	Neurons that synapse with rods/cones on one end and ganglion cells on the other end
Ganglion cell	122.
123.	Connect with bipolar cells and ganglion cells
Horizontal cell	124.

Type of receptor	*Stimuli*	*Function*
125.	126.	Visual image formation
Taste bud	Chemicals	127.
128.	129.	Perception of smell
130.	Electric field	131.
Magnetoreceptor	132.	Directional movement and navigation
133.	134.	Perception of pain

Matching

Match each of the following types of junction with its correct definition.

135.	____	Ocellus	A.	A simple photoreceptor that does not form a visual image
136.	____	Photopigment	B.	The individual visual unit of an invertebrate compound eye
137.	____	Optic chiasm	C.	Specialized for detecting light of low intensity
138.	____	Rod cell	D.	Specialized for detecting light of different wavelengths
139.	____	Cone cells	E.	Type of visual processing that sharpens edges and enhances contrast
140.	____	Ommatidia	F.	Region where portions of each optic nerve cross over
141.	____	Lateral inhibition	G.	Region in the thalamus where optic nerve axons terminate
142.	____	Lateral geniculate nuclei	H.	Hollow sensory bristle that contains taste receptors in insects
143.	____	Sensilla	I.	Light-absorbing complex of retinal and protein
144.	____	Organ of Corti	J.	Contain sensory hair cells that detect vibrations transmitted to the inner ear

Labelling

Identify each numbered part of the following illustration.

145. ____________

146. ____________

147. ____________

148. ____________

149. ____________

150. ____________

151. ____________

152. ____________

153. ____________

154. ____________

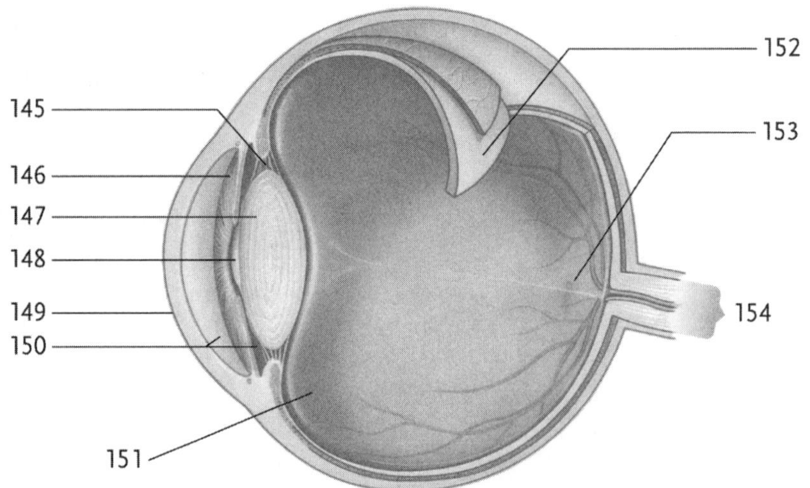

SELF-TEST

1. Sensory transduction involves ____. [p. 820]

   a. One sensory stimulus being converted to another
   b. An increase in the amplitude of an action potential
   c. Stimulus causing a change in membrane potential of a sensory cell
   d. A reduced response of a sensory cell in the face of constant intensity of stimulus

2. Which of the following are sensory structures that help provide information about the position/orientation of a body? [pp. 828–830]

   a. Statocysts
   b. Tympanum
   c. Vestibular apparatus
   d. Organ of Corti
   e. Neuromasts

3. Variations in the quality of sound are recognized by ____. [pp. 831–833]

   a. Number of hair cells stimulated
   b. Pattern of hair cells stimulated
   c. Amplitude of action potential
   d. Frequency of action potential

4. Movement of limbs is detected by ____. [p. 830]

   a. Muscle spindles
   b. Golgi tendon organ
   c. Joint receptors
   d. Barrow receptors
   e. Carotid bodies

5. The membrane in contact with the stapes that transmits sound waves to the inner ear is the _____. [pp. 831–833]

   a. Oval window
   b. Round window
   c. Tympanic membrane
   d. Basilar membrane

6. Visual images can be focused by ______. [pp. 833–840]

   a. Lateral inhibition
   b. Moving the lens back and forth relative to the retina
   c. Altering the number of ommatidia
   d. Changing the shape of the lens

7. _______ is(are) responsible for differences in absorption characteristics that underlie colour vision. [pp. 836–838]

   a. Rod cells
   b. Retinal
   c. Carotine
   d. Opsin
   e. Cone cells

8. _____ are photoreceptive cells that are specialized for detection of light of low intensity. [pp. 836–838]

   a. Rod cells
   b. Ganglion cells
   c. Horizontal cells
   d. Cone cells
   e. Bipolar cells

9. Which of the following chemicals act(s) as a "natural painkiller"? [p. 846]

   a. Capsaicin
   b. Substance P
   c. Insulin
   d. Endorphins

10. ______ is(are) used by animals for the location and capture of food. [pp. 831–833, 844, 847]

   a. Echolocation
   b. Electroreceptors
   c. Pit organs
   d. Cochlea
   e. Pacinian corpuscle

11. A sensory receptor [p. 820]

   a. Detects external and internal information
   b. Converts external and internal information into neural messages
   c. Transduces the messages to the central nervous system
   d. Is decoded and processed by the central nervous system

12. _____ are touch and pressure receptors in human skin. [p. 826]

   a. Stereocilia
   b. Pacinian corpuscles
   c. Ruffin endings
   d. Golgi tendon organs

13. Insects have "ears" on the head, thorax, and abdomen called ______. [p. 831]

   a. Cupula
   b. Pinna
   c. Organ of Corti
   d. Tympanum

14. A portion of each optic nerves crosses over to the opposite side, forming the [p. 839]

   a. Chiasmata
   b. Optic chiasm
   c. Lateral geniculate nuclei
   d. Lateral line

15. Taste receptors respond to which of the following basic tastes? [p. 842]

   a. Sweet
   b. Salty
   c. Umami
   d. Sour

16. Electroreceptors [p. 847]

a. Depolarize in an electric field
b. Depolarize in a magnetic field
c. Hyperpolarize in an electric field
d. Hyperpolarize in a magnetic field

17. Proprioceptors of the tendon [p. 830]

a. Are called Golgi tendon organs
b. Measure stretch and compression of tendons as muscles contract
c. Are called muscle spindles
d. Measure the speed of muscle contractions

18. Information about the intensity and extent of a stimulus can be indicated by [pp. 821–822]

a. The frequency of action potentials
b. The number of afferent neurons sending action potentials
c. The summation of the activities of different types of sensory receptors
d. The duration of the action potentials

19. The vestibular apparatus is responsible for [pp. 828–830]

a. Heat
b. Tasting
c. Equilibrium
d. Vision

20. Nociceptors are sensitive to [p. 847]

a. Heat
b. Pain
c. Touch
d. Vibrations

INTEGRATING AND APPLYING KEY CONCEPTS

1. Despite the independence of sensory quality and information at the receptor level, animals perceive a unified representation of their environment within which information from the entire complement of sensory channels is seamlessly integrated. Why is such integration important? How is the integration accomplished?

2. *Insights from the Molecular Revolution*: Explain the body's responses to eating spicy food in light of capsaicin. Discuss the adaptive significance of temperature-sensitive calcium channels.

OPTIONAL ACTIVITIES

1. In some individuals, a genetic mutation causes them to perceive colours when listening to music. This condition is termed synesthesia. In terms of the sensory systems, explain how this can occur.

2. A common response to being startled is your heart "skipping a beat." In fact, many people use the adage, "I was so scared that my heart stopped." Following such an event, a startled individual's heart begins to beat rapidly. Using your knowledge of the sensory systems, explain why this occurs.

35 The Endocrine System

CHAPTER HIGHLIGHTS

- The endocrine system produces and secretes chemical signals, or hormones, that work with the nervous system to regulate the various processes of animals, including growth, development, reproduction, and metabolism.
- Hormones operate at various levels, including through the blood and locally on neighbouring cells, or even on the cells that secreted them.
- Hormones and local regulators can be grouped into four classes based on their chemical structure: amine, peptide, steroid, and fatty acid derivative.
- The secretion of many hormones is regulated by feedback pathways: negative and positive feedback.
- Hormones may act on cell surface receptors, located on the surface of cell membranes, or in the cytoplasm or nucleus. Hydrophilic hormones bind to surface receptors, usually activating secondary messenger cascade that leads to cell responses. Hydrophobic hormones bind to receptors inside cells, activating or inhibiting genetic regulatory proteins.
- Target cells may respond to more than one hormone, and different target cells may respond differently to the same hormone.
- Many organs in vertebrates produce hormones and serve as endocrine glands, including the brain, pituitary, thyroid, pancreas, adrenal, gonads, kidney, liver, and gastrointestinal tract.

STUDY STRATEGIES AND LEARNING OUTCOMES

- This chapter is extremely long and has many terms. DO NOT try to go through the chapter all in one sitting. Take one section at a time and then work through the companion section(s) in the study guide.

- Through the study of the endocrine system, one quickly recognizes that the concepts can be broken down into a number of steps that can be discussed in a sequence. By understanding a basic diagram (see the topic map) outlining the endocrine system, you can then apply the same basic series of steps to any system and hormone to understand its basic functions.

- Draw diagrams of the various endocrine pathways (source organ, target organ, feedback loops), being sure to label the various parts, noting the chemical nature of the hormone and its actions on the target organ. Some of these pathways are complex and involve intermediate or multiple targets as well as multiple hormone players.

By the end of this chapter, you should be able to

- Describe the functions of the endocrine system, differentiating it from the nervous system
- Describe the four major types of cell signalling
- List and describe the groups of hormones based on their chemical structures and describe their mechanisms of action
- Explain the role of feedback mechanisms in the control of hormone secretion (provide examples)
- List the major human endocrine glands and hormones and provide the functions for each
- Describe the endocrine system in invertebrates

TOPIC MAP

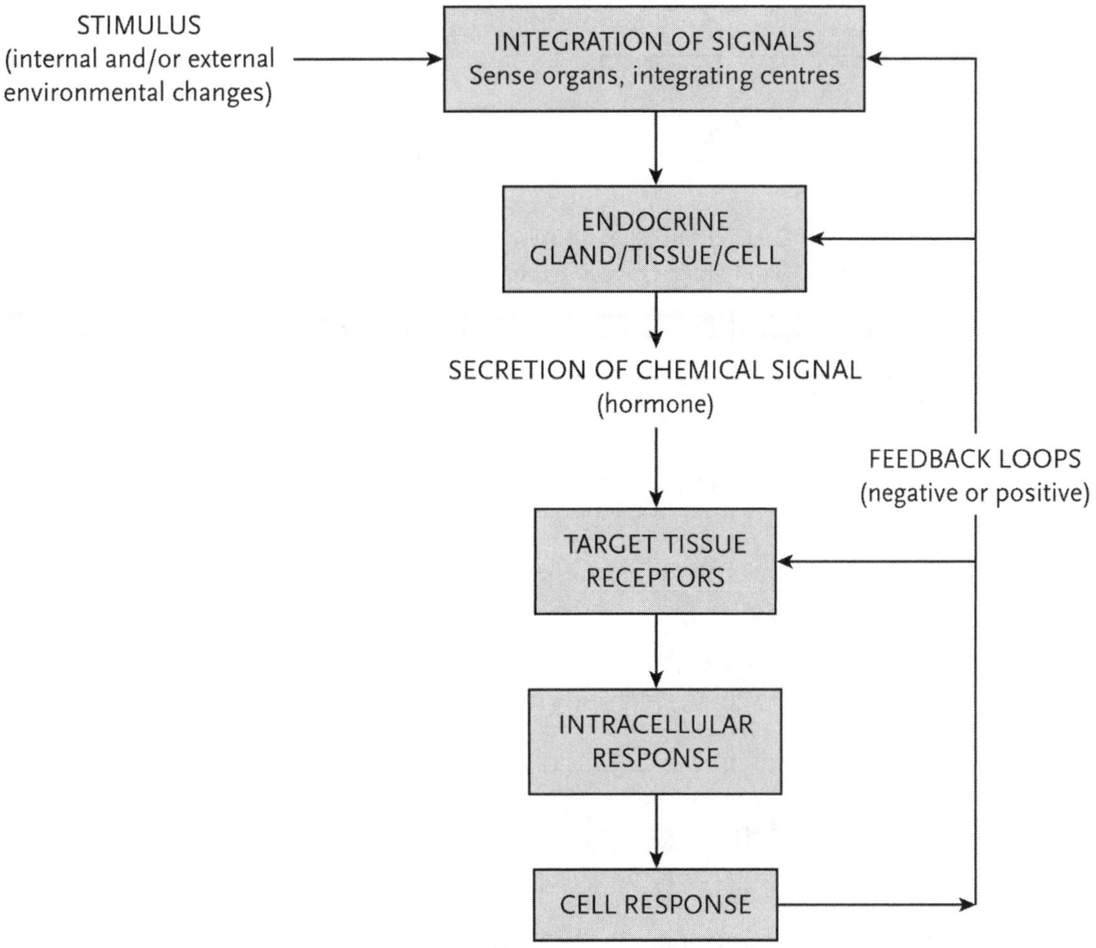
STIMULUS
(internal and/or external environmental changes)
INTEGRATION OF SIGNALS
Sense organs, integrating centres
ENDOCRINE GLAND/TISSUE/CELL
SECRETION OF CHEMICAL SIGNAL
(hormone)
FEEDBACK LOOPS
(negative or positive)
TARGET TISSUE RECEPTORS
INTRACELLULAR RESPONSE
CELL RESPONSE

An example of a topic map for the stress "fight-or-flight response":

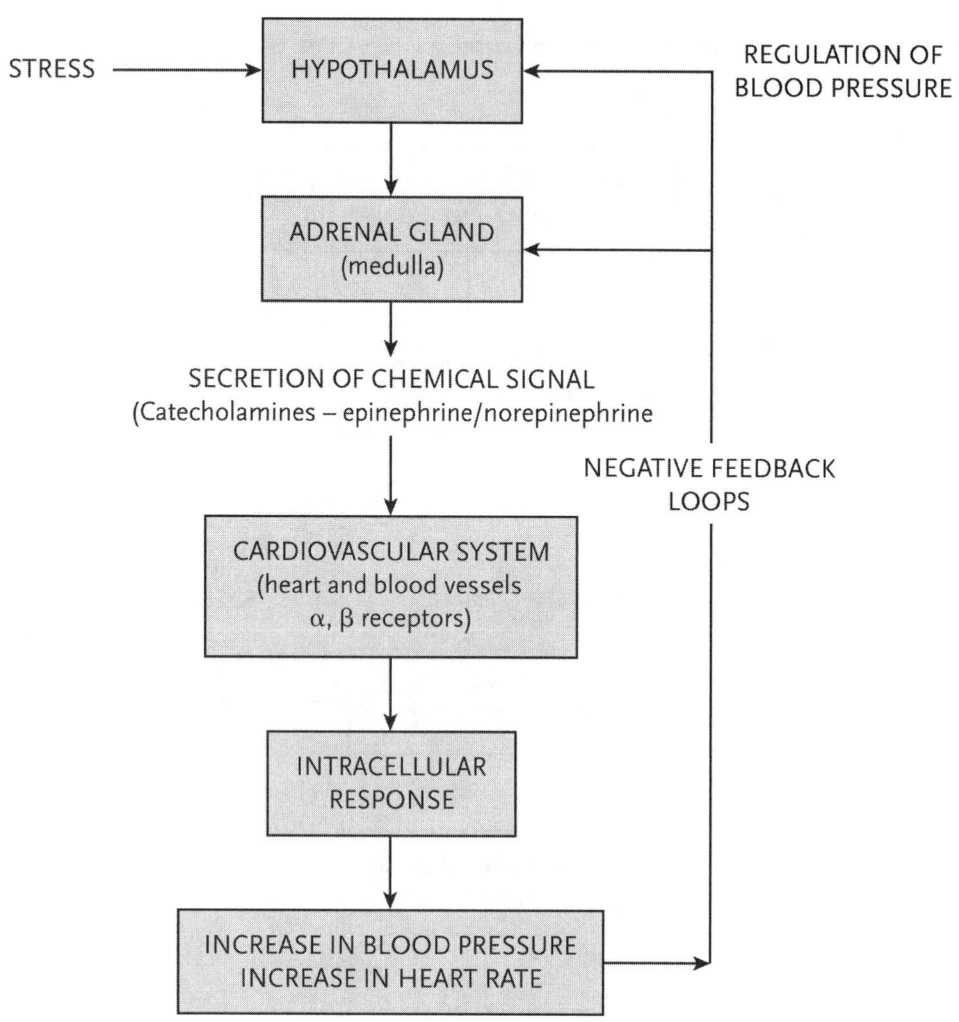

Create your own topic map for each of the endocrine systems listed in Chapter 35 of the textbook.

INTERACTIVE EXERCISES

Why It Matters [pp. 851–852]

Section Review (Fill-in-the-Blanks)

(1) ____ are chemicals released by one cell that affect the activity of another cell. The network of cells that produces hormones is the (2) ____.

35.1 Hormones and Their Secretion [pp. 852–854]

Building Vocabulary

Using the definitions of the prefixes below, assign the appropriate prefix to the appropriate suffix for the following:

Prefix	Meaning
neuro-	of or relating to a nerve
endo-	within, to the inside
hyper-	over
hypo-	under

Suffix	Meaning
-crine	secretion

Prefix	Suffix	Definition
_ ___	-hormone	3. A hormone secreted by a neurosecretory cell
_ ___	-secretion	4. A diminished or undersecretion
____	-secretion	5. An excessive or oversecretion
_ ___	-glycemia	6. An abnormally high level of glucose in the blood
____	-crine	7. A mode of secretion from cells in a ductless gland

Short Answer

8. Outline the differences between the endocrine system and the nervous system.

____

____

9. Define the term “hormone.”

____

____

10. In your own words, what is the role of the endocrine system?

____

____

11. Explain how the secretion of hormones is regulated by feedback pathways. Provide an example of positive and negative feedback pathways. What would be the benefits of such regulatory pathways?

____

____

12. Explain what is meant by “the body processes are regulated by coordinated hormone secretion.”

____

____

Fill-in-the-Blanks

(13) ____ are ductless glands that secrete hormones. (14) ____ are specialized neurons that also secrete hormones. Hormones can be classified into four chemical types: (15) ____, (16) ____, (17) ____, and (18) ____. (19) ____ are examples of peptide hormones, and (20) ____ are examples of a fatty acid derivative. Only (21)____ of a hormone, those with a (22) ____ recognizing and binding that hormone, respond to it.

Complete the Tables

Mode of secretion	*Description*
23.	The release of a hormone from a neuron
Endocrine	24.
Paracrine	25.
26.	The release of a chemical into the extracellular fluid that regulates the activity of the cell that secreted it

Chemical type	*Hormonal example*
Amine	27.
Peptide	28.
Steroid	29.
30.	Prostaglandins

Matching

Match each of the following structures with its correct definition.

31.	____	Growth factor	A.	A chemical signal released from one cell that affects another cell, usually at a distance from its site of secretion
32.	____	Neurosecretory neurons	B.	Network of cells/glands that secrete hormones
33.	____	Endocrine system	C.	A group of peptide hormones that regulate growth and differentiation
34.	____	Neurosecretory neuron	D.	A neuron specialized to release hormones

35.2 Mechanisms of Hormone Action [pp. 854–859]

Short Answer

35. Hormones are not always secreted in an active form. Discuss how hormones may become active. Draw a diagram of hormone activation.

____

____

36. Compare the mechanisms of hormone actions for hydrophilic and hydrophobic hormones. For each, provide a concrete example by drawing a diagram of the mechanisms of hormone action. In each, outline the four features of a hormone mechanism.

____

____

37. Explain why cells may respond to more than one hormone, and different target cells may respond differently to the same hormones, while other cells may not respond to these hormones.

____

____

Predict the Mechanisms of Hormone Action

Determine the mechanism of hormone action for the following chemical signal. Hint: Are these chemical signals hydrophobic or hydrophilic? Would these chemicals bind to surface receptors or intracellular receptors. Justify your answers.

Epinephrine: (38)

____

____

Insulin: (39)

____

____

Testosterone: (40)

____

____

Oxytocin: (41)

____

____

Estrogen: (42)

____

____

Aldosterone: (43)

____

____

Thyroid hormone: (44)

____

____

Prostaglandin: (45)

____

____

35.3 The Hypothalamus and Pituitary [pp. 859–862]

Section Review (Fill-in-the-Blanks)

The (46) ____ consists of two lobes: the (47) ____, which contains axons and nerve endings that originate in the hypothalamus, and the (48) ____. Some of the axons that terminate in the posterior pituitary release (49) ____ that travel through the portal veins and affect the secretions of the anterior pituitary, while others release nontropic hormones that enter the general blood circulation. The two types of tropic hormones are (50) ____ and (51) ____. The two nontropic hormones that enter the general circulation are (52) ___, which helps regulate water balance, and (53) ____, which stimulates milk ejection from the mammary glands of mammals. The anterior pituitary produces and secretes (54) ____, (55) ____, (56) ____, (57) ____, (58) ____, and (59) ____. FSH and LH are referred to as (60) ____ because of their action on gonads. In some species, the anterior pituitary has a distinct intermediate lobe that produces and releases (61) ____ and (62) ____; in those species without an intermediate lobe, these hormones are produced by cells dispersed in the other regions of the anterior pituitary.

Distinguish between Members of the Following Pairs of Terms

63. Anterior pituitary and posterior pituitary

____

____

Complete the Table

Hormone	*Chemical type*	*Site of secretion*	*Function*
Growth hormone	Peptide	64.	65.
66.	67.	68.	Influence reproductive activity, stimulate milk synthesis
69.	70.	Anterior pit	Stimulates thyroid gland to produce T_4
ACTH	Peptide	Anterior pit	71.
FSH	72.	73.	74.
LH	75.	76.	77.
78.	79.	80.	Controls pigmentation/coloration
Endorphin	81.	82.	83.
ADH	84.	85.	86.
87.	88.	89.	Stimulates smooth muscle contraction (includes milk "letdown" or secretion)
Prolactin	Peptide	90.	91.

Labelling

Identify each numbered part of the following illustration.

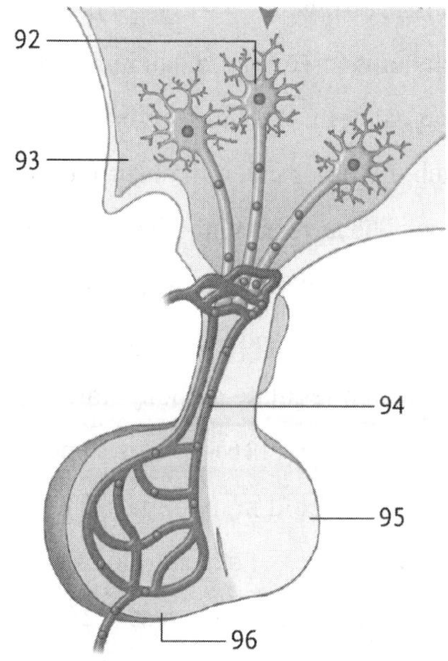

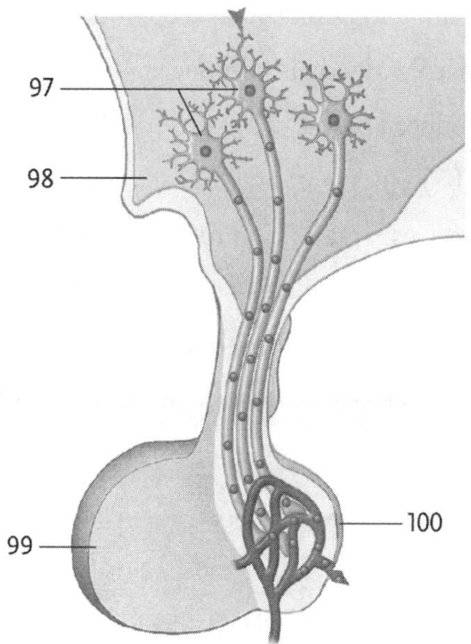

92. ______________

93. ______________

94. ______________

95. ______________

96. ______________

97. ______________

98. ______________

99. ______________

100.______________

35.4 Other Major Endocrine Glands of Vertebrates [pp. 862–867]

Section Review (Fill-in-the-Blanks)

The (101) ____ produces (102) ____, which is converted to its biologically active form, (103) ____, prior to binding to intracellular receptors. In amphibians, T_4 triggers (104) ____. In mammals, the thyroid gland also produces (105) ____, which helps lower blood calcium levels. (106) ____ is produced by the (107) ____ and increases blood calcium. PTH also promotes the conversion of (108) ____ to its active form, the latter of which increases calcium absorption by the gut and synergizes with PTH to release calcium from bone. The adrenal gland has two parts: a centrally located (109) ____ and a peripherally located (110) ____. The medulla releases two (111) ____, hormones derived from the amino acid tyrosine: (112) ____ and (113) ____. The cortex produces two classes of steroid hormones: (114) ____, which regulate carbohydrate metabolism, and (115) ____, which help regulate ion and water balance. (116) ____ is a specific glucocorticoid in mammals and (117) ____ is a specific mineralocorticoid in mammals. The gonads, (118) ____ and (119) ____, are a primary source of three classes of sex steroids: (120) ____, (121) ____, and (122) ____. (123) _____ is a specific androgen, whereas (124) ____ is a specific estrogen, and (125) ____ is a specific progestin. The (126) ____ of the (127) ____ produces several hormones that regulate metabolism. (128) ____ is generally anabolic and stimulates the uptake of nutrients into cells (e.g., glucose) as well as the synthesis of large macromolecules (e.g., glycogen). (129) ____ is generally catabolic and stimulates the breakdown of large macromolecules in tissues and the release of monomers into the blood. (130) ____ can result from several defects, including the inability to produce insulin or the failure of insulin to function properly in targets cells; conditions that generally all lead to abnormally high glucose levels in the blood. The (131) ____ produces (132) ______, which helps maintain daily biorhythms.

Matching

Match each of the following types of junction with its correct definition.

133.	____	Metamorphosis	A.	Change in body form (e.g., tadpole to adult)
134.	____	Catecholamines	B.	Steroidlike molecule that when activated stimulates Ca^{2+} absorption
135.	____	Pancreas	C.	Class of amines derived from tyrosine
136.	____	Gonad	D.	Steroid hormones that stimulate muscle development
137.	____	Estrogen	E.	Typically has both exocrine and endocrine (islets of Langerhans) components
138.	____	Progestin	F.	A disease that typically displays abnormally high glucose levels in the blood
139.	____	Diabetes mellitus	G.	Organ that produces gametes and sex steroids
140.	____	Androgen	H.	Class of sex steroid primarily produced in males
141.	____	Vitamin D	I.	An example is estradiol
142.	____	Anabolic steroid	J.	An example is progesterone

Complete the Table

Hormone	*Chemical type*	*Site of secretion*	*Function*
Thyroxine	143.	Thyroid gland	144.
145.	Peptide	146.	Lowers blood calcium
Parathyroid hormone	Peptide	Parathyroid gland	147.
148.	Amine	149.	Increase blood flow to muscles; stimulates breakdown of macromolecules
150.	151.	Adrenal cortex	Promotes fat and protein breakdown; stimulates gluconeogenesis
Aldosterone	152.	153.	154.
155.	Steroid	Testes	Stimulates sperm maturation/maintenance of secondary sexual characteristics
156.	157.	Ovary	158.
Progesterone	Steroid	Ovary	159.
160.	Peptide	161.	Stimulates secretion of FSH and LH
Insulin	162.	163.	164.
165.	166.	167.	Stimulates glycogen breakdown
168.	169.	Pineal gland	170.

35.5 Endocrine Systems in Invertebrates [pp. 867–869]

Section Review (Fill-in-the-Blanks)

The three hormones that regulate moulting and metamorphosis in insects are (171) ___, (172) ___, and (173) ___. In crustaceans, (174)____ helps regulate moulting by inhibiting ecdysone secretion.

Matching

Match each of the following types of junction with its correct definition.

175.	____	Brain hormone	A.	Peptide hormone that stimulates prothoracic glands
176.	____	Moult-inhibiting hormone	B.	Steroid hormone secreted by prothoracic glands
177.	____	Ecdysone	C.	Peptide hormone secreted by corpora allata
178.	____	Juvenile hormone	D.	Peptide hormone secreted by gland in eye stalk

SELF-TEST

1. ___________ is a chemical released from an epithelial cell in a gland that enters the blood to affect the activity of another cell some distance away. [p. 852]

   a. Hormone
   b. Neurohormone
   c. Pheromone
   d. Bile

2. Which of the following best describes autocrine regulation? [p. 853]

   a. Local regulator acts on the same cell that released it
   b. Local regulator is secreted from a neuron and acts at a distant target organ
   c. Local regulator acts on cells neighbouring the cell that released it
   d. Local regulator acts on a distant organ

3. Epinephrine is [pp. 852–853]

   a. A fatty acid–derived hormone
   b. A steroid hormone
   c. An amine hormone
   d. A peptide hormone

4. Control of hormone secretion by a negative feedback mechanism generally includes [p. 854]

   a. No change in hormone secretion
   b. Decrease in hormone secretion
   c. Change that maintains homeostasis
   d. Increase in hormone secretion

5. Peptide hormones are ___________ and bind to _____________ receptors. [pp. 855–856]

   a. Hydrophilic; intracellular
   b. Hydrophobic; cell surface
   c. Hydrophilic; cell surface
   d. Hydrophobic; intracellular

6. Which of the following chemical signals are small hydrophobic hormones that form hormone-receptor complexes in the cytoplasm of target cells? [pp. 855–856]

   a. Steroids
   b. cAMP
   c. Thyroid hormone
   d. Prolactin
   e. Catecholamines

7. Pituitary dwarfism is caused by [p. 861]

   a. Hypersecretion of cortisol
   b. Hyposecretion of growth hormone
   c. Hyposecretion of thyroid hormone
   d. Hyposecretion of insulinlike growth factor

8. Hormones secreted from the posterior pituitary are [pp. 861–862]

   a. Antidiuretic hormone (ADH) and ACTH
   b. Antidiuretic hormone (ADH) and CRH
   c. Oxytocin and ACTH
   d. Oxytocin and antidiuretic hormone

9. During an acute stress response, catecholamines [p. 864]

   a. Cause an increase in heart rate
   b. Cause an increase in blood pressure
   c. Cause the breakdown in glycogen and fatty acids
   d. Cause dilation of airways

10 ______ are the hormone and target involved in elevating blood glucose by glycogenolysis and gluconeogenesis. [p. 864]

a. Insulin and pancreas
b. Glucagon and liver
c. TSH and thyroid
d. ACTH and adrenal cortex
e. T_4 and skeletal muscle

11. _____ cause(s) the initiation of metamorphosis in lamprey. [p. 862]

a. Low metabolic rate
b. Decreasing T_4 levels
c. Increasing levels of T_4
d. Aldosterone
e. Prolactin

12. Increased skeletal growth results directly and/or indirectly from the activity of [pp. 859–860]

a. Aldosterone
b. Growth hormone
c. Hormones produced in the hypothalamus
d. Progestins
e. Epinephrine

13. The activity of the anterior pituitary is controlled by [p. 859]

a. The hypothalamus
b. ADH
c. Epinephrine
d. Releasing hormones
e. Inhibiting hormones

14. The hormones LH and FSH [p. 865]

a. Are controlled by GnRH
b. Are involved in the control of gonadal function
c. Are involved in the control of adrenal gland function
d. Are involved in osmoregulation
e. Are involved in the control of plasma calcium concentration

15. Prolactin [p. 859]

a. Is a nontropic hormone
b. Stimulates the development of the mammary glands
c. Is a tropic hormone
d. Stimulates mild synthesis
e. Regulates pancreatic secretions

16. The pineal gland [p. 867]

a. Regulates blood glucose levels
b. Regulates biological rhythms
c. Produces the hormone melatonin
d. Synchronizes the biological clock
e. May be considered a third eye

17. Diabetes mellitus is a condition that can be described by which of the following? [p. 866]

a. Lack of insulin
b. Lack of glucagon
c. Lack of antidiuretic hormone (ADH)
d. Lack of cortisol
e. None of the above

18. β-Adrenergic receptors bind which of the following hormones? [p. 856]

a. Thyroxine
b. Glucagon
c. Epinephrine
d. Antidiuretic hormone
e. Insulin

19. Aldosterone [p. 864]

a. Is a mineralocorticoid
b. Increases the amount of Na^+ in the blood
c. Increases the rate of K^+ excretion
d. Is synthesized in the adrenal cortex
e. Is a stress hormone

20. Removing the eye stalk of a lobster would ____. [pp. 868–869]

a. Increase secretion of ecdysone
b. Decrease secretion of ecdysone
c. Accelerate the moult cycle
d. Decelerate the moult cycle
e. Have no effect on ecdysone or the moult cycle

INTEGRATING AND APPLYING KEY CONCEPTS

1. Discuss the validity of the following statement: *The endocrine system and the nervous system are separate and distinct systems that serve to coordinate the function of animals.*

2. The fight-or-flight response leads to the secretion of the stress hormones epinephrine, norepinephrine, and cortisol from the adrenal gland. Describe the regulation of the hypothalamic–pituitary adrenal gland axis (cortisol secretion) and the hypothalamic–adrenal medulla axis (catecholamine secretion).

3. The control of hormone secretion is central for the proper functioning of physiological systems. Explain what effects, if any, disruptions to feedback loops would have to endocrine axes.

OPTIONAL ACTIVITIES

1. In the past number of years, an increasing number of sports figures have been charged and even been found guilty of using performance-enhancing drugs such as steroids, growth hormone, insulinlike growth factors, erythropoietin, and other drugs to bolster their performance. Investigate the benefits of each of these hormones (and others) for performance enhancement. Moreover, what potential side effects would such treatments have on the athlete? Write an essay on the subject matter.

2. Different types of treatments are prescribed for people with high blood pressure. For example, drugs designed to reduce blood pressure or increase urination are strategies for treating hypertension. Investigate the different types of treatments that may be available to treat hypertension. In each case, explain the mechanisms of action of these treatments.

3. What important considerations should be taken into account when one wishes to design drugs for the treatment of different physiological diseases?

36 Muscles, Skeletons, and Body Movements

CHAPTER HIGHLIGHTS

- Skeletal muscles move the joints of the body.
- Skeletal muscle is organized in a hierarchical manner; the functional unit is the sarcomere, which shortens as a result of the relative movement of myosin and actin.
- Muscle contraction is initiated as a result of an action potential. When an action potential arrives at a neuromuscular junction, the axon terminal releases acetylcholine, which triggers an action potential in the muscle fibre. In the muscle, the action potential causes the release of calcium, which initiates the crossbridge cycle that results in the shortening of the sarcomere.
- The response of a muscle fibre to action potentials ranges from twitches to tetanus.
- Muscle fibres differ in their rate of contraction and susceptibility to fatigue.
- Skeletal muscle control is divided among motor units.
- There are three kinds of skeletons in animals: a hydrostatic skeleton, an exoskeleton, and an endoskeleton.
- The bones of an endoskeleton are connected by three types of joints: a synovial joint, a cartilaginous joint, and a fibrous joint.
- Most bones are moved by antagonistic pairs of muscles.

STUDY STRATEGIES AND LEARNING OUTCOMES

- This chapter makes use of the concepts of cell structure and nerve function developed earlier. You should briefly review these concepts to make sure that you understand them.
- While many of the terms in this chapters may be familiar, others probably will not be, so do not be hasty in your study.

- This chapter is divided into sections. DO NOT try to go through them all in one sitting. Take one section at a time and then work through the companion section(s) in the study guide.
- Draw pictures of the functional unit of muscle (sarcomere), being sure to label the various parts and note their role in contraction.

By the end of this chapter, you should be able to

- Describe the internal organization of skeletal muscles
- Explain the sliding filament mechanism of muscle contraction
- List the molecular factors involved in muscle contraction
- Explain the role of calcium and ATP during muscle contraction of relaxation
- List the events of the crossbridge cycle during muscle contraction
- Explain the neural control of skeletal muscle contraction
- Describe the role of skeletal systems and list the different types of skeletal systems
- Contrast between hydrostatic skeleton, exoskeleton, and endoskeleton
- Explain the interactions between muscles and bones

TOPIC MAP

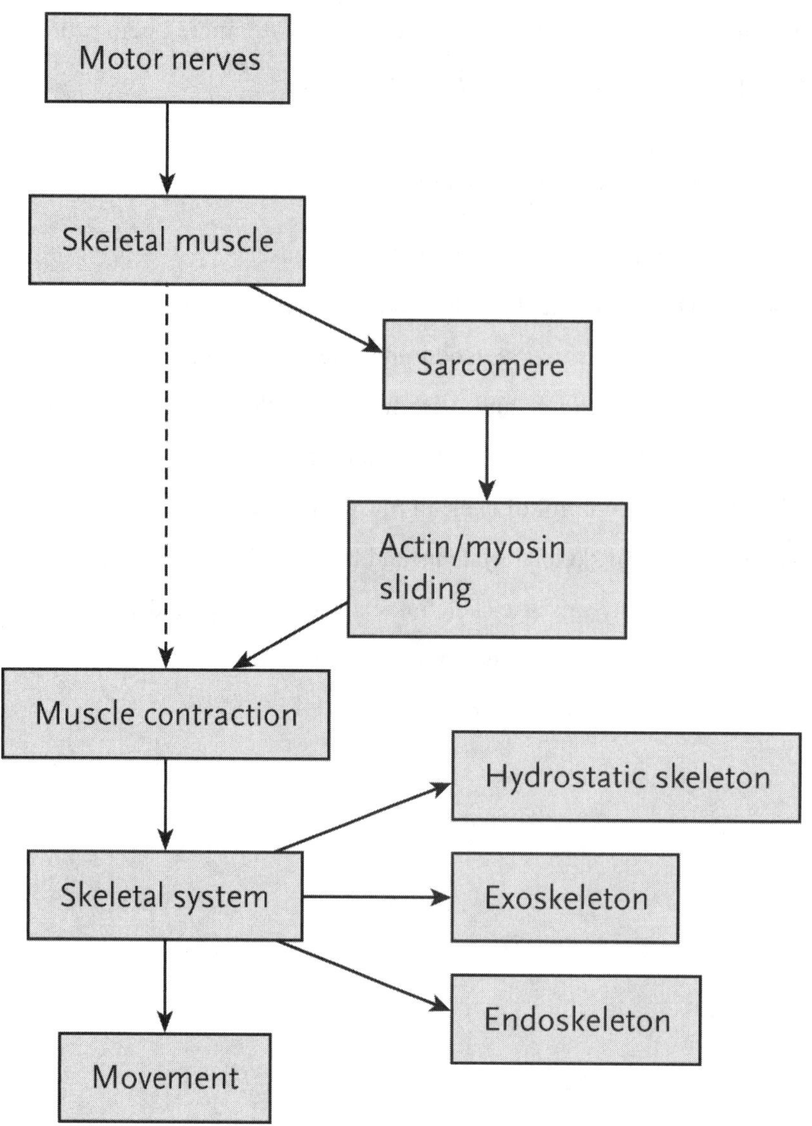

Motor nerves
Skeletal muscle
Sarcomere
Actin/myosin sliding
Muscle contraction
Hydrostatic skeleton
Skeletal system
Exoskeleton
Endoskeleton
Movement

INTERACTIVE EXERCISES

Why It Matters [pp. 873–874]

Section Review (Fill-in-the-Blanks)

There are three types of muscle tissue: (1) ____, (2) ____, and (3) ____. Most (4) ____ are attached by tendons to the skeleton of vertebrates.

36.1 Vertebrate Skeletal Muscle: Structure and Function [pp. 874–880]

Section Review (Fill-in-the-Blanks)

Skeletal muscle consists of elongated, cylindrical cells called (5) ____. Inside the cell, there are contractile elements or (6) ____ that consist of regular arrangements of (7) ______, containing the protein myosin, and (8) ____, containing the protein actin. The functional unit of the myofibril is the (9) ____, which extends Z line to Z line. The plasma membrane of the muscle cell folds into the muscle fibre to form (10) _____, which are in close association with the cell's complex network of endoplasmic reticulum called (11) ____. A neuron comes in contact with a muscle cell to form a(n) (12) ____. The release of (13) ____ from the neuron causes the muscle cell to depolarize and release calcium from the lumen of the sarcoplasmic reticulum and raise cytosolic calcium concentration. Calcium enables the interaction between myosin and actin so that the filaments move relative to one another in a process called the (14) ____; this movement shortens the sarcomere and results in muscle contraction. A single action potential arriving at the neuromuscular junction usually causes a single weak contraction of the muscle cell called a(n) (15) ____. (16) ____ results when muscle fibres cannot relax between rapidly arriving stimuli. (17) ____ contract relatively slowly and the intensity of the contraction is low, whereas (18) ____ contract relatively quickly and powerfully. Controlled contraction of the overall muscle results from organized activation of (19) ____.

Building Vocabulary

Using the definitions of the prefixes below, assign the appropriate prefix to the appropriate suffix for the following:

Prefix	Meaning
sarco-	of or relating to flesh/muscle
myo-	of or relating to muscle

Suffix	Meaning
-mere	unit, part

Prefix	Suffix	Definition
____	____	20. The functional unit of a contractile element in a muscle cell
____	-fibril	21. The contractile element inside of a muscle cell
____	-globin	22. Oxygen-storing molecule in some muscle cells

Choice

For each of the following, choose the most appropriate type of filament from the list below.

a. Thick filament b. Thin filament

23. ____ Parallel bundles of myosin molecules

24. ____ Mostly composed of two linear chains of actin arranged in a double helix

Matching

Match each of the following structures with its correct definition.

25.	____	Skeletal muscle	A.	Connect to bones of the skeleton by tendons
26.	____	Myoglobin	B.	Muscle cell
27.	____	Neuromuscular junction	C.	Cylindrical contractile elements in a muscle cell
28.	____	Myofibrils	D.	The weak contraction of a muscle cell in response to a single action potential
29.	____	Muscle fibres	E.	Process in which there is relative movement of actin and myosin
30.	____	Muscle twitch	F.	Collection of muscle fibres controlled by branch of same neuron
31.	____	Tetanus	G.	Specialized junction between a neuron and a muscle cell
32.	____	Motor units	H.	Oxygen-storing protein in some muscle
33.	____	Sliding filament mechanism	I.	Continuous contraction of a muscle fibre

Labelling

Identify each numbered part of the following illustration.

34. ____________

35. ____________

36. ____________

37. ____________

38. ____________

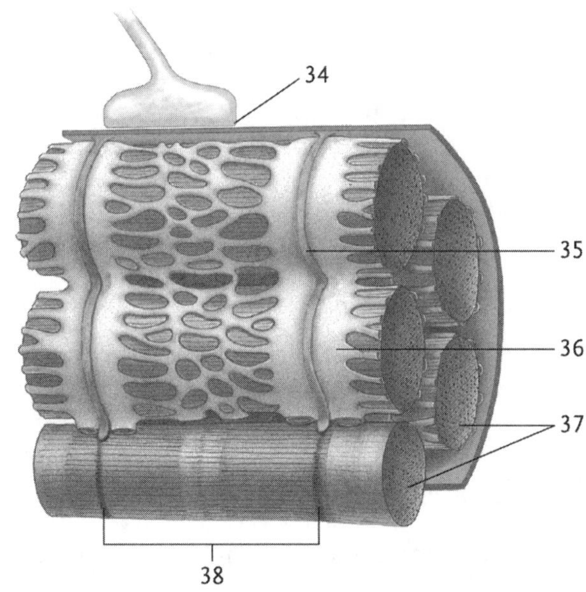

Short Answer

39. Describe how an action potential is propagated along a muscle fibre and initiates a muscle contraction.

____

____

40. Describe the role of calcium and ATP in muscle contraction and relaxation.

____

____

41. Summarize the events of the crossbridge cycle.

____

____

36.2 Skeletal Systems [pp. 881–883]

Section Review (Fill-in-the-Blanks)

A(n) (42) ____ consists of muscles and fluid, with the fluid either within the muscle or in compartments. A(n) (43) ____ is a rigid external covering that provides support, whereas (44) ____ is made up of internal body structures that provide support such as bone. The skeleton of vertebrates is organized into the (45) ____ and the (46) ____.

Short Answer

47. Describe the role of the skeletal system.

____

____

Choice

For each of the following descriptions, choose the most appropriate type of skeleton from the list below.

a. Axial skeleton　　　　b. Appendicular skeleton

48. ____ Contains skull and vertebral column

49. ____ Contains shoulders, hips, and limbs

36.3 Vertebrate Movement: The Interactions between Muscles and Bones [pp. 883–887]

Section Review (Fill-in-the-Blanks)

The bones of vertebrates are connected by three types of joints: (50) ____, (51) ____, and (52) ____. A muscle that causes movement at a joint is a(n) (53) ____. Most bones are moved by muscles in (54) ____: (55) ____, which extend the joint, and (56) ____, which retract the joint.

Choice

For each of the following descriptions, choose the most appropriate type of muscle from the list below.

a. Extensor muscle

b. Flexor muscle

57. ____ Retracts a joint

58. ____ Extends a joint

Matching

Match each of the following types of junction with its correct definition.

59.	____	Agonist	A.	A muscle that causes movement in a joint when it contracts
60.	____	Cartilaginous joint	B.	Arrangement of muscle groups in which one muscle has the opposite effect of the other
61.	____	Fibrous joint	C.	Moveable joint enclosed by a fluid-filled capsule
62.	____	Ligaments	D.	Moveable joint without a fluid-filled capsule
63.	____	Antagonistic pair	E.	Joint connected by stiff connective tissue
64.	____	Synovial joint	F.	Connective tissue that joins bones on either side of a joint

SELF-TEST

1. The sliding filament mechanism states that ____. [p. 875]

   a. Nebulin and titin filaments move relative to one another
   b. Actin and myosin filaments become arranged perpendicular to one another
   c. Actin and myosin filaments move relative to one another
   d. The two actin chains unwind and dissociate into G-actin

2. Myofibrils are composed of ____. [p. 874]

   a. Fibres
   b. Myofibrils
   c. Actin
   d. Myosin
   e. Sarcoplasmic reticulum

3. The human skull is part of the ____. [p.883]

   a. Girdle
   b. Axial skeleton
   c. Atlas
   d. Appendicular skeleton
   e. Hydrostatic skeleton

4. The human axial skeleton includes the [p. 883]

   a. Ulna
   b. Shoulder blades
   c. Centrum
   d. Femur
   e. Sternum

5. Vertebrate appendages are connected to the ____. [p. 883]

   a. Cervical complex
   b. Atlas
   c. Appendicular skeleton
   d. Axial skeleton
   e. Girdles

6. In which of the following animal groups is the primary means of support a hydrostatic skeleton? [p. 881]

   a. Annelids
   b. Cnidarians
   c. Flatworms
   d. Echinoderms
   e. Lobsters and cratfish

7. Internal skeletons are found in ____. [p. 883]

   a. Annelids
   b. Cnidarians
   c. Echinoderms
   d. Lobsters and crayfish
   e. Vertebrates

8. _____ is the primary neurotransmitter released at neuromuscular junctions in vertebrates. [p. 875]

   a. Substance P
   b. Inositol triphosphate
   c. Epinephrine
   d. Acetylcholine
   e. Endorphin

9. _____ facilitates the diffusion of oxygen into tissues from blood and stores oxygen in tissues. [p. 879]

   a. Myoglobin
   b. Hemoglobin
   c. Opsin
   d. Bilirubin
   e. Dystrophin

10. The _____ is(are) connective tissue that joins bones together on either side of a joint. [pp. 883–884]

    a. Periosteum
    b. Stratum corneum
    c. Tendons
    d. Ligaments
    e. Meninges

11. Skeletal muscle contraction is controlled by the [p. 874]

    a. Somatic system
    b. Sensory system
    c. Endocrine system
    d. Skeletal system

12. Skeletal muscles are attached to bones through [p. 874]

a. Ligaments
b. Tendons
c. Z-discs
d. Troponin

13. A sarcomere is delimited by [p. 874]

a. A bands
b. I bands
c. Two M lines
d. Two Z lines

14. During the crossbridge cycle, ATP [pp. 876–877]

a. Causes the detachment of myosin from actin
b. Causes the bending of myosin crossbridges
c. Causes rigor mortis
d. Binds to troponin

15. Ca^2+ that is directly involved in muscle contraction is stored in the [pp. 876–877]

a. Tubules
b. Cytosol
c. Sarcoplasmic reticulum
d. Extracellular fluid

16. Fast anaerobic muscle fibres [p. 879]

a. Contain a high concentration of glycogen
b. Have few mitochondria
c. Have extensive vascularization
d. Are red in colour

17. In skeletal muscles, the force of contraction can be adjusted by [p. 879]

a. Controlling the recruitment of motor units
b. Controlling the speed of contraction
c. Contracting antagonist muscle pairs
d. Tetanic contractions

18. Which of the following would have many motor units in a small area with only a few myofibres in each unit? [p. 879]

a. Eyes
b. Biceps
c. Fingers
d. Back muscles

19. The sliding filament mechanism [p. 874]

a. Causes thick and thin filaments to slide toward the centre of the H band, bringing the Z lines closer together
b. Causes thick and thin filaments to slide toward the centre of the I band, bringing the M lines closer together
c. Causes thick and thin filaments to slide toward the centre of the M line, bringing the Z lines closer together
d. Causes thick and thin filaments to slide toward the centre of the A band, bringing the Z lines closer together

20. Chitin is a compound found in [p. 882]

a. Exoskeleton of vertebrates
b. Exoskeleton of insects and lobsters
c. Ligaments and tendons
d. Hydrostatic skeletons

INTEGRATING AND APPLYING KEY CONCEPTS

1. *Insights from the Molecular Revolution*: Describe the factors that regulate gene expression and identify the possible targets of action for drugs aimed at increasing utropin content in the muscle cells of DMD patients.

2. Coordinated muscle contraction requires neuronal stimulation of motor units. Discuss the roles of calcium in neuromuscular function and the potential consequences of calcium deficiency. (Be sure to consider both the muscle cell and the nerve cell.)

OPTIONAL ACTIVITIES

1. A number of chemicals can interfere with the neuromuscular junction, causing prolonged muscle contraction or preventing muscle contraction. Through literature research, list chemicals that have an ability to interfere with muscle contractions and explain their mechanism of action.

2. Based on your knowledge of muscle contraction, explain what causes rigor mortis (stiffening of the body following death).

37 The Circulatory System

CHAPTER HIGHLIGHTS

- Animal circulatory systems share basic elements to distribute nutrients and gases to tissues and to collect waste products. These elements include blood, vessels, and a heart.
- Some circulatory systems are open (most invertebrates), in which the blood bathes tissues directly. Other systems are closed, in which blood is distributed through a system of vessels and is kept separate from interstitial fluid (vertebrates).
- Vertebrate circulatory systems evolved from a single-circuit system, such as found in sharks and bony fish, in which the gills are in the same circuit as the rest of the blood vessels, to a double-circuit system, in which circulation to the lungs parallels the circulation to the rest of the body.
- The blood of vertebrates is a connective tissue that consists of numerous cell types (e.g., red blood cells, white blood cells, and platelets) and a fluid matrix (plasma). Blood has a number of basic functions: solvent, clotting, nutrient transport, immune responses, gas transport, and waste transport.
- The heart serves as a pump to distribute fluid through the circulatory system. The heartbeat is produced by a cycle of contraction and relaxation of the atria and ventricles and is initiated by action potentials that spread across the muscle cell membranes. Blood pressure is exerted on the walls of the arteries as the heart pumps blood.
- The circulatory system consists of different blood vessel types that begin and end at the heart. Arteries transport blood to tissues, capillaries are the site of exchange between blood and interstitial fluids, and the venules and veins serve as blood reservoirs and conduits to the heart.
- Arterial blood pressure is the principal force moving blood to the tissues. Blood pressure must be regulated to ensure blood flow to the organs. The control of heart rate and strength of contractions, hormones, and the control of arteriole diameter are factors that determine arteriole blood pressure.
- The lymphatic system collects excess interstitial fluid, returns it to the venous blood, and participates in the immune response of vertebrates (covered in more detail in the next chapter).

STUDY STRATEGIES AND LEARNING OUTCOMES

- This chapter is divided into sections. DO NOT try to go through them all in one sitting. Take one section at a time and then work through the companion section(s) in the study guide.
- Draw pictures of the various circulatory systems, being careful to detail the flow of blood, especially the path through the heart.
- Develop a basic understanding of the components and functions of each part of the circulatory system.

By the end of this chapter, you should be able to

- List the basic elements that are shared between circulatory systems of different organisms
- List the functions of the components of the circulatory system
- Explain the difference between an open and a closed circulatory system
- List and explain the functions of the components of blood
- List the components of the heart and describe the flow of blood through the heart of different organisms
- Discuss the heart cycle and how it is controlled
- Explain the changes in blood pressure during the heart cycle and how blood pressure is regulated
- Explain the roles of the lymphatic system

TOPIC MAP

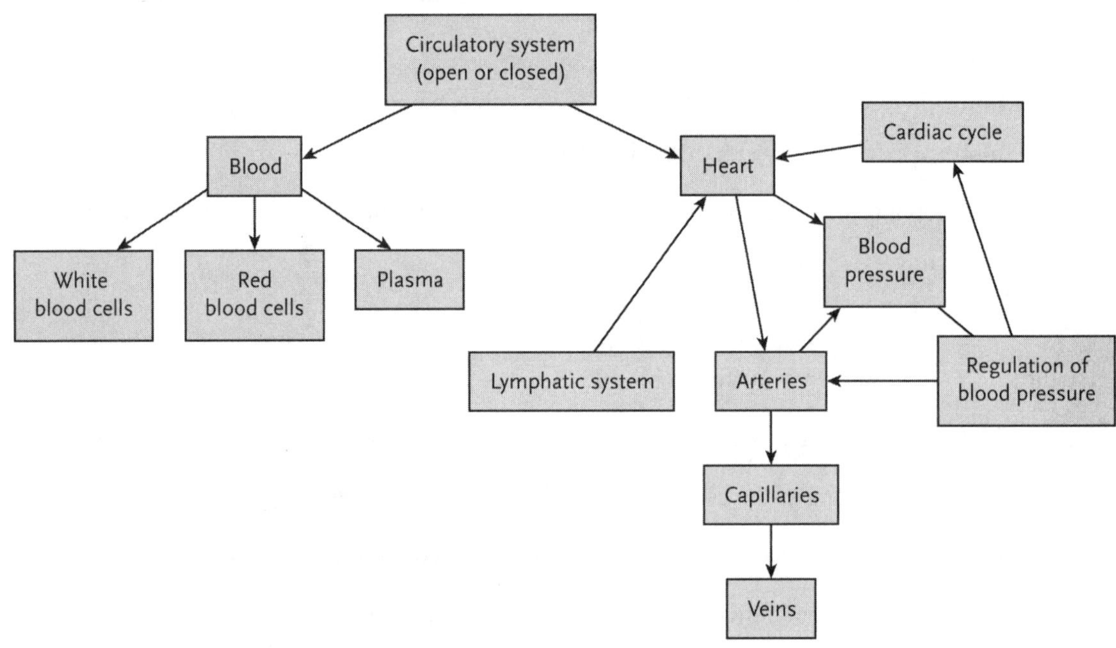

INTERACTIVE EXERCISES

Why It Matters [pp. 891–892]

Section Review (Fill-in-the-Blanks)

A(n) (1) ____ consists of fluid, a heart, and vessels for conducting nutrient, wastes, and gases through the organism. The (2) ____ is an accessory system of vessels and organs that helps balance the fluid content of the blood and participates in the body's (3) _________.

37.1 Animal Circulatory Systems: An Introduction [pp. 892–896]

Section Review (Fill-in-the-Blanks)

In a(n) (4) ____, vessels leaving the heart release (5) ____ directly into body spaces called (6) ____. In a(n) (7) ____, the blood is confined to a network of vessels and is separated from the interstitial fluid. In closed circulatory systems, (8) ____ conduct blood away from the heart and break into highly branched (9) ____ that are specialized for exchange of material between the blood and interstitial fluid. Blood then flows into (10) ____, which return it to the heart. The heart of vertebrates, depending on the species, may consist of one or two (11) ____, which receive blood returning to the heart, as well as one or two (12) ____, which pump blood from the heart. In amphibians, which have a single ventricle, most of the oxygenated blood enters the (13) ____, while deoxygenated blood is directed into the (14) ____. In reptiles, which also have a single ventricle, oxygenated blood enters the systemic circuit, while deoxygenated blood is directed into a(n) (15) ____. Mammals and birds have two atria and two ventricles so that there are two separate circuits: a pulmonary circuit and a systemic circuit.

Short Answer

16. List the common features of circulatory systems in animals.

____

____

17. List the different functions of the circulatory system.

____

____

18. Distinguish between the open and the closed circulatory system.

____

____

Choice

For each of the following, choose the most appropriate type of circulatory system from the list below.

a. Open circulatory system

b. Closed circulatory system

19. ____ Snails

20. ____ Arthropods

21. ____ Annelids

22. _____ Squids

23. _____ Vertebrates

For each of the following, choose the most appropriate circuit of blood from the list below.

a. Systemic circuit b. Pulmonocutaneous circuit c. Pulmonary circuit

24. ____ The circuit that distributes oxygenated blood to the body

25. ____ The circuit in amphibians that distributes deoxygenated blood to the lungs and skin

26. ____ The circuit in reptiles, birds, and mammals that distributes deoxygenated blood to the lungs

For each of the following, choose the most appropriate chamber(s) of the heart from the list below.

a. Atrium/atria b. Ventricle(s)

27. ____ Receive(s) blood returning to the heart

28. ____ Pump(s) blood from the heart

37.2 Blood and Its Components [pp. 896–899]

Section Review (Fill-in-the-Blanks)

Blood is a connective tissue that consists of several cell types and a liquid matrix called (29) ____. There are three classes of protein in the liquid matrix: (30) ____, (31) ____, and (32) ____. (33) ____, commonly called (34) ____, are the cells that carry O_2 to tissues. (35) ____ is a hormone produced by the kidney and stimulates RBC production. White cells, or (36) ____, play a role in the body's defence against invading organisms. (37) ____ are cell fragments that contain factors important for blood clotting, including enzymes that convert fibrinogen into (38) ____.

Building Vocabulary

Using the definitions of the prefixes below, assign the appropriate prefix to the appropriate suffix for the following:

Prefix	Meaning
erythro-	red
leuko-	white (without colour)
hemo-	blood

Suffix	Meaning
-cyte	a cell

Prefix	Suffix	Definition
____	____	39. Red blood cell
____	____	40. White blood cell
____	-globin	41. Protein in red blood cells specialized for binding O_2

Complete the Table

Cell/Component	*Function*
42.	Eliminates dead/dying cells; removes cellular debris
Erythrocytes	43.
44.	Cell fragments produced in bone marrow; play a role in blood clotting

Matching

Match each of the following structures with its correct definition (choices may be repeated more than once).

45.	____	Erythrocytes	A.	Transport of oxygen
46.	____	Leukocytes		
47.	____	Platelets	B.	Involved in immune responses
48.	____	Erythropoietin	C.	Involved in blood clotting
49.	____	Albumins	D.	Red blood cells
50.	____	Hemoglobin	E.	Fluid component of blood that contains nutrients, ions, gases, etc.
51.	____	Globulins	F.	Proteins important for osmotic balance, pH buffering, and transport
52.	____	Plasma	G.	Proteins important in transport and as antibodies
53.	____	Fibrinogen	H.	Hormone that stimulates red blood cell production

Short Answer

54. Explain how hemoglobin transports oxygen.

____

____

55. Discuss the process of blood clotting.

____

____

37.3 The Heart [pp. 899–903]

37.4 Blood Vessels of the Circulatory System [pp. 903–905]

37.5 Maintaining Blood Flow and Pressure [p. 906]

37.6 The Lymphatic System [pp. 907–908]

Section Review (Fill-in-the-Blanks)

Oxygenated blood leaves the heart and enters the (56) ____. The cardiac cycle consists of a period of heart muscle contraction or (57) ____ and a period of relaxation or (58) ____. As the atria contract, blood flows through the (59) ____ and into the ventricle. When the ventricle contract, the AV valves close and blood flows through the semilunar valves in the pulmonary and systemic circuits. Some animals such as crabs and lobsters have a(n) (60) ____ in which the heartbeat is under the control of the nervous system. Other animals, including all vertebrates, have (61) ____ in which the heart has its own endogenous rhythm that does not require outside signals. In mammals, contraction is initiated in the right atrium by the (62) ____, which contains specialized (63) ____. The electrical impulse travels from the atria to the ventricles through the (64) ____. The electrical activity of the heart can be recorded externally to produce a(n) (65) ____.

As blood flows from the heart, it enters progressively smaller arteries until finally reaching the (66) ____. Blood flow into individual capillary beds is controlled by a(n) (67) ____. Capillaries join together to form (68) ____, which merge to form larger veins that return blood to the heart.

Blood pressure is influenced by (69) ____, the degree of constriction of blood vessels (primarily arterioles), and blood volume.

The (70) ____ is a network of vessels that collects excess interstitial fluid or (71) ____ and returns it to the systemic circulation (venous side). (72) ____ act as filters and participate in immune responses.

Choice

For each of the following descriptions, choose the most appropriate type of heart from the list below.

a. Myogenic heart b. Neurogenic heart

73. ____ Possesses endogenous electrical activity that initiates contraction

74. ____ Requires neural signals to initiate contraction

For each of the following descriptions, choose the most appropriate region of the heart from the list below.

a. Sinoatrial node b. Atrioventricular node

75. ____ In the right atrium and possesses pacemaker cells; initiates contraction of atria

76. ____ Conducts electrical impulses from atria to ventricles

For each of the following descriptions, choose the most appropriate phase of the cardiac cycle from the list below.

a. Systole b. Diastole

77. ____ Period of contraction and emptying of heart chambers

78. ____ Period of relaxation and filling of heart chambers

For each of the following descriptions, choose the most appropriate valve from the list below.

a. Atrioventricular valves b. Semilunar valves

79. _____ Valves between atria and ventricles

80. ____ Valves between ventricles and the arteries that leave the heart

Complete the Table

Component	*Function*
Lymphatic system	81.
82.	Filters blood and participates in immune responses
83.	Interstitial fluid that enters lymph vessels

Short Answer

84. What is an electrocardiogram?

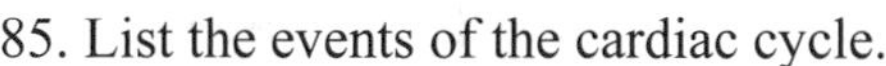

85. List the events of the cardiac cycle.

86. Explain the changes in blood pressure in the aorta, atria, and ventricles during the heart cycle.

87. Using the following diagram, describe the electrical control of the heart cycle.

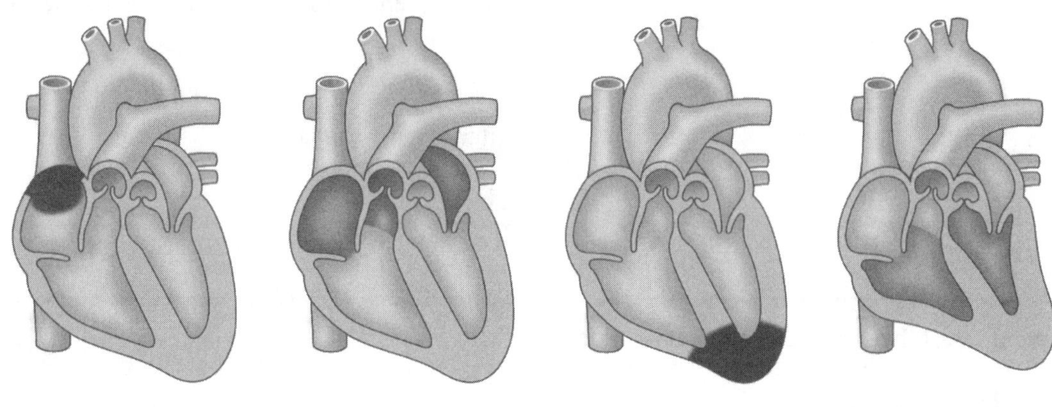

88. Draw a diagram of an electrocardiogram. Relate the characteristics of the electrocardiogram to the events of the heart cycle.

Matching

Match each of the following types of junction with its correct definition.

89.	____	Pacemaker cells	A.	Cells that initiate the endogenous rhythms of myogenic hearts
90.	____	Cardiac cycle	B.	Chronic elevation in blood pressure above normal levels
91.	____	Venules	C.	Small vessels that supply blood to capillaries
92.	____	Aorta	D.	Small vessels that drain blood from capillaries
93.	____	Cardiac output	E.	Control blood flow into capillary
94.	____	Hypertension	F.	Is a function of force of contraction and heart rate
95.	____	Precapillary sphincter	G.	The sequence of heart contraction–relaxation
96.	____	Arterioles	H.	The large vessels into which blood enters after leaving the heart

SELF-TEST

1. Functions of the circulatory system include [p. 892]

   a. Gas transport
   b. Nutrient transport
   c. Transport of waste products
   d. Hormone circulation

2. In a closed circulatory system, [p. 893]

   a. The blood is confined to blood vessels
   b. The blood directly bathes the organs
   c. Substances are exchanged between the blood and the interstitial fluid
   d. The heart releases haemolymph into sinuses

3. _____ is the fluid-filled space that surrounds organs of animals with an open circulatory system. [p. 893]

a. Coelom
b. Lymphocoel
c. Atracoel
d. Sinus

4. Bony fishes have [p. 894]

a. An open circulatory system
b. A single-circuit circulatory system
c. A pulmonary circuit
d. A closed circulatory system

5. Three-chambered hearts generally consist of the following numbers of atria/ventricles: [pp. 895–896]

a. 2/1
b. 1/2
c. 1/1
d. 0/3
e. 3/0

6. Erythrocytes are ____. [pp. 897–898]

a. Also called red blood cells
b. Spherical shaped
c. Produced in bone marrow
d. A kind of white blood cell
e. Specialized to transport O_2

7. Functions of the blood include [pp. 896–897]

a. Gas transport
b. Defence
c. Clotting
d. Stabilizing pH

8. The plasma protein albumin [p. 897]

a. Is important for osmotic balance
b. Plays a role in pH buffering
c. Is involved in lipid transport
d. Participates in blood clotting

9. Which of the following is not a respiratory pigment? [p. 898]

a. Haemoglobin
b. Myoglobin
c. Haemocyanin
d. Haemerythrin
e. Globulins

10. Erythropoietin [p. 899]

a. Is a hormone involved in the activation of the immune responses
b. Is a chemical signal involved in the initiation of blood clotting
c. Is a hormone involved in the stimulation of the production of red blood cells
d. Is a chemical signal involved in the maintenance of extracellular fluid

11. Neutrophils are [pp. 897, 899]

a. Leukocytes
b. Involved in phagocytosis during inflammation
c. Involved in defence against parasitic worms
d. Involved in gas transport

12. Fibrinogen is ____. [p. 899]

a. A plasma lipid
b. A gamma globulin
c. A protein
d. A precursor to fibrin
e. Involved in the clotting mechanism

13. In general, the path of blood in the systemic circulation in a vertebrate occurs in the following order: [pp. 899–900]

a. Veins, venules, capillaries, arterioles, arteries
b. Arterioles, capillaries, arteries, venules, veins
c. Venules, veins, capillaries, arteries, arterioles
d. Arteries, arterioles, capillaries, venules, veins

14. In mammals, blood enters the right atrium from the ____. [p. 900]

a. Right ventricle
b. Superior vena cava
c. Inferior vena cave
d. Pulmonary artery
e. Jugular vein

15. Pulmonary arteries carry blood that is [p. 900]

a. Low in O_2
b. Low in CO_2
c. High in O_2
d. High in CO_2
e. On its way to the lungs

16. In mammals, a semilunar valve is found between a ventricle and ____. [pp. 900–902]

a. An atrium
b. Another ventricle
c. The aorta
d. A pulmonary artery
e. A pulmonary vein

17. Systolic pressure [p. 901]

a. Occurs during the relaxation phase of the heart cycle
b. Is the blood pressure during the contraction of the heart cycle
c. Is approximately 120 mm Hg
d. Is approximately 80 mm Hg

18. The ventricular muscle cells of the mammalian heart receive stimuli to contract from the ____. [pp. 902–903]

a. SA node
b. AV node
c. Atria
d. Purkinje fibres

19. An electrocardiogram [pp. 902–903]

a. Measures the strength of cardiac contractions
b. Measures the electrical activity of the heart
c. Measures the electrical activity of the skeletal muscle
d. Can be measured by attaching electrodes to different points of the body

20. _____ are very small vessels that supply blood to capillary beds. [p. 904]

a. Arterioles
b. Venules
c. Arteries
d. Veins
e. Capillaries

21. Characteristics of veins and venules are [p. 905]

a. Thick walls
b. Large muscle mass in walls
c. Large quantity of elastin in the walls
d. Low blood volume compared with arteries
e. One-way valves to prevent backflow of blood

22. Which of the following aids in the return of venous blood to the heart? [p. 905]

a. Skeletal muscle contraction
b. One-way valves in the veins
c. Respiratory movements
d. Blood pressure in the veins

23. Factors that control blood pressure include which of the following? [p. 906]

a. Cardiac output
b. Degree of constriction of blood vessels
c. Blood volume
d. Hormones such as epinephrine

24. Which of the following is not a role for the lymphatic system? [pp. 907–908]

a. Is a key component of the immune system
b. Returns excess tissue fluid to the blood
c. Produces red blood cells
d. Collects fat absorbed by the intestine

INTEGRATING AND APPLYING KEY CONCEPTS

1. What is the adaptive significance of a closed circulatory system?

2. Explain why some arterioles vasoconstrict in response to epinephrine while others vasodilate. What is the functional significance of the different responses?

OPTIONAL ACTIVITIES

1. Many athletes train at high altitudes (which elevates blood EPO concentrations) or take EPO to boost their performance during competition. What are the benefits of training at high altitude or taking EPO? Discuss the underlying reasons for these benefits. How can one determine if EPO levels in the blood are a result of taking injections or are naturally occurring due to training at a high altitude?

2. Congestive heart failure can be detected using an electrocardiogram. What would an ECG diagram look like for individuals experiencing congestive heart failure? Explain.

3. Discuss a number of adaptations occurring in animals that allow them to be well suited for life at a high altitude and for life as divers.

38 Animal Reproduction

CHAPTER HIGHLIGHTS

- Asexual and sexual reproduction in animals are described.
- Sexual reproduction has the potential to increase species genetic diversity through gamete production and fertilization.
- Mechanisms associated with egg and sperm production and fertilization are described, with a focus on humans.
- Three patterns of embryo development are compared.

STUDY STRATEGIES AND LEARNING OUTCOMES

- First, focus on the purpose and overall processes required to produce offspring by asexual and sexual reproduction.
- Compare the three basic mechanisms of asexual reproduction, noting the special case of parthogenesis.
- Follow the development of the gametes, oogenesis, and spermatogenesis, identifying structures, hormones, and processes involved.
- Examine fertilization and embryo development.
- Next, expand on your knowledge of gamete production, fertilization, and embryo development with a focus on humans.

By the end of this chapter, you should be able to

- Describe and differentiate between the structures and processes involved in asexual and sexual reproduction
- Compare and contrast the cost and benefits of the two modes of reproduction
- Explain gamete production, fertilization, and embryo development in humans, as well as in animals

TOPIC MAP

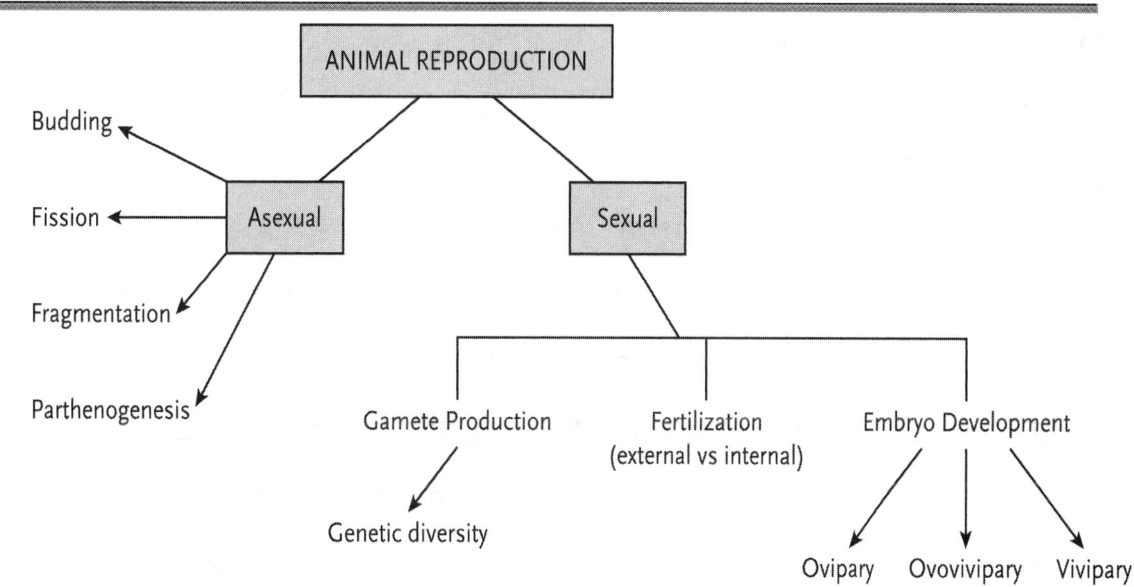
ANIMAL REPRODUCTION
Asexual
Sexual
Budding
Fission
Fragmentation
Parthenogenesis
Gamete Production
Fertilization (external vs internal)
Embryo Development
Genetic diversity
Ovipary
Ovovivipary
Vivipary

INTERACTIVE EXERCISES

Why It Matters [pp. 911–912]

Section Review (Fill-in-the-Blanks)

In many coral species, reproduction requires a mass release of both female (1) ___________ cells and male (2) ___________ cells into the water a few days after a(n) (3) ___________ moon. For (4) ___________ fertilization to be successful, release of these gametes must be (5) ___________. There is evidence that light receptor proteins in these corals (and other animals), called (6) ___________ (CRYs), sense changes in irradiance intensity of (7) ___________ light during a (8) ___________ cycle and trigger the mass spawn. Fertilization can only occur between gametes of the same (9) ___________ due to chemicals in the egg (10) ___________ and sperm (11) ___________.

38.1 The Drive to Reproduce [pp. 912–913]

38.2 Asexual and Sexual Reproduction [pp. 913–914]

Section Review (Fill-in-the-Blanks)

Reproduction is necessary to ensure that an organism's (12) ___________ are passed on to future generations. Aquatic invertebrates and a few vertebrates are able to reproduce by (13) ___________ reproduction (or clonal reproduction), which involves the production of offspring from (14) ___________ parent. There are (15) ___________ major mechanisms of this type of reproduction, and each produces offspring that are genetically (16) ___________ to the parent. A special type of asexual reproduction, called (17) ___________, occurs when a new individual develops from a(n) (18) ___________ egg. Sexual reproduction results in the production of a(n) (19) ___________ from the fertilization of male and female (20) ___________. In comparing the two types of reproduction, (21) ___________ reproduction has greater costs associated with it; however, it provides an overriding advantage: it generates (22) ___________ ___________ among offspring. In an unstable or changing (23) ___________, diversity among individuals will increase the likelihood that some will (24) __________ and reproduce. There are (25) __________ mechanisms during gamete formation that produce genetic diversity, (26) __________ __________ in which alleles within (27) __________ from parents are mixed to form new combinations, and (28) __________ __________ in which chromosomes from parents are (29) __________ combined. (30) __________ is the ultimate source of variation.

Matching

Match each of the following types of asexual reproduction with its correct definition.

31.	____	Parthenogenesis	A.	Parent divides into two offspring
32.	____	Budding	B.	Separate pieces develop into offspring
33.	____	Fission	C.	Egg develops without fertilization
34.	____	Fragmentation	D.	Offspring develops directly off parent

38.3 Cellular Mechanisms of Sexual Reproduction [pp. 914–922]

Section Review (Fill-in-the-Blanks)

During (35) __________, male germ cells in the gamete-producing organ, the (36) _____, first undergo mitosis to produce (37) __________ and then undergo (38) __________ to produce haploid (39) __________ that will mature to become (40) __________ cells. Female germ cells in the (41) __________ produce oogonia from mitosis, which then undergo (42) __________. Uneven division of (43) __________ during this process results in the production of only one (44) __________, which will mature into a functional (45) __________, and two small polar bodies. This process is called (46) __________. Fertilization by most aquatic animals is (47) __________ and requires sperm to (48) __________ in open water until they collide with an egg. Internal fertilization requires close (49) __________ between individuals so that sperm can enter the (50) __________ reproductive tract and swim in a fluid environment to the egg.

Matching

Match each of the following components of an egg or sperm with its correct definition.

51.	____	Zona pellucida	A.	Nonfunctional cells with the correct number of chromosomes but less cytoplasm
52.	____	Oviduct	B.	Embryo develops within mother's body
53.	____	Polar bodies	C.	Contains enzymes necessary for penetration of the surface coating of an egg
54.	____	Amplexus	D.	Gel-like matrix covering the egg in mammals
55.	____	Midpiece of sperm	E.	Egg-laying animals
56.	____	Acrosome	F.	Contains organelles necessary for movement of flagellum
57.	____	Viviparous	G.	More than one sperm fertilizing egg
58.	____	Oviparous	H.	Transport tube for egg
59.	____	Polyspermy	I.	Reflex response in female frogs causing egg release

Sequence

60. Arrange the following steps describing fertilization.

A. Sperm nucleus enters the egg cell and fuses with egg nucleus

B. Enzymes in the acrosome break down jelly

C. Cortical granules are released from the egg cell

D. Sperm head fuses with the egg plasma membrane

E. Sperm reaches jelly layer that coats egg

F. Sperm binds to vitelline coat and bores a hole into it

____ ____ ____ ____ ____ ____

Short Answer

61. Explain the difference between internal and external fertilization.

____

____

62. Explain the importance of the acrosome reaction and the fast/slow block of fertilization.

____

____

63. Distinguish between simultaneous and sequential hermaphroditism.

____

____

38.4 Sexual Reproduction in Humans [pp. 922–931]

38.5 Controlling Reproduction [pp. 931–932]

Section Review (Fill-in-the-Blanks)

In humans, like all vertebrates, the gonads have (64) _____ functions, gamete and (65) _____ production. In males and females, two hormones from the pituitary, (66) ____-_____ hormone and (67) _____ hormone, are under the control of (68) _____-_____ hormone from the hypothalamus. In females, FSH stimulates (69) _____ to undergo meiosis, and one will develop into a mature follicle. The follicular cells secrete (70) _____, which stimulate further development of the follicle. The second hormone from the pituitary, (71) _____, increases just prior to and is thought to be responsible for (72) _____. After ovulation, the follicular cells, as part of the corpus luteum, continue to secrete estrogens but significantly increase the secretion of (73) _____, which further prepares the uterus for implantation. In males, FSH stimulates (74) _____ cells to produce materials that are required for (75) _____, and LH stimulates (76) _____ cells to secrete (77) _____, which also plays a major role in spermatogenesis. Understanding the processes of reproduction in animals, including humans, assists in controlling reproductive (78) _____, maximizing success for animals becoming (79) _____, and reducing the possibility of becoming (80) _____ when desired.

Matching

Match each of the structures with its correct function.

81.	____	Oviduct, near ovary	A.	Entrance for sperm to enter female body
82.	____	Endometrium	B.	Group of external female organs
83.	____	Prostate gland	C.	Site where meiosis I is completed in females
84.	____	Pituitary gland	D.	Produces alkaline secretion that is component of semen
85.	____	Urethra	E.	Produces testosterone
86.	____	Vas deferens	F.	Becomes erect during ejaculation
87.	____	Ovary	G.	Site of spermatogenesis
88.	____	Vulva	H.	Location for sperm storage
89.	____	Leydig cell	I.	Muscular tube that joins with the urethra in males
90.	____	Epididymis	J.	Location where fertilized egg is implanted
91.	____	Vagina	K.	Location of fertilization
92.	____	Greater vestibular gland	L.	Produces lubricant for structures surrounding entrance to vagina
93.	____	Testis	M.	Tube from which semen leaves the male body
94.	____	Penis	N.	Secretes FSH and LH

Choice

For each of the following statements, choose the most appropriate definition from the list below.

A. FSH B. LH C. Androgens D. Estrogens

E. Progesterone F. GnRH G. hCG

95. ____ Leydig cells produce this

96. ____ Corpus luteum produces large amounts of this

97. ____ Without this hormone, FSH and LH will not be released

98. ____ Responsible for development of male secondary sex/reproductive structures

99. ____ Stimulates the Leydig cells

100. ___ Stimulates primary oocytes to undergo development

101. ____ Supportive cells that surround developing spermatocytes are stimulated by this

102. ____ Hormone that is predominant during the luteal phase

103. ____ Progesterone will inhibit this hormone from being released

104. ____ Burst of this hormone results in ovulation

105. ____ Low concentration of testosterone causes this hormone to be secreted

106. ____ Enlarging follicle cells secrete this

107. ____ Secreted by embryo to prevent corpus luteum from breaking down

108. ____ Thickens mucus in uterus that forms a plug

SELF-TEST

1. What does a male acanthocephalan worm do to ensure that his genes are passed on to the next generation? [pp. 912–914]

   a. Mates with female and then seals her gonopore
   b. Mates with female multiple times
   c. Reproduces asexually by budding

2. A sea star was cut into two complete pieces; each developed into a complete sea star. This is an example of _____. [pp. 912–914]

   a. External fertilization
   b. Fission
   c. Budding
   d. Fragmentation

3. Which of the following is an advantage of asexual reproduction? [pp. 912–914]

   a. Independent assortment
   b. Synchronization of gamete release
   c. Genetic diversity
   d. Finding a mate is not necessary
   e. Higher energy cost

4. Which of the following increase(s) genetic diversity of offspring? [pp. 912–914]

   a. Fertilization between genetically different individuals at random
   b. Genetic recombination
   c. Independent assortment
   d. (a), (b), and (c)

5. If the cortical reaction that occurs during the fast block reaction after fertilization were blocked, what might occur? [pp. 914–922]

   a. External fertilization
   b. Fertilized egg would break down
   c. Polyspermy
   d. Greater than diploid number of chromosomes in offspring
   e. Both (c) and (d) are correct

6. The duck-billed platypus is ___. [pp. 914–922]

   a. Placental
   b. Oviparous
   c. Viviparous
   d. Ovoviviparous

7. What is the male cell called that enters meiosis I? [pp. 914–922]

   a. Oocyte
   b. Oogonia
   c. Spermatocyte
   d. Spermagonia

8. What is the chromosome number of a secondary oocyte? [pp. 914–922]

   a. Haploid
   b. Diploid
   c. Triploid

9. What part of the sperm contains the chromosomes that will be passed on to the offspring? [pp. 914–922]

   a. Acrosome
   b. Mitochondria
   c. Tail
   d. Midpiece

10. In a four-year-old human female, at what stage of meiosis are the oocytes in her ovaries? [pp. 914–922]

    a. Beginning of prophase I
    b. End of prophase I
    c. Beginning of metaphase II
    d. End of metaphase II

11. Calcium ion release from the ER to prevent polyspermy is a ___________. [pp. 914–922]

    a. Fast block response that occurs within seconds of fertilization
    b. Fast block response that occurs within minutes of fertilization
    c. Slow block response that occurs within seconds of fertilization
    d. Slow block response that occurs within minutes of fertilization

12. What are hermaphrodites? [pp. 914–922]

    a. Individuals that produce eggs
    b. Individuals that produce sperm
    c. Individuals that produce eggs and sperm
    d. Individuals that produce neither eggs nor sperm

13. In order for progesterone to have its full effect during the luteal phase, which of the following must occur? [pp. 922–932]

    a. Low levels of FSH
    b. Ovulation
    c. The secretory phase
    d. Stimulation of a primary oocyte

14. What is the name of the membrane that human females are born with but that is broken during the first sexual intercourse? [pp. 922–932]

    a. Labia minor
    b. Labia major
    c. Hymen
    d. Clitoris

15. If the levels of FSH and LH were not high enough to inhibit the release of GnRH, what could happen? [pp. 922–932]

    a. Nothing
    b. No fertilization
    c. Possibility of twins
    d. No menstrual phase

16. If the prostate gland secretion were inhibited, other than a decreased volume, what is the effect on the semen? [pp. 922–932]

    a. More basic semen
    b. More acidic semen
    c. More alkaline semen
    d. Both (a) and (c) are correct

17. With respect to fertility, if someone was born without a pituitary gland, what would be the result? [pp. 922–932]

    a. Sterility
    b. They would be female
    c. Multiple births with every pregnancy
    d. GnRH would be extremely low

18. Approximately what percentage of the oocytes in human ovaries at sexual maturity are ovulated during a female human life? [pp. 922–932]

    a. 50%
    b. 20%
    c. 1%
    d. 0.1%
    e. 0.001%

19. As sperm cell develop and mature in human males, they move __________. [pp. 922–932]

    a. Toward the centre of the seminiferous tubule
    b. Toward the wall of the seminiferous tubule
    c. Toward Sertoli cells
    d. Toward Leydig cells

20. Home pregnancy tests evaluate which of the following hormones? [pp. 922–932]

   a. FSH
   b. LH
   c. Relaxin
   d. hCG

INTEGRATING AND APPLYING KEY CONCEPTS

1. Explain why species recognition is important for animals using external fertilization.

2. Identify the components of semen and provide at least one function for each.

3. Assume that pregnancy has occurred. Predict the effect if progesterone levels started to drop instead of increasing.

OPTIONAL ACTIVITIES

1. Some animal species store their sperm for extended periods of time. Identify and compare different techniques that sperm in these species use to maximize their success in sperm competition.

2. Cryptochromes are involved in the regulation of circadian clocks in plants and animals. What other function do they perform in some animals?

39 Animal Development

CHAPTER HIGHLIGHTS

- Major developmental events in animals, from the fertilized egg through organogenesis, are presented.
- Embryonic development of some deuterostomes, with a focus on humans, is described.
- Explanations of underlying cellular processes support development of organ systems.
- Genetic controls regulate all aspects of embryonic development, including cell death.

STUDY STRATEGIES AND LEARNING OUTCOMES

- First, focus on major events of embryological development from fertilization, through cleavage, gastrulation, and organogenesis, to parturition.
- Look for developmental patterns at the different stages, including overall structure of the adult form.
- Next, make comparisons between the embryonic development of humans and other deuterostomes.
- Evaluate the mechanisms of development from a cellular perspective, focusing on genetic and molecular controls.

By the end of this chapter, you should be able to

- Describe the major stages of development from zygote to a whole organism, with a focus on humans
- Explain how genes are able to control processes of embryo development
- Describe how cells change shape, move, and become specialized

TOPIC MAP

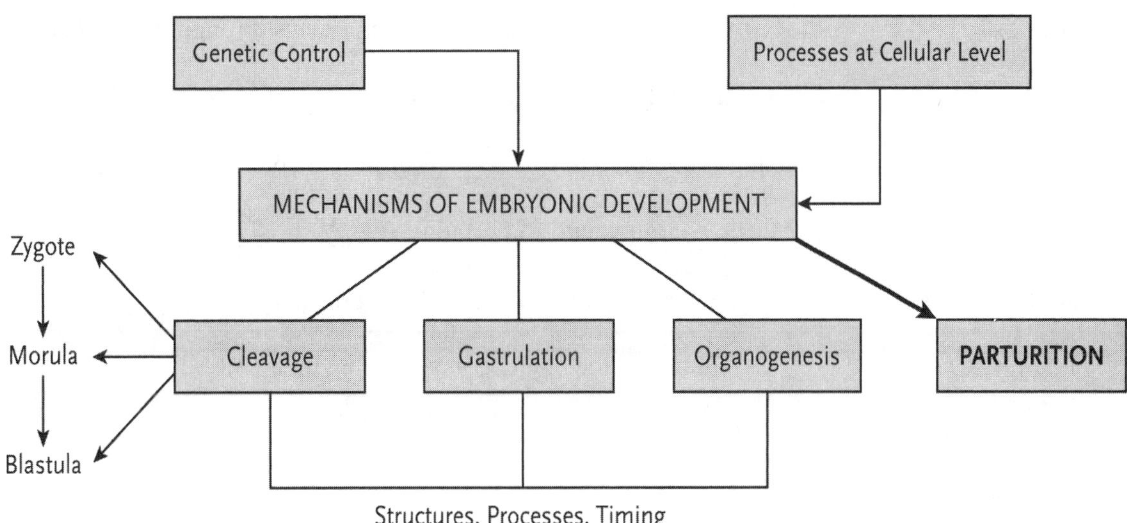

Genetic Control
Processes at Cellular Level
MECHANISMS OF EMBRYONIC DEVELOPMENT
Zygote
Morula
Blastula
Cleavage
Gastrulation
Organogenesis
PARTURITION
Structures, Processes, Timing

INTERACTIVE EXERCISES

Why It Matters [pp. 935–936]

Section Review (Fill-in-the-Blanks)

During (1) __________, the birth process, a mammalian offspring travels through the mother's birth (2) __________ and leaves the body through the opening between bones of the (3) __________ girdle. This opening is often too (4) __________ for the young to pass through, especially in species that give birth to large young. A hormone, (5) __________, which is produced by the (6) __________, assists in parturition by relaxing the (7) __________ fibres of the interpubic (8) __________ surrounding the opening, maximizing their (9) __________ so that the opening is enlarged.

39.1 Housing and Fuelling Developing Young [pp. 936–938]

Section Review (Fill-in-the-Blanks)

The higher the parental investment in caring for offspring, the (10) __________ the likelihood that the parent's (11) __________ will be passed on to future generations. An animal that carries a developing embryo inside its body not only provides (12) __________ against predators but also ensures the best conditions for embryo (13) __________. After parturition, parents that provide nutritious food, such as (14) __________ (primarily mammalian species) or other high-quality food, are investing their energy to promote survival and (15) __________ of young.

39.2 Mechanisms of Embryonic Development [pp. 938–940]

39.3 Major Patterns of Cleavage and Gastrulation [pp. 941–944]

Section Review (Fill-in-the-Blanks)

All instructions for development are located in the nucleus of the (16) __________, the fertilized egg. Since the contribution of the sperm is primarily (17) __________ material, it is the (18) __________ determinants from the (19) __________, consisting of the cytoplasm, all organelles, and mRNA, that play key roles in the initial activities of the zygote. In most species, the nucleus in the zygote is located at one end of the cell, the (20) __________ pole, and the other components are at the (21) __________ pole. Even at this early stage in development, (22) __________ in the egg assists in defining the (23) __________ body axes. During embryo development, there are (24) ___ major development stages. The first stage, (25) __________, progresses through two steps: (i) formation of a solid ball of cells, the (26) __________, resulting from (27) __________ cell divisions, and (ii) the hollowing of the ball of cells to form a (28) __________. Next, the embryo enters the second phase, (29) __________, where indentations and rearrangement of dividing cells result in the formation of three primary tissues, the outer (30)__________ tissue, the inner (31) __________ tissue, and the (32) __________ in between. The cavity that forms in the embryo, the (33) __________ (or gut), has an opening at one end, the (34) __________. If this opening forms a mouth, the animal is classified as a (35) __________, whereas if it forms an anus, it is a(n) (36) __________. The third stage is (37) __________, and it gives rise to all major (38) __________ systems. Six major mechanisms play key roles in these developmental stages: mitotic divisions, cell movements, selective cell (39) __________, induction, determination, and (40) __________ (cell-specific structure and function).

Choice

For each of the following statements, choose the most appropriate gastrulation pattern from the list below. There may be more than one correct answer.

A. Gastrulation with even distribution of yolk

B. Gastrulation with uneven distribution of yolk

C. Gastrulation at one side of yolk

41. ____ Blastodisk is composed of the epiblast and hypoblast

42. ____ Pattern found in amphibians

43. ____ Invagination begins at the vegetal pole of the blastula

44. ____ Blastopore becomes the anus

45. ____ Pattern typically found in amniotes

46. ____ Dorsal lip cells control blastopore formation

47. ____ Primitive streak defines the right and left sides of the embryo

48. ____ Extraembryonic membranes are common

49. ____ Hypoblast cells become germ cell line

50. ____ Involution results in the pigmented cells of the animal half enclosing the vegetal half

Matching

Match each of the following with its correct definition.

51.	____	Chorion	A.	Stores nitrogenous wastes produced by the embryo
52.	____	Primitive groove	B.	Cells of early developmental cleavage
53.	____	Determination	C.	Cell-to-cell contact influencing developmental pathway of group of cells
54.	____	Blastomeres	D.	Layers that gives rise to primary tissues
55.	____	Amnion	E.	Membrane that surrounds the embryo and yolk sac
56.	____	Allantois	F.	Membrane that encloses the embryo
57.	____	Epiblast	G.	Developmental fate of the cell is set
58.	____	Induction	H.	Entrance to archenteron
59.	____	Blastopore	I.	Means for cells to move into the blastocoels

Choice

For each of the following organs, choose the primary tissue type from the list below.

A. Endoderm B. Mesoderm C. Ectoderm

60. ____ Muscles

61. ____ Bone

62. ____ Lining of respiratory tract

63. _____ Spinal cord

64. _____ Heart and kidneys

65. ____ Lining of mouth

66. ____ Lining of digestive tract

67. ____ Cornea of eye

39.4 Organogenesis: Gastrulation to Adult Body Structures [pp. 945–947]

Section Review (Fill-in-the-Blanks)

During the third major developmental stage, (68) ___________ the three primary tissues develop into (69) ___________. This stage involves several major changes that occur in rapid succession. A rod of tissue, the (70) ___________, providing organization along the entire length of the embryo, is derived from (71) ___________. Very rapidly, ectoderm will give rise to the (72) ___________ system in a process called (73) ___________. Above the notochord, cells undergo (74) ___________ and develop into the neural (75) ___________, which gives rise to the central nervous system. At the same time, blocks of (76) ___________ develop into (77) ___________, which will develop into organs and tissues associated with the embryo. In addition to cellular mechanisms used in cleavage and gastrulation, organogenesis has programmed cell death or (78) ___________.

Sequence

79. Put the following steps of eye development in sequence, starting with number 1.

A. ___ Ball of cells forms the lens vesicle

B. ___ Optic vesicles grow outward

C. ___ Outpocket forms optic cup

D. _1_ Neural tube forms

E. ___ Lens placode forms from thickened ectoderm

F. ___ Ectoderm closes over lens and forms the cornea

G. ___ Lens synthesizes crystallin

39.5 Embryonic Development of Humans and Other Mammals [pp. 947–953]

Section Review (Fill-in-the-Blanks)

Human gestation is completed in approximately (80) ___________ and can be divided into three (81) ___________. The first trimester includes three development events (in order): (82) ___________, (83) ___________, and (84) ___________. After the first trimester, the developing embryo is called a(n) (85)___________ until birth. A successful pregnancy requires that (86)

__________ must occur in the first (87) __________ of the oviduct. By the time that (88) __________ in the uterine wall occurs, the embryo is in the form of a (89) __________. This structure contains a fluid-filled cavity, the (90) __________, a(n) (91) __________ __________ __________ that is pushed to one side of the structure, and an outer layer of cells, the (92) __________. The latter secretes (93) __________ that create a hole in the (94) __________, providing a space for implantation of the blastocyst. The inner cell mass divides into a two-layered embryonic (95) __________, in which the inner layer, the (96) __________, develops into the (97) __________ and the outer layer, the (98) __________, assists in the development of extraembryonic (99) __________. All organs and organ systems are formed at (100) ________ weeks, with the remaining weeks providing a time for embryo (101) _______ and development. Gastrulation and organogenesis of a human embryo are similar to the pattern found in (102) __________ or (103) __________.

Matching

Match each of the following with its correct definition.

104.	____	Chorionic villi	A.	Connecting stalk between the embryo and placenta
105.	____	Umbilical cord	B.	Produces the yolk sac
106.	____	Parturition	C.	Develops into female reproductive system
107.	____	Wolffian duct	D.	Extensions that increase surface area
108.	____	Müllerian duct	E.	Develops into male reproductive system
109.	____	Hypoblast	F.	Process involving extracting fluid from cavity containing embryo
110.	____	Amniocentesis	G.	Birth

Choice

For each of the following processes, choose the correct time period for when it would be occurring.

A. First trimester B. Second and third trimesters C. Birth

D. After birth

111.____ Mammary glands are stimulated by prolactin

112.____ Blastocyst breaks out of zona pellucida

113.____ Cervix dilates

114.____ Fetus grows

115.____ Formation of chorionic villi for gas and nutrient exchange between parent and embryo

116.____ Expulsion of the placenta out of vagina

117.____ Amniotic membrane bursts and amniotic fluid is released

118.____ Embryo develops tail and pharyngeal slits

Short Answer

119. Identify and describe the gene that determines whether an embryo develops into a male or a female.

____

____

120. How are the extraembryonic membranes formed?

____

____

39.6 Cellular Basis of Development [pp. 953–958]

39.7 Genetic and Molecular Control of Development [pp. 958–963]

Section Review (Fill-in-the-Blanks)

The final shape and size, as well as the location of organs in the embryo, are determined by the (121) __________, due to the location of cleavage furrows, and (122) __________ of cellular divisions during development. Division rate is determined by the time of the (123) __________ period of interphase. The actual mechanisms for these events are not totally understood; however, scientists do know that various structures and mechanisms influence embryonic development, including cell components (124) __________ and (125) __________, which can lead to cell movement and change in cell shape; the (126)__________ process that uses cells to define the fate of other cells, called (127) __________, and usually results in cell differentiation; and the presence or absence of cell (128) __________ molecules, which allows entire cells to move during embryonic development. At all stages of development, (129) __________ control necessary processes, which are often regulated by master (130) __________ genes. An example is the (131) __________ gene, a homeotic gene that specifies what each embryo segment will become after (132) __________.

True/False

If the statement is true, write a "T" in the blank. If the statement is false, make it correct by changing the underlined word(s) and writing the correct word(s) in the answer blank.

133.____________ Broad regions along the anterior–posterior axis of the embryo are controlled by segment polarity genes.

134.____________ The genes that control the polarity of the zygote and subsequent cleavage events of the embryo are from a maternal source.

135.____________ Fate mapping of embryos allows a clear understanding of cell lineage.

136.____________ Products of segmentation genes divide the embryo into units of two segments each.

137. ____________Homeotic (*Hox*) genes have been identified in animals but not plants.

138.____________ Differentiation leads to determination of a cell's development.

139.____________ A death signal binding to a cell's plasma membrane receptors activates genes that produce killer proteins.

140.____________ The BICOID protein is concentrated at the posterior end of a zygote.

141.____________ Master regulatory genes control the expression of other genes.

142.______________ The MyoD transcription factor converts somite cells into differentiated muscle cells.

143.______________ The sliding movement of microtubules causes column-shaped cells to become wedge-shaped cells.

SELF-TEST

1. Which of the following animals feeds their offspring skin? [pp. 936–938]

   a. Humans
   b. Cockroaches
   c. Caecilians
   d. Pigeons

2. Cytoplasmic determinants have the greatest effect during which stage of development? [pp. 938–944]

   a. Gastrulation
   b. Organogenesis
   c. Cleavage
   d. Second and third trimesters

3. The two major groups of animals are based on the final use of the ____. [pp. 938–944]

   a. Archenteron
   b. Blastopore
   c. Grey crescent
   d. Blastocoel

4. If the endoderm tissue in a developing animal were labelled with a coloured marker, where would you expect to find the marked tissue in the adult? [pp. 938–944]

   a. Nervous system
   b. Muscle and bone
   c. Coverings of the animal or structures
   d. Linings of major organ systems

5. If you were working with a blastodisk of a bird embryo, which portion will develop into the germ cell line in the adult? [pp. 938–944]

   a. Epiblast
   b. Hypoblast
   c. Primitive streak
   d. Primitive groove

6. If the primitive streak were blocked or removed, predict the effect on the embryo. [pp. 938–944]

   a. The adult could not reproduce
   b. Organization, including axes of the embryo, would be lost
   c. Extraembryonic membranes would be lacking
   d. The blastocoel would collapse

7. Which of the extraembryonic membranes surrounds the sac containing nitrogenous wastes? [pp. 938–944]

   a. Amnion
   b. Allantoic membrane
   c. Chorion
   d. Yolk sac

8. What is neurulation? [pp. 945–947]

   a. Development of a blood vessel from the mesoderm
   b. Development of a blood vessel from the endoderm
   c. Development of nervous tissue from the endoderm
   d. Development of nervous tissue from the ectoderm

9. The cranial nerves in an adult vertebrate are derived from _____. [pp. 945–947]

   a. Neural crest cells
   b. Ectoderm
   c. Mesoderm
   d. Somites

10. Which of the following processes does <u>not</u> happen during development of an eye? [pp. 945–947]

    a. Genes coding for keratin are activated
    b. Genes coding for crystallin are activated
    c. Optic vesicles induce lens formation
    d. Ectoderm cells form a cornea

11. A test that evaluates the fluid that surrounds a developing fetus is _____. [p. 947–953]

    a. Chorionic villus sampling
    b. Umbilical cord sampling
    c. Amniocentesis
    d. Blastocoel sampling

12. Why must maternal blood be isolated from embryonic blood? [pp. 947–953]

    a. Maternal blood is toxic to embryonic blood
    b. Prevents an immune reaction if blood cells are seen as “foreign”
    c. Limits adding excessive nutrients to embryo
    d. Prevents maternal wastes from entering into the allantois

13. The embryonic disk consists of _________. [pp. 947–953]

a. Trophoblast cells
b. Epiblast and hypoblast cell layers
c. Amniotic fluid
d. Embryonic blood vessels

14. If the embryo is lacking the SRY protein, the _____ develop into _____. [pp. 947–953]

a. Wolffian ducts, male reproductive structures
b. Wolffian ducts, female reproductive structures
c. Müllerian ducts, male reproductive structures
d. Müllerian ducts, female reproductive structures

15. If you wanted to change the orientation or axes of early cleavage, which of the following would you manipulate? [pp. 947–953]

a. Microtubules and microfilaments
b. Wolffian or Müllerian ducts
c. Blastopore location
d. The chorion

16. Which end of a polar egg undergoes cell division more quickly? [pp. 953–963]

a. Animal pole
b. Vegetal pole
c. End containing the yolk
d. Both ends divide at the same rate

17. Somites, which are derived from mesoderm, are ultimately formed from the products of which genes? [pp. 953–963]

a. Maternal-effect genes
b. Segment polarity genes
c. Segmentation genes
d. Homeotic genes

18. The specialized cells resulting from differentiation ________. [pp. 953–963]

a. Contain more genes than the original cells
b. Contain different genes than the original cells
c. Contain the same genes as the original cells
d. Contain fewer genes than the original cells

19. Why was *C. elegans* able to provide a detailed fate map of every cell during embryonic development? [pp. 953–963]

a. It contains very few cells
b. Its embryonic development is very simple
c. Only one cell lineage is involved in development
d. It is transparent, so markers could be followed

20. *Hox* genes in a human __________. [pp. 953–963]

a. Are in the same order as the *Hox* genes for *Drosophila*
b. Are in a different order from *Hox* genes in *Drosophila*
c. Are a different group of genes than the *Hox* genes in *Drosophila*

INTERACTIVE EXERCISES

1. If each of the primary tissues were assigned a different colour, predict how those colours would be distributed in any given organ or organ system.

2. Address the role and importance of apoptosis in development.

3. Explain how entire cells move during embryonic development.

OPTIONAL ACTIVITIES

1. How is Dr. Hall's laboratory able to track the migration of neural crest cells during embryo development?

2. What is the impact of thyroxine on the metamorphosis of the South African frog (*Hymenochirus*)?

40 Animal Behaviour

CHAPTER HIGHLIGHTS

- Animals must cope with their environment, and behaviour is a way to accomplish this. The behaviours may be instinctive or learned. Most behaviours have both learned and instinctive components.
- An animal's neural circuitry is the basis upon which behaviour occurs, often involving interactions with hormones or the animal's anatomy.
- Different types of behaviours, including communication, space, migration, and mating pattern, have an effect on an animal's ability to survive and reproduce.
- Migration is a way for animals to increase their fitness and avoid stressful environmental conditions, but it is usually stressful and involves risks.
- Social groupings provide both costs and benefits to their members.
- Assisting relatives is not altruistic as the relatedness between the individuals means that there is increased chance that family genes will be passed on to the next generation.
- Understanding the ecological and evolutionary roots of animal behaviour may lead to a better understanding of human behaviour.

STUDY STRATEGIES AND LEARNING OUTCOMES

- First you need to be able to distinguish between the two very different components of animal behaviour, instinctive and learned.
- Since all behaviours are neural based, you need to understand how hormones and anatomy interact with an animal's nervous system.
- After studying the different behaviours, from communication to social behaviours, look at how a combination of behaviours can affect an animal's fitness, an important measure of how successful the animal will be in passing on its genes to the future generation.
- Note that the beginning of this chapter is devoted to how behaviour occurs and the end of the chapter addresses why different types of behaviour occur.

By the end of this chapter, you should be able to

- Differentiate between instinctive and learned behaviour

- Explain how hormones and anatomy each interact with nerves to produce a behaviour
- Describe how communication, space, migration, and mating patterns can increase an animal's fitness
- Differentiate between the different types of mating patterns
- Compare social behaviours among different species, explaining why some species work cooperatively in groups
- Calculate the relatedness between various family members

TOPIC MAP

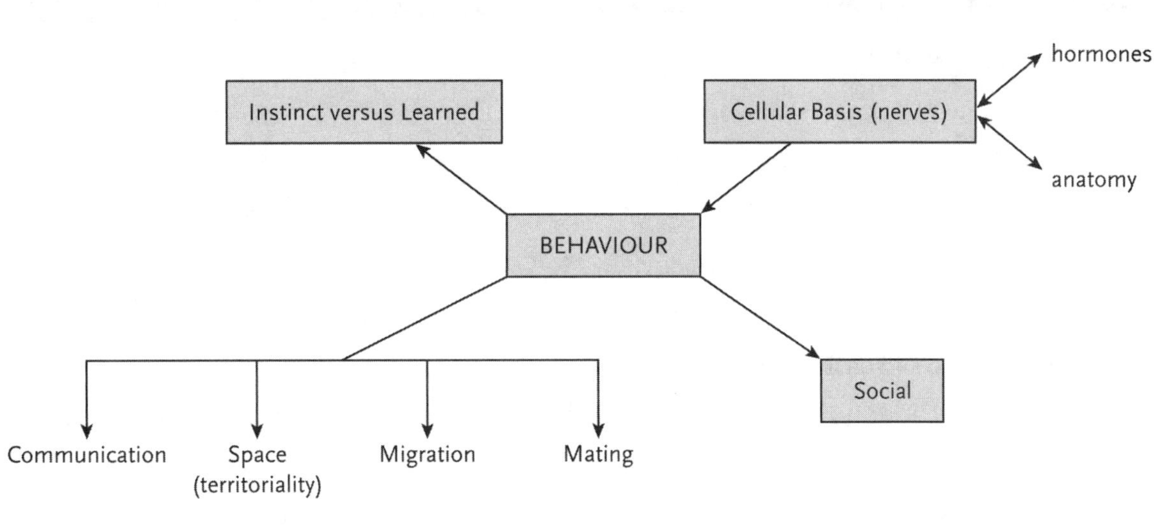

INTERACTIVE EXERCISES

Why It Matters [pp. 967–968]

Section Review (Fill-in-the-Blanks)

A reward, such as food, is often an incentive for an animal to change its (1) ____________. In a changing environment, new opportunities sometimes arise, and if an animal is (2) ____________, it may be able to develop a(n) (3) ____________ action that will provide a positive response. If successful, an animal will then have (4) ____________ a behavioural response that it can continue to use.

40.1 Genes, Environment, and Behaviour [pp. 968–969]

40.2 Instinct [pp. 969–971]

40.3 Learning [p. 972]

Section Review (Fill-in-the-Blanks)

Behavioural scientists have long debated whether an animal's behaviour is (5) ____________, that is, innate and performed correctly the first time it is used, or whether behaviour is (6) ____________, that is, modified by experience. Most attempts to answer this question involved experiments wherein animals were kept in (7) ____________ and not allowed exposure to experiences that could enable them to copy or learn a behavioural response. The consensus today is that instinct and learning are both components of animal behaviour, the result of complex (8) ____________ interactions.

Since instinctive behaviours are (9) ____________ or developmentally programmed, they are often performed in a rather invariant fashion, a (10) ____________ ____________ pattern. Simple stimuli that elicit these behaviours are called (11) ____________ stimuli.

On the other hand, learned behaviours are (12) ____________ variable than instinctive behaviours since they are modified by experiences. Many types of learning have been identified, including (13) ____________, in which an animal learns parents' and/or species' identity when an appropriate stimulus is presented at a (14) ____________ period early in life; (15) ____________ ____________, in which a previously neutral stimulus becomes associated with another stimulus that causes a behavioural response; (16) ____________ conditioning, where behaviour can be modified by linking a stimulus with a positive result called a (17) ____________; (18) ____________ learning, wherein the behavioural solution to a problem appears without trial and (19) ____________; and (20) ____________, in which repeated, unreinforced stimuli are no longer responded to.

Matching

Match each of the following types of behaviour with the appropriate example.

21.	____	Conditioned stimulus	A.	A cat comes running to cat food when it hears the sound of an electric can opener.
22.	____	Habituation	B.	A chick follows a toy car after being exposed to it 24 hours after hatching.
23.	____	Imprinting	C.	A chimp stacks two boxes in order to reach some bananas hanging beyond its reach, thereby solving the problem.
24.	____	Insight learning	D.	A fish exhibits typical courtship behaviour despite being raised from a zygote isolated from other fishes.
25.	____	Instinct	E.	A hungry mouse learns to press a bar (voluntary action) to get a food pellet (favourable consequence).
26.	____	Operant conditioning	F.	A hydra stops contracting when the aquarium it lives in is continually tapped gently.

True/False

If the statement is true, write a "T" in the blank. If the statement is false, make it correct by changing the underlined word(s) and writing the correct word(s) in the answer blank.

27. ____________ In Pavlov's experiments, when dogs started salivating in response to a ringing bell, the food was the conditioned stimulus.

28. ____________ When a mouse presses a bar to obtain food, the act of pressing on the bar is the reinforcement.

29. ____________ A snail's lack of response to repeated light touches is imprinting.

30. ____________ A fixed action pattern is a stereotypical, instinctive behaviour.

31. ____________ Instinctive behavioural differences between animals of different species are a reflection of environmental differences.

32. ____________ Learned behaviour is genetically or developmentally programmed in an animal.

40.4 Neurophysiology and Behaviour [pp. 972–973]

40.5 Hormones and Behaviour [pp. 973–976]

40.6 Neural Anatomy and Behaviour [pp. 976–979]

Section Review (Fill-in-the-Blanks)

Behaviour is a response to physiological activities resulting from (33) ____________ cell biochemistry and structure. Some behaviours are present at birth, as dictated by the animal's (34) ____________ information directing the development of its anatomy and physiology. These behaviours, as well as learned behaviours, may be modified over time through the alteration of nerve cells as a response to an individual's (35) ____________ in its environment. Research has provided many examples of how a behavioural act has a neural basis. For example, songbirds may use a distinctive song to help in defence of their (36) ____________ (an area of land defended by males or couples) or during (37) ____________, the time period when an individual (usually male) tries to attract a mate. Proper development of a song, as well as recognition that other species' songs are different, is controlled by specific nuclei in the (38) ____________. Development of these nuclei to function correctly is usually regulated by (39) ____________.

Matching

Match the species with correct behavioural adaptation.

40.	____	Crickets	A.	Direction of visual stimuli (from above or below) results in different behaviours.
41.	____	Cichlids	B.	Hormonal differences between young males and females results in differences in neural development.
42.	____	Fiddler crabs	C.	Increasing levels of a hormone cause changes in specific social tasks.
43.	____	Honeybees	D.	Most of the cerebral cortex is devoted to processing information from the nose and forelimbs.
44.	____	Bats	E.	Ultrasonic vocalizations of predatory bats cause a neural response resulting in moving away from the source of the sound.
45.	____	Star-nosed moles	F.	Comparison of cell and echo identifies the presence of potential prey.
46.	____	Zebra finches	G.	Ownership of a territory stimulates the increased production of GnRH, resulting in sexual and aggressive behaviour.

40.7 Communication [pp. 979–982]

Section Review (Fill-in-the-Blanks)

Animals use many sensory "channels" or modalities to communicate. Specific sounds or vocalizations are examples of (47) ____________, while movements that convey information are examples of (48) ____________. The use of scents to communicate with others is an example of (49) ____________ signalling, and the compounds used are referred to as (50) ____________. (51) ____________ signalling involves physical contact between individuals. Finally, some fishes use (52) ____________ because they live in murky waters that make other forms of signalling ineffective.

Choice

Choose one or more of the communication channels that would be most appropriate to the situation described.

A. Acoustical B. Chemical C. Electrical D. Tactile E. Visual

53. ____ Animals that live in a cluttered environment

54. ____ Animals that live in dimly lit areas

55. ____ Animals that live in open areas

56. ____ Animals that produce a pheromone to attract a mate

57. ____ Animals that sing to warn intruders entering their territory

58. ____ Animals that send out a mild charge as a threat to invaders

40.8 Space [pp. 982–984]

Section Review (Fill-in-the-Blanks)

No general principles for how animals select a (59) ____________ have been determined. One of the simplest behavioural responses to environmental cues that some animals use to locate suitable habitat is (60)____________, which involves a change in the rate of movement or frequency of turning. Another simple response, but one that involves direct orientation toward or away from a stimulus, is called (61) ____________. Often, after finding a suitable habitat, animals will defend an area from other individuals of the same species, giving it exclusive use of resources in that area, a behaviour known as (62) ____________. Although protecting a territory can be costly, primarily due to the (63) ____________ that must be spent to patrol it, there are (64) ____________, such as having (65) ____________ access to resources and increasing the likelihood of (66) ____________ a female.

Choice

Choose the type of behaviour appropriate to each description.

A. Kinesis B. Taxis C. Territoriality

67. ____ Male yellow-headed blackbirds defending an area against intrusion by other yellow-headed blackbirds

68. ____ Moths flying directly toward a light

69. ____ Wood lice increasing their rate of locomotion in dry areas

70. ____ Cockroaches running to hide in spaces in the dark

40.9 Migration [pp. 984–987]

Section Review (Fill-in-the-Blanks)

The large-scale, predictable movement of animals that occurs on a seasonal basis is (71) ____________. There are a number of mechanisms that enable animals to find their way during these travels. One is (72) ____________, in which animals use landmarks to guide them. Another method is (73) ____________, in which animals use some environmental cue, such as the sun or stars, to enable them to travel in a specific direction and often for a specific period of time. The most complex type of wayfinding is (74) ____________, which requires both a compass and a (75) "____________ map." The costs of (76) ____________ and energy incurred by migrating are offset by benefits such as leaving a northern winter, which requires an (77) ____________ in metabolic rate and (78) ____________ in food supply, and returning when days are (79) ____________, which provides more time for breeding.

Matching

Match each of the animals with the cues it uses for migration. You may choose more than one cue per animal.

80.	____	Digger wasps	A.	Landmarks
81.	____	Homing pigeons	B.	Position of the stars in the sky
82.	____	Indigo buntings	C.	Position of the sun in the sky
83.	____	Salmon	D.	Olfaction (smell)
84.	____	Monarch butterflies		
85.	____	Grey whales		

40.10 Mates as Resources [p. 988]

40.11 Sexual Selection [pp. 988–990]

Section Review (Fill-in-the-Blanks)

A special type of natural selection is called (86) ___________ selection and involves both (87) ___________ for a mate, usually among rival males, and (88) ___________ a mate, which tends to be performed by the female since she often makes the greater (89) ___________ in providing parental care. These activities lead to physical and behavioural differences between males and females, called sexual (90) ___________. Elaborate ornamental structures often evolve in the male, which they display to either attract the attention of females during a (91) ___________ ritual or, in some cases, to (92) ___________ with other males for a female. Sometimes males of a species gather for competition in a common area known as a(n) (93) ___________. Females assess the quality of males and select one for mating. Successful males may mate with numerous females, a mating system known as (94) ___________. Other mating systems include a single female having multiple male mates, called (95) ___________ (rare), a single male pairing with a single female, forming a (96) ___________ relationship, and (97) ___________, where both sexes have multiple mates without any lasting pair bond forming between them.

Matching

Match the mating system with the correct life history pattern.

98.	____	Monogamy	A.	Solitary species with little or no real social structure. Individuals encounter each other randomly.
99.	____	Polyandry	B.	Species in which females are physically drained after egg laying, but resources are temporarily abundant, allowing them to recover quickly and mate with a second male while the first male incubates the eggs
100.	____	Polygyny	C.	Species in which resource availability is so poor that both parents are required to successfully rear offspring
101.	____	Promiscuity	D.	Territorial species in which the quality of territory is highly variable. High-quality territories provide abundant resources, so biparental care is unnecessary.

40.12 Social Behaviour [pp. 990–992]

40.13 Kin Selection and Altruism [pp. 992–993]

40.14 Eusocial Animals [pp. 993–994]

40.15 Human Social Behaviour [pp. 994–997]

Section Review (Fill-in-the-Blanks)

The interactions of animals with members of their own species describe their (102) ____________. Living in groups has costs as well as benefits associated with it. Often groups form a pecking order known as a(n) (103) ____________. Subordinate individuals may stay in the group despite their low status because survival is difficult for solitary individuals. Many social species exhibit (104) ____________, behaviours that help others while putting the performer at somewhat increased risk. This seems to contradict the concept of selfishness inherent in Darwin's view of natural selection, but William Hamilton demonstrated how, by helping relatives, these behaviours could evolve. His theory of (105) ____________ shows how helping a relative can aid in passing on the family's genes into the next generation. (106)____________, the unique genetics of sex determination in ants, bees, and wasps, lends credence to Hamilton's hypothesis since members of any of these (107) ____________ groups show high relatedness and, not surprisingly, are highly cooperative. Under certain circumstances, altruistic behaviour can evolve among nonrelated individuals if there is a good chance that the roles of helper and helped may be reversed in the future. This form of cooperative behaviour is known as (108) ____________.

Short Answer

Calculate the degree of relatedness, r, for the following pairs of individuals.

109. Siblings

____

____

110. Parent and offspring

____

____

111. Uncle and nephew

____

____

112. First cousins

____

____

Matching

Match each of the animals listed below with their social behaviours.

113.	____	Dolphin	A.	Dominants in the group must defend their status
114.	____	Bees	B.	Large brain size provides high social cognitive skill level
115.	____	Grey wolves	C.	A competitive altruist
116.	____	Humans	D.	Sex is determined by haplodiploidy in this eusocial animal

SELF-TEST

1. Peter Marler demonstrated that in the white-crowned sparrow, singing is primarily _____ and song is primarily ________. [pp. 968–972]

   a. Instinctive; instinctive
   b. Instinctive; learned
   c. Learned; instinctive
   d. Learned; learned

2. Birds that lay their eggs in the nests of other species and whose young are raised by the "host" species are called ________. [pp. 968–972]

   a. Brood parasites
   b. Ectoparasites
   c. Endoparasites
   d. Parental hijackers

3. The cues that trigger fixed action patterns are called ________. [pp. 968–972]

   a. Conditioned stimuli
   b. Instinctive stimuli
   c. Sign stimuli
   d. Unconditioned stimulus

4. Stevan Arnold's experiments on newborn garter snakes and banana slugs demonstrated that ________. [pp. 968–972]

   a. Banana slugs exhibit stereotyped avoidance behaviour when they detect a garter snake in their vicinity
   b. Food preferences in garter snakes have a genetic component
   c. Food preferences in garter snakes are learned
   d. The aposematic coloration of banana slugs deters garter snakes from striking

5. A rat presses a bar in its cage and receives a food pellet. Which of the following is true? [pp. 968–972]

   a. Insight learning has occurred
   b. Pressing the bar in exchange for the food pellet is an example of classical conditioning
   c. Pressing the bar is the operant; the food pellet is the reinforcement
   d. Pressing the bar is the reinforcement; the food pellet is the operant

6. A zebra finch that has learned to ignore the songs of neighbours that share his territory is displaying the learning behaviour called __________. [pp. 968–972]

   a. Classical conditioning
   b. Operant conditioning
   c. Habituation
   d. Insight learning

7. Only male zebra finches can sing because __________. [pp. 972–979]

   a. Only males produce estrogen when their brains are developing
   b. Only females produce octopamine when they are adults
   c. Only males have vocal cords
   d. Only females produce increasing concentrations of juvenile hormone as they age

8. Changes in the types of behaviours honeybees perform as they get older are caused by changes in ________. [pp. 972–979]

   a. Diet
   b. Hormone levels
   c. Photoperiod
   d. Savannah

9. The evasive behaviour of crickets when they hear the ultrasonic sounds of a bat is an example of ________. [pp. 972–979]

   a. Habituation to bat vocalizations
   b. Hardwiring between the sensory and motor systems
   c. Hormonal control of behaviour
   d. Imprinting on bat vocalizations during a critical period

10. Star-nosed moles live in dark tunnels, and vision is not a useful sensory modality. Which sense is most important to star-nosed moles? [pp. 972–979]

a. Hearing
b. Smell
c. Taste
d. Touch

11. What produces the acoustical signals that herrings use to communicate with each other? [pp. 979–982]

a. Striped colouring
b. Movement of fins
c. Little bursts of gas (f*rts)
d. Release of pheromones

12. You place several planaria in a pan of water with one side covered so that it is dark while the other side is open and well lit. You notice that their rate of movement slows on the dark side. What phenomenon have you observed? [pp. 982–984]

a. Compass orientation
b. Kinesis
c. Operant conditioning
d. Taxis

13. Food preferences of many vertebrates ________. [pp. 982–984]

a. Are innate
b. Are learned
c. Have both innate and learned components
d. Animals do not show food preferences; diet is determined by the abundance of food only

14. Which resource would likely justify territorial defence? [pp. 982–984]

a. An abundant, localized food supply
b. Food distributed evenly over a large area
c. Oxygen
d. Prey that move over great distances, for example, a school of fish

15. Which of the following is <u>not</u> an environmental cue used by migrating animals to reach their destination? [pp. 984–987]

a. Odours
b. Position of the stars in the sky
c. Position of the sun in the sky
d. Sounds

16. The most favourable reproductive strategies are often different for males and females of a species due to differences in ________. [pp. 988–990]

   a. Body temperature
   b. Emotional makeup
   c. Parental investment
   d. Sex ratio

17. Special ornaments or structures of males that increase their likelihood of attracting females are probably the result of __________. [pp. 988–990]

   a. Artificial selection
   b. Disruptive selection
   c. Sexual selection
   d. Stabilizing selection

18. The mating system in which one female has multiple male mates is ________. [pp. 988–990]

   a. Monogamy
   b. Polyandry
   c. Polygyny
   d. Promiscuity

19. The behaviour of which animals supports the idea of reciprocal altruism? [pp. 990–997]

   a. Honeybees
   b. Musk oxen
   c. Naked mole rats
   d. Dolphins

20. Wilson and Daly found that criminal aggression by an adult toward a child was most common between ________. [pp. 990–997]

   a. Adult and juvenile siblings
   b. Fathers and daughters
   c. Fathers and sons
   d. Stepparents and stepchildren

INTERACTIVE EXERCISES

1. Describe the behaviour an insectivorous bat displays to find its prey and a black field cricket uses to avoid being the prey.
2. Explain why a wandering raven uses the acoustical behaviour of yelling when it finds a carcass (instead of eating it in isolation).
3. Is a male lion who has taken over a pride altruistic? Explain your answer.

OPTIONAL ACTIVITIES

1. Due to inbreeding, members of naked mole rat colonies are more closely related, on average, than siblings. Would individuals enjoy a greater fitness benefit by mating with others in a colony or by assisting a relative? What plausible explanations can you come up with to explain the presence of a sterile "caste" among naked mole rats?
2. Further support for the ability of birds to navigate by the stars might be provided if one could manipulate the position of the stars in the sky. One might be able to "fool" the birds into taking an inappropriate but predictable direction. Can you think of any way Emlen's experiments on stellar navigation could be modified to provide the desired data?
3. Some people have questioned the importance of investing time and money studying the genetics of fruit flies and other nonhuman species. Given the apparent similarity of the *disheveled* gene in *Drosophila* and the gene associated with obsessive–compulsive disorder, Huntington's disease, and schizophrenia (bipolar disorder) in humans, is this argument justified?

41 Plant and Animal Nutrition

CHAPTER HIGHLIGHTS

- Plants and animals require a supply of nutrients; plants acquire main nutrients from photosynthesis and essential minerals from soil or mycorrhizas, whereas animals acquire essential minerals from plants or other animals.
- The size of the particles and the amount of organic matter, air, water, minerals, and living flora in the soil are critical to plant nutrient availability and plant success.
- Nitrogen, phosphorus, and potassium tend to be the most limiting minerals; only nitrates and ammonium forms of nitrogen are accessible by plants.
- Legume plants form a mutualistic relationship with nitrogen-fixing bacteria, forming root nodules that supply ammonium, an accessible nitrogen product for plants.
- Various types of feeding and digestive processes in the animal kingdom are described.
- The digestion of food through a human digestive tract is outlined, indicating the structures and processes involved in mammalian digestion.

STUDY STRATEGIES AND LEARNING OUTCOMES

- Remember that the goals of this chapter are to understand
  - What plants and animals need
  - How plants and animals obtain their nutrients from their environment
- First, focus on the mechanisms used by plants to acquire essential nutrients and then shift your focus to animals and the various feeding mechanisms they use.
- Next, study the structures and processes used in the digestive process in humans, identifying the location and reactions involved in the breakdown of macromolecules and their absorption into the body.
- Make comparisons between the digestive system used by mammals and the various types of feeding, specializations, and nutritional requirements among other animals.

By the end of this chapter, you should be able to

- Differentiate between macronutrients and micronutrients in plants and animals, identifying those nutrients that tend to limit plant success

- Describe the characteristics and conditions of soil that affect plant nutrient uptake, including symbiotic relationships that increase uptake of specific nutrients
- Compare digestive systems of various animals, describing the structures, functions, and reactions occurring in the mammalian digestive system
- Describe how neuron networks and hormones regulate digestion in animals

TOPIC MAP

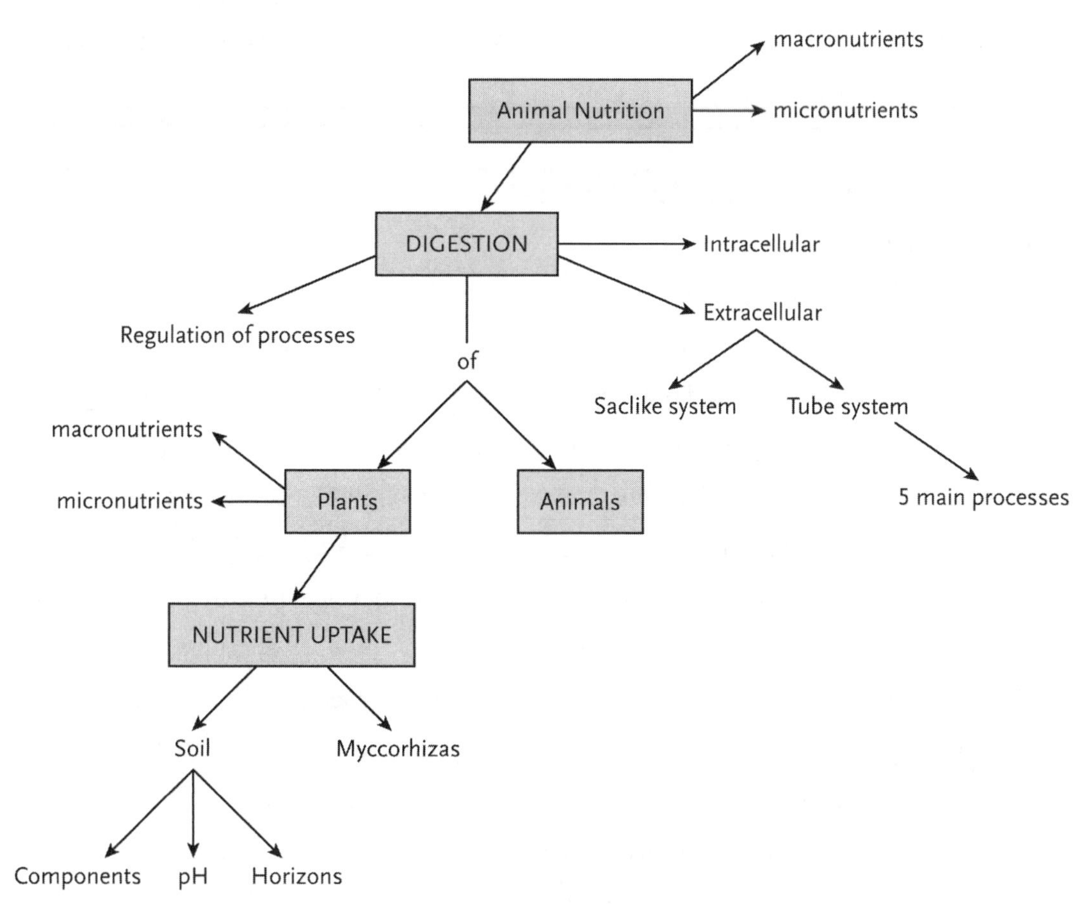

INTERACTIVE EXERCISES

Why It Matters [pp. 1001–1002]

Section Review (Fill-in-the-Blanks)

One of the digestive enzymes, (1) _____ _____, functions in the oral breakdown of starch. This enzyme is produced in the (2) _____ glands and secreted into the oral cavity, where, during the act of (3) _____, it functions to begin breaking down starch into the disaccharide (4) _____. This action on high-starch foods (5) _____ the level of blood glucose immediately. The (6) _____ gene regulates the amount of salivary amylase in saliva. Individuals with a high-starch diet have a (7) _____ number of copies of the gene, and therefore produce (8) _____ salivary amylase, than those with a low-starch diet.

41.1 Nutrition: Essential Materials [pp. 1002–1010]

41.2 Soil [pp. 1010–1012]

Section Review (Fill-in-the-Blanks)

Plants are more than (9) _____% water, with the remainder composing the plant's (10) _____ _____, which is primarily composed of carbon, (11) _____, and oxygen elements. (12) _____ is a technique whereby plants are grown in solutions containing measured amounts of (13) _____, nutrients normally found in the soil. Plants receive most of their nutrients from the process of (14) _____, although many are taken up by the roots (15) _____ (no energy required) with the uptake of water from the soil. Since animals cannot photosynthesize, they acquire energy and nutrients from other living organisms: from plants if they are (16) ___________ consumers and from animals if they are (17) ___________ consumers (or higher level consumers).

(18) ___________ elements are necessary for plants and animals to grow and reproduce. Out of 17 such elements, 9 are required in higher amounts and so are called (19) ___________, while the remainder are required in relatively small amounts and are known as (20) ___________. Deficiency of one or more of these elements in plants can lead to developmental or physiological problems, including (21) ___________ growth and/or (22) ___________of the leaves. In animals, chronic malnutrition or (23) ___________ often results in weight (24) ___________ since animals must use nutrients in their own bodies as fuel. Animals have the added requirement for (25) ___________, organic molecules required in small quantities that they cannot synthesize for themselves. Often these molecules function as coenzymes, assisting (26) ___________ in metabolic activities. In humans, there are (27) ___________ known vitamins required to function normally, some of which are (28) ___________ since they are

hydrophilic and others that are (29) ___________ since they are hydrophobic.

Soils are primarily composed of five components: (a) mineral particles, which determine size and the number of (30) ___________ ___________, characteristics that affect water and nutrient uptake; (b) decomposing dead plants and animals, called the (31) ___________, which forms the organic component of the soil and helps in retaining (32) ___________ in the ground; (c) living organisms, some of which burrow in the soil, thereby (33) ___________ it, and others, such as saprophytic bacteria and fungi, which (34) ___________ organic matter; (d) air, which is located in the pores and provides a source of (35) ___________ to the plant roots; and (e) water, which is also located in pores and forms the (36) ___________ ___________ when mixed with dissolved substances. Minerals can be positively charged, called (37) ___________, or negatively charged, called (38) ___________. Positively charged minerals are less available to plant roots since they tend to bind tightly to (39) ___________ charged soil particles but can be released by the (40) ___________ ___________ mechanism.

Matching

Match each of the following terms with its correct definition.

41.	____	Hydroponics	A.	Layer of soil with distinct texture and composition
42.	____	Essential elements	B.	Cleaning of contaminated soils using plants
43.	____	Macronutrients	C.	Yellowing of the plant tissue due to lack of chlorophyll
44.	____	Micronutrients	D.	Elements that are necessary for normal growth and reproduction of the plants
45.	____	Chlorosis	E.	Elements that are required in very small amounts
46.	____	Leaching	F.	Primarily eats plant material
47.	____	Cation exchange	G.	Primarily eats other animals
48.	____	Horizon	H.	Excessive intake of specific nutrients
49.	____	Primary consumer	I.	Drainage of nutrients from soil by excess water
50.	____	Overnutrition	J.	Growing plants in a solution with known solutes
51.	____	Phytoremediation	K.	Elements that are required in large amounts
52.	____	Secondary consumer	L.	Swap of one positively charged ion (usually H^+) for another

Choice

Choose the element(s) that can be found in each of the listed macromolecules. Elements can be used more than once.

A. Carbon B. Hydrogen C. Nitrogen D. Phosphorus

E. Sulphur F. Oxygen

53. ____ Carbohydrate

54. ____ Protein

55. ____ Lipid

56. ____ Nucleic acid

Complete the Table

57. Soil is a complex mix of important ingredients. Complete the following table by giving the role of each component of the soil.

Soil ingredients	*Value in soil*
Particles	
Humus	
Living organisms	
Minerals	

41.3 Obtaining and Absorbing Nutrients [pp. 1012–1017]

Section Review (Fill-in-the-Blanks)

Plants and animals are adapted to obtain the nutrients required for life. In plants, roots compose between (58) ____________% of plant dry weight, functioning to absorb water and minerals from soil. The greatest absorption by plant roots occurs at root (59) ____________, although the presence of root (60) ____________ increases the surface area of roots, thereby increasing plant water and ion uptake dramatically. Furthermore, cell membranes of some root cells contain (61) ____________ transport proteins, providing channels for the entrance of particular ions, such as potassium ions, into cells. Potassium, (62) ____________, and nitrogen tend to be the most limiting ions to plants. Symbiotic associations between plants and fungi, called (63) ____________, assist in nutrient uptake since the fungi form a network of (64) ____________ filaments, which increase surface area for absorption of water and ions. Nitrogen uptake is often limiting for plants since (65) ___________% of it is in its gaseous form and plants lack the (66) ___________ to convert it into nitrate and ammonium ions, forms that it can absorb. Various types of (67) ___________ are found in the soil that can convert different forms of nitrogen into (68) ___________ and/or ammonium ions. Once plants absorb these ions, they are all converted into (69) ___________, which is then used to make (70) ___________ acids, or other organic molecules.

Some (71) ___________-fixing bacteria live in the soil, but others form a (72) ___________ association with the roots of legume plants, forming swellings in the plant roots called root (73) ___________. These bacteria enter roots through root hairs, attracted into roots by certain chemicals belonging to the (74) ___________. These bacteria then release proteins expressed by their (75) ____________ gene, some of which stimulate root cells to multiply and form root nodules and others that produce the protein (76) ___________, which provides necessary oxygen to the bacteria. Eventually, the enclosed bacteria in a root nodule become larger and immobile, known as (77) ___________, and they produce the enzyme (78) ____________, which converts nitrogen to ammonium.

Animals obtain their nutrients in different ways, depending on their (79) ___________ to their environments. There are (80) ___________ basic groupings of how animals feed, each dependent on the physical state of the food: (81) ___________ feeders have mouthparts that allow animals to ingest foods that are liquid, (82) ___________ feeders filter fluid entering body such that trapped small organisms are ingested, (83) ___________ feeders take in solid material but digest only the edible particles in the sediment, and (84) ___________ feeders ingest relatively large pieces of food, which often requires chewing.

Matching

Match each of the following types of bacteria with their role in providing nitrogen to plants.

85.	____	Nitrogen-fixing bacteria	A.	Bacteria that are able to decompose dead organic matter in the soil to ammonium
86.	____	Ammonifying bacteria	B.	Bacteria that are able to convert atmospheric nitrogen to ammonium
87.	____	Nitrifying bacteria	C.	Bacteria that are able to convert ammonium to nitrates for plants to absorb
88.	____	Bacteroids	D.	Enlarged, immobile bacteria in the legume root nodules

Match each of the following terms with its correct definition.

89. ____ Root nodule

90. ____ Flavenoids

91. ____ Nod gene

92. ____ Leghemoglobin

93. ____ Nitrogenase

A. Bacterial gene that makes proteins to help nitrogen-fixing bacteria enter plant roots

B. Enzyme made by the bacteroids to convert atmospheric nitrogen to ammonium

C. A reddish, heme- and iron-containing protein that is produced by the root nodule cells to increase oxygen supply for the bacteria

D. Chemicals released by plant roots to attract nitrogen-fixing bacteria

E. Swellings in the roots where nitrogen-fixing bacteria reside

Choice

Choose the type of feeding used by each of the animals listed.

94. ____ Alligator

95. ____ Oyster

96. ____ Aphid

97. ____ Earthworm

98. ____ Nectar-feeding butterfly

99. ____ Cougar

A. Suspension feeder

B. Deposit feeder

C. Fluid feeder

D. Bulk feeder

41.4 Digestive Processes in Animals [pp. 1017–1019]

Section Review (Fill-in-the-Blanks)

Digestion is the (100) ____________ of larger particles into subunits that can be (101) ____________ into body fluids and cells. Many of the chemical bonds of larger particles are broken by (102) ____________ ____________. These enzymatic reactions are often very (103) ____________, with a particular type of enzyme breaking down a particular type of macromolecule. These digestive processes can occur within body cells, (104) ____________ digestion, or outside of body cells, (105) ____________ digestion. In summary, digestion requires (106) ____________ main processes: (107) ____________ ____________, which involves tearing large food particles into smaller particles; (108) ____________ of enzymes and other substances, which is required for the third process; (109) ____________ ____________, which involves the further breakdown of food using enzymatic reactions; (110) ___________ so that molecular subunits enter into body fluids and cells; and (111) ___________ to ensure that undigested food leaves the body.

Choice

For each of the following statements, choose the most appropriate enzyme from the list below.

A. Lipases B. Amylases C. Nucleases D. Proteases

112.____ Catalyzes the bonds within carbohydrates, such as starch

113.____ Catalyzes the bonds between amino acids

114.____ Catalyzes the bonds between glycerol and fatty acids

115.____ Catalyzes the bonds within DNA or RNA

Short Answer

116. Explain how endocytosis and exocytosis are involved in intracellular digestion.

____

____

117. Explain how extracellular digestion is advantageous with respect to food source.

____

____

Choice

Choose the appropriate stage of digestion that occurs in each of the following anatomical locations. More than one stage can occur in the various locations.

118.	____	Long intestine of an annelid	A.	Mechanical processing
119.	____	Anus	B.	Secretion of digestive enzymes
120.	____	Gastric ceca of insects	C.	Enzymatic hydrolysis
121.	____	Gizzard	D.	Absorption
122.	____	Intestine	E.	Elimination
123.	____	Liver of bird		
124.	____	Proventriculus of bird		

41.5 Digestion in Mammals [pp. 1019–1027]

Section Review (Fill-in-the-Blanks)

The digestive process in mammals is very similar to other animals, also consisting of the (125) ____________ steps of digestion. The mammalian digestive tract includes (in order) the mouth, pharynx, (126) ____________, stomach, small intestine, large intestine, (127) ____________, and anus. The (128) ____________ and endocrine systems control digestion in all mammals; however, digestive systems in herbivores are (129) ____________ than in carnivores, due to the added digestion of (130) ____________ in plant cell walls.

Matching

Match each of the following structures/components of the digestive system with its correct description.

131.	____	Mucosa	A.	Found in the stomach; fibres are oriented in a diagonal direction
132.	____	Submucosa	B.	Wave of contraction from the circular and longitudinal muscles
133.	____	Muscularis	C.	Outermost layer of the gut; secretes a thin, slippery, lubricating fluid
134.	____	Serosa	D.	Smooth muscle rings that act as valves to ensure one-way flow
135.	____	Sphincter	E.	Muscle oriented perpendicular to the axis of the gut
136.	____	Peristalsis	F.	Innermost layer of the gut; absorption occurs in this layer
137.	____	Circular muscle	G.	Muscle oriented parallel to the axis of the gut
138.	____	Longitudinal muscle	H.	Major muscle coat of the gut—composed of 2–3 layers
139.	____	Oblique muscle	I.	Thick, connective tissue layer of the gut

Consecutive Order

Put the following structures or functions in the proper order, using numbers from 1 to 15. Note: Number 1 has been provided.

140. ____ Gastroesophageal sphincter

141. ____ Esophagus

142. ____ Anal sphincter

143. ____ Large intestine

144. ____ Ileocecal sphincter

145. ____ Stomach

146. ____ Food passes the epiglottis

147. ____ Secretions of pancreas and liver added

148. ____ Gastroesophageal sphincter

149. ____ Pyloric sphincter

150. ____ Rectum

151. 1 Mechanical processing in mouth

152. ____ Secretions of HCl and pepsinogen added

153. ____ Secretions from salivary glands added

154. ____ Anus

155. ____ Chyme enters small intestine

156. ____ Hydrolysis and absorption of nutrients completed

Matching

Match each of the secretions with the location in which it is produced.

157.	____	Pepsinogen	A.	Exocrine cells of pancreas
158.	____	Bile salts	B.	Parietal cells of gastric gland
159.	____	Aminopeptidase, dipeptidases, and mucus	C.	Salivary glands
160.	____	Alkaline mucus for stomach	D.	Chief cells of gastric gland
161.	____	HC1	E.	Microvilli
162.	____	Enzymes, such as trypsin, and bicarbonate ions	F.	Liver cells
163.	____	Amylase, lysozyme, mucus, and bicarbonate ions	G.	Mucus cells of gastric gland

41.6 Regulation of the Digestive Processes [pp. 1027–1030]

Matching

Match each of the hormones with its correct action during digestion.

164.	____	Gastrin	A.	Increases insulin release from the pancreas
165.	____	Cholecystokinin	B.	Inhibits additional HCl release from gastric glands and stimulates HCO_3^- release
166.	____	Secretin	C.	Stimulates secretion of HCl and pepsinogen and gastrointestinal tract motility
167.	____	Leptin	D.	Under control of hypothalamic neurons; thought to inhibit appetite
168.	____	Alpha-MSH	E.	Presence of fat stimulates release; inhibits gastric activity and increases pancreatic enzyme release
169.	___	GIP	F.	Increases when fat deposition increases, reduces appetite

True/False

If the statement is true, write a "T" in the blank. If the statement is false, make it correct by changing the underlined word(s) and writing the correct word(s) in the answer blank.

170.____________ Regulation and control of the digestive process are under local control by the voluntary nervous system.

171.____________ The presence of food in the gastrointestinal tract initiates secretion of mucus and digestive enzymes as well as motility.

172.____________ The neurons that make up the appetite centre are located in the mucosa of the stomach.

173.____________ A(An) decrease in the blood level of leptin will stimulate or increase appetite.

41.7 Variations in Obtaining Nutrients [pp. 1030–1036]

Section Review (Fill-in-the-Blanks)

Different plants and animals have developed adaptations and specializations to assist in obtaining (174) ___________. Examples include plants that are (175) ___________, stealing nutrients from the tissues of other plants, to their detriment; plants that are (176) ___________, growing on other plants but not parasitizing them; animals that are ambushers, (177) ___________ until prey is observed and then attacking it; and animals that use (178) ___________, something that attracts prey to the predator.

Matching

Match each of the following organisms with its specialization for obtaining nutrients.

179.	____	*M. termitophaga* flatworm	A.	Uses bioluminescent bacteria to lure prey
180.	____	Coralroot	B.	Uses roots to access vascular tissues and rob host plant of its nutrients
181.	____	Candiru	C.	Lies in wait on mound, uses eyes to detect movement, and then uses head to capture prey
182.	____	Dodders	D.	Uses its wide mouth to ingest food and its vertebrae as an egg-cracker during digestion
183.	____	Anglerfish	E.	Obtains carbon from other plants using shared mycorrhizal fungi
184.	____	*D. scabra* snake	F.	Attaches to gills of larger fish and feeds on its blood

SELF-TEST

1. Hydroponics is a technique that was developed to __________. [pp. 1002–1010]

   a. Study the role of different minerals in growth of a plant
   b. Replace the agriculture industry
   c. Grow algae in water
   d. Multiply nitrogen-fixing bacteria

2. Some minerals act as cofactors for the plant enzymes. [pp. 1002–1010]

   a. True
   b. False

3. Macronutrients are more important than micronutrients. [pp. 1002–1010]

   a. True
   b. False

4. Which of these macronutrients is not found in proteins? [pp. 1002–1010]

   a. C
   b. N
   c. O
   d. K

5. What fundamental process(es) associated with life are the products of digestion used for? [pp. 1002–1010]

   a. ATP production
   b. Synthesis of biological molecules
   c. Reproduction
   d. Both (a) and (b)

6. Joe has been on antibiotics for 6 weeks due to an infection and is now finding he bleeds easily. Long-term antibiotic use has __________. [pp. 1002–1010]

   a. Decreased the bacteria that produce vitamin D
   b. Decreased the bacteria that produce vitamin K
   c. Increased the level of water-soluble vitamins
   d. Nothing to do with the bleeding problems

7. Which of the following adds organic chemicals to the soil? [pp. 1010–1012]

   a. Soil particles
   b. Humus
   c. Live organisms
   d. Minerals

8. Soil pH does not affect the absorption of minerals by plants. [pp. 1010–1012]

   a. True
   b. False

9. Plants absorb most of the nitrogen in the form of __________. [pp. 1012–1017]

   a. Nitrogen gas
   b. Ammonium
   c. Nitrate
   d. Amino acids

10. Plants use most of the nitrogen in the form of __________. [pp. 1012–1017]

   a. Nitrogen gas
   b. Ammonium
   c. Nitrate
   d. Amino acids

11. A fertilizer bag is labelled 15-30-25. Which minerals, in the correct order, do these numbers represent? [pp. 1012–1017]

   a. Carbon–hydrogen–oxygen
   b. Carbon–nitrogen–oxygen
   c. Nitrogen–phosphorus–potassium
   d. Nitrogen–oxygen–phosphorus

12. What do nitrifying bacteria do? [pp. 1012–1017]

    a. Convert atmospheric nitrogen into ammonium
    b. Decompose organic chemicals to make ammonium
    c. Convert ammonium to nitrates
    d. Convert nitrates into ammonium

13. Bacteroids are __________. [pp. 1012–1017]

    a. Bacteria that live in the soil
    b. Bacteria that live in root nodules
    c. Bacteria that live on dead organic matter
    d. Bacteria that infect plants and parasitize them

14. What is the result of the expression of *nod* genes? [pp. 1012–1017]

    a. Allows plants to access more of the nitrogen in the soil
    b. Stimulates plant cells to make leghemoglobin
    c. Makes the enzyme for converting nitrogen to nitrates

15. An animal with teeth and claws would most likely be a _____ feeder. [pp. 1012–1017]

    a. Fluid
    b. Suspension
    c. Deposit
    d. Bulk

16. Animals that intake food and remove wastes through the same opening have a __________. [pp. 1017–1019]

    a. Saclike digestive system
    b. Tubelike digestive system
    c. Filtering digestive system
    d. Gizzard

17. Which of the following structures are not typically involved with mechanical processing? [pp. 1017–1019]

    a. Gizzard
    b. Stomach
    c. Pancreas
    d. Crop

18. Which of the following are incorrectly matched? [pp. 1019–1027]

    a. Giraffe—long intestinal tract
    b. Tiger—short intestinal tract
    c. Horse—long intestinal tract
    d. Kangaroo—short intestinal tract

19. Where are villi located in mammals? [pp. 1019–1027]

   a. Wall of esophagus
   b. Wall of stomach
   c. Wall of small intestine
   d. Wall of large intestine

20. Essential amino acids and vitamins must be __________. [pp. 1019–1027]

   a. Synthesized before any of the others
   b. Obtained in the diet
   c. Metabolized before others can be synthesized
   d. Obtained from a symbiotic relationship

21. If the pyloric sphincter were blocked, chyme could not move from the __________. [pp. 1019–1027]

   a. Esophagus to the stomach
   b. Large intestine to the rectum
   c. Stomach to the small intestine
   d. Oral cavity to the esophagus

22. One of the hormones produced by the intestine, secretin, was inactive. What effect would this have on the pH of the digestive secretions? [pp. 1027–1030]

   a. Secretions would be more acidic
   b. Secretions would be more basic
   c. Secretions would be neutralized
   d. Secretin has no effect on the pH of digestive juices

23. Orchids that live on the branches of trees are __________. [pp. 1030–1036]

   a. Parasites
   b. Omnivores
   c. Carnivores
   d. Epiphytes

INTERACTIVE EXERCISES

1. Describe differences between clay and sandy soils in terms of their limitations to plant nutrient uptake.

2. Explain the role of *nod* genes in the formation of the root nodule.

3. Why must vegetarians be careful about accessing eight of the essential amino acids?

4. For each of the four macromolecules, identify the enzymes that break down molecules in the small intestine, indicating the molecules that are being digested and the resulting molecules produced.

OPTIONAL ACTIVITIES

1. You eat French fries for lunch. Describe the digestion of one of the macromolecules in a French fry, from its ingestion to its arrival in the circulatory system.

2. Development of a genetically engineered *Arabidopsis thaliana* plant for phytoremediation of contaminated soils appears to be a positive step for helping to clean up the environment. Can you think of any drawbacks to this idea?

42 Gas Exchange: The Respiratory System

CHAPTER HIGHLIGHTS

- All animals exhibit physiological respiration by exchanging O_2 and CO_2 with the environment.
- Gases move by simple diffusion between the respiratory medium of either air or water and the animal across the respiratory surface. The partial pressure of gases across a respiratory surface is the key factor in determining the direction oxygen and carbon dioxide will move during gas exchange. A gas will move down a partial pressure gradient, from a region of high partial pressure to a region of low pressure.
- The rate at which a gas will diffuse across a respiratory surface is dependent on a number of factors: the partial pressure difference of gases, surface area, the thickness of the membrane, and the diffusion coefficient of the gas.
- Adaptations for gas exchange have evolved to allow organisms to exchange gases in water and/or terrestrial habitats. In small animals, the body serves as the respiratory surface. In larger animals, specialized structures such as the tracheal system of insects are responsible for gas exchange.
- Haemoglobin is a respiratory pigment that greatly increases the oxygen-carrying capacity of the blood. In the blood, carbon dioxide is primarily transported in the form of bicarbonate in the plasma but may also be transported as bicarbonate in the red blood cells, bound to haemoglobin, or dissolved as CO_2.

STUDY STRATEGIES AND LEARNING OUTCOMES

- The concepts of this chapter build on elements of diffusion developed earlier. You should briefly review these concepts to make sure that you understand them.
- This chapter contains a lot of new information and new terminology. DO NOT try to go through it all in one sitting. Take one section at a time. Start by skimming the section, writing down the boldface terms, and studying the figures and then go back and read the text, followed by working through the companion section(s) in the study guide. Repeat this process for each section of the text.

- Draw diagrams of the various respiratory systems, carefully labelling each component and noting its function.
- Try to relate Fick's equation of diffusion to make the link between gas exchange and the design of different respiratory systems for breathing in different types of environments.

By the end of this chapter, you should be able to

- Describe the percent composition of air
- Calculate the partial pressure of gases under different atmospheric pressures
- Use Fick's equation of diffusion to explain the factors that influence the rate of gas diffusion across a respiratory surface
- Use Fick's equation of diffusion to explain adaptations for gas exchange in the water and terrestrial habitats
- Discuss the advantages of ventilation and perfusion for gas exchange
- List and explain the advantages and disadvantages of water and air as respiratory media
- Draw and label diagrams of the different respiratory systems: gills, tracheal system of insects, bird lungs, and mammalian lungs
- Compare gas exchange in fish, insects, birds, and mammals
- Draw and label a diagram of a haemoglobin–oxygen dissociation curve and use this diagram to explain oxygen exchange and transport in the blood and the factors that influence oxygen transport
- Explain the mechanisms of CO_2 transport and excretion

TOPIC MAP

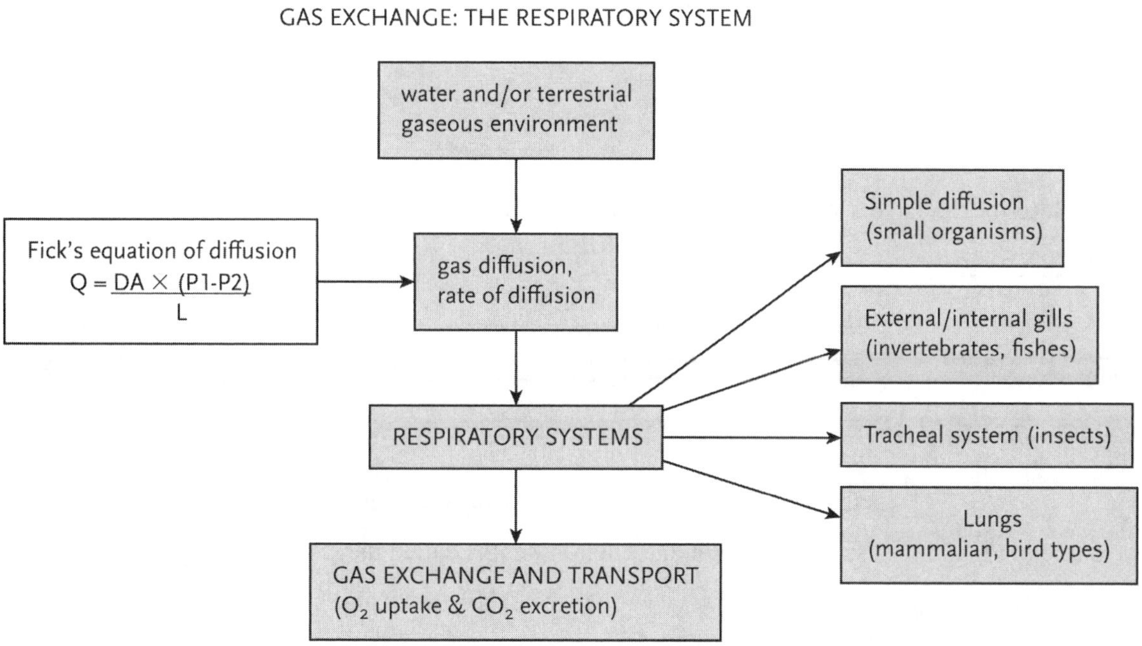

GAS EXCHANGE: THE RESPIRATORY SYSTEM

water and/or terrestrial gaseous environment

Fick's equation of diffusion
$Q = \frac{DA \times (P1-P2)}{L}$

gas diffusion, rate of diffusion

RESPIRATORY SYSTEMS

Simple diffusion (small organisms)

External/internal gills (invertebrates, fishes)

Tracheal system (insects)

Lungs (mammalian, bird types)

GAS EXCHANGE AND TRANSPORT (O_2 uptake & CO_2 excretion)

INTERACTIVE EXERCISES

Why It Matters [pp. 1039–1040]

Section Review (Fill-in-the-Blanks)

All organisms with active metabolism need to exchange gases with their surroundings. (1) ____________ is the process that allows organisms to exchange gases with their environment.

42.1 General Principles of Gas Exchange [pp. 1040–1043]

Section Review (Fill-in-the-Blanks)

Describe the gas composition of air: (2) ____________% N_2; (3) ____________% O_2; and (4) ____________% CO_2.

If atmospheric pressure is the sum of the pressures of all of the gases in a mixture, what would be the partial pressures of N_2, O_2, and CO_2 at the following atmospheric pressures?

P_{atm} = 760 mm Hg (sea level) P_{N_2} = _____ (5) P_{O_2} = _____ (6) P_{CO_2} = _____ (7)

P_{atm} = 500 mm Hg P_{N_2} = _____ (8) P_{O_2} = _____ (9) P_{CO_2} = _____ (10)

P_{atm} = 250 mm Hg (top of Mt. Everest) P_{N_2} = ____ (11) P_{O_2} = ____ (12) P_{CO_2} = _____ (13)

Short Answer

14. Describe the movement of gases.

____

____

15. Describe the factors that influence the rate of diffusion across a membrane.

____

____

Fill-in-the-Blanks

(16) ____________ is the process animals use to exchange gas with their surroundings. The source of O_2 and the "sink" for CO_2 in the environment is called the (17) ____________, which can be either air or water. (18) ____________ is the exchange of gases with the respiratory medium. Gases move by simple diffusion across a(n) (19) ____________. The entire body serves as a respiratory surface in many small animals. Insects possess an extensive (20) ____________ to distribute air to internal organs. In larger animals, (21) ____________ and (22) ____________ are used for gas exchange with water and air, respectively.

Short Answer

23. What roles do ventilation and perfusion play in the rate of gas exchange? (Explain in terms of Fick's equation.)

____

____

Complete the Table

In the following table, list the advantages and disadvantages of water and air as respiratory media.

	Advantages	*Disadvantages*
Water	(24)	(25)
Air	(26)	(27)

42.2 Adaptations for Gas Exchange [pp. 1043–1046]

Section Review (Fill-in-the-Blanks)

Gills are evaginations of the body surface; (28) ____________ extend from the body and do not have protective coverings, whereas (29) ____________ are located within chambers of the body and have a protective cover. Many aquatic vertebrates maximize gas exchange with (30) ____________, a process whereby water flows over the gills in the opposite direction of the flow of blood beneath the respiratory surface. Air enters and exits the tubes or (31) ____________ of the unique respiratory system of insects through (32) ____________. In some air-breathing fish (e.g., lungfish) and amphibians, air is gulped and forced into the lungs by a process called (33) ____________. Other vertebrates, such as reptiles, birds, and mammals, use (34) ____________, in which the air pressure in the lung is lowered by expanding the lung cavity through muscular contraction, resulting in air being pulled inward. The mammalian lung consists of millions of tiny air pockets or (35) ____________. Birds possess nonrespiratory air sacs that ensure that all air in the lungs comes in contact with respiratory surfaces and that create a(n) (36) ____________ flow of air that maximizes gas exchange via cross-current exchange (i.e., air flows perpendicular to blood flow).

Building Vocabulary

Using the definitions of the prefixes below, assign the appropriate prefix to the appropriate suffix for the following:

Prefixes	Meaning
hyper-	over, above
hypo-	under, below
ventila-	to fan

Suffix	Meaning
-tion	the process of

Prefix	Suffix	Definition
____	____	37. The process of moving air or water across a respiratory surface
____	____	38. Excessively rapid breathing
____	____	39. Excessively slow breathing

Describe

40. Describe the process of countercurrent exchange in the gills of fish and its adaptive significance.

____

____

Complete the Table

Animal group	*Principal respiratory medium*	*Principal respiratory structure*
Insects	41.	42.
Most fish	43.	44.
Reptiles	45.	46.
Mammals	47.	48.

Matching

Match each of the following structures with its correct definition.

49.	____	Breathing	A.	The exchange of gases with the respiratory medium
50.	____	Lungs	B.	The air tubes of the unique respiratory system of insects
51.	____	Alveoli	C.	Air hole on insect exoskeleton that controls air flow
52.	____	Tracheae	D.	Small air pocket of mammalian lung
53.	____	Tracheal system	E.	Process by which animals exchange gases with their environment
54.	____	Spiracles	F.	Respiratory structures that represent evaginations of the body
55.	____	Physiological respiration	G.	Respiratory structures that represent invaginations of the body
56.	____	Gills	H.	Extensive system of air tubes in insects

42.3 The Mammalian Respiratory System [pp. 1046–1050]

Section Review (Fill-in-the-Blanks)

Air enters the body through nasal passages and the mouth and enters the (57) ____________, a common pathway leading to the digestive tract and lungs. For respiration, air travels from the pharynx first into the (58) ____________ and then into the (59) ____________, which branches into two (60) ____________, one leading to each lung. The terminal airways or (61) ____________ in the lungs lead into cup-shaped pockets called alveoli. The lungs are covered with (62) ____________, a double layer of epithelial tissue. Inhalation of the lung occurs by contraction of the (63) ____________ and the (64) ____________, actions that expand the thoracic cavity and reduce air pressure in the lung below atmospheric pressure, thereby enabling the lungs to fill passively. Exhalation occurs by relaxation of these muscle groups. With increased activity, air can be forcibly expelled from the lungs by contraction of the (65) ____________. The volume of air entering and leaving the lung is the (66) ____________; the maximum such volume is the (67) ____________. A(n) (68)____________ of air is left in the lungs after exhalation. Respiration is controlled by the medulla oblongata and the pons of the brain stem, which integrate chemosensory information about O_2 and CO_2 levels centrally in the medulla and peripherally from (69) ____________ and (70) ____________.

Choice

For each of the following descriptions, choose the most appropriate type of adaptive immunity from the list below.

a. Carotid bodies b. Aortic bodies

71. ____ Chemoreceptors located in the aortic arch

72. ____ Chemoreceptors located in the carotid arteries

For each of the following descriptions, choose the most appropriate type of response from the list below.

a. Tidal volume b. Vital capacity c. Residual volume

73. ____ The maximum tidal volume of an individual

74. ____ The volume of air entering and leaving the lungs during inhalation and exhalation

75. ____ The volume of air left in the lungs after exhalation

Labelling

Identify each numbered part of the following illustration.

76. ____________

77. ____________

78. ____________

79. ____________

80. ____________

81. ____________

82. ____________

83. ____________

84. ____________

85. ____________

86. ____________

87. ____________

88. ____________

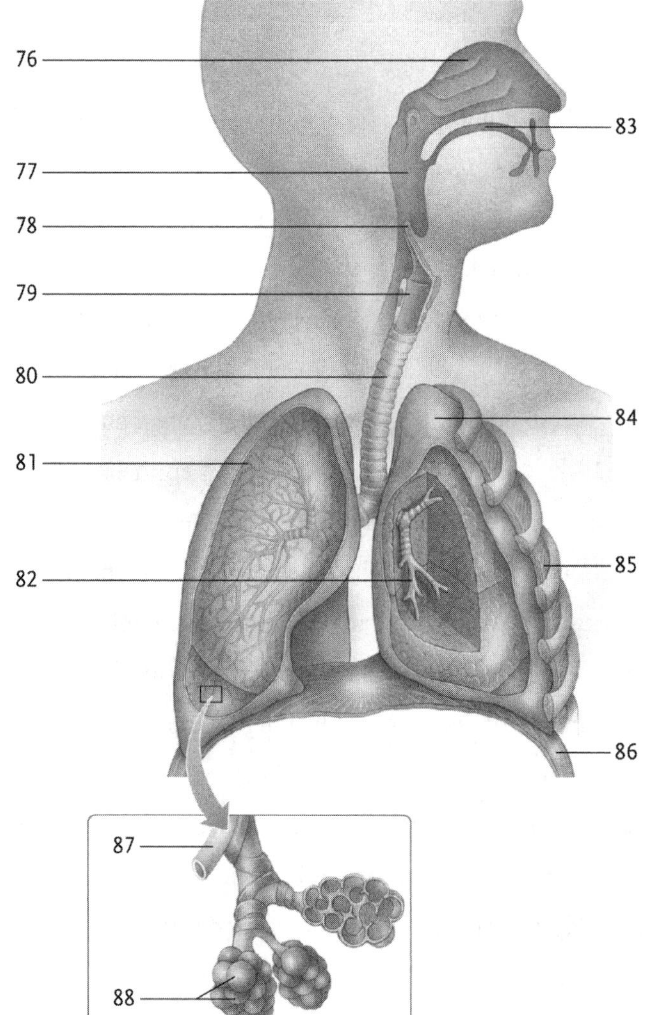

42.4 Mechanisms of Gas Exchange and Transport [pp. 1050–1053]

Section Review (Fill-in-the-Blanks)

The (89) ____________ of O_2 in clean, dry air at sea level is about 160 mm Hg. The carrying capacity of O_2 in the blood of vertebrates is increased by the presence of the respiratory pigment (90) ____________. The binding of O_2 to Hb, a protein with four subunits, can be described as a sigmoid (S)-shaped (91) ____________. The majority of CO_2 is transported in the plasma in the form of bicarbonate (HCO_3^-) produced by the enzyme (92) ____________ in red blood cells and represents an important (93) ____________ to control blood pH.

Discuss

94. What are the reasons for the shape of the haemoglobin–oxygen dissociation curve?

____

____

Short Answer and Label the Diagram

95. Using the following haemoglobin–oxygen dissociation curve, explain oxygen exchange and transport in the blood.

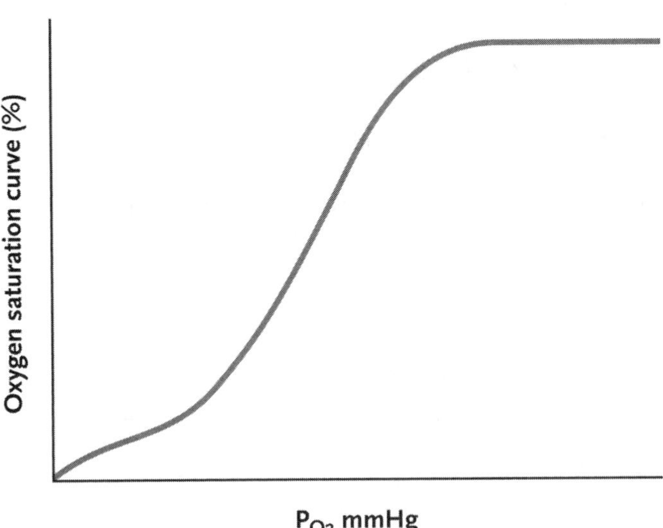

Locate on the diagram where you would expect the P_{O_2} of the blood to be (96) when blood is returning to the heart and (97) when blood has been oxygenated at the lungs.

98. Explain the effects of changes pH on the haemoglobin–oxygen saturation curve. On the diagram, draw the curve if the pH decreases (99). When would you expect this to occur (100)?

101. Explain the significance of the steep part of the curve.

102. List the different forms under which CO_2 is transported in the blood.

103. Explain the mechanisms of CO_2 transport in the blood and excretion at the lungs.

Choice

For each of the following descriptions, choose the most appropriate type of adaptive immunity from the list below.

a. Haemoglobin b. Myoglobin

104.____ Iron-containing respiratory pigment with four subunits

105.____ Iron-containing respiratory pigment with one subunit and a very high affinity for O_2

Matching

Match each of the following types of junction with its correct definition.

106.	____	Buffers	A.	The pressure of an individual gas in a mixture
107.	____	Erythropoietin	B.	Resists changes in pH by accepting or donating electrons
108.	____	Partial pressure	C.	Protein hormone that stimulates RBC production
109.	____	Carbonic anhydrase	D.	Enzyme that catalyzes the reaction CO_2 + H_2O Ξ [H_2CO_3] Ξ H^+ + HCO_3

SELF-TEST

1. The normal composition of air is [p. 1040]

   a. 21 mm Hg O_2; 78 mm Hg N_2; 1 mm Hg CO_2
   b. 160 mm Hg O_2; 590 mm Hg N_2; 7.6 mm Hg CO_2
   c. 21% O_2; 78% N_2; 1% CO_2
   d. 16% O_2; 59% N_2; 7.6% CO_2

2. Which statement is incorrect regarding Fick's equation of diffusion? [p. 1040]

   a. The larger the partial pressure difference, the lower the rate of diffusion
   b. The greater the surface area of exchange, the greater the rate of diffusion
   c. The shorter the length of the barrier, the higher the rate of diffusion
   d. The greater the solubility of a gas, the higher the rate of diffusion

3. Which of the following are disadvantages associated with breathing water? [pp. 1042–1043]

   a. Low oxygen solubility
   b. Dense respiratory medium
   c. Must breathe a greater amount of medium
   d. High oxygen content

4. The tracheal system in insects are characterized by [pp. 1044–1045]

   a. Closed tubes that circulate gases
   b. Uncontrolled diffusion of gases between the atmosphere and the tissues
   c. The transport of respiratory gases directly to every cell
   d. Countercurrent gas exchange

5. Countercurrent gas exchange [pp. 1043–1044]

   a. Allows for the extraction of oxygen from the media along the entire length of the respiratory surface.
   b. Is described as the tidal flow of air in mammalian lungs
   c. Results in the equilibration of partial pressure of gases along the respiratory surface during gas exchange.
   d. Allows establishment of a partial pressure difference that maintains gas exchange along the entire length of the respiratory surface

6. The alternating cycle of contraction and relaxation of the diaphragm and external intercostal muscles underlies ____. [pp. 1048–1049]

   a. Inspiration and expiration
   b. Inhalation and exhalation
   c. Expiration and inspiration
   d. Exhalation and inhalation
   e. Oxygen intake and CO_2 output

7. All respiratory surfaces share which of the following characteristics? [p. 1042]

   a. Moist
   b. Thin
   c. Large surface area
   d. Alveoli
   e. Spiracles

8. A cockroach obtains O_2 for tissues by means of ____. [pp. 1044–1045]

   a. Internal gills
   b. Pseudolungs
   c. Countercurrent exchange
   d. A tracheal system

9. The normal path of air flow through a mammalian respiratory system is ____. [pp. 1046–1047]

   a. Trachea, larynx, pharynx, bronchus, bronchiole
   b. Larynx, pharynx, trachea, bronchus, bronchiole
   c. Pharynx, larynx, trachea, bronchus, bronchiole
   d. Pharynx, trachea, larynx, bronchus, bronchiole

10. Countercurrent exchange ____. [pp. 1043–1044]

   a. Is used by bony fish
   b. Maximizes gas exchange by maintaining diffusion gradient
   c. Is described as the respiratory medium moving in the same direction as blood flow
   d. Is described as the respiratory medium moving in the opposite direction of blood flow

11. ________ detect O_2 levels in blood and provide information to the medulla oblongata. [pp. 1048–1049]

   a. Carotid bodies
   b. Islets of Langerhans
   c. Aortic bodies
   d. Nodes of Ranvier

12. The ability of oxygen to bind to Hb is affected by ____. [pp. 1050–1051]

   a. P_{O_2} in respiratory medium
   b. pH
   c. Temperature
   d. P_{CO_2} in tissues

13. Vital capacity is defined as [p. 1048]
   a. The amount of air taken in the lungs at rest
   b. The maximum amount of air taken up by the lung
   c. The amount of air remaining in the lung following a forced exhalation
   d. The frequency of breathing during exercise

14. In negative pressure breathing, [pp. 1045–1046]

a. The volume of the thoracic cavity increases
b. The pressure of the thoracic cavity increases
c. Air is forced down the trachea by muscular contractions of the mouth and the pharynx
d. All of the above

15. Organisms that rely solely on simple diffusion for gas exchange are [pp. 1040–1041]

a. Mammals
b. Frogs
c. Flatworms
d. Birds

16. The majority of CO_2 in the blood is [pp. 1051–1052]

a. Bound to haemoglobin
b. Transported as HCO_3^-
c. Dissolved as CO_2 in the plasma
d. Dissolved as CO_2 in the red blood cells

17. A haemoglobin–oxygen dissociation curve demonstrates [pp. 1050–1051]

a. The amount of O_2 bound to haemoglobin as a function of atmospheric pressure
b. The amount of O_2 bound to haemoglobin as a function of oxygen partial pressure
c. The amount of O_2 bound to haemoglobin as a function of pH
d. The amount of CO_2 that binds haemoglobin when O_2 is delivered to the tissues

18. If your pet dog lived at atmospheric P_{O_2} = 150 mm Hg and exhibited arterial P_{O_2} = 100 mm Hg and tissues P_{O_2} = 10 mm Hg, you would expect the dog to ____. [pp. 1050–1051]

a. Die
b. Accumulate CO_2
c. Have a serious but nonlethal O_2 deficit
d. Become dizzy from too much O_2
e. Function normally

19. As RBCs circulate into the region of actively metabolizing tissues (e.g., muscle), you would expect [pp. 1050–1052]

a. Increased binding of CO_2 to Hb
b. Increase in reduced Hb
c. Increase in bicarbonate ion concentration in plasma
d. Movement of chloride out of RBC

20. The process of bringing oxygenated water or air into contact with a gas exchange surface is [p. 1042]

    a. Respiration
    b. Inspiration
    c. Exhalation
    d. Ventilation

INTEGRATING AND APPLYING KEY CONCEPTS

1. Carbon monoxide is dangerous because it outcompetes O_2 for binding sites on Hb. The reaction between CO and Hb is reversible (like that between O_2 and Hb), whereas other pollutants, such as NO_x compounds, bind irreversibly to Hb. Discuss the differences in treatment strategies for individuals that you suspect to be suffering from CO and NO_x poisoning.

2. Using Fick's equation of diffusion, explain the adaptations that have allowed for life in water and terrestrial environments.

3. The oxygen partial pressure at sea level is approximately 160 mm Hg. At the top of Mount Everest, the oxygen partial pressure is approximately 53 mm Hg. Explain the challenges of breathing oxygen on top of Mount Everest. Why do some climbers use oxygen tanks to make their ascent? What type of acclimation would allow climbers to avoid the use of oxygen tanks? Explain.

OPTIONAL ACTIVITY

Essay

Create an organism! Design an animal using the principles you have learned in this chapter so that the critter is well suited to the habitat and lifestyle that is imposed upon it. Remember that all natural physical laws apply (i.e., properties of water versus air, etc.). Keep in mind that your animal should be designed to take into the account the requirements of O_2 and CO_2 exchange and transport. To write your essay, remember to first define the presumed gaseous conditions of the habitat and how these may impact gas exchange. Then discuss how your animal has adapted to these conditions.

Description of the Habitat and Lifestyle of the Critter

It is morning, and our hero awakes on the bottom of a shallow, well-mixed, freshwater lake at an altitude of 6500 metres (i.e., Mount Everest) above sea level. The P_{CO_2} of the water is 20 Torr and the pH is 4.0 owing to continual carbonation from seepage from a nearby Diet Cola repository. After acquiring a much-needed caffeine "buzz," he emerges to forage for his favourite food, the nutritious and visually appealing "chocolate cake insects." Sadly, the elusive insects live at an altitude of 500 metres above sea level and are only vulnerable for a brief period of time in the late morning. Thus, the critter must frantically chase the chocolate cake insects in order to gain adequate energy. Although tasty, the chocolate cake insects contain enormous quantities of $NaHCO_3$. To make matters worst, he must return to his home quickly if he is to avoid being eaten by the "Great Predator." Having reached home safely, he "recaffeinates." As usual, the diuretic impact of the caffeine forces him to leave his home once again to seek relief in "the great place." Following meaningful meditation while watching "television," he returns to his underwater dwelling to spend the night alone, yet again.

43 Regulating the Internal Environment

CHAPTER HIGHLIGHTS

- Osmoregulation is defined and comparison of osmoregulatory processes among animals is explored.
- Three forms of excretory wastes differ in their water content, toxicity, and energy requirements.
- Excretion processes and their regulation are presented and compared for invertebrates and vertebrates.
- Mammals use kidneys to produce urine, a structure in which a complex combination of filtration, reabsorption, and secretion ensures maintenance of body solutes at appropriate concentrations.
- Thermoregulation among animals is explained with a focus on ectothermy and endothermy.

STUDY STRATEGIES AND LEARNING OUTCOMES

- First, focus on concepts of osmoregulation and thermoregulation and how they help maintain an organism's internal environment.
- Examine the various types of osmoregulatory mechanisms in the animal kingdom, comparing the overall systems used by invertebrates, mammalian vertebrates, and nonmammalian vertebrates.
- Focus on the mammalian excretory system, both structure and function, making comparisons between mammals and other vertebrates.
- Evaluate thermoregulatory processes in the animal kingdom, with a focus on ectothermy and endothermy.

By the end of this chapter, you should be able to

- Define osmoregulation and thermoregulation, describing the importance of each to survival
- Differentiate between ammonia, urea, and uric acid excretory wastes, as well as describe the advantages and disadvantages of each
- Describe different excretory processes used by invertebrates and vertebrates
- Describe changes in blood and filtrate as molecules and ions are transported through the various structures within a nephron (mammalian)
- Compare mechanisms used by an ectotherm and an endotherm to regulate body temperature

TOPIC MAP

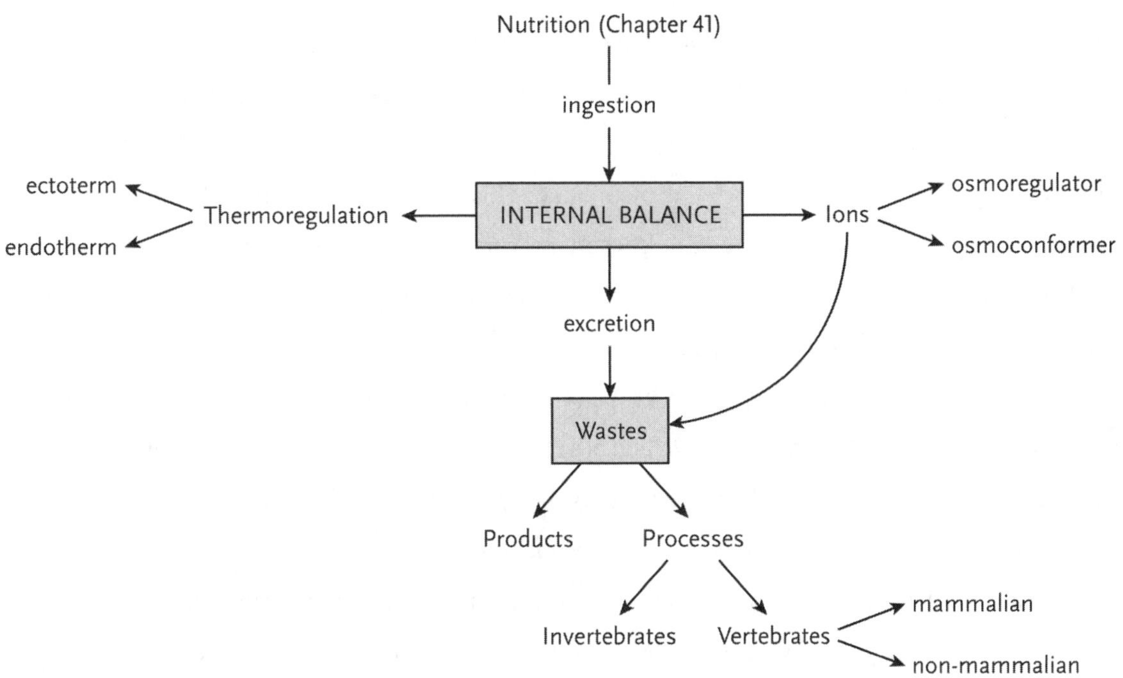

INTERACTIVE EXERCISES

Why It Matters [pp. 1057–1058]

Section Review (Fill-in-the-Blanks)

Organisms have mechanisms that assist them in maintaining their (1) ____________, despite fluctuations in the external environment. Terrestrial organisms are able to survive on land by conserving (2) ____________ in their bodies so that all body (3) ____________ systems are able to function to maintain life. For aquatic organisms living in environments with (4) ____________ concentrations greater than those found in body cells, where the overall movement of water is (5) ____________ the organism, physiological and morphological adaptations counteract this movement and maintain a relatively constant (6) ____________ environment.

43.1 Introduction to Osmoregulation and Excretion [pp. 1058–1061]

Section Review (Fill-in-the-Blanks)

(7) ____________ is the process of water molecules moving from an area of (8) ____________ concentration (having a (9) ____________ concentration of solutes) to an area of (10) ____________ concentration (having a (11) ____________ concentration of solutes), through a(n) (12) ____________ ____________ membrane. This type of movement is (13) ____________ since no energy is required, where movement is down a(n) (14) ____________ gradient. The osmotic concentration in a solution is measured in (15) ____________. When the osmolality of two solutions is compared, specific terms are used: a(n) (16) ____________ solution has the same number of osmoles as the other solution, a(n) (17) ____________ solution has more osmoles than the other solution, and a(n) (18) ____________ solution has fewer osmoles than the other solution. These terms only apply when comparing the osmoles between two solutions on either side of a selectively permeable membrane.

If an animal has control mechanisms that can maintain an internal osmolarity of its body fluids that is significantly different from the external environment, the animal is considered to be a(n) (19) ____________. However, if osmoregulatory processes are lacking or adjust the internal environment to closely match the external environment, the animal is known as a(n) (20) ____________. For animals to maintain the correct osmotic concentration, pH, and ionic balance, (21) ____________ must be eliminated. Wastes produced from (22) ____________ reactions, including the (23) ____________ products from the digestion of proteins and nucleic acids, must also be eliminated. Many wastes are toxic to cells, especially when levels are elevated. The process of (24) ___________ is the removal of metabolic wastes while maintaining water and ion (25) ___________, thus allowing the animal to survive in its environment.

Matching

Match each of the following components of excretion or type of metabolic waste with its correct description.

26.	____	Metabolic water	A.	Solution of relatively low osmolality
27.	____	Urea	B.	Animal that maintains internal osmolality at the same level as its external environment
28.	____	Hypoosmotic	C.	Highly toxic waste only excreted in dilute solutions
29.	____	Osmoconformer	D.	Solution of relatively high osmolality
30.	____	Ammonia	E.	Soluble waste that is relatively nontoxic
31.	____	Hyperosmotic	F.	Solvent for excretion of waste products and other physiological processes
32.	____	Uric acid	G.	Movement of water across a selectively permeable membrane down its concentration gradient
33.	____	Osmosis	H.	Animal that maintains internal osmolality at different level than its external environment
34.	____	Osmoregulator	I.	Nontoxic waste that forms crystals in water

43.2 Osmoregulation and Excretion in Invertebrates [pp. 1061–1063]

Section Review (Fill-in-the-Blanks)

(35) ___________ invertebrates are (36) ___________, since their internal osmolality matches the external environment, while freshwater and (37) ___________ invertebrates are (38) ___________. As osmoconformers, marine invertebrates use very little (39) ___________ on regulating the osmotic concentrations of their body (40) ___________. (41) ___________ wastes, typically in the form of (42) ____________, are directly released into the external

environment. Freshwater and terrestrial invertebrates must expend (43) ____________ to maintain an internal environment, which, in terms of osmolality, is (44) ____________ to their external environment. The types of habitats these animals can inhabit are more (45) ____________ than those of their osmoconformer counterparts.

Matching

Match each of the following types of excretory types with its correct definition.

46.	____	Protonephridia	A.	Closed tubule immersed in haemolymph
47.	____	Metanephridia	B.	Blind ended tubule with a flame cell
48.	____	Malpighian tubules	C.	Funnel-shaped tubule with cilia

Match each of the following animals with the appropriate excretory system type.

49.	____	Insects	A.	Protonephridia
50.	____	Flatworms	B.	Metanephridia
51.	____	Earthworm	C.	Malpighian tubules
52.	____	Larval clam		
53.	____	Adult clam		

43.3 Osmoregulation and Excretion in Mammals [pp. 1063–1069]

Section Review (Fill-in-the-Blanks)

The (54) ____________ is the primary organ of excretion in mammals. The (55) ____________ is the structural and functional unit of the kidney. The outermost region of the kidney is the (56) ____________, while the innermost region is called the (57) ____________. Some nephrons have long loops of Henle, called (58) ____________ nephrons, while the majority of nephrons have short loops, many of which do not cross into the medulla, called (59) ____________nephrons. (60) ____________ is the fluid produced in the nephron that contains water, electrolytes, and (61) ____________ products (by-products of metabolic reactions). After leaving the collecting duct, urine is carried from the (62) ____________ ____________ of the kidney to the (63) ____________ ____________ through a tube, the (64) ____________. In terms of osmolality, mammalian kidneys produce (65) ____________ urine compared to body fluids. The structure and function of the nephrons, the (66) ____________ ____________ that surround the nephrons, and the increasing (67) ____________ of the kidney from the cortex to medulla all play significant roles in producing a water-conserving, (68) ____________ urine.

Sequence

A molecule of nitrogenous waste, in the form of urea, is in a human's afferent arteriole. Identify the pathway this molecule would take through the excretory system, starting with #1, before exiting from the body.

69. ____ Renal pelvis
70. ____ Distal convoluted tubule
71. ____ Urethra
72. __1__ Afferent arteriole
73. ____ Collecting duct
74. ____ Descending loop of Henle
75. ____ Proximal convoluted tubule
76. ____ Bowman's capsule
77. ____ Urinary bladder
78. ____ Ureter
79. ____ Ascending loop of Henle
80. ____ Glomerulus

Matching

Match each of the following regions of the nephron and associated area in the kidney with its primary characteristic. Multiple answers may occur.

81.	____	Glomerulus	A.	Region where some urea moves into interstitial fluid; perrmeable to water
82.	____	Proximal convoluted tubule	B.	Capillaries where filtration occurs
83.	____	Bowman's capsule	C.	Blood vessel that supplies the glomerulus
84.	____	Ascending segment of loop of Henle	D.	Portion of loop that takes filtrate toward the medullary region of the kidney
85.	____	Distal convoluted tubule	E.	Blood vessel that exits the glomerulus
86.	____	Peritubular capillaries	F.	Nephron portion of the filtration area
87.	____	Descending segment of loop of Henle	G.	Blood vessels that surround the convoluted tubule
88.	____	Afferent arteriole	H.	Region of the loop that is impermeable to water
89.	____	Efferent arteriole	I.	Region of the nephrons that reabsorbs electrolytes, nutrients, and 65% of water
90.	____	Collecting ducts	J.	Portion of the nephron located in the cortical region of the kidney where urine is isoosmotic to body fluids

True/False

If the statement is true, write a "T" in the blank. If the statement is false, make it correct by changing the underlined word(s) and writing the correct word(s) in the answer blank.

91. ____________ Essentially all of nutrient molecules, including amino acids and glucose, are reabsorbed from the proximal convoluted tubule.

92. ____________ Movement of water out of the distal convoluted tubule by osmosis is due to overall movement of ions into this structure.

93. ____________ The large number of aquaporins in the wall of the ascending segment of the loop of Henle allows rapid reabsorption of water into interstitial fluid.

94. ____________ The diameter of the afferent arteriole is greater than that of the efferent arteriole, increasing the pressure for movement of fluid into the Bowman's capsule.

95. ____________ Sodium and chloride ions move into the ascending segment of the loop of Henle.

43.4 Kidney Function in Nonmammalian Vertebrates [pp. 1069–1071]

Section Review (Fill-in-the-Blanks)

In nonmammalian vertebrates, the environmental demands placed on the animal determine if (96) ____________ (liquid) or (97) ____ (solid) are conserved or excreted. These vertebrates have a variety of mechanisms to maintain an internal environment that is compatible with life in their external environments. Marine teleosts live in a(n) (98) ____________ environment; therefore, water must be (99) ____________ and salts are (100) ____________for the animals to survive. Sharks and rays retain the metabolic waste (101) ____________ and other (102) ____________ wastes to maintain an internal body fluid that is (103) ____________ to seawater. Freshwater fishes and aquatic amphibians maintain body fluids that are (104) ____________ to their environment by (105) ____________ water and (106) ____________ salts. Terrestrial amphibians must (107) ____________ both water and salts. Reptiles and birds conserve water by excreting water-free (108) ____________ ____________ crystals. In addition to these mechanisms, these animals have a variety of other (109) _____________, some very unique, that play a major role in conservation or excretion of water and salts.

Choice

Choose organism(s) that use the water/salt conservation/excretion mechanisms listed. More than one choice can apply to a mechanism.

110. _____Rectal salt gland	A. Reptiles and birds
111. _____Chloride cells	B. Marine teleosts
112. _____Isoosmotic urine	C. Freshwater fish and amphibians
113. _____Retain urea	D. Sharks and rays
114. _____Hypoosmotic urine	E. Terrestrial amphibians
115. _____Head salt glands	
116. _____Excrete uric acid crystals	
117. _____Excrete ammonia	
118. _____Active uptake of salts by gills	

43.5 Introduction to Thermoregulation [pp. 1071–1074]

43.6 Ectothermy [pp. 1074–1076]

43.7 Endothermy [pp. 1076–1081]

Section Review (Fill-in-the-Blanks)

Temperature regulation is based on (119) ____________ feedback mechanisms and (120) ____________, which detect changes from the internal (121) ____________ ____________. To accomplish these regulating tasks, animals have both heat (122) ____________ (when it is cool) and heat (123) ____________ (when it is hot) adaptations. There are two major strategies to maintain heat gain and loss. (124) ____________ are animals that produce heat internally primarily from physiological sources, while (125) ____________ obtain heat primarily from the external environment.

Ectotherms maintain an internal temperature that is close to the (126) ____________ temperature. Most animals beside (127) ____________ and (128) ____________ are ectotherms. These animals typically use (129) ____________ mechanisms to maintain their internal temperature. During seasonal changes, fish remain in (130) ____ water during the summer and move to the (131) ____________ levels during the winter. Amphibians and reptiles utilize the warming rays from the sun to increase body temperature by (132) ____________. In addition to behavioural mechanisms, these animals often have various physiological mechanisms or (133)

____________ ____________, which ensures survival. One such mechanism is multiple (134) ____________ that can catalyze the same reaction. Each has a different optimal (135) ____________ range of activity.

Endotherms use (136) ____________ reactions and (137) ____________ exchange with the environment to maintain body temperature around an optimal temperature, a (138) ____________ ____________. This is (139) ____________ expensive but allows these animals to be active over a (140) ____________ range of temperatures than ectotherms. In humans, thermoreceptors are located in the hypothalamus, spinal cord, and (141) ____________. The latter assists in regulating temperature through various morphological and physiological adjustments, such as changing blood vessel diameter, (142) ____________ the size to cool the body and (143) ____________ it to raise body temperature; producing (144) ____________ when the body is too hot; and coordinating small muscle contractions or (145) ____________ when body temperature is too low.

Choice

Choose the term that describes or defines the situation.

A. Conduction B. Convection C. Radiation D. Evaporation

146. An animal lying in the sun can either gain or lose heat by ________.

147. When you sweat, you are losing heat through ________.

148. A snake lying on a warm rock during the early morning is gaining heat through ________.

149. Standing in front of a fan allows heat to be lost by ________.

150. If you are naked standing on ice in the winter, you will lose heat through your feet by _________ and through your breath by ________.

Matching

Match the mechanisms with the primary function with respect to thermoregulation.

151.	____	Sweating	A.	Heat-producing mechanism
152.	____	Shivering	B.	Heat loss mechanism
153.	____	Increased thyroid hormone		
154.	____	Decreased epinephrine		
155.	____	Vasodilation		
156.	____	Increased blood flow to the skin		
157.	____	Increased blood flow to the core		

True/False

If the statement is true, write a "T" in the blank. If the statement is false, make it correct by changing the underlined word(s) and writing the correct word(s) in the answer blank.

158.____________ If the external environment is cool, endotherms expend a <u>small</u> amount of energy for heat production.

159.____________ The period of torpor during <u>winter</u> is called estivation, while the period of torpor during <u>summer</u> is hibernation.

160.____________ Gills of fish and the ears of arctic animals have countercurrent exchanges between arterial and venous blood flow to <u>prevent</u> heat loss to the environment.

161.____________ Blubber prevents heat loss due to <u>radiation</u>.

162.____________ The <u>hypothalamus</u> is the primary thermoregulatory centre in birds and mammals.

SELF-TEST

1. Body fluids with an osmolality of 0.5 osmoles/kg are equivalent to body fluids with an osmolality of _____ mOsm/kg. [pp. 1058–1061]

   a. 0.0005
   b. 0.05
   c. 50
   d. 500
   e. 5000

2. If an animal secretes uric acid crystals, you could conclude that the environment in which this animal lives is ______. [pp. 1058–1061]

   a. Freshwater
   b. Marine
   c. Terrestrial and very arid (dry)
   d. Terrestrial and very wet

3. Removal of nitrogenous wastes in salmon at sea would most likely occur by loss of _________. [pp. 1058–1061]

   a. Uric acid crystals
   b. Urea
   c. Trimethylamine oxide
   d. Ammonia

4. If the osmolality of the intracellular fluids is maintained to be isoosmotic to the external environment, the animal is most likely an _______. [pp. 1061–1063]

   a. Osmoregulator
   b. Osmoconformer
   c. Osmoequilibrator
   d. None of these

5. Animals living in freshwater tend to have __________ cells and extracellular fluid. [pp. 1061–1063]

   a. Hypoosmotic
   b. Isoosmotic
   c. Hypoosmotic

6. Invertebrates that excrete wastes using tubules with one open end that empties into a gut use __________. [pp. 1061–1063]

   a. Protonephridia
   b. Malpighian tubules
   c. Nephrons
   d. Metanephridia

7. Cells involved with reabsorption or secretion are involved in active or facilitated transport of materials either into or out of the filtrate. [pp. 1063–1069]

   a. True
   b. False

8. Which of the following segment(s) of nephron is(are) lacking aquaporins? [pp. 1063–1069]

   a. Proximal convoluted tubule
   b. Descending segment of the loop of Henle
   c. Ascending segment of the loop of Henle
   d. Both (a) and (b)

9. In mammals, blood arrives to the kidneys through __________. [pp. 1063–1069]

   a. Renal arteries
   b. Renal pelvis
   c. Renal veins
   d. Ureters

10. What section of a nephron is involved in filtration? [pp. 1063–1069]

   a. Collecting duct
   b. Bowman's capsule
   c. Loop of Henle
   d. Proximal convoluted tubule

11. Osmolality of the filtrate increases as it travels through the __________. [pp. 1063–1069]

   a. Ascending segment of the loop of Henle
   b. Descending segment of the loop of Henle
   c. Bend at the bottom of the loop of Henle
   d. Proximal convoluted tubule

12. Bicarbonate ions are reabsorbed into the peritubular capillaries __________. [pp. 1063–1069]

   a. With H^+ ions to balance acidity in the filtrate
   b. To balance acidity in the blood
   c. Through aquaporins
   d. To increase solute concentration in the interstitial fluid

13. Substances that are <u>not</u> reabsorbed will __________. [pp. 1063–1069]

  a. Be excreted from the body in urine
  b. Remain in the nephron
  c. Leave the kidney through the renal vein
  d. Return to the glomerulus

14. Which of the following organisms engage mechanisms in their bodies to maintain hyperosmotic body fluids? [pp. 1069–1071]

  a. Amphibians
  b. Marine fish
  c. Freshwater fish
  d. Both (a) and (b)

15. How are birds that live by the sea able to survive, even though they drink little or no freshwater? [pp. 1069–1071]

  a. They absorb water from the air through their skin
  b. They use salt glands in their head to excrete salts from their bodies
  c. They are unable to survive in this environment without drinking large quantities of freshwater

16. Boa constrictors will wrap around their eggs and shiver. This is an example of thermoregulation by ______. [pp. 1071–1081]

  a. Evaporation
  b. Conduction
  c. Convection
  d. Radiation

17. Which of the following organisms would you expect to require a constant and high energy demand? [pp. 1071–1081]

  a. Rabbit
  b. Shark
  c. Lizard
  d. Frog

18. Thermal acclimatization __________. [pp. 1071–1081]

  a. Involves regulation of body temperature in response to information provided by a thermoreceptor
  b. Is the change in an animal's physiology that accompanies seasonal changes
  c. Occurs when core body temperature is below normal for an extended period of time
  d. Occurs when core body temperature is above normal for an extended period of time

19. If the set point was increased, which of the following would you expect to occur? [pp. 1071–1081]

    a. Decreased epinephrine
    b. Shivering
    c. Vasodilation
    d. Both (a) and (c)

20. The layer of the skin that contains sweat glands, thermoreceptors, blood vessels, and other thermoregulating structures is the __________. [pp. 1071–1081]

    a. Hypodermis
    b. Epidermis
    c. Dermis

INTEGRATING AND APPLYING KEY CONCEPTS

1. Compare and contrast the structure and function of the various excretory systems of invertebrates with respect to habitat—marine, freshwater, and terrestrial.

2. Explain why osmolality in the filtrate increases as it travels down the descending segment of the loop of Henle.

3. Discuss the advantages and disadvantages of being either an ectotherm or an endotherm.

OPTIONAL ACTIVITIES

1. Trehalose appears to assist midge larvae in surviving in a desiccation state during prolonged dry periods. Provide an explanation of how this sugar is believed to accomplish this task.

2. Describe how the structure of aquaporins assists in permitting H_2O to pass through a membrane.

44 Defences against Disease

CHAPTER HIGHLIGHTS

- Humans and other vertebrates have three lines of defence against pathogens: physical barriers, innate immunity, and adaptive immunity.
- In innate immunity, molecules on the surface of pathogens are recognized by various types of host cells to initiate inflammation and the complement systems that lead to the destruction of the pathogen.
- Adaptive immunity is carried out by B cells and T cells and has two components: an antibody-mediated response and a cell-mediated response.
- In the antibody-mediated response, B cells produce antibodies that bind to the pathogen and mark it for elimination. T cells stimulate B cells in this process.
- In the cell-mediated response, T cells act as killer cells to eliminate infected cells from the body.
- Examples of problems associated with the immune system are presented, as well as interesting mechanisms or techniques some pathogens have developed to circumvent a host's defences.

STUDY STRATEGIES AND LEARNING OUTCOMES

- Be sure to review any concepts of cell biology from other chapters that you do not understand, such as phagocytosis, symbiosis, protein, cell signalling, etc.
- Since this chapter contains a lot of information and terminology and covers complex concepts, read only one section at a time and try not to go through the entire chapter in one sitting. You might want to start by skimming the section, writing down the boldface terms, and studying the figures before going back and reading the text.
- Drawing diagrams or flow charts of the various response pathways may be helpful.

By the end of this chapter, you should be able to

- Differentiate between the three lines of defence used by vertebrates, identifying which is used by invertebrates and plants
- Describe the physical barriers to infection, as well as the different nonspecific responses provided by innate immunity
- Describe the process of both antibody-mediated immunity and cell-mediated immunity, identifying similarities and differences between them
- Explain how malfunction or failure of an immune response can result in disease, or worse
- Provide examples describing how some pathogens are able to overcome an organism's defences

TOPIC MAP

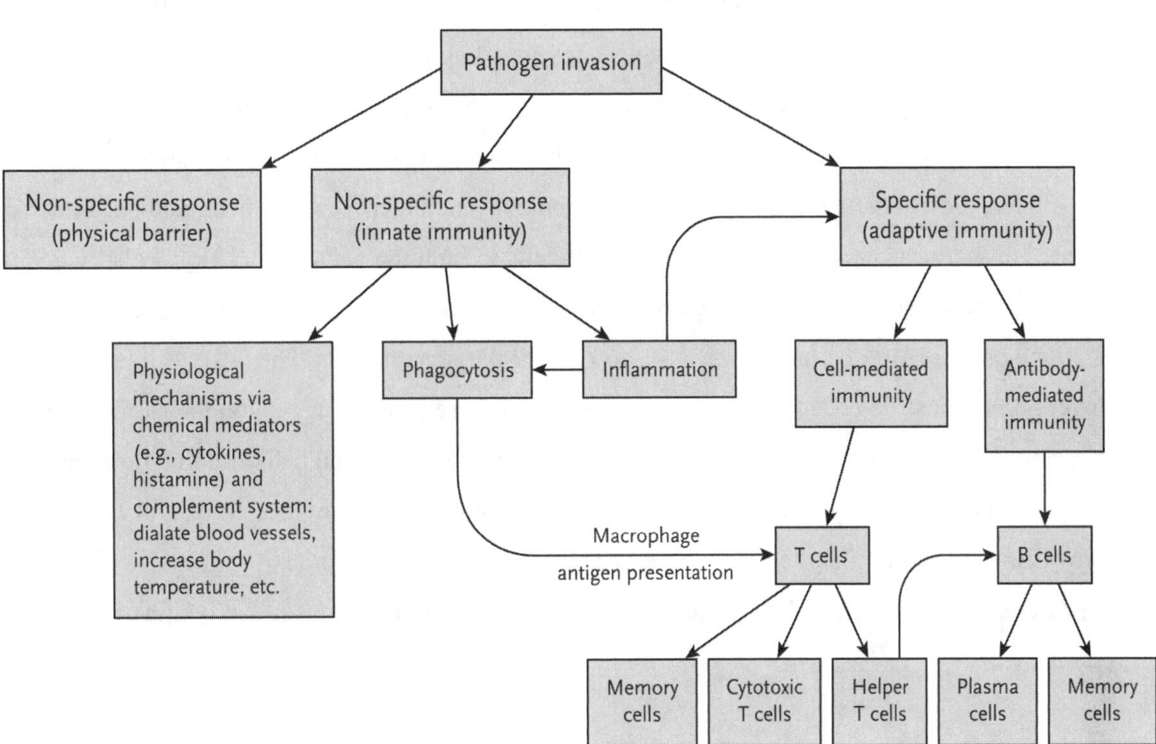

INTERACTIVE EXERCISES

Why It Matters [pp. 1085–1086]

Section Review (Fill-in-the-Blanks)

All types of organisms are infected by (1) ____________, disease-causing organisms. The (2) ____________ system functions as the body's main defence, regulating or (3) ____________ most pathogens. Humans have assisted this organ system by developing (4) ____________, drug treatments that work with the immune system to protect against infection. Some diseases, such as smallpox, have now been (5) ____________ from the human population.

44.1 Three Lines of Defence against Invasion [pp. 1086–1087]

44.2 Nonspecific Defences: Innate Immunity [pp. 1087–1092]

Section Review (Fill-in-the-Blanks)

The three lines of defence against invasion by foreign agents in humans and other vertebrates are physical barriers; (6) ____, a protection against any invading pathogen; and (7) ____, protection against specific pathogens. The (8) ____ is the defensive reaction of the immune system.

(9) ____ is a rapid response to injury that involves heat, pain, redness, and swelling. (10) ____ are phagocytes that are usually the first to recognize pathogens, engulf them, and become activated to secrete (11) ____, which recruit more cells to fight the pathogen. (12) ___ also become activated by foreign substances and release histamine. (13) ____ are attracted to an infection site by (14) ____, as are (15) ____ to help kill large pathogens. The (16) ____, a group of over 30 proteins, also become activated, and some of these form (17) ____ on the surfaces of pathogens. (18) ____, types of cytokines, are produced by viral-infected cells and initiate events that result in RNA degradation and inhibition of protein synthesis in the infected cell. (19) ____ destroy viral-infected cells often by triggering programmed cell death or (20) ____. NK cells are a type of (21) ____, a type of leukocyte that carries out most of its actions in the lymphatic system.

Choice

Choose the defence(s) that could be used by each of the organisms.

22. ____ Grasshopper
23. ____ Corn plant
24. ____ Giraffe
25. ____ Mouse

A. Body surface covering
B. Adaptive immunity
C. Innate immunity

For each of the following descriptions, choose the most appropriate type of defence from the list below.

a. Innate immunity   b. Adaptive immunity   c. Epithelial barrier

26. ____ Immediate, nonspecific response to an invading pathogen by recognizing its nonself molecular pattern
27. ____ Immediate, nonspecific response involving secretion of acid and digestive enzymes
28. ____ Specific response to a particular pathogen
29. ____ Nonspecific response to a pathogenic virus involving shutting down RNA and protein synthesis

Complete the Table

Type of cell	*Function*
Macrophage	30.
31.	Releases histamine upon activation
Neutrophil	32.
33.	Type of leukocyte that helps kill large pathogens by secreting lysozmes and defensins
34.	Destroys viral-infected cells by perforating their plasma membrane

Matching

Match each of the following statements with the most appropriate type of chemical mediator.

a. Cytokines b. Chemokines c. Interferons

35. ____ Proteins produced by infected cells that initiate degradation of RNA and cessation of protein synthesis

36. ____ Released by activated macrophages and initiates heat, redness, and swelling of inflammation

37. ____ Released by activated macrophages and attracts neutrophils to infected site

Match each of the following structures with its correct definition.

38.	____	Apoptosis	A.	The collection of defensive actions of the immune system
39.	____	Histamine	B.	Rapid response that involves swelling
40.	____	Inflammation	C.	Proteins that circulate in the blood and become activated by molecules on the surface of pathogens
41.	____	Complement system	D.	Activated complement proteins that perforate cell membranes
42.	____	Lymphocyte	E.	Type of leukocyte that carries out its actions in the lymphatic system
43.	____	Immune response	F.	Programmed cell death
44.	____	Membrane attack complex	G.	Causes dilation of blood vessels leading to the infected site

44.3 Specific Defences: Adaptive Immunity [pp. 1092–1102]

Section Review (Fill-in-the-Blanks)

A foreign molecule that triggers an adaptive immunity response is a(n) (45) ___________. These molecules are recognized by (46) ___________, which are derived from and mature in bone marrow, and (47) ___________, which are derived from bone marrow and mature in the (48) ___________. The two types of adaptive immunity are (49) ___________ immunity, in which B cells differentiate into plasma cells that secrete (50) ___________, and (51) ___________ immunity, in which a subclass of T cells become activated and, with other cells of the immune system, attach to foreign cells and destroy them. Some activated lymphocytes differentiate into (52) ___________ that circulate and then initiate a rapid response upon re-exposure to the same (53) ___________. B-cell receptors and (54) ___________ ___________ receptors bind to specific regions, or (55) ___________, of an antigen molecule. Antibodies are a large class of proteins known as (56) ___________. An individual protein of this class consists of four subunits: two identical (57) ___________, with one end embedded in the plasma membrane, and two identical (58) ___________. The protein has a (59) ___________ shape, with a(n) (60) ___________ ___________ site at the top end of each of the two arms.

Both types of adaptive immunity involve four basic steps. One or more lymphocytes must encounter, recognize, and then (61) ___________ to an antigen. This interaction with the antigen causes the lymphocyte to become activated and undergo (62) ___________ ___________ to produce identical copies of itself, or (63) ___________. These lymphocytes must then (64) ___________ the host's body of all antigens. Lastly, some of the lymphocytes must differentiate into (65) ___________ cells, which can provide a rapid response should this type of antigen appear in the body again. When exposed to a foreign antigen for the first time, a(n) (66) ___________ immune response results, whereas when a foreign antigen enters the body for a second or subsequent time, a(n) (67) ___________ immune response results. (68) ___________ ___________ is the production of antibodies in response to a foreign antigen, whereas (69) ___________ ___________ is the acquisition of antibodies by direct transfer from another individual.

Sequence

Identify the correct order of events during formation of helpter T cells in antibody-mediated immunity. Note that number 1 is provided.

70. ____ $CD4^+$ T cell begins clonal expansion

71. ____ Lysosomes break down bacterial components

72. ____ $CD4^+$ T cell binds to antigen

73. ____ Infected cell is an antigen-presenting cell

74. _1_ Dendritic cell engulfs bacterium

75. ____ Clones differentiate into helper T cells

76. ____ $CD4^+$ T cell secretes cytokines

77. ____ Antigens bind to class II MHC

78. ____ $CD4^+$ T cell is activated

79. ____ APC secretes interleukin

Identify the correct order of events during B-cell activation in antibody-mediated immunity. Number 1 is provided.

80. ____ Helper T cell secretes interleukin

81. ____ B cell takes in bacterium and breaks it down

82. ____ B cell proliferates

83. ____ B cell activated

84. ____ Linkage between helper T cell and B cell

85. ____ Cloned B cells differentiate into plasma cells and memory cells

86. ____ Antigen binds to class II MHC

87. _1__ B cell binds to bacteria's antigen

88. ____ Plasma cells produce antibodies

89. ____ Antigen presented on cell surface (TCR and BCR for same antigen)

Choice

For each of the following descriptions, choose the most appropriate type of adaptive immunity from the list below.

a. Cell-mediated immunity   b. Antibody-mediated immunity

90. ____ Involves activation of T-cell derivatives, which then attach to foreign cells and kill them

91. ____ Involves production of specific proteins by B-cell derivatives that bind to foreign molecules

92. ____ Involves releasing perforin, which creates pores in membrane of infected cell

Short Answer

93. What is the major histocompatibility complex (MHC)?

____

____

94. Why is it important that B cells and cytotoxic T cells proliferate in antibody-mediated immunity and cell-mediated immunity, respectively?

____

____

Choice

For each of the following descriptions, choose the most appropriate type of response from the list below.

a. Primary immune response   b. Secondary immune response

95. ____ Response mounted by exposure to an antigen for the first time

96. ____ Response mounted by exposure to an antigen for a second or subsequent times

97. ____ Response occurs rapidly

98. ____ Production of a new type of memory B cell

For each of the following descriptions, choose the most appropriate type of immunity from the list below.

a. Passive immunity b. Active immunity

99. ____ The production of antibodies in the body in response to exposure to a foreign antigen

100. ____ The acquisition of antibodies by direct transfer from another individual

101. ____ Blood transfusion between two individuals

102. ____ Resting in bed to recover from a virus

Complete the Table

Type of cell	*Function*
B cell	103.
104.	Lymphocyte that differentiates from stem cells in bone marrow but matures in thymus gland
Plasma cell	105.
Memory cell	106.
Dendritic cell	107.
108.	Derived from activated $CD4^+$ cells and helps stimulate B cells
Helper T cell	109.
110.	A type of T cell that binds to an antigen-presenting cell via a $CD8^+$ receptor and leads to cell-mediated immunity
$CD8^+$ cytotoxic T cell	111.

Matching

Match each of the following structures with its correct definition.

112.	____	Thymus	A.	Organ of lymphatic system involved in maturation of T cells
113.	____	Antigen	B.	Protein that binds to antigens and marks them for elimination
114.	____	Antibodies	C.	Specific region of an antigen molecule
115.	____	Clonal expansion	D.	Family of proteins that serve as antibodies
116.	____	Clonal selection	E.	A cell that displays antigens on its surface
117.	____	Epitopes	F.	The proliferation of a particular clone of cells
118.	____	Immunoglobins	G.	The process by which a lymphocyte is specifically selected for cloning
119.	____	Antigen-presenting cell	H.	Ability to recognize previous antigens and foreign cells
120.	____	Immunological memory	I.	A foreign molecule that triggers an adaptive immunity response

44.4 Malfunctions and Failures of the Immune System [pp. 1103–1104]

44.5 Defences in Other Organisms [pp. 1104–1106]

44.6 How Do Parasites and Pathogens Circumvent Host Responses? [pp. 1106–1108]

Section Review (Fill-in-the-Blanks)

(121) ____________ is a process in which a body's own molecules are protected from attack by the immune system. Failure of this process can lead to an (122) ____________ ____________ in which (123) _____________ antibodies are produced and they attack and eliminate specific body cells. (124) ____________ are a distinct class of antigen that initiate an allergic reaction. A severe reaction can bring on (125) ____________ ____________, an inflammation that blocks airways and interferes with breathing and that could lead to a quick death if a(n) (126) ____________ injection is not administered.

Some pathogens are able to avoid elimination by using various techniques, such as hiding from a host, infecting an area that has a(n) (127) ____________ immune response, or manipulating a response. Some pathogens can constantly change their surface coats, called (128) ____________ ____________. In certain situations, a (129) ____________ relationship between a pair of organisms, both benefiting from the association, allows them to infect a host and then feed and (130) ____________ within the host, before leaving with their offspring to find a(n) (131) ____________ host.

Invertebrates and plants do not have antibodies; they rely on their (132) ____________ immune system to protect them from (133) ____________.

Matching

Match each of the following types of junction with its correct definition.

134.	____	Allergen	A.	Protection of a body's own molecules from immune system attack
135.	____	Antigenic variation	B.	Production of antibodies against molecules of one's own body
136.	____	Immune tolerance	C.	Substance responsible for initiating an allergic reaction
137.	____	Anaphylactic shock	D.	Process that occurs in a host when first exposed to an allergen
138.	____	Autoimmune reaction	E.	Constant changing of the antigen's surface protein to evade a host
139.	____	Sensitization	F.	Extreme inflammatory response that can restrict air passages

SELF-TEST

1. Active immunity can be artificially induced by ____. [pp. 1085–1086]

   a. Transfusions
   b. Injecting vaccines
   c. Passing maternal antibodies to a fetus
   d. Both (a) and (c)

2. Which of the following is a specific defence mechanism in vertebrates? [pp. 1086–1092]

   a. Skin
   b. Acid secretion
   c. Inflammation
   d. Phagocytosis by phagocytes
   e. Antibody production

3. Complement ____ that helps destroy pathogens. [pp. 1086–1092]

   a. Is a system of many proteins
   b. Is an antibody
   c. Is highly antigen specific
   d. Both (a) and (b)

4. Natural killer (NK) cells ________. [pp. 1086–1092]

   a. Secrete interleukin, which activates T cells
   b. Are phagocytes
   c. Are derived from plasma cells
   d. Secrete perforins, which create pores in a cell membrane
   e. Stimulate antibody production

5. How does skin act as a defence? [pp. 1086–1092]

   a. It recognizes the molecular pattern on a pathogen and provides a response
   b. It provides an inflammation response
   c. It forms a physical barrier by connecting cells with tight junctions
   d. It contains cells that have memory of the pathogen

6. What part of the host organism recognizes the molecular pattern of a pathogen? [pp. 1086–1092]

   a. Pathogen-associated molecular pattern
   b. Pathogen effector molecule
   c. PAMP
   d. PRR

7. Why is defensin considered to be a highly conserved antimicrobial peptide? [pp. 1086–1092]

   a. There is only one type of defensin
   b. It is found only in mammals
   c. It is found on all epithelial surfaces
   d. It is found in a wide diversity of organisms, including invertebrates, plants, and vertebrates

8. Why is a different mechanism required to identify viral pathogens than bacterial pathogens? [pp. 1086–1092]

   a. Viruses are often found inside a host cell, hidden away from passing lymphocytes
   b. There is little difference between viral-infected and normal cell surface molecules
   c. Cells become too large to be engulfed by the various leukocytes
   d. Both (a) and (b)

9. During an inflammation response, which leukocyte is killed along with the pathogen, forming pus at the infection site? [pp. 1086–1092]

   a. Macrophage
   b. Neutrophil
   c. Eosinophil
   d. Cytokine

10. Which of the following cells are lymphocytes that target virus-infected hosts with low MHC protein concentrations? [pp. 1086–1092]

    a. Macrophages
    b. Eosinophils
    c. RNAi
    d. Natural killer cells

11. The histocompatibility complex is _____. [pp. 1092–1102]

    a. Found in cell nuclei
    b. Different in each individual
    c. Derived from a large group of genes
    d. Both (a) and (b)
    e. Both (b) and (c)

12. T cells ______. [pp. 1092–1102]

    a. Are lymphocytes
    b. Mature in the thymus gland
    c. Are involved in adaptive immunity
    d. (a), (b), and (c)

13. T-cell receptors __________. [pp. 1092–1102]

   a. Bind antigens
   b. Have no known function
   c. Are identical in all T cells
   d. Stimulate antibody production

14. The __________ site of the antibody determines what immunoglobulin it belongs to. [pp. 1092–1102]

   a. Conservative
   b. Variable
   c. Conservative or variable
   d. Conservative and variable

15. The secondary immune response is due to ______. [pp. 1092–1102]

   a. Killer T cells
   b. Memory cells
   c. Plasma cells
   d. Macrophages
   e. Helper T cells

16. Agglutination occurs when __________. [pp. 1092–1102]

   a. An antigen attaches to one of the arms of an antibody
   b. Each antigen binds to more than one antibody, forming a lattice
   c. An antibody reproduces
   d. An antigen is taken up by an antibody by phagocytosis

17. What method does the AIDS virus use to circumvent a host cell response? [pp. 1103–1108]

   a. Interacts with a symbiont to enter the host cell
   b. Implements antigenic variation
   c. Induces the expression of inhibitors of apoptosis (IAPs)

18. The watery eyes and runny nose resulting from an allergic reaction are due to ________. [pp. 1103–1108]

   a. Signals sent out by mast cells
   b. Lack of histamine production
   c. Production of IgM
   d. Interaction with histocompatibility complex

19. When a nematode and bacteria work together to infect a host, what do the bacteria do to assist in the infection? [pp. 1103–1108]

   a. Penetrate the host
   b. Reproduce and inactivate the host's immune response
   c. Encapsulate the host cells
   d. Change physiology within the nematode to encapsulate the host cells

INTEGRATING AND APPLYING KEY CONCEPTS

1. Describe what happens during an inflammation response.

2. What is the function of cytotoxic T cells in adaptive immunity?

3. What occurs at the cellular level during an allergic reaction?

OPTIONAL ACTIVITIES

1. How can knockout mice be used to study asthma?

2. Devise a treatment strategy for killing melanoma cells based on the *Fas–FasL* system.

3. Describe what occurs when an individual infected with HIV is then infected with the fungus *Pneumocystis carinii*.

45 Population Ecology

CHAPTER HIGHLIGHTS

- Ecologists may study biology from the perspective of the individual, the population, the community, or the ecosystem. This chapter focuses on interactions within populations.
- Populations exhibit unique characteristics, including limited distributions (range), size, density, dispersion, age structure, survivorship, and many others.
- The life history patterns of organisms involve tradeoffs in benefits versus costs. No life history pattern is optimum in all respects.
- Populations can exhibit exponential or logistic growth in numbers. Density-dependent and independent effects play a large role in determining population size.
- Population size is primarily affected by the carrying capacity of the habitat and the intrinsic rate of increase.
- Human population growth has recently been immune to most natural regulatory constraints.

STUDY STRATEGIES AND LEARNING OUTCOMES

- Review the hierarchy of biological investigation. This chapter concentrates on populations, but in upcoming chapters, you will need to know how populations relate to individuals, communities, ecosystems, and the biosphere. You will also need to know that each level of the hierarchy exhibits unique characteristics (emergent properties).
- Review the population concept, noting that interactions among members of a population are intraspecific.
- Be clear on the factors that regulate population growth and final population size.
- Keep in mind that many of the concepts presented in this chapter are presented in a somewhat simplified way to make them easy to understand. In reality, many factors interact with one another, sometimes producing unexpected results.
- Make a table showing the mechanisms that control population size in most organisms and how humans have circumvented those mechanisms.

By the end of this chapter, you should be able to

- Explain how population characteristics affect population growth and size.
- Use exponential and logistic growth curves to predict changes in population size
- Describe how population size and growth are regulated by density-dependent and -independent factors
- Relate your knowledge of population ecology to human populations

TOPIC MAP

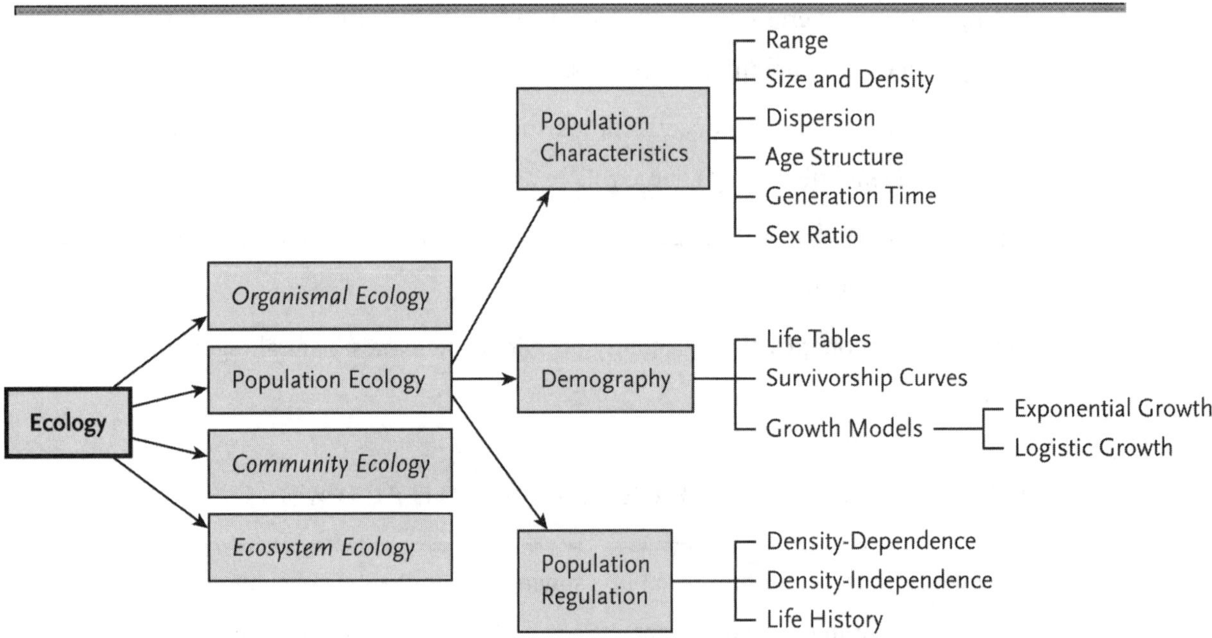

INTERACTIVE EXERCISES

Why It Matters [pp. 1111–1112]

Section Review (Fill-in-the-Blanks)

Ecology provides an understanding of what controls the rate of increase and spread of a population. This applies to bacteria and (1) ____________, as well as plants and animals, and hence has important applications in (2) ____________, the study of the spread of disease. An example is the spread of rabies, a (3) ____________ that has specific characteristics promoting its spread, such as its presence in high densities in an animal's (4) ____________. The rabies eradication program in Ontario consisted of three stages: first was the development of the (5) ____________, second was the vaccination of the primary (6) ____________ of the rabies, (7) ____________, and the third was the measurement of success by (8) ____________ how many foxes were vaccinated. This successful rabies control program was designed to control population (9) ____________ and spread using many different disciplines in biology, including population biology, (10) ____________, immunology, and epidemiology, among others.

45.1 The Science of Ecology [pp. 1112–1113]

Section Review (Fill-in-the-Blanks)

Ecology can be studied on a number of hierarchical levels. (11) ____________ ecology is the study of the various adaptations of individual organisms to the environment. The other levels include (12) ____________ ecology, the study of groups of organisms of the same species living and interacting together; (13) ____________ ecology, which focuses on sympatric populations of different species and their interactions; and (14) ____________ ecology, the study of nutrient cycles and energy flow in the biotic and abiotic environments of the community. Ecologists can address questions at the various hierarchical levels using hypothetical (15) ____________ that provide simulations of natural ecosystems but that should be followed up with exploratory (16) ____________ work and controlled (17) ____________ studies to test predictions generated by their hypotheses.

45.2 Population Characteristics [pp. 1113–1117]

Section Review (Fill-in-the-Blanks)

Populations have characteristics that make them unique, and population ecology focuses on those characteristics. A fundamental characteristic is where you find the organisms, or the populations' geographic (18) ____________. More specifically, the biotic and abiotic features of a population's environment is their (19) ____________. Within a habitat, the population size is simply the number of individuals at any given time, but when this is divided by the volume or area of the habitat, it is the (20) ____________ density. Estimating population size is often difficult, and indirect methods are used, such as (21) ____________, which allows the calculation of population size using a simple ratio equation.

Predicting population growth or decline requires estimates of other population characteristics that relate to their potential for producing young. For example, an age structure that has many (22) ____________ individuals is likely to grow in the near future as they age and reproduce, while populations with a majority of (23) ____________ individuals are likely to decline. Additionally, species with short delays between birth and reproduction have short (24) ____________ times and are capable of rapid population growth. Finally, populations with few females and many males will have limited growth potential; thus, the (25) ____________ ____________ of a population is of interest to population ecologists.

Complete the Table

Fill in the blanks in the following table with the proper dispersal pattern for each described population.

Description of population	*Probable dispersion pattern*
Redwing blackbirds, which defend areas of roughly equal size against entry by others	26.
Dandelions, a hardy species that develop from windblown seeds	27.
Herring off the west coast of Canada, which are found in large schools of tens of thousands of individuals	28.
Crickets, which have very specific habitat requirements and communicate effectively	29.
Marine blue mussels on the coast of Nova Scotia, which settle onto the ocean bottom after about a month of passive floating in the currents	30.

45.3 Demography [pp. 1118–1119]

Section Review (Fill-in-the-Blanks)

The movement of individuals into a preexisting population is called (31) ____________, and the movement of individuals out of a population is referred to as (32) ____________. Both types of movement can change a population's size and density. The process that drives changes in population size and density is a population's (33) ____________, which can be summarized in a life table. A life table follows the lives of a single group of individuals of similar age, known as a (34) ____________, and tabulates age-specific death rates, or (35) ____________, and (36) ____________. Of course, population size and density also depend on the number of offspring produced and when they are produced; thus, life tables also typically give age-specific (37) ____________, the average number of offspring produced by females of a given age.

Calculations

Fill in the blanks in the following partial life table and then, using the information in the table, answer the question below.

Age interval (years)	*Number alive at the start of the interval*	*Number dying during interval*	*Age-specific mortality rate*	*Age-specific survivorship rate*
0–2	1000	800	(38) _____	(39) _____
2–4	(40) _____	140	0.700	0.300
4–6	40	25	0.625	0.375
6–8	15	(41) _____	0.533	0.467
8–10	7	(42) _____	(43) _____	(44) _____
10–12	0	—	—	—

45. What type of survivorship curve does this population most closely resemble? Explain your answer.

____

____

Matching

Match each of the following populations with the survivorship curve below that best describes it.

46. ____ Humans. Infant survivorship is relatively high, and individuals surviving past their first 6 months typically survive for many years until mortality increases sharply in old individuals.

47. ____ Starfishes. A species with few predators whose numerous larvae are highly vulnerable and early mortality is high. Those that survive and metamorphose into adults typically live for a long time.

48. ____ *Hydra*. A small, stationary freshwater organism often eaten by larger organisms. The chances of being preyed upon are fairly constant throughout their life span.

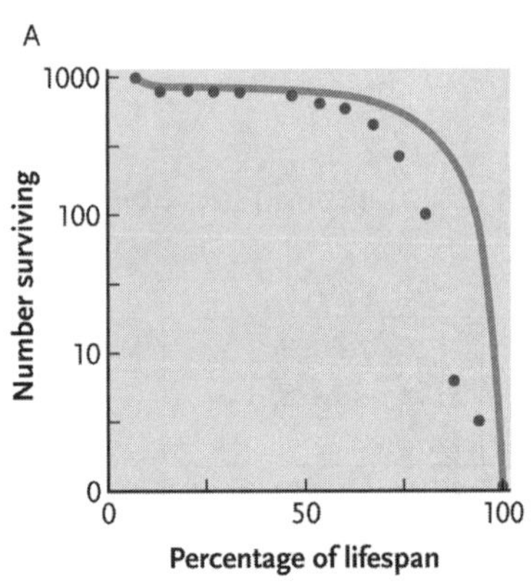

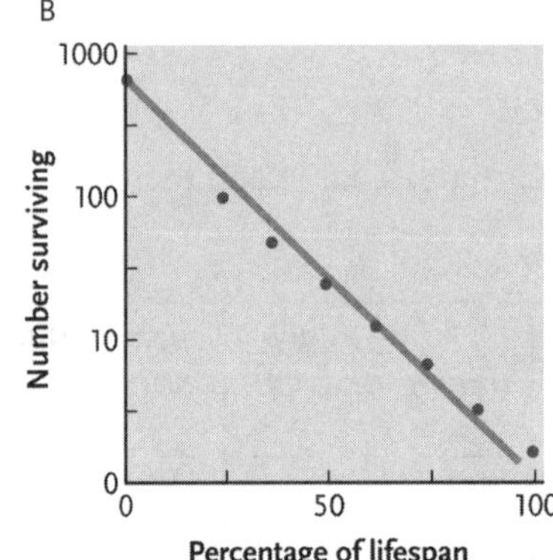

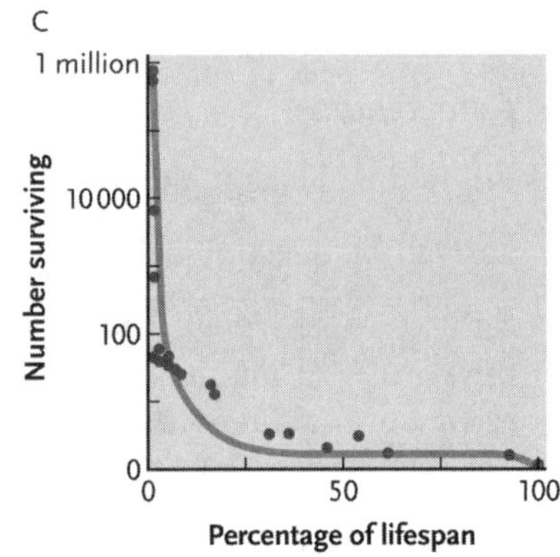

45.4 The Evolution of Life Histories [pp. 1120–1121]

Section Review (Fill-in-the-Blanks)

A life history is the lifetime pattern of development, or (49) ____________, reaching reproductive age at (50) ____________, and producing offspring, or (51) ____________. An organism's life history is constrained by the total energy accumulated (referred to as its (52) ____________ ____________), so organisms must make (53) ____________ between the components of their life history. The energetically expensive functions of most organisms can be placed into three categories: the energy required for basic functioning, or (54) ____________; the energy required to increase body size, or (55) ____________; and the energy needed to produce offspring through (56) ____________. There is a wide diversity in the way species allocate their energy among the categories, especially for reproduction: energy invested in young before birth or hatching is referred to as (57) ____________ parental care, while energy spent after birth or hatching to raise offspring is referred to as (58) ____________ parental care.

True or False

If the statement is true, write a "T" in the blank. If the statement is false, make it correct by changing the underlined word(s) and writing the correct word(s) in the answer blank.

59. ____________ Generally, species that produce large numbers of relatively small offspring exhibit <u>active</u> parental care.

60. ____________ In many organisms that get older and larger, their fecundity <u>increases</u>, and they may delay reproduction.

61. ____________ If an organism has a low likelihood of surviving past reproduction (such as coho salmon), it <u>will likely reproduce many times</u>.

62. ____________ <u>High</u> levels of predation may select for early reproduction.

63. ____________ <u>Natural selection</u> probably shapes most species' life history strategies.

Choice

For each of the following population characteristics, choose the more likely time of reproduction.

a. Early reproduction   b. Delayed reproduction

64. _____ High survivorship among adults

65. _____ Long life span

66. _____ Low fecundity of larger individuals

67. _____ Size of individuals increases with age

45.5 Models of Population Growth [pp. 1121–1127]

Section Review (Fill-in-the-Blanks)

The (68) ____________ ____________ ____________ ____________, or "r," is an important component of the two common models of population growth. The (69) ____________ model of population growth results in a population growing at an unrestricted and increasing rate over time. Under perfect conditions, r attains its highest possible value and is referred to as r_{max}, or the population's (70) ____________ ____________ ____________ ____________. In the (71) ____________ model of population growth, r becomes smaller as the population approaches its

(72) ____________ ____________, or "*K*," the number of individuals that the environment can sustain indefinitely. This model takes into account the effect of (73) ____________ ____________, the competition for a limited resource by two or more members of the same species. One of the flaws of this model is that it assumes that fecundity and survivorship respond immediately to changes in population density. In reality, there is a delay or (74) ____________ ____________ involved.

Matching

Match the correct change to a population associated with the description of a model. Use each answer only once.

75.	____	N is small (logistic model)	A.	Population size will decrease
76.	____	B = D (exponential model)	B.	Population size will remain constant
77.	____	N > K (logistic model)	C.	Population size will increase exponentially
78.	____	$r > 0$ (exponential model)	D.	Population will go extinct
79.	____	B = 0 (exponential model)	E.	r will approach r_{max}

True/False

If the statement is true, write a "T" in the blank. If the statement is false, make it correct by changing the underlined word(s) and writing the correct word(s) in the answer blank.

80. ______________ In the logistic model of population growth, when dN/dt = 0, the population <u>has gone extinct</u>.

81. ______________ Populations <u>cannot</u> exhibit exponential growth indefinitely.

82. ______________ The logistic model of population growth predicts that *r* <u>decreases</u> with population size.

83. ______________ The logistic model of population growth accounts for increasing <u>interspecific</u> competition as population size increases.

45.6 Population Regulation [pp. 1127–1132]

Section Review (Fill-in-the-Blanks)

As population density increases, it becomes easier for parasites to spread from host to host. This is an example of a density- (84) ____________ effect on population size. Natural disasters, on the other hand, kill a fixed percentage of a population regardless of density. This is an example of a density- (85) ____________ effect.

The type of population growth a species exhibits is often correlated with other life history strategies. Species that show exponential growth when conditions are favourable are said to be (86) ____________. Species exhibiting logistic growth and utilizing density-dependent mechanisms to maintain population size near carrying capacity are referred to as (87) ____________; however, that distinction is relative, so it can be difficult to unambiguously categorize a population. The difference between (88) ____________ (internal) and (89) ____________ (external) control of population size is that the first depends primarily on changes in the population or organisms, while the latter depends primarily on changes in the environment (including other species). Thus, population (90) ____________ may be the result of (91) ____________ ____________ in the response of the population to changes in complex interactions between the species and their environment.

Choice

For each of the following population characteristics, choose the most likely life history pattern.

a. *r*-selected b. *K*-selected

92. _____ Small body size

93. _____ Long life span and large body size

94. _____ Type III survivorship curve

95. _____ Repeated and frequent reproductive events in an individual's lifetime

96. _____ Substantial active parental care of offspring

True or False

If the statement is true, write a "T" in the blank. If the statement is false, make it correct by changing the underlined word(s) and writing the correct word(s) in the answer blank.

97. ____________ In density-dependent populations, as density increases, body size often decreases.

98. ____________ In density-dependent populations, the probability of survival increases as population density increases.

99. ____________ In density-independent populations, very low population density reduces reproductive success.

100. ____________ In density-independent populations, high population density often results in higher predation mortality

45.7 Human Population Growth [pp. 1132–1136]

Section Review (Fill-in-the-Blanks)

Over the last 200 years, human population growth has been (101) ____________. Humans have avoided limits to population growth in three ways: (i) humans have expanded their (102) ____________ ____________ and colonized virtually all terrestrial (103) ____________, (ii) the shift from hunting and gathering has increased the (104) ____________ ____________ of human habitats, and (iii) modern medicine and improved hygiene have served to reduce the (105) ____________ rate.

The economic status of a country can have dramatic impacts on life history factors that drive population growth. A (106) ____________ ____________ model depicts the expected impact of preindustrial through postindustrial development on birth and death rates, ultimately determining expected population growth rates. Many countries with large and growing populations are attempting to control population growth through (107) ____________ ____________, although in some countries that is culturally unacceptable and even unlawful.

True or False

If the statement is true, write a "T" in the blank. If the statement is false, make it correct by changing the underlined word(s) and writing the correct word(s) in the answer blank.

108. __________ The preindustrial stage is characterized by <u>low</u> death rates and high birth rates.

109. __________ The postindustrial stage is characterized by birth rates <u>lower</u> than death rates and population decline.

110. __________ Women delaying having their first child will <u>increase</u> population growth.

111. __________ Mexico will have faster population growth than Canada because it has <u>fewer</u> prereproductive individuals.

Matching

Match each age structure pyramid in the figure below to its description.

112. ____________ A. Many young males and females soon to reproduce: population "bomb"

113. ____________ B. Many older males and females: declining population numbers

114. ____________ C. Equal numbers of reproductive and prereproductive people: zero growth

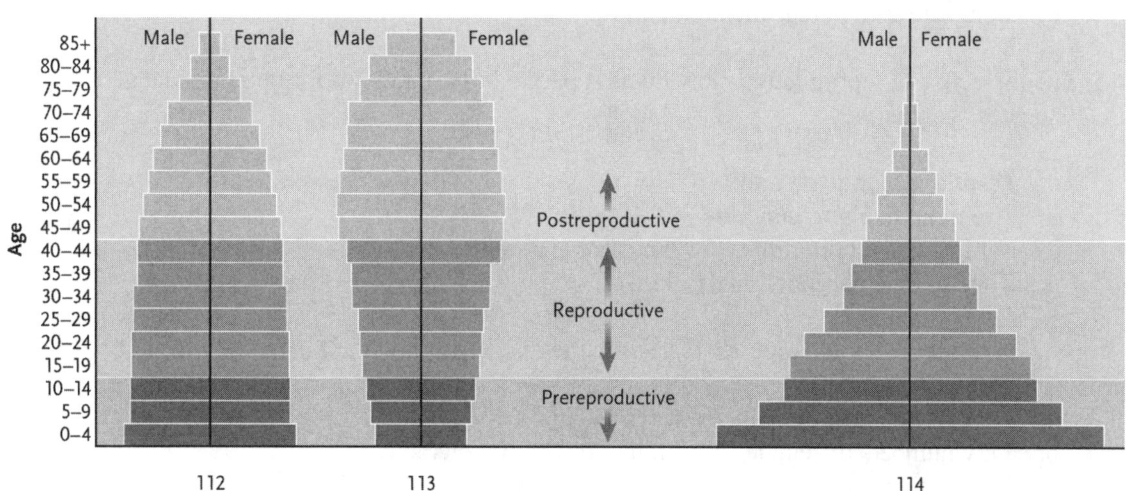

SELF-TEST

1. Which of the following is <u>not</u> a branch of the science of ecology? [pp. 1112–1113]

   a. Abiotic ecology
   b. Community ecology
   c. Organismal ecology
   d. Population ecology

2. Which dispersion pattern best describes the distribution of humans in Canada? [pp. 1114–1117]

   a. Clumped
   b. Even
   c. Random
   d. Uniform

3. The statistical study of the factors that affect population size and density is called ________. [p. 1118]

   a. Biogeography
   b. Demography
   c. Organismal ecology
   d. Logistics

4. Which of the following pairs represents tradeoffs in life history strategies? [pp. 1120–1121]

   a. High fecundity versus intense parental care
   b. Geographic range versus habitat
   c. Intraspecific competition versus population density
   d. Long life span versus metabolic rate

5. The incidence of reproductive individuals is particularly important for rare or endangered species because they _______. [p. 1117]

   a. Have a higher death rate
   b. Are more likely to emigrate
   c. Will have the opportunity for rapid population growth
   d. Tend to be older individuals

6. Why is sex ratio a population characteristic that can affect population growth? [p. 1117]

   a. A 50:50 sex ratio maximizes survivorship
   b. Low numbers of females can limit population growth
   c. Low numbers of males will block reproduction
   d. High numbers of females lead to male emigration

7. The exponential model of population growth assumes ________. [pp. 1121–1125]

   a. That birth rate is equal to death rate
   b. That density-dependent factors are involved
   c. Ideal conditions with no limits to growth
   d. That species are *K*-selected

8. Which of the following is a key component of the logistic model of population growth? [pp. 1125–1127]

   a. Species range
   b. Body size
   c. Density-independent factors
   d. Carrying capacity

9. Which of the following is not a density-dependent factor affecting population size? [pp. 1127–1129]

   a. Crowding
   b. Intraspecific competition
   c. Nighttime air temperature
   d. Spread of disease

10. The rabies virus primarily extends its range by _______. [pp. 1111–1112]

    a. Density-independent factors
    b. Transmission through biting
    c. Altering its dispersion pattern
    d. Insect carriers

11. All species have geographic range boundaries; this is driven primarily by _____. [p. 1113]

    a. *K*-selection
    b. The logistic growth curve
    c. Birth versus death rates
    d. Availability of suitable habitat

12. Extrinsic cycles in population size may be due to _______________. [pp. 1131–1132]

    a. Predation
    b. Hormonal suppression of maturity
    c. Large habitat size
    d. Temperature tolerance

13. It took ________ years for the human population of the world to grow from 5 to 6 billion. [p. 1132]

    a. 1.5 million
    b. 40 000
    c. 500
    d. 12

14. Which population characteristic would be most important in determining whether a community should spend money on schools versus nursing homes? [pp. 1133–1135]

    a. Age structure
    b. Carrying capacity
    c. Population density
    d. Sex ratio

15. If a population decreases from 500 individuals to 300 in their first year of life, what are the age 0–1 specific mortality and survivorship rates, respectively? [pp. 1118–1119]

    a. 0.400 and 0.600
    b. 0.600 and 0.400
    c. 0.375 and 0.625
    d. 60% and 40%

16. If a population of salmon with fecundity that increases with age and body size experience a change in their predation rate such that they have a much higher survivorship late in life, what should happen to their age of first reproduction? [pp. 1120–1121]

    a. It will not change
    b. It will increase
    c. It will decrease
    d. The population will decline

17. What causes differences in life history patterns among populations of a single species? [p. 1121]

    a. Habitat differences
    b. Natural selection
    c. Intrinsic rate of increase
    d. Both (a) and (b)

18. In the logistic model of population growth, what is expected to happen when K > N? [pp. 1125–1127]

    a. The population will decline
    b. The population will continue to grow
    c. The population will be eliminated
    d. The population will expand its range

19. Which of the following are likely population responses to overcrowding? [pp. 1127–1129]

    a. Increased mortality rates
    b. Decreased individual condition
    c. Increased emigration
    d. All of the above

20. In r-selected populations, juvenile survival is expected to be ______. [p. 1130]

a. Higher than in K-selected populations
b. About the same as in K-selected populations
c. Lower than in K-selected populations
d. Exponentially increasing with density

INTEGRATING AND APPLYING KEY CONCEPTS

1. Compare the impact of commercial fishing pressure on fish populations to the effect of pike cichlid predators on guppy populations in Trinidad. Explain similarities or differences you would expect. Would you expect similar impacts from the effect of pollutants on fish?

2. Some of the countries in the world with the worst overcrowding exhibit the fastest population growth, yet we know from our study of density-dependent factors that overcrowded populations usually suffer higher death rates and lower birth rates. Why are human populations not following that "rule"?

3. The Ontario rabies eradication program included three phases for ultimately eliminating rabies in southern Ontario, but despite success, rabies still persists at low levels. Using your knowledge of density-dependent factors that act on the rabies virus, explain why it would be very difficult to eradicate (eliminate) rabies from southern Ontario.

OPTIONAL ACTIVITIES

1. *Focus on Research:* Population fluctuations are common, even in long-established and presumably stable populations. We know time lags contribute to population cycling, but what would contribute to random population size fluctuations? List four possible factors and check the Internet for an example of a species that has been shown to respond to your proposed factor.

2. The Canadian Great Lakes are being rapidly invaded by a host of non-native aquatic species. In some cases (such as round gobies and fishhook waterfleas), the speed of the population increase and range expansion is faster than expected even under exponential growth. Can you think of a mechanism that may be driving such a remarkably rapid invasion process? How would you test your hypothesis?

46 Population Interactions and Community Ecology

CHAPTER HIGHLIGHTS

- Populations adapt to the physical characteristics of their environment, as well as to other organisms they encounter (coevolution).
- Feeding is key to survival; obvious species interactions in ecological communities are predation and herbivory, with prey using a variety of mechanisms for defence.
- Ecological communities are characterized by complex species interactions such as interspecific competition, trophic interactions, and symbioses (i.e., commensalism, mutualism, and parasitism).
- Communities differ in the number of species present (diversity) and the relative abundance of each species (evenness).
- Species richness in a community may actually be increased by predation and/or moderate levels of disturbance. The theory of island biogeography is a useful model to explain variations in species richness.
- Communities undergo a characteristic sequence of changes through time, a process known as ecological succession.

STUDY STRATEGIES AND LEARNING OUTCOMES

- Since the emphasis of this chapter is on interactions among species and the effect of those interactions on populations and communities, be sure to become familiar with the various types of interactions as they may be confusing and then try describing the types of interactions in your own words.
- Next, review the variety of predation/herbivory defences that have evolved, noting which apply to plants and which to animals.
- Review interspecific competition in terms of mechanisms that maximize the likelihood of "winning" versus those that minimize competition.
- To become familiar with measurements of diversity, make up a data set of species numbers in a simple community (say five or six species), calculate Shannon indices of

diversity (H′) and evenness (E_H) and then change the numbers in the way you think will increase and/or decrease diversity. Did diversity change as you expected when you recalculated H′? Try this with E_H as well.

- After listing the various factors that are thought to affect climax community composition, think of examples where each factor may be in play.

By the end of this chapter, you should be able to

- Explain how coevolution can ultimately shape species interactions in ecological communities
- Describe the various forms of species interactions that contribute to community characteristics
- List and provide examples of the types of defences that animals and plants use to avoid becoming prey
- Define, calculate, and interpret indices of species diversity and evenness
- Predict how competition may affect species diversity
- Explain trophic interactions and draw a simple food web
- Describe how the succession process determines the climax community
- Explain how the island theory of biogeography predicts species diversity in a variety of communities

TOPIC MAP

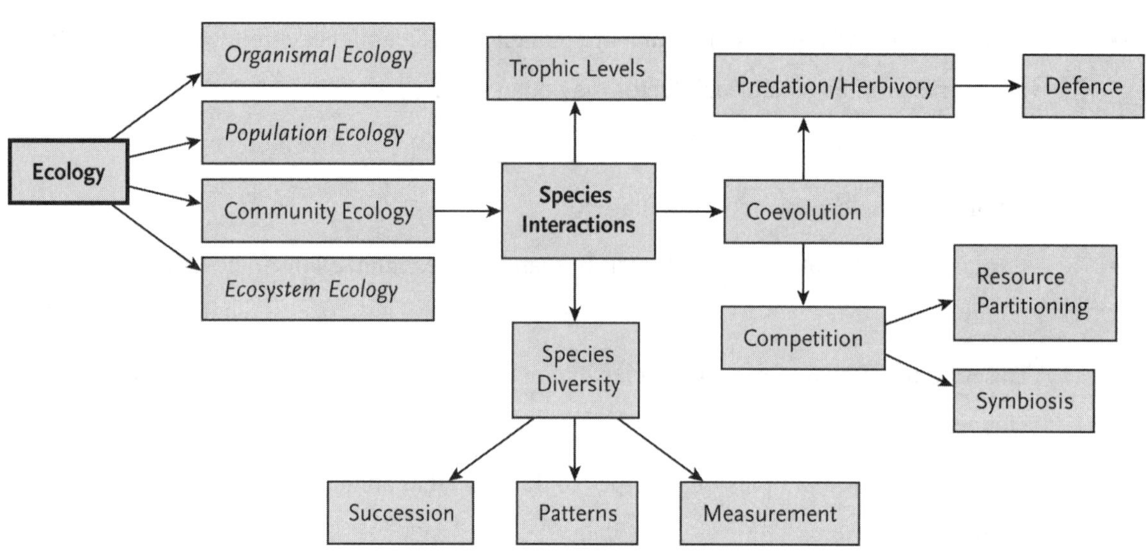

INTERACTIVE EXERCISES

Why It Matters [pp. 1141–1142]

Section Review (Fill-in-the-Blanks)

The natural variation in the species composition and abundance in various ecosystems is apparent to even the casual observer; however, the study of (1) ____________ seeks to explain differences and patterns in species diversity and abundance. For example, the diet of tropical caterpillars is more (2) ____________ than the species found in temperate forests. This difference is thought to allow more ecological (3) ____________ for the caterpillars in the tropical forest and thus more "room" for species of butterflies. One of the basic approaches in community ecology is to explain differences in ecological community diversity through detailed study of species (4) ____________.

46.1 Interspecific Interactions [pp. 1142–1143]

Section Review (Fill-in-the-Blanks)

Often species interactions affect one or more of the species involved positively or negatively. If a particular interaction affects survival or reproduction in one of the species, then we expect (5) ____________ through natural selection. However, if two (or more) species are involved, both interacting species will experience natural selection, thus potentially driving (6) ___________, or genetically based reciprocal adaptation. A simple example is predator–prey adaptation, where, for example, the speed of the antelope prey forces the cheetah to adapt by evolving, even (7) ___________ chasing speeds to survive. In reality, that interaction is much more complex since predators (cheetahs) can have multiple potential (8) ___________, and prey (antelopes) have many potential (9) ___________.

46.2 Getting Food [pp. 1143–1144]

Section Review (Fill-in-the-Blanks)

The process of gaining nutrients by consuming other organisms is herbivory if eating (10) ___________ and predation if eating (11) ___________. Predators and herbivores have evolved effective mechanisms to locate and feed upon their prey: for example, rattlesnakes use (12) ___________ found in pits on their heads to detect warm bodies, while (13) ___________ use chemical sensors on their legs to find food. Predators also differ in the range of food they can tolerate: those with a broad range are (14) ___________, while those with specific needs are (15) ___________. The choice of a predator's (or herbivore's) food is based on maximizing the ratio of (16) ___________, and the models of how predators choose the best food to maximize this ratio are identified by the (17) ___________ ___________ theory.

True or False

If the statement is true, write a "T" in the blank. If the statement is false, make it correct by changing the underlined word(s) and writing the correct word(s) in the answer blank.

18. ___________ Fish that feed on only one species of coral in tropical coral reef ecosystems are <u>generalist</u> feeders.

19. ___________ Optimal foraging theory predicts that predators choose prey that provide the <u>most</u> energy for the least effort and risk.

20. ___________ If the encounter rate for a specific type of prey is <u>low</u>, the predator will often concentrate on that prey.

21. ___________ Omnivores are very good examples of generalist feeders.

46.3 Defence [pp. 1144–1151]

Section Review (Fill-in-the-Blanks)

Most species are prey to another species, so prey species develop elaborate mechanisms of (22) _____ to avoid being eaten. Some of these mechanisms are simple, such as having a (23) _____ body size, which may make the prey not worth capturing, or (24) _____ body size, which may make the prey difficult to capture and kill. Some prey defend against predators through physiological and behavioural specializations, such as (25) _____ predators prior to attack, or through cryptic or (26) _____ coloration or behaviour. Alternatively, prey species can avoid attack through more aggressive means such as defensive (27) _____ (as on a porcupine or some plants) or through (28) _____ defence (e.g., a skunk or a rattlesnake). When an animal uses a chemical defence, it may advertise the fact through bright colours or distinctive patterns; such species are referred to as (29) _____. However, such an elegant adaptation is a target for species that (30) _____ the well-defended species. (31) _____ _____ is when a harmless species looks similar to the poisonous species, while (32) _____ _____ is when two poisonous species resemble each other to increase the likelihood that they will be recognized by predators.

Matching

Match the prey species on the left to the type of defence it would use on the right.

33.	____	Arrowhead frog	A.	Predator detection
34.	____	Leopard tortoise	B.	Chemical defence
35.	____	Meerkats	C.	Large size
36.	____	Cowhorn euphorb	D.	Armour defence
37.	____	Elephants	E.	Active defence (spines/thorns)

Choice

For most forms of prey defence, predators have developed mechanisms to overcome the defence. For each prey defence listed below, choose the best potential adaptation for the predator to overcome the defence. Choices may be used more than once.

38.	____	Evasive action	A.	Detect prey by characteristic smell
39.	____	Camouflage	B.	Capture prey by stealth or speed
40.	____	Vigilance	C.	Evolve resistance
41.	____	Poisonous	D.	Switch to different prey
42.	____	Batesian mimicry	E.	A or C

46.4 Competition [pp. 1151–1155]

Section Review (Fill-in-the-Blanks)

Two or more individuals of competing species for the same limited resource are interacting by (43) ____________ competition. This phenomenon takes two forms: (44) ____________ competition, in which individuals interact directly and attempt to harm members of the other species, and (45) ____________ competition, where use of the resource by one species reduces its availability to the other. The experiments of G.F. Gause indicated that when two species are competing for the same limited resource in the same way, the more efficient species eliminates the other from the system. Based on these results, he proposed the more general (46) ____________ ____________ principle. Resource utilization and the environmental conditions an organism requires are major components of an organism's ecological (47) ____________. The fact that two or more species use the same resource does not necessarily imply that competition is occurring since (48) ____________ ____________ can minimize interspecific competition. Often closely related species living in the same area will specialize on different resources and evolve structures to help them exploit the specific resource, a phenomenon known as (49) ____________ ____________.

There are three main forms of symbiosis: (50) ____________, where both species benefit from an interaction; (51) ____________, where one species benefits and the other is unaffected; and (52) ____________, where one species benefits and the other is harmed by the association. True parasites usually do not kill their hosts. Since (53) ____________ are animals that lay eggs within the larvae of host insects, eventually consuming and killing their host, they lie somewhere between a parasite and a predator.

Choice

For the species interactions listed, choose the correct pair of effects for the two species involved. Answers may be used more than once.

54. ____ Commensalism
55. ____ Competition
56. ____ Herbivory
57. ____ Mutualism
58. ____ Parasitism
59. ____ Predation

a. Species A benefits; species B is harmed
b. Species A benefits; species B is unaffected
c. Both species benefit
d. Both species are harmed

46.5 The Nature of Ecological Communities [pp. 1155–1156]

Section Review (Fill-in-the-Blanks)

The American ecologist Frederic Clements hypothesized that a mature community was at (60) ____________ and that if the community was disturbed, it would tend to return to its original state. On the other hand, Henry Gleason proposed that a community is an assemblage of species individually adapted to similar conditions and that no (61) ____________ would be attained; rather, the community would change in response to (62) ____________ or environmental change. Robert Whittaker suggested a test of these two hypotheses by examining species composition of communities over environmental (63) ____________. Most communities showed gradual change, thus supporting (64) ____________ view. The transition zone between adjacent communities is an (65) ____________ and is often characterized by elevated levels of species (66) ____________.

True or False

If the statement is true, write a "T" in the blank. If the statement is false, make it correct by changing the underlined word(s) and writing the correct word(s) in the answer blank.

67. ____________ Ecotones are <u>broad</u> when there is an abrupt change in critical resources.

68. ____________ Ecotones generally have <u>high</u> species diversity because they include species from both bordering communities and species adapted to intermediate conditions.

69. ____________ The <u>interactive</u> view of communities is that it will return to its original state after disturbance.

70. ____________ The <u>nonequilibrium</u> view of community structure predicts stable structure over time.

46.6 Community Characteristics [pp. 1156–1161]

Section Review (Fill-in-the-Blanks)

The number of species found in a community comprises that community's species (71) ____________. The relative commonness or rarity of a species in a community is its species (72) ____________. Combining these two community characteristics allows ecologists to objectively quantify the species (73) ____________ of a community.

All ecological communities have (74) ____________ structure that includes all plant–herbivore, prey–predator, and host– (75) ____________ interactions. At the base of this hierarchy are photosynthetic organisms or (76) ____________ producers, which are photoautotrophs. Animals and other nonphotosynthesizing consumers are (77) ____________. Herbivores occupy the second level and are called (78) ____________ consumers. Secondary consumers are carnivores that feed on the herbivores and occupy the (79) ____________ level in the trophic structure. Carnivores on the fourth level are called tertiary consumers, and so on. Animals that feed on organisms from the first and higher trophic levels are called (80) ____________. Organisms that extract energy from dead animals or organic matter are called (81) ____________, while decomposers are smaller organisms such as (82) ____________ and fungi.

The trophic structure of a community is often depicted in the form of a (83) ____________ ____________, a simple, linear diagram. But a more realistic depiction is the (84) ____________ ____________ since it allows for organisms to feed on different foods and from different trophic levels. Species diversity, (85) ____________ complexity, and community (86) ____________ are interlinked, such that more links in the food web will increase community stability.

Complete the Table

In the table below, column 2 shows the species distribution for a simple community (Community 1). By performing the calculations and filling in the blanks, you will calculate Shannon's index of diversity (H′). First, calculate p_i (the proportion of each species to the total number of individuals) in the third column. Next, calculate the term $p_i(\ln(p_i))$ in column 4 using data from column 3. Calculate the sum of column 4 for the total.

Community 1	*Number of individuals*	p_i	$p_i(\ln(p_i))$
Species A	6	(87) ____	(88) ____
Species B	10	0.22	–0.33
Species C	3	0.07	–0.18
Species D	1	(89) ____	–0.08
Species E	25	0.56	(90) ____
TOTALS	**45**	**1.00**	(91) ____

92. What is the Shannon's index of species diversity (H′) for this community? Is this community more or less diverse than a community with H′ = 0.50?

Matching

Using the information in the following diagram, match each of the organisms with one or more appropriate terms.

93.	____	Autotroph	A.	Crabeater seal
94.	____	Carnivore	B.	Emperor penguin
95.	____	Herbivore	C.	Orca
96.	____	Primary consumer	D.	Phytoplankton
97.	____	Primary producer	E.	Zooplankton

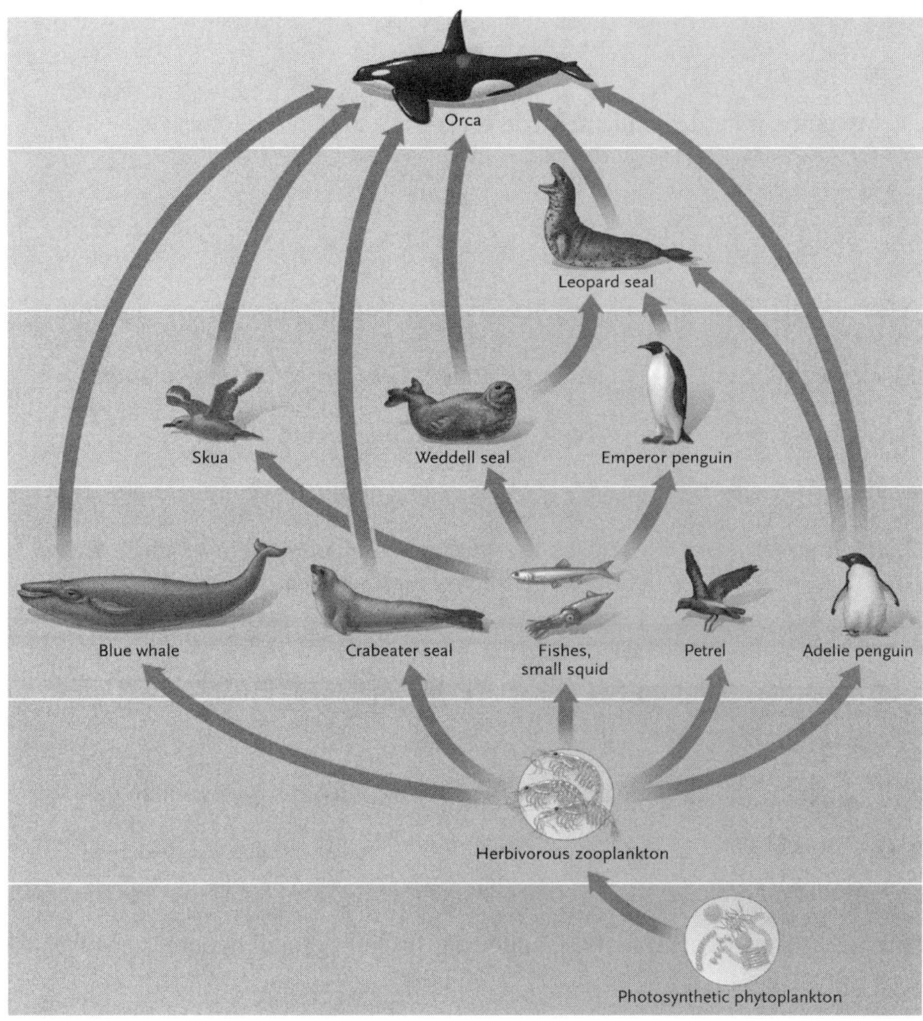

46.7 Effects of Population Interactions on Community Structure [pp. 1161–1162]

Section Review (Fill-in-the-Blanks)

Interspecific competition can limit species diversity since the inferior competitor may go locally (98) ___________, or competition may exclude new species from becoming established. Experiments designed to test the effect of adding or removing species from a community have resulted in complex outcomes that are not completely understood by ecologists. However, in surveys of published competitive interactions, researchers have estimated that more than (99) ___________% of the species in ecological communities are sensitive to changes in competition. There is variation between species since ecologists working with plant or (100) ___________ vertebrate species both conclude that competition is a critical factor in community structure, while insect ecologists working with (101) ___________ species argue that competition is not a major factor. Interestingly, predators can act to (102) ___________ species richness, as is the case for the marine mussel predator the sea star (*Pisaster*). When *Pisaster* is excluded, the marine mussel (*Mytilis*) dominates the near-shore community and species (103) ___________ declines. Because of the wide range of ecological roles *Pisaster* plays in the marine near-shore community, it has been identified as a (104) ____________ species.

Matching

Match the species description on the left with its interactions within a community on the right.

105.	____	*K*-selected species	A.	Relatively few competitive interactions
106.	____	Keystone species	B.	A dominant competitor
107.	____	*r*-selected species	C.	Many strong competitive interactions
108.	____	*Mytilis* mussel	D.	Larger impact on community than expected based on numbers

46.8 Effects of Disturbance on Community Characteristics [pp. 1163–1165]

Section Review (Fill-in-the-Blanks)

Recent work generally supports the individualistic view of community structure, where the species composition is not at (109) ____________, because physical (110) ____________ such as storms, floods, fires, etc. often eliminate some species, while making way for new ones. Based on analyses of long-term monitoring of coral reef communities, Joseph Connell hypothesized that communities that are subject to disturbances of moderate frequency and intensity show greater species diversity than those that are severely disturbed and/or with great frequency or those that are rarely disturbed. The (111) ____________ ____________ hypothesis predicts that communities experiencing moderate levels of disturbance will include both (112) ____________ species that colonize the disturbed areas and (113) ____________ species that remain in the undisturbed areas. Communities that have high species (114) ________ richness were proposed to be more stable over time, which is supported by a study of (115) ____________ plots recovering from fire.

True or False

If the statement is true, write a "T" in the blank. If the statement is false, make it correct by changing the underlined word(s) and writing the correct word(s) in the answer blank.

116. ____________ Frequent and very large storms can foster <u>very high</u> species diversity in coral reef communities.

117. ____________ <u>Physical</u> disturbances include storms, fires, floods, avalanches, and earthquakes.

118. ____________ High species diversity buffers change in future species composition.

119. ____________ Recruitment is the growth and survival of new individuals arising from larvae that have settled in coral reefs.

120. ____________ Oceanographic models are not useful for predicting the impact of disturbances on marine communities.

46.9 Succession [pp. 1165–1169]

Section Review (Fill-in-the-Blanks)

When a community is disturbed, it does not immediately return to its equilibrium state but rather goes through a sequence of changes known as (121) ____________. If an area has never supported a community before, for example, a volcanic island, the sequence of changes is referred to as (122) ____________ ____________. The final, relatively stable community in the sequence is the (123) ____________ ____________. Secondary succession is the (124) ____________ of a community after it has been disturbed. The process of ecosystem development is not restricted to terrestrial communities. If the process occurs in a water body, such as a lake ecosystem, it is referred to as (125) ____________ ____________.

Several hypotheses have been proposed to explain the mechanisms of ecological succession. The (126) ____________ hypothesis states that earlier species alter conditions in ways that make it easier for later species to get established. The (127) ____________ hypothesis is based on the notion that earlier communities resist colonization by new species, but disturbances and life span of species ultimately lead to changes in community structure. The (128) ____________ hypothesis states that new species colonize according to their superior competitive ability and that the earlier species neither facilitate nor inhibit new species colonization. The three models are not mutually exclusive, and all three may play a role in succession. Sometimes biotic or abiotic disturbance disrupts succession from proceeding to a true climax community; the resulting community is a (129) ____________ community.

Choice

For each statement, choose the term from those below. Answers may be used more than once.

a. Not predictable
b. *K*-selected species
c. Dispersal
d. Primary succession
e. Disturbance

130. The development of a new rocky habitat that alters the environment to allow subsequent colonizers is ______.
131. The final equilibrium community usually includes many ______.
132. A climax community includes both established and newly colonized species due to new, open habitat created by ______.
133. According to the inhibition hypothesis, the order of species colonization during succession is _______.
134. High species richness may be a result of succession coupled with moderate _______.

46.10 Variations in Species Richness among Communities [pp. 1169–1173]

Section Review (Fill-in-the-Blanks)

Community ecologists try to identify patterns in the variation in species richness among communities. Two large-scale patterns that have been documented are (135) ___________ that reflect community differences from the tropics to the polar regions and (136) ___________ ___________, which occur in small isolated patches of habitat.

A number of hypotheses exist to explain the generally observed higher species richness in tropical communities. One possibility is that tropical ecosystems may have higher speciation rates due to a higher number of (137) ___________ per year, the result of a benign climate, warmer temperatures, and a higher level of local isolation due to reduced (138) ___________ in the stable climate of the tropics. Other hypotheses focus on the year-round availability of abundant resources that lead to feeding (139) ___________ and hence high species richness.

MacArthur and Wilson's (140) ___________ ___________ of island biogeography predicts species richness on islands (or isolated habitats such as mountain tops and lakes) based on island size, distance from the mainland, and immigration and extinction rates. Under this theory, the mainland acts as a (141) ___________ ___________, providing colonizers for the islands, and islands closer to the mainland will receive more colonizers. (142) ___________ islands also receive more colonizers than (143) ___________ islands. The combined rate of (144) ___________ of new species will be balanced by the rate of species (145) ___________ at equilibrium to produce a characteristic species richness that depends on a balance between the arrival of new species versus the loss of the existing species.

Matching

Match each island with the appropriate characteristics.

146.	____	Large, distant island	A.	High immigration rate, high extinction rate
147.	____	Large island, close to mainland	B.	High immigration rate, low extinction rate
148.	____	Small, distant island	C.	Low immigration rate, high extinction rate
149.	____	Small island, close to mainland	D.	Low immigration rate, low extinction rate

SELF-TEST

1. Which of the following is most likely a result of coevolution? [pp. 1142–1143]

   a. Ability to tolerate extreme low temperatures
   b. Batesian mimicry
   c. Specialized digestive organs
   d. Territoriality

2. Which of the following is <u>not</u> a symbiosis? [pp. 1153–1155]

   a. Commensalism
   b. Intraspecific competition
   c. Mutualism
   d. Predation

3. Toxins used for defence may be synthesized by the user or they may be from ______. [p. 1146]

   a. Other plants or animals
   b. Photosynthesis
   c. The organisms' mother
   d. Water or air

4. Interference competition is characterized by _______. [p. 1151]

   a. One species that benefits; the other is unaffected
   b. Individuals that directly harm their competitors
   c. Individuals that use the same limiting resource
   d. Both species benefitting

5. Under intense competition for the same resources, __________ may occur. [p. 1152]

   a. Realized niche
   b. Facilitation
   c. Mimicry

d. Resource partitioning

6. Which of the following is(are) true of commensalism? [p. 1153]

   a. It is rare in nature
   b. One species benefits, the other is unaffected
   c. Both species are harmed
   d. Both (a) and (b)

7. Parasitoids are organisms that parasitize other individuals, but they are different from true parasites because they ______. [pp. 1154–1155]

   a. Kill their host
   b. Benefit their host
   c. Have no effect on their host
   d. Do not have a host

8. The transition zone between adjacent communities that often has a high species diversity is called a(n) ________. [pp. 1155–1156]

   a. Climax community
   b. Ecotone
   c. Ecotype
   d. Trophic zone

9. Which aspect of species diversity is Shannon's evenness index (E_H) designed to reflect? [p. 1158]

   a. Species richness
   b. Competition
   c. Species number
   d. Relative abundance

10. Organisms can be classified as ________ based on how they obtain the energy necessary for life. [p. 1159]

   a. Consumers
   b. Decomposers
   c. Detritivores
   d. All of the above

11. Complex trophic relationships are typically shown as ______. [pp. 1159–1160]

   a. Trophic levels
   b. Diet diagrams
   c. Food webs
   d. Food chains

12. MacArthur proposed that species-rich communities with complex food webs would be _________ over time. [pp. 1159–1161]

   a. Extinct
   b. A climax community
   c. Stable
   d. Unstable

13. Predators may affect communities in many ways, such as they may _______. [pp. 1161–1162]

   a. Increase species richness
   b. Lead to local extinction of specific species
   c. Prevent new species from colonizing
   d. All of the above

14. A ________ is an organism that has a greater impact on community structure than its numbers might suggest. [pp. 1161–1162]

   a. Keystone species
   b. Müllerian mimic
   c. Mutualist
   d. Primary producer

15. Which of the following community characteristics would be predicted by the intermediate disturbance hypothesis? [pp. 1164–1165]

   a. A preponderance of *K*-selected species
   b. A preponderance of *r*-selected species
   c. Extreme sensitivity to slight disturbances
   d. High species diversity

16. In which of the following areas would you most likely observe primary succession? [p. 1165]

   a. A clear-cut forest with all trees removed
   b. A prairie after a fire
   c. An abandoned farm
   d. Rocky outcrops created by a receding shore

17. Which of the following is <u>not</u> a hypothesis that attempts to explain the processes that drive succession? [pp. 1165–1169]

   a. Island equilibrium hypothesis
   b. Facilitation hypothesis
   c. Inhibition hypothesis
   d. Tolerance hypothesis

18. Disturbances in a climax community tend to favour what kind of colonist? [pp. 1168–1169]

   a. *K*-selected
   b. Aquatic
   c. Generalist
   d. *r*-selected

19. Latitudinal clines are global patterns in species diversity where ______. [p. 1169]

   a. There are no species in the Arctic
   b. There are fewer species with increasing latitude
   c. There are more species with increasing latitude
   d. There is greater evenness with increasing latitude

20. Which assumption is necessary in MacArthur's and Wilson's equilibrium theory of island biogeography? [pp. 1171–1172]

   a. Islands close to the mainland have relatively low immigration rates
   b. Large islands have relatively low extinction rates
   c. Large islands have relatively low immigration rates
   d. Small islands have relatively low extinction rates

INTEGRATING AND APPLYING KEY CONCEPTS

1. Imagine four species: A and B are Müllerian mimics, C is a noxious model, and D is its Batesian mimic. Do you think each of the following scenarios would be beneficial or detrimental to each species? (i) A greatly outnumbers B; (ii) C greatly outnumbers D; and (iii) D greatly outnumbers C. Explain your answer.

2. Why would *K*-selected species be more likely to show strong competitive interactions than *r*-selected species?

3. The equilibrium theory of biogeography was developed for true islands, but it also applies to "islands" of suitable habitat in a "sea" of unsuitable habitat. List three such scenarios, explain how they are "islands," and describe what you think the "mainland" sources of new species might be.

OPTIONAL ACTIVITIES

1. The plasmid in the *Rhizobium* bacterium that promotes its mutualistic association with various legumes is similar to the plasmid in another bacterium that causes crown gall disease in deciduous trees. Speculate on the evolutionary relationship between the two bacteria.

2. Review Wilson and Simberloff's test of the equilibrium theory of island biogeography in terms of island size and distance from the mainland. Which predictions of the theory were supported by their results? Besides testing the theory, how else was this research important to field ecologists?

3. Go into your backyard and count the number of different species of plant (you do not have to identify them; just group them into the same species). Calculate Shannon's indices of species richness and evenness and repeat this for a more "wild" tract of land. How do you think richness and evenness may differ?

47 Ecosystems

CHAPTER HIGHLIGHTS

- The interaction of the biologic community with the physical, nonliving part of the environment is an ecosystem.
- Ultimately, the energy to maintain ecosystem structure and function comes from the sun. Ecosystems are usually open systems with respect to energy, with energy entering an ecosystem as solar radiation and leaving as entropy. Energy flows through ecosystems.
- Unlike energy, the materials that are used for ecosystem structure are in finite supply and must be recycled between the living and nonliving parts of the system. This is the basis of nutrient cycling.
- Ecosystems are extremely complex systems, with many components. Ecologists often attempt to understand ecosystem function by cutting through the noise in the system and creating models that incorporate only those variables that have the greatest impact on ecosystem function.

STUDY STRATEGIES AND LEARNING OUTCOMES

- It is important to understand that energy FLOWS THROUGH ecosystems, which can be tracked and quantified using food webs and ecological pyramids, whereas elements that make up the organisms in an ecosystem are in finite supply, so material CYCLES BETWEEN the living and nonliving parts of the system.
- Using this basic information, you can then interpret models that attempt to reduce a complex system to what is believed to be its most important components. It is important to realize that factors not incorporated into the model can have an impact.

By the end of this chapter, you should be able to

- Define and describe the flow of energy through ecosystems, from solar radiation to top predators
- Define and estimate various types of efficiencies of energy transfer in an ecosystem
- Interpret ecological pyramids in terms of efficiencies and relative availability of energy at each level

- Describe hydrogeologic, carbon, nitrogen, and phosphorus cycles and draw schematic diagrams to illustrate the cycles
- Explain how ecosystem models are constructed and what they are used for

TOPIC MAP

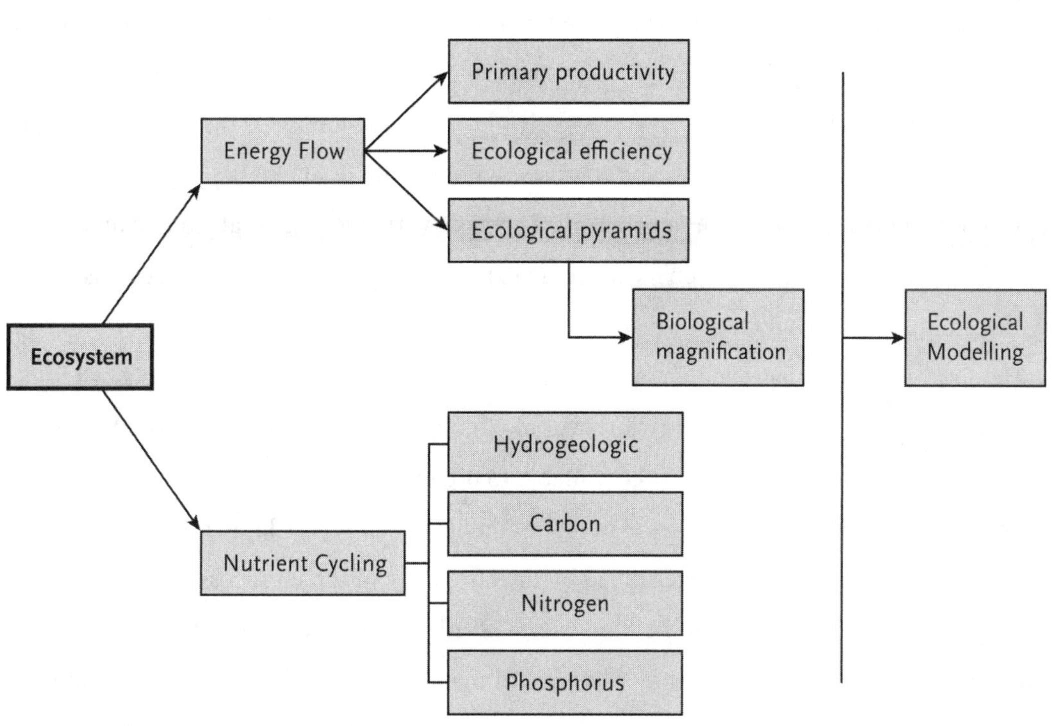

INTERACTIVE EXERCISES

Why It Matters [pp. 1179–1180]

Section Review (Fill-in-the-Blanks)

(1) ____________ ecosystems are the fastest growing habitat on earth today. Such ecosystems are very different from the way they were before human development; the runoff of (2) ____________ over the asphalt and concrete and the reflectance of energy from the (3) ____________ fundamentally change the nature of the ecosystem. These ecosystems are not recent; one of the earliest, Tell Brak, existed in Syria about (4) ____________ years ago. Although urban sprawl may not be new, it represents a serious threat to the conservation of (5) ____________.

47.1 Energy Flow and Ecosystem Energetics [pp. 1180–1189]

Section Review (Fill-in-the-Blanks)

The conversion of solar energy into chemical energy by autotrophs is known as the (6) ____________ primary productivity. Primary producers are, however, complex organisms, and the energy left after deducting the amount spent on maintenance is the (7) ____________ primary productivity. Generally, producers use between (8) ____________% and (9) ____________% of the energy they capture for respiration. There are several ways to estimate these different forms of energy. One is by measuring (10) ____________, the dry weight of organic matter per unit area or volume of a habitat. The (11) ____________ ____________ of biomass is the total amount of organic matter in an area, which is not to be confused with the rate of (12) ____________ (or production) of that matter. Like all organisms, plants need a variety of materials to live and function. The element in shortest supply is called a (13) ____________ nutrient because its scarcity determines productivity.

Heterotrophic organisms acquire some of the energy accumulated by autotrophs when they consume the autotrophs. Any energy left over after consumer maintenance is stored as (14) ____________ productivity. That energy is available to organisms at the next trophic level. The ratio of net productivity at one trophic level to that of the level below it is known as (15) ____________ efficiency, which is determined by the efficiency of three processes: (16) ____________ efficiency (the ratio of the food energy available to the energy in the food actually consumed), (17) ____________ efficiency (the ratio of the energy gained by consumption to the energy contained in the food consumed), and (18) ____________ efficiency (the ratio of the energy content of new biomass produced to the energy gained by consumption). The flow of energy through a food web is often represented diagrammatically in the form of ecological (19) ____________, of which there are three types. The (20) ____________ ____________ ____________ depicts the amount of energy temporarily stored in a trophic level as dry weight. This pyramid is often inverted in aquatic systems if the producers reproduce rapidly; that is, they have high (21) ____________ rates. The amount of energy at a trophic level often affects the population size of the organisms at the next higher level; thus, the (22) ____________ ____________ ____________ uses the number of individuals at a particular trophic level to represent energy. The (23) ____________ ____________ ____________ is the only one that actually measures energy flow, and the second law of thermodynamics dictates that it must show a reduction between each trophic level. Some interactions between two trophic levels have indirect impacts on lower trophic levels, a phenomenon known as a trophic (24) ____________.

The interaction among trophic levels and the interconnectedness of organisms in an ecosystem is dramatically demonstrated when (25) ____________ are introduced into the environment (e.g., DDT). The process whereby the concentration of the contaminant increases in higher trophic levels is (26) ____________ ____________.

Matching

Match each of the following ecological efficiencies with the correct definition.

27.	____	Assimilation efficiency	A.	Ratio of energy absorbed from food consumed to total energy content of food
28.	____	Harvesting efficiency	B.	Ratio of net productivity at one trophic level to the net productivity at the trophic level below
29.	____	Production efficiency	C.	Ratio of energy content of new tissue produced to energy assimilated from food
30.	____	Ecological efficiency	D.	Ratio of energy content of food consumed to energy content of available food

True/False

If the statement is true, write a "T" in the blank. If the statement is false, make it correct by changing the underlined word(s) and writing the correct word(s) in the answer blank.

31. _____________ A pyramid of biomass is sometimes inverted because it overestimates the importance of high turnover rates.

32. _____________ A pyramid of energy can never be inverted because the transfer of energy from one trophic level is always less than 100% efficient.

33. _____________ Individuals at the top of a pyramid of numbers are often of considerable conservation concern.

34. _____________ Photosynthesis converts approximately 25% of the solar energy hitting Earth's surface into chemical energy

47.2 Nutrient Cycling in Ecosystems [pp. 1189–1200]

Section Review (Fill-in-the-Blanks)

The circulation of nutrients from the living to the nonliving part of an ecosystem and back again is referred to as a (35) ____________ cycle. Unlike the flow of energy through an ecosystem, which is eventually lost, matter is (36) ____________ in the various cycles. The (37) ____________ ____________ model describes nutrient cycling based on whether molecules and ions are available or unavailable for assimilation and whether they are in organic or inorganic form.

Although not technically a nutrient, water cycles through ecosystems much as minerals do. This cycling is referred to as the (38) ____________ cycle. It is based on evaporation and (39) ____________ of water as rain or snow.

Carbon enters the food web when producers convert (40) ____________ ____________ into carbohydrates. Although most carbon is stored as sedimentary rock, most (41) ____________ carbon is present as bicarbonate ions in the ocean. Human activities, such as burning (42) ____________ ____________, are transferring carbon into the atmosphere at an unprecedented rate, causing a general (43) ____________ of the climate.

The (44) ____________ cycle is primarily based on the activity of prokaryotic organisms since they are the only organisms capable of converting atmospheric nitrogen (N_2) into organic form that can be utilized by eukaryotes; a process known as nitrogen (45) ____________. The breakdown of organic nitrogen in detritus into ammonia by various bacteria and fungi is (46) ____________. This ammonia can be assimilated by plants. The ammonia may also be converted to nitrites and nitrates by other bacteria in a process called (47) ____________. Still other bacteria convert unused nitrites and nitrates into N_2 through the process of (48) ____________. N_2 can also be fixed through an artificial process (the Haber–Bosch process), which is how (49) ____________ are produced.

Phosphorus compounds lack a (50) ____________ phase since the phosphorus cycle consists of movement between the terrestrial and (51) ________ aquatic ecosystems in a (52) ____________ cycle. Within a community, phosphorus cycles rapidly, with plants (53) ____________ it and heterotrophs (54) ____________ it.

Choice

Choose the biogeochemical cycle associated with each item in the following list. Answers can be used more than once.

a. Carbon cycle b. Hydrologic cycle c. Nitrogen cycle d. Phosphorus cycle

55. ____ Ammonification

56. ____ Fixation into organic form mainly by photosynthesis

57. ____ Fixation into organic form mainly by prokaryotes

58. ____ Largest available reservoir is bicarbonate ions in the oceans

59. ____ Largest available reservoir is the atmosphere

60. ____ Largest available reservoir is Earth's crust

61. ____ Largest available reservoir is Earth's oceans

47.3 Ecosystem Modelling [pp. 1200–1201]

Ecosystem modelling helps in the understanding of ecosystem dynamics. Conceptual models tend to be more general in scope than (62) ____________ models, which, if well constructed, can give precise answers to specific questions about ecosystems without disturbing them. It can also be used to (63) ____________ how future disturbances (such as climate change) can affect ecosystem function.

True/False

If the statement is true, write a "T" in the blank. If the statement is false, make it correct by changing the underlined word(s) and writing the correct word(s) in the answer blank.

64. ______________ Ecologists use ecosystem modelling to make predictions about how a species will respond to specific changes in the environment.

65. ______________ Ecosystem simulation models rely on detailed data concerning the interaction and relationships in a specific ecosystem.

66. ______________ Ecosystem modelling allows us to explore potential impacts of change without environmentally damaging experimentation.

67. ______________ Ecosystem models do not rely on assumption.

SELF-TEST

1. Urban ecosystems are similar to natural ecosystem in which way? [pp. 1179–1180]

   a. Both have similar rainfall runoff patterns
   b. Both have similar solar reflectance patterns
   c. Both have similar biodiversity levels
   d. Both have animal and plant communities

2. Energy in ecosystems usually flows through which two types of food webs? [pp. 1180–1182]

   a. Detrital and decomposer food webs
   b. Grazing and detrital food webs
   c. Predator and prey food webs
   d. Producer and consumer food webs

3. The contribution of various ecosytems to overall net productivity depends on ______. [p. 1182]

   a. The nitrogen cycle
   b. The size of the ecosystem
   c. Their productivity per unit area or volume
   d. Both (b) and (c)

4. The rate of conversion of solar energy into chemical energy by autotrophs is called ________. [pp. 1181–1182]

   a. Gross primary productivity
   b. Net primary productivity
   c. Respiration
   d. Standing crop biomass

5. The element in shortest supply in an ecosystem is called a(n) _________. [pp. 1181–1182]

   a. Essential nutrient
   b. Primary nutrient
   c. Basal nutrient
   d. Limiting nutrient

6. Which of the following are forms of new biomass? [pp. 1181–1182]

   a. Growth of existing biomass
   b. Reproduction to make new producers
   c. Storage of carbohydrates
   d. All of the above

7. Energy stored by consumers, as it is transferred to them from producers, is ________. [p. 1183]

   a. Gross primary productivity
   b. Net primary productivity
   c. Respiration
   d. Secondary productivity

8. The ratio of net productivity at one trophic level to that at the trophic level below is _______. [pp. 1183–1185]

   a. Harvesting efficiency
   b. Production efficiency
   c. Ecological efficiency
   d. Assimilation efficiency

9. Ectothermic animals generally convert more than ______ of their assimilated energy into new biomass, which is usually _______ than endothermic animals. [pp. 1183–1185]

   a. 50%; more
   b. 10%; more
   c. 50%; less
   d. 10%; less

10. Energy pyramids usually have very broad bases relative to the top of the pyramid. This is because of ________. [pp. 1185–1186]

   a. Low turnover rates
   b. Low ecological efficiency
   c. Reduced numbers of organisms
   d. Increasing biomass

11. The two types of nutrient cycles are ________. [pp. 1189–1190]

   a. Atmospheric and sedimentary
   b. Biotic and abiotic
   c. Organic and inorganic
   d. Terrestrial and aquatic

12. The source of energy driving the hydrogeologic cycle is _______. [pp. 1190–1191]

   a. Photosynthesis
   b. Hydrolysis
   c. Solar
   d. Hydrothermal

13. Carbon enters the biotic (living) component of the ecosystem via ________. [pp. 1191–1192]

   a. Ammonification
   b. Photosynthesis
   c. Precipitation
   d. Respiration

14. Which of the following would not cause levels of atmospheric carbon dioxide to rise? [pp. 1191–1192]

   a. Burning fossil fuels
   b. Destruction of terrestrial forests
   c. Volcanic eruptions
   d. Increased rates of photosynthesis

15. Which of the following processes returns N_2 to the atmosphere? [pp. 1192–1195]

   a. Ammonification
   b. Denitrification
   c. Nitrification
   d. Nitrogen fixation

16. Which of the following can convert N_2 into ammonium and nitrate ions? [pp. 1192–1195]

   a. Microorganisms
   b. Lightning
   c. Volcanic activity
   d. All of the above

17. Which of the following statements relating to the phosphorus cycle is correct? [pp. 1195–1200]

   a. Excess phosphorus from fertilizers is a pollutant of lakes and ponds
   b. Phosphorus cycles slowly through terrestrial ecosystems
   c. Phosphorus is made available to plants through fixation by prokaryotes
   d. The atmosphere is the largest reservoir of phosphorus

18. The phosphorus cycle is different from many of the other biogeochemical cycles because it _____. [pp. 1195–1200]

   a. Includes artificial fertilizers
   b. Includes freshwater and marine components
   c. Does not include an atmospheric component
   d. Has the atmosphere as the main reservoir

19. Which of the following is not true of ecosystem modelling? [pp. 1200–1201]

    a. It attempts to identify the most important factors that are involved in ecosystem function
    b. It describes ecosystem processes but has no predictive value
    c. It simplifies events that occur in nature
    d. It uses mathematical equations to define relationships between populations and the environment

20. Which of the following would not likely be included in an ecosystem simulation model? [pp. 1200–1201]

    a. Temperature
    b. Limiting nutrient levels
    c. Energy flow
    d. Time of day

INTEGRATING AND APPLYING KEY CONCEPTS

1. This chapter is focused on two ecosystem processes: ecosystem energetics and nutrient cycles. What are the fundamental differences between these two processes, and how are they related?

2. Compare and contrast gross and net primary productivity. Explain how production efficiency plays a role in the determination of net primary productivity.

3. Describe how net primary productivity could vary depending on the ecosystem being considered. Give examples.

4. This chapter describes four biogeochemical cycles. Identify (a) which two rely on solar radiation to drive part of the cycle, (b) which cycle does not include the atmosphere as part of the cycle, and (c) where the main reservoirs for each cycle are located.

OPTIONAL ACTIVITIES

1. *Unanswered Questions:* If human activity is increasing CO_2 levels in the atmosphere and causing a rise in temperature, what might be the effect on ecosystem processes such as carbon fixation, decomposition, productivity, and nutrient cycling?

2. *Insights from the Molecular Revolution:* Based on what you have learned about the use of mtDNA to determine migration routes of loggerhead turtles, can you think of any other interesting or important questions regarding the movement of other animals (including humans) that might be solved using this methodology?

3. *Focus on Research:* Summarize the environmental impact of deforestation based on what you have learned about the carbon cycle in this chapter and the Bormann and Likens experiments in the Hubbard Brook Experimental Forest watershed.

4. Which biogeochemical cycle do you think humans have impacted the most? Explain why.

48 Conservation of Biodiversity

CHAPTER HIGHLIGHTS

- Various reasons have been suggested for extinction of different species, as well as mass extinctions, in the past, but the largest mass extinction is occurring presently and is primarily the result of human activities.
- Examples of extinctions and reasons for the loss of these species are presented.
- It is necessary to identify species that are at risk and then categorize them according to data-based criteria, which is undertaken by various governing bodies.
- Conserving biodiversity requires addressing the growth of human populations, the root problem of declining biodiversity.

STUDY STRATEGIES AND LEARNING OUTCOMES

- After learning about the negative impact humans have had on populations, you can then look at what humans are doing to try to protect those species at risk.
- Be sure to differentiate between the different governing bodies, as well as understand how these bodies determine which group of organisms should be protected.
- You can now address the conflicting needs of humans versus nature, noting the challenges that must be addressed.

By the end of this chapter, you should be able to

- Define and provide examples of species extinction
- Discuss why species have become extinct or are threatened to become extinct
- Explain how species are categorized by a governing body in terms of their status (number and distribution) in nature
- Identify conservation efforts explaining why they have or have not been successful, especially in terms of the human population

TOPIC MAP

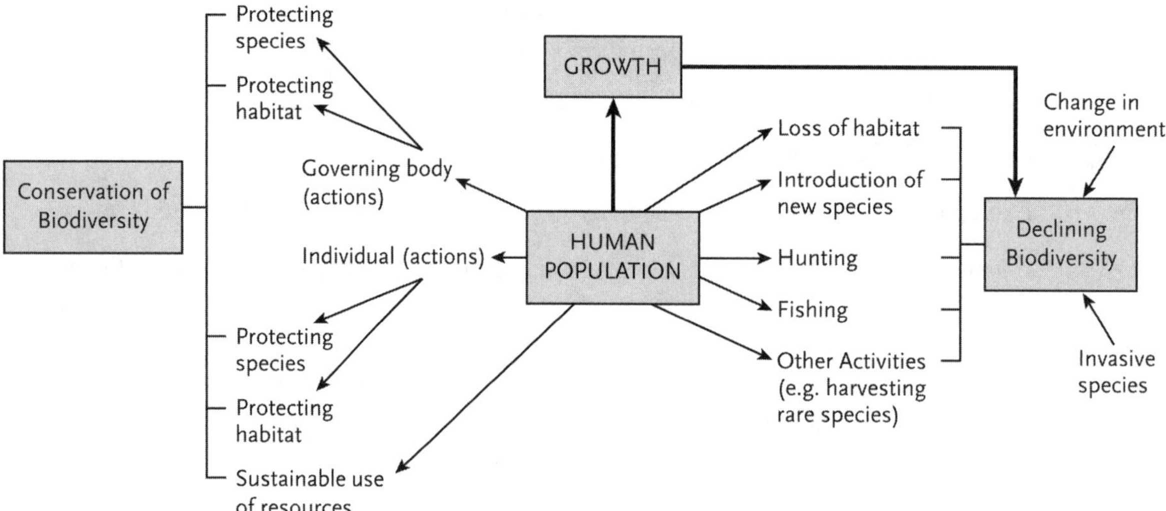
Protecting species
Protecting habitat
GROWTH
Change in environment
Loss of habitat
Conservation of Biodiversity
Governing body (actions)
Introduction of new species
Declining Biodiversity
Individual (actions)
HUMAN POPULATION
Hunting
Fishing
Protecting species
Other Activities (e.g. harvesting rare species)
Invasive species
Protecting habitat
Sustainable use of resources

INTERACTIVE EXERCISES

Why It Matters [pp. 1205–1206]

Section Review (Fill-in-the-Blanks)

To maintain biodiversity, it will be necessary for humans to change their attitudes about nature from a (1) ____________ view to a view in which each animal and plant, as well as other living organisms, is as important in nature as (2) ____________ believe that they are. Historically, some religious beliefs have resulted in the (3) ____________ of humans from nature, thereby justifying increasing productivity, at the (4) ____________ of biodiversity, to supply the needs of humans. This has included development of (5) ____________ farming (to produce large amounts of one food species) and genetically modified organisms, as well as increased use of (6) ____________ (to enhance the soil's nutrient level), water, and energy to increase food production. It is hoped that the (7) ____________ ____________ ____________ project will aid in identifying and naming species since species number is an important indicator of (8) ____________, a measure of ecosystem stability and productivity.

48.1 Extinction [pp. 1206–1209]

Section Review (Fill-in-the-Blanks)

It is expected that over time some species become extinct. This low rate of species loss is known as the (9) ____________ ____________ rate, and it usually results from climate change and the inability of some organisms to adapt. Consequently, there is a continual (10) ____________ of species living on earth over time. For species that are closely related, usually due to having a common (11) ____________, some (12) ____________ may be found in more than one taxa, making it difficult to differentiate between the taxa and therefore to determine if a particular species has become (13) ____________, that no individuals of the species are alive. However, in many situations, it is clear when a species has become extinct, although it can be difficult to ascertain why the species has become extinct. For example, it is surmised that the demise of multituberculates in early Oligocene resulted from being (14) ____________ by rodent species, newly evolved due to (15) ____________ ____________; however, information from fossil records is limited.

There are at least (16) ____________ times in history when mass extinctions have occurred. The cause of the mass extinction at the end of the Cretaceous period has provoked extensive debate. Although it is generally agreed that a(n) (17) ____________ hit the earth at that time, as evidenced by the presence of (18) ____________, a substance often found in asteroids, in rocks from that time period and the 180 km diameter (19) ____________ off a peninsula in Mexico, it is not clear if this event resulted in the mass extinction. It has been suggested that changes in the (20) ____________ from the asteroid resulted in the extinction of many birds, mammals, and other organisms.

Choice

Choose the correct period for each of the events or results relating to a mass extinction.

A. End of Ordovician B. End of Devonian C. End of Permian

D. End of Triassic E. End of Cretaceous F. Present day

21. ____ Glaciations due to formation of Pangea
22. ____ Extinction of dinosaurs
23. ____ Asteroid hitting Earth
24. ____ Extinction of trilobites and trees of coal swamp forests
25. ____ Human degradation of environment
26. ____ Most severe; > 85% of species became extinct
27. ____ Glaciation from move of Gondwana toward South Pole

48.2 The Impact of Humans [p. 1210]

48.3 Introduced and Invasive Species [pp. 1211–1213]

48.4 How We Got/Get There [pp. 1213–1216]

Section Review (Fill-in-the-Blanks)

Human actions have been responsible, whether directly or indirectly, for the (28) ____________ in individual numbers of various species, sometimes resulting in the loss of the species, its (29) ____________. These actions usually involve overharvesting the species, (30) ____________ a new species to an area, or destroying a species (31) ____________ so that it can no longer access what it needs for survival. For example, the extinction of the Dodos on Mauritius Island around (32) ____________ resulted from extensive hunting by new settlers and introduced (33) ____________, such as dogs, cats, and rats, as well as from habitat loss due to the expanding settler population on the island. Many introduced or invasive species are not successful in their new environment; however, it has been found that invaders are usually more successful if the new habitat is (34) ____________, allowing the invader to use (35) ____________ growth and reproduction rates to outcompete native species. Surprisingly, a recent study on plants indicated that even in nutrient-poor sites, invasive species had a higher (36) ____________ ____________ ____________, assimilating more carbon (biomass) per unit of resource, which indicated that limited resources may not limit the success of invaders. Effects of human activities seldom affect a species in isolation since a species interacts with its environment at many levels, all the way up to the (37) ____________ level. For example, it is hypothesized that the dramatic (38) ____________ in bay scallops is not directly due to overfishing but rather is an indirect effect of overharvesting various species of (39) ____________. Decreased numbers of these animals allowed their prey, skates and rays, to increase in numbers, and since they are (40) ____________of bay scallops, the number of bay scallops decreased.

Matching

Match the name of the invader with each of the species known to be negatively affected by its presence.

41.	____	Flatworms	A.	Flightless wrens
42.	____	Starlings	B.	Chestnut tree
43.	____	Zebra mussels	C.	Native bird species
44.	____	*Endothia parasitica*	D.	Earthworms
45.	____	Cats	E.	Eastern pond mussels

Choice

Choose the appropriate action by humans that has caused a large reduction in numbers of individuals belonging to each of the following organisms.

A. Reduction in animal numbers (by humans) of species that feed on predators of this species

B. Caught as bycatch by fishers

C. Hunted and shot by trophy hunters

46. ____ Sea turtles

47. ____ Cape buffalo

48. ____ Bay scallops

49. ____ Barndoor skate

50. ____ African elephant

Short Answer

51. How is the extinction of the Dodos, which occurred more than 300 years ago, related to the approaching extinction of the Mauritian calvania tree?

____

____

52. What is it about the horns of black rhinos that has made this species so attractive to hunters, resulting in its dramatic drop in numbers?

____

____

48.5 Protecting Species [pp. 1216–1219]

48.6 Protecting What? [pp. 1219–1221]

48.7 The Downside of Being Rare [pp. 1221–1224]

48.8 Protecting Habitat [p. 1224]

Section Review (Fill-in-the-Blanks)

Before organisms can be protected, it is necessary to ascertain whether they are (53) ____________ for consideration, as regulated by a government agency. A group of organisms being considered must be identified as one or more (54) ____________ units, such as a species, population, ecosystem, etc. If this group is identified as at risk, it must then be categorized using (55) ____________ criteria for assessing the risk posed. These data include number of individuals, fecundity, (56) ____________ (rate at which they are dying), intrinsic rate of increase, and (57) ____________ ____________ (maximum number of individuals in a population without negatively affecting their habitat). Based on the criteria and assessment procedures identified by the International Union for the Conservation of Nature, a Canadian body called (58) ____________ (acronym) assesses the data collected and categorizes the group of organism as (59) ____________, if it no longer exists; (60) ____________, if it no longer exists in a particular area but does live elsewhere; (61) ____________, if it is expected to become extinct or extirpated; (62) ____________, if it is expected to become endangered; of (63) ____________ ____________, if there are threats or it has biological characteristics that indicate it may become threatened or endangered; or (64) ____________ ____________, if there is insufficient information available to determine if it is eligible and/or at risk. When some species are identified to be at risk, their value tends to (65) ____________ due to increased competition to acquire the last individuals.

In some situations, identifying species at risk does not ensure (66) ____________ of the group of organisms or its habitat. The expansion of (67) ____________ population size and associated needs reduces the area available for the habitat and needs of other species.

Matching

Match each of the acronyms associated with conservation regulatory bodies to the description of its purpose.

68.	____	IUCN	A.	Categorizes potential species at risk in Canada based on data-based criteria
69.	____	CITES	B.	Developed criteria and assessment procedures for identifying species at risk at an international level
70.	____	COSEWIC	C.	Identifies animals or animal parts that should be banned from importation

Choice

Choose the correct category that each of the animals belong to. Choices can be used more than once.

A. Extinct B. Extirpated C. Endangered

D. Threatened E. Special concern F. Not at risk

71. ____ Chinese bahaba

72. ____ White-coloured moose

73. ____ Banff Springs snail

74. ____ Pacific water shrew

48.9 Effecting Conservation [pp. 1224–1228]

48.10 Taking Action [pp. 1228–1229]

Section Review (Fill-in-the-Blanks)

The root problem of declining biodiversity is the increasing (75) ____________ ____________, and if it continues to grow at its present rate, it will double in (76) ____________ years. Although the human population growth rate is (77) ____________ globally, some countries, such as Afghanistan, continue to maintain relatively high population growth rates. It is hoped that with continued education on family planning, more women will have the knowledge to control their own (78) ____________.

It will be up to humans to take action to maintain global (79) ____________. Both ecosystems and biodiversity are very (80) ____________ systems, and a disruption, such as a species going extinct or introduction of a new species, tends to lead to (81) ____________ biodiversity. If humans continue to try to achieve (82) ____________ use of resources, as well as taking action in (83) ____________ species and habitats locally, these conservation efforts could help maintain biodiversity.

True/False

If the statement is true, write a "T" in the blank. If the statement is false, make it correct by changing the underlined word(s) and writing the correct word(s) in the answer blank.

84. ____________ Human population growth rate in Afghanistan is <u>less</u> than in Sri Lanka.

85. ____________ Bird flu can spread from birds, such as chickens, to <u>humans</u>.

86. ____________ Reducing the human growth rate requires empowering women to control <u>their own fertility</u>.

87. ____________ <u>Mammals and birds</u> dominate the list of threatened species both internationally and in Canada.

SELF-TEST

---

1. Which of the following reasons explain(s) why humans drive species to extinction? [pp. 1206–1209]

   a. Harvesting organisms in order to survive
   b. Harvesting organisms that have medicinal or magical value
   c. Harvesting organisms for industrial development
   d. (a), (b), and (c)

2. All dinosaurs were large, dominating all other creatures living during this time period. [pp. 1206–1209]

   a. True
   b. False

3. Background extinction rate is a measure of __________. [pp. 1206–1209]

   a. The number of species becoming extinct due to human activities
   b. The number of species becoming extinct due to environmental change
   c. The number of species becoming extinct in terrestrial ecosystems only
   d. The number of species becoming extinct in aquatic ecosystems only

4. The survivorship curve for *Albertosaurus* most closely follows a __________ survivorship curve. [pp. 1206–1209]

   a. Type I
   b. Type II
   c. Type III
   d. Type IV

5. The cartwheel flower is an introduced species to __________. [pp. 1210–1216]

   a. British Isles
   b. British Columbia
   c. Hawaii
   d. Great Lakes

6. Ballast water __________. [pp. 1210–1216]

   a. Can contain many species of many different phyla
   b. Is water that is taken in and released by ships in the same port
   c. Ensures that species remain in their own waters
   d. Both (a) and (b)

7. Which of the following is an introduced species that is now established in Hawaii? [pp. 1210–1216]

   a. Flatworms
   b. Chestnut blight
   c. Monterey pine
   d. Zebra mussels

8. Which RUE (umol CO_2*m^{-2}*s^{-1}) would provide an invading species with the greatest chance of success? [pp. 1210–1216]

   a. 1
   b. 5
   c. 10
   d. 15

9. Which of the following have been hunted to extinction? [pp. 1210–1216]

   a. Black rhinos
   b. Passenger pidgeons
   c. Starlings
   d. Barndoor skates

10. Which of the following characteristics of a potential species at risk would members of COSEWIC <u>not</u> consider? [pp. 1216–1224]

    a. Morphology of individuals
    b. Population information
    c. Generation time
    d. Risks to species survival

11. What category does the Bali starling fall under? [pp. 1216–1224]

    a. Extinct
    b. Extirpated
    c. Endangered
    d. Threatened

12. Why are some organisms that appear to be at risk not protected? [pp. 1216–1224]

    a. They are a subspecies (not a species)
    b. The species is not at risk
    c. They are morphologically distinct individuals but belong to a not-at-risk population
    d. (a), (b), and (c)

13. On what basis were killer whales on the west coast of Canada divided into four designatable units? [pp. 1216–1224]

    a. Different species
    b. Different subspecies
    c. Different behaviour and geography
    d. Different populations

14. Why is it difficult to protect the Chinese bahaba? [pp. 1216–1224]

    a. They are easy to catch
    b. Their swim bladders have medicinal properties
    c. Their fins have medicinal properties
    d. They are valued as a pet

15. Only a group of organisms classified as a species can be considered at risk, not a population. [pp. 1216–1224]

    a. True
    b. False

16. What percentage of drylands are suffering from severe land degradation? [pp. 1224–1229]

    a. 1–5%
    b. 5–10%
    c. 10–20%
    d. 25–50%

17. What is the root problem of declining biodiversity? [pp. 1224–1229]

    a. Introduced species
    b. Human population growth
    c. Trophy hunting
    d. Changing climate

18. Historically, which of the following causes have humans <u>not</u> managed to achieve? [pp. 1224–1229]

    a. Abolition of slavery
    b. Emancipation of women
    c. Maintenance of biodiversity

INTEGRATING AND APPLYING KEY CONCEPTS

1. If you had a time machine and travelled back to the end of the Cretaceous, what animals would you expect to see and what would you expect to happen to life on earth after an imminent collision with a large asteroid occurred?

2. American ginseng is categorized as endangered, but it is difficult to protect. Explain why.

3. Why is the number of birds decreasing globally?

OPTIONAL ACTIVITIES

1. Name an endangered species found in Canada and then identify why it is endangered and what is being done to protect it.

2. Discuss how successful you think nations are likely to be in meeting the goals outlined by the United Nations Millennium Development Goals in 2000 (see page 1226).

49 Domestication

CHAPTER HIGHLIGHTS

- Domestication requires selective breeding to promote characteristics desirable to humans.
- Domestication was developed historically to increase the amount of food available.
- Over time, different species have been domesticated to provide many resources, including food, materials for clothing, a source of labour, and proteins and other molecules with medicinal value.
- Technology has greatly increased yield, offsetting the increased food requirement for an increasing human population.
- Costs, such as increased input of fertilizers, water, and energy, and increased potential for pest infestations are associated with farming.

STUDY STRATEGIES AND LEARNING OUTCOMES

- After noting the importance of domestication through history, focus on the costs and benefits of associated technological development.

By the end of this chapter, you should be able to

- Define domestication and describe its development over time
- Provide examples of plants and animals that have been domesticated and identify their uses
- Describe the benefits and costs of advanced technology relating to domestication

TOPIC MAP

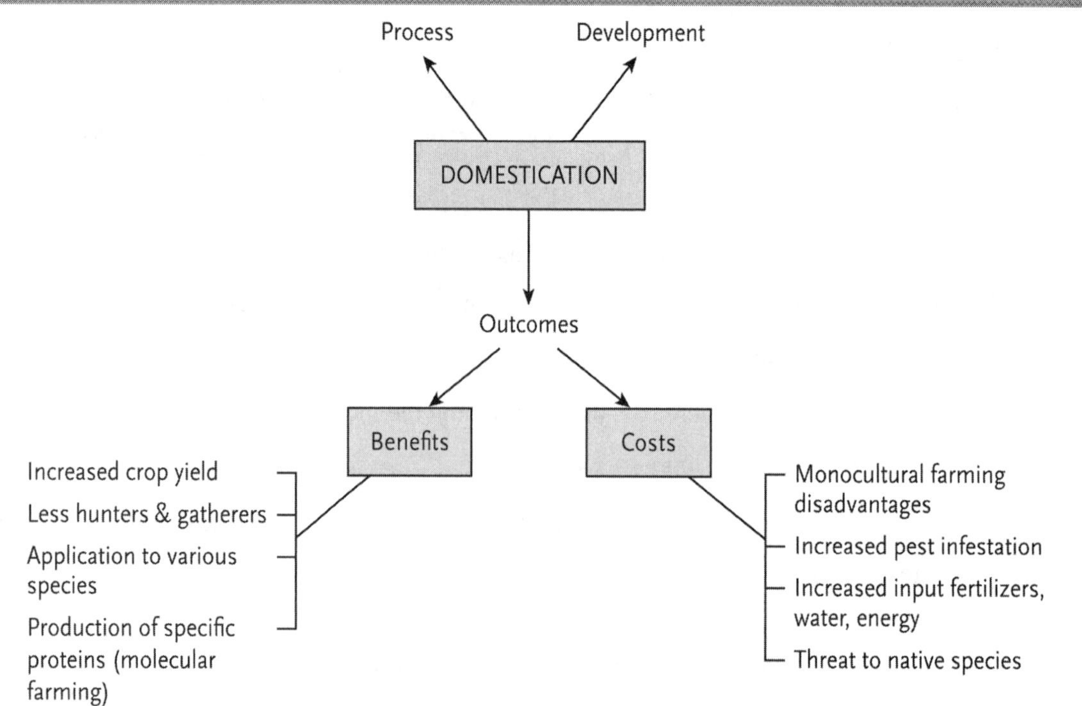
Process
Development
DOMESTICATION
Outcomes
Benefits
Costs
Increased crop yield
Less hunters & gatherers
Application to various species
Production of specific proteins (molecular farming)
Monocultural farming disadvantages
Increased pest infestation
Increased input fertilizers, water, energy
Threat to native species

INTERACTIVE EXERCISES

Why It Matters [pp. 1233–1234]

Section Review (Fill-in-the-Blanks)

Improvements in agricultural techniques, including combining new genetic strains, better (1) ____________ (to promote plant growth), and more efficient harvesting and processing, have (2) ____________ yields. The advent of chemical fertilizers and other agricultural techniques is known as "the (3) ____________ ____________." (4) ____________ also plays an important role, as seen when comparing areas with different amounts of rainfall, such as Zimbabwe, Africa, with its (5) ____________ annual precipitation, relative to southwestern Ontario, Canada, with its relatively (6) ____________ annual precipitation. Even with technological advancements, the food available for children to eat differs between countries, as indicated by the statistics that one in three children in Zimbabwe are (7) ____________, whereas an increasing number of children in Ontario suffer from (8) ____________.

49.1 Domesticate [pp. 1234–1238]

Section Review (Fill-in-the-Blanks)

Originally, humans acquired food by hunting and (9) ____________ resources from the environment, but, over time, they progressed to (10) ____________ seeds and growing their own plants, a process known as (11) ____________. Eventually, humans began to selectively breed individuals of a species, or (12) ____________ them, thereby promoting characteristics that benefited humans. Many species of (13) ____________, yeast, and animals have been domesticated at different times and places throughout history. The use of (14) ____________ and molecular genetics has provided evidence of the development of cultivation and domestication of many organisms. For example, genetic evidence indicates that the domestication of pigs occurred at (15) ____________ different locations throughout Eurasia. Archaeological evidence of fires indicates the possible emergence of cultivation and domestication by humans as far back as (16) ____________ years ago (in Mossel Bay, South Africa).

True/False

If the statement is true, write a "T" in the blank. If the statement is false, make it correct by changing the underlined word(s) and writing the correct word(s) in the answer blank.

17. ____________ Humans were the first species to manipulate organisms for their own benefit.

18. ____________ Parthenogenic figs can be domesticated rapidly due to their use of asexual reproduction.

19. ____________ Early cultivation of rice in China required clearing of land using fire.

20. ____________ The birth of agriculture arose from domestication.

21. ____________ Aquatic species were the first organisms to be domesticated.

22. ____________ The change from hunting and gathering to cultivation and domestication provides an opportunity for increased population size.

Short Answer

23. Explain how the appearance of indehiscent wheat grains promoted cultivation and domestication of wheat.

____

____

24. Why is fire indicative of cultivation?

____

____

____

49.2 Why Some Organisms Were Domesticated [pp. 1238–1246]

Section Review (Fill-in-the-Blanks)

Although the primary motivation for cultivation and domestication of plants and animals was to provide a supply of (25) ____________, they have provided many other uses.

Choice

For each domesticated organism, choose one or more ways that it is useful or substances it helps produce.

A. Food or food source　　B. Beer　　C. Labour

D. Pollination　　E. Wine

26. ____ Corn

27. ____ Grapes

28. ____ Honeybees

29. ____ Squash

30. ____ Yeast

31. ____ Cattle

32. ____ Wheat

Matching

Match each of the following organisms with the latest date at which domestication is believed to have occurred. Note that BP = before present.

33.	____	Cotton	A.	4400 years BP
34.	____	Dogs	B.	6000 years BP
35.	____	Cattle	C.	6250 years BP
36.	____	Squash	D.	9000 years BP
37.	____	Yeast	E.	10 000 years BP
38.	____	Corn	F.	14 000 years BP

49.3 Yields [pp. 1246–1247]

49.4 Complications [pp. 1247–1251]

49.5 Chemicals, Good and Bad [pp. 1251–1252]

Section Review (Fill-in-the-Blanks)

Farmers must consider many aspects of crop production when trying to determine which crop provides the (39) ____________ greatest yield on a particular piece of land and, overall, which crop will bring in the (40) ____________highest income, after (41) ____________ deducting all costs. Increasing yield is possible if the farmer is able to take advantage of (42) ___________ technology, such as a tractor or bailer, or improve the land with the addition of (43) __________ fertilizer (if low nutrient soil) or water (if little or seasonal rainfall), but all of these add to the cost of producing the crop.

Choice

Choose whether each of the following farming conditions will have a positive or negative impact on yield.

A. Positive B. Negative

44. ____ Milkweed mixed in with barley

45. ____ Addition of manure to soil

46. ____ Use of cisterns to capture rainwater during a relatively dry growing season

47. ____ Infestation of fungus in a grain storage tank

48. ____ Increase in fuel prices

49. ____ Purchase of a tractor

50. ____ Introduction of cats to control a rodent problem

True/False

If the statement is true, write a "T" in the blank. If the statement is false, make it correct by changing the underlined word(s) and writing the correct word(s) in the answer blank.

51. ____________ Ginseng is a <u>toxin</u> that stimulates the immune system.

52. ____________ The yield on terrain that is terraced will be <u>lower</u> than on terrain that is flat.

53. ____________ The domestication of cats is believed to have occurred <u>before</u> the development of agriculture.

54. ____________ In 2 A.D., Rome was able to produce a <u>sufficient</u> amount of wheat to feed all of its people.

49.6 Molecular Farming [pp. 1252–1253]

49.7 The Future [pp. 1253–1255]

Section Review (Fill-in-the-Blanks)

The domestication of plants, along with the ability to (55) ___________ modify plants, has provided the opportunity to mass produce specific (56) ___________ at a relatively low cost. This type of agriculture is called (57) ___________ ___________. One example of molecular farming is the planting of tobacco plants that contain the inserted human (58) ___________ gene. These plants produce the protein interleukin-10, which, after harvest, can be used by humans to treat (59) ___________ ___________ disease. Although there is concern that the products of these genetically modified plants might enter an ecosystem's (60) ___________ ___________ and (61) ___________ affect existing crops, tobacco is a (62) ___________ plant, and research has indicated that the protein does not enter the soil or enter insects that infect the plant.

Molecular farming, as well as other forms of domestication, is allowing us to (63) ___________ other species. The (64) ___________ for humans are many, including an increased food supply, labour resource, production of chemicals and other materials, and pest control. With these benefits, there are costs, so it is imperative that innovations are scrutinized to ensure that implementation of any development will result in benefits that (65) ___________ costs.

Matching

Match each of the following descriptions with how tobacco is used or has been used in the past.

66.	____	In molecular farming	A.	Smoking a cigarette
67.	____	Traditional medicine	B.	Applied to plants to control or prevent pest infestation
68.	____	An insecticide	C.	Producing the protein that treats irritable bowel disease
69.	____	Recreational use		Reducing the pain of childbirth

Short Answer

70. Explain why using corn to produce the biofuel ethanol is not environmentally friendly.

____

____

SELF-TEST

1. Domestication of wheat was possible due to __________. [pp. 1234–1238]

    a. A mutation resulting in production of indehiscent wheat grains
    b. A mutation resulting in production of dehiscent wheat grains
    c. Ability to asexually reproduce
    d. Growth of wheat in marine environments

2. Freshwater animals, as well as marine plants and animals, have been domesticated over the past __________. [pp. 1234–1238]

    a. 100 000 years
    b. 10 000 years
    c. 1000 years
    d. 100 years

3. Why is it sometimes difficult for archaeologists to determine when an organism has undergone domestication? [pp. 1234–1238]

    a. The DNA has degraded over time
    b. Domestication of some organisms originated from more than one location
    c. Problem distinguishing between domestic and wild-type organisms using bone or plant material

4. What is niche construction? [pp. 1234–1238]

    a. Ability of an organism to build its own shelter
    b. Destruction of habitat that occurs when an undomesticated animal moves to a new location
    c. Modification of the environment in preparation for domestication
    d. Breeding of a species for its specific characteristics

5. Which of the following changes assisted in the emergence of domestication? [pp. 1234–1238]

    a. Climate
    b. Human behaviour
    c. Human settlement

d. (a), (b), and (c)

6. Which of the following has <u>not</u> been domesticated? [pp. 1238–1246]
   a. Dogs
   b. Honeybees
   c. Yeast
   d. Atlantic salmon
   e. Mushrooms

7. Which of the following domesticated plants originated in more than one location? [pp. 1238–1246]
   a. Corn
   b. Barley
   c. Rice
   d. Wheat

8. The formation of indehiscent reproductive structures aided in the domestication of __________. [pp. 1238–1246]
   a. Rice
   b. Cotton
   c. Lentils
   d. Eggplant

9. Which of these plants do not belong to the *Solonaceae* family? [pp. 1238–1246]
   a. Tomatoes
   b. Locoweed
   c. Squash
   d. *Capsicum* species of chili peppers

10. The First Nations people believed that there were “three sisters” that needed to be present each year. They were referring to the staple foods _____________. [pp. 1238–1246]
    a. Tomatoes, potatoes, and corn
    b. Potatoes, corn, and beans
    c. Corn, beans, and squash
    d. Beans, squash, and wheat

11. Honeybees are important pollinators in the production of __________. [pp. 1238–1246]
    a. Apples
    b. Broccoli
    c. Cotton
    d. Both (a) and (b)

12. Which of these substances does not contain a phenolic compound? [pp. 1246–1252]
    a. Cinnamon
    b. Coffee

c. Nutmeg
d. Conine

13. Renal failure can result from eating a crop infected with __________. [pp. 1246–1252]

    a. Milkweed
    b. *Aristolochea clematitis*
    c. *Panax quinquefolius*
    d. Queen Anne's lace

14. Domestic cats were the result of a single domestication event. [pp. 1246–1252]

    a. True
    b. False

15. EROI __________. [pp. 1252–1255]

    a. Stands for "energy return on investment"
    b. Is a ratio that relates energy input to energy output
    c. Is a high value for corn-produced ethanol (~17:1)
    d. (a) and (b)
    e. (a), (b), and (c)

16. Where does the *IL-10* gene inserted into IL-10 protein–producing tobacco plants come from? [pp. 1252–1255]

    a. Mouse genome
    b. Human genome
    c. Tobacco genome
    d. *Solonaceae* genome

17. By advancing from being hunters and gatherers to using cultivation and domestication techniques, fewer individuals need to produce food, allowing a majority of people to contribute to society in other ways. [pp. 1252–1255]

    a. True
    b. False

INTEGRATING AND APPLYING KEY CONCEPTS

1. Differentiate between cultivation and domestication.

2. Describe the effect of cultivation on the Abu Hureyra settlement from 12 000 years BP to 9400 years BP.

3. To become domesticated, some species must undergo one or more specific genetic changes. Contrast the genetic changes required to promote domestication in lentils and in rice.

4. You have just inherited a 20-hectare parcel of agricultural land in central Canada. What do you need to consider if you want to be successful at farming?

OPTIONAL ACTIVITIES

1. Why would canola seeds that travelled into space (in a space shuttle) germinate faster and grow more quickly than canola seeds that remained on earth (control seeds)?

2. How is the single species *Brassica oleracea* able to produce such a variety of plants (broccoli, cauliflower, kale, Brussels sprouts, cabbage)?

Case Studies

The case studies included in this Study Guide are a compilation of short cases based on scientific investigations and stories experienced by everyday people. Each case can be used in conjunction with material from the textbook *Biology: Exploring the Diversity of Life*, First Canadian Edition.

The cases found within are intended to provide you with opportunities to apply and integrate concepts and research in biology from the classroom to real situations. To this end, I have tried to cover a wide spectrum of concepts you may encounter within your biology course. Each case begins with an objective, which provides you with the context and intended application of the material. This is followed by a series of *Learning Outcomes* and *Suggested Readings*, which link to material found within the textbook to further your understanding. Each case varies in length and is followed by a series of questions to consider. These questions should assist you in connecting with the content of the case and achieving the Learning Outcomes set out at the beginning of the chapter.

You may notice that the answers to the case study questions are not provided in this study guide. It is my hope that you will come to your own conclusions through research and application of your knowledge and will not feel limited by any packaged answers I could have provided. Instructors may choose to use them as assigned homework, integrated into a lecture, or as part of laboratory exercises. The case studies can be used for group work or independently.

The use of case studies in biology for teaching and learning is somewhat of a new (and exciting!) approach, and, as such, we are all learning together. I invite any and all prospective users to learn more about teaching and learning with case studies through research into the various literature and Internet sources that exist. I welcome students and instructors alike to modify the cases and/or associated questions for their own purposes and course objectives. I sincerely hope that these cases will provide instructors with an alternative to the standard lecture format to engage their students and for students, in turn, to benefit from that engagement as they come to realize how biology impacts the world around them. As an instructor, some of these cases have already been used in my classes (i.e., Genetics, Cellular Physiology, Animal Physiology, and Environmental Physiology), and it is my intention to use the new case studies I have developed in future courses.

I would like to thank those involved at Nelson Education Ltd. for recognizing the benefit of case studies in teaching biology and allowing me to bring this exciting approach to all of you. My thanks go as well to Clyde F. Herried* of SUNY, Buffalo, whose case study workshop provided me with the tools and the foundation I required to begin my journey into developing and teaching with cases. I was pleased to be able to take part in Nelson's Case Study Workshop, run by Dr. Herreid, as a presenter, demonstrating to fellow Canadian biology instructors how my students and I have benefited from this approach. Finally, I would like to thank two of my senior students, Gurjot Jassy and Eric DiMeo, who participated in and contributed cases to this project.

I welcome any feedback and hope that with the coming years, through your input and my own personal assessment of the cases, these will evolve and improve and more case studies will be added. Enjoy, and stay tuned.

Colin Montpetit
University of Ottawa

*For more information on Clyde F. Herreid and the National Center for Case Study Teaching in Science, go to http://ublib.buffalo.edu/libraries/projects/cases/case.html.

Table of Cases

1. A New Strain on the System C-4
2. A Swimmer's Journey C-7
3. Firish Fries C-11
4. The Case of the Two-Toned Lobster C-14
5. Microbes in the Ocean's Depths C-17
6. A Case of Natural Doping C-20
7. The Case of Bees C-24

A New Strain on the System

CASE OBJECTIVE

The goal of the following case is to gain an understanding of the concepts of viruses, infectious biological particles, and defences against disease.

LEARNING OUTCOMES

By the end of this case, you should be able to

1. Describe the characteristics of viruses
2. Explain how viruses infect cells and replicate in hosts
3. Explain the treatment of viral infections
4. Explain why some treatments of viral infections are ineffective
5. List and describe the immune responses to viral and other infections
6. Discuss how organisms may combat pathogenic viruses

Suggested Readings: Chapters 10, 22, and 44; independent studies

A New Strain on the System

By Gurjot Jassy and Colin J. Montpetit, Department of Biology, University of Ottawa

"*Health officials are becoming increasingly concerned with the growing number of confirmed swine influenza virus cases being reported in North America. Normally, the flu outbreak is seasonal, and thus health departments are able to take the necessary precautions in order to minimize the outbreak and repercussions on the provided health services. However, this isn't the case this time around . . . .*"—Reported in a recent telecast from a local news station

Everett, a volunteer in the emergency room at the local hospital, did not understand what the big fuss was about. There had only been a few reported cases to date, and he thought that there could not have been rapid transmission from person to person. He felt safe. However, just to be sure, he went to talk to the resident on call.

The resident was quick to respond: "The most recent strand, H1N1, is a new strain that scientists have yet to completely investigate. Commonly, if the strain is known, it can be attenuated and administered via flu shots. However, the fact remains that individuals have yet to acquire natural immunity. Also, the symptoms of this new flu virus are similar to the ones of the seasonal flu, which include fever, cough, sore throat, headache, chills, and fatigue. This is making it especially difficult to accurately diagnose H1N1 cases. Based on the symptoms, you may think you have the seasonal flu; however, you may have contracted a more virulent form of the virus, which your body may have difficulty fighting off. Also, if you have it and do not know it, there is a greater risk of transmitting the virus to others. Even though there were a few reported cases, this can turn into a pandemic, a massive outbreak. One has to take the precautionary measures not to transmit this virus, or any other viruses for that matter."

Everett was finally beginning to grasp the seriousness of the situation and decided he was going to take such news a little bit more seriously from now on.

QUESTIONS

1. What is a virus? Outline the characteristics of viruses.
2. Explain how viruses infect and replicate.
3. How are viruses transmitted from host to host?
4. Why are doctors worried that this new H1N1 strain may result in a pandemic? (Define pandemic.)
5. What is the H1N1 virus? What symptoms does this virus cause in infected hosts?
6. Why are a number of treatments for the flu not effective in individuals diagnosed with the H1N1 strain? Why are antibiotics ineffective against viruses?
7. Explain the significance of genetic variation in relation to the influenza virus. What factors are responsible for this variation? Why do new vaccines against viruses need to be developed each year?
8. If the new strain of H1N1 causes symptoms similar to those of seasonal flus, how can doctors effectively diagnose cases of infections by the new H1N1 strain? What do scientists need to know about this strain, and what types of tests do they need to develop to help doctors make accurate diagnoses?
9. In which season is the flu most prevalent and why?
10. Explain how organisms recognize pathogens.
11. List and outline the types of defences organisms have against invasions.
12. Why would the body have more difficulty in fighting off this new strain of H1N1?
13. How can we combat infections by pathogenic viruses?
14. What needs to be known about the strain to attenuate and create flu shots? What is the purpose of the flu shot in the first place? How will this help patients acquire immunity to this strain?
15. What precautionary measures can you take to prevent transmission of the influenza virus?

A Swimmer's Journey

CASE OBJECTIVE

The objective of this case is to gain an understanding of the role and functioning of the body systems and cellular processes involved in physiological function during activities.

LEARNING OUTCOMES

By the end of this case, you should be able to

1. Discuss the significance of diets during training
2. Discuss the roles and functioning of body systems and cellular processes involved in the physiology to perform activities
3. Discuss the regulation of energy mobilization and metabolism
4. Discuss the steps involved in carbohydrate and lipid metabolism
5. List the advantages and disadvantages to switch from carbohydrate metabolism to lipid metabolism
6. Discuss the benefits and biological effects of training to enhance performance

Suggested Readings: Chapters 4, 5, 6, 8, 35, 36, 37, 41, 42, and 43; independent studies

A Swimmer's Journey

By Eric DiMeo and Colin J. Montpetit, Department of Biology, University of Ottawa

Months before his much-anticipated swim across the English Channel, Paul Hopfensperger is thinking about the 34 kilometres he must swim to complete this feat. This will be the longest swim he has ever undertaken. At this time of year, the waters are cold and frigid, and he will have to swim through choppy waters.

The success of the mission will depend on how well he manages his diet and training. His diet consisted of bananas, pastries, and pastas and any foods that provide for slow energy release. Since the beginning of his training, he has gained 7 kilograms of fat and his arms and legs have thickened. Paul's cardiac output improved from 20 to about 35 litres/minute.

On the day of the swim, Paul slowly walks down the beach. He blankly stares across the channel, seemingly fixing the point where he will finish the swim and meet the people gathered on the French coast to welcome him. He now focuses on the task, the things he must do to complete the journey, blocking out any negative thoughts about doing this. He gives a final nod to his team manager, slowly enters the water, and begins to swim.

Stroke after stroke, Paul remembers to take a breath of fresh air at constant intervals. Carbohydrates from Paul's last meal stored in the liver and muscles quickly convert to glucose and combine with oxygen to provide energy for each stroke that will carry him closer to the French coast. Paul's muscles are burning 3000 calories/hour. But after only a few hours and 5 kilometres later, glucose is running out, and Paul is facing a fuel crisis. Paul's muscles are now aching and burning. His desire to stop and turn back is growing strong. His trainer and team manager encourage him from a nearby boat. Paul knows his training will help him overcome this energy crisis through the mobilization of a new fuel source, lipids.

To get this fuel, lipolysis in fat cells and muscles is stimulated to provide free fatty acids in the blood. Converting fat to fuel, however, demands extra oxygen. Paul is taking in 76 litres of air/minute. This energy powers Paul through 19 more kilometres across the Channel during the first 6 hours. To succeed, he will have to keep providing fuels until the end. Each minute, Paul's heart pumps blood and fuel throughout his body, delivering the much-needed oxygen and fuel to his muscles to prevent fatigue.

After 12 hours, Paul covered almost 29 kilometres. Within the 5 kilometres of the French coast, the swim has cost Paul almost 6 kilograms of body fat. He is excited about the finish but must stay calm and focused to save his energy to finish the crossing. He knows it will be over soon!

Despite the fatigue, Paul prepares his heart and muscles for the final push. After 14 exhausting hours, Paul reaches the opposite side of the Channel on the French coast, to the applause of his awaiting team and supporters.

QUESTIONS

Before the Swim

1. Discuss the importance of Paul's diet for his training.
2. Explain cardiac output. What is the significance of increasing cardiac output during training?
3. What changes are occurring to Paul's muscles during training? Explain.

During the Swim

4. How do carbohydrates provide energy to the muscles during the beginning minutes of the race?
5. Why do Paul's muscles ache and burn after a few hours into the swim?
6. After a few hours of swimming, Paul is facing a fuel crisis. How does his body respond to the crisis?
7. Compare lipid and glucose metabolism. What are the advantages and disadvantages to switching from carbohydrate metabolism to lipid metabolism? Why would Paul have to slow down a little to conserve his energies to finish the race?
8. What tissues are responsible to provide energy to Paul's working muscles? Explain the control of carbohydrate and lipid mobilization during the swim.
9. Explain the importance of Paul keeping a good breathing rhythm.
10. Explain how Paul delivers oxygen to his muscles. How has he been able to maximize this delivery?
11. How does Paul regulate his body temperature in the cold and frigid waters of the English Channel?

12. What role do the endocrine and autonomic nervous systems play during the swim?
13. Explain how Paul is able to generate contractions and force to propel him forward during the swim.
14. Explain why Paul lost nearly 6 kilograms during the race.

Firish Fries

CASE OBJECTIVES

The objective of the following case is to gain an understanding of the concepts of biodiversity and plant growth and the issues regarding the use of chemicals to control the spread of crop pathogenic diseases.

LEARNING OUTCOMES

By the end of this case, you should be able to

1. Explain the importance of genetic diversity
2. Discuss the biological advantages and disadvantages of using insecticides, herbicides, and fungicides to protect crops from pathogens
3. Discuss how pathogens damage crop plants
4. Discuss strategies that would enable farmers to protect crop plants from pathogens
5. Explain factors responsible for plant growth
6. Discuss the life cycle of potatoes
7. Discuss the importance of crop plants to communities

Suggested Readings: Chapters 16, 25, 28, 29, 30, 31, 48, and 49; independent studies

Firish Fries

By Gurjot Jassy and Colin J. Montpetit, Department of Biology, University of Ottawa

St. Patrick's Day's was reason enough for Madeleine and her friends to head out to the local pub after class. Murphy was already there and had ordered his usual side dish of French fries.

"You know, they really shouldn't be called French fries," he said, as they all sat down. Everybody looked confused. "Why not?" asked Madeleine. "Well, I think it would be nice if we called them Irish fries to pay tribute to the potato famine in Ireland," Murphy replied. Everybody nodded their heads in agreement.

Madeleine, being of Irish descent, had heard the stories from her grandparents many times and, as a biology major, understood exactly what had caused the famine. "I agree, Murphy. Back in those times, people were unaware of the need to conserve and maintain the biodiversity of plants. Ireland was only using a few varieties of potatoes for almost all its food, none of which were resistant to the fungus that was quickly spreading. What they should have done is used more varieties of potatoes and sprayed the crops with chemicals."

"Wait a minute; haven't you been listening to the local news recently? The government has banned using certain chemical products that include insecticides, herbicides, and fungicides," exclaimed Murphy. "Some of those products contain harmful chemical products. For example, Killex, used to kill weeds growing on lawns, produces a breakdown product, 2,4-dichlorophenol (2,4-DCP), which is a carcinogen. Also, given the way that these chemical products are synthesized these days, it is highly likely that they are laced with chlorinated dioxins. This is bad since these toxic substances have been linked to cancer."

Murphy then added, "Maybe spraying the crops with similar compounds would have averted the potato famine, but it could have caused other problems—such as polluting the environment, killing other organisms!" Murphy then asked, "You're a biology major, Madeleine. What could we do to avoid such problems and maintain crops in the presence of pathogens?"

"I'll get back you on this—I'm hungry right now! Let's focus on this day and dig in to our fries. I guess the best way is simply to find an appropriate balance, and for now we can call them Firish fries," replied Madeleine. With that, everybody focused their attention on the hockey game.

QUESTIONS

1. What types of plants are potatoes?
2. Explain the life cycle of potatoes.
3. What factors regulate plant growth? What factors regulate potato plant growth?
4. Why were potatoes so important to the Irish community? Explain the importance of crop plants to communities in general.
5. How do pathogens inflict damage on crop plants? What tissues are usually damaged? What plant processes are affected by the pathogens?
6. What caused the potato famine in Ireland? How did the pathogen inflict the damage in this case?
7. Explain why the lack of genetic diversity of potatoes used in Europe exacerbated the potato famine. What could have been done to avoid the potato famine?
8. Compare insecticides, herbicides, and fungicides. What are their targets?
9. Outline the advantages and disadvantages of using such products to control pathogens. What ecological risks would they have?
10. Why do you think the Government of Canada is now banning the use of specific compounds to control the spread of pathogens and unwanted plants? Could this decision lead to another famine or other problems in the future? Explain.
11. Describe different strategies that scientists and farmers can use to increase biodiversity and to decrease the reliance on chemicals to control pathogens and unwanted plants. Why would these strategies be safer and better for biodiversity?
12. As a biology major, what types of experiments could Madeleine perform to develop such strategies?
13. Based on our knowledge of the potato famine, explain the significance of conserving biodiversity.

The Case of the Two-Toned Lobster

CASE OBJECTIVE

The goal of the following case is to cover the following concepts associated with genetics, the cell cycle, DNA technology and genomics, and animal development:

1. Genes and approaches to genetic studies
2. Gene expression, gene mutations
3. Cell cycle and embryonic development

LEARNING OUTCOMES

By the end of this case, you should be able to

1. List different types of mutations and discuss the importance of mutations of genetic studies
2. Differentiate between genotype and phenotype, genes and alleles
3. Discuss different approaches for genetic studies
4. Discuss the role of model organisms in research
5. Discuss the role of mitosis during embryonic development
6. List the events of mitosis and the cell cycle.

Suggested Readings: Chapters 9, 11, 12, 13, 14, 15, 16, and 38; independent studies

The Case of the Two-Toned Lobster

By Colin J. Montpetit, Department of Biology, University of Ottawa

On a bright, sunny day, Jay (10 years old) decided to accompany his father (Edward Pothier, a fisherman) on his daily trips on the ocean, just off the coast of Nova Scotia. But on this day, Jay noticed a rare catch in one of his father's traps. Jay could not believe his eyes and quickly called his father. Edward and the crew members were very excited when they saw this odd-looking lobster. "You never know what's going to be in there!" exclaimed Edward. "Considering the day's haul, this catch was the highlight of the day!" Edward decided to donate the lobster to the Bedford Institute of Oceanography in Dartmouth, N.S. Doug Pezzack, a biologist from the Bedford Institute, explained that 20 to 30 billion lobsters are trapped around these waters every year, but we only hear about one or two of these two-toned lobsters a year. "What is even more fascinating is that this lobster is a female. It is unheard of for such a lobster to be all female. Two-toned lobsters such as this one are usually hermaphrodites—half male and half female." Doug also added that it would be even more interesting to know what the outcome would be if the lobster breeds. Doug went on to explain that lobsters are typically a mix of greens and blues, but strange lobsters such as this are sometimes discovered. "We have even seen lobsters that are bright blue, orange, and even albino!" said Doug. Doug believes that the lobster has a genetic mutation. "Something happened during the development when the egg split in half. The genetics on one side stays normal, but the other side turns red. Edward admitted that he had caught a number of yellow-toned lobsters, which turned red when cooked.

QUESTIONS

1. What is meant by genetic mutation? Why does Doug believe that the two-toned phenotype is caused by a genetic mutation? List and briefly explain different types of mutations.
2. Define the following terms: genotype, phenotype, alleles.
3. What type of mutation do you believe Jay's lobster is exhibiting? Justify your answer.
4. As a biologist, what approaches would you take to test whether or not this is caused by a gene mutation?
5. What is hermaphroditism? Why are two-toned lobsters usually hermaphrodites?
6. Lobsters of different colours (blue, yellow, albino, etc.) have been detected in the past. What test could you conduct to find out if the colours are controlled by multiple alleles for a given gene?
7. Many scientists are interested in studying these types of phenotypes (two-toned lobsters, hermaphroditism, etc.). Is the lobster an appropriate model organism to study the genetics of these phenotypes? Why or why not? (Hint: First define what a model research organism is and why it is used.)
8. Explain how a fertilized egg develops into an embryo. Explain the steps of embryonic development.
9. What role does mitosis play during embryonic development? Explain the steps of mitosis.
10. Explain the structure of genes and how genes are expressed. How might this explain the lobster's phenotype? Explain.

Microbes in the Ocean's Depths

CASE OBJECTIVE

The objective of this case is to gain an understanding of the characteristics of hydrothermal vent ecosystems and how these influence the evolution and diversity of organisms present in that ecosystem.

LEARNING OUTCOMES

By the end of this case, you should be able to

1. Outline differences between prokaryotes and eukaryotes
2. Discuss how bacteria may live in communities
3. List the different modes of obtaining energy and carbon in prokaryotes
4. Discuss different ways in which prokaryotes differ in whether oxygen can be used in their metabolism
5. Define the key features that group prokaryotes in different domains
6. List and discuss approaches and criteria to identify species of bacteria
7. Explain different mechanisms of adaptations for life near hydrothermal vents
8. Discuss the importance of bacteria to humans

Suggested Readings: Chapters 16, 18, 19, 21, 46, and 47; independent studies

Reference: Stennhuysen J., Reuters. (2007, October). *Thousands of new microbes discovered.* Retrieved from http://www2.canada.com/montrealgazette/features/greenlife/story.html?id=67866b02-a941-4b23-99c0-375d923ca67d&k=97074

Microbes in the Ocean's Depths

By Colin J. Montpetit, Department of Biology, University of Ottawa

Recently, Dr. Huber and her team discovered thousands of marine microbes living in the deep sea (1500 m) next to the hydrothermal vents of an underwater volcano off the Oregon coast. As many as 37 000 types of bacteria living on two separate hydrothermal vents have been identified using DNA sequencing techniques. Dr. Huber's research interests are to describe and understand the evolutionary and community dynamics of microbial groups in the subseafloor habitat.

Hydrothermal vents typically release hot fluids containing hydrogen and sulphur gases, which are compounds you would not normally find in large quantities in the sea. These are mixing with cold and oxygenated seawater, creating different niches for life in these environments.

The team found many epsilon-proteobacteria in their samples. Epsilon-proteobacteria are found in many different places, including the human gut and hydrothermal vents. Although the samples were taken from different sites on the same volcano, the epsilon-proteobacteria from the two sites had totally different chemistries and population structures. Dr. Huber thinks that this is a result of the different geochemistries of the vents and the adaptations to the different environments.

Microbes make living on Earth possible. They produce oxygen and are also involved in carbon, nitrogen, and sulphur cycling. Dr. Huber is quick to point out the need to understand the organisms that live near hydrothermal vents and to understand what they are doing.

QUESTIONS

1. List the general features of prokaryotes. In what ways do they differ from eukaryotes and viruses?
2. Describe the ecosystem at the hydrothermal vents.
3. Near hydrothermal vents, oxygen levels may be low to nonexistent. Explain how prokaryotes may produce ATP in the absence of oxygen. What are the different modes of obtaining energy and carbon in prokaryotes?
4. In nature, how do bacteria live in communities? Explain how studying the community dynamics of microbes is of importance.
5. How are bacteria able to live near hydrothermal vents? What types of adaptations enable them to live in an environment that offers elevated temperatures and low levels of oxygen?
6. In the case description, no discussion is provided as to the methods employed to identify the "37 000" different kinds of bacteria found at the two hydrothermal vents. What criteria may have been used to identify the species of bacteria near the hydrothermal vents?
7. In the case, it is also mentioned that the prokaryotes were identified using new DNA technology. Explain how prokaryotic species may be identified using DNA information.
8. In your own words, explain why there would be a need to understand the organisms that live near hydrothermal vents and to understand what they are doing.
9. What reasons may account for the differences in species observed on the two hydrothermal vents, respectively? What are the community characteristics that allow supporting the diversity of prokaryotes at the hydrothermal vents?
10. List different characteristics of these environments. How do you think these influence microbe adaptations to these environments?
11. Explain in which ways bacteria are important to humans.
12. What are the broader implications of this study?

A Case of Natural Doping

CASE OBJECTIVE

The goal of the following case is to gain an understanding of concepts related to cell membranes, membrane transport, energy metabolism and mobilization, animal nutrition, and the control of gene expression

LEARNING OUTCOMES

By the end of this case, you should be able to

1. Explain the purpose of using model organisms in research
2. Outline the components of cell membranes and their roles
3. Discuss how membrane fluidity can be influenced by changes in components of the cell membranes
4. Discuss the transport of molecules in and out of cells
5. Explain the significance of homeoviscous adjustment to transport across membranes and aerobic capacity
6. Compare glucose and lipid metabolism, aerobic and anaerobic metabolism
7. Explain why lipids are the preferred energy source for activities such as migration
8. Discuss how fatty acids may control gene expression

Suggested Readings: Chapters 4, 5, 6, 15, and 41

References

Maillet, D., & Weber, J.-M. (2006). Performance-enhancing role of dietary fatty acids in a long-distance migrant: The semipalmated sandpiper. *Journal of Experimental Biology, 209*, 2686–2695.

Maillet, D., & Weber, J.-M. (2007). Relationship between n-3 PUFA content and energy metabolism in the flight muscles of a migrating shorebird: Evidence for natural doping. *Journal of Experimental Biology, 210,* 413–420.

Vaillancourt, E., & Weber, J.-M. (2007). Lipid mobilization of long-distance migrant birds *in vivo*: The high lipolytic rate of ruff sandpipers is not stimulated during shivering. *Journal of Experimental Biology, 210,* 1161–1169.

A Case of Natural Doping

By Colin J. Montpetit, Department of Biology, University of Ottawa

For the past several years, many athletes have resorted to cheating techniques to outperform their competitors. Treatments such as anabolic steroids, insulinlike growth factors, and erythropoietin (EPO), for example, have all been used to help "cheating" athletes enhance their performance. Have athletes now reached their physiological limits and knowledge on how to naturally improve muscular performance so that they now have to depend on unnatural means to increase muscular activity? Are their examples in nature that could teach us otherwise? Professor Weber, from the Department of Biology at the University of Ottawa, thinks so: "Studies have shown that long-distance migratory birds, such as semipalmated sandpipers, depend on lipids as their main source for flight; however, different roles for this type of fuel have not been considered."

According to Dr. Weber, dietary lipids could also be used as performance-enhancing substances to prepare animals for migration. Prior to their migration, which involves a nonstop flight lasting three to four days to South America, semipalmated sandpipers stop briefly in the Bay of Fundy (New Brunswick, Canada) to feed on burrowing shrimps (*Corophium volutor*). Over a two-week period, sandpipers may double their body mass by feeding on these shrimps. These marine invertebrates are a rich source of omega-3 polyunsaturated fatty acids (PUFAs), eicosapentaenoic acid (EPA), and docosahexaenoic acid (DHA). Studies have shown that the omega-3 PUFAs are known to boost aerobic capacity and to have beneficial health effects in humans.

To investigate the role of lipids as a performance-enhancing substance, Dr. Weber's research team set out to study semipalmated sandpipers (*Calidris pusilla*). "Semipalmated sandpipers provide a unique model to study the effects of nutrition on the metabolism of long-distance migrants and on their exercise capacity," states Dr. Weber. He adds, "We have used refuelling sandpipers as a natural experiment to test the idea that lipids could serve to enhance muscle performance."

It turns out that during refuelling, in addition to increasing lipid content in adipose tissue, omega-3 PUFA from *Corophium* are rapidly incorporated into the cell membranes of the bird's flight muscles. During this feeding period, sandpiper flight muscle cells also demonstrate an increase in the density of mitochondria and activities of key enzymes associated with oxidative metabolism.

According to Dr. Weber, "The significance of these findings suggests that the diet and the incorporation of omega-3 PUFA make their flight muscle cell membranes more fluid and improve their ability for fatty acid transport and their aerobic capacity. In fact, omega-3 PUFA appears to act as a potent metabolic signal to improve muscle performance in these birds."

QUESTIONS

1. What is meant by performance enhancing and muscular performance? How can one enhance muscular performance? What physiological processes can be changed, improved, or influenced to enhance muscular performance? Explain.
2. What are lipids? Explain the difference between unsaturated and polyunsaturated fatty acid. What type of lipids are EPA and DHA? List the benefits they have for humans.
3. What was the hypothesis of the study? What observations led the investigators to formulate their hypothesis?
4. Why are model organisms used in research? Give different examples. In this study, why was the semipalmated sandpiper an appropriate model organism?
5. Why are lipids a better fuel for migration than carbohydrates? Discuss the metabolic pathways involved in the metabolism of carbohydrate and lipids. What are the benefits of storing fats instead of carbohydrates (glucose and glycogen)?
6. What effects would changes in membrane fluidity have on the functions of membrane proteins (i.e., transporters etc.)?
7. Explain the effects of incorporating omega-3 PUFA in cell membranes. How would these affect membrane functions?
8. Based on the information presented in the case study and your knowledge of membrane fluidity, how would a diet rich in fatty acids increase aerobic capacity and metabolic performance in these birds? Explain.
9. In addition to changing membrane fluidity, explain how EPA and DHA could act as signals to boost metabolism and increase aerobic capacity.
10. What significant implications does this study have for humans, athletes, and research on obesity? Explain.

The Case of Bees

CASE OBJECTIVE

The goal of the case is to gain an understanding of the link between physiological function and the design of metabolic pathways. Topics covered in this case include energy, metabolism, metabolic pathways, and cell respiration.

LEARNING OUTCOMES

By the end of this case, you should be able to

1. Draw plots of enzyme assays and gain different measures of enzyme characteristics from the plots
2. Describe the dynamics of the regulation of the glycolytic pathway
3. Outline the major pathways involved in oxidation of carbohydrates, lipids, and proteins
4. Compare and contrast different ways to measure metabolic rate
5. Provide an in-depth background to understand the case study
6. Define key words

Suggested Readings: Chapters 4 and 5; independent studies

Reference: Darveau, C. A., Hochachka, P. W., Welch, K. C., Jr., Roubik, D. W., & Suarez, R. K. (2005). Allometric scaling of flight energetics in Panamanian orchid bees: A comparative phylogenetic approach. *Journal of Experimental Biology, 208,* 3581–3591.

The Case of Bees

Dr. Charles Darveau is a professor in the Department of Biology at the University of Ottawa. His research interests are focused on understanding how metabolic rates scale with body size across species. To this end, Darveau uses different species of bees to understand the underlying factors responsible for the variations observed in metabolic rates in organisms. Based on previous reports demonstrating relationships between body size and metabolic rate and wingbeat frequencies in insects, Darveau and his collaborators, Raul Suarez and Kenneth Welch, set out to investigate the effects of body mass on the design of the metabolic pathways.

For these purposes, the team selected a number of related orchard bee species. They measured the bees' wingbeat frequencies, metabolic rates, body sizes, wing size, and wing loading (a measure of the body mass supported by a given wing area). When the team examined the data, they discovered that wing loading explained most of the variation they saw in wingbeat frequency, and this was highly correlated with metabolic rate. As Darveau explains, "Wing morphology explains more of the differences in metabolic rate in these bees than body size alone."

Having already determined that orchid bees fuelled their flight entirely with glucose, the team then investigated whether the metabolic pathway for glycolysis may be a contributing factor to the variances observed in metabolic rates in these organisms. This was accomplished by plotting flight muscle enzymatic activity against metabolic rate. The results demonstrated that of the 10 enzymes involved in the catabolism of glucose, only hexokinase activity (the enzyme responsible for converting glucose into glucose 6-phosphate) correlated with the variation in metabolic rate in the orchid bees.

Originally, the research team believed that the control of the glycolytic flux was distributed among the 10 enzymes involved in glycolysis and was not restricted to one "rate-limiting" step near the start of the pathway. Overall, the team's results suggest that the evolutions of orchid bee size, flight metabolic rate, and hexokinase are all closely correlated. Darveau and his collaborators report that smaller bees need to beat their wings 250 per second in order to hover; however, wingbeat frequency is almost four times lower in bigger bees to perform the same task. Surprisingly, the team observed that hexokinase activity follows exactly the same scaling pattern as that of wingbeat frequency.

The ultimate goal, explains Dr. Darveau, notwithstanding the energetics of flight, is to understand the evolution of biological design.

QUESTIONS

PART I

Understanding the Case: Defining Key Terms and Concepts

1. What are enzymes? What factors determine their activity? How can their activity be measured? What information regarding their characteristics can be obtained (i.e., catalytic characteristics under different conditions)?
2. Describe the chemical basis for cellular respiration.
3. Discuss the relationship between metabolism and metabolic rate.
4. How can flux through metabolic pathways be regulated?
5. Summarize the reactions of glycolysis. What enzymes are involved in these reactions, and which reactions do they catalyze? Explain the significance of the first step in glycolysis (hexokinase).
6. Summarize the ATP production from the complete oxidation of glucose (i.e., glycolysis, Krebs cycle, etc.).
7. Discuss the relationship between body size and metabolic rate. Draw a figure showing your understanding of how metabolic rates scale with body size across species. What is the effect of body mass on metabolic rate? What factors might explain these observations?

PART II

Analysis of the Study

8. What was the hypothesis of this study, and how was it tested?
9. Why was the glycolytic pathway studied in this case? What was the reasoning behind this choice?
10. How might hexokinase characteristics differ in different orchid bee species? What would enzyme assay plots demonstrate? Support your answer by drawing the plots of enzyme activity versus time for bees of different sizes. Label your figures. How might this support the scaling between metabolic rate and body size?

11. What is meant by "flux" in this case? How may flux through a metabolic pathway increase or decrease?
12. How do changes in hexokinase activity influence "flux" through the glycolytic pathway?
13. Based on the answers provided above, provide an explanation as to why evolution might have chosen hexokinase activity, and not other enzymes, as a mechanism to scale metabolic rate in orchid bees.
14. How are changes in hexokinase activity beneficial to the hovering abilities of bees?
15. What new hypotheses can be drawn from this study? How would you test them? What variables would you measure to test your hypothesis? Propose an experiment.

PART III

Broader Implications

16. What would be the broader implications of this study? What would be the importance of gaining such an understanding as to the energetics of insect flight muscle and the design of energy production pathways?
17. What additional adaptations can you predict that may explain the observation of the scaling of metabolic rate with body size?

Answers

Chapter 1 Light and Life

1.1 The Physical Nature of Light [pp. 2-5]

1. energy; 2. information; 3. electromagnetic radiation; 4. electrical/ magnetic; 5. magnetic/electrical; 6. photons; 7. inversely; 8. reflected; 9. transmitted; 10. absorbed; 11. pigments; 12. conjugated; 13. electron; 14. ground; 15. equal/match; 16. C, B, D, A; 17. A. Photon; B. Electron at ground state; C. Electron at excited state ; D. Low energy level; E. High energy level; 18. False: the energy of light is inversely related to its wavelength; 19. False: it is the result of the fusion of hydrogen nuclei into helium; 20. True; 21. False: chlorophyll absorbs red and blue light meaning the differences are equal the energy of red and blue wavelengths; 22. C; 23. E; 24. F; 25. B; 26. A; 27. D.

1.3 Light as a Source of Information [pp. 6- 10]

28. potential; 29. ATP/NADPH; 30. NADPH/ATP ; 31. carbohydrates; 32. bacteriorhodopsin; 33. ATP; 34. information; 35. rhodospin; 36. retinal; 37. phototaxis; 38. phytochrome; 39. photomorphogenesis; 40. eye-spot; 41. compound; 42. single lens; 43. B; 44. J; 45. F; 46. I; 47. C; 48. H; 49. E; 50. A; 51. D; 52. G; 53. A: non-image-forming, senses light intensity and direction; B: Single lens; C.arthropods such as insects and crustaceans; 54. Eyes contain the photoreceptors but also require a brain or simple nervous system to interpret and respond to the information whereas eyespots trigger a change in flagellar movement through a simple signal transduction response; 55. They both allow the same type of response (movement in response to light) however the ocellus transmits information via nerves to a cerebral ganglion; 56. The compound eye is extremely sensitive to movement; 57. Delocalized electrons are promoted to their excited states and this causes a conformational change in the pigment molecule.

1.4 Light Can Damage Biological Molecules [pp. 11- 13]

58. 400; 59. 700; 60. absorbed; 61. Ozone/O_3; 62. water; 63. CO_2; 64. Ultraviolet/UV; 65. infrared; 66. electrons; 67. water; 68. damage; 69. repair; 70. carotenoids; 71. heat; 72. thymine dimers; 73. melanin; 74. False: light of wavelengths less than 400nm (i.e. in the UV range) are referred to as ionizing radiation because their high energy levels breaks bonds, oxidizing molecules so that they release ions; 75. False: it absorbs UV-C but not the longer wavelengths of UV-A and UV-B; 76. Generally true except in the case of tthe Inuit (natives of the Arctic) whose traditional diet provides high levels of vitamin D.

1.5 Role of Light in Ecology and Behaviour [pp. 13 - 18]

77. circadian; 78. daily; 79. biological; 80. brain; 81. optic; 82. melatonin; 83. seasonal; 84. flowering; 85. hibernation; 86. colour; 87. Camouflage; 88. pattern/behaviour; 89. behaviour/pattern; 90. appearance; 91. colour; 92. colour; 93. pollinators; 94. light pollution ; 95. nocturnal; 96. A. Lens; B. Retina; C. Optic nerve; D. Suprachiasmatic nuleus; E. Pineal gland; F. Melatonin; G. Brain; 97. False: all forms of life display circadian rhthym; 98. True; 99. True; 100. False: they are able to absorb light of the blue region of the spectrum which is the only type of light that can penetrate that deep; 101. Pollinators are attracted to shape, smell and colour and the colour that is favoured depends upon the abilities of the pollinator to perceive light of specific wavelengths; 102. The Industrial Revolution caused an increase in air pollution and a darkening of the bark on the trees in their habitat, making the dominant light-coloured moths more visible to predators. The dark-coloured variants thus had a selective advantage and became more common.

1.7 Organisms Making Their Own Light: Bioluminescence [pp. 18-20]

103. visual; 104. mole rat; 105. darkness; 106. photoreceptors; 107.light; 108. biological; 109. circadian; 110. bioluminescent; 111. electron; 112. excited; 113. photon; 114. mate; 115. prey; 116. defense; 117. C; 118. D; 119. A; 120. B.

SELF-TEST

1. C [400nm is in the blue region of the spectrum, closest to the UV region]
2. B [The colour is the result of transmitted light]

3. D [Chlorophyll pigments absorb blue and red light but not green, giving plants their green colour]
4. C [Plants release oxygen as a by-product and use the absorbed energy to convert CO_2 to the organic molecules that allow the plant to grow]
5. D [Bacteriorhodopsin is a membrane-embedded light-activated proton pump that generates a proton gradient which is then used to make ATP]
6. D [Retinal is the light-absorbing pigment in bacteriorhodopsin as well as rhodopsin]
7. C [The *C. reinhartii* eyespot responds directly to light and does not require the nervous system that is an essential component of vision]
8. C [Rods and cones are the light-absorbing cells of the retina]
9. A [Phytochrome is the light-sensing photoreceptor in the cytoplasm of plants that is essential for the developmental changes associated with exposure of seedlings to light]
10. C [Light in the UV region of the spectrum is shorter wavelength, higher energy and capable of breaking bonds causing oxidation of the molecule and release of ions]
11. C [UV light causes thymine dimers however cells have repair systems that are able to repair damage to DNA]
12. A [A critical characteristic of circadian rhythms is that they are NOT dependent on daily changes in exposure to light]
13. B [Melanin is the pigment produced by the skin's melanocytes to protect skin cells from UV-induced damage]
14. A [Carotenoids are accessory pigments that absorb excess light and dissipate it as heat, thereby protecting the photosynthetic apparatus from light-induced damage]
15. B [The common form is light coloured allowing it to blend with the light colour of lichen-covered trees in its normal habitat]
16. D [Phototaxis is the ability to swim in a light-dependent direction; it is based simply of light absorption and a signal transduction resulting in altered flagellar movement]
17. C [Phytoerythrin preferentially absorbs blue light which, because of the absorption of longer wavelength light by water, is the only type of light that penetrates to the greater depths]
18. D [The eyes of mole rats have been overgrown by layers of tissue making them blind however their retinas can still respond to light and this information is transmitted via the optic nerve to the biological clock]
19. A [Excess night-time light in urban areas is called ecological light pollution]
20. A [Phototaxis is a light-dependent behavior seen in simple organisms such as *C. reinhartii*]

Integrating and Applying Key Concepts

1. Read section 1.3d and the figure legend of Fig. 1.17.

Chapter 2 Origins of Life

2.2 The Chemical Origins of Life [pp. 25-29]

1. atoms; 2. molecules; 3. chemistry; 4. physics; 5. order; 6. energy; 7. stimuli; 8. development; 9. evolve; 10. cell; 11. 4.6; 12. 500; 13. CO_2, NH_3, CH_4, H_2S and H_2; 14. Oparin/Haldane; 15. Haldane/Oparin; 16. spontaneously; 17. Miller-Urey; 18. clay; 19. protobiont; 20. T; 21. F – although the meteorite has been dated at ~4.5 billion years old, the calculated age of Earth and the solar system, and may contain evidence of past life, it arrived on Earth only ~13 000 years ago, long after life began on Earth; 22. F – a reducing environment was critical to the abiotic synthesis of organics and the lack of oxygen in the atmosphere of early Earth meant no ozone was present to block a critical energy source for these synthetic reactions: ultraviolet light; 23. T; 24. D; 25. C; 26. B; 27.E; 28. A; 29. D, B, C, A; 30. A. Display order, B. Energy comes from the environment and is required to maintain order, C. Reproduction, D. Respond to stimuli, E. The ability to maintain constant internal conditions, F. The increase in size and number of cells and for some organisms, the change in form with time, G. Evolution;31. Oxygen is a strong oxidizing agent and its presence would prevent the spontaneous formation of the large, complex electron and energy-rich organic molecules; 32. Viruses contain nucleic acids, reproduce and evolve ; 33. The original experiment used H2, CH4, NH3, water vapour and sparking electrodes and yielded many organic compounds. A subsequent modification that included hydrogen cyanide and formaldehyde yielded all of the essential building blocks of biological molecules.

2.3 The Origins of Information and Metabolism [pp. 29-31]

34. DNA; 35. RNA; 36. protein; 37. Proteins/enzymes; 38. ribozymes; 39. shapes; 40. RNA; 41. information; 42. catalyst; 43. proteins; 44. selective; 45. enzymes; 46. diversity; 47. DNA; 48. stable; 49. thymine; 50. double; 51. complementary; 52. metabolism; 53. reduction; 54. oxidation; 55. Czech discovered that certain modern-day RNA molecules, "ribozymes" can function as biological catalysts. This discovery lead to the proposal that the earliest life forms used RNA for both information storage and catalysis then slowly evolved the proteins and DNA components typical of modern cellular information flow; 56. Three reasons: 1. Enzymes have far greater catalytic power than any known ribozyme, 2. Because there are 20 different building blocks (amino acids) for proteins, but only four possible bases for RNA molecules means protein structures exhibit far more diversity and 3. The various amino acids can interact with eachother in more diverse ways than can the four RNA bases, giving rise to even more structural diversity through complex folding; 57. DNA is double-stranded, DNA bases contain the more stable sugar deoxyribose and include thymine whereas RNA is single-stranded, its bases possess the sugar ribose and include uracil in place of thymine; 58. Changes in DNA sequences result in changes in the organism from one generation to another, and these may be maintained if they provide a selective advantage to the organism; 59. T – the ribosome is a large complex of RNA molecules and proteins but it is one of the ribosomal RNA molecules that actually catalyzes the polymerization of amino acids into proteins; 60. F – the flow of information in modern cells is DNA to RNA to protein and the end product, proteins, are required to catalyze each step; 61. T – they can only fold into the specific shapes required for catalytic activity because of their single-stranded nature; 62. F – the first types of metabolic reactions probably involved oxidation of food molecules and direct, one-step transfers of the electrons in reduction (biosynthetic) reactions, leading to significant losses of energy; 63. T.

2.4 Early Life [pp. 32-35]

64. carbon; 65. 3.9; 66. stromatolites; 67. cyanobacteria; 68. 3.5; 69. panspermia; 70. cooled; 71. spores; 72. dormancy; 73. prokaryotes; 74. nucleus; 75. plasma membrane; 76. cytoplasm; 77. oxidation; 78. photosynthesis; 79. organelles; 80.chromosomes; 81. transcription; 82. translation; 83. oxygen; 84. banded iron; 85. 2.5; 86. cyanobacteria; 87. oxygenic; 88. abundant; 89. eukaryotes; 90. E; 91. D; 92. A; 93. F; 94. C; 95. B; 96. Organisms preferentially incorporate the carbon-12 isotope over other isotopes so demonstration of a depletion of other carbon isotopes (e.g. ^{13}C) is indicative of a biological source for the carbon; 97. The stromatolites in Shark Bay are approximately 2000 years old and are mineral deposits formed by the action of cyanobacteria. Fossilized stromatolites are highly similar in structure, are dated to ~3.5 billion years ago and contain fossils that resemble modern-day cyanobacteria; 98. F – primitive forms of metabolism probably included anaerobic respiration, fermentation and simple photosynthesis using the more easily oxidized H_2S and Fe^{2+} as electron sources; 99. F – geological evidence indicates that oxygen only started to appear in Earth's atmosphere around 2.5 billion years ago, long after the first life forms appeared; 100. T; 101. T; 102. bacterial flagellum; 103. plasma membrane; 104. cell wall; 105. nucleoid.

2.5 Eukaryotic Cells [pp. 35-43]

106. nuclear; 107. membrane; 108. organelles; 109. motor/contractile; 110. infolding; 111. plasma; 112. nuclear; 113. reticulum; 114. Golgi; 115. mitochondrion/chloroplast; 116. chloproplast/mitochondrion; 117. Endosymbiosis; 118. prokaryotes/bacteria; 119. prokaryotic; 120. beneficial; 121. aerobic; 122. mitochondria; 123. oxygenic; 124. chloroplasts; 125. mitochondria; 126. chloroplasts; 127. stages; 128. cytoskeleton; 129. fibres/tubes; 130. tubes/fibres; 131. movement; 132. multicellular; 133. fossil; 134. physical/chemical; 135. chemical/physical; 136. The chloroplast and mitochondrion share a similar size and morphology with bacterial cells, reproduce by binary fission as do prokaryotic cells, possess their own DNA that encodes essential functions of the organelle, exhibit similar transcription and translational processes and generate ATP using their own electron transport chains; 137. Prokaryotic cells are believed to have engulfed smaller oxygenic photosynthetic bacteria and bacteria capable of aerobic respiration to give rise to the eukaryotic cells; 138. They are both structures on the surface of the cell, are both surrounded by the plasma membrane and both possess microtubules arranged in a 9 + 2 configuration. However cilia are shorter and more numerous than flagella, move by beating and function to move material over the surface of the cell, whereas flagella provide locomotion through a whipping action; 139. Motor proteins interact with microtubules and microfilaments and move by "walking: over the surface of these structures, using

ATP to provide the power for movement; 140. F – they are believed to have developed from in-folding of the plasma membrane; 141. F – the fungal, animal and plant lines are believed to have evolved independently; 142. T; 143. F – although they serve the same function they are not evolutionarily related so are considered analogous structures; 144. T – some float free in the cytoplasm for the synthesis of proteins that are required in the cytosol or cytoplasmic structures or in the nucleus; 145. T; 146. A; 147. E; 148. D; 149. C; 150. H; 151. J; 152. K; 153. B; 154. G; 155. F; 156. I; 156. Endoplasmic reticulum; 157. Rough ER; 158. Smooth ER; 159. Golgi complex; 160. mitochondrion; 161. nucleus; 162. A – The process of releasing material to the outside of the cell. B -Endocytosis. C - Formed by budding from the Golgi membranes and travel to the plasma membrane for release by exocytosis. D. Part of the cytoskeleton that is involved in cytoplasmic streaming, amoeboid movement, contraction of muscles and dividing the cytoplasm. E.Microtubules. F. Intermediate filaments.

SELF-TEST

1. c [viruses also contain nucleic acids but lack the cellular machinery to replicate, transcribe and translate the information on their nucleic acids, so are not considered to be life forms]
2. b [viruses also contain nucleic acids but are not cells]
3. a [a critical aspect of the Oparin-Haldane hypothesis was the absence of oxygen and a resulting reducing atmosphere and the spontaneous assembly of complex electron-rich molecules]
4. a [the absence of oxygen gave rise to a resulting reducing atmosphere, allowing for the spontaneous assembly of complex electron-rich molecules.]
5. e [they combined methane, ammonia, hydrogen gas, water vapour and continuously sparking electrodes]
6. d [scientists debate whether the methane and ammonia concentrations of primitive Earth's atmosphere were comparable to those used by Miller and Urey, however deep sea vents do have the conditions analogous to those used by Miller and Urey]
7. d [deep sea vents have methane and ammonia concentrations similar to those used by Miller and Urey and the heat from the vents would supply the necessary energy]
8. b [life is believed to have begun in an RNA world]
9. b [modern stromatolites are formed by cyanobacteria and have similar structures to fossilized cells in 3.5 billion year old stromatolites]
10. c [the chromosome of bacterial cells is found in a compact structure called the nucleoid; the chromosomes of eukaryotes are bound by a nuclear membrane]
11. c [aerobic respiration involves the oxidation of food molecules for energy and the use of oxygen as a final electron acceptor]
12. b [the endomembrane system is believed to have developed from in-folding of the plasma membrane and includes the nuclear membrane, the ER and the Golgi complex]
13. c [the endomembrane system is believed to have developed from in-folding of the plasma membrane to give rise to the nuclear membrane, the ER and the Golgi complex]
14. a [the Golgi receives proteins synthesized in the ER, chemically modifies them and regulates the movement of proteins within the cell]
15. c [the ancestor of the mitochondrion is believed to have be a bacterium that performed aerobic respiration while an endosymbiotic oxygenic photosynthetic bacterium is believed to have given rise to the chloroplast]
16. b [virtually all eukaryotic cells possess mitochondria but only plants and algae possess both mitochondria and chloroplasts]
17. e.[similarities between these organelles and prokaryotes include morphology and size, method of reproduction, presence of DNA that encodes essential genes, transcription and translation machinery including ribosomes of similar structure, and energy generation involving electron transport chains and ATP]
18. e [the cytoskeleton includes protein fibres and tubules, gives the cell shape and uses motor proteins to facilitate intracellular movement as well as movement of the entire cell]
19. c [as a cell grows the surface area increases as a square of its length but the volume increases at a greater rate, as a cube of its length]
20. c [the oldest known fossil eukaryotes are 2.2 billion years old]

Integrating and Applying Key Concepts

1. Consider the by-product of this type of photosynthesis, the energetic advantages of aerobic respiration, as well as the fact that endosymbiosis must provide an advantage to the endosymbiont and the host. So what does that suggest about the type of benefit(s) the endosymbiosis of an aerobically respiring bacterium provided to the host?
2. Consider the size difference between typical eukaryotic cells and typical prokaryotic cells, how surface area and volume change with an increase in size, and then look at the structural differences that have evolved in eukaryotic cells. For these unique structures, think about their function within the cell and whether this same function (but not the actual structure) occurs in prokaryotic cells. If the answer is yes, what would be the explanation for the evolution of this structure?

Chapter 3 Selection, Diversity and Biosphere

3.2 Selection [pp. 50-52]

1. floor; 2. atmosphere; 3. different; 4. carbon; 5. energy; 6. autotrophs; 7. heterotrophs; 8. phototrophs; 9. chemotrophs; 10. photoheterotrophs/chemoautotrophs; 11. chemoautotrophs/photoheterotophs; 12. metabolic; 13. selection; 14. lethal; 15. selective; 16. genetic; 17. antibiotic-resistant; 18. syphilis; 19. whales; 20. climbing; 21. E; 22. C; 23. G; 24. F; 25. I; 26. H; 27. A; 28. B; 29. D; 30. A – only certain prokaryotes, B – CO_2, C – Organics, D - Chemoautotrophs; 31. D, H, A, E, F, B, G, C.

3.3 Evolution [pp. 52-55]

32. Evolution; 33. unity; 34. diversity; 35. common; 36. ATP; 37. DNA; 38. bilayers; 39. natural selection; 40. variation; 41. reproduce; 42. offspring; 43. incidence; 44. adaptive; 45. oxygenic; 46. aerobic; 47. ozone; 48. terrestrial; 49. T; 50. F – air is less dense and viscous than water so oxygen is more readily available; 51. T; 52. Because air is less viscous and dense than water, land animals developed skeletons and muscles for locomotion and support while land plants developed structures like woody stems and roots for support and fluid transport, respectively; 53. The accumulation of oxygen in the atmosphere led to the evolution of the more energy-efficient aerobic respiration in some prokaryotes, the evolution of eukaryotes (through the endosymbiotic generation of mitochondria and chloroplasts as discussed in Ch. 2) and the increase in atmospheric ozone levels increased causing a reduction in UV radiation and allowing the movement of organisms, including plants and animals, onto land; 54. A –origin of life, B - time, C – present, D - common ancestor, E – origin of a new lineage, F – extinct group.

3.4 The Biosphere [pp. 56-61]

Section Review (Fill-in-the-Blanks)

55. abiotic; 56. climate; 57. diversity; 58. latitudinal; 59. solar; 60. axis; 61. sun; 62. 90; 63. atmosphere; 64. oblique; 65. area; 66. tilt; 67. sun; 68. duration/intensity; 69. intensity/duration; 70. Northern; 71. Southern; 72. poles; 73. equator; 74. heating; 75. air; 76. winds; 77. latitudinal; 78. rotation; 79. deflection; 80. east-west; 81. rainfall; 82. sunlight, temperature, humidity, wind speed, cloud cover and rainfall; 83. Altitude, air pressure and topography; 84. It is defined as a decrease in temperature without a loss of heat energy and results when warm air masses rise and expand so that their heat energy occupies a larger volume; 85. T; 86. F – ocean currents in coastal regions exchange heat with overlying air masses, moderating the temperature over the nearby land; 87.F – they are the result of adiabatic cooling.

3.6 Cumulative Effect on Biotic and Abiotic Factors [pp. 63-64]

88. E; 89. G; 90. A; 91. B; 92. H; 93. C; 94. F; 95. D; 96. F – it is the genetic sequence of mitochondrial cytochrome c oxidase 1; 97. T.

Self-Test

1. a [Based on easily observed organisms and features, insects are the group of organisms with the biggest number of species]
2. c [only certain prokaryotes exhibit the photoheterotrophic or chemoautotrophic modes of nutrition]
3. d [Autotrophs use CO_2 for carbon and are often photoautotrophs, such as most plants]

4. c [A recently discovered fossilized ungulate from an extinct, raccoon-sized animal that lived a hippo-like existence is believed to be the ancestor of whales]
5. d [Darwin's theory of evolution states that all life on Earth descended from the same common ancestor and therefore share certain critical traits such as DNA as the hereditary material]
6. a [an adaptive radiation occurs when an evolutionary breakthrough allows diversification of life]
7. b [the increase in atmospheric oxygen levels lead to the development of an ozone layer which blocks UV rays]
8. b [Ammonia is a toxic waste product excreted by aquatic animals but, due to the lack of water for dilution, terrestrial animals had to evolve the ability to excrete non-toxic wastes such as urea and uric acid]
9. a [the root of the tree of life represents the origin of life, the common ancestor from which all life is believed to have evolved]
10. b [abiotic factors are the physical and chemical conditions of the environment whereas parasitism involves a harmful symbiosis with another organism]
11. c [symbiotic interactions are interactions between different organisms and these do not exhibit a global pattern; Earth's rotation, orbit and spherical shape hence latitudinal variations in solar radiation all determine global temperatures, season, winds and precipitation which affect diversity]
12. a [because of the Coriolis effect, the north-south winds are deflected to give rise to east-west winds]
13. d [Latitudinal variation in the speed of Earth's rotation about its axis causes a deflection of the north-south winds in equatorial regions to the predominantly east-west flow of the Trade Winds]
14. d [Latitudinal variation in the intensity of solar radiation causes uneven temperatures and latitudinal variation in air circulation patterns
15. a [as the air in equatorial regions warms it absorbs ocean waters, rises and expands to occupy a larger volume without losing heat energy, it starts to cool adiabatically and releases rain]
16. c [latitudinal variation in solar radiation causes uneven heating of ocean waters which results in changes in water density, causing water to sink towards the poles or rise towards the equator, depending upon the temperature and density and Earth's gravity. Surface water flows under the influence of surface winds and land masses also influences the movement of currents]
17. b [Microclimate refers to the abiotic conditions immediately surrounding organisms and has the greatest effect on survival and reproduction]
18. d [Sea slugs co-opt the chloroplasts of their algal prey to become photosynthetic, they are animals that are otherwise hetertrophic]
19. d [although most plants are photoautotrophs and therefore producers, some lack chloroplasts and therefore must obtain their energy and carbon by parasitism or mutualism with fungi]
20. a [parasitism is a symbiotic relationship where one organism benefits (the parasite) while the host is harmed; ectoparasites live on the surface of their host whereas endoparasites live within the host]

INTEGRATING AND APPLYING KEY CONCEPTS

1. Read section 3.4 then do the corresponding section of the study guide in order to focus on the basic interactions. It may help to derive an idea map for these factors.
2. Begin by defining adaptive radiation and making a list of the abiotic factors that influence life, then determine which factors would be extreme or limiting in each of the three environments. Finally, think about what abiotic conditions are likely to exert the most selective pressure on animals and on plants and what kind of adaptations would be required to survive these pressures.

Chapter 4 Energy and Enzymes

Why It Matters [71-72]

1. enzymes; 2. millions; 3. life

4.1 Energy and the Laws of Thermodynamics [pp. 72-75]

4. Energy; 5. energy; 6. radiation energy; 7. Kinetic; 8. Potential; 9. Thermodynamics; 10. system; 11. surroundings; 12. isolated system; 13. closed system; 14. open systems; 15. conservation; 16. created; 17. destroyed; 18. potential; 19. kinetic; 20. heat; 21. sound; 22. electricity; 23. potential; 24. randomness; 25. entropy; 26. entropy; 27. surroundings; 28. disintegrate; 29. energy; 30. entropy; 31. D; 32. G; 33. F; 34. H; 35. E; 36. A; 37. B; 38. C. 39. Electrons; 40. Molecules; 41. Photons; 42. Photons; 43. There is energy stored in the arrangement of atoms of the molecules making up the cereal; 44. There is potential energy due to the position of the skier relative to Earth's gravitational field.

4.2 Free Energy and Spontaneous Reactions [pp. 75-78] ***Note that in section 4.2a of the Fenton text, exothermic and endothermic are used incorrectly (these terms refer to release of heat or absorption of heat); the correct terms are exergonic and endergonic, respectively. Also in section 4.2c, at equilibrium it is ΔG and <u>not</u> G which is equal to zero.***

45. Spontaneous; 46. exergonic; 47. endergonic; 48. potential; 49. disorder; 50. H; 51. entropy; 52. G; 53. free; 54. enthalpy; 55. entropy; 56. less; 57. exergonic; 58. free energy; 59. equilibrium; 60. equal; 61. ΔG; 62. closed; 63. negative; 64. catabolic; 65. anabolic; 66. positive; 67. B; 68. G; 69. H; 70. A; 71. D; 72. C; 73. E; 74. F; 75. False – It is catabolic because it involves the break-down of large molecules into smaller ones and the release of energy; 76.True; 77. False – water is more disordered than ice so the ice melts spontaneously and the reaction is exergonic (-ve ΔG); 78. True.

4.3 The Energy Currency of the Cell: ATP [pp.79-81]

79. positive; 80. hydrolysis; 81. ribose; 82. adenine; 83. phosphate; 84.replusion; 85. free; 86. exergonic; 87. phosphate group; 88. stable; 89. exergonic; 90. exergonic; 91. energy coupling; 92. exergonic; 93. carbohydrates or fats or proteins; 94. ATP cycle; 95. D; 96. E; 97. B; 98. A; 99. C; 100. Reactions that are connected with one reaction requiring energy (endergonic) and the other reaction releasing energy (exergonic); 101. They make possible such endergonic processes as growth, reproduction, movement and responses to stimuli; 102. ATP is made of ribose sugar to which adenine and 3 phosphates are attached. It helps in transfer of energy from exergonic to endergonic reactions; 103. Shivering causes the spontaneous hydrolysis of ATP and the release of the free energy as heat; 104. Approximately 10 million per second.

4.4 The Role of Enzymes in Biological Reactions [pp. 81-85]

105. spontaneously; 106.Enzymes; 107. strained; 108. broken; 109. activation; 110. transition; 111. kinetic; 112. kinetic; 113. activation; 114. catalysts; 115. free energy; 116. ΔG; 117. substrate; 118. enzyme specificity; 119. lock-and-key; 120. induced-fit; 121. conformational; 122. thousands; 123. unchanged; 124. Catalytic/enzyme; 125. 10 million; 126. active; 127. reacting molecules or reactants; 128. charged; 129. shape; 130. transition; 131. E; 132. B; 133. F; 134. G; 135. A; 136. D; 137. C.

A. Enzymes use their active site to bring the reactants closer together.
B. Enzymes provide the reactants the appropriate charge environment for forming the transition state.
C. Enzymes use their active site to change the conformation of the substrate to one mimicing the transition state.

139. False – propane has a high amount of free energy however requires the activation energy provided by the spark to initiate its combustion; 140. True.

4.5 Conditions and Factors that Affect Enzyme Activity [pp. 85-89]

141. collisions; 142. saturated; 143. cycling; 144. inhibitors; 145. shape; 146. compete; 147. competitive inhibitors; 148. active; 149. non-competitive; 150. Conformation or shape; 151. irreversibly; 152. reversible; 153. allosteric; 154. allosteric; 155. conformational; 156. activators; 157. end-products; 158. feedback; 159. optimal; 160. charged groups; 161. collisions; 162. three; 163. unfolding; 164. A molecule that competes with the reactants for the active site; 165. A molecule that binds to sites other than the active site of the enzyme, causing a change in 3D structure of the enzyme; 166. The reversible binding of an activator or inhibitor to an enzyme, changing its 3D structure, and thereby activating or inhibiting its activity; 167. Where the end product of a metabolic pathway inhibits the activity of the first enzyme in the pathway; 168. True; 169. False—Competitive inhibitor binds to the active site while noncompetitive inhibitor works by binding to site other than the active site; 170. False – allosteric inhibitors convert allosteric enzymes to the low-affinity conformation whereas activators convert them to the high-affinity conformation.

SELF-TEST

1. b [Enzymes are proteins that act as chemical catalysts]
2. c [Heat is a form of kinetic energy]
3. a [Glucose has potential or chemical energy]
4. d [Entropy is defined as the state of disorder in the system]
5. c [Endergonic reactions have a positive ΔG]
6. c [Active site is where the reactant/s fit]
7. d [Enzymes decrease the activation energy—i.e. the energy required to start a reaction]
8. a [Competitive inhibitors inhibit by fitting into the active site—preventing the reactants from entering the active site]
9. c [Cofactors are inorganic chemicals—such as minerals—that help enzymes]
10. a [coenzymes are organic—often vitamins—that help enzymes]
11. b [reactions tend to be spontaneous when the products are less ordered than the reactants]
12. c [catalysts (i.e. enzymes) speed up chemical reactions]
13. d [Enzymes are affected by temperature, pH, and substrate concentration]
14. b [Coupled reactions have an overall negative ΔG]
15. b [this control mechanism involves the product of a reaction being the regulator of the reaction]

Chapter 5 Membranes and Transport

5.1 An Overview of the Structure of Membranes [pp. 94-95]

1.mutations; 2.membrane; 3. chloride; 4. electrical; 5. selectively; 6. Fluid; 7. move; 8. proteins; 9. transport; 10. enzymes; 11. transport; 12. bilayer; 13. asymmetry; 14.
Membranes are lipid bilayers that are fluid in nature and contain a wide assortment, or mosaic, of proteins. The consistency of membranes is fluid allowing the lipids and proteins within the bilayer; 15. The membrane proteins of human and mouse cells were labeled with differently-coloured dye molecules and then fused. Within less than 1h, the two different types of protein were evenly distributed over the surface of the hybrid cells, providing a demonstration of the fluid nature of the membrane; 16. F – Freeze fracture and electron microscopy allow the separation and visualization of the inner and outer layer of the membrane. It therefore demonstrates the asymmetry of the membrane.

5.2 The Lipid Fabric of a Membrane [pp. 96-99]

17. phospholipids; 18. fatty acid; 19. alcohols; 20. phosphate; 21. hydrophobic; 22. hydrophilic; 23. amphipathic; 24. bilayer; 25. composition; 26. tightly; 27. gel; 28. fluid; 29. leakage; 30. Saturation/unsaturation; 31. unsaturated; 32. double bond; 33. space; 34. desaturases; 35. double bonds; 36. carbon; 37. saturation/unsaturation; 38. temperatures; 39. increase; 40. buffers; 41. gelling; 42. fluidity; 43. C; 44. E; 45. F; 46. B; 47. G; 48. A; 49. D; 50. T; 51. F – they remove two hydrogen atoms from neighbouring carbon atoms in fatty acid tails; 52. F – at high temperatures they help restrain movement between neighbouring fatty acid tails. At low

temperatures they disrupt the interactions between neighbouring fatty acid tails, thus slowing the transition to the nonfluid gel state; 53. Double bonds introduce bends in the fatty acid chains meaning that the more double bonds within the fatty acid chain, the lower the degree of packing of the membrane and the higher the fluidity; 54. Detergents are amphipathic like lipids so can dissolve oil (lipid) stains.

5.3 Membrane Proteins [pp. 99-101]

55. proteins; 56. Transport; 57. enzymatic; 58. electron; 59. signal transduction; 60. attachment; 61. recognition; 62. integral; 63. peripheral; 64. A - Made up of a mixture of polar and non-polar amino acids like the majority of proteins and associated with one of the membrane surfaces through hydrogen or ionic bonding with lipid molecules or integral membrane proteins, B – Composed of domains of approximately 20 nonpolar/hydrophobic amino acids, giving rise to membrane-spanning alpha-heces, connected by flexible loops of hydrophilic/polar amino acids that are exposed to the aqueous environment on either side of the membrane; 65. Transport, enzymatic activity, signal transduction and attachment/recognition; 66. A – cholesterol, B – integral protein (glycoprotein), C – peripheral protein, D – carbohydrate groups (glycoprotein), E – lipid bilayer, F – polar (phosphate) head group.

5.4 Passive Membrane Transport [pp. 101 - 104]

67. hydrophobic; 68. diffuse; 69. transport ; 70. Diffusion; 71. energy; 72. entropy; 73. size/charge; 74. charge/size; 75. nonpolar; 76. hydrophobic; 77. hydration; 78. charge; 79. facilitated; 80. channel; 81. carrier; 82. channels ; 83. ions; 84. bind; 85. conformational; 86. solute; 87. gated; 88. uniport; 89. specificity; 90. saturated; 91. osmosis; 92. aquaporins; 93. G; 94. D; 95. A; 96. B; 97. C; 98. F; 99. E; 100. F – Diffusion is entropy-driven with maximum entropy being attained at equilibrium; 101. F – Glycerol is small and polar but uncharged and can cross membranes by simple diffusion through the hydrophobic interior; 102. Both involve the movement of solutes areas of high concentration to lower concentration without the expenditure of energy but facilitated diffusion uses a transport protein specific for a particular type of substrate, has a faster initial rate but displays saturation at high concentration of substrate molecules; 103. The channel contains positive charges which repel protons and has such a narrow diameter that water molecules must pass through single file; 104. In the absence of active transport of ions to balance the osmotic pressure, water would rush into the cell causing them to swell and burst; 105. A - simple diffusion, B – facilitated diffusion, C – all transporters are occupied; 106. A - membrane, B – solute, C – into the cell, D – hypertonic, E – no net movement (equilibrium), F – isotonic, G – out of the cell, H - hypotonic.

5.5 Active Membrane Transport [pp. 104 - 107]

107. energy; 108. 25; 109. primary; 110. secondary; 111. ions; 112. sodium; 113. potassium; 114. voltage; 115. potential; 116. symport; 117. antiport; 118. Both involve transport proteins that are substrate-specific, saturable and undergo conformational changes to move the solute from one side of the membrane to the other; 119. There is a concentration gradient of the two chemicals on either side of the membrane and because the two different ions are not transported in equal numbers, there is also a charge or electrical gradient across the membrane; 120. Muscle cells require a calcium gradient across the membrane of approximately 30 000, so the calcium pump would be present in extremely high levels and be extremely active in these cells; 121. A – active transport, B – symport, C – antiport.

5.6 Exocytosis and Endocytosis [pp. 107 - 109]

122. endocytosis; 123. exocytosis; 124. exocytic; 125. plasma membrane; 126. endocytosis; 127. pinocytosis; 128. receptor; 129. integral; 130. clathrin; 131. endocytic; 132. clathrin; 133. lysosomes; 134. transport; 135. phagocytosis; 136. cytoplasmic; 137. vesicle; 138. C; 139. D; 140. B; 141. A; 142. B; 143. A; 144. A; 145. D; 146. There is no initial receptor binding so extracellular water is taken up along with any other molecules in solution; 147. The vesicles fuse with lysosomal vesicles and lysosomal enzymes digest the substrate into smaller molecules which, if of value to the cell, are transported into the cytoplasm by transport proteins.

SELF-TEST

1. b [the phosphate groups are hydrophilic whereas the fatty acid chains are hydrophobic making the entire molecule amphipathic. In aqueous solution a bilayer forms so that the hydrophilic region of the molecules is facing the aqueous environment while the hydrophobic regions are sequestered from water inside the bilayer]
2. b [the spontaneous assembly means that the bilayer end product has a lower energy state than the initial state (individual lipids in aqueous solution]
3. a [the fatty acid chains are hydrophobic and sequestered from the aqueous environments on either side of the membrane by facing each other in the middle]
4. d [unsaturated fatty acids introduce bends in the chain so that there is more space between adjacent lipid molecules. Sterols such as cholesterol interrupt the interactions between adjacent lipid molecules]
5. a [desaturases are used to create double bonds in fatty acid tails of lipids and therefore create unsaturated lipids; these are required with decreasing temperatures]
6. d [membrane proteins function in transport, as enzymes and as receptors to mediate recognition and attachment]
7. c [diffusion is the movement of molecules from a region of high concentration to one of low concentration]
8. c [at equilibrium, maximum entropy is reached and there is no net movement of molecules]
9. d [because phosphate ions are charged and have a hydration shell, the membrane is virtually impermeable to these molecules]
10. c [aquaporins are extremely narrow channels that are filled with positive charges so that water passes through single-file and protons are prevented from entering]
11. b [active transport involves the transport of molecules against their concentration gradient using energy to drive the process]
12. c [The concentration of the solutes Na^+ and Cl^- are both twice as high as in compartment A]
13. c [Because the concentration of the solutes Na^+ and Cl^- are higher in compartment B as compared to A and the membrane is permeable to both, they will move down their concentration gradient into compartment A until equilibrium is reached]
14. b [water moves across a membrane by osmosis from a region of low solute concentration to a region of high concentration, so from compartment A to B]
15. d [Primary active transport is generally used to transport ions]
16. a [primary active transporters bind the solute and ATP, hydrolyzing ATP to provide the energy for transport]
17. b [calcium transporters are widely distributed primary active transporters in eukaryotic cells and generally create a high concentration of calcium outside the cell and inside the lumen of the ER. They are essential for muscle cell function.]
18. c [receptor-mediated endocytosis is the method of transport for molecules or substrates that are too large to be transported by primary or secondary active transport (using symporters or antiporters)]
19. d [clathrin is the name for the proteins that form a network on the cytoplasmic side of the membrane and are part of receptor-mediated endocytic vesicles]
20. a {Lap-Chee Tsui and his team discovered the CF gene while working in the Dept. of Genetics in Toronto's Hospital for Sick Children]

INTEGRATING AND APPLYING KEY CONCEPTS

1. Read the section on membrane lipids (5.2) and consider the differences between membrane structure, gene expression and enzyme activity.
2. Read section 5.6 and think about where cholesterol is needed within cells and what process would likely allow the cholesterol to reach that destination. What gene product might play a critical role in this trafficking? Now use your imagination (or the internet) to figure out how cholesterol-lowering medications such as Lipitor might work.
3. Read section 5.6 and then section 5.4. Think about the major differences in body structure between plants and animals. What organ in human might be used to control regulate the volume and composition of body fluids?

Chapter 6 Cellular Respiration

6.1 The Chemical Basis of Cellular Respiration [pp. 116-119]

1. respiration; 2. electron; 3. ATP; 4. diseases; 5. diabetes; 6. photosynthesis; 7. CO_2 ; 8. sugar/organic; 9. respiration; 10. oxidation; 11. ATP; 12. CO_2; 13. C - H; 14. equidistant; 15. energy; 16. electronegative; 17. oxidation; 18. removal; 19. reduction; 20. redox; 21. enzymatic; 22. dehydrogenases; 23. NAD^+; 24. F; 25. A; 26. G; 27. C; 28. E; 29. B; 30. D; 31. A – makes/breaks, B – releases/uses, C – uses/releases, D – stores/releases; 32. T; 33. F – as atoms move further from the atom's nucleus, they gain energy; 34. F – Sometimes other molecules are used and sometimes molecules are only partially oxidized or reduced; 35. Methane is partially reduced to carbon dioxide because the carbon electrons move from being equally shared in the C –H bonds to being much closer to the electronegative oxygen; oxygen is partially reduced to water because the electrons in the O-H bond of water are held closer to the oxygen atom than in the O_2 bonds.

6.4 Pyruvate Oxidation and the Citric Acid Cycle [pp. 122-123]

36. three; 37. ATP; 38. glycolysis; 39. glucose; 40. pyruvate; 41. ATP; 42. four; 43. NADH; 44. groups; 45. ADP; 46. Substrate-level; 47. cytosol; 48. cytosol; 49. matrix; 50. porees; 51. carrier; 52. acetate; 53. Acetyl-coA; 54. Citric acid; 55. NADH; 56. $FADH_2$/ATP; 57. ATP/$FADH_2$; 58.CO_2; 59. oxidation; 60. C; 61. E; 62. D; 63. B; 64. A; 65. F – Organisms from all branches of the tree of life carry out this process; 66. F- while this is true of prokaryotic cells, the citric acid cycle happens in the mitochondrial matrix in eukaryotes; 67. A – glycolysis, B – glucose, C – 2 ATP, 2 NADH, 2 pyruvate, D – pyruvate oxidation, citric acid cycle, E – pyruvate then acetryl coA, F – for the first reaction, 1 acetyl-coA, 1 NADH, 1CO_2, for the citric acid cycle, 3 NADH, 2 CO_2, 1 ATP and 1 $FADH_2$; 68. It is widespread among life forms, does not require oxygen, which appeared later in the evolution of life and is a series of cytoplasmic enzymatic reactions without the requirement of electron transport chains or organelles; 69. On entering, the acetate from acetyl coA combines with the oxaloacetate and the product is then oxidized and decarboxylated in a series of reactions until oxaloacetate is again produced, letting the process cycle again; 70. Glycolysis has a 6 carbon stage then a 3-C stage, the citric acid cycle starts with a 6-C stage then loses a CO2, yielding a 5 – C stage, and this is followed by another decarboxylation to give a four carbon stage. With the addition of acetate from acetyl CoA, the 6C stage begins again.

6.5 Electron Transport and Chemiosmosis [pp. 123 - 128]

71. NADH/$FADH_2$; 72. $FADH_2$/NADH; 73. acceptor; 74. complexes; 75. Peripheral; 76. Inner membrane; 77. prokaryotes; 78. I; 79. III; 80. IV; 81. prosthetic; 82. $Fe3^+$; 83. Fe^{2+}; 84. NADH; 85. $FADH_2$; 86. ubiquinone; 87. cytochrome c; 88. complex IV ; 89. exergonic; 90. spontaneous; 91. proton; 92. work; 93. impermeable; 94. I; 95. IV; 96. ubiquinone; 97. same; 98. electrical; 99. motive; 100. chemiosmosis; 101. transport; 102. synthase; 103. down; 104. ADP; 105. $PO_4{}^{3-}$; 106. G; 107. A; 108. F; 109. D; 110. B; 111. C; 112. E; 113. The former uses the ATP synthase and inorganic phosphate for phosphorylation, whereas the latter uses soluble enzymes and other phosphorylated molecules as donor for the phosphorylation reaction; 114. NADH and $FADH_2$ both possess free energy and this is extracted after they donate their electrons to the electron transport chain which converts the chemical energy to the proton motive force; 115. In eukaryotes it is in the inner membrane of the mitochondrion whereas in the prokaryotes it is in the plasma membrane; 116. It is a complex of protein molecules and part of the complex rotates with the flow of protons through it, allowing for the conversion of the potential energy of the gradient into ATP; 117. A - outer membrane, B - intermembrane space, C – inner membrane, D - matrix; 118.E; 119. D; 120. D; 121. C; 122. C; 123. C, D, A, B.

6.6 The Efficiency and Regulation of Cellular Respiration [pp. 128 - 131]

124. 9-10; 125. 3; 126. 3; 127. $FADH_2$; 128. 38; 129. 32; 130. regulated; 131. matches; 132. phosphofructokinase; 133. ATP; 134. NADH; 135. citrate; 136. ADP; 137. ATP; 138. proteins; 139. lipids/fats; 140. other sugars; 141. points; 142. flexibility 143. Biosynthesis/anabolism; 144. F – some is made in the cytoplasm through glycolysis and substrate-level phosphorylation; 145. F – some of the intermediates are not fully oxidized but are diverted to biosynthetic reactions; 146. A – cytosol, 2, 1, 0, no, no, B – matrix, 0, 1, no, yes, C – matrix, 1, 3, 1, no, yes, D – inner membrane, 3, 0, 0, yes, no; 147. In glycolysis, phosphofructokinase binds ATP to become inhibited and as levels of ATP drop this inhibition is relived and the enzyme is activated by ADP. The enzyme is also

inhibited by two components of the citric acid cycle, NADH and citrate. Various enzymes in the citric acid cycle are inhibited by ATP; 148. Proteins are hydrolyzed to amino acids, the amino group is removed then the molecules enter as pyruvate, acetyl coA or at various points in the citric acid cycle. Fats such as triglycerides are broken down to glycerol and fatty acids. The glycerol is converted to a 3C glycolytic intermediate to enter the pathway while the fatty acids are broken down to 2C units and enter as acetyl- CoA; 149. The enzymes within the pathways are reversible, allowing the organism to direct the flow of material and electrons in whichever direction is necessary.

6.7 Oxygen and Cellular Respiration [pp. 131 - 136]

150. anaerobic; 151. nitrate; 152. sulfate; 153. Fe^{3+}; 154. Fermentation; 155. pyruvate; 156. lactate; 157. oxidized; 158. NADH; 159. alcohol; 160. decarboxylation; 161. NADH; 162.glycolytic; 163. respiration; 164. reactive; 165. anaerobes; 166. aerobes; 167. brain; 168. C; 169. E; 170.D; 171. B; 172. F; 173. A; 174. The enzymes superoxide dismutase and catalase act together to convert superoxide then hydrogen peroxide to water by catalyzing their reduction. Molecules such as vitamin C and E do the same thing; 175. The reduction of oxygen to water requires four electrons. Complex IV fills with all four electrons before simultaneously transferring all four to the electron acceptor; 176. A – cytosol, 2, 2, 0, no, no, B – cytosol, 0, 0, 0, no, sometimes; 177. T; 178. F – it happens in the cytosol.

SELF-TEST

1. a [During the first half of the pathway ATP donates two phosphate groups; 4 are then generated in the second half]
2. b [this 3 carbon molecule is then oxidized and decarboxylated to enter the citric acid cycle as acetyl-CoA]
3. a [this is true for all respiratory organisms]
4. b [the glycerol enters the 3-C stage of glycolysis and the fatty acids are broken down to 2-C units and combined with acetyl-CoA to enter the citric acid cycle]
5. b [the products of this reaction are CO_2, NADH and acetyl-CoA]
6. b [this is based on the second law of thermodynamics]
7. b [when pyruvate starts to accumulate in the cytoplasm it is reduced through fermentation in order to oxidize the NADH of glycolysis and allow the glycolytic pathway to continue functioning, generating ATP by substrate-level phosphorylation]
8. a [using the end product of glycolysis: pyruvate]
9. d [the sour taste is due to the accumulation of lactate or lactic acid as a result of lactic acid fermentation]
10. b [yeast such as *Saccharomyces cerevisiae* are anaerobic and in the presence of excess sugar will ferment the sugar to CO_2 and ethanol in an alcohol fermentation reaction. The ethanol evaporates during backing and the CO_2 allows the dough to rise]
11. a [oxygen is the most electronegative electron donor this reduction of this molecule is as exergonic as possible]
12. d [defective mitochondria have been implicated in many age-related diseases including these three and others and even including the aging process itself]
13. c [because the electrons in the C-H bonds are equally shared between the two atoms, they have more energy. The C – O bonds in sugars are polar because the electrons are closer to the oxygen atom so the carbon is already partly reduced]
14. b [heat is only used for combustion; in respiration it is provided by the various enzymes in the pathways]
15. a [he was a biochemist in the U.K. who determined that the proposed "high energy intermediate" used by the ATP synthase to make ATP was, in fact, the proton motive force]
16. b.[their electron transport chains are in the plasma membrane and protons are pumped from the cytosol to the outside then flow back in through the ATP synthase, so that oxidative phosphorylation happens in the cytosol where ATP is required]
17. b [electron transport chains function to generate a proton motive force, regardless of the source of free energy (and electrons) that they use. Once a proton motive force is created, the proton flow through the membrane-bound ATP synthase triggers the catalytic activity and oxidative phosphorylation]
18. a [present in virtually all forms of life suggesting it evolved very early in the evolution of life on Earth]
19. c [some are proteins that insert into the membrane to allow the unregulated flow of protons down their concentration gradient. They do not prevent electron transport through the chain but the free energy that is released cannot be harnessed as proton motive

force so is dissipated as heat. In certain cells such as those of brown adipose fat cells, this is accomplished by the synthesis of natural uncouplers in order to regulate body heat]

20. d [facultatively anaerobic prokaryotes as well as the muscle cells of animals will use oxygen if present but can perform anaerobic fermentation or respiration if not. Given that they can function aerobically they must possess this complex since it is universal among aerobic respiratory electron transport chain]

INTEGRATING AND APPLYING KEY CONCEPTS

1. Read section 6.2, 6.5d and 6.7 in order to determine the relative localization of the pathways and transfer reactions in the mitochondrion and aerobic bacteria, as well as anaerobic bacteria.
2. Read section 6.2d to confirm the location of the two electron transport chains then sections 2.5 c and c on the endosymbiotic origin of the mitochondrion then the discussion in sect. 5.6 on phagocytosis.
3. Read the section on the metabolic effects of anaerobic conditions (sect. 6.7).

Chapter 7 Photosynthesis

Why it matters [pp. 139-140]

1. photosynthetic; 2. aquatic; 3. CO_2; 4. 11 x 10^{13}; 5. microorganisms/phytoplankton; 6. half; 7. temperate/eqautorial; 8. Artic.

7.1 Photosynthesis: An Overview [pp. 140-142]

9. organic; 10. producers; 11. consumers; 12. decomposers; 13. fungi; 14. light-dependent; 15. Calvin; 16. NADPH; 17. ATP; 18. carbohydrate; 19. fixation; 20. organic; 21. chloroplasts; 22. plasma; 23. A – chloroplast inner and outer membranes, B – thylakoid/thylakoid membrane, C – stroma, D - thylakoid lumen; 24. B; 25. D; 26. F – it comes from the light-dependent phase of photosynthesis; 27. F – it also happens in certain prokaryotes such as cyanobacteria; 28. C; 29. G; 30. I; 31. H; 32. A; 33. B; 34. D; 35. F; 36. E; 37. A – makes/breaks, B – releases/uses, C – uses/releases, D – stores/releases; 38. 6 CO_2 + $6H_2O \rightarrow 6C_6H_{12}O_6$ + 6H2O + 6O2, the light-dependent reaction occurs in the thylakoids and the CO_2 is fixed in the stroma of the chloroplast.

7.2 The Photosynthetic apparatus [pp. 142- 145]

39. pigment; 40. electrons; 41. ground; 42. excited; 43. excited; 44. primary; 45. chlorophyll; 46. a; 47. b; 48. oxidized; 49. carotenoids; 50. accessory; 51. inductive resonance; 52. photosystems; 53. thylakoid; 54. photosystems; 55. reaction center; 56. antenna; 57. reaction; 58. primary; 59. F; 60. G; 61. A; 62. K; 63. H; 64. C; 65. E; 66. I; 67. L; 68. M; 69. D; 70. B; 71. J; 72. This phase traps light energy and uses it to create NADPH and ATP which are then used in the Calvin cycle to convert CO_2 to carbohydrates and other organic molecules for cell structures; 73. These pigments have a different absorption spectrum and so extend the range of wavelengths that can be used for photosynthesis. Upon absorption, they transfer the absorbed energy to neighbouring molecules until the energy if funneled to the reaction center chlorophyll; 74. This pigment of chlorophyll a is found in the PSI complex and absorbs light with a wavelength of 700nm; 75. F – this is the technique used to measure the activity spectrum. The absorption spectrum is measured using a spectrophotometer which provides relative units of absorption as a function of wavelength; 76. F – they are both chlorophyll a molecules. The difference in absorbed wavelength is the result of the various molecules that surround them in their respective photosystems.

7.3 Photosynthetic Electron Transport [pp. 145 - 150]

77. oxygenic; 78. water; 79. atmosphere; 80. II; 81. P680; 82. primary; 83. electron; 84. oxidant; 85. water; 86. respiration; 87. proton; 88. ATP; 89. chemiosmosis; 90. Oxidation; 91. electron; 92. reductase; 93. cyclic; 94. ATP; 95. NADPH; 96. E; 97. G; 98. B; 99. A; 100. H; 101. I; 102. J; 103. F; 104. C; 105. D; 106. B, D, A, D, C ; 107. F – it is to augment the ATP production; 108. T; 109. F – protons are pumped across thylakoid membranes into the thylakoid lumen; 110. A - To make ATP and NADPH, B – The reaction center chlorophyll of PSII once excited by absorption of light, C – this is where the photosystems and electron transport chains are located and where the ATP synthase is located; 111. The water-splitting complex is associated with PSII which is involved in non-cyclic electron flow, so that once oxidized, the P700 reaction center chlorophyll cannot undergo another oxidation reaction until it has

been reduced. The electrons for the come from water; 112. The carbon fixation process requires 18 ATP for every molecule of glucose made and only 12 NADPH. Since non-cyclic electron flow makes one of each, the additional ATP required is supplied by cyclic flow, which only generates ATP; 113. To get one electron from H_2O/PSII to NADPH, two photons of light must be absorbed, one for each PS. Each water molecule releases 2 electrons when split and the formation of oxygen requires two oxygen atoms, i.e. two water molecules. Therefore 2x2x2 = 8 photons are required to generate one molecule of O_2.

7.5 Photorespiration and CO_2 – Concentrating Mechanisms [pp. 152-157]

114. carbohydrate; 115. stroma; 116. cytosol; 117. three; 118. ribulose 1,5 –bisphosphate/ RuBP; 119. rubisco; 120. 3PGA; 121. ATP; 122. NADPH; 123. RuBP; 124. enzymatic; 125. 3; 126. G3P; 127. organic; 128. rubisco; 129. oxygenase; 130. RuBP; 131. toxic; 132. CO_2; 133. photorespiration; 134. temperature; 135. CO_2; 136. PEP carboxylase; 137. PEP; 138. oxaloacetate; 139. C4; 140. Energy/ATP; 141. rubisco; 142. CAM; 143. oxygen; 144. hydrolyzed; 145. CO_2; 146. E; 147. G; 148. H; 149. C; 150. B; 151. D; 152. I; 153. A; 154. J; 155. F; 156. A - acceptor molecule for CO_2 fixation by Calvin cycle, B – the product of the Calvin cycle which is then used as a building bock to make other organic molecules, C – the immediate product of CO_2 fixation in the Calvin cycle, D – the carboxylase enzyme of the carbon cycle, E – the aqueous environment of the chloroplast in which the Calvin cycle occurs; 157. A – thylakoid, produced, produced, produced, no, no, B – stroma, used, used, n.a., used, produced; 158. F – it produces one molecule of G3P per cycle but requires three cycles to generate a G3P not required to continue the cycle. G3P is a 3-carbon molecule and two of these are required to make glucose; 159. F – as with all plants, much of the glucose is used for biosynthesis or respiration; 160. T; 161. T; 162. The widespread nature of the enzyme is believed to be due to its having evolved very early in the development of life on Earth. In addition, this would have been at a time when there was no oxygen in Earth's atmosphere so no problem with oxygen as a competitive inhibitor. Even in the presence of equal amount of CO_2 and O_2, the enzyme uses CO_2 far more efficiently, however the modern-day atmosphere consists of 21% O_2 and less than 1% CO_2, giving rise to the problem of photorespiration; 163. The aquatic microorganisms actively pump HCO_3^-, the dissolved across the membrane so that once enzymatically converted to CO_2 in the cytosol it diffuses into the chloroplast where it is at much higher levels than O_2. C4 plants use a supplemental enzyme for carbon fixation and normally use this method in the mesophyll leaf cells near the surface, where O_2 concentrations are high, produce the storage product malate and this diffuses into deeper leaf cells, the bundle sheath cells, where O_2 is less concentrated, and in these cells the malate is hydrolyzed, releasing high quantities of CO_2 for the Calvin cycle. In CAM cells, carbon fixation by the C4 enzyme is done at night when the stomata can be opened without risking loss of water. During the day as the heat rises, these close and the stored malate is released from storage vacuoles and hydrolyzed, providing the CO_2 for the Calvin cycle. 164. C4 plants are generally ones in hot climates where the opening of stomata for the diffusion of gases in and out has the problem of a consequent water loss. CAM plants are found in regions with hot days and cool nights so that they open their stomata at night when the temperature is lower, closing them in the day to stop gas diffusion and water loss.

SELF-TEST

1. b [fungi are heterotrophs and decomposers]
2. d [water is split to release electrons and oxygen]
3. b [chlorophyll P700 and P680 are both chlorophyll a molecules]
4. d [chlorophyll a and b as well as the accessory pigments carotenoids are in the antenna as well as chlorophyll a forming the reactin center chlorophylls]
5. a [water is split to supply the electrons to PSII and allow non-cyclic photosynthesis]
6. d [the thylakoid membranes hold the photosystems, electron transport chains and the ATP synthase]
7. c [this involves PSII funneling electrons to PSI which in turn funnels electrons to NADPH]
8. a [chlorophyll is required for the light-dependent part of photosynthesis]
9. c [after three cycles, 6 G3P are made, 5 of which are required to regenerate RuBP]
10. b [RuBP is a 5C molecule which, when combined with CO_2 gives a total of 6 carbons divided into two 3PGA molecules]
11. c [this is the 5C acceptor molecule that is the acceptor for the carboxylation reaction catalyzed by rubisco]

12. a [photo – because they are photosynthetic, and autotroph refers to the ability to fix carbon dioxide, literally meaning "self-feeding"]
13. d [these are the pores that allow diffusion of water and gases into and out of the cell. They are regulated so that the plant can at least partially control these processes. The cuticle is a waxy covering which inhibits the diffusion of gases and water except through the stomata]
14. a [C3 plants do not have the C4 enzyme and pathway so when temperatures rise, CO_2 diffusion rates in the stroma decline at a much greater rate than O_2 and the result is photorespiration]
15. d [malate is formed from oxaloacetate, the product of the carboxylation reaction catalyzed by the PEP carboxylase enzyme. This storage product may diffuse from mesophyll cells to bundle sheath cells or may be stored in vacuoles until conditions change so that photorespiration is not a problem]
16. b [cyclic photosynthesis involve PSI only, with the electron flow generating a proton motive force for use by the ATP snthase, but no NADPH]
17. a [the Calvin cycle generates G3P which, by combining with a second molecule of G3P and reversing the process of glycolysis, is converted to glucose]
18. c [when present in equal concentrations, the enzyme still performs the carboxylation reaction ~80x faster than the oxygenation reaction. However since there is almost 20 times more oxygen in the atmosphere than CO_2, the enzyme becomes very inefficient]
19. c [this pump allows them to bring in the dissolved form of CO_2 to high concentrations, and once in the cytosol, carbonic anhydrase converts the bicarbonate anion to CO_2 which enters the chloroplast for fixation]
20. b [During the day when it is hot, the stomata of CAM plants are closed and stored malate is hydrolyzed to liberate intracellular CO_2 which is used by rubisco. At the same time, water absorbed at night is cleaved while photosynthesis is occurring. At night, the temperature drops, the stomata open, O_2 is released, CO_2 and water diffuse into the leaves and the C4 cycle for the synthesis of malate occurs]

INTEGRATING AND APPLYING KEY CONCEPTS

1. Read section 7.5 over again, playing particular attention to Table 7.1
2. This question is asking you to be able to reproduce the information in Fig. 7.10 without referring to the diagram for guidance.
3. Read sections 7.4a and 7.4b then 7.3f. Remember that glucose is a 6-carbon molecule and that it is made from 3 molecules of G3P.

Chapter 8 Cell Communication

8.1 Cell Communication: An Overview [pp. 162 - 164]

1. multicellular; 2. plants; 3. channels; 4. contact; 5. intracellular; 6. ions; 7. gap junctions ; 8. plasmodesmata; 9. cell adhesion molecules; 10. intracellular; 11. signal; 12. recognition; 13. receptors; 14. reception; 15. transduction; 16. response; 17. cascade; 18. transduction; 19. D; 20. E; 21. F; 22. A; 23. B; 24. C; 25. receptor; 26. transduction; 27. response; 28. T; 29. T; 30. F – they are integral membrane proteins with domains extending beyond the plasma membrane surface and into the cytosol; 31. He discovered the hormone epinephrine and determined its mechanism of action laying the foundation for our understanding of hormones and signal transduction.

8.6 Integration of Cell Communication Pathways [pp. 174 - 175]

32. vertebrates; 33. steroids; 34. hormones; 35. extracellular; 36. pathways; 37. conformational; 38. activation; 39. phosphorylation; 40. kinases; 41. cascade; 42. phasphatases; 43. endocytosis; 44. amplification; 45. enzymes; 46. phosphate; 47. tyrosine; 48. target; 49. dimers; 50. kinase; 51. binding; 52. phsophate; 53. first; 54. conformational; 55. GTP; 56. effector; 57. second; 58. ATP; 59. cAMP; 60. kinases; 61. phospholipase; 62. IP_3; 63. DAG; 64. Ca^{2+}; 65. kinase; 66. talk; 67. kinase; 68. network; 69. Ras; 70. kinase; 71. Plasma membrane; 72. Steroid nuclear; 73. nuclear; 74. nucleus; 75. DNA; 76. control; 77. D; 78. E; 79. G; 80. H; 81. B; 82. C; 83. A; 84. K; 85. M; 86. L; 87. I ; 88. F; 89. J; 90. T; 91. F – They are either broken down at a steady rate by enzymes or in their target cells or in organs such as the liver, or excreted by the

kidneys; 92. F –They exist in Drosophila, higher plants and yeast indicating that they evolved prior to the split leading to the fungi, animals and plants; 93. F – the activated G protein in turn activates the effector protein phospholipase C or adenylyl cyclase, which then make the second messengers; 94. F – the second messengers are IP_3 and DAG. IP_3 activates release of Ca^{2+} from the lumen of the ER and either the Ca^{2+} itself or Ca^{2+} in combination with DAG then activate protein kinases; 95. T; 96. T; 97. D, H, F, A, G, C, E, B; 98. The signal molecules are bound by surface or intracellular receptor molecules that specifically fit the molecule, much like a substrate in an enzyme. So whether a cell responds to a particular signal molecule that is present depends upon whether the "matching" receptor is present on or in the cell; 99. Both are surface receptor proteins involved in cell communication and both ultimately result in regulating intracellular protein kinases to transducer a signal. However the receptor proteins have different responses to signal molecule binding, with the former autophosphorylating and directly triggering a phosphorylation cascade and the latter instead responding by activating an associated protein, the G protein, through conformational changes rather than phosphorylation; 100. The Ras protein is a special kind of G protein that has an associated adaptor protein which recognizes and binds the phosphoryl groups on specific tyrosine kinase receptors. Once bound, these activate Ras which in turn is activated by binding GTP and initiates in a phosphorylation cascade involving MAP proteins. The last MAP protein in the cascade enters the nucleus and phosphorylates proteins which alter gene expression; 101. A - The second messenger in one type of G protein-coupled receptor response pathways. It activates cAMP-dependent protein kinases to transducer a signal, B – IP_3, C – second messenger which functions with Ca_{2+} to activate protein kinases of the pathway, D – steroid hormone receptor/steroid nuclear receptor, E – autophosphorylation followed by phosphorylation of target molecules, F – G protein-coupled receptor, G - steroid hormone receptor/steroid nuclear receptor; 102. A - phosphatases + endocytosis, B - phosphodiesterase + endocytosis + phosphatases, C - endocytosis + phosphatases, D - phosphatases + endocytosis, E – phosphatases.

SELF-TEST

1. d [The signal molecule, having reached its target cell initiates transduction of the signal. This starts with receptor binding, transduction to an intracellular response, then the final cellular response]
2. d [the surface receptor proteins for many signal molecules are glycoproteins]
3. a [steroid hormone nuclear receptors work by binding directly to DNA to trigger changes in gene expression]
4. a [these are internal receptors that, once bound with the signal molecule, enter the nucleus and bind the appropriate control regions to alter gene expression]
5. d [this enzyme catalyzes the formation of cAMP from ATP]
6. d [the phospholipase C activation triggers the IP3/DAG pathways, MAP kinases are activated by the Ras G protein and adenylyl cyclase activation triggers the cAMP-dependent G protein pathways]
7. b [these are internal protein receptors with two domains: one that binds the steroid signal molecule and one that gets activated to bind control sequences associated with specific genes]
8. a [these are surface receptor proteins that transduce a signal by autophosphorylating a tyrosine residue on the cytoplasmic domain of the protein. This then binds target proteins and phosphorylates them to trigger a cellular response or a phosphorylation cascade to trigger the cellular response]
9. e [phsophatases remove phosphate groups resulting from a phosphorylation cascade. Phosphodiesterase degrades the second messenger cAMP, cross talk involves the interaction of different signaling pathways, usually those that involve protein kinases – these can trigger activation or inhibition, which means that one pathway can inhibit another pathway. Endocytosis of a surface receptor-signal molecule complex will eliminate the signal transduction mediated by the activated surface receptor]
10. d [G proteins are peripheral membrane proteins plus are too big to pass through gap junctions]
11. d [this is a surface glycoprotein that can interact with other cell surfaces or extracellular matices and trigger a cell reponse that is capable of cross-talk with cAMP and IP_3/DAG pathways]
12. c [the same steroid hormone may interact with internal receptors and trigger one type of response and with a surface receptor and trigger another type of response]

13. a [amplification is the result of the catalytic activity of enzymes which can very rapidly bind substrates, catalyze a response then release and repeat]
14. a [The cAMP-dependent pathway uses ATP as the substrate for generation of the cAMP second messenger and GTP to activate the G protein]
15. b [approximately 1000 different G protein-coupled receptors have been identified but only about 50 receptor tyrosine kinases]
16. d [these bacteria all produce a toxin that is an enzyme which chemically modifies the G protein so that it is always active and the pathway therefore cannot be turned off]
17. c [hormones are released by specialized gland cells]
18. e [this plant hormone has a diverse array of effects. It is a gas at physiological temperatures and as such, is soluble in aqueous and lipid phases. Its receptor is found within the transmembrane portion of the membrane of the endoplasmic reticulum]
19. d [second messengers are part of the G protein-coupled pathways. They are the molecules that are generated by the effector enzymes, normally either adenylyl cyclase or phospholipase C which in turn, trigger the activation of the protein kinases which ultimately generate the final cellular response]
20. c[steroid hormone nuclear receptor proteins are internal receptor proteins, this hydrophilic and soluble. They have two domains: one that binds the steroid signal molecule and one that gets activated to bind control sequences associated with specific genes]

Integrating and Applying Key Concepts

1. Think about what structural features of plant cells allow communication between neighbouring cells, the kinds of molecules that might pass through these structures and the types of pathway molecules that might be able to move from one cell to another. Finally, read the "*Molecule behind biology*" box.
2. Think about the nature of the insulin receptor, which is described in the textbook. Given that it is a protein, it is encoded by a gene and therefore heritable. With both of these facts established, you should be able to think about what structural defects would lead to the symptoms of this disease.
3. Read the description of steroid hormones in section 8.5 then describe two different types of signal transduction pathways.

Chapter 9 Cell Cycles

9.1 The Cycle of Cell Growth and Division: An Overview [pp. 180-181]

1. 3: 2. DNA replication; 3. DNA; 4. cytokinesis; 5. daughter; 6. parental; 7. genetically; 8. mitosis; 9. chromosomes; 10. proteins; 11. chromosomes; 12. parent; 13. diploid; 14. haploid; 15. ploidy; 16. chromatids; 17. E; 18. F; 19. B; 20. G; 21. A; 22. C; 23. D; 24. T; 25. F – All of the cells that make up a multicellular organism arose by mitosis from a single cell, the zygote, so they all have the same genetic material.

9.2 The Mitotic Cell Cycle [pp. 182-187]

26. interphase; 27. mitosis; 28. G_1; 29. S; 30. G_2; 31. prophase; 32. condense; 33. centrosomes; 34. spindle; 35. prometaphase; 36. nuclear; 37. spindle; 38. spindle; 39. kinetochore; 40. metaphase; 41. spindle; 42. metaphase; 43. anaphase; 44. sister chromatids; 45. spindle; 46. segregation; 47. daughter; 48. telophase; 49. decondense; 50. spindle; 51. nuclear; 52. cytokinesis; 53. furrowing; 54. plate; 55. N; 56. D; 57. E; 58. A; 59. B; 60. H; 61. M; 62. K; 63. L; 64. G; 65. I; 66. O; 67. F; 68. C; 69. F – DNA is synthesized during the S phase; 70. T; 71. F – they are synthesizing RNA and proteins in preparation for mitosis; 72. C – G_2, D - 4*n*, E – prophase, F - 4*n*,G – prometaphase, H -4*n*, I – metaphase, J – 4*n*, K – anaphase, L – 4*n*, M – telophase, N – 4*n* ; O – G_1, P – 2*n*; 73. By undergoing this transition, cells commit themselves to cell division. It is subject to molecule controls and loss of these controls is a hallmark of cancer; 74. Sister chromatids are the replicated chromosome pairs that exist in a cell prior to cytokinesis; 75. Cell furrowing is the process used by animals in cytokinesis and involves the formation of a band of microfilaments just inside the plasma membrane which gradually contracts. This gives rise to a furrow which expands until the daughter cells are separated. Cell plate formation is used by plants to separate the two daughter cells. It involves the vesicular deposition of cell wall material at the former spindle midpoint until the daughter cells are separated.

9.3 Formation and Action of the Mitotic Spindle [pp. 187-189]

76. mitotic spindle; 77. centrosome; 78. organizing; 79. microtubules; 80. nucleus; 81. kinetochores; 82. kinetochore; 83. nonkinetochore; 84. midpoint; 85. kinetochore; 86. disassembles; 87. nonkinetochore; 88. poles; 89. D; 90. C; 91. A; 92. B; 93. F – they are responsible for generating the microtubules for cilia and flagella; 94 T.

9.4 Cell Cycle Regulation [pp. 190-194]

95. checkpoints; 96. reactions; 97. cyclin; 98. cycle; 99. cyclin-dependent; 100. phosphorylation; 101. cycle; 102. cyclins; 103. surface receptors; 104. peptide; 105. growth; 106. contact; 107. phosphate; 108. contact inhibition; 109. dividing; 110. tumours; 111. metastasis; 112. body; 113. death; 114. apoptosis; 115. surplus; 116. infected; 117. caspases; 118. B; 119. D; 120. E; 121. G; 122. C; 123. D; 124. A; 125. F – the cyclin proteins vary with the cell cycle, rising in concentrations until a critical level where they are able to bind and activate their cognate CDK kinase activity. After that transition has passed, the cyclin is degraded; 126. T; 127. This is the loss of replicative ability in "older" cells and it is thought to be caused, at least in part, by progressive damage to DNA sequences and to the shortening of chromosomal telomeres; 128. Cancerous cells often have suffered mutations to genes that encode the proteins and molecules involved in the internal or external regulation of cell division. These mutated genes are called oncogenes.

9.5 Cell Division in Prokaryotes [pp. 194-195]

129. binary fission; 130. chromosome ; 131. DNA replication; 132. origin; 133. origins; 134. Cytokinesis; 135. wall; 136. . F – the mechanism of segregation of bacterial chromosomes is unknown; 137. This means the splitting or dividing into two parts and is the process by which prokaryotic cells undergo cell division.

SELF-TEST

1. b [Mitosis is the process by which eukaryotes grow, ensuring that the new cells are exactly the same as the parental cells]
2. a [G_1 is the most variable in length; cells make the transition to S only when they need to divide]
3. a [The envelope breaks down in prophase, the chromatids begin to condense and the duplicated centrosomes begin to make the mitotic spindle. The nuclear envelope has disappeared by prometaphase]
4. c [G_o is the phase that some cells enter from G_1 and is a state of division arrest]
5. c [Each chromatid has a single kinetochore through which the molecule attaches to and eventually moves along a kinetochore microtubule]
6. c [Flowering plants and conifers lack a centrosome and instead have multiple MTOCs]
7. a [Cyclin B is the internal signal molecule that, at a critical concentration, binds to and activates CDK1, which in turn initiates a signal transduction pathway that causes the cell to enter the M phase]
8. b [Contact inhibition is a normal property of healthy eukaryotic cells but is lost in cancerous cells, allow them to overgrow each other, forming dense masses of cells called tumours.]
9. c [The method of segregation of the daughter chromosomes in prokaryotic cells is unknown.]
10. b [Only prokaryotic cells, with their much smaller chromosomes, have an origin of replication, that is a unique DNA sequence at which replication begins]
11. a [Apoptosis is a process of programmed cell death that in required for normal development of multicelluair organisms whereas senescence is a process that is not understood but that leads to the gradual loss of replicative ability over time]
12. d [Cellular senescence was discovered by Haylick and Moorrhead and scientists have been unable to confirm the underlying explanation or "Haylick factor" although shortening of the telomeres and accumulation of DNA damage are both strong possibilities]
13. d [This is the technique of growing living cells in the laboratory and can be done with many different types of prokaryotic or eukaryotic cell types, including various human cells in cancer-testing laboratories]
14. b [Karyotypes are done using fully condensed metaphase chromosomes]

15. c [Cancer is a disease of multicellular organisms that involves the loss of normal regulatory control of the cell cycle so that tumours develop]
16. b [Peptide hormones are signal molecules that bind to surface receptors on target cells and give rise to a cellular response that can modify the cyclin:CDK phosphorylation state]
17. b [Although exhibiting similarities with the process of division in bacteria, the process in yeast is believed to have evolved separately from that of animals and higher plants]
18. a [In plants, the layer of microtubules that persists at the spindle midpoint serves as an organizing center for vesicles derived from the ER and Golgi and which gradually coalesce to give cytokinesis and a new cell wall – the cell plate]
19. c [Eukaryotes cannot replicate the linear ends of their chromosomes so during replication they employ this enzyme to enzymatically add repetitive DNA sequences or telomeres, to the ends, however as cells "age" this enzyme does not maintain the length of the ends, resulting in shortening of these sequences and possibly contributing to senescence]
20. d [This is the first cancer cell line developed in tissue culture; it was first grown in 1951 and remains in use today. Cancer cell lines are, by their nature, naturally immortal]

INTEGRATING AND APPLYING KEY CONCEPTS

1. Read the section on cell division in prokaryotes and pay particular attention to the description of the genome in prokaryotic cells.
2. Remember that all cells in a multicellular organism are derived by mitosis from a single parental cell, the zygote. Pay attention to the descriptions in this chapter of external factors that control the cell cycle, the resulting intracellular processes, and take note of the fact that the blastema is described as being a temporary structure. Once regenerated, are the cells likely to continue growing and dividing?
3. After rereading the relevant sections, think about the reasons for a cell entering each of these three phases, what is happening or not happening in cells in each of the three phases/states and the end result for each.

Chapter 10 Genetic Recombination

10.1 Mechanism of Genetic Recombination [pp. 200-201]

1. sameness; 2. genetically; 3. population; 4. selection; 5. different; 6. heritable; 7. mutation; 8. recombination; 9. meiosis; 10. differ; 11. proximity; 12 backbones; 13. exchange; 14. paste; 15. Homologous; 16. line up; 17. backbone; 18. exchanged; 19. four; 20. four; 21. single; 22. circular; 23. fusing; 24. C; 25. D; 26. A; 27. B; 28. A single crossover event between two molecules involves cutting the two backbones of each molecule, then joining the cut ends from one molecule to the other. If the two molecules are circular then the result is fusion of the two together to give a single recombinant molecule whereas with two linear molecules, the cut ends are "swapped" giving two recombinant linear molecules.

10.2 Genetic Recombination in Bacteria [pp. 201-208]

29. recombination; 30. organisms; 31. offspring; 32. E. coli; 33. conjugation; 34. donor; 35. pilus; 36. F plasmid; 37. F^+; 38. integrated; 39. Hfr ; 40. chromosomal; 41. plasmid; 42. nutritional; 43. entry/transfer; 44. map; 45. dead; 46. transformation; 47. transformed; 48. genetic; 49. transduction; 50. bacteriophages; 51. host; 52. lytic; 53. random; 54. generalized; 55. specialized; 56. λ; 57. lysogenic; 58. E; 59. C; 60. A; 61. F; 62. B; 63. D; 64. J; 65. D; 66. E; 67. G; 68. K; 69. B; 70. C; 71. F; 72. O; 73. I; 74. M; 75. A; 76. N; 77. H; 78. L; 79. A – the absorption of free DNA across the plasma membrane, B – the one-way transfer of DNA, usually a conjugative plasmid, through a pilus from a donor cell. When the plasmid integrates into the chromosome, conjugation results in transfer of chromosomal genes, C – Occurs when a virus accidentally packages host DNA then transfers that DNA when it infects the next host cell; 80. A – The bacteriophage injects its DNA, the DNA is transcribed, viral products are translated, viral particles assemble and are released by rupture or lysis of the host cell, and the cycle is therefore called a lytic lifecycle, B - The bacteriophage injects its DNA and the DNA becomes integrated into the host's chromosome where it is replicated as part of the host's DNA. If conditions within the cell change the virus can excise from the host's chromosome, and revert to a lytic lifestyle; 81. T; 82. F – only certain species are able to undergo transformation unless "forced" in the lab; 83. F – it depends upon whether the donor is F^+, in which case the recipient will become F^+, but if the donor is an Hfr then the entire F sequence will not get transferred so the recipient remains

F⁻; 84. T; 85. P22 encodes an enzyme that degrades the host DNA into short fragments which can become integrated in a random fashion when the phage particles assemble; in bacteriophage λ, the phage chromosome becomes integrated into a specific region of the host's chromosome; when triggered to excise the phage may accidentally remove a host gene on either side of the prophage (*gal* or *bio*); 86. A bacteriophage is the infectious form of a virus whereas a prophage is the form when the phage DNA is integrated into the host's chromosome.

10.3 Genetic Recombination in Eukaryotes: Meiosis [pp. 209-218]

87. meiosis; 88. half; 89. recombination; 90. three; 91. haploid; 92. diploid; 93. mitotic; 94. diploid; 95. gametes; 96. mitosis; 97. fertilization; 98. meiosis; 99. four; 100. haploid; 101. different; 102. interphase; 103. I; 104. prophase I; 105. chromatids; 106. synapsis; 107. recombination; 108. prometaphase I; 109. kinetochore; 110. opposite; 111. I; 112. metaphase; 113. anaphase I; 114. haploid; 115. disassembly; 116. decondensation; 117. prophase II; 118. prometaphase II; 119. each; 120. metaphase II; 121. anaphase II; 122. telophase II; 123. reproductive; 124. gametes; 125. zygote; 126. I; 127. I; 128. maternal; 129. paternal; 130. recombinant; 131. fertilization; 132. I; 133. I; 134. II; 135. I; 136. II; 137. I; 138. C; 139. D; 140. A; 141. E; 142. B; 143. A, E, D, F, C, B; 144. It is a condition that results from nondisjunction of chromosome 21 so that babies are born with 3 copies of this chromosome instead of 2; 145. It is the brief interlude between meiosis I and II during which no DNA replication occurs; 146. Tetrads are the homologous pairs of sister chromatids which form through the process of synapsis; 147. F – it is a rare event and generally zygotes that form from cells that have suffered nondisjunction rarely survive; 148. T; 149. A -2, diploid, B - occurred, C – crossovers, D – homologous chromosomes, random, no, E – sister chromatids, haploid; 150. i - E, F, ii – D, iii – B, iv – C, v – G, vi – A.

10.4 Mobile Elements [pp. 219-224]

151. transposable elements; 152. transposition; 153. transposase; 154. pasting; 155. replicating; 156. inactivate; 157. expression; 158. cancer; 159. mutagens; 160. insertion sequences; 161. transposons; 162. transposase; 163. resistance; 164. F plasmid; 165. conjugative; 166. retrotransposons; 167. retroviruses; 168. RNA; 169. transcribing ; 170. gametes; 171. I; 172. C; 173. E; 174. F; 175. A; 176. B; 177. D; 178. G; 179. H; 180. F – insertion sequences are short sequences with usually only the gene for the transposases; transposons carry extra genes; 181. T; 182. The different colours in the kernels are the result of the movement of transposons into and out of genes controlling pigment production; 183. Plasticity refers to changes in the genome, the raw material for evolution. Stability refers to the fact that genetic change must occur at very low levels and that as the organism grows and develops and the somatic cells are replicated, the daughter cells that are produced must be genetically identical to the parental cell.

SELF-TEST

1. a [binary fission is the method of cell division and growth and since bacteria are haploid, there are no homologous alleles to undergo recombinatin]
2. d [auxotrophs are nutritional mutants that arise from exposure to mutagens such as X-rays and UV]
3. c [a sex pilus is a tube-like structure that connects the cytoplasm of donor cell to recipient cell]
4. c [Hfr stands for high frequency recombination, so named because the F plasmid has been integrated into the chromosome resulting in the transfer of chromosomal genes to recipient cells and the generatin recombinants]
5. a [In an Hfr, the F plasmid is integrated into the chromosome and transfer originates from a gene in the middle of the plasmid so that chromosomal genes are transferred but the entire plasmid sequence does not get transferred because mating pairs are unstable. Those genes that are transferred make the recipient a partial diploid and the new sequences can undergo homologous recombination with the recipient cell's DNA]
6. b [This can happen through generalized or specialized transduction and involves the transfer of chromosomal DNA from one host cell to the next via a phage head containing host DNA]
7. b [Only certain strains, such as *Streptococcus* are naturally transformable although most strains can be "forced" to undergo transformation in the lab]
8. b [Lysogenic bacteriophage insert their chromosome into the host's chromosome and reside there until conditions change and they excise from the chromosome and revert to the lytic life cycle]

9. a {In the lytic life cycle, bacteriophage particles are released by rupturing or lysis of the host cell]
10. c [Bacteriophage λ in *E. coli* and retroviruses in humans both integrate their chromosomes into the host chromosome]
11. d [Transposable elements insert by non-homologous recombination so whatever DNA is present in the cell could be a target]
12. c [Retrotransposons are only found in eukaryotic cells]
13. b [In a diploid organism, i.e. $2n$, the haploid number is n. So $2 \times n = 16$, $n = 8$ chromosomes of maternal origin and 8 of paternal origin]
14. c [Alleles are variants of a particular gene and depending upon the alleles in the mother and father and whether these particular genes were recombined during formation of the two gametes giving rise to the individual, they could be different or the same]
15. c [After replicating the chromosomes in the S phase the sister chromatids enter meiosis I where they undergo synapsis, are separated in anaphase I so that at the end, there are two haploid daughter cells which then go on to meiosis II, during which the sister chromatids are separated]
16. a [Anaphase I is the phase of meisos I where the homologous pairs are separated; during anaphase II the sister chromatids are separated to give, by the end, four haploid daughter cells]
17. d [In nondisjunction, a homologous pair fails to separate and moves to one pole of the cell so that if the the diploid number is 6 then there are three homologous pairs and after nondisjunction one cell will be missing one of the homologous chromosomes and the other will have both]
18. b [Females are XX and males are XY and during meiosis, the X and Y chromosomes in the male behave as a homologous pair so gametes are either X or Y]
19. a [Cross-overs do not alter the amount of DNA unless nondisjunction happens. So if the diploid number is X then when homologous pairs are separated during meiosis I, the haploid pair will be half of that, 0.5X]
20. d [As long as there is sexual reproduction and alternating haploid and diploid phases with fertilization of gametes, the four sources of genetic diversity are active]
21. c [Cross-overs during meiosis I followed by random segregation of homologous pairs and recombinant chromatids all give rise to genetic diversity so that the gametes are not alike nor are they the same as the parental cell]
22. d [Insertion sequences are only found in bacteria]
23. c [Retroviruses are animal viruses]

INTEGRATING AND APPLYING KEY CONCEPTS

1. Remember the purposes of meiosis: to reduce the diploid number to haploid by separating the homologous pairs of chromosomes and generating genetic variability through recombination of homologous sequences. Think about what role synapsis plays in both of these.
2. The textbook starts and ends with a discussion of the tension between having to replicate cells of the individual without introducing change vs. the needs of the population for genetic variability in order for evolution to function. These needs must be met through introducing change in the DNA in the gametes.
3. Think about what "homologous" means and the consequences of non-disjunction, which might be analogous to this hypothetical situation.
4. This question may seem complicated but simply requires rereading the section on bacterial conjugation and paying attention to the details and the figures. In terms of the control(s), you want to be able to ensure that the clones you had growing on your "diagnostic" media were the results of recombination and not mutation of the parent's mutant allele back to functional. In terms of the units of measurement, it may help to know the alternate name for this type of experiment: "time-of-entry" experiments.

Chapter 11 Mendel, Genes and Inheritance

11.1 The Beginnings of Genetics: Mendel's Garden Peas [pp. 230-239]

1. Gregor Mendel; 2. blending; 3. generation; 4. garden pea; 5. true; 6. pollination; 7. self-fertilized; 8. alleles; 9. P; 10. F_1; 11. self; 12. F_2; 13. meiosis; 14. characters; 15. phenotype; 16. genotype; 17. monohybrid; 18. pairs; 19. dominant; 20. separately; 21. segregation; 22. test-crosses; 23. true-; 24. genotype; 25. dihybrid; 26. independently; 27. assortment; 28. inheritance; 29. F; 30. K; 31. D; 32. H;

33. G; 34. A; 35. L; 36. B; 37. E; 38. M; 39. I; 40. C; 41. J; 42. A; 43. B; 44. B; 45. A; 46. A; 47. A- 2, B- R, C- r, D- T, E- T, F- R, G- r, H- 4, I- T, J- t, K- R, L- r; 48. A- Rr, B- red, C- phenotype, D- tall, E- red, F- TT, G- Rr, H- rr, I- tt, J- dominant, K- recessive, L- test; 49. This refers to the fact that two alleles for a given character will segregate or separate and enter the gametes singly; 50. This refers to the fact that during gamete formation, the alleles for two different traits segregate independently from each other; 51. A test cross is performed to determine if a phenotype is the product of a homozygous genotype or a heterozygous dominant genotype. To do the test you cross your organism of interest with a true-breeding homozygous recessive parent and examine the F_1 progeny. They will have a specific ratio of phenotypes depending upon the genotype of the organism in question; 52. Genes and their alleles are carried on chromosomes; 53. A- T, t; t, t; B. ½ Tt, ½ tt, C- ½ tall, ½ dwarf, D- A test cross because a monohybrid cross involves crossing two individuals that are heterozygous for the same pair of alleles (e.g. Tt x Tt); 54. A- TS, TS, TS, TS and ts, ts, ts, ts, B- 1 TtSs, C- 1 tall, strong, D- 1/16TTSS, 2/16TTSs, 2/16TtSS, 4/16TtSs, 2/16Ttss, 1/16ttSS, 2/16ttSs, 1/16ttss, 1/16TTss, E- 9/16 tall strong, 3/16 tall weak, 3/16 dwarf strong, 1/16 dwarf weak.

11.2 Later Modifications and Additions to Mendel's Hypotheses [pp. 239-246]

55. dominant; 56. recessive; 57. incomplete; 58. pink; 59. sickle cell anemia; 60. co- ; 61. MN; 62. allele; 63. organ; 64. epistasis; 65. deposition; 66. deposition; 67. polygenic; 68. continuous; 69. polygenic; 70. bell; 71. pleiotropy; 72. P could be a true dominant, masking the recessive p allele or P and p could be co-dominant so that the effect(s) of P and p would both be visible or P might be incompletely dominant in that the effect(s) of p are not completely masked so if they represented two colours the heterozygote would be a blended colour; 73. In the case of a pleiotropic character, a single gene would give rise to multiple traits whereas in polygenic inheritance and single character (e.g. the height of a human adult) is determined by the combined effects of multiple genes.

SELF-TEST

1. d [Meiosis is the process of segregating the maternal and paternal chromosomes so that haploid gametes are formed. The separation of any pair of homologous chromosomes is independent of the separation of any other two pairs]
2. c [Looking at the product of the first cross, the offspring would all be heterozygous and therefore have the phenotype T; with the second cross there would be two possible genotypes in the offspring: TT and Tt, which would both have the T phenotype]
3. a [The phenotype is the expression, i.e. blood type and can arise, in the case of dominance and multiple alleles, from different possible genotypes in the parental cross, for example I^Ai or I^AI^A. If either or both of the parents was heterozygous then the children would have a different blood type, i.e. *ii* or type O.]
4. a [The a and b alleles segregate or assort independently so the Aa pair would give gametes of A and a; the BB pair would only give gametes that are B. Then the various possible combinations for these are: AB, AB, aB and aB, so only two types and two of each]
5. d [A cross of this type would generate all heterozygote offspring so in the case of complete dominant, they should all look like the AA parent. Since they don't look like either, there are two possible explanations – A is not completely dominant over a so that the heterozygote has a blended phenotype or A and a are equally or co-co-dominant so that each contributes to a phenotype that differs from AA and aa]
6. c [Bb x bb yields the F_1 offspring 50% Bb and 50% bb. Since B is dominant, the heterozygotes would be blue and the homozygous recessive would be white]
7. c [Bb x bb yields the F_1 offspring 50% Bb and 50% bb.]
8. b [Type A blood (father) can result from either I^AI^A or I^Ai; Type B blood (mother) results from either I^BI^B or I^Bi. So if both parents were homozygous, the offspring would have type AB blood (which results from I^AI^B only) with no other possibilities; if the father was heterozygous but the mother was homozygous, the offspring would either have AB or B type blood; if the father was homozygous and the mother was heterozygous, then the offspring would have either AB or A-type blood; finally if both parents were heterozygous, the offspring would have either A, B or O type blood (which results from *ii* only)]
9. a [co-dominance occurs when neither allele dominates; this is the case with the AB blood type I^AI^B, however the *i* allele is a normal recessive allele as it is masked by either of the other two alleles]

10. d [Yellow labs are the result of a homozygous recessive epistasis allele (*ee*) which prevents the pigment or product of the B allele from being deposited. So it doesn't matter if the parents were BB (black) or Bb (black) it only matters that each had one *e* allele (if either had had two *e* alleles, they would have been yellow rather than black.]
11. c [The *E* gene is epistatic, that is it has an effect on another trait: colour (B); the determination of colour is not just dependent upon the ability to synthesize the colour but the ability to then deposit it in the hair. Retrievers that are homozygous for the recessive allele *e* are yellow regardless of how much pigment they produce]
12. b [Adult human height is a polygenic trait or the product of multiple genes so that one sees a continuous gradient of height distribution in the population]
13. d [Sickle cell disease results from a single mutation in the gene for a hemoglobin subunit and this mutation has multiple effects, as depicted in Fig. 11.14]
14. c [Co-dominance means that neither allele is dominant over the other AND they are both expressed whereas in incomplete dominance neither allele is dominant however one is able to mask the expression of the other to a limited degree]
15. b [Looking at the product of the first cross, the offspring would all be heterozygous and therefore have a single phenotype of Tt; with the second cross there would be two possible genotypes in the offspring: TT and Tt, which the same phenotypes]
16. b [Looking at the product of the first cross, the offspring would all be heterozygous and therefore have a single blended phenotype of Tt; with the second cross there would be two possible genotypes in the offspring: TT and Tt, where the first would display the pure phenotype of T while the Tt offspring would have a blended phenotype
17. c [Mendel performed his analyses with seven distinct traits all of which were true-breeding and which were the product of two alleles that were either dominant or recessive. Although flower colour in snap dragons displays incomplete dominance so that his results based strictly on those experiments would have supported the prevailing blending theory of inheritance, other snap dragon traits may display the same two allele, dominant/recessive traits that Mendel relied on in his studies]
18. c [Cystic Fibrosis is a genetic disorder that is the result of a mutation in a single membrane transporter]
19. a [Although the result of a single genetic mutation, individuals who are heterozygous for this mutation display a milder form of the disease as a result of incomplete dominance]

Integrating and Applying Key Concepts

1. Assume complete dominance (given the good health of the heterozygous individuals) and use a Punnett square to determine the genotypes of the F_1 offspring. Now assuming you could do this with multiple lines of mice (to prevent in-breeding), which offspring would you mate?
2. Read the section on co-dominance and take note of the description of the antigens that each blood type has on the surface – which of these would match Suzie and her two parents? Keep in mind that the basis of rejection or adverse reactions is based on antibody recognition of "non-self" antigens
3. Read the section on co-dominance and take note of the description of the antigens that each blood type has on the surface. Now design a Punnett square for the mother's blood type alleles and Joe's. Examine the offspring for the AB blood type.

Chapter 12 Genes, Chromosomes and Human Genetics

12.1 Genetic Linkage and Recombination [pp. 250-254]

1. linked; 2. chromosomes; 3. independently; 4. linked; 5. recombination; 6. recombination; 7. cross-over; 8. homologous; 9. proportional; 10. frequency; 11. map unit; 12. each other; 13. map; 14. frequencies; 15. relative; 16. *AaBb*; 17. aabb; 18. unlinked; 19. double; 20. not linked; 21. D; 22. E; 23. G; 24. A; 25. C; 26. B; 27. The number of recombinants over the total number of progeny multiplied by 100 ; 28. In the cross *TtRr* x *ttrr*, based on the principle of independent assortment of two unlinked genes, the expected progeny would be 4 *TtRr*, 4 *Ttrr*, 4 *ttRr* and 4 *ttrr* that is a ratio of 1:1:1:1 . If the *T* and *R* genes are linked, then the ratio you actually obtain would differ from that.; 29. F – the genes for flower colour and seed colour are on the same chromosome however they are so far apart they behave as if they were not linked; 30. T.

12.2 Sex-Linked Chromosomes [pp. 254-260]

31. pairs ; 32. sex chromosomes; 33. autosomal chromosomes; 34. fruit flies; 35. X; 36. Y; 37. inheritance; 38. reciprocal; 39. X-linked; 40. parents; 41. sex; 42. X-linked; 43. inactivation; 44. condensation; 45. Barr body; 46. G; 47. F; 48. A; 49. H; 50. B; 51. C; 52. E; 53. F – it is an X-linked trait so mothers are generally carriers and male offspring that inherit the recessive gene will have the disease; 54. T; 55. F – inactivation of one of the X-chromosomes occurs in cells of the developing multicellular embryo, and inactivation is random so that all decendents of a given cell (for example in a a particular developing tissue type) will have the same chromosome inactivated but this is independent of which chromosome is inactivated in another cell; 56. If she was homozygous recessive she would display the trait; 57. The SRY gene is a Y-linked gene that is te master switch for sex determination. In its absence the fetus will develop female physical features and sexual organs but will be genetically male; 58. There is a small region of homology between the X and Y chromosomes so that in males these two chromosomes pair during meiosis I; 59. If the mutation is recessive and mothers are healthy then they must be carriers, in which case their father may have had the disease. There is a 50% chance that any male offspring of the carrier mother would acquire the defective allele and exhibit the same disease their grandfather had; 60. A- Suzie would be a heterozygous carrier but phenotypically normal due to the dominant wild type allele, B- There would be a 50% chance that a male child will be phenotypically and genotypically SQ, however any female offspring would be phenotypically normal although a 50% chance that they will be heterozygous carriers, C- $X^{SQ}Y$ and X^{+} Y, D- X^{+} X^{SQ} and X^{+} X^{+} .

12.4 Human Genetics and Genetic Counselling [pp. 264-266]

61. inheritance; 62. radiation; 63. viruses; 64. duplications, deletions, inversions; 65. translocations; 66. homologous; 67. evolution; 68. germ; 69. nondisjunction; 70. homologous; 71. sister chromatids; 72. polyploid; 73. mitosis; 74. three; 75. autosomal; 76. carriers; 77. unaffected; 78. dominant; 79. recessive; 80. X-linked; 81. heterozygous; 82. counselors; 83. pedigree; 84. offspring; 85. F; 86. D; 87. A; 88. I; 89. B; 90. H; 91. E; 92. G; 93. C; 94. T; 95. F – in plants it can give rise to hardier plants with increased growth and production; 96. For an autosomal recessive trait, both male and female heterozygous individuals are healthy carriers but homozygous recessive individuals exhibit the trait. It is the opposite for autosomal dominant defects where heterozygous and homozygous dominant individuals are both affected but homozygous recessive individuals are not.

12.5 Nontraditional Patterns of Inheritance [pp. 266-268]

97. Cytoplasmic; 98. imprinting; 99. mitochondrial; 100. Maternal; 101. energy generation; 102. silenced; 103. imprinting; 104. methylation; 105. gametes; 106. imprinted; 107. imprint; 108. imprinting; 109. gene; 110. F – loss of imprinting means that an allele which should be silenced by methylation, i.e. imprinted, is not. As a result, the individual has an effective gene dosage twice what it should be; 111. This term is used to describe the phenomenon where all of the progeny, regardless of sex, inherit the genotype of one of the parents. It is usually the result of cytoplasmic inheritance; 112. This would indicate that a trait is cytoplasmically inherited, that is associated with the maternal mitochondria.

SELF-TEST

1. b [If the two genes are tightly linked then they will behave as if they were a single gene and be inherited together]
2. a [Since both parents share the same phenotype, 50% of the male offspring vary from the phenotype while all of the females have the same phenotype as the parents one can conclude that the parents are $X^{+}X^{-}$ and $X^{+}Y$ where X^{+} represents the X-linked gene dominant gene for short wings and X^{-} represents the recessive gene for long wings.]
3. b [Red-green colour blindness is an X-linked recessive trait and so a carrier would have two copies, one normal and one defective and would therefore have to be female (XX)]
4. c [Barr bodies are the condensed, inactivated second X chromosome that forms in females in order to reduce the X-linked gene dosage to one, as it is in males (XY)]
5. d [the sequence GFED is an inversion from the normal sequence and appears twice before it appears that the rest of the normal sequence reappears]

6. d [assuming the normal diploid chromosome content is 50, if there was a failure of the spindle during mitosis of a germ line cell, giving rise to 100 chromosomes (4*n*) and this was followed by meiosis, the content would be reduced to 2*n* rather than the normal 1*n*; fertilization with a normal gamete would give 3*n* that is a triploid individual]
7. b [A karyotype would clearly demonstrate three copies of chromosome 21 (and two of every other homologous pair)]
8. a [failure of chromosome 21 homologous pairs during meisos I or of the sister chromoatids during meiosis II would both lead to a single extra copy of the chromosome in one gamete (i.e. 2) and a missing copy in another gamete. Upon fertilization those numbers would increase by 1]
9. c [since 50% of the offspring would be expected to be male one can conclude this is not X-linked (i.e. sex-linked). If the trait were autosomal dominant and if the parents were Aa and Aa and one assumed the A allele is dominant, conferring the disease, the offspring would be Aa, Aa, Aa, aa and only the aa individual (25%) would not exhibit the disease]
10. b [PKU is an enzymatic deficiency that can be assayed following birth. When identified early, diet modification will prevent the development of symptoms]
11. b [SRY is the master switch that controls the sex of an individual and it is located on the Y chromosome]
12. c [it is due to inactivation of one of the X chromosomes in the female]
13. c [The calico colour is the result of epistasis, X-linkage and inactivation of one of the X-chromosomes in females]
14. d [Sharks, which evolved earlier than humans, only have one gene for hemoglobin however humans have a few suggesting that these arose subsequently by duplication]
15. b [Aneuploidy is usual fatal and may be responsible for ~ 70% of miscarriages]
16. a [Cystic is a genetic disease caused by a gene that encodes a mutant transport protein. Only individuals that are homozygous for this trait will develop the disease]
17. d [either amniocentesis or chorionic villus sampling would provide the cells of the embryo which could then be screened by karyotyping for trisomy of chromosome 21]
18. e [X-linked recessive traits are manifested in homozygous females but not heterozygous and males from a phenotypically normal mother will exhibit the trait if she is heterozygous; in contrast, if a mother has the trait and it is cytoplasmically-inherited then all offspring will also have the trait because of the much greater volume of cytoplasm in the female egg than the sperm that fertilizes it]
19. c [Although the ratio suggest the genes are not linked and assort independently, when genes are linked but separated by a wide distance on the chromosome they beahev as if they were not linked]
20. c [Determining the distance from one outside gene to one in the middle and then from the other outside gene to that same one in the middle will allow you to add up the two measurements to get the distance between the two outside genes]

INTEGRATING AND APPLYING KEY CONCEPTS

1. Use the topic map at the beginning of this study guide chapter and incorporate the concepts and ideas from the textbook. Or start from scratch and focus on heredity patterns and chromosomal alterations.
2. Think about the differences in complexity and developmental requirements.
3. Read the section on aneuploids and study table 12.1
4. Study figure 12.8

Chapter 13 DNA Structure, Replication and Organization

13.1 Establishing DNA as the Hereditary Molecule [pp. 272-275]

1. DNA; 2. deoxyribonucleic acid; 3. acidic; 4. phosphorus; 5. 20th ; 6. hereditary; 7. proteins; 8. amino acids; 9. transformation; 10. killed; 11. transforming; 12. Avery; 13. heat- killed; 14. 3; 15. enzymatic; 16. transform; 17. *E. coli*; 18. bacteriophage; 19. not; 20. DNA; 21. labeled; 22. E; 23. A; 24. D; 25. F; 26. C; 27. B.

13.2 DNA Structure [pp. 276-278]

28. nucleotides; 29. adenine; 30. guanine; 31. pyrimidines; 32. thymines; 33. guanines; 34. Chargaff's; 35. nucleotides; 36. phosphodiester; 37. 5' ; 38. sugar; 39. hydroxyl; 40. 5'; 41. 3'; 42. Watson and Crick; 43. 2; 44. parallel; 45. bases; 46. complementary; 47. purine; 48. thymine; 49. guanine; 50. hydrogen; 51. antiparallel; 52. hereditary; 53. sequences; 54. E; 55. F; 56. B; 57. A; 58. D; 59. C; 60. F-the AT base pairs are held together by two hydrogen bonds while the CG pairs are held together by three hydrogen bonds; 61. The DNA consists of a double helix of nucleotides linked together by phosphodiester linkages and held together by hydrogen bonds between complementary base pairs. The two strands are antiparallel with a 5' phosphate group on one end of one strand opposite a 3' hydroxyl on the other strand; 62. A - 5' phosphate, B – 5' carbon, C – 4' carbon, D – 3' carbon, E – 1'carbon, F – 2' carbon, G – phosphate, H – deoxyribose sugar.

13.3 DNA Replication [pp. 278-287]

63. semi-conservative; 64. template; 65. complimentary; 66. helicase; 67. replication origin; 68. RNA primer; 69. primase; 70. DNA polymerase; 71. 3' → 5'; 72. 5' → 3'; 73. hydroxyl; 74. 3'; 75. origin; 76. polymerase; 77. fork; 78. leading; 79. lagging strand; 80. Okazaki fragments; 81. RNA; 82. DNA ligase; 83. 3' → 5'; 84. 5'; 85. shorter; 86. replication; 87. genes; 88. telomeres; 89. thousands; 90. shorter; 91. telomerase; 92. divisions; 93. dies; 94. aging; 95. cancer; 96. telomere; 97. telomerase; 98. B; 99. E; 100. C; 101. D; 102. D; 103. A; 104. F – Meselson and Stahl determined the semi-conservative nature of DNA replication; 105. F – reactivated telomerase is correlated with cancer; 106. A – leading strand, B – lagging strand; 107. The leading strand (A) has continuous replication because the DNA polymerase is synthesizing the DNA in the 5' to 3' direction, the only direction it can move in, and this is following immediately behind the helicase. The lagging strand has discontinuous replication because of the antiparallel nature of the DNA strands: the orientation of the template strand is such that the polymerase on the lagging strand template cannot following behind the helicase while polymerizing in the 5' to 3' durection; 108. Semiconservative replication means that each strand of the DNA molecule serves as a template for the synthesis of a complementary strand yielding two molecules which contain one strand of the original molecule and one strand that is new.

13.4 Mechanisms that Correct Replication Errors [pp. 287-289]

109. mismatches; 110. proofreading; 111. mismatched; 112. forward; 113. distortion; 114. mismatch; 115. mismatched; 116. DNA polymerase; 117. ligase; 118. accuracy; 119. mutations; 120. evolutionary; 121. In humans, the same repair mechanisms detect and correct changes to the DNA arising from exposure to UV radiation and certain chemicals. People who have a genetic defect in their DNA repair mechanisms must avoid all exposure to sunlight as they have an extremely high incidence of skin cancer upon exposure to sunlight.

13.5 DNA Organization in Eukaryotes and prokaryotes [pp. 289-292]

122. nucleus; 123. histones; 124. positively; 125. phosphate; 126. H2A, H2B, H3 and H4; 127. nucleosome; 128. H1; 129. solenoid; 130. chemical; 131. chromatin; 132. euchromatin; 133. heterochromatin; 134. turned off; 135. Barr; 136. heterochromatin; 137. nonhistone; 138. genes; 139. histone; 140. packing; 141. circular; 142. loops; 143. positively; 144. nucleoid; 145. plasmids; 146. circular; 147. origin of replication; 148. E; 149. G; 150. F; 151. B; 152. A; 153. C; 154. D; 155. A – DNA double helix, B - nucleosome, C - linker, D - H1, E - solenoid; 156. Euchromatin is loosely condensed DNA in which genes are relatively accessible to the proteins involved in gene expression. Heterochromatin is more tightly packed and the genes found within it are not expressed; 157. The most basic DNA structure is the double helix. This is then condensed through the winding of the DNA strand around the nucleosomes and further organized into the solenoid structure by the binding of the H1 histone protein to the linker region between the nucleosomes and to the DNA entering the nucleosome.

SELF-TEST

1. d [Proteins were thought to the hereditary material because they were thought to have a greater coding capacity with 20 amino acid building blocks than DNA with only 4 nucleotide building blocks]
2. b [Transformation is the process that some bacteria can undergo where they absorb DNA directly from the environment]

3. b [Some amino acids in proteins have sulfur; phosphate groups make up the backbone of DNA. The separate labeling revealed that the progeny viruses from infected bacteria possessed radiolabeled DNA but not proteins proving that DNA was the hereditary material]
4. c [The purines and pyrimidines are the bases that base pair together and hold the two strands of the double helix together]
5. c [The X- ray diffraction studies of Franklin and Wilkins revealed a pattern that Franklin correctly interpreted as a helix}
6. a [The semi-conservative mechanism of replication was discovered through the experiments of Meselson and Stahl]
7. a [The 3' end of the double helix has the 3' hydroxyl group required for formation of the phosphodiester bonds that make up the backbone of the DNA]
8. d [DNA helicase unwinds the double helix, separating the two strands so that the DNA polymerase can replicate the DNA]
9. a [The polymerase synthesizes short fragments moving in the opposite direction to the movement of the helicase and replication fork. These fragments, the Okazaki fragments, are then linked into a continuous chain through the action of DNA ligase]
10. a [The helicase unwinds the DNA helix, allowing the replication fork to progress]
11. b [The proofreading combined with mismatch repair maintains an extremely low rate of error however the errors that do persist may lead to mutations and if these occur in the germ cells they can be heritable. This is an essential component leading to evolution]
12. d [The euchromatin is more loosely packed and proteins can access the DNA within it, allowing the proteins required for gene expression to transcribe the genes. The heterochromatin is too densely packed and its formation is believed to be a mechanism of gene inactivation]
13. c [They heat-killed S-type *Streptococcus* and treated the cell extracts with enzymes to degrade the proteins and RNA molecules, leaving the DNA. Upon exposure to this extract, the previously rough bacteria were transformed to grow as smooth colonies on growth media]
14. d [The phosphate groups of DNA contained the ^{32}P radioisotope and because the DNA was the hereditary material and the phage injects its DNA when it infects bacteria, this label was found inside the cells]
15. d [The purines adenine and guanine are double ring structures that base pair with the single rig-structured thymines and cytosines, respectively]
16. b [A conservative mode of replication would mean that the new DNA molecules contain two new strands of DNA and the original, intact DNA molecule. Therefore the original molecule would be uniformly labeled with the heavy nitrogen while the new DNA, replicated after transfer to the "light" medium, would be entirely consisting of light nitrogen]
17. b [the inner phosphate group is integrated into the phosphodiester backbone of the DNA and the bond linking it to the terminal two phosphates is hydrolyzed with the energy used for polymerization]
18. c [Primase does not require 3' hydroxyl groups unlike DNA polymerase however it synthesizes RNA strands which must later be degraded and replaced with DNA]
19. a [Because the bacterial chromosome is circular the replication forks simply progress from the origin of replication bi-directionally until the two forms meet at the opposite side of the circle. This means that there are no gaps that can't be filled by DNA polymerase]
20. b [Rolling circle replication is used during conjugative transport of the F plasmid. Once transfer starts DNA replication of the leading strand fills in behind the strand being transferred while the lagging strand is replicated once the single strand of the plasmid enters the recipient cell}

Integrating and Applying Key Concepts

1. Begin by drawing a circular double-stranded molecule as depicted in Fig. 13.19. Arbitrarily assign one parental strand as having a 5' to 3' direction in the clockwise direction and the other parental strand being antiparallel. Now remember the antiparallel nature of the DNA double helix to determine the orientation of the newly replictated DNA and whether this follows the direction of the advancing replication fork.

2. Reread the section "*Why it matters*" then the section on DNA organization in eukaryotes paying attention to the effects of having these proteins in living cells and the difference in the susceptibility of DNA that is not associated with these proteins compared to those that are.
3. Remember the effects of shortening telomeres on the viability of the cell and that for traditional drug therapy you need to target the abnormal (or pathogenic) cells.

Chapter 14 Gene Structure and Expression

14.1 The Connection between DNA, RNA and Protein [pp. 296-300]

1. protein; 2. DNA; 3. proteins; 4. genes; 5. RNA; 6. ribosomes; 7. proteins; 8. Beadle; 9. Tatum; 10. enzymes; 11. minimal; 12. nutrients; 13. auxotrophs; 14. gene; 15. nutrient; 16. gene; 17. protein; 18. gene; 19. polypeptide; 20. proteins; 21. transcription; 22. RNA; 23. translated; 24. protein; 25. Crick; 26. Central dogma; 27. rRNA, tRNA and snRNA; 28. template; 29. RNA; 30. RNA polymerase; 31. ribosome; 32. amino acids; 33. polypeptide; 34. prokaryotic; 35. simultaneously; 36. mRNA; 37. nucleus; 38. cytoplasm; 39. three; 40. universal; 41. mRNA; 42. peptide; 43. 64; 44. 20; 45. degeneracy; 46. codons; 47. nonsense; 48. stop; 49. methionine; 50. start/initiator; 51. C; 52. E; 53. G; 54. J; 55. M; 56. K; 57. A; 58. L; 59. F; 60. B; 61. D; 62. I; 63. H; 64. A – Transcription, B – Processing, C - Translation; 65. F – some proteins comprise multiple subunits or polypeptides; 66. This is a heritable defect in a metabolic pathway; 67. Uracil is a nucleotide that replaces thymine in RNA.

14.2 Transcription: DNA-Directed RNA Synthesis [pp. 301-304]

68. transcription; 69. uracil; 70. template; 71. single; 72. RNA; 73. 3'; 74. primer; 75. promoter; 76. 3' ; 77. 5' ; 78. RNA; 79. 5' ; 80. 3' ; 81. DNA; 82. mRNA; 83. DNA; 84. all; 85. 3; 86. RNA II; 87. I; 88. III; 89. rRNA; 90. tRNA; 91. promoter; 92. transcription; 93. TATA; 94. termination; 95. terminators; 96. mRNA; 97. C; 98. F; 99. B; 100. H; 101. G; 102. A; 103. J; 104. D; 105. K; 106. I; 107. L; 108. E; 109. B, E, G, D, A, C, F; 110. A – DNA: C-G, A-T, RNA: C-G, A-U, B – DNA:2, RNA:3, C-DNA:2, RNA:1, D-DNA: DNA polymerase, RNA:RNA polymerase, E-DNA:yes, RNA:no; 111. A-prokaryotes:RNA polymerase, eukaryotes:RNA polymerase I, II, III, B-prokaryotes:direct binding, eukaryotes:require transcription factors, C-prokaryotes:two types of termination sequences, eukaryotes:no termination sequences; 112.Prokaryotes have one type of RNA polymerase that transcribe all types of genes whereas eukaryotic cells have three RNA polymerases. RNA polymerase I and III transcribe non-protein-encoding genes while RNA polymerase II transcribes protein-encoding genes; 113. It is a sequence found in the promoter that is recognized by transcription factors. These bind to the sequence then recruit the appropriate polymerase; 114. Morer RNA polymerases are following along immediately behind; 115. T.

14.3 Processing of mRNAs in Eukaryotes [pp. 304-307]

116. polypeptide; 117. processing; 118. nucleus; 119. guanine; 120. 5' - cap; 121. ribosome; 122. enzyme; 123. termination; 124. polyadenylation; 125. Poly (A) polymerase; 126. poly A-tail; 127.nucleus; 128. cytoplasm; 129. exons; 130. introns; 131. mRNA splicing; 132. snRNPs; 133. introns; 134. exon; 135. exons; 136. exons; 137. genes; 138. exon shuffling; 139. B; 140. M; 141. A; 142. K; 143. F; 144. J 145. D; 146. L; 147. C; 148. E; 149. H; 150. I; 151.G; 152. F-no one knows how they evolved as they are not present in prokaryotes; 153. T; 154. Through alternate splicing it provides the ability to make more functionally distinct proteins without having to increase the size of the genome and through exon shuffling it provides a mechanism for evolving new proteins; 155. A - Contains introns interspersed with exons and their splicing prior to exiting the nucleus provides an opportunity to make new proteins without having to have additional genes, B – provides a site for the ribosome to first bind, C - protects the mRNA from enzymatic degradation as it migrates from the nucleus to the ribosomes in the cytoplasm, D – generates the mature mRNA for translating.

14.4 Translation: mRNA-Directed Polypeptide Synthesis [pp. 307-317]

156. ribosomes ; 157. proteins; 158. amino acid; 159. cytoplasm; 160. RNA polymerase; 161. cytoplasm; 162. chloroplast; 163. nucleus; 164. ribosome; 165. transfer; 166. amino acids; 167. four; 168. clover leaf; 169. anticodon; 170. mRNA; 171. amino acids; 172. Aminoacyl-tRNA; 173. translation; 174. small; 175. large; 176. function; 177. antibiotics; 178. three; 179. A; 180. P; 181. E; 182.

initiation; 183. AUG; 184. 5'; 185. methionine; 186. P; 187. elongation; 188. tRNAs; 189. peptide; 190. peptidyl; 191. rRNA; 192. GTP; 193. E; 194. P; 195. aminoacyl; 196. 3' ; 197. stop; 198. termination; 199. release; 200. A; 201. mRNA; 202. processing; 203.amino acids ; 204. Carbohydrates, lipids; 205. chaperones; 206. free; 207. signal; 208. signal; 209. SRP; 210. DNA; 211. function; 212. mutations; 213. codon; 214. missense; 215. nonsense; 216. termination; 217. nucleotide; 218. reading; 219. sequence; 220. C; 221. G; 222. A; 223. K; 224. B; 225. O; 226. D; 227. L; 228. E; 229. F; 230. H; 231. I; 232. N; 233. J; 234. M; 235. T; 236. F – they are interchangeable. If a eukaryotic signal sequence for targeting to the RER is attached to a prokaryotic protein then the recombinant protein is targeted to the plasma membrane, the membrane structure that fulfills most of the functions of the RER in eukaryotic cells. The opposite is also true; 237. A - Mutation where a nucleotide is replaced by another type of nucleotide, B – Missense mutation, C – nonsense mutation, D - Mutation where a nucleotide is deleted or added that changes all the codons on mRNA; 238. Prokaryotes do not have a nucleus like eukaryotes and their mRNA molecules do not require processing. In eukaryotes, the processing of the mRNA and the physical separation of transcription and translation make simultaneous transcription and translation impossible; 239. Because of the redundancy in the genetic code, more than one codon can encode the same amino acid; 240. A-codon, B-anticodon, C-aminoacyl tRNA, D-peptide chain, E-released/uncharged tRNA, F-5' end of the mRNA, G-3' end of mRNA.

SELF-TEST

1. a [RNA polymerase makes the mRNA required by the ribosome]
2. b [The codon is the three base code in the mRNA that threads through the ribosome and base pairs with the complimentary anticodon in the appropriate tRNA]
3. c [AUG is the universal start codon, encoding methionine; the uracil is the base in RNA that replaces the thymine base in the DNA]
4. b [AUG encodes methionine and is a universal start codon]
5. d [the number is determined by the formula 4^3, which is derived from the fact that there are four different bases and they are combined in a three base code]
6. a [Prokaryotes do not have membrane-bound organelles so transcription and translation happen in the same cellular "compartment" and occur simultaneously]
7. b [Eukaryotic mRNA is generally made as a larger pre-mRNA which is then processed by removing introns]
8. a [Introns are removed by the snRNPs so that the final transcript has the 5' cap, 3' polyA tail and the sequence comprises the exons only]
9. a [The snRNPs are small ribonucleoprotein particles with ribozme activity for recognizing, binding and splicing mRNA to remove the introns]
10. c [The anticodon is the complementary sequence to the codon(s) for whichever amino acid is carried by the tRNA]
11. a [The A stands for aminoacyl and is the site where the charged tRNA binds. The exception is the initiator tRNA carrying methionine. This binds in the P site and the next charged tRNA binds in the A site. Upon formation of the peptide bond between this amino acid and the peptide chain on the tRNA in the P site, the ribosome translocates by three bases on the mRNA moving the peptidyl tRNA that is now in the A site over to the P site, allowing the next aminoacyl-tRNA to enter and bind the next codon]
12. b [The P or peptidyl site binds the tRNA carrying the growing peptide chain]
13. d [One of the rRNA molecules in the ribosome has the peptidyl transferase activity so is a ribozyme]
14. c [Frameshift mutations result when a base is inserted or deleted so that from the point of insertion/deletion, the reading frame of the protein is altered, the amino acid sequence of the remaining polypeptide chain is altered and the structure and function of the protein is lost]
15. b [Multiple ribosomes attach to the mRNA giving rise to a structure called a polysome]
16. b [Alternative splicing allows for alterations in the way introns are spliced out of the pre-mRNA; exon shuffling is believed to be a method of creating or evolving new proteins however is likely not a routine method of mRNA processing]
17. c [The process is unique to eukaryotes and allows for an increase in protein-coding capacity without having to increase their genome size through the mechanism of alternative splicing]

18. d [There are 20 different tRNAs, one for each amino acid however most amino acids can be encoded by more than one codon. The base pairing of the particular tRNA for a particular amino acid "wobbles" generally in the third position so that one tRNA can base pair with any of the codons that encodes their particular amino acid]
19. a [Their work with auxotrophic mutants of *Neurospora crasa* led Beadle and Tatum to propose the one gene-one protein hypothesis]
20. e [In prokaryotes as the mRNA is still being transcribed, ribosomes can attach and follow along behind the advancing RNA polymerase – the two processes are not separated in space or time and there is no need for processing of the mRNA or, in general, the proteins after polymerization]

INTEGRATING AND APPLYING KEY CONCEPTS

1. Read the section 14.1a describing Beadle and Tatum's work to establish the one gene-one enzyme hypothesis. You might assume that you have some experience with biochemistry, such as in the case of Archibald Garrod (sect. 14.1a) as well as ability to perform the appropriate microbiological and genetic experiments performed by Beadle, Tatum and William Bateson.
2. Remember that the genetic code is a triplet code and that the reading frame is established by the presence of the start codon with no subsequent "commas" in the sequence.
3. Think about the structure of a simple protein vs. a complex protein like hemoglobin and about the processing that eukaryotic pre-mRNA must undergo to generate the translatable mRNA.
4. Reread the section on mRNA sorting signals. Remember from previous chapters that the nuclear membrane breaks down during cell division cycles of eukaryotic cells, reforming at the end of the division processes.

Chapter 15 Control of Gene Expression

Why it matters [pp. 321-322]

1. inactive; 2. sperm; 3. active; 4. differentiate; 5. genes; 6. off; 7. short; 8. short; 9. long.

15.1 Regulation of Gene Expression in Prokaryotes [pp. 322-327]

10. reversibly; 11. environment; 12. off; 13. available; 14. on; 15. available; 16. operon; 17. Jacob; 18. Monod; 19. single; 20. transcriptional; 21. operator; 22. repression; 23. operon; 24. corepressor; 25. active; 26. metabolize; 27. absent; 28. operator; 29. repression; 30. RNA polymerase; 31. inducible; 32. inducer; 33. available; 34. inactive; 35. active; 36. corepressor; 37. repressible; 38. activator; 39. inactive; 40. glucose; 41. *lac* operon; 42. adenylate; 43. cAMP; 44. catabolite activator; 45. activator; 46. adenylate; 47. cAMP; 48. CAP; 49. C; 50. F; 51. G; 52. A; 53. J; 54. K; 55. H; 56. L; 57. B; 58. I; 59. D; 60. E; 61. N; 62. M; 63. A – An inducible operon is one that is normally repressed and is turned on or induced when a molecule, i.e. the inducer, binds and inactivates the repressor, B – A repressible operon is one that is normally on and the repressor is normally inactive but in the presence of a corepressor, usually the end product of the genes in the cluster, a complex is formed between the two to give the active repressor, C – Negative gene regulation is one that involves a repressor binding when it is in its active form, D – positive gene regulation is the opposite, that is expression is subject to activation by an functional activator protein; 64. The *lac* genes would be expressed; 65. The strain would be unable to make a functional activator complex so would not be able to activate transcription of the *lac* operon; 66. The *trp* operon would be expressed regardless of whether tryptophan was present; 67. A-*lacI.* B-promoter, C-Lac repressor, D- β-galactosidase, E-*lacY,* F-*lacA,* G-CAP sequence.

15.2 Regulation of Transcription in Eukaryotes [pp. 327-334]

68. operons; 69. regulated; 70. short; 71. long; 72. post; 73. post; 74. chromatin; 75. chromatin; 76. activator; 77. remodeling; 78. acetylation; 79. histone; 80. accessible; 81. transcription; 82. TATA; 83. activator; 84. proximal; 85. enhancer; 86. coactivator; 87. promoter; 88. activators; 89. coactivator; 90.) histone; 91. chromatin; 92. enhancer; 93. strengths; 94. combining; 95. few; 96. combinatorial; 97. coordinates; 98. tissues; 99. plasma membrane; 100. receptors; 101. tissues; 102. active; 103. response element; 104. methylation; 105. silencing; 106. imprinting; 107. gene; 108. C; 109. E; 110. D ; 111. A; 112. B; 113. K; 114. H; 115. F; 116. J; 117. M; 118. L; 119. G; 120. I; 121. A-Chromatin in its condensed state prevents access of transcriptional proteins to genes within the

chromatin. This condensed structure can be loosened around the promoter regions of genes that need to be expressed, either through binding of an activator protein upstream of a promoter, and the subsequent recruitment of either a remodeling complex which displaces the nucleosome in the region of that promoter or the recruitment of acetylation enzymes that then act on the histone proteins, causing them to loosen their interaction with the chromosome, B-this has three levels: transcription factors bound to the promoter and a resulting recruitment of the RNA polymerase has minimal activation which can be further increased by binding of activator proteins at the promoter proximal region. Full activation adds to this the binding of an activator to the enhancer region further upstream and its interaction with a coactivator which loops out the intervening DNA so that it can simultaneously bind the proteins in the promoter region, C-the use of specific combinations of regulators for a given gene allows the specific regulation of specific genes without having to have a gene for a specific regulator but rather gene expression can be effected by a small number of regulator proteins, D-Methylation of whole chromosomes or segments of chromosomes allows for the complete inactivation of that region/chromosome an is the basis for silencing. Methylation of specific genes inactivates them and is the basis for genomic imprinting; 122. F – long-term regulation is the basis for cellular differentiation in multicellular organisms; 123. T; 124. The zygote is the cell produced by fertilization; a deletion would result in the loss of steroid X regulation in all target cells and tissues. In the embryo, cells begin to differentiate into different tissue types so a deletion in one cell undergoing differentiation would result in the loss of steroid X regulation in that particular tissue but not in other steroid-regulated tissues.

15.3 Post-Transcriptional, Translational and Post-Translational Regulation [pp. 334-337]

125. quantities; 126. pre-mRNA; 127. translation; 128. degradation; 129. Pre-mRNA; 130. splicing; 131. exons; 132. masking; 133. unfertilized; 134. fertilized; 135. degradation; 136. half-life; 137. higher; 138. micro-RNAs; 139. Development; 140. precursor; 141. stem-loop; 142. Dicer; 143. mi-RNA; 144. cleavage; 145. translation; 146. 3'poly(A) tail; 147. Increase / decrease; 148. modification; 149. precursor; 150. degradation; 151. ubiquitin; 152. proteasome; 153. acetylation; 154. transduction; 155. cycle; 156. insulin; 157. precursor; 158. H; 159. D; 160. A; 161. E; 162. G; 163. B; 164. F; 165. C; 166. A-posttranscriptional control may involve alternate splicing to control which proteins are produced, a masking protein which blocks access of the ribosome, modulating the steps involved in the degradation of mRNA to increase or decrease the half-life, or using RNA interference to either prevent transcription of cause the mRNA to be cleaved, B – The most common mechanism of translational control is through alterations to the 3'poly(A) tail length. The length is proportional to the rate of translation.C –Posttranslational control may be accomplished through chemical modification of the protein, e.g. acetylation or phosphorylation, through targeting the protein for degradation by the addition of ubiquitin or through control of the processing of preproteins; 167. T; 168. F – the 5' UTR seems to be involved in determining the half-life of the mRNA molecule; 169. Researchers are able to synthesize double-stranded RNA molecules of a particular sequence that can be targeted by Dicer and the other proteins of the RNA interference pathway and then base pair through complementary sequences with a gene being studied. Through the RNAi mechanism, this gene will be silenced.

15.4 The loss of Regulatory Controls in Cancer [pp. 337-339]

170. genes; 171. tumours; 172. dedifferentiation; 173. benign; 174. malignant; 175. cancer; 176. metastasis; 177. proto-oncogenes; 178. oncogenes; 179. tumour-suppressor; 180. inhibitory; 181. tumour-suppressor; 182. multi-step; 183. C; 184. D; 185. B; 186. G; 187. I; 188. H; 189. E; 190. A; 191. F; 192. A, B – benign tumours vs. malignant: both arise from cells displaying uncontrolled growth and division however benign tumours show no evidence of the cells breaking loose and spreading, while malignant invade and disrupt surrounding tissue and may be metastatic, C, D- proto-oncogenes are normal genes that encode functions that promote cell division whereas oncogenes are mutant versions of these where they cause cells to become cancerous, E,F-Differentiation is the process whereby cells become specialized with some genes being turned on and others being turned off. This is usually accompanied by a slowing or cessation of cell growth. Dedifferentiation is something that happens when differentiated cells become cancerous and the genes that were turned off are turned back on, the cell reverts partially or completely to the embryonic state and the cell starts growing rapidly. G, H-Tumour-suppressor genes are normal cellular genes that encode proteins that inhibit cell growth and division. p53 is a protein encoded by a mutant tumour-suppressor gene, TP53, which no longer inhibits cell growth and division. It is associated with many different types of cancer; 193. Spontaneous mutations to the promoter region can alter the normal transcriptional controls for the gene; alternatively, spontaneous mutations within the coding portion of the gene can cause the protein to become highly active.

Translocations of the gene to another chromosome such that the gene is under the control of different promoters can have the same effect. Viruses can insert in the promoter region so that, again, the normal controls are lost and highly active viral promoters start to drive expression of the gene. Alternatively viruses that insert into the host genome can introduce genes for proteins that alter the normal controls.

SELF-TEST

1. c [An operon is the name proposed by Jacob and Monod when studying the cluster of genes for lactose metabolism in *E. coli*]
2. b [In prokaryotes, the RNA polymerase can bind directly to the promoter in order to transcribe a gene or operon]
3. d [An inducer is a molecule that interacts with a repressor to cause the repressor to become inactivated, relieving repression of the gene or operon]
4. b [The *trp* operon is under negative control with a repressor that is produced in an inactive form; when the end-product of the pathway starts to accumulate, that is the amino acid tryptophan, the amino acid binds to the repressor, and the complex then becomes a functional repressor. Tryptophan is therefore called the corepressor.]
5. a [Transcription factors are required to bind the TATA box of the promoter and recruit one of the RNA polymerases; for protein-encoding genes, this is RNA polymerase II.]
6. b [Although regulation if gene expression in eukaryotes is far more complex and occurs at many different levels, regulation of transcription is the most important.]
7. c [Histones are the positively-charged proteins that are responsible for compacting the eukaryotic chromosomes into chromatin. These include H2A, H2B, H3, H4 and H1.]
8. a [Steroid hormones enter the cell by diffusing through the plasma membrane, combining with specific steroid hormone receptor molecules in the cytoplasm, then enter the nucleus and bind regulatory regions (promoter proximal) to enhance transcription]
9. a [Methyl groups are added to cytosines to inactivate whole regions of a chromosome or whole chromosomes to "silence" the genes within that region or, in the case of the Barr body, the whole chromosome. They can also be added to the cytosines in a particular gene, to silence just that gene, as is the case with genomic imprinting.]
10. b [It is the TATA box that is recognized and bound by transcription factors in order that the RNA polymerase can then be recruited to transcribe the downstream gene.]
11. b [Masking proteins are used as a method of post-transcriptional regulation, binding to the mRNA and preventing the ribosomes from gaining access to translate the message. When eggs are fertilized, these proteins are displaced, allowing translation to occur.]
12. a [The poly(A)-tail is the untranslated string of adenosines that is added to eukaryotic mRNA molecules. The length can be regulated, through an unknown mechanism, with the length being directly related to the efficiency of translation of the message by the ribosome.]
13. b [A malignant tumour is one where the cells, in addition to exhibiting uncontrolled growth and division, invades neighbouring tissues and from which cells may break free and spread to another part of the body via the circulatory system.]
14. b [Proto-oncogenes are normal genes that encode proteins that promote cell division. When they become abnormal, they are referred to as oncogenes. The products of oncogenes cause cell growth and division to become unregulated.]
15. c [Metastasis means "change of state" and is the term used to describe the process by which cells from a malignant tumour break free from that growth and travel through the body's circulatory system to another part of the body.]
16. b [TP53 is a normal tumour-suppresor gene that encodes a protein involved in repressing cell growth and division. When it is mutant, i.e. $TP53^-$ it loses its inhibitory ability. The mutant product, p53, is associated with many different types of cancer.]
17. d [Prokaryotes are single-celled organisms that lack a nuclear membrane and perform transcription and translation simultaneously. It is therefore able to adapt quickly to changes in the environment, allowing them to have generation times on the order of minutes, for example in the case of *E. coli*, where they can still maintain a rapid generation time even if their environment quickly changes from the intestinal tract of a cow to water treatment facility.]
18. c [Methylation of an entire chromosome is the underlying mechanism for inactivation of one of the X chromosomes to give rise to the Barr body. When restricted to a specific gene, provides the explanation for genomic imprinting, another aspect of gene silencing.]

19. a [miRNA or microRNA molecules are short single stranded RNA molecules that are generated from a double-stranded precursor by the Dicer enzyme to then take part in the posttranscriptional regulation of gene expression by gene silencing]
20. d [Prokaryotes do not have a nuclear membrane nor do they synthesize pre-mRNA molecules that require processing before they can be translated. The result of both of these is that transcription and translation happen simultaneously. In addition, given their changing environmental conditions and their lack of control over those conditions, the fact that their mRNA has an average life-span of 3 minutes means that while the appropriate adjustments in gene expression are made, the mRNA for now inappropriate proteins is being rapidly degraded]

INTEGRATING AND APPLYING KEY CONCEPTS

1. Read sections 15.1d and 15.2c, paying attention to the regulatory molecules, whether a partner is required in both cases, the scope of the gene(s) that are targeted, the mechanism of regulation (negative or positive? transcriptional/translational?). Identify as many similarities as possible and critical differences.
2. Think about whether this is a short-term or long-term type of regulation and, having determined that, read through the sections on regulation of eukaryotic gene expression looking for mechanisms that are specifically used for long-term vs. short-term regulation. Remember that blood cells and every other cell in the body are derived from the zygote.
3. Read the section in 15.3a on regulation of gene expression by small RNAs. While reading this section, pay attention to the specific types of genes that are regulated by miRNA and think about the structural/physiological differences between worms and humans.

Chapter 16 DNA Technologies and Genomics

Why it matters [pp. 343-344]

1.Biotechnology; 2. technology; 3. engineering; 4. metagenomics; 5. grow; 6. Genes.

16.1 DNA cloning [pp. 344-352]

7. cloning; 8. function; 9. *mutations*; 10. protein; 11.gene ; 12. restiction; 13. *sequences* 14. *sticky*; 15. *ligase* ; 16. *recombinant*; 17. *transformation ;* 18. *replicated*; 19. *recombinant*; 20. *antibiotics*; 21. *transformed*; 22. *loss*; 23. *blue*; 24. recombinant; 25. hybridization; 26. protein; 27. library; 28. *mRNA*; 29. introns; 30. *expressed*; 31. reverse transcriptase; 32. mRNA; 33. polymererase; 34. complementary; 35. polymerase chain reaction; 36. primers; 37. *denaturation*; 38. *annealing*; 39. *extending*; 40. *double*; 41. Gel electrophoresis; 42. *C*; 43. *K*; 44. *L*; 45. *A*; 46. *G*; 47. *D*; 48. *M*; 49. *B*; 50. *H*; 51. *E*; 52. *O*; 53. *J*; 54. *F*; 55. *I*; 56. *N*; 57. Restriction enzymes vary in the specific DNA sequence that they recognize and cleave; 58. Reverse transcriptase is an enzyme made by the RNA viruses retroviruses. It is an RNA-dependent DNA polymerase that allows for the synthesis of a complementary DNA strand from an RNA template; 59. DNA hybridization is a technique used to identify the presence of DNA sequences of interest based on complementary base-pairing to a labeled nucleic acid primer; 60. In order to perform PCR one must have the sequence information that allows for the synthesis of the two primers that will be complementary to and hybridize to either end of the gene or sequence of interest; 61. F – metagenomics is the technique of deriving sequence data, including genomic sequence data from a mixture of DNA purified straight from an environmental sample. It does not require the ability to culture or grow the source organism; 62. T; 63. F – they were awarded the Nobel prize for their development of the first DNA cloning techniques.

16.2 Applications of DNA Technologies [pp. 352-362]

64. DNA; 65. genetic; 66. mutant; 67. restriction; 68. restriction fragment length polymorphisms; 69. electrophoresis; 70. DNA hybridization; 71. identify; 72. DNA; 73. short tandem repeats; 74. identical twins; 75. PCR; 76.gel electrophoresis; 77. length; 78. engineering; 79. plants; 80. domestic; 81. crop; 82. proteins; 83. somatic; 84. hereditary; 85. genetic engineering; 86. genetically modified organisms; 87. biosafety; 88. C; 89. D; 90. E; 91. G; 92. A; 93. B; 94. F; 95. The genetic material is isolated and digested with a restriction enzyme that gives different-sized restriction fragments depending on whether the allele is normal or mutant. The restriction digest fragments are separated by gel electrophoresis then DNA hybridization is performed using a labeled nucleic acid probe for the gene. If a person has sickle cell trait then they will have the normal allele and the mutant allele and this can be clearly

seen in the results of the hybridization analysis; 96. This terms denotes an organism that contains, in its genome, a gene or genes from another species; 97. One can clone the gene for a protein of interest into the fertilized egg of a mammal, for example a sheep, under the control of the promoter for the milk protein gene, so that the foreign protein is secreted in the milk; 98. F – in oder for the change to be heritable, it must be made to the germ-line or a fertilized egg; 99. T; 100. F – although it has been used successfully with animals, it is illegal with humans, to use germ-line gene therapy.

16.3 Genome Analysis [pp. 362-369]

101. structural; 102. fucntional; 103. bioinformatics; 104. shotgun; 105. fragments; 106. cloned; 107. termination; 108. polymerase; 109. dideoxyribonucleotides; 110. polymerization; 111. tag; 112. algorithms; 113. genome; 114. protein; 115. codon; 116. three; 117. Human Genome Project; 118. base pairs; 119. protein-coding; 120. 100,000; 121. noncoding; 122. repeated; 123. active; 124. microarray; 125. fluorescently; 126. protein-coding; 127. levels; 128. proteins; 129. proteomics; 130. T; 131. F – it was predicted to have that many genes based on the number of different human proteins, however it has ¼ - 1/5 that amount; 132.F-this is illegal. Somatic gene therapy however has had some success and holds promise for the future.

SELF-TEST

1. d [insulin is a human peptide hormone that is not produced in bacteria unless they are genetically engineered to do so]
2. a [Reverse transcriptase is used to make a DNA copy (complementary) of an mRNA so that either the gene can be cloned without need for the eukaryotic machinery for processing pre-mRNA or so that one can clone specifically the genes that are being expressed]
3. b [A DNA primer provides the starting point for the DNA polymerase to synthesize a complementary strand to the template to which the primer binds]
4. d [DNA polymerase replicates more copies of the template DNA using successive rounds of denaturation, annealing of the primers and polymerization of new DNA]
5. d [It separates molecules based on shape and charge and is used for nucleic acids or proteins]
6. c [If DNA have single-stranded overhands they will base pair with complementary, single-stranded sequences, that is, hey would behave as if they were sticky, if provided with the appropriate partner molecules. Many restriction enzymes generate sticky ends that can then base pair with foreign DNA with similar sticky ends]
7. a [Current research is aimed at discovering the role of these sequences, however this is a phenomenon that seems to exist through the eukaryotic world, based on those organisms whose genome has been sequenced]
8. a [The rest includes intro sequences, intergenic sequences that may be involved in regulation of gene expression, as well as a significant amount of non-coding sequences of unknown function]
9. a [A diploid cell derived from the mammary gland of a sheep was fused with an enucleated unfertilized egg]
10. a [This enzyme is an RNA-dependent DNA polymerase produced by retroviruses, which are RNA viruses that convert their chromosome to DNA]
11. b [retroviruses are a class of RNA viruses that convert their RNA chromosome backwards to DNA using an enzyme that they bring into the host upon infection. The DNA copy of their chromosome is then integrated into the host's DNA chromosome]
12. a {Kary Mullis won the Nobel Prize in 1993 for this]
13. a [Some good examples are found in work with mice, for example in the case of the use of germ-line gene therapy to give rise to normal-sized mouse pups from the cells of mutant dwarf mice]
14. a [The need to process pre-mRNA with eukaryotic genes precludes using genomic sources of eukaryotic genes to clone into bacteria. Mature mRNA contains the processed gene sequence and can be reverse transcribed to a cDNA copy which then can be cloned into a bacterial host cell for large-scale production]
15. c [germ-line gene therapy has been successfully used with animals but is not allowed with humans]
16. d [They are particularly useful and of interest for their ability to produce and express, through milk secretion, high levels of easily purified proteins with pharmaceutical applications]

17. d [It is not uncommon and in fact a champion milk-producing dairy cow has been successfully cloned and a commercial enterprise is now selling these animals]
18. d [The Ti factor of the Ti plasmid from *Rhizobium radiobacter* is one common tool for introduction DNA into a plant's genome. This sequence inserts itself and gives rise to a tumour or gall, however in using this for the genetic engineering of plants, the gene you want to introduce is cloned into the middle of the T- or tumour-inducing sequence so that it is no longer able to cause disease]
19. c [That is these are projects directed by pharmaceutical companies for the biological, large-scale production of proteins of interest to pharmaceutical companies]
20. d [Craig Venter and his colleagues were the first to sequence the genome of an organism, the bacterium *Haemophilus influenza*, it was his genome that was first sequenced in the Human Genome Project and he and colleagues traveled by ship through the Atlantic, Panama Canal and Pacific to collect sea water from which was isolated and sequenced DNA from microbial organisms.]

Integrating and Applying Key Concepts

1. Read section on gene therapy paying particular attention to the problems that have arisen with somatic gene therapy. Now consider the differences between the process of somatic gene therapy and germ-line gene therapy and the differences in outcomes.
2. Read section 16.2d to review the discussion of GMOs, genetically modified organisms, for a discussion of the concerns, possible problems and oversight governing GMO production and commercialization.
3. Reread the section on the features of the human genome sequences and, if necessary, the relevant sections in 14.3 and 15.3a on processing of human mRNA.

Chapter 17 Microevolution: Genetic Changes Within Populations

Why it matters [pp. 373-374]

1. Penicillin; 2. resistant; 3. microevolution; 4. Species.

17.1 Variation in Natural Populations [pp. 374-377]

5. phenotypic; 6. quantitative; 7. qualitative; 8. polymorphism; 9. bar; 10. curve; 11. variation; 12. frequency; 13. The hydrangea produces flowers that are pink if grown in alkaline soil and blue flowers when grown in acidic soil; 14. Researchers investigating the basis for increased wheel running behavior and increased running speed in certain house mice, were able to show, by selective breeding, that the trait had a genetic basis; 15. It is important to determine if differences such as the flower colour differences in hydrangeas have a genetic basis – if they do, then the genes for the trait (flower colour) will be passed to the offspring and evolution can act select for or against the trait. If the observed differences are based on the environment then mutations or shuffling of alleles will not result in any natural selection because these do not control the flower colour; 16. Changes in the codons for a particular amino acid may not change the amino acid encoded due to the redundancy of the genetic code and the fact that there are generally multiple codons that will specify the same amino acid; 17. If the height of a population of people were plotted on a graph against the number of people at each height, depending upon the size of the population, a bar graph showing a bell curve or a bell curve will result with the peak representing the average height of the population and width representing the degree of variation in height; 18. This refers to the existence of discrete variants of a character (e.g. Mendel's garden pea flower colours); 19. This is a heritable change in the genetics of a population; 20. Gel electrophoresis; 21. This is defined as all members of the same species that live together in the same place and time.

17.2 Population Genetics [pp. 377-379]

22. variation; 23. gene pool; 24. genotype frequency ; 25. allele frequency; 26. *p*; 27. *q*; 28. Hardy-Weinberg; 29. equilibrium; 30. observational; 31. null; 32. No mutations are occurring; 33. The population is closed to migration of individuals of other populations (gene flow); 34. The population is infinite in size; 35. All genotypes in the population survive and reproduce equally well; 36. Individuals within the population mate randomly with respect to genotype.

17.3 The Agents of Microevolution [pp. 379-388]

37. microevolution; 38. mutation; 39. gene flow; 40. genetic drift; 41. variability; 42. bottlenecks; 43. founder; 44. alleles; 45. endangered; 46. population; 47. breeding; 48. drift; 49. microevolution; 50. heritable; 51. fitness; 52. stabilizing; 53. directional; 54. disruptive; 55. sexual; 56. intersexual; 57. Intrasexual; 58. dimorphism; 59. B; 60. C; 61. D; 62. E; 63. A.

17.5 Adaptation and Evolutionary Constraints [pp. 391-392]

64. genetic; 65. phenotypic; 66. diploid; 67. heterozygous; 68. polymorphism; 69. fitness; 70. heterozygote advantage; 71. malaria; 72. variable; 73. fitness; 74. frequency – dependent; 75. neutral variation; 76. selectively neutral; 77. bottleneck; 78. natural selection; 79. adaptive; 80. adaptation; 81. changing; 82. existing; 83. B; 84. E; 85. D; 86. A; 87. C.

SELF-TEST

1. b [A recessive trait is, by definition, masked in heterozygotes, so natural selection cannot act on these individuals. In the case of an undesirable allele, this means that it will not be eliminated from the population]
2. d [In the case of colouration, for example, genetic factors may determine the colour (e.g. Mendel's garden peas) or environmental factors may be involved, as is the case with hydrangeas and the determination of flower colour based on soil pH]
3. c [For example crops that grow quickly and taller may do so because of genetic traits, and these can be selected for in selective breeding programs, or may simply be the result of growth conditions (nutrient availability, moisture levels]
4. d [Alleles are different variants of a genetic trait and so changes arise by changes in the DNA sequence, i.e. mutations. Sexual reproduction results in the fusing of the male and female gamete, thus giving new combinations of alleles, based on those inherited from the mother and the father]
5. b [Genetic drift is the change in allele frequencies due to chance events and it can be seen in a population that has undergone a population bottleneck, where there has been a massive die-off within a population and alleles are lost]
6. d [This is a measure of reproductive success, which is the essence of natural selection]
7. d [Nonrandom mating will alter genotype frequency but not allele frequency]
8. c [Stabilizing selection is seen when individuals expressing intermediate traits have the highest relative fitness; this therefore eliminates phenotypic extremes]
9. c [This is specifically the result of intersexual selection where extreme traits in males help attract mates and therefore results in a higher relative fitness]
10. a [Some mutations may have no measureable fitness consequences; selection does not produce perfect organisms regardless of human intervention; many traits that evolved for one purpose in an ancestor get co-opted for other purposes in descendents]
11. c [Sexual reproduction only shuffles existing alleles into new combinations]
12. b [The bacterial population is genetically diverse and exposure to antibiotics represents a selective pressure. Within the population will be individuals with varying levels of resistance and those with the highest level of resistance will have the greatest relative fitness.]
13. b [Because of the redundancy of the genetic code, some base changes will not cause a change in the amino acid sequence of the encoded protein]
14. a [The use of antibiotics is a selective pressure that selects for individuals within a population that have a greater relative fitness. It does not alter the genes]
15. a [The gene pool is dangerously low in populations that have undergone a population bottleneck or in cases where a small number of individuals have created a new population – for example in the case of endangered species and captive breeding programs. This has undesired effects]
16. b [As in the case of the sickle cell trait where the heterozygotes have greater relative fitness in areas where malaria is endemic. The result is the maintenance of an otherwise harmful trait, so that both the normal and mutant allele are maintained in those populations]]
17. a [Qualitative traits are those that give rise to a continuum of variation rather than having discrete phenotypes]

18. c [Dwarfism is the result of a mutation so is not represented by a continuum of phenotypes but rather numbers/frequencies of affected individuals]
19. c [These individuals will have already mated and produced offspring]
20. a [Population bottlenecks are an example of genetic drift where there is a high rate of mortality among individuals of a small population; this can eliminate certain alleles or reduce genetic variation with the loss of so many individuals. It is not, however, stabilizing selection because the individuals that succumb are not being selected based on a genotype or phenotype]

INTEGRATING AND APPLYING KEY CONCEPTS

1. Given the relative differences in life-span, think about the relative differences in the reproductive periods of each. Reread section 17.1c.
2. Reread section 17.2b and determine what experimental methods would be used to test each of the 5 criteria for satisfying the Hardy-Weinberg principle. Reading sections 17.1 again may also help.
3. Reread sections 17.1 and 17.2a.

Chapter 18 Speciation

Why it matters [pp. 373-374]

18.1 What's in a Name? [p. 396]

18.2 Definition of "Species" [pp. 396-397]

1. communication; 2. information; 3. species; 4. scientific; 5. signals; 6. classification; 7. Biological; 8. Phylogenetic; 9. Ecological; 10. asexually; 11. fertile; 12. females; 13. hermaphrodites; 14. Gynogenetic; 15. Recombination; 16. mutation; 17. C; 18. A; 19. B; 20. F-although bacteria and archaea reproduce asexually, they are given species names; this is critical when distinguishing between pathogenic and nonpathogenic species; 21. T; 22. F-they can interbreed however their offspring are sterile; 23. Hybridization occurs when members of different species mate and produce fertile offspring; 24. This refers to hybrid offspring that are not only fertile but strong competitors, that is their relative fitness is high.

18.4 Gene Flow: Four Examples [pp. 399-401]

25. organisms; 26. evolution; 27. cohesiveness; 28. flow; 29. isolated; 30. speciation; 31. morphological; 32. Morphological; 33. morphological; 34. fossils; 35. flow; 36. gene flow; 37. gene flow; 38. gene flow; 39. DNA; 40. distribution; 41. hitchhike; 42. genetic; 43. DNA; 44. DNA; 45. social; 46. interbreeding; 47. gene flow; 48. interbreed; 49. Geographic separation prevents gene flow while an increase in habitat continuity allows an increase in gene flow which may then result in a reduction if biodiversity; 50. In the case of house mice, males are less easily assimilated into another population, limiting gene flow. However females are readily assimilated and can interbreed with males of other populations; 51. Gene flow between discontinuous habitats can be accomplished if one organism can hitchhike on another, for example on migrating birds.

18.6 Geographic Variation [pp. 402-404]

52. region; 53. phenotypic; 54. subspecies; 55. phenotype; 56. gene flow; 57. ring; 58. cline; 59. When an uninhabitable region is surrounded by members of a particular species, they may diverge into subspecies with individuals readily interbreeding in the zones of overlap to give hybrids of intermediate phenotypes. If a ring does not completely close, then the subspecies at each end may diverge to the extent that they are no longer able to interbreed; 60. Clinal variation arises when there is continuous variation in environmental conditions over a geographic gradient. These changing environment conditions select for differing alleles and phenotypes which also show a gradual but continuous change over the geographic gradient; 61. F-they exist as different subspecies based on geographically isolated habitats.

18.7 Reproductive Isolation [pp. 404-407]

62. Reproductive isolating; 63. prezygotic; 64. postzygotic; 65. ecological; 66. Temporal; 67. behavioural; 68. mates; 69. mechanic al; 70. gametic; 71. interspecific; 72. intraspecific; 73. inviability; 74. sterility; 75. breakdown; 76. recombinations; 77. T; 78. F-sympatric

species are ones that occupy the same geographic area but are reproductively isolated; 79. This is thought to arise through mate choice by females and sexual selection so that they do not mate with and raise offspring that will be inviable, sterile or exhibit hybrid breakdown. This would therefore be the product of natural selection, so that the relative fitness of the females is maximized; 80. This arises when the male gamete must recognize and bind to a specific receptor molecule on the surface of the female gamete prior to fertilization. Incompatible gametes are unable to do this due to the presence of molecularly distinct gamete receptors; 81. Mechanical isolation in flowers may physically prevent access by inappropriate pollinators or may have colours which are not recognized by inappropriate pollinators; 82. A-ecological, B-mating times differ, for example if one species mates during a different month than another, or at a different time of day, C-Different species recognize appropriate mates based on species-specific courtship displays, D-when reproductive structures are incompatible, E-gametic isolation; 83. A-hybrid inviability, B-hybrid sterility, C-the offspring of the hybrids have reduced viability due to the accumulation of chromosomal abnormalities or genetic recombinations.

18.10 Back to the species Concept in Biology [p. 416]

84. allopatric; 85. secondary; 86. hybrid; 87. reinforcement; 88. Parapatric; 89. phenotypes; 90. Sympatric; 91. host; 92. host; 93. flow; 94. genetic; 95. contact; 96. inviability; 97. reinforce; 98. mechanical; 99. plants; 100. polyploidy; 101. meiosis; 102. autopolyploidy; 103. tetraploids; 104. reproductively; 105. sterile; 106. allopolyploidy; 107. double; 108. diploid; 109. tetraploid; 110. species; 111. Para; 112. Auto; 113. Allo; 114. Sympatric; 115. Allo; 116. F- natural selection cannot select for reproductive isolation between allopatric populations as they must be able to together in the same location to allow interbreeding and selection to select against hybrid offspring and select for mechanisms that prevent fertilization. Selection will therefore occur upon secondary contact; 117. T; 118. Genetic changes such as inversions may cause speciation because of failure of chromosomes to pair (e.g. if the region that suffered the inversion includes the centromere); 119. Species clusters occur when founding populations from the same parent species repeatedly colonize an isolated land mass (e.g. an island) and each evolves reproductive isolation mechanisms to give discrete, but closely related species; 120. A tetraploid organism will produce $2n$ gametes while a normal diploid organism will produce haploid gametes. When the gametes from these two organisms fuse, the zygote will be $3n$ and while mitosis of somatic cells may occur normally, meiosis cannot reduce $3n$ germ line cells: there will be an odd number of chromosomes.

SELF-TEST

1. a [The biological species concept is based on an individual's ability to interbreed in nature; the morphological species concept is based on anatomical traits; the phylogenetic species concept is based on analysis of a phylogenetic tree]
2. b [The morphological species concept is based on anatomical, i.e. morphological traits and although an old concept, is still used by biologists in the field and by paleontologists]
3. d [Some biologists use "race" and "subspecies" synonymously]
4. a [This is a pattern of smooth variation along a geographic gradient]
5. c [Examples are the vocalizations, colour and behaviour used by song birds to attract females of their species]
6. d [Another example would be the opposite coiling of some species of snails, preventing them from getting close enough to mate]
7. a [A prezygotic isolating mechanism would prevent fertilization of the egg]
8. a [Allopatric speciation requires geographical isolation of populations. The other choices do not]
9. d [Sympatric populations are those found in the same geographic area; because polyploidy results in the reproductive isolation of the organism (usually a plant), they can mate by self-fertilization or with other plants of the same ploidy number]
10. a [b, c and d are true. Autopolyploidy is rare in animals because most animals are incapable of self-fertilization]
11. d [Habitat can be discontinuous and interfere with gene flow; the physical conditions of a habitat may vary sharply and influence gene flow; behavioural differences such as species-specific courtship displays will discourage gene flow between the different species as females of one species will not recognize the courtship display of a male of a different species]
12. b [A cline is a smooth variation in a particular trait such as the length of appendages, as a result of a gradual variation in some environmental condition (such as average temperature)]
13. c [An example of this situation is seen with the different subspecies of the salamander *Ensatina eschscholtzi* that surround California's Central Valley]

14. a [A cline is a smooth variation in a particular trait such as the length of appendages, as a result of a gradual variation in some environmental condition (such as average temperature)]
15. a [An example of this would be the various closely-related species of duck where they can breed and give fertile, viable hybrids but in the wild sexual selection has given rise to morphologically distinct males and females breed exclusively with males of their own species, exhibiting behavioural isolation]
16. b [Polyploid plants often display hybrid vigour and can be created through the use of chemicals by plant breeders]
17. c [The use of Giemsa stain revealed chromosomal alterations between humans, chimps, orangutans and gorillas with inversions and chromosomal fusions but with preservation of nearly all of the 1000 different chromosomal bands in each]
18. a [Giemsa stains reveal distinctive banding patterns that are reflective of chromosomal structures, as in the case of the comparison of chromosomal similarities and alterations between humans, chimps, gorillas and orangutans. Chromosomal inversions seem to correlate with increased divergence in the genes within the inverted region compared to those outside of the inversion – these can trigger speciation. Although humans and chimps share approximately 99% genetic similarity, the phenotypic/physiological differences between them greatly exceeds the 1% genetic difference between them]
19. a [Sterile hybrids often are more robust than their parent strains, growing faster and taller. They do not, however have a higher relative fitness than their parents: if they are sterile, they have a relative fitness (which is a measure of their ability to produce healthy, viable offspring) of zero. Coffee, plantains, cotton, potatoes, sugar cane, tobacco and wheat are all polyploid crop plants. The wheat strain used for bread may have evolved by hybridization and allopolyploidy of three ancestral wild wheat species thousands of years ago]
20. b [Autopolyploidy and allopolyploidy can result in speciation very rapidly and without the requirement of geographic isolation. In addition, chromosomal alterations such as those seen among the primates can cause speciation without geographic isolation]

Integrating and Applying Key Concepts

1. Reread section 18.7 and make sure you are clear on the underlying mechanisms of prezygotic and postzygotic isolation; also reread section 18.9. Think about what kind(s) of technique(s) you would need to use to identify genetic differences between two species and how you would determine if the reproductive isolation was prezygotic vs. postzygotic.
2. Reread the section on allopatric speciation (18.8a) and examine fig. 18.20, then reread the section on genetic divergence (18.9a). Again, be aware of the molecular or physical basis of prezygotic isolating mechanisms and what sexual selection generates.
3. Based on the biological species concept, distinct species are able to breed with members of their own species to give fertile and viable offspring. Although there are many examples of organisms that do not exactly match these criteria (e.g. gynogenetic organisms) think about what conclusions you would draw if: a) the different populations were able to give fertile fully viable hybrids, and b) if they were not?

Chapter 19 Evolution and Classification

Why it matters [pp. 419-421]

1. conditions; 2. parallel; 3. convergent; 4. deadly.

19.4 From Classification to Phylogeny [p. 424]

5. Systematic; 6. phylogeny; 7. phylogenetic trees; 8. independently; 9. taxonomy; 10. classification; 11. evolutionary; 12. Linnaean; 13. taxonomic; 14. taxon; 15. domain; 16. morphology; 17. subcellular; 18. organ; 19. sequences; 20. E; 21. G; 22. B; 23. H; 24. D; 25. A; 26. I; 27. F; 28. I; 29. H; 30. J; 31. K; 32. C; 33. G; 34. A; 35. D; 36. B; 37. F; 38. E; 39. T; 40. F-life on Earth is classified into 3 domains, which includes the *Eukarya*; 41. Phylogenetic trees are pictorial representations of the proposed evolutionary relationship between the organisms on the tree. These relationships cannot be proved but they can be tested and disproved and revised based on available data.

19.6 Phylogenetic Inference and Classification [pp. 427-429]

42. genetic; 43. independent; 44. Homologous; 45. genetic; 46. analogous; 47. genetic; 48. mosaic; 49. ancestral; 50. derived; 51. lineage; 52. outgroup; 53. monophyletic; 54. polyphyletic; 55. paraphyletic; 56. assumption of parsimony; 57. derived; 58. evolutionary; 59. morphological; 60. Cladistics; 61. clade; 62. cladogram; 63. parsimony; 64. evolutionary relationships; 65. G; 66. K; 67. M; 68. A; 69. C; 70. J; 71. D; 72. E; 73. L; 74. H; 75. B; 76. I; 77. F; 78. F-it considers only those traits that are shared among the group or organisms, that is derived traits. Traditional evolutionary systematics is based on shared ancestral and derived traits; 79. T; 80. F – they evolved separately in all three so they are analogous structures, however because they evolved form the same ancestral structure, certain aspects, e.g. the bones, are homologous; 81. F – it does not use any of these but classifies based solely on clades ; 82. Homologous characters emerge from comparable embryonic structures and grow in similar ways during development, presumably because evolution has conserved the pattern of embryonic development; 83. Derived traits are those that are new and are heritable therefore shared by all subsequent descendents so serves as a marker for an entire evolutionary lineage whereas ancestral traits do not provide this kind of information unless one were to move backwards in evolutionary time and determine at which point that particular trait evolved, thus determining when it was "derived"; 84. A- derived, B- ancestral, C- ancestral, D- ancestral, E-insects.

19.9 Putting it Together [pp. 435-436]

85. protein; 86. characters; 87. shared; 88. environmental; 89. quickly; 90. slowly; 91. mosaic; 92. constant; 93. divergence; 94. distant; 95. recent; 96. clocks; 97. fossil; 98. divergence; 99. align; 100. phylogenetic trees; 101. maximum likelihood; 102. derived; 103. four; 104. tree of life; 105. Carl Woese; 106. domains; 107. Protista; 108. F-these sequences evolve quickly so are more useful for determining evolutionary divergence that has arisen in the last few million years; 109. F-while the prokaryotes were assigned to a polyphyletic kingdom (Monera), this problem was not recognized at the time. It was, however, always recognized that the kingdom Protista was polyphyletic; 110. Because of its critical role in the very complex process of translation and the fact that all life forms perform this process, this molecule allows the comparison of evolutionary relatedness among even the most diverse organisms and, because it evolves very slowly, any sequence differences can be inferred as being derived traits with the degree of sequence divergence providing an accurate measure of evolutionary time since the organisms being compared diverged; 111. This refers to differences in DNA sequences that provide a time of evolutionary divergence.

SELF-TEST

1. c [The taxon above the species level is Genus, which is more inclusive than species but less inclusive than the next higher level, Family]
2. a [The two major goals of systematics is to determine the phylogeny (i.e. evolutionary relatedness of organisms) and to identify and name organisms so that they can be placed within a classification scheme]
3. d [Analogous characters would be such things as wings, which have evolved four separate times, with bats, birds, insects and pterosaurs]
4. d [Fossil evidence indicates that the earliest animals lacked backbones whereas the comparison of the Monarch butterfly, with its four walking legs to the majority of other insects suggests that 6 walking legs is the ancestral character and four walking legs is the derived (more recently-evolved) trait]
5. b [Traditional phylogeny as been criticized as lacking clarity because it classifies based on two separate phenomena: branching evoluation and morphological divergence. Cladistics is a more recent approach for classification that is based solely on evolutionary relationships, using derived characters as a basis for classification]
6. d [Because DNA is inherited, either the genes or DNA sequences themselves can be used, or the RNA or proteins that the genes encode]
7. b [If comparing gene X from two organisms, if the base at position 200 is the same between the two organisms and a third presumed outgroup, based on that alone, one cannot determine if the change represents a shared derived trait or if it evolved separately as there are only four possible bases that could occupy that position meaning that the chances are high for independent evolution of the same change]

8. c [The traditional tree of life consisted of five kingdoms: Monera (thought to be a monophyletic taxon containing the bacteria), Animalia, Fungi, Plantae and Protista (known to be polyphyletic) whereas the tree of life determined by Carl Woese using rRNA sequence data consisted of three domains: Bacteria, Archaea and Eukarya with the latter containing monophyletic branches for the animals, plants, fungi and a variety of new lineages representing organisms that had previously been characterized as Protista]
9. c [The stapes of the inner ear of four-legged vertebrates evolved from the hyomandibula, a bone that supports the jaw joint of most fishes]
10. b [A monophyletic taxon is one that includes an ancestral species and all of its descendants]
11. c [A paraphyletic taxon is one that includes the ancestral species and some but not all of its descendants]
12. a [The node where the two lineages branch towards A and B represents the common ancestor]
13. d [Species I is the most distantly related so would be the best choice for the outgroup, sharing only ancestral characters]
14. d [Cladistic phylogenies are based only on evolutionary relatedness, not morphological divergence, therefore contains only monophyletic clades. Some scientists that are supporters of cladistics maintain that we should dispense with traditional nomenclature and use a Phylocode, which is based strictly on clades]
15. a [They share the trait of trapping and consuming insects for a source of nitrogen however they use a variety of methods for trapping and, based on molecular data, have evolved multiple times with some more closely related than others]
16. d [These three aquatic vertebrates have evolved separately, despite having the same fishlike body. They are all vertebrates but dolphins and tuna have bony skeletons, sharks have cartilaginous skeletons and dolphins are mammals. Depending upon how distantly related they are, one could consider these analogous body styles as being the product of parallel or convergent evolution]
17. b [While this fossil had a flattened vertebral tail, the skull is quite distinct from that of the Canadian beaver, suggesting that the name and flattened tail notwith (most specific), is: Domain, Kingdom, Phylum, Class, Order, Family, Genus, Species. The molecular analyses of Carol Woese revealed that the prokaryotes are subdivided into two domains: Bacteria and Archaea]
19. d [Homologous characters are those shared traits that indicate a common genetic basis, inherited from a common ancestor and generally evident during embryonic development due to the conservation, by evolution, of embryonic developmental patterns. Since these have a genetic basis, they are not subject to variation based on environmental conditions]
20. a [This is based on the Principle of Parsimony, that is the least evolutionary changes to give rise to a particular lineage is the most likely explanation]

Integrating and Applying Key Concepts

1. Think about the morphological characteristics of prokaryotes compared to plants, animals and even fungi. If this doesn't help, reread section 19.8 and think about how phylogenetic classifications were determined from the time of Linnaeus to the time of Carol Woese.
2. Think about the requirements of a molecular clock, the intricacies of translation and the source of, and variation of mRNA (especially among eukaryotes). You might also think about the nature of the genetic code (how many codons are there compared to how many amino acids?) You may need to reread sect. 19.7 and even sect. 14.3 and 14.4.
3. Reread the section on p. 430 about *Constructing a Cladogram*. You may also need to refresh your memory on what an outgroup is (p. 427) and the differences between ancestral and derived traits between a specific group of organisms (p. 426).

Chapter 20 Darwin, Fossils and Developmental Biology

20.3 Charles Darwin [pp. 442-445]

1. Aristotle; 2. observation ; 3. experimentation; 4. inductive; 5. biogeography; 6. geology; 7. Le Comte de Buffon; 8. vestigial; 9. fossils; 10. environments; 11. use and disuse; 12. acquired characteristics; 13. change; 14. environment; 15. mechanisms; 16. gradualism; 17. Charles Lyell; 18. uniformitarianism; 19. glaciers; 20. Charles Darwin; 21. populations; 22. reproduce; 23. generations; 24. Alfred Russell Wallace; 25. On the Origin of Species by Means of Natural Selection; 26. A - Thomas Malthus, B- Proposed that theory of uniformitarianism, C- Presented a parallel theory to that of Darwin July 1, 1958 at the Linnaean Society of London, D- Jean Baptiste de Lamarck, E- Proposed the theory of gradualism, F- Le Comte de Buffon, G- Aristotle, H- Proposed the theory of catastophism; 27. Darwin grew up in the country and was aware of the practice of artificial selection whereby breeders could

breed individuals with desirable traits and enhance those traits in future generations. After reading Malthus's essay on the inevitable starvation of individuals in England as a result of food shortages, he proposed that in nature, those animals with traits which provided a better advantage in a given environment would surviv and reproduce while others would not; 28. Darwin argued that all organisms that have ever lived arose through descent with modification, that is evolved, from a common ancestor.

20.4 The Fossil Record [pp. 445-447]

29. Paleontology; 30. mineralization; 31. oxygen; 32. acidity; 33. teeth; 34. leaves; 35. minerals; 36. moulds; 37. decomposition; 38. incomplete; 39. radiometric dating; 40. half-life; 41. ^{14}C: ^{12}C; 42. deeper/older; 43. shallower/younger; 44. F – they do not form in mountain forests as they require flowing water, including rain or run-off that carries fine rocks or soil downstream to a swamp, lake or the sea whereupon the particles settle as sedminets; 45. F-because many types of structures or organisms are soft-bodied and normally do not form fossils, and because fossil-formation requires specific conditions and only occurs in some areas, the fossil record is significantly incomplete; 46. T; 47. Soft-bodied organisms generally undergo microbial decay when they die however this requires the presence of oxygen, which is not present in peat bogs, tar pits and glaciers; 48. Amber is the fossilized resin of coniferous trees and upon formation, it can trap insects, tiny lizards and frogs; these do not decay because of the lack of oxygen.

20.5 Earth History, Biogeography, and Convergent Evolution [pp. 447-450]

49. environmental conditions; 50. tectonics; 51. mantle; 52. continental drift; 53. asteroid impacts; 54. Biogeography; 55. historical; 56. continuous; 57. disjunct; 58. dispersal; 59. vicariance; 60. Pangaea 61. biota; 62. realms; 63. endemic; 64. morphological; 65. convergent; 66. Biogeography; 67. T; 68. T; 69. T; 70. continuous; 71. The common torpedo shape of many fish and aquatic mammals is the result of convergent evolution and is due to the fact that this form facilitates movement in water.

20.7 Macroevolutionary Trends [pp. 455-457]

72. linear; 73. extinct; 74. intermediate; 75. anagenesis; 76. cladogenesis; 77. gradualist 78. punctuated equilibrium; 79. cladogenesis; 80. fossil; 81. complexity; 82. structures; 83. Edward Drinker Cope; 84. preadaptation; 85. allometric; 86. heterochrony; 87. neoteny; 88. pedomorphosis; 89. reproduce; 90. juvenile traits; 91. pedo; 92. hetero ; 93. allo; 94. clado; 95. D; 96. E; 97. B; 98. A; 99. C; 100. T.

20.9 Evolutionary Biology since Darwin [p. 462]

101. Evolutionary developmental biology; 102. sequence; 103. morphological; 104. Homeotic; 105. form; 106. genetic tool-kit; 107. *Hox*; 108. homeobox; 109. homeodomain; 110. head-to-tail; 111. tool-kit; 112. light-sensing; 113. levels; 114. times; 115. development; 116. switches; 117. allometric growth; 118. heterochrony; 119. F – it regulates genes that control the development of spines on the pelvic fins of sticklebacks; 120. T; 121. T; 122. F –tool-kit genes have been preserved for at least hundreds of millions of years, and differ in their activities and time of activation to give rise to substantial differences in body form; 123. Evo-devo is the study of the effects of changes in genes regulating embryonic development and these studies are revealing that such changes often result in changes in body plans which may give rise to adaptive radiations if the new body plan allows the organism to move into a new adaptive zone; 124. Marine populations of three-spined sticklebacks have bony armour along their sides and prominent spines whereas the freshwater descendents of this fish have greatly reduced armour and may lack pelvic spines altogether. In the latter, deactivation of the homeotic gene *Ptx1*during embryonic development of fin buds provides the explanation. This change in expression appears to have been maintained through natural selection because the lack of long spines helps protect the freshwater fish from attack by dragonfly larvae. The gene is active in other areas of the developing fish, suggesting an unidentified mutation is specifically blocking expression of the gene in the fin buds; 125. When fruit flies were genetically engineered to express the *pax-6* gene of a squid or a mouse, its eyes developed normally, indicating that the gene has been maintained unchanged since the three lineages leading to squid, mice and fruit flies diverged.

SELF-TEST

1. a [By the 14th century, Europeans had merged Aristotle's classification with the biblical account of creation, and sought to name and catalogue all of God's creations]
2. c [He proposed that organisms were conceived by Nature and produced by Time]
3. a [Georges Cuvier proposed that abruptly changing strata were the result of a dramatic shift in ancient environments and that the fossils within those strata were ones that died in a local catastrophe, such as a flood]
4. b [Although he proposed the idea of use and disuse, the idea behind this was that organisms change in response to their environment, an underpinning for Darwin's theory of natural selection]
5. c [Malthus wrote "Essays on the Principles of Population" that discussed the fact that the rate of increase in the England's population was outpacing its agricultural capacity and predicted that with limiting food some people would inevitably starve]
6. b [Fossils are associated primarily with sedimentary rock, formed when fine rocks and sediment are carried by rain or run-off into lakes, rivers, bogs or oceans, where they settle and gradually get buried by the increasing deposition of sediment. They do not form in the presence of oxygen as decomposition is increased with exposure to air, especially for soft-bodied organisms. Geological processes such as volcanic eruptions cause erosion and tend to destroy fossils]
7. a [Radioactive isotopes are unsteady and decay at a fixed rate that is characteristic of the particular isotope, therefore the amount of a particular isotope in a stratum gives a measure of the age of the stratum and the fossils within that stratum. The amount of carbon 14 is constant while an organism is living but starts to decrease once it dies. Therefore the ratio of $^{14}C:^{12}C$ gives a measure of the age of a fossil as long as there is still organic material associated with it]
8. b [This kind of distribution is characterized by the existence of widely separated populations and is explained by the fact that the populations were previously part of one big land mass, e.g. Pangea, which then separated, separating with them, the population}
9. d [The other realms are the Nearctic, Palearctic, and Oriental]
10. c [Tracing the evolution of horses reveals a complex pattern of branching with most branches becoming extinct]
11. d [This hypothesis suggests that most species experience long periods of stasis or morphological equilibrium punctuated by brief periods of rapid cladogenesis]
12. c [This observation is most applicable with vertebrates which concomitantly show an increase in morphological complexity]
13. a [Preadaptation is the phenomenon where a structure that has one adaptive role in one environment becomes coopted for a different role in a different habitat. The best example is that of wings and feathers which are believed to have existed prior to the evolution and adaptive radiation of birds, in certain dinosaurs for capturing prey and maintaining body heat, respectively]
14. b [In humans, the head, torso and limbs all grow at a different rates.]
15. c [For example the *Hox* genes are universal among animals and are responsible for the development of the body form.]
16. c [The first two are inferences he made based on observations while the last is his hypothesis of evolutionary change based on natural selection]
17. d [Earth's mantle is semisolid and flows in currents, causing the large rigid plates of Earth's crust and the continents embedded in them, to drift. The merging of these masses into the supercontinent Pangea followed by its separation into smaller continents provides the vicariance the underlies disjunct distributions]
18. c [For example many of the species on the continent of Australia, which has been geographically isolated for approximately 55 million years]
19. b [Their lineages arose independently long after their respective continents had separated]
20. a [This process is the evolutionary transformation of an existing species so does not give rise to an increase in the number of species however if morphological changes are large, the organisms may be given different names at different times in their history]

INTEGRATING AND APPLYING KEY CONCEPTS

1. Study Fig. 20.21 and reread section 20.8a.
2. Reread sections 20.5a and 20.5b.

3. Reread section 20.8a, paying particular attention to the experiments with genetically-engineered fruit flies.

Chapter 21 Prokaryotes

21.2 Prokaryote Structure and Function [pp. 466-475]

1. bacteria; 2. disease; 3. antibiotics; 4. prokaryotes; 5. domains; 6. diverse; 7. diversity; 8. organelles; 9. circular; 10. nucleoid; 11. plasmid; 12. horizontal; 13. smaller; 14. antibiotics; 15. cell wall; 16. cell wall; 17. peptidoglycan; 18. positive; 19. peptidoglycan; 20. negative; 21. peptidoglycan; 22. membrane; 23. Gram; 24. capsule; 25. pathogens; 26. flagella; 27. pili; 28. plasmids; 29. electricity; 30. energy; 31. chemoheterotrophy; 32. photoautotrophy; 33. chemoautotrophs; 34. photoheterotrophs; 35. requirement; 36. aerobes; 37. anaerobes; 38. obligate; 39. facultative anaerobes; 40. biogeochemical; 41. fixation; 42. nitrification; 43. proteins; 44. binary; 45. mutation; 46. horizontal; 47. toxin; 48. endotoxin; 49. lipopolysaccharide; 50. H; 51. A; 52. D; 53. E; 54. G; 55. I; 56. B; 57. F; 58. N; 59. J; 60. L; 61. C; 62. O; 63. K; 64. M; 65. A- Mutation, B- When DNA is absorbed by bacterial cells from their environment to cause genetic recombination, C- When DNA is passed from one bacteria to another through a physical contact to cause genetic recombination, D- Transduction; 66. Chromosomes; 67. plasmids; 68. T; 69. 1%; 70. The prokaryotes existed on Earth for 3 billion years before the eukaryotes evolved; 71. It protects against the entry of certain toxic chemicals such as the penicillins; 72. They can be used for bioremediation and sewage treatment but they may also be detrimental to human health such as the biofilms that form on teeth as plaque or on pacemakers and artificial joints; 73. They may experience a mutation so that the antibiotic no longer binds and inhibits its target; they may actively pump the antibiotic out of the cell or they may enzymatically alter the structure of the antibiotic so that it loses its effect; 74. A- Chemoautotroph/Lithotroph, B- Obtain energy by oxidizing organic molecules and use already prepared organic chemicals as their carbon source, C- Use light as their energy source and CO_2 as their carbon source, D- Photoheterotrophs, E- Facultative anaerobes, F- Obligate anaerobes, G- Use oxygen as the electron acceptor for cellular respiration and cannot grow in its absence.

21.4 The Domain Archaea [pp. 478-480]

75. evolutionary; 76. proteobacteria; 77. purple; 78. photoheterotrophs; 79. chlorophyll; 80. pathogens; 81. proteobacteria; 82. green bacteria; 83. photoheterotrophs; 84. oxygen; 85. cyanobacteria; 86. chloroplast; 87. oxygen; 88. nitrogen; 89. Gram $^+$ bacteria; 90. *Lactobacillus*; 91. peptidoglycan; 92. intracellular; 93. extremophiles; 94. eukaryotes; 95. lipids; 96. Korarchaeota; 97. Euryarchaeota; 98. halophiles; 99. methanogens; 100. thermophiles; 101. Crenarchaeota; 102. psychrophiles; 103. thermophiles; 104. mesophilic; 105. plankton; 106. C; 107. B; 108. F; 109. D; 110. A; 111. E; 112. F-They have a different kind of photosynthetic pigment than plants and do not produce oxygen. Some of them are even photoheterotrophs; 113. It is a spiral-shaped bacterium that causes syphilis; 114. T; 115. They are the same size, both smaller than eukaryotic ribosomes however some of the components of the archaeal ribosome are the same as found in eukaryotic ribosomes so that antibiotics do not target archaeal ribosomes; 116. PCR/the polymerase chain reaction was made possible by the isolation of the thermophilic archaea *Thermus aquaticus* and its thermostable DNA polymerase.

SELF-TEST

1. d [The vast majority of prokaryotes cannot grown using currently available laboratory conditions; in addition, prokaryotes constitute the majority of biomass in the world's oceans, which represents approximately 70% of the Earth's surface and much of this is inaccessible]
2. c [Prokaryotes existed for approximately 3 billion years before the appearance of the eukaryotes]
3. d [The chromosome is condensed in the region known as the nucleoid but they do not have any internal, membrane bound organelles]
4. c [The have a single, generally circular chromosome and many strains will also have multiple copies of small, circular, plasmid DNA]
5. d [Using the Gram staining technique, Gram negative bacteria stain with the second stain, that is they stain pink or red]
6. d [The capsule is a polysaccharide layer that may surround some strains of bacteria and archaea and which may help in pathogenesis or protect against dessication, bacterial viruses or the penetration of antibiotics]

7. d [Prokaryotic viruses are made of a helical arrangement of a protein called flagellin and they move by rotation, much like a propeller]
8. d [Conjugation, that is the movement of plasmids between bacterial cells occurs through the hollow passage of the sex pili that join two cells during conjugation. Pili are also important in attachment, either during biofilm formation or, in the case of *Neisseria gonorrhoeae*, to host genitourinary tract cells. Finally, in the bacterium *Geobacter*, the pili can be used to pass electrons to another cell or conducting surface, generating a current]
9. a [Facultative anaerobes may function aerobically when oxygen is present but if it is not, they generate energy anaerobically]
10. c [It is a biological process that is strictly limited to small numbers of prokaryotic species and is an essential component of the biogeochemical cycling of nitrogen in the environment. This is because nitrogen gas is a large component of the atmosphere however its conversion to ammonia, through nitrogen fixing bacteria is what provides bioavailable nitrogen for all other organisms on Earth]
11. a [Binary fission is the process of replication in prokaryotes where the cell grows to twice its initial size then divides in half to give two genetically identical daughter cells. Their tremendous genetic variability is due to an extremely high reproductive rate combined with an very high mutation rate, and to a lesser extent, horizontal gene transfer. Because they represent a huge proportion of the biomass on Earth, colonizing virtually every niche on the planet, their numbers alone provide a basis for their genetic diversity compared to other life forms]
12. c [Lipopolysaccharide is the molecule of the Gram negative outer membrane that contains endotoxin. Specifically the Lipid A portion of this complex molecule causes severe symptoms when the lipid is exposed to human immune cells, giving rise to an overwhelming immune response that can be deadly]
13. a [Chemoautotrophs are also known as lithotrophs, literal translation, "rock-eaters". They are the critical base of deep sea vent food chains]
14. b[For example the green bacteria and the purple proteobacteria both contain photosynthetic bacteria that are photoheterotrophic, rather than photoautotrophic]
15. e [Bacteria have evolved multiple ways to resist the effects of antibiotics, some of which are encoded by genes on conjugative or transmissible plasmids. The overuse or inappropriate use of antibiotics provides increasing selective pressure for the growth and spread of resistant bacteria]
16. a [They can be used in bioremediation of toxic areas, removing toxin from an environment and they are a critical component of sewage treatment facilities]
17. c [They may be pathogenic or non-pathogenic, are named for their spiral shape however it is the cyanobacteria that form a symbiotic relationship with fungi to give rise to lichens]
18. a [The Korarchaeota have not yet been cultured in the lab]
19. b [Halophiles are associated with places like Great Salt lake, in Utah, and the Dead Sea. They can grow in fully saturated salt solutions]
20. a [Psychrophilic archaea are a dominant component of the cold depths of the world's oceans and the Antarctic, especially the Antarctic lakes]

Integrating and Applying Key Concepts

1. Think about the relationship between the size of an organism's genome and its coding capacity. If these organisms evolved to live permanently-associated with animal hosts, knowing that prokaryotes have a very high mutation rate, what kinds of mutations might have occurred that would have very little impact given the niche these bacteria occupy? Would there have been a benefit to undergoing such mutations?
2. Remembering that most microbes live in moist environments and looking at the cyanobacteria depicted in Fig. 21.17, would organisms that do not produce oxygen and are photosynthetic be likely to require oxygen? If not, relative to the cyanobacteria in this figure, where might purple and green bacteria live so that there is minimal oxygen, water and sunlight?

3. Think about the cell wall structure of bacteria, especially Gram positive bacteria such as *Streptococcus pneumonia*. This cell wall is critical to the survival of these bacteria. Now think about the cellular structure of animal (human) cells. You may want to check Table 21.1 to help you if the answer is still not clear. If this antibiotic is lethal to Streptococcus and other Gram positive bacteria, how might this be an advantage to other microorganisms in nature, such as specific fungi or antibiotic-producing bacterial species?

Chapter 22 Viruses

22.4 Viruses May Have Evolved from Fragments of Cellular DNA or RNA [p. 492]

1. *nucleic acid*; 2. *capsid*; 3. *envelope*; 4. *energy*; 5. *living*; 6. *RNA*; 7. *DNA*; 8. *single*; 9. *double*; 10. *recognition*; 11. *helical*; 12. *polyhedral:* 13. *plant*; 14. *animal*; 15. *envelope*; 16. *orders*; 17. *genera*; 18. *species*; 19. *diversity*; 20. *animals*; 21. *plants*; 22. *bacteria*; 23. *bacteriophages*; 24. *tail*; 25. *virulent*; 26. *temperate*; 27. *virulent*; 28. *lysis*; 29. *lytic*; 30. *lambda*; 31. *lysogenic*; 32. *prophage*; 33. *lytic*; 34. *latent*; 35. *fusion*; 36. *endocytosis*; 37. *lysis*; 38. *plasma membrane*; 39. *release;* 40. *expression*; 41. *envelope*; 42. *injuries*; 43. *insects*; 44. *seeds*; 45. *plasmodesmata*; 46. *bacterial*; 47. *release*; 48. *reverse transcriptase*; 49. *mutation*; 50. *E*; 51. *M*; 52. *F*; 53. *A*; 54. *K*; 55. *B*; 56. *J*; 57. *L*; 58. *N*; 59. *I*; 60. *C*; 61. *H*; 62. G; 63. D; 64. T; 65. F – they rely on the host's cellular machinery for their replication and growth; 66. F-there are very few drugs for the treatment of viral infections and these are all specific for viral structures or processes whereas antibiotics target specific bacterial structures or processes; 67. Generalized transduction is done by bacteriophage that are lytic viruses; they sometimes randomly incorporate pieces of the host's chromosome when they assemble and then are released by lysis. Specialized transduction is done by certain bacteriophage that have a lysogenic life cycle; they become incorporated into the host's chromosome at a specific locus and when they excise in order to revert to the lytic life cycle, the excision may be imprecise so that the virus incorporates one or two genes from either side of the site of insertion; 68. The virus has 8 chromosomes and when an animal host is infected by more than one strain, the chromosomes may randomly assort into new strains with new combinations of the parental chromosomes. In addition, they have a very high rate of mutation so that, between the two processes, their coat proteins vary and look different to the human immune system; 69. They may have emerged from the primordial soup before life evolved or they may have evolved from random "escaped' fragments from living cells.

22.5 Viroids and Prions Are Infective Agents Even Simpler in Structure than Viruses [pp. 492-494]

70. *RNA*; 71. *crop plants*; 72. *RNA*; 73. *proteins*; 74. *spongiform encephalopathies*; 75. *BSE*; 76. *scrapie*; 77. *kuru*; 78. *Creutzfeldt-Jakob disease*; 79. *misfolding*; 80. Virus - RNA, DNA, single-stranded or double-stranded, proteins, +/- envelope, viroid – single-stranded RNA, no protein, no envelope, prions – no genetic material, single, misfolded protein, no envelope; 81. BSE arose in cattle, possibly through the consumption of scrapie-contaminated rendered cattle feed. vCJD arose in humans through the consumption of BSE-contaminated beef.

SELF-TEST

1. a [Bacteria are infectious life forms, the others are infectious but not life forms]
2. d [The Spanish 'flu was a pandemic in 1918 that killed approximately 1 in 20 people and the basis for its extreme virulence is still not well understood]
3. a [Viruses contain protein coats, genetic material and may be surrounded by an envelope but they do not contain any of the structures typical of cells, like the ribosomes that are used for protein synthesis]
4. e [Most viruses infect only one or a few closely related host cell types or organisms, they acquire their envelope as they pass through the host's membrane. They are essential in the cyanobacterial regions of the world's oceans and they are believed to keep the growth of bacteria in check so that the world is not overrun by bacteria. Because viruses are so prevalent, can insert into the host's chromosome and may not give rise to symptoms or even be active]
5. b [They are virulent, infecting *E. coli*, and emerging by lysis. They do not have an envelope and they do not have a lysogenic phase]
6. a [Certain lytic viruses may accidentally package a random fragment of the host's chromosome and transfer the genes from the first host cell to the next host cell]

7. c [Bacteriophage lambda inserts into a specific site on the *E. coli* chromosome and occasionally excises inaccurately so that it packages a host gene located on one side or the other of the site of insertion, taking that gene from the original host to the next host cell]
8. d [It is an enveloped virus so picks up the envelope when it passes through the host's membrane. It does, however, indirectly cause lysis of host cells by rapid release giving rise to damage to the host's membrane]
9. a [It has two single-stranded RNA chromosomes]
10. c [zanamavir is an anti-flu antiviral drug that is commercially available]
11. d [These are the two prevailing theories for the evolution of the viruses; viruses have been suggested by some researches to have infected the ancestor to the eukaryotic cells to give rise the the eukaryotic nucleus]
12. c [They are single-stranded RNA particles that infect plants and are currently thought to act through complementary base-pairing with target host RNA molecules and interfering with RNA processing]
13. d [All of these routes of infection have been documented. vCJD arises from consumption of contaminated meat while CJD has been shown to be transmitted through contaminated surgical instruments and corneal transplants]
14. b [Cows had been fed meal from rendered cow and sheep carcasses for many years however changes in the rendering process, reducing the heating time, is thought to have allowed for entry of active prion proteins into the cows food supply]
15. c [The name prion is based on the descriptive term proteinaceous infectious particles]

INTEGRATING AND APPLYING KEY CONCEPTS

1. Reread the section on "Infection of Animal Cells".
2. Think about the structures of animal viruses and compare this to bacterial cells. Are there any obvious shared structures? If you are thinking possibly a DNA chromosome or a membrane, think about the nature of the host cell structures.
3. Reread section 22.3. and the section that describes the Spanish 'flu pandemic of 1918 in "Why it Matters".

Chapter 23 Protists

23.3 Protists' Diversity is Reflected in Their Metabolism, Reproduction, Structure, and Habitat [pp. 500-502]

1. *Giardia lamblia*; 2. related; 3. Protista; 4. ancestor; 5. 1.5 to 2 billion; 6. mitochondria; 7. chloroplasts; 8. mitochondrion; 9. Animals, plants and fungi; 10. fungi; 11. cellulose; 12. stem, leaves, roots, and seeds; 13. unicellular; 14. diverse; 15. habitat; 16. photoautotrophs; 17. aquatic; 18. host organisms; 19. colonial; 20. cytoplasmic; 21. contractile; 22. flagella; 23. pseudopodia; 24. sexual; 25. F; 26. E; 27. A; 28. B; 29. G; 30. D; 31. C; 32. A – aquatic or moist terrestrial environments or inside host organisms, B-unicellular, colonial or multicellular, motile using flagella, cilia or pseudopodia, unicellular contain unusual structures such as the contractile vacuole, C-heterotrophic, photoautotrophic or a combination of both, D-asexual or sexual; 33. F – they do not all share a common lineage but are grouped together for convenience. Their classification and phylogeny has been extremely problematic and there are many current classification schemes all of which are likely to change as more data (particularly molecular0 becomes available; 34. They may be unicellular or multicellular and range in size from microscopic to the largest organisms on Earth; 35. Sea weeds to not have leaves, stems, roots and seeds.

23.4 The Protist Groups [p. 502-518]

36. common ancestor; 37. ultrastructural; 38. mitochondria; 39. glycolysis; 40. feeding; 41. *Giardia lamblia*; 42. *Trichomonas vaginalis*; 43. heterotrophic; 44. cristae; 45. *Euglena*; 46. *Trypanosoma*; 47. photosynthetic; 48. alveoli; 49. Apicomplexa; 50. Ciliaphora; 51. Dinoflagellates; 52. cellulose; 53. apical complex; 54. flagella; 55. Oomycota; 56. diatoms; 57. Chrysophyta; 58. Phaeophyta; 59. fungi; 60. hyphae; 61. potato blights; 62. silica; 63. fucoxanthin; 64. blades; 65. stipes; 66. holdfasts; 67. giant kelp; 68. pseudopods; 69. tests; 70. axopods; 71. perforations; 72. projections; 73. pseudopodia; 74. plasmodial; 75. pseudopods; 76. binary fission; 77. *Dictyostelium*; 78. multinucleate; 79. plasma membrane; 80. Rhodophyta; 81. Chlorophyta; 82. Plantae; 83. phycobilins; 84. charophytes; 85. flagellum; 86. animals; 87. D; 88. O; 89. A; 90. L; 91. B; 92. G; 93. I; 94. M. 95. K; 96. N; 97. C; 98. F; 99. E; 100. J; 101. H; 102. A-*Trypanosoma brucei*, B- *Giardia lamblia*, C- Sexually-transmitted diseases that are asymptomatic in men but if untreated may cause infertility in women, D- Malaria, E- Dinoflagellates, F- *Toxoplasma,* G- *Phytophthora infestans,* H- *Entamoeba*

histolytica; 103. Diplo-; 104. Proto-; 105. Dino-; 106. Hetero- ; 107. Foramin-; 108. Rhodo-; 109. Chloro-; 110. Opistho-; 111. Choano-.

23.5 Some Protist Groups Arose from Primary Endosymbiosis and Others from Secondary Endosymbiosis [pp. 518-520]

112. cyanobacterium; 113. Red and green; 114. euglenoids; 115. dinoflagellates; 116. heterokonts; 117. eukaryote; 118. additional.

SELF-TEST

1. b [Protists are a heterogeneous collection of eukaryotes]
2. c [They have been difficult to classify because of their complex evolutionary history. Some are closely related to animals, others to fungi and yet others to land plants]
3. b [The animal-like protists are unicellular and may have complex cytoplasmic structures but do not have complex differentiated structures]
4. c [Many of the photosynthetic protists may switch to the heterotrophic mode of nutrition]
5. b [The contractile vacuole is found in protists like *Paramecium* that live in dilute environments. They do not have cell walls so the excess water is retained in the contractile vacuole, and when it swells to a certain size, is expelled to the outside through the movement of the vacuole to the plasma membrane and its contraction]
6. b [The excavates lack typical mitochondria and are restricted to making ATP through the anaerobic process of glycolysis}
7. a [The name *Euglena* roughly translates as "eyeball organism"]
8. d [A typical example is *Paramecium*: these have cilia that they use for swimming, to guide prey into their mouth-like gullet and they have one macronucleus and one or more micronuclei that contain the diploid genome]
9. a [They are a major primary producer in the world's oceans, i.e. they are a major component of the phytoplankton]
10. c [They are all non-motile]
11. c [They were originally thought to be fungi because they grow by forming hyphae and mycelia however molecular data indicates they are not part of the fungi]
12. a [They are all unicellular with intricately-formed silica cell walls in two sections that enclose the organism like two halves of a Petri dish]
13. b [Fucoxanthin is the accessory pigment in brown and golden algae that masks the colour of the chlorophyll]
14. b [This is an extract from red algae. All of the other structures are associated with the kelps formed by many of the brown algae]
15. b [The amoeba in this group form rigid pseudopods, for example the radiolarians and forams, the latter being a major component of the White Cliff of Dover; when they die their hard shells eventuall become part of the sedimentary rock]
16. a [The water moulds, or oomycetes (which form hyphae and mycelia) are not related to the slime moulds and plasmodial slime moulds, which, in addition to the amoebas that are part of this group, are "shape-shifters" that use flexible pseudopods for feeding and locomotion, for at least part of their life cycle]
17. d [The ability of the cellular slime moulds as well as the plasmodial slime moulds, both members of the group Amoebozoa, to differentiate into stalks and fruiting bodies under certain environmental conditions has been extensively studies]
18. a [The phycobilins are the accessory pigments that mask the colour of the chlorophyll in these photosynthetic organisms]
19. c [This group includes the diplomonads and the parabasalids, both of which lack typical mitochondria although they may possess some remnant of a mitochondrion, and they rely on glycolysis for generating ATP]
20. b [This group includes the green algae, the red algae and the land plants which evolved from the green algae]
21. a [This group includes choanoflagellates, believed to be the ancestor to the animals and fungi]
22. b [The ancestor of the chloroplast in these organisms is related to the modern day cyanobacteria]
23. d [This appears to have happened at least three times giving rise to the photosynthetic protists other than the green and red algae. The chloroplasts in these organisms have additional membranes derived from the new host]
24. a [The amoeba only reproduce asexually and do so by binary fission]

Integrating and Applying Key Concepts

1. Reread section 23.2 and remember the characteristics of the prokaryotes from Ch. 21 (what is their basic structure?)
2. Think about how classifications have traditionally been done using basic morphological and metabolic similarities and differences. Also remember from Ch. 21 how, until the advent of molecular phylogeny, the prokaryotes were thought to all be one domain/Kingdom, based on a similar structure. Now think about the diversities you have seen among the protists as well as the smiliarities between some of the protists and other life forms (kelps and plants, slime moulds and fungi) as well as the complex evolutionary tree depicted in Fig. 23.2.
3. Reread section 23.5 and pay particular attention to Fig. 23.31

Chapter 24 Fungi

24.1 What is a Fungus? General Characteristics of Fungi. [pp. 525-527]

1. Cellulose; 2. enzymes; 3. break down; 4. absorb; 5. saprotrophs; 6. symbionts; 7. mutualistic; 8. parasites; 9. yeasts; 10. hyphae; 11. mycelia; 12. chitin; 13. septa; 14. organelles; 15. tips; 16. streaming; 17. spores; 18. fragmentation; 19. plasmogamy; 20. karyogamy; 21. diploid; 22. haploid; 23. dikaryon; 24. F; 25. K; 26. I; 27. C; 28. G; 29. J; 30. L; 31. E; 32. B; 33. A; 34. D; 35. H; 36. A-Organisms that live on the organic material of dead matter, producing degradative enzymes to break them down, B- Organisms that live on another living organism (the host), causing harm to the host, C- A symbiotic relationship where both partners derive benefit; 37. A-Eukaryotic heterotrophs, B-Chitin, C-unicellular/yeasts or multicellular through hyphae and mycelia, D-Absorptive nutrition, they secrete degradative enzymes from the growing hyphal tips, the smaller breakdown products are transported into the cell and are disseminated to the rest of the organism by cytoplasmic streaming, E-Sexual or asexual involves fusion of spores or hyphae, with plasmogamy, karyogamy (sometimes delayed leading to a dikaryon stage) and meiosis; 38. F-it is a symbiotic bacterium growing on the surface of the ants that is producing the antibiotics. This is therefore a symbiotic relationship that involves three partners; 39. T; 40. They produce a number of important secondary metabolites including antibiotics and steroids; 41. Fossil evidence of possible arbuscular mycorrhyzae suggests the fungi may have played a critical role in facilitating the colonization of land by plants at least 500 million years ago; 42. These are metabolites or cellular products which are not required for the day-to-day survival of the organism but which do provide some benefit to the organism.

24.2 Evolution and Diversity of Fungi. [pp. 527-538]

43. sexual reproduction; 44. five; 45. paraphyletic; 46. Deuteromycota; 47. flagellated; 48. aquatic; 49. saprotrophs; 50. mutualists; 51. parasites; 52. amphibians; 53. rhizoids; 54. sporangium; 55. saprotrophs; 56. moulds; 57. commercial; 58. aseptate; 59. hyphae; 60. gametangia; 61. zygospore; 62. dormant; 63. Sporangium; 64. meiosis; 65. asexually; 66. mitosis; 67. arbuscular; 68. roots; 69. wheat/maize; 70. hypha; 71. spore; 72. genetic; 73. *Saccharomyces cerevisiae*; 74. lichens; 75. *Candida albicans*; 76. budding; 77. mycelia; 78. conidia; 79. plasmogamy; 80. Karyogamy; 81. dikaryotic; 82. asci; 83. 4-8 ; 84. ascocarp; 85. lignin; 86. DDT and PCBs; 87. invertebrates; 88. rusts and smuts; 89. Mushrooms; 90. toxic/poisonous; 91. yeasts; 92. karyogamy; 93. dikaryons; 94. favourable; 95. basidiocarp; 96. basidia; 97. meiosis; 98. basidiospores; 99. conidia; 100. *Penicillium*; 101. O; 102. E; 103. G; 104. H; 105. M; 106. K; 107. B; 108. D; 109. L; 110. N; 111. C; 112. A; 113. F; 114. I; 115. J; 116. A-form chains of haploid spores called conidia, and in sexual reproduction, form haploid ascospores inside sacs called asci which may be housed inside fruiting bodies, often sup-shaped, called ascocarps, B- forms aseptate hyphae, asexual reproduction by spores formed in sporangia, sexual reproduction involves gametangia forming at the end of specialized + and - hyphae which fuse, and the nuclei then fuse, functioning as gametes to give rise to a thick-walled zygospore which may remain dormant for years. When conditions improve, these germinate with a sporangium emerging from the zygospore to release haploid spores of the two mating types, C-Reproduce primarily sexually through the formation of a large, club-shaped fruiting body called a basidiocarp (mushroom) which has a stalk and a cap. The underside of the cap has basidia inside of which are haploid basidiospores which may be dispersed and germinate to form haploid mycelia. D-These are the only fungi that produce flagellated spores and generally live as unicellular organisms however they may form chains of cells and rhizoids that anchor them to a surface. They are normally haploid and most species reproduce asexually through the formation of sporangia from which emerge the spores. E-These form the arbuscular mycorrhizae and reproduce asexually by walling off a fragment of hypha to form a spore; 117. A-*Saccharomyces cerevisiae* – brewer's and bread-maker's yeast, *Candida albicans* causes thrush in

humans, B- these are the moulds that grow on breads, and rot fruits, vegetables and grains, C-smuts and rusts cause blights of many crops and other plants, they are also the mushrooms, both edible and poisonous, D-symbionts of cows and other herbivores, it is these fungi that break down the cellulose in the animal's diet, they also cause chytridomycosis, a fungal disease that is contributing to the world-wide die-off of amphibians, E-these are most important as mycorrhyzae, they are the only kind of fungus to form arbuscular mycorrhizae; 118. T; 119. Close biochemical relationship between fungi and animals may explain why these infections are so difficult to treat ; 120. T; 121. F-they reproduce sexually during adverse conditions and when conditions improve, switch to asexual reproduction; 122. The fungi in this group are assigned to this group based on the inability to observe sexual reproduction, which is the typical means of classifying fungi. The organisms in this group are therefore not grouped together based on evolutionary relatedness. Classification to this group is assumed to be temporary, until such time as molecular data or experimentation allows the organism to be properly classified; 123. Molecular analyses plus flagellated spores and aquatic lifestyle suggest that the fungal and animal lineages may have diverged from the choanoflagellate protist lineage around 945 million years ago. However, fungi do not appear in the fossil record until approximately 500 million years ago.

24.3 Fungal Lifestyles [pp. 538-542]

124. saprotrophs; 125. symbionts; 126. leaf litter/detritis; 127. CO_2 ; 128. compound; 129. green algae; 130. cyanobacteria; 131. mycobiont; 132. photobiont; 133. thallus; 134. mycobiont; 135. soredia; 136. phycobiont; 137. hyphae; 138. fungal hyphae; 139. carbon; 140. mutualistic; 141. desiccation; 142. Glomeromycota; 143. mutualistic; 144. surface area; 145. mineral ions; 146. arbuscular; 147. abundant; 148. colonization of land; 149. basidiomycetes; 150. ascomycetes; 151. Boreal; 152. rain forests; 153. endophytic; 154. mutualistic; 155. herbivores; 156. H; 157. A; 158. G; 159. F; 160. I; 161. C; 162. D; 163. B; 164. E; 165. T; 166. Some mycorrhizal relationships increase the plant's defenses against pathogens.

SELF-TEST

1. b [they are eukaryotic heterotrophs that obtain their nutrients by absorptive nutrition]
2. b [they obtain organic nutrients living as saprotrophs, degrading dead material or through symbiotic relationships with other living organisms]
3. a [Spores may be produced both sexually and asexually and some fungi, although not all, may produce either type, depending upon the stage of its life cycle]
4. b [Most are multicellular, growing as hyphae and mycelia however some may be unicellular, growing as yeasts]
5. c [They are not autotrophic as their source of carbon is also the organic material from which they obtain their energy. They also do not ingest their nutrients using phagocytosis/endocytosis, rather they secrete degradative enzymes to liberate small enough molecules to transport them across their plasma membrane. It is because of these enzymes that they are such prolific saprotrophs]
6. a [This is thought to be the oldest lineage and the one that evolved from the choanoflagellates due to the similar types of flagellated spores and molecular data]
7. c [The name basidium means club, and is given to these fungi due to the shape of the fruiting bodies which include the mushrooms]

8/. d [The Zygomycota may also be used commercially for their products (secondary metabolites) however they are well known spoilage organisms for fruits, vegetables and grains]

9. b [These fungi take their name from the sac-like structures or asci in which the sexual spores form]
10. e [The Glomeromycota are dominant members of the soil and form mycorrhizal structures inside of root cells that resemble little trees, the appearance referred to by the designation "arbuscule"]
11. b [lichens are a compound organism that is made of two or sometimes three organisms, living in a symbiotic relationship. While not well understood, the partnership appears to be a parasitic one in many cases, with the phytobiont being "enslaved" by the mycobiont as a sole source of carbon, however, in some cases, for example when some green algae are the phytobiont, the relationship may be mutualistic, with the phytobiont also deriving some benefit (protection against desiccation and harmful UV rays. The relationship does not, however, involve any decomposing reactions]

12. c [The Deuteromycota is a temporary grouping for organisms which have not yet been determined to have a sexual reproductive cycle and for which molecular data does not allow classification to any of the 5 evolutionary lineages. The members of this group are therefore not evolutionarily related]
13. d [These two lineages are not monophyletic as they contain subgroups that are not all descended from a common ancestor]
14. d [Organisms in this group are assigned here temporarily until such time as sexual reproduction, the typical means of classifying the fungi, is observed, or until molecular data allows classification to the appropriate evolutionary lineage]
15. a [The fungus, or mycobiont, makes up the majority of the lichen body and it is based on the identity of the mycobiont that the lichen is given a species designation]
16. c [These are the reproductive structures and are clusters of the algae or cyanobacteria wraped up in fungal hyphae. These are the means of dispersal for the lichen]
17. c [This ascomycete produces a psychrotropic toxin called ergot, from which LSD is made. It is thought to be the explanation for the bizarre "bewitched" behaviour of young girls and cows documented during the Salem Witch Trials]
18. d [Basidiomycota can grow for most of their lives as dikaryon mycelia, and this may be an advantage given the dispersive abilities of the basidiospores emitted from the biodiocarp structures. In fact, one of the world's largest organisms is thought to be a basidiomycete in Oregon where the mycelia of a single organism have spread over almost 9 km^2 of land]
19. d [Fungi are either saprotrophs or symbionts, aand those that can produces enzymes to break down the complex ligen polymers of woody plants often use those same enzymes to break down DDT and PCBs. Those same fungi must obtain nitrogen from a source other than the woody plants, which are poor in nitrogen, so they are also carnivorous, trapping and degrading invertebrates/insects]
20. d [It is not known exactly what the nature of the endophytic relationship is however in at least a few cases it seems to be mutualistic. Taxol, a new anticancer agent, is produced by fungal endophytes]

Integrating and Applying Key Concepts

1. Reread sect. 24.3a to clarify the importance and sect. 24.1to clarify the nature of nutrient acquisition.
2. Reread sect. 24.1 but also try to find the information on the dispersive abilities of spores of *Rhizopus* and the basidiomycetes which have a prolonged dikaryon stage and a somewhat bizarre, very large, fruiting body.
3. Reread sect. 24.2a and the section "Why it Matters".

Chapter 25 Plants

25.1 Defining Characteristics of Land Plants. [p. 548]

1. cellulose; 2. photoautotrophs; 3. achlorophyllous; 4. heterotroph; 5. roots; 6. generations; 7. dominant; 8. sporophytes; 9. gametophyte; 10. sporophyte; 11. meiosis; 12. mitosis; 13. embryos; 14. gametophyte; 15. A-having a single set of chromosomes, B-having two sets of chromosomes, one from each gamete/parent, C-the multicellular structure that gives rise to the spores, D-the multicellular structure that gives rise to the gametes, E-fusion of the female gamete by the male gamete, F-haploid cells that arise by the process of meiosis in the sporophyte, G-haploid cells arising by mitosis in the gametophyte, H-the immediate product of fertilization; 16. T; 17. Plants alternate between the diploid sporophyte and the haploid gametophyte generations.

25.2 The Transition to Life on Land. [pp. 548-536]

18. charophyte; 19. bryophytes; 20. poikilohydric; 21. diffusion; 22. cuticle; 23. stomata; 24. vascular; 25. xylem 26. phloem; 27. sunlight; 28. lignin; 29. apical meristem; 30. Roots; 31. mycorrhizas; 32. nutrients; 33. land plants; 34. sporophyte; 35. sporangia; 36. meiosis; 37. homosporous; 38. gametophyte; 39. heterosporous; 40. male; 41. female; 42. sporophyte; 43. gametangia; 44. sporangium; 45. gametophyte; 46. sporophyte; 47. homosporous; 48. heterosporous; 49. L; 50. O; 51. J; 52. A; 53. M; 54. B; 55. D; 56. C; 57. F; 58. E; 59. I; 60. G; 61. K; 62. L; 63. H.

25.4 Seedless Vascular Plants [pp. 559-564]

64. bryophytes; 65. roots; 66. liverworts; 67. hornworts; 68. mosses; 69. gametophyte; 70. gametangium; 71. archegonium; 72. antheridium; 73. fertilization; 74. germinates; 75. protonema; 76. gametophytes; 77. rhizoids; 78. antheridia; 79. archegonia; 80. archegonium; 81. egg; 82. sporophyte; 83. sporangium; 84. water; 85. sugar; 86. phloem; 87. seeds; 88. lycophytes; 89. ferns; 90. swimming; 91. vascular; 92. leaves; 93. sporophyte; 94. sporangia; 95. gametophytes; 96. archegonium; 97. antheridium; 98. sporophyte ; 99. nutritionally; 100. heterosporous; 101. male; 102. megaspores; 103. F; 104. H; 105. J; 106. G; 107. I; 108. A; 109. D; 110. B; 111. E; 112. C; 113. A-often poikilohydric, gametophyte dominant, motile sperm, B-sporophyte dominant, motile sperm, vascular tissues, root and shoot system/apical meristem, C-sporophytes retained attached to tips of gametophytes, spores are released and once they land and germinate they give rise to a protonema from which eventually arise the male and female gametophytes, D-sporangia form on the underside of fern leaves, homosporous –only one type of spore. Male and female gametophytes form on the underside of the sporangium leaves, male and female on the same leaf, after fertilization, the embryo gives rise to the sporopophyte which remains attached to the gametophyte until it becomes nutritionally independent, at which point the gametophyte dies; 114. T; 115. T; 116. F-the various bryophyte lineages evolved separately in parallel with vascular plants; 117. If a bisexual gametophyte is surrounded by mainly male gametophytes and many of these are from other spores, this increases the chances that the eggs will be fertilized by the sperm of other plants, increasing genetic diversity; 118. They lack vascular tissue and they did not evolve from the same structures as leaves and stems did.

25.6 Angiosperms: Flowering Plants [pp. 568-574]

119. Gymnosperms; 120. angiosperms; 121. pollination; 122. seed; 123. naked; 124. male; 125. female; 126. pollen; 127. megagametophyte; 128. pollen tube; 129. pollen tube; 130. zygote; 131. pine nut; 132. cycads; 133. Gingkophyta; 134. gnetophytes; 135. angiosperms; 136. vascular; 137. ovary; 138. endosperm; 139. pollinators; 140. fruits; 141. Anthophyta; 142. monocots; 143. eudicots; 144. gametophyte; 145. endosperm; 146. fruit; 147. dispersal; 148. G; 149. J; 150. F; 151. K; 152. M; 153. C; 154. A; 155. O; 156. E; 157. B; 158. D; 159. N; 160. I; 161. L; 162. H; 163. A-Confined to subtropics and tropics, look like small palm trees, may have giant cones, B-Gingkophta, C-Generally form woody cones in which the gametangia form, are evergreen with needlelike leaves and thick cuticle, fibrous epidermis and sunken stomata for moisture retention, D- Gnetophytes, E-lilies, grasses and orchids, F-most flowering shrubs and trees, herbs and cacti; 164. F – the coconut is the fruit that protects and disseminates the seeds within it; 165. F – it is a mix of organic compounds that are a by-product of metabolism. It flows in resin ducts and serves as a deterrent to wood-boring insects and certain microbes; 166. T; 167. The ovule represents the female gametophyte which develops inside the megaspore wall inside the megasporangium, remaining physically attached to the sporophyte. These multiple layers and physical attachment means the female gamete is protected from predation and environmental conditions. Once fertilized, this becomes the seed; 168. This forms in heterosporous plants from the microspore. In these plants themale gametophyte develops inside the microspore wall and is very much smaller than non-seed plants, comprising only a few cells: this is the pollen. The result is a structure that is extremely light weight and is easily carried by air currents; 169. The evolution of flowers with their distinctive shapes, colours, nectar and odours evolved in parallel with the animal pollinators (e.g. insects, birds and bats) to maximize attraction of the appropriate pollinators and efficient transfer of the male pollen to the female reproductive parts.

SELF-TEST

1. d [Angiosperms evolved more a more efficient vascular system, their embryos are in seeds which are protected and dispersed by fruits. Their sperm form in a pollen tube after pollination and the pollen grains are dispersed by specific animal pollinators}
2. a [Double fertilization is associated with the angiosperms and is what gives rise to the fruit]
3. e [Gymnosperms are heterosporous: the microsporocytes gives rise to haploid male microspores by meiosis and these undergo mitosis to generate the pollen grains, which are immature male gametophytes]
4. d [The strobilus is a structure which is associated with the sporophyte generation ($2n$) of lycophytes or club mosses. It is a cone-like cluster of sporophylls, which are the specialized leaves that give rise to the sporangia]

5. a [The sporophyte generation is one of the two generations of the alternation of generations typical of plants. It is the diploid generation, and ultimately gives rise to haploid spores through meiosis; these spores then give rise to gametophytes. Both the sporophyte and gametophyte generation are multicellular]
6. c [The sorus is the name for the cluster of sporangia that forms on the underside of the leaves of the fern (sporophyte or diploid generation). The fiddlehead is the newly growing sporophyte generation which is generated by fertilization of the female gamete by the male gamete]
7. a [The charophytes are a lineage of green algae and they share a number of characteristics (ancestral traits) with the least evolved of the land plants, the liverwort lineage of the bryophytes]
8. b [The gametophyte generation is the generation that gives rise to the gametes. It arises by development from the spores of the sporophyte generation. The spores are produced by meiosis and the gametophyte and gametes all arise by mitosis]
9. d [The antheridia is the name given to a particular type of male gametangium. It is a multicellular structure made of haploid cells and gives rise to sperm in the bryophytes and seedless vascular plants]
10. b [Spores form in a sporangium from meiosis. The sporangium is part of the sporophyte – the diploid structure/generation]
11. e [To grow up and away from the ground, plants needed strengthening tissue, lignin, an internal water and sugar circulating system (xylem, phloem and roots) and an anchoring structure (roots). Evolution of the apical meristem allowed roots and shoots which then allowed for the acquisition of water and nutrients from the soil and increased solar capture]
12. c [pollen represents the immature male gametophyte which, after landing on a female reproductive structure, produces a pollen tube which invades the female gametophyte tissue and which, at the same time, produces sperm for delivery to the egg and the subsequent formation of a zygote]
13. a [Archegonia is the name given to certain female gametangia formed by heterosporous plants. Eggs form within this structure and remain inside after fertilization, giving rise to the embryo. It does not aid in dissemination but has a protective function]
14. a [Plants are the only organisms that have an alternation of generations. In the case of the algae, including the green algae (such as charophytes), the haploid stage dominates the life cycle and a diploid, single-celled zygote undergoes meiosis to generate the haploid phase.]
15. e [They appear suddenly in the fossil record without a fossil sequence that links them to any other plant group]
16. b [Pine, spruce and fir trees are conifers, which make up the majority of the gymnosperms; cacti are eudicot angiosperms]
17. b [Rhizomes are horizontal, modified stems that have the ability to penetrate substrates, providing support for plants. They do not absorb nutrients so they are not homologous with roots, are not part of a vascular system and since they do not absorb water and nutrients, they do not form symbiotic relationships with mycorrhizas]
18. c [Apical meristems are constantly dividing, undifferentiated cells at the growth sites of roots and shoots. They give rise to all of the tissues of the plant body and are present in vascular plants only]
19. e [Mycorrhizas are fungal symbionts that are associated with the root systems of most land plants. They are highly branched, often penetrate the root cells and extend well beyond the zone of the roots, efficiently absorbing water and nutrients and exchanging these with the plants]
20. a [This is a process unique to angiosperms. Two sperm are delivered to the ovule, one fertilizing the egg and the other fertilizing the endosperm-forming cell to give rise to a binucleate cell which gives rise to a $3n$ endosperm which eventually forms part of the fruit]
21. d [Many floral features correlate with the morphology and behaviour of certain animal pollinators which coevolved with the flowers. For example, the bats pollinate intensely sweet-smelling flowers with white or pale petals which are easier to see at night than coloured petals.

Integrating and Applying Key Concepts

1. Think about what challenges faced early plants as they colonized land then reread the section that describes how the evolution of pollen, pollination, ovules and seeds provided an adaptive advantage (sect. 25.5a). Then think about the further evolutionary changes in the angiosperms (compared to the gymnosperms) –sect. 25.6 and how the diversity of the flowering plants compares to that of the gymnosperms.

2. Reread the introductory section of sect. 25.2, the discussions of figs. 25.6, 25.10 and sect. 25.6a. pay attention to the description of fig. 26.10: is there agreement and confidence in the relationships presented? What kind of role has molecular data played in determining relationships? Are all of the genomes sequenced? Are there questions remaining that might be answered using molecular data?
3. Have a look at the topic map at the beginning of this chapter and see if all of the major points make sense to you. Can you identify the most ancient lineage and derive a time-line starting through the various lineages?

Chapter 26 Protostomes

Why It Matters [pp. 577-578]

1. Cambrian; 2. Species; 3. Mudslide; 4. Burgess Shale; 5. Sponges; 6. extinct

26.2 Key Innovations in Animal Evolution [pp. 579-583]

7. monophyletic; 8. eukaryotic; 9. lack; 10. energy; 11. animals; 12. heterotrophs; 13. motile; 14. sessile; 15. sexual; 16. sperm; 17. asexual; 18. body; 19. metazoans; 20. diploblastic; 21. ectoderm; 22. endoderm; 23. triploblastic; 24. mesoderm; 25. asymmetrical; 26. bilateral; 27. radial; 28. central; 29. acoelomate; 30. pseudocoelomate; 31. mesoderm; 32. coelomate; 33. K; 34. G; 35. M; 36. N; 37. H; 38. A; 39. O; 40. B; 41. F; 42. C; 43. P; 44. L; 45. D; 46. J; 47. E; 48. I

26.3 An Overview of Animal Phylogeny and Classification [pp. 583-585]

49. morphology; 50. embryology; 51. molecular; 52. nucleotide; 53. mitochondrial; 54. genes; 55. lineages; 56. Radiata; 57. three; 58. phylogenetic

26.5 Metazoans with Radial Symmetry [pp. 586-590]

59. Porifera; 60. Metazoan; 61. filter; 62. pores; 63. sessile; 64. larvae; 65. Cnidaria; 66. Ctenophora; 67. two; 68. gastrovascular; 69. mouth; 70. polyp; 71. medusa; 72. cnidocytes; 73. nematocysts; 74. sticky; 75. cilia; 76. B; 77. D; 78. A; 79. A; 80. D; 81. B; 82. E; 83. C; 84. B; 85. C; 86. B; 87. A; 88. B; 89. A; 90. F

26.6 Lophotrochozoan Protostomes [pp. 590-603]

91. bilateral; 92. two; 93. Lophotrochozoan; 94. Ecdysozoan; 95. three; 96. lophophore; 97. capture; 98. gas; 99. Ectoprocta; 100. Brachiopoda; 101. calcium; 102. Phoronida; 103. Platyhelminthes; 104. flatworms; 105. digestive; 106. nervous; 107. ganglion; 108. reproductive; 109. excretory; 110. flame; 111. four; 112. Turbellaria; 113. parasitic; 114. Rotifera; 115. cilia; 116. parthenogenesis;117. unfertilized; 118. complete; 119. mouth; 120. anus; 121. proboscis; 122. everted; 123. three; 124. visceral mass; 125. head-foot; 126. mantle; 127. eight; 128. Gastropoda; 129. Cephalopoda; 130. Annelida; 131. three; 132. segments; 133. septa; 134. metanephridia; 135. setae; 136. Polychaeta; 137. Oligochaeta; 138. Hirudinea; 139. P; 140. B; 141. O; 142. H; 143. N; 144. D; 145. K; 146. J; 147. L; 148. M; 149. C; 150. A; 151. I; 152. E; 153. G; 154. F; 155. Platyhelminthes; 156. ribbon worm; 157. Annelida; 158. Brachiopoda; 159. Rotifer; 160. Mollusca

26.7 Ecdysozoan Protostomes [pp. 603-614]

161. exoskeleton; 162. three; 163. Ecdysozoan; 164. Nematoda; 165. parasites; 166. Onychophora; 167. southern; 168. Arthropoda; 169. three-quarters; 170. ecdysis; 171. grows; 172. three; 173. abdomen; 174. cephalothorax. 175. A significant disadvantage of shedding the exoskeleton is loss of protection and support of the exoskeleton. An animal is extremely vulnerable to predation during this molting period.; 176. Nematodes can reproduce sexually (as well as by parthenogensis). The large number of eggs a fertilized female produces, as well as the short gestation period, plays a major role in the success of these worms; 177. B; 178. G; 179. E; 180. F; 181. A; 182. H; 183. D; 184. C; 185. C; 186. B; 187. C; 188. B; 189. D; 190. E; 191. D

Self-Test

1. c [other selections are associated with plants; being heterotrophic is most unique to animals]
2. d [Platyhelminthes do not have a body cavity, they are Acoelomates]
3. b [the origin of mesoderm is different between protostomes and deuterostomes]
4. a [these are characteristics that are associated with protostomes]
5. a [blastophore is an opening linking the gut to the environment and becomes a mouth in protostomes]
6. a [the innermost tissue, the endoderm, forms the lining of the gut in animals]
7. c [cephalization is the formation of a head end, which can only occur for organisms that are bilaterally symmetrical]
8. b [new information that affects a developed phylogenetic tree can be added at any time]
9. b [when one individual can produce both eggs and sperm, it is monoecious]
10. b [nematocyst and polyp apply to Cnidarians and a porocyte is a cell lining the pore of a sponge; gemmule is a cluster of cells awaiting appropriate conditions to germinate]
11. d [*Obelia* metamorphose from a polyp to a medusa, and sponges and coral metamorphose from ciliated larvae to adult forms; only *Hydra* does not change its form during its life]
12. b [a radula is a scraping device, mollusks typically have a radula and are often bottom- or rock-dwelling animals]
13. d [cephalopods move rapidly and require a high level of oxygen to support the increased mobility, thus a closed circulatory system will increase delivery of oxygen delivery to tissues.]
14. c [each organism of Phylum Rotifera has a corona and a mastax]
15. d [rotifers make up a large portion of fresh water zooplankton]
16. b [selection a is an incorrect statement, and selections c and d are associated with insects]
17. b [animals with repeating units of internal systems are typically in Phylum Annelida, the segmented worms]
18. c [millipedes and centipedes do not have a thorax and shrimps and spiders have a cephalothorax]
19. c [occupying different habitats and eating different food make insects that undergo complete metamorphosis very successful]

Integrating and Applying Key Concepts

1. An organism is an animal if it is eukaryotic and multicellular, with plasma membranes of neighbouring cells in contact with each other. All animals are heterotrophs, feeding on other living organisms, as well as being motile at some time during their life cycle. They reproduce either sexually or asexually.
2. Tapeworms, round worms, and earthworms are classified in three different phyla because, although they each are commonly called 'worms', this is a generic term and is not based on phylogenetic characteristics. Earthworms and tapeworms have characteristics that makes them Lophotrochozoans but within this group, earthworms belong in the Phylum Annelida because they produce trocophore larvae whereas tapeworms belong in the Phylum Platyhelminthes because they do not produce trocophore larvae and they have no coelom. Round worms have a cuticle that they shed (ecdysis) so they belong in the group Ecdysozoa (not Lophotrochozoa) and since they are not segmented, they belong to the Phylum Nematoda.
3. Segmentation is found in animals in the Phyla Annelida, Onychophora, and Arthropoda within the protostomes. Since Phylum Annelida is in a separate and independent lineage from the other two phyla, segmentation must be a result of convergent evolution in response to environmental conditions (not from a common ancestor).

---

27 Deuterostomes: Vertebrates and Their Closest Relatives

Why It Matters [pp. 617-619]

1. molecular data; 2. classification; 3. phylogenetic; 4. *Xenoturbella*; 5. Mollusca; 6. deuterostome; 7. deuterostome; 8. morphological; 9. eaten

27.3 Phylum Hemichordata [p. 623]

10. anus; 11. blastopore; 12. mouth; 13. mouth; 14. arms; 15. cord; 16. adult; 17. bilaterally; 18. radially; 19. 5; 20. water; 21. tube; 22. Hemichordata; 23. acorn; 24. proboscis; 25. into; 26. out of; 27. E; 28. C; 29. F; 30. A; 31. B; 32. D; 33. D; 34. I; 35. B; 36. J; 37. H; 38. C; 39. E; 40. G; 41. F; 42. A

27.5 The Origin and Diversification of Vertebrates [pp. 625-628]

43. notochord; 44. gill slits; 45. 3; 46. vertebral; 47. invertebrates; 48. backbone; 49. bone; 50. Vertebrata; 51. cranium; 52. neural; 53. duplications; 54. Hox; 55. complex; 56. location; 57. shape; 58. complex; 59. more; 60. jaw; 61. Agnathostomata; 62. terrestrial; 63. tetrapods; 64. four; 65. locomotion; 66. Amniota; 67. terrestrial; 68. C; 69. H; 70. G; 71. D; 72. E; 73. A; 74. F; 75. B; 76. B; 77. D; 78. A; 79. B; 80. C; 81. A; 82. B; 83. A; 84. A; 85. B; 86. D

27.7 Jawed Fishes: Jaws Expanded the Feeding Opportunities for Vertebrates [pp. 629–635]

87. Agnathans; 88. hagfishes; 89. lamprey; 90. sucking; 91. arches; 92. larger; 93. Chondrichthyes; 94. dorsoventrally; 95. sharks; 96. Bony; 97. endoskeleton; 98. scales; 99. mucus; 100. bony; 101. Teleosteii; 102. operculum; 103. lateral; 104. chemoreceptive; 105. diet; 106. locomotion; 107. G; 108. H; 109. M; 110. F; 111. J; 112. K; 113. L; 114. A; 115. C; 116. D; 117. E; 118. B; 119. N; 120. I

27.8 Early Tetrapods and Modern Amphibians [pp. 635-637]

121. air; 122. skeletal; 123. pectoral; 124. tympanum; 125. sound; 126. skin; 127. moist; 128. larval; 129. aquatic; 130. both; 131. Anura; 132. toads; 133. Urodela; 134. Gymnophonia; 135. E; 136. G; 137. A; 138. F; 139. B; 140. D; 141. C

27.9 The Origin and Mesozoic Radiations of Amniotes [pp. 638-642]

142. sac; 143. embryo; 144. amnion; 145. terrestrial or dry; 146. skin; 147. keratin; 148. waterproof; 149. amniotic; 150. shell; 151. membranes; 152. Dessication; 153. Gases; 154. yolk; 155. albumin; 156. Uric acid; 157 ammonium; 158. synapsids; 159. anapsids; 160. Diapsids; 161. Skull; 162. F; 163. D; 164. G; 165. A; 166. C; 167. E; 168. B

27.12 Aves: Birds [pp. 645–649]

169. anapsids; 170. Shell; 171. Rib cage; 172. Head; 173. Legs; 174. Diapsids; 175. Two; 176. Tuatara; 177. Snakes; 178. Alligators; 179. Birds; 180. Bill; 181. Wing; 182. D; 183. B; 184. E; 185. B; 186. E; 187. C; 188. A; 189. B; 190. A

27.15 The Evolution of Humans [pp. 657-662]

191. mammals; 192. synapsids; 193. temporal; 194. furry; 195. homeothermic; 196. diaphragm; 197. occipital; 198. palate; 199. cortex; 200. heterodont; 201. diphyodont; 202. milk or deciduous; 203. adult; 204. milk; 205. reproduction; 206. monotremes; 207. placental; 208. marsupials; 209. teat; 210. marsupium or pouch; 211. convergence; 212. dorsoventrally; 213. diversity; 214. Hominids; 215. bipedal; 216. erect; 217. power; 218. precision; 219. hypotheses; 220. Multiregional; 221. African Emergence; 222. 200,000; 223. H; 224. L; 225. F; 226. G; 227. K; 228. A; 229. J; 230. D; 231. C; 232. B; 233. I; 234. E

SELF-TEST

1. c [larvae are radially symmetrical, adults are bilaterally symmetrical, and sexual reproduction is external; tube feet are the only correct characteristic]
2. b [the bilaterally symmetrical larval form is characteristic of the phylum Echinodermata, pedicellariae are found in the asteroidea]
3. d [only acorn worms have a proboscis and gill slits; dorsal hollow nerve chord is not found in acorn worms nor echinoderms]
4. c [pearl fish live in the tubes of other animals
5. c [description is a tunicate, which belongs to the Urochordata phylum]
6. d [seahorses are teleosts, a more complex and diverse group of fish, and therefore expected to have the greatest number of Hox gene complexes]
7. b [class Chondrichthyes includes sharks, rays, and chimeras, all of which have a cartilaginous skeleton]

8. a [gill arches; some evolved into forming a jaw]
9. d [all except for d are Agnathans]
10. d [the swim bladder is a hydrostatic organ which increases buoyancy; if destroyed, the fish would not be as buoyant]
11. a [of the selections, dehydration is a major problem in the terrestrial environment]
12. b [gas exchange occurs across the skin so blood vessels must carry oxygen and carbon dioxide to it]
13. a [of the selections, only keratin and lipid in skin will be advantageous to terrestrial life]
14. d [4-chambered heart is a striking adaptation of crocodiles and is homologous to bird heart]
15. c [hollow limb bones significantly reduces the weight of the skeleton of birds]
16. e [biomolecules captured by the tongue enter the mouth and are identified by receptors in the roof of a snake's mouth]
17. d [all selections are either other names or examples of prototheria]
18. d [teeth are an example of diversity; only the flattened tail and spines are examples of convergence
19. c [mammals can replace their teeth once (milk teeth replaced by adult teeth]
20. b [the precision grip allows us to manipulate objects with fine movements]

Integrating and Applying Key Concepts

1. Both echinoderms and humans are deuterostomes because during embryonic development the blastopore forms an anus. However, in terms of other characteristics, individuals within the deuterstomes can vary greatly. An adult echinoderm has no 'head', no excretory system, no respiratory system and only a minimal circulatory system, whereas a human is cephalized (has a head), as well as a complete excretory system, respiratory system, and circulatory system.
2. Movement from an aquatic to a terrestrial environment was possible with the evolution of four limbs for locomotion, lungs to breathe atmospheric oxygen, and specialized eggs which had coverings to protect them from dehydration.

Chapter 28 The Plant Body

Why It Matters [pp. 667-668]

1. root system; 2. water; 3. shoot; 4. photosynthesis; 5. bark; 6. roots; 7. shoots

28.1 Plant Structure and Growth: An Overview [pp. 668-673]

8. organs; 9. shoot system; 10. cells; 11. chloroplasts; 12. vacuoles; 13. cell walls; 14. lignin; 15. determinate growth; 16. indeterminate growth; 17. meristem; 18. apical meristems; 19. primary tissues; 20. primary plant body; 21. secondary tissues; 22. lateral meristems; 23. secondary growth; 24. monocots; 25. eudicots; 26. annuals; 27. biennials; 28. perennials; 29. f—able; 30. t; 31. f—shoot apical meristem; 32. f—secondary ; 33. t; 34. Multiples of 3; 35. Arranged randomly; 36. Network; 37. Taproot

28.2 The Three Plant Tissue Systems [pp. 673-678**]**

38. ground tissue system; 39. vascular tissue system; 40. dermal tissue system; 41. parenchyma; 42. collenchyma; 43. sclerenchyma; 44. end-to-end; 45. Xylem; 46. tracheids; 47. vessel members; 48. phloem; 49. sieve tube members; 50. companion cells; 51. Epidermis; 52. cuticle; 53. guard cells; 54. stomata; 55. trichomes; 56. c; 57. f; 58. a; 59. e; 60. b; 61. d

28.3 Primary Shoot Systems [pp. 678-682]

62. petiole; 63. node; 64. internode; 65. terminal buds; 66. lateral buds; 67. apical dominance; 68. primary meristems; 69. protoderm; 70. ground meristem; 71. procambium; 72. vascular bundles; 73. circle; 74. cortex; 75. pith; 76. leaf primordial; 77. mesophyll; 78. veins; 79. b; 80. e; 81. a; 82. f; 83. d; 84. c

28.5 Secondary Growth [pp.685-691]

85. taproot; 86. lateral roots; 87. fibrous root; 88. adventitious; 89. root cap; 90. zone of cell division; 91. zone of elongation; 92. zone of maturation; 93. exodermis; 94. endodermis; 95. Vascular; 96. pericycle; 97. lateral; 98. vascular cambium; 99. cork cambium; 100.

cork; 101. epidermis; 102. Xylem; 103. Phloem; 104. wood; 105. heartwood; 106. sapwood; 107. bark; 108. F--are not; 109. T; 110. F—zone of maturation 111. T

Self-Test

1. c [Morphology is defined as an organism's external form.]
2. a [If a secondary cell wall is present, it is laid down <u>inside</u> the primary wall.]
3. b [Woody plants contain secondary tissues.]
4. a [Shoot and root apical meristems are responsible for growth in length.]
5. b [Secondary cell wall is laid down <u>inside</u> the primary wall.]
6. c [It is constantly undergoing cell division.]
7. a [Apical meristems are responsible for growth in length.]
8. b [Lilies are monocots.]
9. b [Sclerenchyma cells produce thick secondary cell walls.]
10. b [Parenchyma cells have the greatest number of chloroplasts.]
11. e [Angiosperms have tracheids and vessel members, as well as sieve tube members.]
12. a [Companion cells assist in providing sugars to sieve tube member cells.]
13. b [Leave attach to the stem at the node. The distance between two nodes in the internode.]
14. d [Vascular cambium functions in secondary growth.]
15. b [Bamboo is a hard-walled monocot plant that develops a hollow stem.]
16. d [Trichomes in shoots take various forms.]
17. c [Palisade mesophyll cells contain the greatest number of chloroplasts.]
18. d [The zone of cell division produces new cells, not the root cap.]
19. d [Xylem cells are larger in the spring than summer so reflect light differently.]
20. b [Size of tree rings indicates amount of growth; narrow rings means little growth.]

Integrating and Applying Key Concepts

1. Dendritic means to have a spreading or branching form. Plant shoot systems need to have a branched form for leaves to be spread out to reduce shading thereby maximizing the light absorbed by the plant and optimizing photosynthesis for energy (sugar) production. Plant root systems also need to be dendritic so that soil water and mineral access and absorption can be maximized.
2. Within a vegetative bud, cells must differentiate to form each of the necessary plant tissue systems. Identify what tissue system that each of the primary meristems develops into and give their resulting functions.
   Within an apical meristem (bud), three primary meristems develop: protoderm, procambium, and ground meristem. Over time, the protoderm will develop into the plant epidermis, which is a tissue of the dermal tissue system. This tissue system functions in protection of the internal parts of the plant. The procambium develops into primary vascular tissues (primary xylem and primary phloem), which are the main components of the vascular tissue system. This system functions in transport of water and solutes throughout the plant. Lastly, the ground meristem will form the ground tissue system (parenchyma, collenchyma, and sclerenchyma tissues), which make up the majority of the plant body. This tissue system has many functions within a plant including storage, secretion, photosynthesis, flexibility, support, and protection.
3. You planted a red oak tree in your backyard three years ago. Explain how growth this summer will be different from the growth occurred in the tree's first summer.
   During the first year of the tree's (seedling's) growth, only primary growth (from apical meristems) would have occurred to increase the tree's length. In the third year of growth (this summer), both primary and secondary growth will occur. Apical meristems in the shoot (terminal and lateral buds) and root systems (root tips) will occur resulting in a further increase in tree length but there will also be growth in girth as the vascular cambium (lateral meristem) produces secondary xylem and secondary phloem cells. The secondary xylem tissue produced this summer will appear as a ring in the tree's wood (if it is cut down).

Chapter 29 Transport in Plants

Why It Matters [pp. 695-696]

1. water; 2. pump; 3. cohesion; 4. evaporation; 5. gravity

29.1 Principles of Water and Solute Movement in Plants [pp. 696-700]

6. passive transport; 7. active transport; 8. transport proteins; 9. membrane potential; 10. electrochemical; 11. down; 12. symport; 13. antiport; 14. bulk flow; 15. xylem sap; 16. osmosis; 17. water potential; 18. zero; 19. solution potential; 20. into; 21. turgor pressure; 22. pressure potential; 23. zero; 24. turgid; 25. central vacuole; 26. tonoplast; 27. aquaporins; 28. wilt; 29. T; 30. F—Symport; 31. T; 32. T; 33. T; 34. B; 35. C; 36. B; 37. A

29.3 Transport of Water and Minerals in the Xylem [pp. 703-708]

38. apoplast pathway; 39. symplast pathway; 40. transmembrane pathway; 41. endodermis; 42. radially; 43. casparian strip; 44. cohesion- tension model;45. cohesion; 46. adhesion; 47. transpiration; 48. stomata; 49. guard cells; 50. root pressure; 51. endodermis; 52. guttation; 53. crassulacean acid metabolism (cam); 54. evening; 55. day; 56. G; 57. E; 58. A; 59. B; 60. D; 61. F; 62. C

29.4 Transport of Organic Substances in the Phloem [pp. 709-711]

63. translocation; 64. phloem sap; 65. source; 66. sink; 67. pressure flow; 68. pressure; 69. decreases ; 70. osmosis; 71. bulk flow; 72. transfer cells; 73. F—sucrose; 74. F—in any direction; 75. T; 76. T; 77. F—only at the end of

SELF-TEST

1. c [Transport proteins include carrier proteins, which do not provide pores.]
2. d [Symport and antiport require use of energy that can be released after primary active transport has occurred.]
3. c
4. b [Water potential is the sum of the solute and pressure potentials.]
5. d
6. d
7. c [Soil water does not enter cortexcells in the apoplastic pathway and uses diffusion (not use plasmodesmata) to move between cells in transmembrane pathway.]
8. b
9. d [Sieve tube member cells are alive even though they do not contain a nucleus.]
10. c
11. c [Cohesion is the attraction of water molecules to each other. Viscosity is the friction of liquid molecules as the flow past each other or over a solid surface.]
12. b
13. c
14. a [When expanded, the inner walls of the guard cells do not expand as much as the outer cells, resulting in formation of a pore (stoma) between them.]
15. c [low CO_2 concentration, daylight, and moist soil all encourage stomata to open.]
16. c [CAM photosynthesis occurs most often in hot, dry habitats like those where *Sedum* is found.]
17. d [Phloem sap is primarily composed of sugars but other solutes, including amino acids and fatty acids, are present as well.]
18. c [It was phloem sap pressure that caused the phloem sap to travel through the gut and out the anus of an aphid.]
19. a [As long as a pressure gradient exists, phloem sap will move toward the region of lower pressure.]
20. a [Only in the early spring, before leaves are functional, are roots and buds acting as sources.]

Integrating and Applying Key Concepts

1. For a stoma to open, first H^+ must be actively transported out of the two surrounding guard cells. As the H^+ move back down the concentration gradient, energy is released and used to actively transport K^+ into the guard cells. The higher solute concentration in the guard cells causes water to enter by osmosis, which results in guard cells expanding and bending away from each other, resulting in the opening of the central pore (stoma).
2. Transport in xylem is unidirectional since water and minerals are taken up in the roots only and therefore can only move upwards in the plant (by cohesion-tension model and/or root pressure model mechanisms). However, the organic solutes that enter the phloem from all different areas of the shoot system may be needed in any region of a plant (roots or shoots) and therefore travel upwards, downwards, or sideways in phloem tissue.
3. Adaptations in root systems could include:
   1) increased length of roots to access greater soil water resources,
   2) increased root hair production to increase surface area for water uptake.
   Adaptations in shoot systems could include:
   1) decreased leaf surface area (e.g. spines) (to reduce water loss from surface of epidermis and stomata)
   2) decreased number of stomata (less water loss due to transpiration when stomata open)
   3) greater number of stomata on underside of leaf (reduces temperature and relative humidity conditions that promotes water loss during carbon dioxide uptake)
   4) recessed stomata or trichomes around stomata (reduces air movement in area of stomata so reduces water loss)
   5) CAM (stomata open at night when less water loss due to temperature, humidity, etc)
   Note that there are many other examples.

Chapter 30 Reproduction and Development in Flowering Plants

TOPIC MAP

1. dedifferentiation; 2. differentiation; 3. maturation; 4. meiosis; 5. meiosis; 6. pollination; 7. double fertilization; 8. mitosis; 9. germination; 10. maturation

Why It Matters [pp. 715-716]

11. flowers; 12. pollination; 13. bees; 14. colony collapse disorder; 15. fruits

30.2 The Formation of Flowers and Gametes [pp. 718-721]

16. sporophyte; 17. haploid; 18. pollen grain; 19. embryo sac; 20. gametes; 21. diploid; 22. mitosis; 23. floral shoot; 24. influorescence; 25. sepals; 26. petals; 27. stamen; 28. filament; 29. anther; 30. pollen; 31. carpels; 32. ovary; 33. style; 34. stigma; 35. microsporocyte; 36. microspores; 37. pollen grains; 38. sperm; 39. pollen tube; 40. ovary; 41. megasporocyte; 42. one; 43. embryo; 44. one; 45. two; 46. one; 47. central; 48. micropyle; 49. E; 50. C; 51. D; 52. F; 53. A; 54. G; 55. B; 56. H; 57. E; 58. A; 59. G; 60. B; 61. I; 62. L; 63. C; 64. J; 65. D; 66. K; 67. F; 68. B; 69. C; 70. A; 71. F; 72. D; 73. E

30.3 Pollination, Fertilization, and Germination [pp. 721-728]

74. pollination; 75. anther; 76. stigma; 77. allele; 78. s allele; 79. pollen; 80. style; 81. sperm; 82. fertilization; 83. zygote; 84. endosperm; 85. apical; 86. suspensor; 87. cotyledons; 88. cotyledon; 89. endosperm; 90. radical; 91. plumule; 92. coat; 93. coleorhizae; 94. coleoptiles; 95. ovule; 96. ovary; 97. pollen; 98. pericarp; 99. protection; 100. dispersal; 101. dormancy; 102. germination; 103. imbibitions; 104. coat; 105. enzymes; 106. mitosis; 107. radical; 108. root; 109. plumule

Complete the Table

110.Complete the following table by describing each process.

Events	*Description*
A. Pollination	The transfer of pollen from an anther to the stigma of the carpel
B. Fertilization	The fusion of male sperm cell and the female egg cell

C. Germination	The seed imbibes water to come out of dormancy and grow into a seedling

111.Complete the following table by describing each fruit type.

Fruits	*Description*
A. Simple fruits	A fruit that develops from a single ovary of a flower
B. Aggregate fruits	A fruit that develops from multiple ovaries of a single flower
C. Multiple fruits	A fruit that develops from several ovaries of multiple flowers

112. E; 113. A; 114. B; 115. F; 116. C; 117. G; 118. D; 119. Endosperm is not produced when fertilization does not occur whereas gametophyte tissue would be produced whether or not fertilization occurred (wasted energy cost); 120. Cotyledons in an eudicot absorb and store the nutrients absorbed from the endosperm (no endosperm in mature seed) whereas in a monocot, nutrients are stored in the endosperm.

30.5 Early Development of Plant Form and Function [pp. 731-731]

121. vegetative; 122. totipotency; 123. fragmentation; 124. unfertilized; 125. apomixis; 126. callus; 127. hormones; 128. identical; 129. root-shoot; 130. apical; 131. basal; 132. transcription factors; 133. differentiation; 134. C; 135. D; 136. A; 137. G; 138. H; 139. B; 140. E; 141. F; 142. I; 143. Totipotent describes cells that can develop into a complete new organism (in plants), pluripotent describes cells that can only develop into many kinds of cells but not a new organism (in animals), and multipotent describes cells that can only develop into specific kinds of cells.

SELF-TEST

1. b [A three celled pollen grain is a male gametophyte because it produces the male gametes (sperm cells).]
2. d [A mature pollen grain is composed of one pollen tube cell and two sperm cells.]
3. c [The ovule of the flower transforms into a seed after fertilization.]
4. d [The (diploid) central cell of the embryo sac fuses with the (haploid) sperm cell to form triploid endosperm.
5. c [The type of fruit that is formed from ovaries from different flowers in an influorescence, such as a pineapple produces, is a multiple fruit.]
6. a[All the cells in a callus are genetically alike since they arise from one or more cells of a single plant.]
7. d [A radicle forms roots since the radicle is an embryonic root.]
8. b [Specialized cells dedifferentiate when they become unspecialized
9. c [The pathway the growing pollen tube travels is through stigma, then style, then micropyle, the latter being the entrance into the ovule.]
10. a [The correct order of processes occurring during germination is imbibition (absorption of water), seed swelling (due to hydration), seed coat rupturing (due to swelling), and radicle growing (root growth).]
11. c [A developing eudicot embryo becomes heart-shaped because some of the embryo cells will differentiate and form the two cotyledons.]
12. b [Pollinators function to transfer pollen grains from one plant to a stigma on a different plant.]
13. b [An advantage of reproducing asexually is that there is a lower energy cost for producing a new plant than for sexual reproduction. There are a number of disadvantages to reproducing asexually.]
14. a [In monoecious species, all flowers contain both male and female reproductive structures. Monoe
15. a [Self-fertilization is reduced due to the incompatibility of a pollen grain and a stigma having the same S gene allele. This prevents inbreeding and promotes genetic variation.]
16. c [A somatic embryo can develop following differentiation of totipotent cells, which results in the formation of unspecialized cells that can then to develop to form a new embryo.]
17. e [For seeds requiring a dormancy period, dormancy ends when germination begins. During dormancy, no biological activity is

occurring however, during germination, many reactions are occurring.]

18. c[An endosperm contains three nuclei, two it received from the diploid central cell in the ovule and one it received from the sperm cell.]
19. d [Different plants have different cues to promote germination and increased day length, increased soil moisture, and increased temperature are all possible cues.]
20. b [In a developing seed, the suspensor transfers nutrients from the parent plant to the embryo, ensuring that the developing embryo has the energy it needs to reach maturity.]

Integrating and Applying Key Concepts

1. Totipotency is demonstrated in plants. Explain the concept by giving an example.
   Totipotency is a characteristic of virtually all cells in a plant and it is the ability for a cell to be able to form an entire new organism. If a branch (or root, or leaf, etc.) was cut off a plant and provided the correct conditions (water and nutrients), the cells at the cut would dedifferentiate and be able to form root cells so that the branch became an entire new plant.
2. Describe the S gene concept to explain incompatibility between pollen and female tissues.
   The S gene concept is that there are multiple alleles for the S gene so that individual plants of the same species will have different alleles. If the S gene allele in the pollen grain is different from the S gene allele in the stigma, then the pollen grain and stigma are compatible (came from different plants) and the processes for sexual reproduction will occur. However, if the S gene alleles in the pollen grain and the stigma are the same, they are incompatible (and possibly could have come from the same plant), fertilization will not occur (no seed production). This prevents inbreeding, which reduces the variation (and therefore success) in a plant population.
3. What is CCD and why is it a problem for farmers growing apple trees and/or berry plants?
   Colony Collapse Disorder (CCD) is the mysterious loss of honeybee colonies that have been noted globally. Since pollination of apples and berries require honeybee pollinators, continued loss of further colonies will have an impact on these plant species, as well as on all of the other plant species that rely on these pollinators (about 1/3 of North American plants). Since apple and berry farmers rely on bees to pollinate their orchards and crops, if the cause of CCD cannot be found and eliminated, there will be a drastic reduction in fruit production resulting in the loss of their livelihood.

Chapter 31 Control of Plant Growth and Development

Why It Matters [pp. 735-736]

1. sessile; 2. defense; 3. germination; 4. reproductive; 5. senescence; 6. environment; 7.volatile

31.1 Plant Hormones [pp. 736-746]

8. chemical; 9. environmental; 10. hormones; 11. small; 12. vascular; 13. meristems; 14. stimulate; 15. elongation; 16. light; 17. bend; 18.phototropism; 19. promote; 20. tips; 21. stimulate; 22. internodes; 23. bolting; 24. dormancy; 25. tips; 26. division; 27. xylem; 28. gas; 29. division; 30. fruit; 31. senescence; 32. asbscission zones; 33. steroid; 34. shoot; 35. elongation; 36. inhibits; 37. carotenoids; 38. inhibits; 39. environmental 40. germinate; 41. fatty; 42. carbohydrates; 43. regulate; 44. pathogens; 45. F; 46. A; 47. G; 48. B; 49. D; 50. C; 51. E; 52. D; 53. A; 54. B; 55. C;

56.

Hormone	*Action*
Auxins	Promote growth of plant primarily through cell elongation; promote fruit development; help plant in responding to light and gravity
Gibberellins	Promote cell division and elongation, seed germination, and bolting
Cytokinins	Promote cell division and inhibit senescence
Ethylene	Regulate many physiological processes; promote senescence, abscission, and fruit ripening
Brassinosteroids	Promote stem cell division and elongation, inhibits root elongation;

	promotes vascular development and growth of pollen tube
Abscisic acid	Inhibits growth and promotes dormancy
Jasmonates and Oligosaccharins	Protect plants from pathogens; regulates growth and seed germination

31.3 Plant Chemical Defenses [pp. 748-752]

57. response; 58. chemical message; 59. growth; 60. receptor; 61. cytoplasm; 62. protein; 63. signal; 64. secondary; 65. target; 66. bacteria, viruses, fungi and insects; 67. pathway; 68. jasmonates; 69. physical; 70. inhibitor; 71. proteins; 72. hypersensitive; 73. oxygen-containing; 74. nutrient; 75. salicylic acid; 76. pathogenesis-related; 77. secondary; 78. phytoalexins; 79. volatile; 80. predators; 81. specific; 82. resistance; 83. receptor; 84. avirulence; 85. salicylic acid; 86. pr proteins; 87. long-term; 88. A secondary messenger is produced by a primary messenger and functions as an intermediary to cause a change in a cell's activity whereas a secondary metabolite is a molecule that is not usually a product of metabolism but is produced when needed as a defense response.; 89. Pathogenesis-response proteins produce hydrolytic enzymes that break down a pathogen's cell walls. This kills pathogen cells, wounding and killing the pathogen.; 90. C; 91. B; 92. A; 93. E; 94. D

31.6 Plant Responses to the Environment: Responses to Temperature Extremes [pp. 760]

95. tropism; 96. phototropism; 97. cryptochrome; 98. Gravitropism; 99. downward; 100. upward; 101. elongation; 102. Thigmotropism; 103. tendrils; 104. support; 105. nastic; 106. turgor; 107. biological clock; 108. circadian; 109. photoperiodism; 110. phytochrome; 111. long-day; 112. short-day; 113. darkness; 114. vernalization; 115. dormancy; 116. long; 117. E; 118. F; 119. D; 120. G; 121. H; 122. B; 123. C; 124. A

Self-Test

1. a [Auxins were the first plant hormone to be identified.]
2. b [Went grew his seedlings in the dark so that he could observe the effect of the agar block on plant growth.]
3. a [Auxins exhibits polar transport and move away from unidirectional light.]
4. a [The breaking of bonds between microfibrils, increasing the plasticity of the cell wall.]
5. b [Gibberellins are involved in breaking of seed and bud dormancy.]
6. b [Gibberellins are involved in bolting seen in rosette plants.]
7. c [Cytokinins coordinate growth of roots and shoots in concert with the auxins.]
8. d [Ethylene stimulates the ripening of fruit.]
9. c [Abscisic acid promotes dormancy in plants.]
10. b [Abscission is the process of dropping of flowers, fruits, and leaves.]
11. d [Bolting is used for extension of the floral stem in rosette plants.]
12. e [Systemin, first peptide hormone, provides a defense response when it binds to a receptor.]
13. b [Shoots grow upwards, so show negative gravitropism.]
14. c [Thigmotropism is movement or growth of a plant in response to contact with an object.]
15. d [Nastic movement refers to temporary, reversible response to a non-directional stimulus.]
16. e [Photoperiodism refers to response of a plant due to changes in the length of light and dark periods during each 24-hour period.]
17. e [Vernalization is low temperature stimulation of flowering.]
18. a [Senescence is aging in plants.]
19. e [Phytochrome is involved in photoperiodism.]
20. a [Gibberellins are made by plants and fungi.]

Integrating and Applying Key Concepts

1. Fruit growers are able to pick and transport their fruit before it has completely ripened, which reduces the damage caused during transport. Once at the destination, ethylene can be applied to the fruit, which will cause it to ripen quickly for sale. Also, since

fruit produces its own ethylene, fruit growers can apply a chemical to inhibit ethylene production by fruit, thereby allowing storage of fruit for a long period of time.

2. A phytochrome pigment occurs in two forms, P_r which absorbs red light and P_{fr} in which it absorbs far red light. As P_f absorbs the red light, it is converted to P_{fr} and as P_{fr} absorbs far-red light, it is converted to P_r. A high concentration of P_{fr} tells a plant that there is lots of sunlight (middle of the summer) but this concentration will decrease as the plant approaches the end of the season.
3. An action potential is a reversal to the membrane charge across a plasma membrane, relative to the membrane potential at rest. Resulting from this charge reversal, potassium ion channels in the plasma membrane open and potassium ions will move out of a cell, reverting the membrane charge to its original state. The increased ion concentration outside of the cell causes water to move out of the cell by osmosis, resulting in decreased turgor pressure in the cell. When pulvinar cells become flaccid, associated leaves will move closer together (and when these cells become turgid again, associated leaves will spread apart).

Chapter 32 Introduction to Animal Organization and Physiology

Why It Matters [pp. 765-766]

1. homeostasis; 2. internal; 3. organ systems; 4. tissue; 5. function; 6. physiology; 7. anatomy

32.1 Organization of the Animal Body [pp. 766-767]

8. multicellular; 9. interstitial fluid; 10. molecules; 11. waste; 12. cells; 13. osmosis; 14. function; 15. tissue; 16. organ; 17. different; 18. organ system; 19. organs

32.2 Animal Tissues [pp. 767-775]

20. cytoskeleton; 21. extracellular; 22. junction; 23. epithelial; 24. cells; 25. shape; 26. extracellular; 27. fibers; 28. gel-like; 29. three; 30. muscle; 31. smooth; 32. nervous; 33. neurons; 34. glial; 35. D; 36. A; 37. B; 38. C; 39. D; 40. A; 41. B; 42. E; 43. C; 44. C; 45. G; 46. A; 47. F; 48. B; 49. E; 50. D; 51. B; 52. D; 53. E; 54A. Absorption, secretion, protection, diffusion; 54B. Lines body cavities, covers surfaces; 54C. Connective; 54D. Contraction (shorten); 54E. Body muscles, walls of organs and tubes, heart; 54F. Communication and control between body parts, conducts electrical signals; 55. Tight junctions hold epithelial cells together to line the urinary bladder. This junction is necessary since only this junction seals the spaces between cells, so waste molecules and ions cannot leak out of the bladder into other body tissues. 56. An osteoblast produces the collagen and mineral deposits of the extracellular matrix (building bone tissue) whereas osteoclasts remove the mineral deposits (break down bone tissue). 57. Fat is able to store more chemical energy per weight than carbohydrates.

32.3 Coordination of Tissues in Organs and Organ Systems [pp. 775-776]

58. cell; 59. Survive; 60. Organ system; 61. 11; 62. Nutrients, 63. Wastes; 64. Responding; 65. Reproducing; 66. C; 67. D; 68. F; 69. E; 70. K; 71. A; 72. I; 73. J; 74. B; 75. E; 76. G; 77. F; 78. A, I; 79. I; 80. C; 81. G; 82. H

32.4 Homeostasis [pp. 776-779]

83. Homeostasis is a dynamic process that ensures animals maintain a relatively constant internal environment (within an acceptable range). If a body function, such as body temperature, changes due to a change in the external (or internal) environment, the body's control mechanisms are able to detect any changes and make modifications accordingly to keep (or return) the body function to within its acceptable range. 84. A positive feedback mechanism is an amplification process, intensifying the change and therefore producing even more of the product, whereas a negative feedback mechanism will respond to this excess product by compensating for this change and reducing the amount of product being produced, returning the amount of product to within an acceptable range.; 85. C; 86. F; 87. E. 88. B; 89. A; 90. D

SELF-TEST

1. a [a cell must maintain homeostasis whether a change comes from internal or external environments]
2. b [an organ, like a muscle, is composed of two or more different tissues]

3. a [the liver (organ) contains epithelium tissue on its surface and contains hepatic cells, each of which contains mitochondria (organelles)]
4. d [interstitial fluid is found outside of cells, providing various functions to ensure cell survival]
5. c [anatomy is the study of structure]
6. d [epithelial tissue is the lining of body cavities and has little or no extracellular matrix]
7. b [only gap junctions allow molecules and ions to move between cells using open channels]
8. c [these characteristics are of blood, a connective tissue; red blood cells are involved in oxygen delivery to cells; and white blood cells are involved in immunity]
9. c [connective tissue is the most varied major tissue, being composed fibers, extracellular matrix, and a specific cell type associated with the particular type of connective tissue]
10. b [an osteon a unit of bone consisting of osteocytes, porous calcium phosphate minerals, a blood vessel, and nerve endings (sensory)]
11. b [movement of ions across gap junctions ensures that smooth muscle cells contract together as a unit]
12. c [erythrocytes are red blood cells and leukocytes are white blood cells; both are components of blood]
13. a [an electric signal travels through a dendrite to the cell body and then through the axon within a neuron]
14. d [chemical signal arising from one neuron can travel to a gland cell (secretes), a muscle cell (contracts), or another nerve cell (continue transmitting message)]
15. b [the excretory system is critical in maintaining osmotic balances, ions or electrolytes, and pH of body fluids]
16. b [sweat glands are derivatives of the skin (integument) which is a major protective coat between the internal and external environment of organisms]
17. e [the digestive system takes in food (containing proteins and carbohydrates) and breaks them down into aminoa acids and sugars respectively]
18. c [the integrator is a control centre that compares a set point to a detected change]
19. b [positive feedback systems are characterized by an amplification of the response or product]
20. a [increasing sweat gland activity is a way to reduce body temperature since heat is removed when sweat evaporates from the skin's surface]

Integrating and Applying Key Concepts

1. Sweat is produced in epithelial tissue that has formed an exocrine gland and therefore the sweat secretions produced by these secreting epithelial cells are emptied onto the epithelial surface via ducts.
2. Bone, cartilage, and adipose are all connective tissues and therefore are composed of cells in extracellular matrices. Bone is composed of bone cells, called osteocytes, and a surrounding porous extracellular matrix composed of collagen fibres and glycoproteins impregnated with calcium phosphate minerals. This tissue functions to form the skeleton, supports the body, and protects softer tissues. Cartilage is composed of cells called chondrocytes which are surrounded by collagen fibres in an elastic and resilient matrix of glycoprotein. It functions in various ways including as a support for body parts such as larynx, trachea, and small air passages, a cushion between vertebrae, and a precursor to bone during embryonic development. Adipose tissue is composed of cells called adipocytes (store fat) which is surrounded by little extracellular matrix. It cushions the body and may help insulate it.
3. The three types of muscle tissue are skeletal, cardiac, and smooth muscle tissue. All three function to contract, shortening to provide movement. Skeletal tissue is composed of long muscle cells (muscle fibres) that have striations due to the orderly actin and myosin filaments with each of the cells. It is attached to bones by tendons and forms muscle organs, such as the biceps. Cardiac tissue is found in the walls of the heart and is composed of relatively short striated cells, like skeletal tissue, but unlike skeletal tissue, the cells are connected to neighbouring cells by intercalated disks in branches and form a network. This allows contractions in many directions (unlike the other tissues), allowing for the pumping action required to produce a heartbeat. Smooth tissue, like the other two tissues, contains actin and myosin filaments but they are arranged so loosely that no visible striations are apparent in their small, spindle-shaped cells. Contraction is relatively slow but can be maintained for a longer period

than the other two tissues. It is found in the wall of the digestive tract organs, wall of many blood vessels, and the wall of the uterus.

Chapter 33 Information Flow: Nerves, Ganglia, and Brains

Why It Matters [pp. 783]

1. nervous system

33.1 Neurons and their Organization in Nervous System: An Overview [pp. 784-787]

2. neural signaling; 3. neurons; 4. reception; 5. transmission; 6. integration; 7. response; 8. afferent neurons; 9. sensory neurons; 10. interneurons; 11. efferent neurons; 12. effectors; 13. cell body; 14. dendrites; 15. axons; 16. axon hillock; 17. axon terminals; 18. neural circuits; 19. glial cells; 20. astrocytes; 21. oligodendrocytes; 22. Schwann cells; 23. nodes of Ranvier; 24. synapse; 25. presynaptic cell; 26. postsynaptic cell; 27. electrical synapse; 28. chemical synapse; 29. neurotransmitter; 30. synaptic cleft; 31. pre-; 32. post-; 33. interneuron; 34. b; 35. a; 36. a; 37. b; 38. b; 39. c; 40. a; 41. Efferent neurons carry impulses away from the interneuron/network to effectors in general (could be a gland or muscle). Motor neurons are specific efferent neurons that carry impulses to muscle (could be smooth muscle, skeletal muscle, or cardiac muscle); 42. b; 43. d; 44. a; 45. c; 46. neural support cells; 47. astrocytes; 48. wrap around axons of neurons in the central nervous system; 49. wrap around axons of neurons in a peripheral nervous system; 50. neurons; 51. e; 52. c; 53. f; 54. d; 55. b; 56. g; 57. a; 58. dendrites; 59. axon; 60. axon hillock; 61. axon terminal; 62. cell body (soma)

33.4 Integration of Incoming Signals by Neurons [pp. 798-799]

63. membrane potential; 64. resting potential; 65. polarized; 66. action potential; 67. depolarized; 68. threshold potential; 69. hyperpolarized; 70. all-or-nothing principle; 71. refractory period; 72. voltage-gated ion channels; 73. propagation; 74. salutatory conduction; 75. presynaptic membrane; 76. ligand-gated ion channels; 77. postsynaptic membrane; 78. synaptic vesicles; 79. exocytosis; 80. direct neurotransmitters; 81. indirect neurotransmitters; 82. excitatory postsynaptic potential (EPSP); 83. inhibitory postsynaptic potential (IPSP); 84. graded potential; 85. temporal summation; 86. spatial summation; 87. neuron-; 88. de-; 89. hyper-; 90. a; 91. b; 92. b; 93. a; 94. a; 95. b; 96. b; 97. a; 98. b; 99. a; 100. All cells display a separation of positive and negative charges across their membrane, but this potential (membrane potential) remains unchanged. Some cells, such as neurons and muscles, possess membranes (excitable membranes) that are capable of changing their potential. The resting potential is the membrane potential of an unstimulated nerve or muscle cell; 101. e; 102. b; 103. a; 104. d; 105. c; 106. f;
107. resting potential; 108. threshold potential; 109. depolarization; 110. repolarization; 111. refractory period;
112. hyperpolarization (undershoot).

33.6 Vertebrates Have the Most Complex Nervous Systems [pp. 802-804]

113. nerve nets; 114. ganglia; 115. brain; 116. nerve cords; 117. central nervous system (CNS); 118. peripheral nervous system (PNS); 119. neural tube; 120. spinal cord; 121. ventricles; 122. central canal; 123. forebrain; 124. midbrain; 125. hindbrain; 126. hypothalamus; 127. postganglionic; 128. preganglionic; 129. a; 130. b; 131. loose meshes of neurons in certain animal groups with radial symmetry; 132. ganglion; 133. bundle of nerves that extend from a central ganglion.

33.8 The Peripheral Nervous System [pp. 810-812]

134. somatic nervous system; 135. autonomic nervous system; 136. sympathetic division; 137. parasympathetic division; 138. meninges; 139. cerebral spinal fluid; 140. gray matter; 141. white matter; 142. reflex; 143. brainstem; 144. cerebral cortex; 145. blood-brain barrier; 146. reticular formation; 147. cerebellum; 148. thalamus; 149. hypothalamus; 150. basal nuclei; 151. limbic system; 152. amygdale; 153. hippocampus; 154. olfactory bulbs; 155. corpus callosum; 156. primary somatosensory area; 157. association areas; 158. primary motor area; 159. lateralization; 160. b; 161. a; 162. a; 163. b; 164. a; 165. b; 166. control body movement (mostly voluntary); 167. autonomic; 168. sympathetic; 169. para-sympathetic; 170. connects the two cerebral hemispheres; 171. primary somatosensory area; 172. primary motor area; 173. integrates sensory information and formulates responses; 174. a; 175. j; 176. l;

177. e; 178. g; 179. c; 180. d; 181. i; 182. k; 183. n; 184. h; 185. m; 186. b; 187. f

33.9 Memory, Learning, and Consciousness [pp. 812-814]

188. memory; 189. learning; 190. consciousness; 191. short-term memory; 192. long-term memory; 193. long-term potentiation; 194. electroencephalogram; 195. rapid eye-movement (REM) sleep; 196. a; 197. b; 198. a; 199. d; 200. f; 201. c; 202. b

SELF-TEST

1. b, c, d, e
2. a, c
3. a, c, d
4. a, c, e
5. a, c
6. a, c
7. b, e
8. a, d [a is correct because the membrane cannot be stimulated if the influx of sodium is blocked; therefore, the inactivation of voltage-gated sodium channels marks the onset of the refractory period. The membrane will remain refractory until the inactivation of the voltage-gated sodium channel is lifted; this corresponds with the reestablishment of the resting potential, making d also correct.]
9. c, d [c is correct because as sodium influx occurs and the membrane becomes depolarized in a specific region, adjacent voltage-gated sodium channels are prompted to open; however, because voltage-gated sodium channels become inactivated at the peak of an action potential, only those channels that are in front of the wave (which have not recently opened and are not inactive) are capable of responding and opening, while those behind the wave (which have just recently opened) are inactive and not capable of responding. Thus, the impulse travels only in one direction in a wavelike manner; d also is correct because of the all-or-nothing nature of action potentials; once threshold is reached, the voltage-gated sodium channels open, allowing rapid (initially) sodium influx and depolarization of the membrane in that specific region. Because the amplitude of each action potential is the same, the impulse is constantly "refreshed" as the wave of depolarization moves along the axon, assuring that it will reach the axon terminus. (This is important because some motor neurons can be several meters in length!)]
10. a, c, d
11. b, d [b is correct because an inhibitory neuron releases neurotransmitters that bind to the postsynaptic cell and can open chloride and potassium channels. Because of the asymmetric distribution of ions across the membrane (high sodium and chloride outside, high potassium inside), the opening of chloride channels causes chloride influx (down its concentration gradient), while the opening of potassium channels results in potassium efflux (down its concentration gradient). The net result is that the inside becomes more negative relative to the outside than when the membrane was at resting potential; therefore, the membrane is said to be hyperpolarized; d also is correct because the flux of potassium and chloride is proportion to the strength of the inhibitory signal (based on how many neurotransmitters where bound and how many channels were affected).]
12. a
13. d [d is correct because the hypothalamus is responsible for coordination various growth, development, reproductive, osmoregulatory and other processes; neurons in the hypothalamus release hormones that affect these processes (see chapter 40); a is not correct because this region is involved in sensory integration and motor control; b is not correct because this region is involved in high functions such as emotions, memory, etc.; c is not correct because this region is the region that receives sensory information; and e is not correct because this region integrates and sorts sensory information]
14. b, d [b is correct because the autonomic nervous system has primary responsibility for the viscera and blood vascular system; d is correct because most arteriole are innervated only with branc
15. b, c [b is correct because the autonomic nervous system has primary responsibility for the viscera and blood vascular system; d is correct because the parasympathetic division is responsible for coordinated most processes associated with feeding and digestion (e.g., saliva release, gut peristalysis, etc.)]

16. a, b, c, e
17. b
18. a, b
19. b
20. a, c
21. c

Chapter 34 Sensory Systems

Why It Matters [pp. 819-820]

1. sensory systems

34.3 Mechanoreception and Hearing [pp. 831-833]

2. sensory receptors; 3. sensory transduction; 4. mechanoreceptors; 5. photoreceptors; 6. chemoreceptors; 7. thermoreceptors; 8. nociceptors; 9. frequency of action potentials; 10. number of neurons activated; 11. sensory adaptation; 12. Pacinian corpuscles; 13. proprioceptors; 14. statocysts; 15. statoliths; 16. sensory hair cells; 17. lateral line system; 18. neuromasts; 19. stereocilia; 20. cupula; 21. vestibular apparatus; 22. semicircular canals; 23. utricle; 24. saccule; 25. otoliths; 26. stretch receptors; 27. muscle spindles; 28. Golgi tendon organ; 29. tympanum; 30. pinna; 31. outer ear; 32. tympanic membrane; 33. middle ear; 34. malleus; 35. incus; 36. stapes; 37. oval window; 38. inner ear; 39. semicircular canal; 40 utricle; 41. saccule; 42. cochlea; 43. organ of Corti; 44. round window; 45. echolocation; 46. chemoreceptor; 47. thermoreceptor; 48. otolith; 49. proprioceptor; 50. photoreceptor; 51. nociceptor; 52. lateral line system; 53. detect position and orientation; used for equilibrium in invertebrates; 54. perceives position and motion of head; 55. muscle spindle; 56. organ of Corti; 57. Proprioceptors that detect stretch and compression of tendon; 58. neuromast; 59. a; 60. h; 61. i; 62. e; 63. c; 64. g; 65. b; 66. f; 67. d; 68. pinna; 69. Eustachian tube; 70. stapes; 71. incus; 72. malleus; 73. semicircular canals; 74. oval window; 75. auditory canal; 76. tympanic membrane; 77. round window; 78. cochlea; 79. outer ear; 80. middle ear; 81. inner ear

34.7 Electroreceptors and Magnoreceptors [pp. 847]

82. ocellus; 83. compound eye; 84. ommatidia; 85. cornea; 86. photopigment; 87. single-lens eye; 88. lens; 89. retina; 90. iris; 91 pupil; 92. accommodation; 93. aqueous humor; 94. vitreous humor; 95. ciliary body; 96. rods; 97. cones; 98. fovea; 99. peripheral vision; 100. photopigments; 101. retinal; 102. opsins; 103. rhodopsin; 104. bipolar cells; 105. ganglion cells; 106. horizontal cells; 107. amacrine cells; 108. lateral inhibition; 109. photopsin; 110. optic chiasm; 111. lateral geniculate nuclei; 112. sensilla; 113. taste buds; 114. olfactory hairs; 115. transient receptor potential (TRP); 116. magnoreceptors; 117. A compound eye contain 100s to 1000s of individual visual units. A single-lens eye has one lens and operates like a camera; 118. Photo-pigments consist of a covalent complex of retinal and one of several different proteins know as opsins. The photopigment in rod cells is rhodopsin. Cone cells contain different types of photopsins based upon different opsin forms; humans have three photopsins; 119. Accommodation is the process of focusing and image by moving the lens back and forth relative to the retina; 120. Electroceptors detect electric fields. Electroceptors depolarize in an electric field. Magnetoceptors allow animals to detect and use Earth's magnetic field as a source of directional information. 121. bipolar cell; 122. extend over entire retina, and axons come together to form optic nerve; 123. amacrine cell; 124. connect with different photoreceptor cells and bipolar cells; 125. photoreceptor; 126. radiant energy; 127. distinguish tastes of sweet, sour, salty, and umami; 128. olfactory hair; 129. chemicals; 130. electroreceptor; 131. communication; locate objects (including prey); 132. magnetic field; 133. nociceptor; 134. tissue damage; noxious chemicals; 135. a; 136. i; 137. f; 138. c; 139. d; 140. b; 141. e; 142. g; 143. h; 144. j;145. ciliary body; 146. iris; 147. lens; 148. pupil; 149. cornea; 150. aqueous humor; 151. vitreous humor; 152. retina; 153. fovea; 154. optic nerve

SELF-TEST

1. c
2. a, c, e [a is correct because these are used by invertebrates to detect position; c is correct because this structure is used for maintaining equilibrium in vertebrates; e is correct because some fish have these structures to provide information about orientation; b and d are not correct because these structures detect vibration (sound)]
3. b
4. a, b,
5. a
6. b, d [invertebrates use b to focus images, whereas vertebrates use d; a is not correct because this process sharpens images by enhancing contrast; c is not correct because this structure is the visual unit of a compound eye]
7. d
8. a
9. d [d is correct because these compounds bind to opioid receptors and block the release of substance P, which makes b not correct because this compound is released from axons and conveys the sensation of pain to the CNS; a is not correct because this chemical gives the "hot" taste in food; c is not correct because insulin stimulates the uptake of glucose and other nutrients into cells and does not affect pain perception]
10. a, b, c
11. a, b, c, d
12. b, c
13. d
14. b
15. a, b, c, d
16. a
17. a, b
18. a, b
19. c
20. b

---

35 The Endocrine System

Why It Matters [pp. 851-852]

1. hormones; 2. endocrine systems

35.1 Hormones and Their Secretions [pp. 852-854]

3. neurohormone; 4. hyposecretion; 5. hypersecretion; 6. hyperglycemic; 7. endocrine;

8. The nervous system and the endocrine system regulates and coordinates bodily functions. The two systems are structurally, chemically, and functionally related, but they control different types of activities. The nervous system communicates through neurons and primarily by chemical synapses and involve rapid communication. The nervous system directs highly specific localized target. It is a "private mode" of communication. The endocrine system typically involves the secretion of hormones in the blood circulation that act at distant target sites. The endocrine system controls activities that involve slower, longer-lasting responses. The endocrine system is more "public", often affecting several tissues or organs.

9. a signaling molecule secreted by a cell that can alter the activities if any cell with receptors for it. Hormones are typically transported in the bloodstream to alter its physiological activity of target cells/tissues.

10. System of glands that release their secretions (hormones) directly into the circulatory system to alter the physiological activity of target tissues.

11. Hormones within an endocrine axis can interact with the cells that secreted the hormones within the axes to either inhibit or activate further secretion of hormones within that axes. In the case of negative feedback, an example includes the ability of cortisol to

inhibit cortisol secretion from the adrenal cortex, inhibit the secretion of ACTH from the anterior pituitary, and inhibit CRH secretion from the hypothalamus. Overall, this inhibition controls the secretion of cortisol from the adrenal cortex. An example of positive feedback control includes the continuous secretion of prolactin as an infant suckles the breast of a lactating mother. The suckling reflex causes additional secretion of prolactin from the pituitary which promotes milk secretion. In this way, the mother continuously provides milk to her offspring. The benefits of such regulatory pathways to organisms is to regulate hormone concentrations in the blood and to regulate cell/tissue responses to the hormones. 12. At any given time, a number of hormones are secreted into the bloodstream to elicit a number of physiological responses. The hormones secreted at any given time will interact with a number of cells/tissues/organs to coordinate the integration of a number of physiological responses that will ultimately controls a given process. 13. endocrine glands; 14. neurosecretory cells; 15. amine; 16. peptide; 17. steroid; 18. fatty acid derivative; 19. growth factors; 20. prostaglandins; 21. the target cells; 22. target proteins; 23. neuronsecretion; 24. the release of a hormone from an epithelial cell in a gland that is transported in the blood and generally effective at a distance from its site of secretion; 25. the release of a chemical signal into extracellular fluid that regulated the activity of a neighboring cell; 26. autocrine (or autoregulation); 27. epinephrine; 28. insulin; 29. cortisol; 30. fatty acid derivative; 31. c; 32. d; 33. b; 34. a

35.2 Mechanisms of Hormone Action [pp. 854-859]

35. Many hormones are secreted in an inactive or less active form (a prohormone) and converted by the target cells or enzymes in the blood or other tissue to the active form; 36. Hydrophilic hormones bind cell surface receptors and activate a signal transduction cascade, involving enzymatic reactions, inside the cell that leads to a cell response. Hydrophobic hormones, are usually transported bound to a protein and bind to receptors inside cells activating or inhibiting genetic regulatory proteins. Both cases include reception of the signal (cell surface or intracellular receptor), transduction and amplification of the signal, and a cellular response; 37. Only cells that express receptors to a given hormone will respond to that hormone. Cells may express a number of different types of receptors and therefore respond to different hormones. Different cells may respond differently to the same hormone owing to the different mechanisms that are activated by the receptors in those cell types. In addition, the actions of one hormone may modify the responses to another hormone in different cell types; 38., epinephrine is an amine and a hydrophilic compound, it interacts with cell surface receptors; 39. insulin is a protein and is hydrophilic, it interacts with cell surface receptors; 40. testosterone is a steroid and is hydrophobic, it interacts with intracellular receptors to regulate gene expression; 41. oxytocine is a peptide and is hydrophilic, and interacts with cell surface receptors; 42. estrogen is a steroid and is hydrophobic, it interacts with intracellular receptors to regulate gene expression; 43. Aldosterone is a steroid and is hydrophobic, it interacts with intracellular receptors to regulate gene expression; 44. Thyroid hormone is an amine but is hydrophobic, it interacts with intracellular receptors to regulate gene expression; 45. Prostaglandin is derived from fatty acids and is hydrophobic. Prostaglandins, however, interacts with cell surface receptors.

35.3 The Hypothalamus and Pituitary [859-862]

46. pituitary gland; 47. posterior pituitary; 48. anterior pituitary; 49. tropic hormone; 50. releasing hormones; 51. inhibiting hormones; 52. antidiuretic hormone (ADH); 53. oxytocin; 54. prolactin; 55. growth hormone; 56. thyroid stimulating hormone (TSH); 57. adrenocorticotropic hormone (ACTH); 58. follicle stimulating hormone (FSH); 59. luteinizing hormone (LH); 60. gonadotropins; 61. melanocyte stimulating hormone (MSH); 62. endorphins; 63. The anterior pituitary is a distinct lobe that produces and secretes many hormones into the general circulation (e.g., growth hormone). The posterior pituitary does not produce any hormone. It is the site where axons from the hypothalamus terminate to release tropic hormones into the portal system that regulate the anterior pituitary or to release peptides such as ADH or oxytocin directly into the general circulation; 64. anterior pituitary; 65. stimulates growth and regulates metabolism; 66. prolactin; 67. peptide; 68. anterior pituitary; 69. TSH; 70. peptide; 71. stimulates adrenal cortex to produce cortisol and aldosterone; 72. peptide; 73. anterior pituitary; 74. promotes gamete development; 75. peptide; 76. anterior pituitary; 77. promotes sex steroid production; 78. MSH; 79. peptide; 80. anterior pituitary (intermediate lobe, when present); 81. peptide; 82. anterior pituitary (intermediate lobe, when present); 83. inhibit perception of pain; 84. peptide; 85. posterior pituitary (produced in hypothalamus); 86. stimulates water reabsorption (conservation); 87. Oxytocin; 88. peptide; 89. posterior pituitary (produced in hypothalamus); 90. anterior pituitary; 91. breast development and milk secretion. 92. neurosecretory neuron (some produce releasing hormones; some produce inhibiting hormones); 93. hypothalamus; 94. protein vein; 95. posterior pituitary; 96. anterior pituitary; 97.

neurosecretory neuron (some produce ADH, some produce oxytocin); 98. hypothalamus; 99. anterior pituitary; 100. posterior pituitary;

35.4 Other Major Endocrine Glands of Vertebrates [pp. 862-867]

101. thyroid gland; 102. thyroxin (T4); 103. triiodothyronine (T3); 104. metamorphosis; 105. calcitonin; 106. parathyroid hormone; 107. parathyroid gland; 108. vitamin D; 109. adrenal medulla; 110. adrenal cortex; 111. catecholamines; 112. epinephrine; 113. norepinephrine; 114. glucocorticoids; 115. mineralocorticoids; 116. cortisol; 117. aldosterone; 118. testes; 119. ovaries; 120. androgens; 121. estrogens; 122. progestins; 123. testosterone; 124. 17β-estradiol; 125. progesterone; 126. Islets of Langerhans; 127. pancreas; 128. insulin; 129. glucagon; 130. diabetes mellitus; 131. pineal gland; 132. melatonin; 133. a; 134. c; 135. e; 136. g; 137. i; 138. j; 139. f; 140. h; 141. b; 142. d; 143. amine; 144. regulate basal metabolic rate; triggers metamorphosis in amphibians; 145. calcitonin; 146. thyroid gland; 147. increases blood calcium; 148. epinephrine; 149. adrenal medulla; 150. cortisol; 151. steroid; 152. steroid; 153. adrenal cortex; 154. increase sodium and water reabsorption; 155. testosterone; 156. estradiol; 157. steroid; 158. promotes development and maintenance of secondary sex characteristics; 159. prepares and maintains uterus for implantation; 160. gonadotropin releasing hormone; 161. posterior pituitary (produced in hypothalamus); 162. peptide; 163. islets of Langerhans; 164. anabolic; stimulates nutrient uptake into cells and macromolecule synthesis; 165. glucagon; 166. peptide; 167. islets of Langerhans; 168. melatonin; 169. peptide; 170. helps maintain daily biorhythms.

35.5 Endocrine Systems in Invertebrates [p. 867-869]

171. brain hormone; 172. ecdysone; 173. juvenile hormone; 174. molt-inhibiting hormone; 175. a; 176. d; 177. b; 178. c

SELF-TEST

1. a
2. a
3. c
4. b, c [b is correct because the secretion of the hormone from the source cell is reduced in negative feedback; c is correct because homeostasis is maintained with negative feedback]
5. c
6. a, c, e [a and c are correct because steroids and thyroid hormones are soluble in the cell membrane and enter the cytoplasm where they bind to receptors inside the cell; e is correct because the hormone-receptor complexes formed after hormone binding affect the transcription of genes; b and d are not correct because these hormones bind to membrane-associated receptors and initiate rapid responses inside cells without altering gene expression]
7. b
8. d
9. a, b, c, d
10. b [b is correct because glucagon acts on the liver to promote the breakdown of stored glycogen to glucose as well as the formation of glucose (i.e., gluconeogenesis) from noncarbohydrate sources (e.g., amino acids); these actions result in the elevation of glucose levels in the blood]
11. b
12. b, c
13. a, d, e
14. a, b
15. a, b, d
16. b, c, d, e
17. a
18. c
19. a, b, c, d

20. a, c

Chapter 36 Muscle, Bones, and Body Movements

Why It Matters [pp. 933–934]

1. skeletal; 2. cardiac; 3. smooth; 4. skeletal muscle

36.1 Vertebrate Skeletal Muscle: Structure and Function [pp. 874-880]

5. muscle fibers; 6. myofibrils; 7. thick filaments; 8. thin filaments; 9. sarcomere; 10. T-tubules; 11. sarcoplasmic reticulum; 12. neuromuscular junction; 13. acetylcholine; 14. sliding filament mechanism; 15. muscle twitch; 16. tetanus; 17. slow-muscle twitch; 18. fast-muscle twitch; 19. motor units; 20. sarcomere; 21. myofibril; 22. myoglobin; 23. a; 24. b; 25. a; 26. h; 27. g; 28. c; 29. b; 30. d; 31. I; 32. f; 33. e; 34. neuromuscular junction; 35. T-tubule; 36. sarcoplasmic reticulum; 37. myofibril; 38. sarcomere; 39. When an action potential arrives at a neuromuscular junction, the axon terminal releases a neurotransmitter, acetylcholine, which triggers an action potential in the muscle fibre. The action potential travels in all direction over the muscle fibre's surface membrane and penetrates into the interior of the fibre through the T tubules. When an action potential reaches the end of a T tubule, it opens ion channels in the sarcoplasmic reticulum that allow Ca^{2+} to flow out into the cytosol. When Ca^{2+} flows out into the cytosol, the troponin molecules of the thin filament bind the calcium and undergo a conformational change that causes the tropomyosin fibres to slip into the grooves of the atin double helix. The slippage uncovers the actin's binding site for the myosin crossbridge, allows myosin to bind to actin. Bending of the myosin molecule allows the muscle to then contract. 40. During contraction, calcium is responsible to bind to troponin so that troponin can undergo a conformational change that causes the tropomyosin fibres to slip into the grooves of the actin double helix and allow myosin to bind actin and initiate a contraction. During relaxation, calcium must be pumped back into the sarcoplasmic reticulum. The decrease in cytosolic calcium will cause troponin to cover the actin's binding site for myosin (myosin can no longer bind actin) thus relaxing the muscle. ATP, on the other hand, provides the energy for the bending of the myosin molecule to cause the movement of actin filaments relative to the myosin filaments, thus causing contraction. ATP is also needed to cause the detachment of myosin from the actin molecule. In the absence of ATP, myosin remains attached to actin and cannot proceed with contraction or relaxation (an example is *rigor mortis*). 41. Following the arrival of the action potential at the muscle fibre and the release of calcium from the sarcoplasmic reticulum, the crossbridge events include: a) calcium binding to troponin on actin filaments causing tropomyosin to be displaced into the grooves, this uncovers the actin's binding site for the myosin crossbridge. B) ATP is hydrolyzed and the myosin crossbridge bends and binds to a binding site on an actin molecule. C) The binding triggers the crossbridge to snap back toward the tail, pulling the thin filament over the thick filament (the power stroke). ADP is released. d) ATP binds to the crossbridge, causing myosin to detach from actin. The cycle then repeats itself. When action potentials stop, calcium is taken up by the sarcoplasmic reticulum and the contraction stops. The thin filaments slide back to their original relaxed positions.

36.2 Skeletal Systems [pp. 881-883]

42. hydrostatic skeleton; 43. exoskeleton; 44. endoskeleton; 45. axial skeleton; 46. appendicular skeleton; 47. The skeletal system provide physical support for the body and protection for the soft tissues. It also acts as a framework against which muscles work to move parts of the body or the entire organism; 48. a; 49. b;

36.3 Vertebrate Movement: The Interactions between Muscles and Bones [pp. 943–946]

50. synovial; 51. cartilaginous; 52. fibrous; 53. agonist; 54. antagonistic pairs; 55. extensor muscles; 56. flexor muscles; 57. b; 58. a; 59. a; 60. d; 61. e; 62. f; 63. b; 64. c

SELF-TEST

1. c [c is correct because only actin and myosin filaments move relative to one another; b is not correct because actin and myosin are arranged parallel to one another; d is not correct because actin does not normally dissociate during sarcomere shortening]
2. c, d
3. b

4. c, e
5. d, e
6. a, b, c
7. c, e
8. d
9. a
10. d
11. a
12. b
13. d
14. a, b
15. c
16. a, b
17. a
18. a, c
19. c
20. b

Chapter 37 The Circulatory System

Why It Matters [pp. 891-892]

1. circulatory system; 2. lymphatic system; 3. body's defences.

37.1 Animal Circulatory Systems: An Introduction [pp. 892-896]

4. open circulatory system; 5. haemolymph; 6. sinuses; 7. closed circulatory system; 8. arteries; 9. capillaries; 10. veins; 11. atria; 12. ventricles; 13. systemic circuit; 14. pulmonocutaneous circuit; 15. pulmonary circuit; 16. specialized fluid medium contain some cells (carries nutrients, wastes, gases), tubular vessels for the specialized fluid, and a muscular heart that pumps the fluid through the circulatory system; 17. Functions: circulation of respiratory gases, transport of nutrients, transport of products of metabolism, transport of wastes, maintaining blood pressure, immune defences, blood clotting; 18. in an open circulatory system , vessels leaving the heart release fluid, haemolymph, directly into the body spaces or into sinuses surrounding organs. The haemolymph re-enters the heart through valves in the heart wall that close each time the heart pumps. In a closed circulatory system, the blood is confined to blood vessels and is distinct from the interstitial fluid. Substances are exchanged between the blood and the interstitial fluid, and then between the interstitial fluid and cells; 19. a; 20. a; 21. b; 22. b; 23. b; 24. a; 25. b; 26. c; 27. a; 28. b

37.2 Blood and Its Components [pp. 896-899]

29. plasma; 30. albumins; 31. globulins; 32. fibrinogen; 33. erythrocytes; 34. red blood cells (RBC); 34. erythrocytes; 35. erythropoietin; 36. leukocytes; 37. platelets; 38. fibrin; 39. erythrocyte; 40. leukocyte; 41. haemoglobin; 42. leukocytes; 43. specialized to transport O_2; 44. platelets; 45. d; 46. b; 47. c; 48. h; 49. f; 50. a; 51. g; 52. e; 53. c; 54. Haemoglobin consists of four polypeptides, each linked to a nonprotein haem group that contains an iron atom in its centre. The iron atom binds O_2 molecules as the blood circulates through the lungs and releases the O_2 as the blood flows through other body tissues.; 55. When blood vessels are damaged, collagen fibres in the extracellular matrix are exposed to the leaking blood. Platelets in the blood stick to the collagen fibres and release signalling molecules that induce additional platelets to stick to them. The process continues, forming a plug that helps seal off the damaged site. As the plug forms, platelets release other factors that convert the soluble plasma protein, fibrinogen into insoluble threads of fibrin. Crosslinks between the fibrin threads form a meshlike network that traps blood cells and platelets and further seals the damaged area.

37.3 The Heart [pp. 899–903] **37.4 Blood Vessels of the Circulatory System** [pp. 903-905] **37.5 Maintaining Blood Flow and Pressure** [pp. 906] **37.6 The Lymphatic System** [pp. 907-908]

56. aorta; 57. systole; 58. diastole; 59. atrioventricular valves; 60. neurogenic heart; 61. myogenic heart; 62. sinoatrial node; 63. pacemaker cells; 64. atrioventricular node; 65. electrocardiogram; 66. arterioles; 67. precapillary sphincter; 68. venules; 69. cardiac output; 70. lymphatic system; 71. lymph; 72. lymph nodes; 73. a; 74. b; 75. b; 76. a; 77. b; 78. a; 79. a; 80. b; 81. extensive network of vessels that collects excess interstitial fluid; 82. lymph node; 83. lymph; 84. a record reflecting the electrical activity of the heart by attaching electrodes to the surface of the body; 85. a) heart is relaxed, atria begins to fill with blood, b) blood fills atria and pushes AV valves open, ventricles begin to fill with blood, c) atria contract filling ventricles completely, d) ventricles begin to contract, forcing AV valves closed, e) ventricles contract fully, forcing semilunar valves open and ejecting blood into arteries.

86. see figure 37.11; 87. Pacemakers generates a wave of signals (action potentials) in the SA node which spreads into both atria. This wave of signals cause the atria to contract. The propagation of the wave reaches the AV node. The AV node cells are stimulated to produce a signal, which spreads along Purkinje fibres to the bottom of the heart. The signals then spread from the bottom of the heart upward, causing the ventricles to contract.

88.

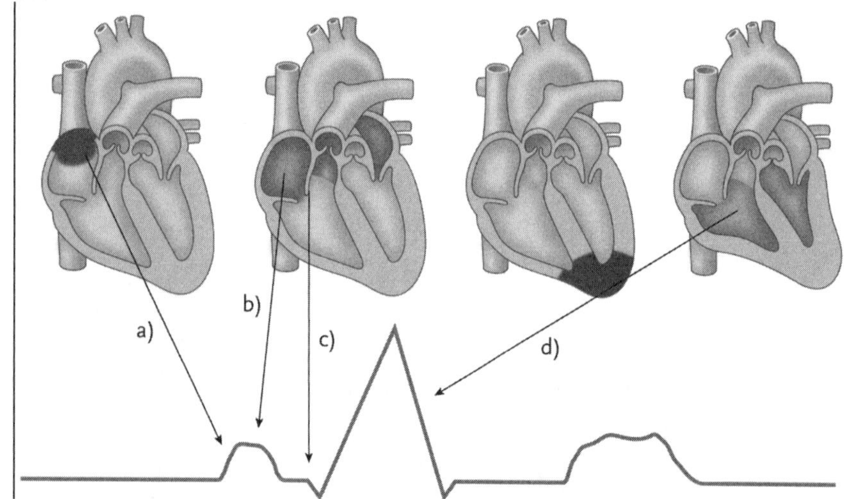

a) Prior to the generation of waves from the SA pacemakers, the atria are filling with blood. This beginning of the first wave of the ECG appears (depolarization of the pacemakers). b) Once the action potentials are produced and spread to the both atria, the atria will contract to pump the blood into the ventricle (peak of the first wave of the ECG). c) The spread of the wave reaches the AV node. The delay in the spread of the wave allows the ventricles to fill with blood (delay between depolarization of the atria and the depolarization wave of the ventricles (d)). d) Once the AV node are stimulated to produce a signal, the signal spreads to the Purkinje fibres and upwards the heart, causing the ventricles to contract and pump the blood out of the heart (depolarization wave of the ventricles on the ECG). The last wave is related to the repolarization of the ventricles. During this time the heart is filling with blood once again.; 89. a; 90. g; 91. d; 92. h; 93. f; 94. b; 95. e; 96. c

Self-Test

1. a, b, c, d
2. a, c
3. d
4. b, d
5. a
6. a, b, c, e
7. a, b, c, d
8. a, b
9. e
10. c

11. b
12. c, d, e
13. d
14. b, c
15. a, d, e [a and e are correct because the pulmonary artery carries deoxygenated blood from the right ventricle to the lungs to get oxygenated; d is correct because this blood also carries relatively high levels of CO_2 to the lungs where it can be released into the air]
16. c, d
17. b, c
18. d
19. b, d
20. a
21. e
22. a, b, c, d
23. a, b, c, d
24. c

Chapter 38 Animal Reproduction

Why It Matters [pp. 911-912]

1. egg; 2. sperm; 3. full; 4. external; 5. synchronized; 6. cryptochromes; 7. blue; 8. lunar; 9. species; 10. coatings; 11. acrosome

38.2 Asexual and Sexual Reproduction [pp. 913-914]

12. genes; 13. asexual; 14. one; 15. three; 16. identical; 17. parthenogenesis; 18. unfertilized; 19. zygote; 20. gametes; 21. sexual; 22. genetic diversity; 23. environment; 24. survive; 25. two; 26. genetic recombination; 27. chromosomes; 28. independent assortment; 29. randomly; 30. mutation; 31. C; 32. D; 33. A; 34. B

38.3 Cellular Mechanisms of Sexual Reproduction [pp. 914-922]

35. spermatogenesis; 36. testes; 37. spermagonia; 38. meiosis; 39. spermatids; 40. sperm; 41. ovaries; 42. meiosis; 43. cytoplasm; 44. ootid; 45. egg or ovum; 46. oogenesis; 47. external; 48. swim; 49. contact; 50. female; 51. D; 52. H; 53. A; 54. I; 55. F; 56. C; 57. B; 58. E; 59. G; 60. E, B, F, D, A, C; 61. External fertilization occurs outside of the body, typically in aquatic species; synchronization of female and male gametes is critical to success. Internal fertilization occurs within the female reproductive tract in many terrestrial species; typically synchronization of gamete release is not necessary.; 62. The acrosome reaction involves release of enzymes that breakdown egg-coating material. Fast block is a wave of depolarization at the egg plasma membrane that occurs within seconds of fusion of the sperm nucleus into the cytoplasm of the egg, while the slow block involves the release of calcium ions from ER into cytosol causing release of enzymes from cortical granules outside egg (barrier to sperm penetration). These are mechanisms to prevent multiple sperm from fusing with the egg.; 63. Hermaphrodites are individuals that can produce eggs and sperm. Simultaneous hermaphroditism occurs in organisms that develop functional ovaries and testes at the same time whereas sequential hermaphroditism is when organism changes from one sex to the other.

38.5 Controlling Reproduction [pp. 931-932]

64. two; 65. hormone; 66. follicle-stimulating (FSH); 67. luteinizing (LH); 68. gonadotropin-releasing (GnRH); 69. oocytes; 70. estrogen; 71. LH; 72. ovulation; 73. progesterone; 74. Sertoli; 75. spermatogenesis; 76. Leydig; 77. testosterone`; 78. output; 79. extinct; 80. pregnant; 81. K; 82. J; 83. D; 84. N; 85. M; 86. I; 87. C; 88. B; 89. E; 90. H; 91. A; 92. L; 93. G; 94. F; 95. C; 96. E; 97. F; 98. C; 99. B; 100. A; 101. A; 102. E; 103. F; 104. B; 105. F; 106. D; 107. G; 108. E

SELF-TEST

1. a [male acanthocephalan worms inserts cement secretions in female gonopore after copulating]
2. d [fragmentation occurs when there are separate pieces of the parent]
3. d [The only advantage of asexual reproduction listed would be that it is not necessary to find a mate; it has lower energy cost and provides lower genetic diversity]
4. d [all selections increase genetic diversity of offspring]
5. e [if the fast block were inhibited, then polyspermy could occur; if this happened, the number of chromosomes would be greater than the diploid number]
6. b [this animal is a monotreme, which means that it is a mammal but it lays eggs]
7. c [during the meiosis process the immature sperm cell is called a spermatocyte]
8. a [a secondary oocyte is haploid since it is the result of meiosis I, after pairs of chromosomes have separated]
9. a [sperm nucleus (containing chromosomes) is located in the acrosome
10. b [human females produce oocytes as embryos but their development is arrested at the end of prophase I until they reach sexual maturity]
11. d [calcium ion release is a slow block response occurring within minutes of fertilization, a slower response than a fast block]
12. c [hermaphrodites develop functional ovaries and testes, producing eggs and sperm]
13. b [after ovulation, the follicular cells become the corpus luteum, which produce progesterone]
14. c [hymen partially covers the opening to the vagina so will be ruptured during first sexual intercourse, if not already broken]
15. c [if GnRH were not inhibited, another cycle could start prior to the end of the first, thus additional FSH and LH would be released from the pituitary and another oocyte could be stimulated]
16. b [the prostate produces a very alkaline (basic) fluid, thus if this were inhibited, the semen would be more acidic]
17. a [without a pituitary, the individual would be sterile, there would be no FSH or LH]
18. e [of the 380,000 oocytes present at maturity, only about 380 are ovulated during a human life time]
19. a [as sperm develop through the stages of meiosis, they are located towards the lumen at the centre of the seminiferous tubule]
20. d [human chorionic gonadotrophin (hCG) is secreted by embryonic cells after implantation]

INTEGRATING AND APPLYING KEY CONCEPTS

1. For external fertilization to occur, mass spawning of egg and sperm into open water must be synchronized. Furthermore, successful fertilization requires that the egg and sperm belong to the same species. Since it is likely that gametes from more than one species will be present in the water during a mating season, it is important for reproductive success that gametes can recognize whether a potential mate belongs to the same species. This is not important for species that use internal fertilization since sperm are released close to or inside a female's reproductive tract.
2. Semen is composed of a) sperm, which carry a haploid set of chromosomes to contribute to the offspring, b) secretion from seminal vesicles, viscous liquid which contains prostaglandins that trigger contractions in the female reproductive tract, c) secretion from the prostate gland, milky alkaline fluid that raises the pH of the semen to pH=6 and also provides an enzyme that changes semen into a gel at ejaculation to prevent it from draining out of the vagina, and e) secretion from bulbourethral glands, mucus-rich fluid that lubricates tip of penis and neutralizes any residual acidity in urethra.
3. If, after pregnancy, progesterone levels started to drop, the arteries supplying blood to the uterine lining would contract, removing the blood supply and causing the endometrium to break down. A menstrual flow would occur, removing the degenerating endometrium and the dying embryo from the female's body. A miscarriage has occurred so the pregnancy is terminated.

Chapter 39 Animal Development

Why It Matters [pp. 935-936]

1. parturition; 2. canal; 3. pelvic; 4. small; 5. relaxin; 6. ovaries; 7. elastic; 8. ligaments; 9. stretchabiity

39.1 Housing and Fuelling developing Young [pp. 936-938]

10. greater or higher; 11. genes; 12. protection; 13. development; 14. milk; 15. growth

39.3 Major Patterns of Cleavage and Gastrulation [pp. 941-944]

16. zygote; 17. genetic (nuclear); 18. Cytoplasmic 19. egg; 20. animal; 21. vegetal; 22. polarity; 23. three; 24. Three; 25. cleavage; 26. morula; 27. Mitotic; 28. blastula; 29. gastrulation; 30. ectoderm; 31. endoderm; 32. mesoderm; 33. archenteron; 34. blastopore; 35. protostome; 36. Deuterostome; 37. Gastrulation; 38. Organ; 39. adhesions; 40. Differentiation; 41. C; 42. B; 43. A; 44. A, B, C; 45. C; 46. B; 47. C; 48. C; 49. C; 50. B; 51. E; 52. I; 53. G; 54. B; 55. F; 56. A; 57. D; 58. C; 59. H; 60. B; 61. B; 62. A; 63. C; 64. B; 65. C; 66. A; 67. C

39.4 Organogenesis: Gastrulation to Adult Body Structures [pp. 945-947]

68. organogenesis; 69. Organs; 70. notochord; 71. mesoderm; 72. nervous; 73. neurulation; 74. induction; 75. neural plate; 76. mesoderm; 77. somites; 78. Apoptosis; 79A. 5; 79B. 2; 79C. 3; 79D. 1; 79E. 4; 79F. 7; 79G. 6

39.5 Embryonic Development of Humans and Other Mammals [pp. 947-953]

80. 38; 81. trimesters; 82. cleavage; 83. gastrulation; 84. organogenesis; 85. fetus; 86. fertilization; 87. third; 88. implantation; 89. blastocyst; 90. blastocoel; 91. inner cell mass; 92. embryo; 93. trophoblast; 94. trophoblast; 95. proteases; 96. endometrium (uterine wall); 97. disc; 98. epiblast; 99. embryo; 100. hypoblast; 101. membranes; 101. eight; 102. birds; 103. reptiles; 104. D; 105. A; 106. G; 107. E; 108. C; 109. B; 110. F; 111. E; 112. A; 113. C; 114. B; 115. A; 116. E; 117. C; 118. A; 119. The presence of SRY gene (sex-determining region of the Y chromosome) determines that the embryo will become male (if absent embryo will become female). This gene encodes a protein that initiates development of fetal testes which secrete testosterone and anti-Mullerian hormone. These two hormones cause the Wolffian ducts to become male reproductive organs and inhibit the development of Mullerian ducts.; 120. The extra-embryonic membranes, the chorion, amnion, yolk sac, and allantoic membrane are primarily derived from the hypoblast cell layer of the embryonic disc and the trophoblast layer of cells.

39.7 Genetic and Molecular Control of Development [pp. 958-963]

121. orientation; 122. rate; 123. G1; 124. microtubules; 125. microfilaments; 126. induction; 127. determination; 128. adhesion; 129. genes; 130. regulatory; 131. hox; 132. metamorphosis; 133. False, gap; 134. True; 135. True; 136. False, pair-rule; 137. False, and; 138. False, induction; 139. True; 140. False, anterior; 141. True; 142. False, undifferentiated; 143. False, microfilaments

SELF-TEST

1. c [all other selections feed on milk or milk-like products]
2. c [cytoplasmic determinants are from the egg having their effect throughout development but primarily during early cleavage of the zygote]
3. b [the blastopore becomes the mouth in protostomes and the anus in deuterostomes]
4. d [endoderm is the innermost layer and will form the linings of major organ systems]
5. b [a few cells from the hypoblast develop into the germ cells, which migrate to the developing gonads of the embryo]
6. b [the primitive streak defines the axis of the embryo, providing organizational cues for right and left or bilateral symmetry]
7. b [nitrogenous wastes are stored in the allantois, which is surrounded by the allantoic membrane]
8. d [neurulation is the beginning of organogenesis which begins with the formation of a neural plate from the ectoderm]
9. a [neural crest cells are unique to vertebrates; they form when the neural tube closes; cranial nerves are derived from these cells]
10. a [all cells contain all genes but genes coding for keratin are activated in skin (and other) cell, not in the eye]
11. c [amniocentesis is evaluation of amniotic fluid that surrounds the fetus]
12. b [maternal and embryonic blood must be isolated from each other since components in blood could react to foreign bodies and cause blood clotting]
13. b [epiblast cell layer of the embryonic disk forms the embryo and the hypoblast layer aids in forming extraembyonic membranes]

14. d [the SRY protein is a product of a gene on the Y chromosome, so without this protein, the Mullerian ducts would develop into female reproductive structures]
15. a [microtubules and microfilaments appear to play a significant role in the orientation or axes of the development of cleavage furrows that determine cell orientation in the embryo]
16. a [yolk is found at the vegetal pole and causes these cells to divide more slowly]
17. c [segmentation genes subdivide the embryo into regions or segments (somites) of the embryo]
18. c [all cells contain the same genes but different genes are expressed, which results in specialized cells]
19. d [because *C. elegans* is transparent, it was possible to follow the cell lineage for all cells produced from zygote to the end of embryonic development]
20. a [sequence of the Hox genes is highly conserved so mammal Hox genes are in virtually the same order as in *Drosophila*]

Integrating and Applying Key Concepts

1. Each organ or organ system arises from one of the three primary tissues, ectoderm, mesoderm, or endoderm. The organs of the nervous system and integumentary system arise from the ectoderm, the muscular, skeletal, circulatory, reproductive, and excretory systems primarily arise from the mesoderm, and the lining of many organs, including those in the digestive tract and respiratory tract arise from the endoderm.
2. Apoptosis functions to breakdown tissues no longer required as newly formed adult tissues are formed. This allows developing organisms to change structures in preparation for its adult form; for example, removal of tissue between fingers and toes, changing paddle-shaped structures into freely moving fingers and toes.
3. The movement of a whole cell involves attachment, stretching, and contraction steps. If a cell is attached to a substrate, it will elongate itself (using its microtubules and microfilaments) and attach itself to the substrate at the advancing tip. The cell will then contract itself at the back end of the cell so that it eventually breaks free from its previous attachment point. These steps will be repeated until the cell reaches its destination.

Chapter 40 Animal Behaviour

Why It Matters [pp. 967-968]

1. behaviour; 2. versatile; 3. novel; 4. learned

40.3 Learning [p. 972]

5. instinctive; 6. learned; 7. isolation; 8. gene-environment; 9. genetically; 10. fixed action; 11. sign; 12. more; 13. imprinting; 14. critical; 15. classical conditioning; 16. operant; 17. reinforcement; 18. insight; 19. error; 20. habituation; 21. A; 22. F; 23. B; 24. C; 25. D; 26. E; 24. C; 25. D; 26. E; 27. F, unconditioned; 28. F, operant; 29. F, habituation; 30. T; 31. F, genetic; 32. F, Instinctive

40.6 Neural Anatomy and Behaviour [pp. 976-979]

33. nerve; 34. Genetic; 35. Experiences; 36. territory; 37. Courtship; 38. Brain; 39. Hormones; 40. E; 41. G; 42. A; 43. C; 44. F; 45. D; 46. B

40.7 Communication [pp. 979-982]

47. acoustic signaling; 48. visual signaling; 49. chemical; 50. pheromones; 51. tactile; 52. electrical; 53. b (and possibly a or d); 54. c (and possibly a, b, d); 55. e (and possibly a); 56. b; 57. a; 58. c

40.8 Space [pp. 982-984]

59. habitat; 60. kinesis; 61. taxis; 62. territoriality; 63. energy; 64. costs; 65. exclusive; 66. attracting; 67. c; 68. b; 69. a; 70. b

40.9 Migration [pp. 984-987]

71. migration; 72. piloting; 73. compass orientation; 74. navigation; 75. mental; 76. time; 77. increase; 78. decrease; 79. longer; 80. A;

81. C, D; 82. B; 83. D; 84. D; 85. A

40.11 Sexual Selection [pp. 988-990]

86.sexual; 87. competition; 88. choosing; 89. investment; 90. dimorphism; 91. courting (or courtship); 92. compete; 93. lek; 94. polygyny; 95. polyandry; 96. monogamous; 97. promiscuity; 98. C; 99. B; 100. D; 101. A

40.15 Human Social Behaviour [pp. 994-997]

102. social behavior; 103. dominance hierarchy; 104. altruism; 105. kin selection; 106. haplodiploidy; 107. eusocial; 108. reciprocal altruism; 109. 0.5; 110. 0.5; 111. 0.25; 112. 0.125; 113. C; 114. D; 115. A; 116. B

SELF-TEST

1. b [sparrows can make singing sounds instinctively but a song must be learned]
2. a [species that insert their eggs into another species nest are brood parasites]
3. c [sign stimuli cue fixed action patterns]
4. b [newborn garter snakes instinctively were interested in eating slugs, so their food preferences must have a genetic component]
5. c [The operant is the desired behavioral response from the subject. The reinforcement is the reward for performing the operant.]
6. c [habituation is the learned response involving loss of responsiveness to a repeated stimulus, such as song of a neighbouring bird species]
7. a [only males produce estrogen, which increases the nerve cells in the higher vocal centre, allowing the bird to sing (females do not produce estrogen)]
8. b [honeybees produce increasing concentrations of the hormone octopamine as they age]
9. b [crickets have ears (sensory system) that 'hear' a predator and initiates an immediate motor response]
10. d [moles live in dark places, using the tactile (or touch) sensors on their mouth tentacles]
11. c [herring communicate using fast repetitive transient signals (f*rts)]
12. b [kinesis is the change in rate of movement in response to environmental stimuli, as displayed by the planaria]
13. c [some food preferences in vertebrates are instinctive, others are learned]
14. a [The other choices all involve resources that would be difficult to defend.]
15. d [migrating animals do not use sound cues, only visual and olfactory cues]
16. c [females usually have a greater investment in parental care than males, which explains their different reproductive strategies]
17. c [male ornamentation or structures are used by females who choose their mates, one of the components of sexual selection]
18. b [polyandry is rare since females rarely mate with more than one male]
19. d [dolphins assist other group members, as well as members of other species]
20. d [research indicated that abuse was more likely between adults and children that were not genetically related]

INTEGRATING AND APPLYING KEY CONCEPTS

1. An insectivorous bat uses echolocation calls to "hear" if and where a potential prey is located. This involves comparing the original call with the returning echo. The black field cricket is one of the bat's prey, however, this animal has ears on its legs which sense the direction of the sound (stronger stimulus on the leg closer to the call) and causes an automatic response. The response includes a jerking motion by the hindleg farthest away from the sound which blocks wing movement on that side of the body, causing the cricket to swerve away from the sound and downwards towards the ground (away from the bat).
2. A wandering raven is an intruder into another raven's territory. The wandering raven will yell when it finds a carcass because, although it will have to share its food, it will need the assistance of other ravens to assist in fighting off the resident raven.
3. The new head of a pride is not altruistic because when takes over the pride, he kills all of the nursing young (infanticide) so that the female lions in the pride will go into estrus. This allows the new male to produce offspring of his own relatively quickly, passing on <u>his</u> genes to the next generation.

Chapter 41 Plant and Animal Nutrition

Why It Matters [pp. 1001-1002]

1. salivary amylase; 2. saliva; 3. chewing; 4. maltose; 5. increases; 6. amy1; 7. greater or higher; 8. more

41.2 Soil [pp.1010-1012]

9. 90; 10. dry matter; 11. hydrogen; 12. Hydroponics; 13. minerals; 14. photosynthesis; 15. passively; 16. primary; 17. secondary; 18. Essential; 19. macronutrients; 20. micronutrients; 21. stunted; 22. yellowing; 23. undernutrition; 24. loss; 25. vitamins; 26. enzymes; 27. 13; 28. water-soluble; 29. fat-soluble; 30. air spaces; 31. humus; 32. water; 33. aerating; 34. decompose; 35. oxygen; 36. soil solution; 37. cations; 38. anions;. 39. negatively; 40. cation exchange; 41. J; 42. D; 43. K; 44. E; 45. C; 46. I; 47. L; 48. A; 49. F; 50. H; 51. B; 52. G; 53. A,B,F; 54. A,B,C,F, often E; 55. A,B,F, sometimes D; 56. A,B,C,D,F

57.

Soil Ingredients	*Value in Soil*
Particles	Hold water and air
Humus	Source of nutrients, holds water and air
Living organisms	Provide organic chemicals and aerates the soil
Minerals	Nutrients for the plant

41.3 Obtaining and Absorbing Nutrients [pp. 1012-1017]

58. 20-50; 59. tips; 60. hairs; 61. ion-specific; 62. phosphorous; 63. mycorrhizae; 64. hyphal; 65. 80; 66. enzymes; 67. bacteria; 68. nitrate; 69. ammonium; 70. amino; 71. nitrogen; 72. symbiotic; 73. nodules; 74. flavinoids; 75. nod; 76. leghemoglobin; 77. bacteroids; 78. nitrogenase; 79. adaptations; 80. 4; 81. fluid; 82. suspension; 83. deposit; 84. bulk; 85. B; 86. A; 87. C; 88. D; 89. E; 90. D; 91. A; 92. C; 93. B; 94. D; 95. A; 96. C; 97. B; 98. C; 99. D

41.4 Digestive Processes [pp. 1017-1019]

100. breakdown; 101. absorbed; 102. enzymatic hydrolysis; 103. specific; 104. intracellular; 105. extracellular; 106. 5; 107. mechanical processing; 108. secretion; 109. hydrolytic enzymes; 110. absorption; 111. elimination; 112. B; 113. D; 114. A; 115. C; 116. Intracellular digestion occurs within the cell. In order for the cell not be broken down by digestive enzymes, particles are contained within vacuoles that will fuse with lysosomes containing digestive enzymes. Food particles are taken into the cell by endocytosis. After digestion has occurred and materials absorbed from the vacuole, waste products are eliminated from the cell by exocytosis; 117. Extracellular digestion occurs in a tube or saclike structure, actually outside of the organism. The potential for different types of food sources is increased, since size is not a limiting factor. The primary limiting factor will be the available enzymes to breakdown the food material; 118. B; 119. E; 120. C, D; 121. A; 122. C, D; 123. B; 124. B, C

41.5 Digestion in Mammals [pp. 1019-1027]

125. 5; 126. esophagus; 127. rectum; 128. nervous; 129. longer; 130. Cellulose; 131. F; 132. I; 133. H; 134. C; 135. D; 136. B; 137. E; 138. G; 139. A; 140. 6; 141. 4; 142. 16; 143. 14; 144. 13; 145. 7; 146. 3; 147. 11; 148. 5; 149. 9; 150. 15; 151. 1; 152. 8; 153. 2; 154. 17; 155. 10; 156. 12; 157. D; 158. F; 159. E; 160. G; 161. B; 162. A; 163. C

41.6 Regulation of the Digestive Process [pp. 1027-1030]

164. C; 165. E; 166. B; 167. F; 168. D; 169. A; 170. F, autonomic; 171. T; 172. F, hypothalamus; 173. T

41.7 Variations in Obtaining Nutrients [pp. 1030–1036]

174. nutrients; 175. Parasitic; 176. Epiphytic; 177. Hiding; 178. Lures; 179. C; 180. E; 181. F; 182. B; 183. A; 184. D

Self-Test

1. a [Hydroponics was developed to learn about the mineral requirements of plants]
2. a [Some of the minerals act as cofactors for the plant enzymes]
3. b [They are both important, just the amount needed is different]
4. d [Proteins contain C, H, O, and N but do not contain the macronutrient K (potassium)]
5. d [ATP production as well as biological molecule production is dependent on the digestion of organic molecules]
6. b [long term antibiotic treatment will decrease the bacteria that normally inhabit the gut; these bacteria synthesize vitamin K which is necessary for synthesis of clotting factors]
7. b [Humus adds organic chemicals to the soil]
8. b [pH directly affects the absorption of minerals by the plants]
9. c [Plants absorb nitrogen mostly in the nitrate form]
10. b [Plants mostly use nitrogen in the ammonium form]
11. c [nitrogen-phosphorus-potassium]
12. c [Nitrifying bacteria convert ammonium to nitrates, a form that plants absorb]
13. b [Bacteroids are bacteria that live in root nodules]
14. b [Nod genes stimulate plant root cells to make leghemoglobin]
15. d [bulk feeders consume chunks or particles of food; teeth and claws would be advantageous adaptations]
16. a [saclike body plans have one opening for both intake of food and removal of waste products]
17. c [all selections except the pancreas are involved in mechanical processing or mixing]
18. d [Kangaroos are herbivores which have long intestinal tracts, only the tiger, a carnivore, has a short intestinal tract]
19. c [microvilli are on villi in wall of small intestine; increases surface area for nutrient absorption]
20. b [essential amino acids, fatty acids, vitamins and minerals must be obtained in the diet]
21. c [the pyloric sphincter is a ring of smooth muscle that acts as a valve between the stomach and small intestine; if this valve were blocked, chyme would not be able to move out of the stomach into the small intestine]
22. a [Secretin reduces gastric acid release as well as stimulates the pancreas to release a rich bicarbonate solution; reduction in this hormone would result in a more acidic environment]
23. d [Most orchids are epiphytes and anchor on the tree to reach light and catch raindrops]

Integrating and Applying Key Concepts

1. Plants grown in clay soils tend to be more limited in nutrient uptake than plants grown in sandy soils. Since clay soils are composed of small particles, usually <0.002 mm in diameter, they are closely packed together. The resulting small air spaces, which have negatively charged surfaces, bind strongly to polar water molecules, reducing the uptake of water and dissolved minerals into plant roots. Furthermore, when water fills the small air spaces, the lack of air in the spaces reduces the uptake of oxygen into plant roots. This is quite different from sandy soils, which are composed of large particles (0.02 – 2.0 mm diameter) and, therefore, large air spaces. When water is present in the soil, water, minerals, and oxygen can be taken up by plant roots. However, due to the large air spaces water drains down rapidly from horizon layers containing plant roots.
2. The expression of nod genes in the bacteria, in response to flavenoids released by the legume's roots, cause a root hair to curl towards the bacteria and the bacteria to release enzymes that break down the cell wall of the root hair so that the bacteria can enter the cell. As the bacteria multiply, an infection thread from the root cell forms, acting like a tube, so that bacteria can invade root cortical cells. The bacteria become bacteroids, large and immobile cells, which express more nod genes, causing the cortical cells to divide; this will form the root nodule. In the bacteroids, nitrogenase converts nitrogen gas to ammonium, which leaves the bacteroids and is converted to other products, such as amino acids, in the nodule cells.
3. The eight essential amino acids must be acquired through diet. Vegetarians must be careful to eat the correct combination of foods to acquire all eight of the essential amino acids because, although many meats, fish, and dairy products provide all of these amino acids, many plants are deficient in one or more of the amino acids. For example, corn contains little lysine and isoleucine and

lentils contain little methionine and tryptophan. If both of these food are eaten, all of the essential amino acids will be provided, separately they will not.

4. In the small intestine, carbohydrates are digested by pancreatic amylase (secreted by the pancreas) and disaccharidases (secreted by the epithelial cells of the small intestine wall), resulting in the production of monosaccharides; proteins are digested by proteases, such as trypsin and carboxypeptidases (secreted by the pancreas) and aminopeptidase and dipeptidase (secreted by the epithelial cells of the small intestine wall), resulting in the production of amino acids; lipids are digested by bile salts (secreted by the liver) and lipase (secreted by the pancreas), resulting in the production of fatty acids and glycerol; and nucleic acids are digested by nucleases (secreted by the pancreas) and nucleotidases, nucleosidases, and phosphatases (secreted by the epithelial cells of the small intestine wall), resulting in the production of 5-carbon sugars, nitrogenous bases, and phosphates.

Chapter 42 Gas Exchange: The Respiratory System

Why It Matters [pp. 1039-1040]

1. respiration

42.1 The Function of Gas Exchange [pp. 998–1000]

2. 78%; 3. 21%; 4. less than 1%; 5. 592.8 mmHg; 6. 159.6 mmHg; 7. 7.6 mmHg; 8. 390 mmHg; 9. 105 mmHg; 10. 5 mmHg; 11. 195 mmHg; 12. 52.5 mmHg; 13. 2.5 mmHg; 14. Gases will diffuse down their partial pressure gradient (from a region of high partial pressure to a region of low partial pressure).; 15. The factors that influence the rate of diffusion are the diffusion coefficient, the surface area involved in gas exchange, the partial pressure difference of the gas being exchanged, and the distance for diffusion; 16. physiological respiration; 17. respiratory medium; 18. breathing; 19. respiratory surface; 20. tracheal system; 21. gills; 22. lungs; 23. Ventilation and perfusion of respiratory surfaces are essential to bring oxygenated water or air to the exchange surface and to circulate blood high in oxygen and low in CO_2 (gases that were exchanged during respiration) away from the exchange surface, respectively. This allows the respiratory system to maintain partial pressure differences to allow the gases to diffuse through the exchange surface. By maintaining a partial pressure difference, the rate of diffusion of the gases are maintained. Otherwise, equilibrium between the partial pressure difference of the gases could be reached and reduce the rate of gas diffusion to zero (no gas exchange would occur under these conditions . 24. aquatic animals have no problems keeping respiratory surfaces wet (minimizes evaporative loss of water; this occurs in air breathing organisms). 25. Diffusion coefficient is low; less oxygen dissolved and therefore a lower partial pressure of oxygen in water, makes oxygen less available and accessible; aquatic animals must pass a greater amount of medium to obtain required oxygen; water is more dense and viscous than air, therefore much energy is required to breath water; higher temperature or salinity causes decreased amount of dissolved gases in the water. Therefore changing environmental conditions such as temperature and salinity will have an impact on oxygen concentration and partial pressure; 26. diffusion coefficient higher than water; more oxygen available in air; air is less dense and viscous than water therefore energy required to ventilate is lower. Allows air breathing animals to breath in and out, reversing the direction of flow of the respiratory medium, without a large energy penalty; 27. constant evaporative loss of water from the respiratory surface. Animal lose water through breathing that must be replaced to keep the respiratory surface from drying.

42.2 Adaptations for Respiration [pp. 1043-1046]

28. external gills; 29. internal gills; 30. counter-current exchange; 31. tracheae; 32. spiracles; 33. positive pressure breathing; 34. negative pressure breathing; 35. alveoli; 36. one-way; 37. ventilation; 38. hyperventilation; 39. hypoventilation; 40. Counter-current exchange occurs when the respiratory medium (in this case, water) flows in the direction opposite to that which the respiratory surface (in this case, lamellae) is perfused with blood. The adaptive significance is that the diffusion gradient is maintained across the entire length of the respiratory surface, increasing the O_2 extraction efficiency; 41. air; 42. tracheal system; 43. water; 44. gills; 45. air; 46. lungs; 47. air; 48. lungs; 49. a; 50. g; 51. d; 52. b; 53. h; 54. c; 55. e; 56. f

42.3 The Mammalian Respiratory System [pp. 1046-1050]

57. pharynx; 58. larynx; 59. trachea; 60. bronchi; 61. bronchioles; 62. pleura; 63. diaphragm; 64. external intercostals muscles; 65. internal intercostals muscles; 66; tidal volume; 67. vital capacity; 68. residual volume; 69. carotid bodies; 70. aortic bodies; 71. b; 72 a; 73. b; 74. a; 75. c; 76. nasal passages; 77. pharynx; 78. epiglottis; 79. larynx; 80. trachea; 81. lung; 82. bronchi; 83. mouth; 84. pleura; 85. intercostals muscles; 86. diaphragm; 87. bronchiole; 88. alveoli.

42.4 Mechanisms of Gas Exchange and Transport [pp. 1050-1053]

89. partial pressure; 90. hemoglobin; 91. oxygen dissociation curve; 92. carbonic anhydrase; 93. buffer; 94. The sigmoidal shape of the haemoglobin-oxygen saturation curve reflects the cooperate binding characteristics of the four subunits of the Hb molecule. Initially, O_2 binds to the first subunit with some difficulty (lag phase). After O_2 binds to the first subunit, it changes the shape of the Hb molecule such that the binding affinity of the second Hb subunit is increased and O_2 binds with greater ease; similarly, binding of the second O_2 changes the shape of the Hb molecule further and results in heightened affinity of the 3rd subunit for O_2 (exponential phase). Finally, as the 4th O_2 binds, the Hb molecule becomes saturated; 95. At the lungs, PO_2 in the lung is higher than in the blood. The partial pressure difference causes oxygen in the lung to diffuse into the blood (and binds haemoglobin in the RBCs). When the oxygenated blood arrives at tissues, the PO_2 of the blood is higher than that of tissues. The partial pressure difference, once again, causes oxygen to detach from the haemoglobin and to diffuse into cells where the partial pressure of oxygen is low. Overall, the haemoglobin-saturation curve provides a representation of the amount of oxygen that is transferred across exchange surfaces.

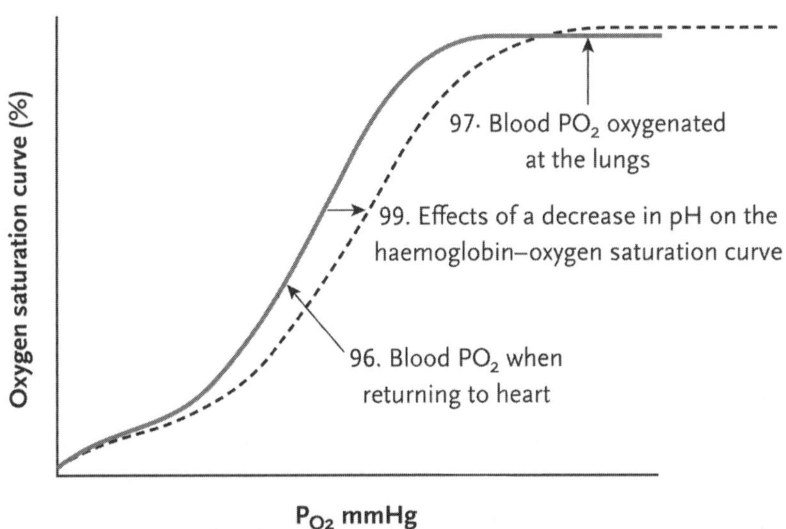

98. pH can cause a right shift or left shift in the haemoglobin-oxygen saturation curve. A decrease in pH causes the curve to right shift, whereas an increase in pH causes the curve to left shift. 100. This would occur when blood reaches tissues. A working tissue has a lower pH as a result of metabolic wastes and a region of elevated CO_2 levels. The decrease in pH promotes the delivery of oxygen to the tissues (the lower pH reduces the affinity of Hb for oxygen); 101. The significance of the steep part of the curve is related to the fact that there will be a large change in haemoglobin oxygen saturation over small changes in oxygen partial pressure. This is the part of the curve where large changes in saturation can occur as a result of small changes in oxygen partial pressure. When one considers oxygen transfer at the lungs or at the tissues, the oxygen partial pressure difference is such that it occurs in this part of the curve, thus underlying the large amount of oxygen that can be delivered during the transfer across the exchange surface; 102. CO_2 may be transported as CO_2 dissolved in the blood or in the red blood cell, it may be transported bound to haemoglobin, or transported as HCO_3 in the blood and red blood cells (CO_2 converted to HCO_3 by carbonic anhydrase); 103. In body tissues, some of the CO_2 released into the blood combines with water in the blood plasma to form HCO_3 and H^+. However, most of the CO_2 diffuses into

erythrocytes, where some combines directly with hemoglobin and some combines with water to form HCO_3 and H^+. The H^+ formed by this reaction combines with hemoglobin; the HCO_3 is transported out of erythrocytes to add to the HCO_3 in the blood plasma. In the lungs, the reactions are reversed. Some of the HCO_3 in the blood plasma combines with H^+ to form CO_2 and water. However, most of the HCO_3 is transported into erythrocytes, where it combines with H^+ released from hemoglobin to form CO_2 and water. CO_2 is released from hemoglobin. The CO_2 diffuses from the erythrocytes and, with the CO_2 in the blood plasma, diffuses from the blood into the alveolar air; 104. b; 105. a; 106, b; 107. c; 108. a; 109. d

SELF-TEST

1. c
2. b, c
3. a, b, c
4. c
5. a, d
6. a, b, e
7. a, b, c
8. d
9. c
10. a, b, d [a is correct because counter-current exchange is used in bony fish; d is correct because in counter-current circulation, the respiratory medium (in this case, water) flows in the direction opposite to that which the respiratory surface (in this case, lamellae) is perfused with blood. B is correct because the counter-current circulation results in the pO_2 of water always being higher than the pO_2 of blood, thus favoring the diffusion of O_2 from water to blood across the entire length of the respiratory surface]
11. a, c
12. a, b, c, d
13. b
14. a
15. c
16. b
17. b
18. e [e is correct because atmospheric pO_2 is typically 150 mmHg near sea level; normal arterial pO_2 is generally 100 mmHg (after extraction from atmospheric air) and tissue pO_2 levels are typically 10 mmHg (where O_2 is being used)]
19. a, b, c [In tissues, CO_2 is generated as a by-product of metabolism and its levels increase; the more CO_2 produced by metabolism, the more diffuses from the tissues into the blood and, into RBCs (remember the cell membrane is permeable to small molecules like CO_2); a is correct because the more CO_2 that enters the RBC, the more CO_2 is available to react with Hb; b is correct because the higher the CO_2 concentration, the more CO_2 there is to react with water to shift the equilibrium of the reaction of $CO_2 + H_2O$ Ξ $[H_2CO_3]$ Ξ H^++HCO_3^- to the right (as written); therefore, the more CO_2 that enters the RBC, the more H^+ there is produced and the more H^+ that can react with Hb-O_2, forcing the dissociation of O_2 from Hb, thereby forming HHb (reduced Hb); c is correct because the RBCs exchange Cl^- with HCO_3^-; with higher levels of HCO_3^- being produced in the RBC (from above carbonic acid reaction), the more Cl^- will be taken into the cell in exchange for HCO_3^-, resulting in higher levels of HCO_3^- in the plasma (making HCO_3^- in the plasma the major form in which CO_2 is transported in the blood).]
20. d

Chapter 43 Regulating The Internal Environment

Why It Matters [pp. 1057–1058],

1. homeostasis; 2. water; 3. organ; 4. solute; 5. out of; 6. internal

43.1 Introduction to Osmoregulation and Excretion [pp. 1058–1061]

7. Osmosis; 8. high; 9. low; 10. low; 11. high; 12. selectively permeable; 13. passive; 14. concentration; 15. osmoles; 16. isoosmotic; 17. hyperosmotic; 18.hypoosmotic; 19. osmoregulater; 20. osmoconformer; 21. ions; 22. metabolic; 23. nitrogenous; 24. excretion; 25. balance; 26. F; 27. E; 28. A; 29. B; 30. C; 31. D; 32. I; 33. G; 34. H

43.2 Osmoregulation and Excretion in Invertebrates [pp. 1061–1063]

35. Marine; 36. osmoconformers; 37. terrestrial; 38. osmoregulators; 39. energy; 40. cells; 41. Nitrogenous or toxic metabolic; 42. ammonia; 43. energy; 44. hyperosmotic; 45. varied; 46. B; 47. C; 48. A; 49. C; 50. A; 51. B; 52. A; 53. B

43.3 Osmoregulation and Excretion in Mammals [pp. 1063–1069]

54. kidney; 55. nephron; 56. cortex; 57. medulla; 58. juxtamedullary; 59. cortical; 60. urine; 61. waste; 62. renal pelvis; 63. urinary bladder; 64. ureter; 65. hyperosmotic; 66. pertiubular capillaries; 67. osmolality; 68. hyperosmotic;69. 9; 70. 7; 71. 12; 72. 1; 73. 8; 74. 5; 75. 4; 76. 3; 77. 11; 78. 10; 79. 6; 80. 2; 81. B; 82. I; 83. F; 84. H; 85. J; 86. G; 87. D; 88. C; 89. E; 90. A; 91. T; 92. F, out of; 93. F, descending; 94. T; 95. F, out of

43.4 Kidney Function in Nonmammalian Vertebrates [pp. 1069–1071]

96. water; 97. salts; 98. hyperosmotic; 99. conserved; 100. excreted; 101. urea; 102. nitrogenous; 103. isoosmotic; 104. hyperosmotic; 105. excreting; 106. conserving; 107. conserve; 108. uric acid; 109. specializations; 110. D; 111. B; 112. B; 113. D; 114. C, E; 115. A; 116. A; 117.B,C; 118. C

43.7 Endothermy [pp. 1076–1081]

119. negative; 120. thermoreceptors; 121. set point; 122. gain; 123. loss; 124. Endotherms; 125. ectotherms; 126. external; 127. birds; 128. mammals; 129. behavioral; 130. deep; 131. upper; 132. radiating; 133. thermal acclimatization; 134. enzymes; 135. temperature 136. Oxidative; 137. Heat; 138. Set point; 139. Energetically; 140. Broader; 141. Skin; 142. Increasing; 143. Decreasing; 144. Sweat; 145. Shivering;146. C; 147. D; 148. A; 149. B; 150. A, B; 151. B; 152. A; 153. A; 154. B; 155. B; 156. B; 157. A; 158. F, large; 159. F, summer, winter; 160. T; 161. F, conduction; 162. T

SELF-TEST

1. d [there is 1000 milliosmoles in one osmole]

2. c [removal of nitrogenous wastes in the form of uric acid crystal would indicate a conservation of water, the environment is most likely terrestrial and very dry]

3. d [salmon are marine teleosts, and nitrogenous waste removal would be by secretion of ammonia from the gills]

4. b [if the osmolarity of the intracellular and external environment were equal (isoosmotic), the animal would be an osmoconformer]

5. c [freshwater animals are hyperosmotic since their osmolality is greater than their environment]

6. b [both protonephridian and Malpighian tubules have only one open end but only Malpighian tubules empty into a gut]

7. a [movement of many ions and molecules into and out of the convoluted tubule often requires energy (active transport) or membrane proteins (facilitated transport)

8. c [aquaporins assist in water movement across a membrane; lack of aquaporins in ascending segment of loop of Henle prevents reabsorption of water at that location

9. a [renal arteries carry blood (oxygenated) to kidneys, renal veins carry blood (deoxygenated) away from kidneys]

10. b [Bowman's capsule surrounds glomerulus; pressure in glomerulus forces substances from blood into the Bowman's capsule (filtration)]

11. b [as water is reabsorbed from the descending loop of Henle, osmolality of filtrate increases]

12. b [metabolic reactions in body increases acidity in blood, which negatively charged bicarbonate ion helps offset]

13. a [substances in filtrate that are not reabsorbed, remain in the nephrons and exit the body as a component of urine]

14. c [freshwater fish tend to be hyperosmotic, a situation that must be maintained to survive in their environment] 15. b [sea birds primarily take in salt water with food, then use salt glands to collect excess salt and excrete it through their nostrils, eliminated the need for freshwater]

16. b [shivering would produce heat ; heat transfer is by conduction because the snake is in direct contact with eggs]

17. a [rabbit is the only endotherm in the list; it provides optimal body temperature and activity across wide range of temperatures, but this requires a constant energy supply]

18. b [thermal acclimatization adjusts the range of temperatures over which it can optimally function to coincide with the external temperature]

19. b [shivering would increase heat production, decrease epinephrine, and vasodilation would decrease heat production]

20. c [only the dermis layer of skin contains sweat glands, thermoreceptors, and blood vessels, all of which assist in thermoregulation]

INTEGRATING AND APPLYING KEY CONCEPTS

1. Most marine invertebrates are osmoconformers so they have few, if any, structures or mechanisms for osmoregulation. The marine invertebrates that do osmoregulate, are able to use behavioural mechanisms, such as closing shells or retreating into burrows to prevent dessication, often at low tide. Freshwater invertebrates must maintain their hyperosmocity, which requires using mechanisms to remove excess water and take up salt ions (in food or by active transport from environment). Osmoregulation in terrestrial invertebrates involves constantly taking up an adequate supply of water and salts from the environment, often through food. Water loss for animals on land occurs primarily through evaporation and excretion, which must continually be replaced. Some invertebrates have specialized tubule osmoregulatory systems: freshwater organisms such as flatworms and adult mollusks use protonephridian (cilia move filtrate through flame cells and tubule), terrestrial invertebrates such as annelids use metanephridia (cilia move filtrate into funnel-shaped opening and tubule), and invertebrates such as insects use Malpighian tubules (ultrafiltrate moves into tubule that empties into gut).
2. Osmolality in the filtrate of the descending segment of the loop of Henle increases as it moves downward due to the reabsorption of water from the filtrate into interstitial fluid as it travels through this segment. Increasing solute concentrations in the interstitial fluid as the tubule descends into the medulla, causes water to move by osmosis into the interstitial fluid. Furthermore, the presence of aquaporins in the wall of this segment of tubule fosters rapid movement of water molecules out of the tubule. Increasing solute concentrations in the interstitial fluid during descent in the medulla is the result of a) Na^+ and Cl^- reabsorption, primarily by active transport, to interstitial fluid surrounding the ascending segment of the loop of Henle in the medulla and b) urea reabsorption as urine travels down the collecting duct in the medulla.
3. Ectotherms and endotherms use different mechanisms to thermoregulate. Endotherms are able to be active in a wider range of temperatures than ectotherms, but the cost of this higher metabolic rate is that they have greater energy (food) requirements and so must put themselves into danger as they forage for food. Ectotherms activities are limited by cool temperatures, when they are sluggish and therefore less likely to escape from a predator. However, because of their lower metabolic rates, they have a relatively low energy (food) requirement and consequently spend more time in the safety of their homes and less time exposed to outside dangers while foraging for food.

Chapter 44 Defenses against Disease

Why It Matters [pp. 1085-1086]

1. pathogens; 2. immune; 3. eliminating or killing; 4. vaccines; 5. eliminated

44.2 Nonspecific Defences: Innate Immunity[pp. 1087-1092]

6. innate immune system; 7. adaptive immune system; 8. immune response; 9. Inflammation; 10. Macrophages; 11. cytokines; 12. Mast cells; 13. neutophils; 14. chemokines; 15. eosinophils; 16. complement system; 17. membrane attach complexes; 18. Interferons; 19. Natural killer cells; 20. apoptosis; 21. lymphocyte; 22. A,B; 23. A,B; 24. A,B,C; 25. A,B,C; 26. a; 27. c; 28. b; 29. a; 30. first to recognize pathogens; engulfs pathogen and kills it; secretes signal to initiate other immune reponses; 31. mast cell; 32. attracted to infected site by chemokines; engulfs pathogen and kills it; usually dies itself afterward; 33. eosinophils; 34. natural killer cells; 35. c;

36. a; 37. b; 38. F; 39. G; 40. B; 41. C; 42. E; 43. A; 44. D

44.3 Specific Defences: Adaptive Immunity [pp. 1092-1102]

45. antigen; 46. b cells; 47. t cells; 48. thymus gland; 49. antibody-mediated; 50. antibodies; 51. cell-mediated; 52. memory cells; 53. antigen; 54. t cell; 55. epitopes; 56. immunoglobins; 57. heavy chains; 58. light chains; 59. y; 60. antigen-binding; 61. bind; 62. cell division; 63. clones; 64. clear; 65. memory; 66. primary; 67. secondary; 68. Active immunity; 69. passive immunity; 70. 9; 71. 2; 72. 5; 73. 4; 74. 1; 75. 10; 76. 8; 77. 3; 78. 7; 79. 6; 80. 6; 81. 2; 82. 8; 83. 7; 84. 5; 85. 9; 86. 3; 87. 1; 88. 10; 89. 4; 90. A; 91. B. 92. A; 93. The MHC is derived from a large cluster of genes expressed in a few immune cells types (dendritic cells; macrophages, B cells). MHC proteins bind to antigen molecules inside the cell, then translocates it to the surface of the cell, making the cell an antigen-presenting cell.; 94. To combat the large number of pathogens that tend to be present at an infection site, it is important that both B and T cells undergo cell division. B cells must proliferate to form plasma cells, which produce antibodies specific for the infecting pathogen that can function in various ways including neutralizing toxins produced by the pathogens, immobilize pathogens through agglutination, and enhance phagocytosis of pathogens. Each of the cloned cytotoxic T cells function in producing perforins that form pores in the membrane of the infected cell, eventually resulting in its death.; 95. a; 96. b; 97. b; 98. a; 99. b; 100. a; 101. a; 102. b; 103. lymphocytes that arise and mature in the bone marrow; derivatives contribute to antibody-mediated immunity; 104. T cells; 105. B cell derivative that produces antibodies; 106. cell types derived from T cells and B cells and are responsible for initiating a rapid immune response upon reexposure to an antigen; 107. phagocytic cell that initiates adaptive immunity by engulfing foreign cell; 108. $CD4^+$ cell; 109. derived from activated CD4+ cells and leads to antibody-mediated immunity; 110. $CD8^+$ cell; 111. Derived from activated $CD8^+$ cells and destroys infected cells; 112. A; 113. I; 114. B; 115. F; 116. G; 117. C; 118. D; 119. E; 120. H

44.6 How Do Parasites and Pathogens Circumvent Host Responses? [pp. 1106-1108]

121. Immunological tolerance; 122. autoimmune reaction; 123. antiself; 124. Allergens; 125. anaphylactic shock; 126. Epinephrine (or adrenaline); 127. Low; 128. Antigenic variation; 129 symbiotic; 130. Reproduce; 131. New; 132. Innate; 133. Pathogens; 134. C; 135. E; 136. A; 137. F; 138. B; 139. D

SELF-TEST

1. b [vaccine causes host to provide an immune response to a mild infection so that host can quickly respond to a major infection]
2. e [producing an antibody combats a *specific* antigen; all other choices are general immune responses]
3. a [complement is a complex of proteins in the plasma which, when activated, attach to the surface of pathogens (general response), resulting in the perforation of their membrane and destruction]
4. d
5. c [skin is the first line of defence, the cells and tight junctions assist in preventing pathogens from entering the host]
6. d [PRR, pattern recognition receptors, are molecules that recognize groups of pathogenic organisms]
7. d [defensin is a type of antimicrobial peptide that is found in many types of organisms, indicating that it has had an important role throughout evolution]
8. a [because viruses tend to hide inside a host cell, or have surface molecules very similar to host surface molecules, they are more difficult to identify and other mechanisms must be implemented in the innate immune system]
9. b
10. d [natural killer cells are a type of lymphocyte that use MHC concentration to identify infected cells]
11. d [a large number of genes code for the histocompatibility complex and they are different for each vertebrate individual]
12. d
13. a [TCRs are used to identify specific antigens and therefore differ among T cells]
14. a [the conservative site of the antibody determines the class of immunoglobulin; the variable site results in differently shaped antigen-binding sites]
15. b [memory cells allow organisms to provide a rapid response if an antigen appears in the organism in future]
16. b [bacteria agglutinate or clump as a result of binding to more than one antibody, which prevents the bacteria from infecting cells]

17. b [antigenic variation is the ability of some pathogens, such as the AIDS virus, to continually change surface proteins produced]
18. a [the allergic response , which is due an inflammation response, results from IgE antibodies producing histamine in response to signals sent by mast cells]
19. b [after the nematode penetrates the host, bacteria enter, become phagocytosed, reproducing within the host cell and inactivating the insect's immune response]

Integrating and Applying Key Concepts

1. During an inflammation response, monocytes (one type of leukocyte) differentiate into macrophages which recognize and engulf the invading pathogens. If there are too many pathogens, the activated macrophages secrete cytokines to recruit more immune cells, as well as chemokines to attract neutrophils. As host cells die, they activate mast cells to release histamine which, with cytokines, dilate blood vessels and make them more permeable to movement of fluid, including neutrophils, from blood vessels into body tissues. Neutrophils also recognize and phagocytose pathogens. The movement of fluid into body cells results in the heat, redness, and swelling of an inflamed area. Both macrophages and neutrophils break down engulfed pathogens by various methods including enzymes, defensins, and toxic compounds.
2. During cell-mediated immunity (a type of adaptive immunity), cytotoxic T cells are produced from activated T cells and the receptors on their surfaces function to recognize antigens on antigen-presenting cells, which causes the cytotoxic T cells to destroy the pathogen. These cells have various methods to kill pathogens including releasing the protein perforin and various proteases. Perforin causes pores to form in the pathogen's surface membrane, allowing ions and molecules to cross the membrane which causes the infected cell to rupture. Proteases that enter the infected cell cause it to undergo apoptosis.
3. During an allergic reaction, allergen antigens cause B cells to over-secrete IgE antibodies which induce mast cells and basophil leukocytes to secrete histamine. Histamine produces inflammation in the infected area. Mast cells also release signals causing mucosal cells to secrete lots of mucus (runny nose) and smooth muscle cells to contract (constricting airways).

Chapter 45 Population Ecology

Why It Matters [pp. 1111-1112]

1. viruses; 2. epidemiology; 3. Virus; 4. saliva; 5.vaccine; 6. Vector; 7. foxes; 8. Estimating; 9. Size; 10. behaviour

45.1 The Science of Ecology [pp. 1112-1113]

11. organismal; 12. population; 13. community; 14. ecosystem; 15. models; 16. field; 17. laboratory

45.2 Population Characteristics [pp. 1113-1117]

18. range; 19. habitat; 20. population density; 21. mark-release-recapture; 22. pre-reproductive; 23. post-reproductive; 24. generation; 25. sex ratio; 26. uniform; 27. random; 28. clumped; 29. clumped; 30. random

45.3 Demography [pp. 1118-1119]

31. immigration; 32. emigration; 33. demography; 34. cohort; 35.mortality; 36. survivorship; 37. fecundity; 38. 0.800; 39. 0.200; 40. 200; 41. 8; 42. 7; 43. 1.000; 44. 0.000; 45. Type III. This type of survivorship curve describes a population in which there is high juvenile mortality but low mortality once individuals reach approximately 4 years of age.; 46. A; 47. C; 48. B

45.4 Evolution of Life Histories [pp. 1120-1121]

49. growth; 50. maturation; 51. reproduction; 52. energy budget; 53. tradeoffs; 54. maintenance; 55. growth; 56. reproduction; 57. passive; 58. active; 59. F, passive; 60. T; 61. F, may reproduce once and die; 62. T; 63. T; 64. b; 65. b; 66. a; 67. b

45.5 Models of Population Growth [pp. 1121-1127]

68. per capita growth rate; 69. exponential; 70. intrinsic rate of increase; 71. logistic; 72. carrying capacity; 73. intraspecific competition; 74. time lag; 75. e; 76. b; 77. a; 78. c; 79. d; 80. F, size remains constant; 81. T; 82. T; 83. F, intraspecific

45.6 Population Regulation [pp. 1127-1132]

84. dependent; 85. independent; 86. r-selected; 87. K-selected; 88. intrinsic; 89. extrinsic; 90. cycles; 91. time lags; 92. a; 93. b; 94. a; 95. a; 96. b; 97. T; 98. F, decreases; 99. F, has no effect on; 100. F, no effect on

45.7 Human Population Growth [pp. 1132-1136]

101. exponential; 102. geographical range; 103. habitats; 104. carrying capacity (K); 105. death; 106. demographic transition; 107. family planning; 108. F, high; 109. T; 110. F, decrease; 111. F, more; 112. c; 113. b; 114. a

SELF-TEST

1. a [The study of ecology involves living things, "abiotic" refers to non-living things]
2. a [Canadian are clustered in cities, with many areas having very few people]
3. b [Demography is the study of population size, density]
4. a [High fecundity produces lots of low value offspring, which the parent generally does not provide care for]
5. c [high numbers of reproductive means the population can produce many offspring in the current generation]
6. b [Females are more limiting for producing offspring, since a single male can fertilize many females]
7. c [If there is any limit to population growth, the population will cease growing exponentially]
8. d [Carrying capacity (K) is one of the variables in the equation that defines logistic growth]
9. c [Nighttime air temperature is not related to population density]
10. b [The rabies virus concentrates in the saliva, which is then transferred through biting]
11. d [Populations inhabit suitable habitat, thus areas without suitable habitat form range boundaries]
12. a [Extrinsic cycles are controlled by external factors, predation is an external factor]
13. d [Examination of Figure 45.19 shows the human population increased from 5 to 6 billion in 12 years]
14. a [Age structure would show if there would be more schoolchildren or seniors in the next few years]
15. a [Mortality = 200/500; Survivorship = 300/500]
16. b [lower mortality at older and bigger ages would make it more profitable to delay reproduction]
17. d [Habitat differences can lead to different survivorship curves; natural selection drives life history variation]
18. b [When N is less than K, the population is below the carrying capacity, and continues to grow]
19. d [Populations that are overcrowded suffer higher death rates, a loss of condition (due to starvation), and increased emigration with individuals looking for better habitat]
20. c [r-selected populations the young generally have little parental care and suffer high mortality]

INTEGRATING AND APPLYING KEY CONCEPTS

1. Commercial fishing pressure is more selective than the guppy predators (fishers generally target larger individuals by law). Also commercial fishing does not select based on behaviour or colouration. However, since commercial fishing does shorten the average lifespan, it will tend to drive fish towards early reproduction. Pollution is non-selective and not density dependent, so its effects will be quite different from either commercial fishing or natural predators. Fish may evolve natural resistance to the contaminants, but no life history changes are expected.
2. Human populations (even in developing countries) have reduced many of the density-dependent factors that may limit other species' population growth. Thus technology, medicine, civic hygiene, and very high birth rates contribute to allow very fast population growth. Secondly, more developed countries have the resources for very fast population growth, but family planning limits population growth to provide better standards of living.
3. As the population density of the rabies virus declines, the likelihood of vaccinating the few foxes still carrying the virus becomes very low indeed. Furthermore, the natural density-dependent factors that can limit the growth of the rabies population will not help the government's efforts. Furthermore, if the foxes are eliminated as a carrier species, the virus may evolve its' life history to utilize a new host (e.g. skunks or bats). Finally, infected carriers may immigrate into Ontario, bring new sources for population

growth.

Chapter 46 Population Interactions and Community Ecology

Why It Matters [pp.1141-1142]

1. community ecology; 2. specialized; 3. niches; 4) interactions

46.1 Interspecific Interactions [pp. 1142-1143]

5) evolution; 6) coevolution; 7) faster; 8) prey; 9) predators

46.2 Getting Food [pp. 1143-1144]

10) plants; 11) animals; 12) heat sensors; 13) insects; 14) generalists; 15) specialists; 16) cost:benefit; 17) optimal foraging; 18. F, specialist; 19. T; 20. F, high; 21. T

46.3 Defence [pp. 1144-1151]

22) defence; 23) small; 24) large; 25) detecting; 26) camouflage; 27) spines; 28) chemical; 29) aposematic; 30) mimic; 31) Batesian mimicry; 32) Műllerian mimicry; 33. B; 34. D; 35. A; 36. E; 37. C; 38. b; 39. a; 40. b; 41. c; 42. e.

46.4 Competition [pp. 1151-1155]

43) interspecific; 44) interference; 45) exploitative; 46) competitive exclusion; 47) niche; 48) resource partitioning; 49) character displacement; 50) mutualism; 51) commensalism; 52) parasitism; 53) parasitoids; 54. b; 55. d; 56. a; 57. c; 58. a; 59. a.

46.5 The Nature of Ecological Communities [pp. 1155 – 1156]

60) equilibrium; 61) equilibrium; 62) disturbance; 63) gradients; 64) Gleason's; 65) ecotone; 66) diversity; 67. F, narrow; 68. T; 69. T; 70. F, equilibrium

46.6 Community Characteristics [pp. 1156 – 1161]

71) richness; 72) evenness; 73) diversity; 74) trophic; 75) parasite; 76) primary; 77) heterotrophs; 78) primary; 79) third; 80) omnivores; 81) detritivores; 82) bacteria; 83) food chain; 84) food web; 85) trophic; 86) stability; 87) 0.13; 88) -0.27; 89) 0.02; 90) -0.33; 91) -1.19; 92. 1.19, more diverse since as H' increases more species diversity is present in the community ; 93. D; 94. A, B or C; 95. E; 96. E; 97. D

46.7 Effects of Population Interactions on Community Structure [pp. 1161-1162]

98) extinct; 99) 50-75; 100) K-selected; 101) r-selected; 102) increase; 103) diversity; 104) keystone; 105. C; 106. D; 107. A; 108. B.

46.8 Effects of Disturbance on Community Characteristics [pp. 1163-1165]

109) equilibrium; 110) disturbances; 111) intermediate disturbance; 112) r-selected; 113) K-selected; 115) grassland; 116. F, low (large disturbances lead to loss of species, intermediate disturbances lead to high diversity); 117. T; 118. T; 119. T; 120. F, useful

46.9 Succession [pp. 1165 – 1169]

121) succession; 122) primary succession; 123) climax community; 124) recovery; 125) aquatic succession; 126) facilitation; 127) inhibition; 128) tolerance; 129) disclimax; 130. d; 131. b; 132. e; 133. a; 134. e.

46.10 Variations in Species Richness among Communities [pp. 1169 – 1173]

135) latitudinal trends; 136) island patterns; 137) generations; 138) migration; 139) specialization; 140) equilibrium theory; 141) species pool; 142) Larger; 143) small; 144) immigration; 145) extinction; 146. D; 147. B; 148. C; 149. A.

Self-Test

1.b [a and c are simply adaptations that do not necessarily result from interaction with other organisms; d results from intraspecific interactions; coevolution is based on interactions between different species]

2. b [Symbioses are interactions between individuals of different species; intraspecific refers to 'within' species]

3. a

4. b [a is commensalism, c is exploitative competition, and d is mutualism]

5. d

6. d [commensalim is rarely found in nature since species interactions usually affect both species]

7. a [parasitoids kill their "hosts" making them more like predators]

8. b

9. d [E_H is designed to reflect the variation in the relative number of each species]

10. d

11. c [a food web is used where there is more than one predator eating the same prey, and each predator may eat more than one prey type]

12. c [complex communities are expected to be more stable since the loss of one species is unlikely to make a big difference]

13. d [the effect of predators can be highly varied, depending on the specific circumstances]

14. a

15. d [such a community should have a mix of K- and r-selected species, and should be no more sensitive to disturbance than any other community]

16. d [a, b and c had established communities before the disturbance, therefore the recovery is secondary succession]

17. a [the island equilibrium hypothesis relates to the species composition at equilibrium on isolated habitat patches]

18. d [a disturbance will allow fast-growing high fecundity r-selected species to colonize]

19. b

20. b [The assumption is that extinction rates are inversely correlated with island size; larger islands have more potential niches and competition between immigrants is less likely.]

Integrating and Applying Key Concepts

1. (i) Probably beneficial to both species. The greater the number of distasteful individuals of either species, the faster predators will learn to avoid them. (ii) Probably beneficial to both species. Since the majority of individuals are models, predators are most likely to sample distasteful individuals; (iii) Detrimental to both species. Predators are most likely to encounter palatable mimics and will be less apt to avoid both species.
2. K-selected species are generally slow-growing, high parental contribution species that are usually associated with climax communities. These communities tend to have populations in which growth is limited by carrying capacity, hence there tends to be strong competitive interactions. r-selected species, on the other hand, are adapted to fast population growth with few resource restrictions, so there is low competition pressure between individuals in the community.
3. 1) Lakes: lakes are islands of freshwater aquatic habitat surrounded by dry land. The probable source of immigrants for lakes is larger lakes or inland seas connected by rivers. 2) Mountaintops: mountaintops are islands of cold, dry alpine habitat surrounded by lowland habitat. The source of immigrants would be other mountaintop habitat on mountains close by. It is unlikely that isolated mountains would follow the equilibrium theory of island biogeography. 3) Coral reefs: coral reefs are islands of complex coral habitat surrounded by low productivity, deep ocean waters. Sources of immigrants for coral reefs would be other coral reefs or shore-line shallow coral habitat. Other possible examples exist.

Chapter 47 Ecosystems

Why It Matters [pp.1179-1180]

1. Urban; 2. rain; 3. sun; 4. 6,000; 5. biodiversity

47.1 Energy Flow and Ecosystem Energetics [pp. 1180 – 1189]

6. gross; 7. net; 8. 10; 9. 50; 10. biomass; 11. standing crop; 12. synthesis; 13. limiting; 14. secondary; 15. ecological; 16. harvesting; 17. assimilation; 18. production; 19. pyramids; 20. pyramid of biomass; 21. turnover; 22. pyramid of numbers; 23. pyramid of energy; 24. cascade; 25. contaminants; 26. biological magnification; 27. A; 28. D; 29. C; 30. B; 31. F, underestimates; 32. T; 33. T; 34. F, less than 1%.

47.2 Nutrient Cycling in Ecosystems [pp. 1189 - 1200]

35. biogeochemical; 36. conserved; 37. generalized compartment; 38. hydrogeologic; 39. precipitation; 40. carbon dioxide; 41. available; 42. fossil fuels; 43. warming; 44. nitrogen; 45. fixation; 46. ammonification; 47. nitrification; 48. denitrification; 49. fertilizers; 50. gaseous; 51. aquatic; 52. sedimentary; 53. absorbing; 54. excreting; 55.c; 56. a; 57. c; 58. a; 59. c; 60. d; 61. b.

47.3 Ecosystem Modelling [pp. 1200 – 1201]

62. simulation; 63. predict; 64. F, ecosystems; 65. T; 66. T; 67. F, do

SELF-TEST

1. d [Urban ecosystems do have plant and animal communities, but they are very simplified]
2. b
3. d [the bigger the ecosystem, the more primary productivity possible, and the more productivity per unit ecosystem size, the greater the total productivity]
4. a
5. d
6. d [New biomass is created when existing organisms grow, reproduce, or store energy]
7. d
8. c
9. a
10. b [low ecological efficiency means that there is high energy loss between steps of the pyramid]
11. a
12. c [solar energy drives evaporation, a prime driver of the hydrogeologic cycle]
13. b
14. d [Photosynthesis is the only one the four choices that removes carbon from the atmosphere.]
15. b [denitrification is the only process that produces N_2]
16. d
17. a
18. c
19. b [ecosystem modeling is designed to be predictive]
20. d [all except time of day are critical ecosystem factors]

Integrating and Applying Key Concepts

1. Ecosystem energetics rely on an external source of energy (primarily solar radiation) and energy is not conserved, but flows through the ecosystem. On the other hand, nutrients are conserved and they cycle among available and non-available forms repeatedly. Both processes limit ecosystem productivity, and hence function; both are needed for complex and dynamic ecosystems to function normally.
2. Gross primary productivity is the total proportion of the sun's energy primary producers convert into energy, for their own use (e.g. maintenance) as well as for the production of new biomass. Net primary productivity, is the amount of energy available for consumers to access (gross primary productivity minus the primary producers maintenance costs). If a particular species of primary producer has high production efficiency (i.e. it uses a low proportion of the energy it collects for its own maintenance), it will contribute to a high net primary productivity.
3. Low water and nutrient availability could limit net primary productivity, for example desert ecosystems get plenty of sunshine, but are limited by water and nutrient availability.
4. The hydrogeologic and carbon cycles rely directly on solar radiation for evaporation and photosynthesis. The phosphorus cycle has no gaseous component, hence it does not include the atmosphere as part of the cycle. The hydrogeologic cycle has the oceans as the main reservoir, the nitrogen cycle's main reservoir is the atmosphere, the phosphorus cycle's reservoir is the earth's crust, and the carbon cycle's main reservoir is sedimentary rock (although the ocean is the largest reservoir of available carbon).

Chapter 48 Conservation of Biodiversity

Why It Matters [pp.1205-1206]

1. human-centric; 2. humans; 3. separation; 4. cost; 5. monocultural; 6. fertilizers; 7. Barcode of Life; 8. biodiversity;

48.1 Extinction [pp. 1206 – 1209]

9. background extinction; 10. turnover; 11. ancestor; 12. genes; 13. extinct; 14. outcompeted; 15. adaptive radiation; 16. five; 17. asteroid; 18. iridium; 19. crater; 20. atmosphere; 21. C; 22. E; 23.E; 24. C; 25. F; 26. C; 27.A

48.4 How We Got/Get There [pp. 1213-1216]

28. reduction; 29. extinction. 30. introducing; 31. habitat; 32. 1690; 33. predators; 34. nutrient-rich; 35. higher; 36. resource use efficiency; 37. ecosystem; 38. decrease; 39. sharks; 40. predators; 41. D; 42. C; 43. E; 44. B; 45. A; 46. B; 47. C; 48. A; 49. B; 50. C; 51. Mauritian calvaria tree life cycle includes travel through a Dodo's digestive tract to initiate seed germination. Presently, these trees are dying of old age and cannot produce offspring since Dodo extinction has removed the possibility of seed germination.; 52. The horns are not only collected as trophies but also to make bowls that some cultures believe to have magical properties, to make handles of ceremonial daggers called jambiya, and to breakdown into powder form for medicinal use as a fever suppressant.

48.8 Protecting Habitat [p. 1224]

53. eligible; 54. designatable; 55. data-based; 56. mortaility; 57. carrying capacity; 58. COSEWIC; 59. extinct; 60. extirpated; 61. endangered; 62. threatened; 63. special concern; 64. data deficient; 65. increase; 66. protection; 67. human; 68. B; 69. C; 70. A; 71. C; 72. F; 73. C; 74. D

48.10 Taking Action [pp. 1228-1229]

75. human population; 76. 40; 77. slowing; 78. Fertility; 79. Biodiversity; 80. Complex; 81. Decreased; 82. Sustainable; 83. Protecting; 84. F, more; 85. F, other birds; 86. T; 87. T

Self-Test

1. d [humans drive species to extinction for reasons ranging from survival to greed]
2. b [some dinosaurs were small and delicate]
3. b

4. a [similar to a human's survivorship curve]

5. c

6. a [taken in at one port and released in another so potential introduction of many aquatic species]

7. c

8. d [the greater the resource use efficiency, the better a competitor the invading species will be]

9. b

10. a [morphology alone does not determine differences between species (e.g. white moose)]

11. c

12. d [the group of organisms must be eligible for categorization]

13. c

14. b

15. b [populations, and even ecosystems, can be identified as 'at risk']

16. c

17. b

18. c

Integrating and Applying Key Concepts

1. You would find yourself among various species of dinosaurs, as well as many other organisms of that time period. After the asteroid hits the earth, a large amount of dust would arise into the atmosphere from the collision and block out sunlight. This would reduce photosynthesis by autotrophs, resulting in the death of many species from producers up to high level consumers; a mass extinction.
2. American ginseng grows in areas, such as in Sugar Maple forests, that are easily accessible to humans. Since it is believed to have medicinal properties, its value has increased as its numbers have decreased. This is another species which pays a cost for being rare.
3. Birds are showing signs of stress due to loss of habitat, especially near expanding urban areas, as well as areas in which agricultural operations are expanding. Migrating birds are affected by changes occurring anywhere between tropical to temperate areas. Also, some birds, especially chickens and other domesticated birds, are raised in high densities, increasing the chance of spreading disease.

Chapter 49 Domestication

Why It Matters [pp.1233-1234]

1. fertilizers; 2. increased; 3. green revolution; 4. Climate; 5. low; 6. high; 7. malnourished; 8. obesity

49.1 Domesticate [pp. 1234 – 1238]

9. gathering; 10. collecting; 11. cultivation; 12. domesticating; 13. plants; 14. archeology; 15. 14; 16. 55,000; 17. F, ant farmers; 18. T; 19. T; 20. T; 21. F, terrestrial plants and animals; 22. T; 23. Indehiscent wheat grains are grains that stick to the plant when they mature so they are easy to collect for cultivation. The dehiscent wheat grains scatter when they mature, so they are very difficult to collect.; 24. Fire was used by various historic settlements for 'niche construction', to clear land and prepare them for cultivation of plant species.

49.2 Why Some Organisms Were Domesticated [pp. 1238-1246]

25. food; 26. A; 27. A,E; 28. A,D; 29. A; 30. A,B,E; 31. A,C; 32. A; 33. a; 34. f; 35. d; 36. e; 37. b; 38. c

49.5 Chemicals, Good and Bad [pp. 1251-1252]

39. greatest or highest; 40. highest; 41. deducting; 42; technology; 43. fertilizer; 44. B; 45. A; 46 A; 47. B; 48. B; 49. A; 50. A; 51. F, medicinal plant; 52. T; 53. F, with; 54. F, insufficient

49.7 The Future [pp. 1253-1255]

55. genetically; 56. proteins; 57. molecular farming; 58. il-10 (or interleukin-10); 59. irritable bowel; 60. food webs; 61. negatively; 62. non-food; 63. exploit; 64. advantages; 65. outweigh; 66. C; 67. D; 68. B; 69. A; 70. Production of corn-produced ethanol has an EROI of ~1:1 (1 L ethanol requires 1 L petroleum), which is a high price in terms of petroleum fuel. Also, the natural land cleared to grow the corn will result in much higher carbon dioxide generation than the original natural habitat produced.

Self-Test

1. a [collecting grain would have been difficult to do if they were dehiscent]
2. d
3. c
4. c [another term for niche construction is ecosystem engineering]
5. d [any or all of the choices could have assisted in the emergence of domestication]
6. e [mushrooms are cultivated by they have not been domesticated]
7. c [it is believed that rice was probably domesticated in India, Myanmar, Thailand, and southern China]
8. a
9. c
10. c
11. d [reduction in honeybees will greatly affect apple and broccoli production but will not affect cotton production]
12. d [conine is a poison, not a phenolic compound]
13. b
14. b [there were at least five founder populations of domestic cats and they were transported where needed to control the rodent population in developing agricultural areas]
15. d [the EROI (energy return on investment) is less than 4:1 for corn-produced ethanol]
16. b [human IL-10 gene is inserted into plants]
17. a

Integrating and Applying Key Concepts

1. Domestication involves selectively breeding individuals of a plant or animal species for desireable characteristics, whereas cultivation does not require selective breeding since it involves just the systematic sowing of wild plant seeds.
2. The Abu Hureyra settlement changed between 12000 – 9400 years BP from a small population of less than 200 people who lived in semi-subterranean pit dwellings and lived on the fruits and seeds of local plants and animals collected from a nearby woodlot to a large population of greater than 4000 people who lived in mud and brick dwellings and lived on cultivated plants (and relatively few wild plants). Climate change was believed to have triggered this change to cultivation, resulting in extensive development of the land.
3. Two genetic mutations occurred that made it possible to domesticate lentils: a mutation that removed dormancy as part of the lentil's life cycle and a mutation that increased the number of seeds that a lentil plant produced. The former reduced the time for seed production each generation and the latter benefitted the crop yield. Rice required a mutation that prevented the abscission layer connecting the flower to the pedicel from completely breaking down, thereby causing them to remain attached, even at maturity, allowing for easier collection of the indehiscent rice product.
4. You would need to consider many factors including (but not limited to): crop prices, technology available to increase yield, (and therefore any limitations of terrain) crop yield for different species, climate, costs associated with producing a particular crop, benefits/costs of fertilizer use, possible pests and pathogens in area, etc.